Methods in Enzymology

Volume 244
PROTEOLYTIC ENZYMES:
SERINE AND CYSTEINE PEPTIDASES

METHODS IN ENZYMOLOGY

EDITORS-IN-CHIEF

John N. Abelson   Melvin I. Simon

DIVISION OF BIOLOGY
CALIFORNIA INSTITUTE OF TECHNOLOGY
PASADENA, CALIFORNIA

FOUNDING EDITORS

Sidney P. Colowick and Nathan O. Kaplan

Methods in Enzymology

Volume 244

Proteolytic Enzymes: Serine and Cysteine Peptidases

EDITED BY

Alan J. Barrett

BIOCHEMISTRY DEPARTMENT
STRANGEWAYS RESEARCH LABORATORY
CAMBRIDGE, ENGLAND

ACADEMIC PRESS

San Diego   New York   Boston   London   Sydney   Tokyo   Toronto

This book is printed on acid-free paper. ∞

Academic Press, Inc.
A Division of Harcourt Brace & Company
525 B Street, Suite 1900, San Diego, California 92101-4495

United Kingdom Edition published by
Academic Press Limited
24-28 Oval Road, London NW1 7DX

International Standard Serial Number: 0076-6879

International Standard Book Number: 0-12-182145-5

PRINTED IN THE UNITED STATES OF AMERICA
94   95   96   97   98   99   MM   9   8   7   6   5   4   3   2   1

Table of Contents

Section I. Serine Peptidases

Section II. Cysteine Peptidases

Contributors to Volume 244

Article numbers are in parentheses following the names of contributors.
Affiliations listed are current.

MAGNUS ABRAHAMSON (49), *Department of Clinical Chemistry, University of Lund, University Hospital, S-221 85 Lund, Sweden*

MAGGY ADAM (19), *Centre d'Ingénierie de Protéines, Université de Liège, Institut de Chimie, B6, B-4000 Sart Tilman, Liège 1, Belgium*

WILLIAM W. BACHOVCHIN (10), *Department of Biochemistry, Tufts University School of Medicine, Boston, Massachusetts 02111*

ALAN J. BARRETT (1, 2, 32), *Department of Biochemistry, Strangeways Research Laboratory, Cambridge CB1 4RN, United Kingdom*

GERALD W. BECKER (29), *Biotechnology Division, Lilley Research Laboratories, Lilley Corporate Center, Eli Lilley and Company, Indianapolis, Indiana 46285*

KERRY BEMIS (29), *Statistical and Mathematical Sciences Division, Lilley Research Laboratories, Lilley Corporate Center, Eli Lilley and Company, Indianapolis, Indiana 46285*

ALISON BEVAN (11), *Department of Biochemistry, Stanford University School of Medicine, Stanford, California 94305*

JENS J. BIRKTOFT (8), *Roche Research Center, Hoffmann-La Roche Inc., Nutley, New Jersey 07110*

KLAUS BREDDAM (8, 18), *Department of Chemistry, Carlsberg Research Laboratory, DK-2500 Valby, Copenhagen, Denmark*

CHARLES BRENNER (11), *Rosenstiel Center, Brandeis University, Waltham, Massachusetts 02254*

DIETER BRÖMME (48), *Khepri Pharmaceuticals, Inc., South San Francisco, California 94080*

MARK T. BROWN (27), *Biology Department, Brookhaven National Laboratory, Upton, New York 11973*

DAVID J. BUTTLE (37, 38, 45), *Department of Human Metabolism and Clinical Biochemistry, University of Sheffield Medical School, Sheffield S10 2RX, United Kingdom*

MICHEL CHRÉTIEN (13), *Department of Molecular Neuroendocrinology, Clinical Research Institute of Montreal, Montreal, Quebec, Canada H2Q 1R7*

LÉON CHRISTIAENS (19), *Centre d'Ingénierie de Protéines, Université de Liège, Institut de Chimie, B6, B-4000 Sart Tilman, Liège 1, Belgium*

CHIN HA CHUNG (25), *Department of Molecular Biology, Seoul National University, Seoul, Korea*

JAMES A. COOK (29), *Virology Research Division, Lilley Research Laboratories, Lilley Corporate Center, Eli Lilley and Company, Indianapolis, Indiana 46285*

JACQUES COYETTE (19), *Centre d'Ingénierie de Protéines, Université de Liège, Institut de Chimie, B6, B-4000 Sart Tilman, Liège 1, Belgium*

ROSS E. DALBEY (21), *Department of Chemistry, Ohio State University, Columbus, Ohio 43210*

HANS-ULRICH DEMUTH (48), *Department of Biochemistry, Martin-Luther-University, D-06099 Halle (Saale), Germany*

HAKIM DJABALLAH (24), *Department of Biochemistry, University of Leicester, Leicester LE1 7RH, United Kingdom*

COLETTE DUEZ (19), *Centre d'Ingénierie de Protéines, Université de Liège, Institut de Chimie, B6, B-4000 Sart Tilman, Liège 1, Belgium*

ix

JOHN J. DUNN (27), *Biology Department, Brookhaven National Laboratory, Upton, New York 11973*

JEAN DUSART (19), *Centre d'Ingénierie de Protéines, Université de Liège, Institut de Chimie, B6, B-4000 Sart Tilman, Liège 1, Belgium*

CLAUDINE FRAIPONT (19), *Centre d'Ingénierie de Protéines, Université de Liège, Institut de Chimie, B6, B-4000 Sart Tilman, Liège 1, Belgium*

JEAN-MARIE FRÈRE (19), *Centre d'Ingénierie de Protéines, Université de Liège, Institut de Chimie, B6, B-4000 Sart Tilman, Liège 1, Belgium*

ROBERT S. FULLER (11), *Department of Biochemistry, Stanford University School of Medicine, Stanford, California 94305*

MORENO GALLENI (19), *Centre d'Ingénierie de Protéines, Université de Liège, Institut de Chimie, B6, B-4000 Sart Tilman, Liège 1, Belgium*

JÉAN-MARIE GHUYSEN (19), *Centre d'Ingenierie de Protéines, Université de Liège, Institut de Chimie, B6, B-4000 Sart Tilman, Liège 1, Belgium*

WADE GIBSON (28), *Virology Laboratories, The Johns Hopkins School of Medicine, Baltimore, Maryland 21205*

JOANNA GIORDANO (29), *Virology Research Division, Lilley Research Laboratories, Lilley Corporate Center, Eli Lilley and Company, Indianapolis, Indiana 46285*

ALFRED L. GOLDBERG (25, 26), *Department of Cell Biology, Harvard Medical School, Boston, Massachusetts 02115*

STUART G. GORDON (39), *Department of Pathology, University of Colorado, Health Sciences Center, and the Colorado Cancer Center, Denver, Colorado 80262*

JACQUELINE GRANDCHAMPS (19), *Centre d'Ingénierie de Protéines, Université de Liège, Institut de Chimie, B6, B-4000 Sart Tilman, Liège 1, Belgium*

BENOÎT GRANIER (19), *Centre d'Ingénierie de Protéines, Université de Liège, Institut de Chimie, B6, B-4000 Sart Tilman, Liège 1, Belgium*

BEULAH GRAY (3), *Department of Microbiology, The University of Minnesota, Minneapolis, Minnesota 55455*

JOHN R. HOIDAL (3), *Pulmonary Division, University of Utah School of Medicine, Salt Lake City, Utah 84132*

YUKIO IKEHARA (16), *Department of Biochemistry, Fukuoka University School of Medicine, Fukuoka 814-01, Japan*

SHIN-ICHI ISHII (42), *Faculty of Pharmaceutical Sciences, Hokkaido University, Sapporo 060, Japan*

MARC JAMIN (19), *Centre d'Ingénierie de Protéines, Université de Liège, Institut de Chimie, B6, B-4000 Sart Tilman, Liège 1, Belgium*

WANDA M. JONES (17), *The Rockefeller University, New York, New York 10021*

BERNARD JORIS (19), *Centre d'Ingénierie de Protéines, Université de Liège, Institut de Chimie, B6, B-4000 Sart Tilman, Liège 1, Belgium*

CHIH-MIN KAM (31), *School of Chemistry and Biochemistry, Georgia Institute of Technology, Atlanta, Georgia 30332*

BAEK KIM (20), *Laboratory of Cell Biology, National Cancer Institute, National Institutes of Health, Bethesda, Maryland 20892*

SEUNG-HO KIM (23), *Laboratory of Cell Biology, National Cancer Institute, National Institutes of Health, Bethesda, Maryland 20892*

HEIDRUN KIRSCHKE (34), *Institute of Biochemistry, Medical Faculty, Martin-Luther University, D-06097 Halle, Saale, Germany*

MICHAEL D. KRAMER (4), *Laboratory for Immunopathology, Institut für Immunologie, D-59120 Heidelberg, Germany*

ALLEN KRANTZ (47), *Syntex Discovery Research, Palo Alto, California 94303*

KOTOKU KURACHI (7), *Department of Human Genetics, University of Michigan Medical School, Ann Arbor, Michigan 48109*

STEFAN KUZELA*(26), *Institute of Molecular Biology, Slovak Academy of Sciences, Bratislava, Slovak Republic*

JEAN LABUS (29), *Biochemical Pharmacology Research Division, Lilley Research Laboratories, Lilley Corporate Center, Eli Lilley and Company, Indianapolis, Indiana 46285*

BERNARD LAKAYE (19), *Centre d'Ingénierie de Protéines, Université de Liège, Institut de Chimie, B6, B-4000 Sart Tilman, Liège 1, Belgium*

MIJIN LEE (27), *Biology Department, Brookhaven National Laboratory, Upton, New York 11973*

MÉLINA LEYH-BOUILLE (19), *Centre d'Ingénierie de Protéines, Université de Liège, Institut de Chimie, B6, B-4000 Sart Tilman, Liège 1, Belgium*

HANS-DIETER LIEBIG (40), *Department of Biochemistry, Medical Faculty, University to Vienna, A1030 Vienna, Austria*

LIH-LING LIN (20), *Genetics Institute, Cambridge, Massachusetts 02140*

JOHN W. LITTLE (20), *Department of Biochemistry and Molecular and Cellular Biology, Universiy of Arizona, Tucson, Arizona 85721*

MARK O. LIVELY (22), *Department of Biochemistry, Bowman Gray School of Medicine, Wake Forest University, Winston-Salem, NC 27157*

JENNIFER LUDFORD (28), *Virology Laboratories, The Johns Hopkins School of Medicine, Baltimore, Maryland 21205*

YU-TING MA (44), *Department of Biological Chemistry, and Molecular Pharmacology, Harvard Medical School, Boston, Massachusetts 02115*

JOSEPH S. MANETTA (29), *Biochemical Pharmacology Research Division, Lilley Research Laboratories, Lilley Corporate Center, Eli Lilley and Company, Indianapolis, Indiana 46285*

WALTER F. MANGEL (27), *Biology Department, Brookhaven National Laboratory, Upton, New York 11973*

JAMES M. MANNING (17), *The Rockefeller University, New York, New York 10021*

TAKEHARU MASAKI (9), *Faculty of Agriculture, Ibaraki University, Ibaraki 300-03, Japan*

MICHAEL R. MAURIZI (23, 25), *Laboratory of Cell Biology, National Cancer Institute, National Institutes of Health, Bethesda, Maryland 20892*

ROBERT MÉNARD (33), *Biotechnology Research Institute, National Research Council of Canada, Montréal, Quebec, Canada H4P 2R2*

YOSHIO MISUMI (16), *Department of Biochemistry, Fukuoka University School of Medicine, Fukuoka 814-01, Japan*

RICHARD P. MOERSCHELL (25), *Department of Cell Biology, Harvard Medical School, Boston, Massachusetts 02115*

KAZUHISA NAKAYAMA (12), *Institute of Biological Sciences and Gene Experiment Center, University of Tsukuba, Ibaraki 305, Japan*

ANN L. NEWSOME (22), *Department of Neurobiology and Anatomy, Bowman Gray School of Medicine, Wake Forest University, Winston-Salem, North Carolina 27157*

MARTINE NGUYEN-DISTÈCHE (19), *Centre d'Ingénierie de Protéines, Université de Liège, Institut de Chimie, B6, B-4000 Sart Tilman, Liège 1, Belgium*

MICHAEL J. NORTH (36), *Department of Biological and Molecular Sciences, University of Stirling, Stirling FK9 4LA, Germany*

MOHAMAD NUSIER (23), *Department of Neurobiology and Anatomy, Bowman Gray School of Medicine, Wake Forest University, Winston-Salem, North Carolina 27157*

* deceased

SHIGENORI OGATA (16), *Department of Biochemistry, Fukuoka University School of Medicine, Fukuoka 814-01, Japan*

JOZEF OLEKSYSZYN (30), *OsteoArthritis Sciences, Inc., Cambridge, Massachusetts 02139*

MANUEL C. PEITSCH (5), *University of Lausanne, Institute de Biochemie, CH-1066 Epalinges, Switzerland*

ANDREW G. PLAUT (10), *Gastroenterology Division, Department of Medicine, Tufts University School of Medicine, and New England Medical Center Hospital, Boston, Massachusetts 02111*

LÁSZLÓ POLGÁR (14), *Institute of Enzymology, Biological Research Center, Hungarian Academy of Sciences, H-1518 Budapest, Hungary*

JAMES C. POWERS (30, 31), *School of Chemistry and Biochemistry, Georgia Institute of Technology, Atlanta, Georgia 30332*

ROBERT R. RANDO (44), *Department of Biological Chemistry, and Molecular Pharmacology, Harvard Medical School, Boston, Massachusetts 02115*

N. V. RAO (3), *Pulmonary Division, University of Utah School of Medicine, Salt Lake City, Utah 84132*

NEIL D. RAWLINGS (2, 32), *Strangeways Research Laboratory, Cambridge CB1 4RN, United Kingdom*

S. JAMES REMINGTON (18), *Institute of Molecular Biology, University of Oregon, Eugene, Oregon 97403*

A. JENNIFER RIVETT (24), *Department of Biochemistry, University of Leicester, Leicester LE1 7RH, United Kingdom*

KENNETH L. ROLAND (20), *Department of Microbiology and Immunology, Emory University, Atlanta, Georgia 30322*

ANDREW D. ROWAN (38), *Pharmacia Biotech Ltd., St. Albans, Herts AL1 3AW, United Kingdom*

FUMIO SAKIYAMA (9), *Division of Protein Chemistry, and Research Center for Protein Engineering, Institute for Protein Research, Osaka University, Osaka 565, Japan*

PETER J. SAVORY (24), *Department of Biochemistry, University of Leicester, Leicester LE1 7RH, United Kingdom*

ANDREA SCALONI (17), *The Rockefeller University, New York, New York 10021*

HENNING SCHOLZE (35), *Department of Biology/Chemistry, Biochemistry, University of Osnabrueak, D-49069 Osnabrueck, Germany*

LAWRENCE B. SCHWARTZ (6), *Department of Internal Medicine, Virginia Commonwealth University, Richmond, Virginia 23298*

NABIL G. SEIDAH (13), *Biochemical Laboratory, J.A. DeSève Laboratory of Biochemical Neuroendocrinology, Clinical Research Institute of Montreal, Montreal, Quebec, Canada H2W 1R7*

ELLIOTT SHAW (46), *Friedrich Miescher-Institut, CH-4002 Basel, Switzerland*

MARKUS M. SIMON (4), *Max-Planck-Institute for Immunobiology, D-79108 Freiburg, Germany*

SATYENDRA K. SINGH (23), *Laboratory of Cell Biology, National Cancer Institute, National Institutes of Health, Bethesda, Maryland 20892*

TIM SKERN (40), *Department of Biochemistry, Medical Faculty, University of Vienna, A1030 Vienna, Austria*

STEVE N. SLILATY (20), *Quantum Biotechnologies, Montreal, Quebec, Canada H4P 2R2*

MARGARET H. SMITH (20), *Department of Biochemistry, University of Arizona, Tucson, Arizona 85721*

MICHELE C. SMITH (29), *Virology Research Division, Lilley Research Laboratories, Lilley Corporate Center, Eli Lilley and Company, Indianapolis, Indiana 46285*

ANDREW C. STORER (33), *Biotechnology Research Institute, National Research Council of Canada, Montréal, Quebec, Canada H4P 2R2*

EGBERT TANNICH (35), *Department of Molecular Biology, Bernhard Nocht Institute for Tropical Medicine, D-20359 Hamburg, Germany*

MARK W. THOMPSON (23), *Laboratory of Cell Biology, National Cancer Institute, National Institutes of Health, Bethesda, Maryland 20892*

NANCY A. THORNBERRY (43), *Department of Biochemistry, Merck Research Laboratories, Rahway, New Jersey 07065*

KAROLY TIHANYI (41), *Department of Microbiology CHUS, University of Sherbrooke, Sherbrooke, Quebec, Canada J1H 5N4*

DIANA L. TOLEDO (27), *Biology Department, Brookhaven National Laboratory, Upton, New York 11973*

ADRIAN TORRES-ROSADO (7), *Department of Human Genetics, University of Michigan Medical School, Ann Arbor, Michigan 48109*

WILLIAM R. TSCHANTZ (21), *Department of Chemistry, Ohio State University, Columbus, Ohio 43210*

JUERG TSCHOPP (5), *University of Lausanne, Institute de Biochemie, CH-1066 Epalinges, Switzerland*

AKIHIKO TSUJI (7), *Department of Human Genetics, University of Michigan Medical School, Ann Arbor, Michigan 48109*

DAISUKE TSURU (15), *School of Pharmaceutical Sciences, Nagasaki University, Nagasaki 852, Japan*

ELCIRA C. VILLARREAL (29), *Virology Research Division, Lilley Research Laboratories, Lilley Corporate Center, Eli Lilley and Company, Indianapolis, Indiana 46285*

MARK WAKULCHIK (29), *Virology Research Division, Lilley Research Laboratories, Lilley Corporate Center, Eli Lilley and Company, Indianapolis, Indiana 46285*

JOSEPH M. WEBER (41), *Department of Microbiology CHUS, University of Sherbrooke, Sherbrooke, Quebec, Canada J1H 5N4*

ANTHONY R. WELCH (28), *Virology Laboratories, The Johns Hopkins School of Medicine, Baltimore, Maryland 21205*

BERND WIEDERANDERS (34), *Institut für Biochemie, Freidrich-Schiller-Universität Jena, D-07740 Jena, Germany*

JEÁN-MARC WILKIN (19), *Centre d'Ingenierie de Protéines, Université de Liège, Institut de Chimie, B6, B-4000 Sart Tilman, Liège 1, Belgium*

KIMBERLY WORZALLA (27), *Biology Department, Brookhaven National Laboratory, Upton, New York 11973*

TADASHI YOSHIMOTO (15), *School of Pharmaceutical Sciences, Nagasaki University, Nagasaki 852, Japan*

WILLY ZORZI (19), *Centre d'Ingénierie de Protéines, Université de Liège, Institut de Chimie, B6, B-4000 Sart Tilman, Liège 1, Belgium*

Preface

Through the earlier, general volumes on proteolytic enzymes (Volumes 19, 45, and 80), *Methods in Enzymology* made available over 200 authoritative articles on these enzymes and their inhibitors. Since the appearance of the latest of these volumes, however, there have been many profound advances in this field of study. The biomedical importance of proteolytic enzymes, suspected for so long, has been established beyond reasonable doubt for a number of groups, including the matrix metalloproteinases, the viral polyprotein-processing enzymes, and the prohormone-processing peptidases. The more recent, specialized Volumes 222, 223, and 241 have dealt with some of these areas, but others have remained to be covered.

The resurgence of excitement about proteolytic enzymes has inevitably resulted in an information explosion, but some of the new understanding has also helped us develop novel approaches to the management of the mass of data. As a result, we can now "see the forest for the trees" a little more clearly. Like other proteins, the proteolytic enzymes have benefited from the recent advances in molecular biology, and amino acid sequences are now available for many hundreds of them. These can be used to group the enzymes in families of evolutionarily related members. Also, there has been a major overhaul of the recommended nomenclature for peptidases by the International Union of Biochemistry and Molecular Biology. In Volumes 244 and its companion Volume 248 on aspartic and metallo peptidases, the chapters on specific methods, enzymes, and inhibitors are organized within the rational framework of the new systems for classification and nomenclature.

In the peptidases dealt with in this volume, the nucleophilic character of a serine or cysteine residue is at the heart of the catalytic mechanism, whereas the peptidases of the aspartic and metallo types described in Volume 248 depend for their activity on an ionized water molecule. A wide variety of specificities of peptide bond hydrolysis is represented in each set of peptidases, together with an equally wide range of biological functions.

ALAN J. BARRETT

METHODS IN ENZYMOLOGY

VOLUME XVII. Metabolism of Amino Acids and Amines (Parts A and B)
Edited by HERBERT TABOR AND CELIA WHITE TABOR

VOLUME XVIII. Vitamins and Coenzymes (Parts A, B, and C)
Edited by DONALD B. MCCORMICK AND LEMUEL D. WRIGHT

VOLUME XIX. Proteolytic Enzymes
Edited by GERTRUDE E. PERLMANN AND LASZLO LORAND

VOLUME XX. Nucleic Acids and Protein Synthesis (Part C)
Edited by KIVIE MOLDAVE AND LAWRENCE GROSSMAN

VOLUME XXI. Nucleic Acids (Part D)
Edited by LAWRENCE GROSSMAN AND KIVIE MOLDAVE

VOLUME XXII. Enzyme Purification and Related Techniques
Edited by WILLIAM B. JAKOBY

VOLUME XXIII. Photosynthesis (Part A)
Edited by ANTHONY SAN PIETRO

VOLUME XXIV. Photosynthesis and Nitrogen Fixation (Part B)
Edited by ANTHONY SAN PIETRO

VOLUME XXV. Enzyme Structure (Part B)
Edited by C. H. W. HIRS AND SERGE N. TIMASHEFF

VOLUME XXVI. Enzyme Structure (Part C)
Edited by C. H. W. HIRS AND SERGE N. TIMASHEFF

VOLUME XXVII. Enzyme Structure (Part D)
Edited by C. H. W. HIRS AND SERGE N. TIMASHEFF

VOLUME XXVIII. Complex Carbohydrates (Part B)
Edited by VICTOR GINSBURG

VOLUME XXIX. Nucleic Acids and Protein Synthesis (Part E)
Edited by LAWRENCE GROSSMAN AND KIVIE MOLDAVE

VOLUME XXX. Nucleic Acids and Protein Synthesis (Part F)
Edited by KIVIE MOLDAVE AND LAWRENCE GROSSMAN

VOLUME XXXI. Biomembranes (Part A)
Edited by SIDNEY FLEISCHER AND LESTER PACKER

VOLUME XXXII. Biomembranes (Part B)
Edited by SIDNEY FLEISCHER AND LESTER PACKER

VOLUME XXXIII. Cumulative Subject Index Volumes I–XXX
Edited by MARTHA G. DENNIS AND EDWARD A. DENNIS

VOLUME XXXIV. Affinity Techniques (Enzyme Purification: Part B)
Edited by WILLIAM B. JAKOBY AND MEIR WILCHEK

VOLUME XXXV. Lipids (Part B)
Edited by JOHN M. LOWENSTEIN

VOLUME 195. Adenylyl Cyclase, G Proteins, and Guanylyl Cyclase
Edited by ROGER A. JOHNSON AND JACKIE D. CORBIN

VOLUME 196. Molecular Motors and the Cytoskeleton
Edited by RICHARD B. VALLEE

VOLUME 197. Phospholipases
Edited by EDWARD A. DENNIS

VOLUME 198. Peptide Growth Factors (Part C)
Edited by DAVID BARNES, J. P. MATHER, AND GORDON H. SATO

VOLUME 199. Cumulative Subject Index Volumes 168–174, 176–194 (in preparation)

VOLUME 200. Protein Phosphorylation (Part A: Protein Kinases: Assays, Purification, Antibodies, Functional Analysis, Cloning, and Expression)
Edited by TONY HUNTER AND BARTHOLOMEW M. SEFTON

VOLUME 201. Protein Phosphorylation (Part B: Analysis of Protein Phosphorylation, Protein Kinase Inhibitors, and Protein Phosphatases)
Edited by TONY HUNTER AND BARTHOLOMEW M. SEFTON

VOLUME 202. Molecular Design and Modeling: Concepts and Applications (Part A: Proteins, Peptides, and Enzymes)
Edited by JOHN J. LANGONE

VOLUME 203. Molecular Design and Modeling: Concepts and Applications (Part B: Antibodies and Antigens, Nucleic Acids, Polysaccharides, and Drugs)
Edited by JOHN J. LANGONE

VOLUME 204. Bacterial Genetic Systems
Edited by JEFFREY H. MILLER

VOLUME 205. Metallobiochemistry (Part B: Metallothionein and Related Molecules)
Edited by JAMES F. RIORDAN AND BERT L. VALLEE

VOLUME 206. Cytochrome P450
Edited by MICHAEL R. WATERMAN AND ERIC F. JOHNSON

VOLUME 207. Ion Channels
Edited by BERNARDO RUDY AND LINDA E. IVERSON

VOLUME 208. Protein–DNA Interactions
Edited by ROBERT T. SAUER

VOLUME 209. Phospholipid Biosynthesis
Edited by EDWARD A. DENNIS AND DENNIS E. VANCE

VOLUME 210. Numerical Computer Methods
Edited by LUDWIG BRAND AND MICHAEL L. JOHNSON

VOLUME 211. DNA Structures (Part A: Synthesis and Physical Analysis of DNA)
Edited by DAVID M. J. LILLEY AND JAMES E. DAHLBERG

[1] Classification of Peptidases

By ALAN J. BARRETT

Introduction

The establishing of a system for classification and nomenclature can be seen as a clear indication of the "coming of age" of a branch of science. The introduction of the Linnaean system for naming and classifying organisms in the eighteenth century and the invention of a system of nomenclature for enzymes in the 1950s were such landmarks, and their value has been obvious. However, the peptide hydrolases (peptidases) have never fitted comfortably into the system of classification and nomenclature introduced for enzymes in general, in which the characteristics of the reaction catalyzed are all important, and this has stimulated efforts to develop other forms of classification for them.

Three major criteria are currently in use for the classification of the peptidases: (1) the reaction catalyzed, (2) the chemical nature of the catalytic site, and (3) the evolutionary relationship, as revealed by structure. The purpose of this chapter is to describe how these criteria are used, and to consider their strengths and weaknesses.

Classification by Reaction Catalyzed

The classification and naming of enzymes by reference to the reactions they catalyze has become the underlying principle of the enzyme nomenclature of the International Union of Biochemistry and Molecular Biology (see Ref. 1, pp. 5–8). In this method, knowledge of the systematic chemical name of the substrate and the type of reaction that it undergoes allows one to derive a systematic name for the enzyme and to decide how it should be classified. In *Enzyme Nomenclature 1992*,[1] class 3 contains the hydrolases, and subclass 3.4 contains all of the enzymes that we are concerned with here, the peptide hydrolases or peptidases. Subclass 3.4 is further divided primarily on the basis of the reactions catalyzed (Table I).

The sub-subclasses of peptidases fall into two sets, comprising the exopeptidases and the endopeptidases. The exopeptidases act only near the ends of polypeptide chains. Those acting at a free N terminus liberate a single amino acid residue, a dipeptide or a tripeptide (aminopeptidases,

---

[1] Nomenclature Committee of the International Union of Biochemistry and Molecular Biology, "Enzyme Nomenclature 1992." Academic Press, Orlando, Florida, 1992.

TABLE I
CLASSIFICATION OF PEPTIDASES BY TYPE OF REACTION CATALYZED[a]

Action	Group	EC subsection
Exopeptidases		
	Aminopeptidases	3.4.11
	Dipeptidyl-peptidases, tripeptidyl-peptidases	3.4.14
	Carboxypeptidases	3.4.16–18
	Peptidyl-dipeptidases	3.4.15
	Dipeptidases	3.4.13
	Omega peptidases	3.4.19
	Endopeptidases	3.4.21–24 and 99

[a] Open circles represent amino acid residues and filled circles are the residues comprising the blocks of one, two, or three terminal amino acids that are cleaved off by these enzymes. The triangles indicate the blocked termini that provide substrates for some of the omega peptidases. Further subdivisions of the carboxypeptidases and the endopeptidases have been made on the basis of catalytic type, as shown in Table III.

dipeptidyl-peptidases, and tripeptidyl-peptidases, respectively). The exopeptidases acting at a free C terminus liberate a single residue or a dipeptide (carboxypeptidases and peptidyl-dipeptidases, respectively). Other exopeptidases are specific for dipeptides (dipeptidases), or remove terminal residues that are substituted, cyclized, or linked by isopeptide bonds (peptide linkages other than those of α-carboxyl to α-amino groups) (omega peptidases).

Endopeptidases act preferentially in the inner regions of peptide chains, away from the termini, and the presence of free α-amino or α-carboxyl groups has a negative effect on the activity of the enzyme. The oligopeptidases are a subset of the endopeptidases that are confined to action on oligopeptide or polypeptide substrates smaller than proteins, although the size ranges of the substrates differ between the individual enzymes.[2]

[2] A. J. Barrett and N. D. Rawlings, *Biol. Chem. Hoppe-Seyler* **373,** 353 (1992).

Methods

The classification of peptidases according to their mode of action on substrates has proved more useful for the exopeptidases than for the endopeptidases, as may be judged from the fact that sub-subclasses only of exopeptidases are distinguished in this way in *Enzyme Nomenclature 1992* (see also Ref. 3).

Synthetic substrates have become increasingly important in both assay and characterization of peptidases in recent years. Factors responsible for this shift away from dependence on natural substrates for many purposes include the discovery of convenient leaving groups, such as 7-amino-4-methylcoumarin, especially for serine and cysteine peptidases, and the recognition of the ability of dipeptide and tripeptide sequences to confer high sensitivity as well as selectivity on the substrates (illustrated by Ref. 4). The use of larger peptide substrates has been facilitated by the increased availability of high-performance, reversed-phase chromatography and the development of quenched fluorescence substrates (reviewed by Knight[5]). As a result of improvements in the methods of peptide synthesis it is commonly possible to assemble a range of artificial substrates, allowing the specificity of a newly discovered peptidase to be explored in some detail.*

The omega peptidases are best thought of as exopeptidases, because they hydrolyze terminal residues, but some of them act on atypical residues that do not bear free α-carboxyl or α-amino groups. Examples would be peptidyl-glycinamidase (EC 3.4.19.2), the pyroglutamyl-peptidases (EC 3.4.19.3; EC 3.4.19.6), and *N*-formylmethionyl-peptidase (EC 3.4.19.7).

The extended substrate-binding sites of many endopeptidases tend to

---

[3] J. K. McDonald and A. J. Barrett, "Mammalian Proteases: a Glossary and Bibliography. Volume 2: Exopeptidases," pp. 1–6. Academic Press, London, 1986.

[4] A. J. Barrett and H. Kirschke, this series, Vol. 80, p. 535.

[5] C. G. Knight, this series, Vol. 248, Chapter 2.

* In describing the specificity of peptidases, use is made of a model in which the catalytic site is considered to be flanked on one or both sides by specificity subsites, each able to accommodate the side chain of a single amino acid residue. These sites are numbered from the catalytic site, S1⋯Sn toward the N terminus of the structure, and S1′⋯Sn′ toward the C terminus. The residues they accommodate are numbered P1⋯Pn, and P1′⋯Pn′, respectively, as follows:

Substrate: –P3–P2–P1 $+$ P1′–P2′–P3′–
Enzyme: –S3–S2–S1 * S1′–S2′–S3′–

This scheme is based on that initially described by I. Schechter and A. Berger, *Biochem. Biophys. Res. Commun.* **27,** 157 (1967) and slightly modified in "Enzyme Nomenclature 1992," p. 372. Academic Press, Orlando, 1992. The peptide bond cleaved (the scissile bond) is indicated by the crossbar symbol ($+$), and the catalytic site of the enzyme is denoted by an asterisk.

require larger synthetic substrates for their characterization than do those of the exopeptidases. The binding site of an endopeptidase normally will not accommodate a free N or C terminus, whereas one or both is essential for a typical exopeptidase.

Errors in the characterization of peptidases with substrates have arisen primarily from the use of impure enzyme preparations, so that the first set of products was rapidly converted into others. This can give the impression of an enzyme quite different from any that is actually present. For example, there was a case in which Z-Ala-Ala-Phe-NHMec [Z, benzyloxycarbonyl; NHMec, 7-(4-methyl)coumarylamide] was cleaved to release the free fluorescent aminomethylcoumarin. This is what chymotrypsin does, and the natural conclusion was that a chymotrypsin-like endopeptidase was present. However, more detailed analysis showed that an initial cleavage at the -Ala+Phe- bond (probably by neprilysin, EC 3.4.24.11) was followed by hydrolysis of the Phe-NHMec product by an aminopeptidase. It is a help toward avoiding such pitfalls if efforts are made to identify the first products of cleavage of the substrate, as can be done by a timed series of high-performance liquid chromatograms (HPLC) or thin-layer chromatograms of the products as the reaction proceeds. A second type of problem has come with the technology that now facilitates the synthesis of oligopeptide substrates. Initially, it was natural to assume that the endopeptidases that cleave these substrates would also act on the same sequences when they occurred in proteins. This has proved not always to be the case, however, because the members of the oligopeptidase subset of endopeptidases will act only on the small substrates.

Information about the reaction catalyzed by an individual peptidase can form a basis for naming as well as classifying it, when the specificity clearly depends on the identity of one or, at most, two amino acid residues. An exopeptidase is commonly named by reference to the type of reaction it catalyzes (aminopeptidase, etc.; see Table I). When the enzyme shows a marked preference for a particular amino acid residue in the P1 or P1' position, the name of this may form a qualifier (e.g., "prolyl aminopeptidase") (Table II). The names currently recommended[1] for many aminopeptidases, dipeptidases, carboxypeptidases, and omega peptidases are derived in this way. For other exopeptidases the specificity is too complex to form a qualifier for the name, however, and then alphabetical or numerical serial names such as "dipeptidyl-peptidase I," "dipeptidyl-peptidase II," "peptidyl-dipeptidase A," and "peptidyl-dipeptidase B" are used.

Endopeptidases that show a clear preference for a particular amino acid residue in the P1 or P1' position also can be named with reference to this ("glycyl endopeptidase," "peptidyl-Lys endopeptidase"). It is important to note that such a term is a name for the reaction catalyzed,

TABLE II
SEMISYSTEMATIC NAMES FOR EXOPEPTIDASES[a]

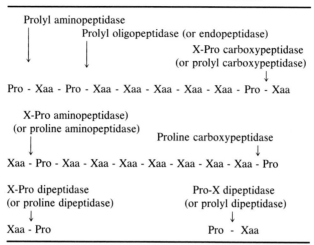

Prolyl aminopeptidase
Prolyl oligopeptidase (or endopeptidase)
X-Pro carboxypeptidase
(or prolyl carboxypeptidase)

Pro - Xaa - Pro - Xaa - Xaa - Xaa - Xaa - Xaa - Pro - Xaa

X-Pro aminopeptidase)
(or proline aminopeptidase)
Proline carboxypeptidase

Xaa - Pro - Xaa - Xaa - Xaa - Xaa - Xaa - Xaa - Xaa - Pro

X-Pro dipeptidase   Pro-X dipeptidase
(or proline dipeptidase)   (or prolyl dipeptidase)

Xaa - Pro   Pro - Xaa

[a] Two ways in which semisystematic names for exopeptidases have been based on the nature of preferred P1 or P1′ residues are illustrated by the example of enzymes cleaving bonds adjacent to proline. The term "prolyl" is appropriate when the prolyl bond is cleaved, whereas "X-Pro" or "proline" has been used when it is the bond on the imino side of proline that is hydrolyzed. Adapted from J. K. McDonald and A. J. Barrett, "Mammalian Proteases: a Glossary and Bibliography. Volume 2: Exopeptidases." Academic Press, London, 1986.

and not for any individual enzyme protein, so that there could be several quite different proteins, each of which could correctly be described as "glutamyl endopeptidase," say. To distinguish between these separate enzyme proteins one may again need to use serial names such as "glutamyl endoeptidase I" or "glutamyl endopeptidase II." For the majority of endopeptidases, the specificity is too complex to provide the basis for a name. In such cases, trivial names such as "trypsin," "papain," "thermopsin," and "pitrilysin" may serve very well. Although these convey little or no information in themselves, they relate to the enzyme protein, rather than just the enzymatic reaction, which can be useful, and they are well suited to the storage and retrieval of information by use of conventional or computerized indexes.

Classification by Catalytic Type

The difficulties associated with the classification and naming of many peptidases on the basis of reaction catalyzed have been recognized for

TABLE III
SUBDIVISION OF CARBOXYPEPTIDASES AND ENDOPEPTIDASES ACCORDING
TO CATALYTIC TYPE[a]

Group of Peptidases	EC sub-subclass
Carboxypeptidases	
Serine-type carboxypeptidases	3.4.16
Metallocarboxypeptidases	3.4.17
Cysteine-type carboxypeptidases	3.4.18
Endopeptidases	
Serine endopeptidases	3.4.21
Cysteine endopeptidases	3.4.22
Aspartic endopeptidases	3.4.23
Metalloendopeptidases	3.4.24
Endopeptidases of unknown catalytic mechanism	3.4.99

[a] Nomenclature Committee of the International Union of Biochemistry and Molecular Biology, "Enzyme Nomenclature 1992," Academic Press, Orlando, Florida, 1992.

many years, and other options have been sought. Hartley[6] pointed out that four distinct types of catalytic mechanism were used by peptidases, and this observation has been developed into a practical system for classification.[7,8] In the enzyme list,[1] the carboxypeptidases and the endopeptidases are divided into sub-subclasses on the basis of catalytic mechanism (Table III). The serine-type peptidases have an active center serine involved in the catalytic process, the cysteine-type peptidases have a cysteine residue in the active center, the aspartic-type endopeptidases depend on two aspartic acid residues for their catalytic activity, and the metallopeptidases use a metal ion (commonly zinc) in the catalytic mechanism. A number of endopeptidases cannot yet be assigned to any of the sub-subclasses EC 3.4.21–24, and these form sub-subclass EC 3.4.99 of the enzyme list.

Methods

The effects of inhibitors give the most reliable information as to the catalytic type of a peptidase, although few if any of the available inhibitors are perfect diagnostic reagents, and false-negative or false-positive reactions do occur. There are wide variations in the susceptibility of the

[6] B. S. Hartley, *Annu. Rev. Biochem.* **29,** 45 (1960).
[7] A. J. Barrett, *Ciba Found. Symp.* **75,** 1. (1980).
[8] A. J. Barrett, *in* "Proteinase Inhibitors" (A. J. Barrett and G. Salvesen, eds.), p. 3. Elsevier, Amsterdam, 1986.

peptidases to the inhibitors, even within a catalytic type. Sometimes these variations reflect the fact that the enzymes are members of different evolutionary families, and thus are structurally quite dissimilar, although they use the same type of catalytic mechanism. When partial inhibition is detected, the effect of time of exposure of the enzyme to the inhibitor on the degree of inhibition can be informative. The covalently reacting, irreversible inhibitors such as 3,4-dichloroisocoumarin (3,4-DCI) and L-3-carboxy-*trans*-2,3-epoxypropyl-leucylamido(4-guanidino)butane (E64) are expected to inactivate their target enzymes progressively, but eventually reaching total inhibition.[9] Any significant percentage of activity that persists after prolonged exposure to an excess of such a reagent indicates that the system is anomalous, and further tests will be necessary before conclusions can safely be reached. The inactivation of metallopeptidases by chelating agents also is a time-dependent process, for different reasons, as is explained elsewhere.[10] In contrast, time dependence of inhibition is generally not seen with the reversible inhibitors.[9]

Earlier accounts of the use of inhibitors in the classification particularly of endopeptidases include those in Refs. 11 and 12. A simple protocol for the identification of the catalytic type of peptidase by use of inhibitors is given in Table IV.

Serine-Type Peptidases. The reagent of choice for the recognition of a serine peptidase is 3,4-DCI. This is safe and convenient to use, and reacts rapidly and irreversibly with a wide range of serine peptidases (see [31] in this volume). Serine peptidases that have little reactivity with other standard reagents, such as diisopropyl fluorophosphate (DFP) and phenylmethylsulfonyl fluoride(PMSF), but are inhibited by 3,4-DCI, include the glutamyl endopeptidase of *Staphylococcus*. The inhibition of some of the activities of the multicatalytic endopeptidase complex by 3,4-DCI has been taken as an indication that this is a serine peptidase (see [24] in this volume). In contrast, however, prolyl oligopeptidase is inhibited by DPF, but reportedly not by 3,4-DCI or PMSF (Ref. 13; see also [14] in this volume).

The other reagents that have proved valuable for serine peptidases are DFP and PMSF. The disadvantage of DFP is its toxicity, which is due to inhibition of acetylcholinesterase, and is exacerbated by its volatility, but once DFP has been diluted to a 100 m*M* working stock solution in dry

---

[9] J. G. Bieth, this series, Vol. 248, Chapter 5.
[10] D. S. Auld, this series, Vol. 248, Chapter 14.
[11] A. J. Barrett and G. S. Salvesen, "Proteinase Inhibitors." Elsevier, Amsterdam, 1986.
[12] J. S. Bond and R. J. Beynon (eds.), "Proteolytic enzymes: A Practical Approach." IRL Press, Oxford, 1989.
[13] S. Kalwant and A. G. Porter, *Biochem. J.* **276,** 237 (1991).

TABLE IV
USE OF INHIBITORS TO DETERMINE CATALYTIC TYPE OF PEPTIDASE[a]

1. Prepare stock solutions of inhibitors:
   Serine peptidase inhibitor: 3,4-DCI—100 mM in dimethyl sulfoxide (DMSO) (fresh)
   Cysteine peptidase inhibitor: E64—100 mM in water (stable at $-20°$)
   Aspartic peptidase inhibitor: Pepstatin—100 μg/ml in DMSO (stable at $-20°$)
   Metallopeptidase inhibitor: 1,10-Phenanthroline—100 mM in DMSO (stable at $-20°$)
2. Prepare usual assay buffer for the peptidase, and stock solution (100-fold) of the enzyme. Store the latter at 4°.
3. To six tubes, each containing 0.98 ml of assay buffer, add 10 μl of enzyme solution and incubate at 30° for 5 min.
4. To separate tubes, add 10 μl of each inhibitor (tubes 1–4), 10 μl of DMSO (tube 5), and 10 μl of buffer (tube 6).
5. Incubate all six tubes at 30° for 10 min to 1 hr, to allow irreversible inhibitors to react.
6. Add substrate to start the peptidase assay and monitor this as usual.
7. Interpret results:
   (a) Tube 6 should show at least 90% of the activity of a sample of the enzyme that is diluted and assayed immediately. If not, then the enzyme is unstable during the test and results are unreliable. Find conditions under which the enzyme is more stable (e.g., lower temperature, presence of 20% (v/v) glycerol, 0.1% Brij 35, 0.01% serum albumin), and repeat the test.
   (b) Tube 5 shows the effect of DMSO on enzymatic activity. If there is inhibition, find a more suitable solvent (e.g., methanol, 2-propanol) and repeat the test. Try to avoid final solvent concentrations above 1%.
   (c) Essentially complete inhibition in just one of tubes 1–4 is required for a clear result. If there is only partial inhibition, repeat the test with longer incubation or higher inhibitor concentrations.
   (d) If inhibition is obtained with 1,10-phenanthroline, compare the effects of the nonchelating analogs, 1,7-phenanthroline and 4,7-phenanthroline. See the text for further comments.

[a] Adapted from B. M. Dunn, in "Proteolytic Enzymes. A Practical Approach" (R. J. Beynon and J. S. Bond, eds.), p. 57. IRL Press, Oxford, 1989, wherein sources of the inhibitors are also given.

2-propanol for storage at $-20°$ it is not difficult to use safely. Among the peptidases, DFP shows good selectivity for the serine-dependent enzymes, and wide reactivity with them.[14] Serine peptidases known to be inhibited by DFP, in addition to those of the chymotrypsin family, include prolyl oligopeptidase, dipeptidyl-peptidase IV, and serine-type carboxypeptidases.

[14] J. C. Powers and J. W. Harper, in "Proteinase Inhibitors" (A. J. Barrett and G. Salvesen, eds.), p. 55. Elsevier, Amsterdam, 1986.

PMSF is somewhat more convenient to use than DFP, being a solid with only moderate toxicity, but tends to react more slowly,[14] so ample time should be allowed. Both DFP and PMSF are stable in stock solutions in dry 2-propanol, but have quite short half-lives once diluted into water, e.g., $t_{0.5}$ is 55 min for PMSF at pH 7, 25°.[15] DFP and PMSF may inhibit cysteine peptidases, but the effect is reversed by thiol compounds such as dithiothreitol. Synthetic inhibitors of serine peptidases have been comprehensively reviewed by Powers and colleagues (Ref. 14; see [30] and [31] in this volume).

Such protein inhibitors of serine peptidases as soybean trypsin inhibitor, lima bean trypsin inhibitor, and aprotinin cannot be regarded as general reagents for the serine peptidases because they are mostly selective for endopeptidases of the chymotrypsin family. Several of the inhibitors are easy to obtain and use, however, and seldom give false-positive reactions, so they deserve consideration.

Cysteine-Type Peptidases. Many of the cysteine peptidases that are encountered are cysteine endopeptidases of the papain or calpain families, and these are mostly susceptible to rapid, specific, and irreversible inactivation by compound E64.[16,17] An important characteristic of E64 and the related epoxide inhibitors is that they are almost unreactive with low molecular mass thiols such as cysteine and dithiothreitol, so these may safely be used in the buffer that activates the cysteine peptidase.

Many cysteine peptidases of other families, such as clostripain and streptopain, are poorly inhibited by E64, and general reagents for thiol groups are needed to recognize these. Iodoacetate, iodoacetamide, and N-ethylmaleimide (1 mM) are generally suitable, but they are less convenient to use because they react rapidly with the low molecular mass thiol activators of cysteine peptidases. The cysteine peptidase must be fully activated before it is exposed to the inhibitor, but then the activator must be removed or diluted so as not to destroy the thiol-blocking reagent. Rapid gel chromatography has commonly been used for this, and the "spin columns" advocated by Salvesen and Nagase[18] for use with metallopeptidases may be well suited for this purpose also.

In using general reagents for thiol groups as peptidase inhibitors, one must remember that there are many serine and metallo-type peptidases that show significant thiol dependence. That is to say, they are activated by

[15] G. T. James, *Anal. Biochem.* **86,** 574 (1978).

[16] A. J. Barrett, A. A. Kembhavi, M. A. Brown, H. Kirschke, C. G. Knight, M. Tamai, and K. Hanada, *Biochem. J.* **201,** 189 (1982).

[17] C. Parkes, A. A. Kembhavi, and A. J. Barrett, *Biochem. J.* **230,** 509 (1985).

[18] G. Salvesen and H. Nagase, *in* "Proteolytic Enzymes: A Practical Approach" (J. S. Bond and R. J. Beynon, eds.), p. 83. IRL Press, Oxford, 1989.

thiol compounds such as dithiothreitol and/or inhibited by thiol-blocking reagents. Among such serine peptidases are yeast carboxypeptidase Y together with other members of the carboxypeptidase C family (see [18] in this volume), prolyl oligopeptidase (see [14] in this volume), and an important subset of the subtilisin family (see [2] and [13] in this volume). Thiol-dependent metallopeptidases include thimet oligopeptidase and other members of this family,[19] as well as insulysin[20] and aminopeptidase P.[21] Several of the thiol-dependent serine peptidases and metallopeptidases contain a cysteine residue close to the catalytic site (see [2] and [18] in this volume; Ref. 22). It may well be that molecules reacting with the side chains of these cysteine residues sterically hinder the catalytic activities of the enzymes. Clearly, these effects can complicate the assignment of a peptidase to its catalytic type, and indeed they have been responsible for many erroneous assignments. Organomercurial compounds are especially prone to cause nonspecific inhibition of peptidases, and should not be used in attempts to identify catalytic types.

As a class, the peptidyldiazomethanes display a useful degree of selectivity for the inactivation of cysteine peptidases (see [46] in this volume), but no broadly reactive diazomethane has been described that can be used as a diagnostic reagent for this type of peptidase. Like E64, the diazomethanes can be used in the presence of thiol compounds, but a few serine peptidases, including prolyl oligopeptidase, have been found to be inhibited.[23] The peptide aldehyde, leupeptin, has sometimes been treated as if it were a type-specific inhibitor for cysteine peptidases, but in fact it is a potent inhibitor of many serine peptidases also.

Reviews of the inhibitors of cysteine peptidases have been provided by Rich[24] and Shaw.[25]

Aspartic-Type Peptidases. The aspartic-type peptidases (all known examples of which are endopeptidases) are reversibly inhibited by pepstatin (K_i values in the range 10^{-8}–10^{-12} M). This is a highly specific reagent seldom affecting peptidases of other types. However, there are a number of endopeptidases maximally active at low pH, and therefore

[19] A. J. Barrett, this series, Vol. 248, Chapter 32.
[20] A. B. Becker and R. A. Roth, this series, Vol. 248, Chapter 44.
[21] G. Vanhoof, I. De Meester, F. Goossens, D. Hendriks, S. Scharpé, and A. Yaron, *Biochem. Pharmacol.* **44**, 479 (1992).
[22] C. Betzel, A. V. Teplyakov, E. H. Harutyunyan, W. Saenger, and K. S. Wilson, *Protein Eng.* **3**, 161 (1990).
[23] S. R. Stone, D. Rennex, P. Wikstrom, E. Shaw, and J. Hofsteenge, *Biochem. J.* **283**, 871 (1992).
[24] D. H. Rich, *in* "Proteinase Inhibitors" (A. J. Barrett and G. Salvesen, eds.), p. 153. Elsevier, Amsterdam, 1986.
[25] E. Shaw, *Adv. Enzymol.* **63**, 271 (1990).

probably dependent on carboxyl groups, that are not detectably homologous with pepsin and are not much affected by pepstatin; these have to be treated as of unknown catalytic type, for the present. Examples are thermopsin[26] and *Aspergillus* proteinase A.[27]

Metallopeptidases. 1,10-Phenanthroline is the most useful inhibitor with which to recognize metallopeptidases, the very great majority of which are zinc-containing enzymes. 1,10-Phenanthroline has a vastly higher affinity for zinc than for calcium (stability constants $2.5 \times 10^6 \ M^{-1}$ and $3.2 \ M^{-1}$, respectively), so that it can be used in the presence of 10 mM Ca^{2+}, which is required for the stability or activity of a number of metallo- and other peptidases. The planar molecule of phenanthroline binds nonspecifically in the active sites of some enzymes, however, so controls are necessary. The nonchelating analogs, 1,7- and 4,7-phenanthroline, are expected to be noninhibitory, when inhibition is truly due to metal chelation. Moreover, it is commonly possible to demonstrate recovery of activity when phenanthroline is removed and zinc is restored. Salvesen and Nagase[18] have recommended a convenient "spin-column" system for removal of the chelator. The presumed apoenzyme should be treated with only 10–100 μM Zn^{2+}, however, because excess zinc is normally inhibitory, and many erroneous negative results have been reported in the literature following the use of millimolar concentrations of metal ion.

The inhibition of a metallopeptidase by a chelating agent is not necessarily due to removal of the metal ion from the protein. The stability of binding of zinc by metalloenzymes varies enormously, but is often so high that loss of the metal ion to the chelator (by equilibrium dissociation, independent of the presence of the chelator) would not occur quantitatively during the time course of a normal experiment. It seems that an agent such as 1,10-phenanthroline commonly binds to the metal ion while it is still associated with the enzyme, forming a ternary complex that is inactive, and may dissociate more rapidly than the binary enzyme–metal complex.[10] Because inhibition of metallopeptidases by chelating agents often involves slow equilibrations, the process may be time dependent.

Inhibition of a peptidase by an unspecific chelating agent such as EDTA cannot be taken as a reliable indication that the enzyme is a metallopeptidase, because many peptidases of other types are activated by cations, notably Ca^{2+}. Such enzymes include members of the subtilisin family (see [13] in this volume), the calpains and clostripain.

Phosphoramidon [N-(α-L-rhamnopyranosyl-oxyhydroxyphosphinyl)-Leu-Trp] is an effective inhibitor of many bacterial metallopeptidases of

[26] X. Lin and J. Tang, this series, Vol. 248, Chapter 11.
[27] K. Takahashi, this series, Vol. 248, Chapter 10.

the thermolysin family, but most mammalian metallopeptidases other than neprilysin are unaffected by this compound, so this is not a diagnostic agent for metallopeptidases.

Increasingly, a deduced amino acid sequence becomes available early in the study of a peptidase, and this may allow the enzyme to be assigned to an evolutionary family. If this is done by use of rigorous criteria such as those described below, it seems safe to assume that the peptidase is of the same catalytic type as other members of the family, and further experimental verification may not be needed.

Classification by Evolutionary Relationship

Now that amino acid sequences have been determined for many of the known peptidases, much can be learned about the evolutionary and structural relationships among the enzymes from the comparison of the sequences. This provides a third approach to the classification of the peptidases, and it is a powerful one, because the structural similarities within a family of peptidases commonly reflect important similarities in catalytic mechanism and other properties, even extending to biological functions.

Methods

Rawlings and Barrett[28] used the term "family" to describe a group of peptidases that are shown by their primary structures to be evolutionarily related. That usage is retained in the present volume. All the members of a family defined in this way are believed to have arisen by divergent evolution from a single ancestral protein, and are thus homologous as defined by Reeck et al.[29]

Operationally, the members of families are recognized by the fact that each shows a statistically significant relationship in amino acid sequence to at least one other member, either throughout the whole sequence or at least in the domain responsible for catalytic activity. The programs FASTP,[30] FASTA, and TFASTA[31] are used to detect similarities between sequences, and on the basis of these, provisional assignments to the system of families are made. These assignments are refined by construction of

[28] N. D. Rawlings and A. J. Barrett, *Biochem. J.* **290**, 205 (1993).
[29] G. R. Reeck, C. de Haën, D. C. Teller, R. F. Doolittle, W. M. Fitch, R. E. Dickerson, P. Chambon, A. D. McLachlan, E. Margoliash, T. H. Jukes, and E. Zuckerkandl, *Cell (Cambridge, Mass.)* **50**, 667 (1987).
[30] D. J. Lipman and W. R. Pearson, *Science* **227**, 1435 (1985).
[31] W. R. Pearson and D. J. Lipman, *Proc. Natl. Acad. Sci. U.S.A.* **85**, 2444 (1988).

optimized alignments. In many cases the similarities between the sequences are so close that no further analysis is necessary, but whenever the similarity is questionable, the RDF program[30] is applied. This tests the significance of a similarity by comparing the actual score for the alignment with the mean and standard deviation of scores for random shuffles of the sequences. A score of at least six standard deviation units above the mean for the random shuffles is required for addition of a peptidase to an already recognized family. It is generally agreed that similarity at this level of significance reflects common ancestry rather than chance or convergent evolution.[32]

The present usage of "family" for groups of peptidases is not very different from that of "superfamily" by Dayhoff et al.[32] for proteins generally, except that the present definition of family takes account of the need to deal with chimeric (mosaic) proteins by requiring that the relationship be demonstrable in the catalytically active domain. The necessity for this is illustrated by the example of bone morphogenetic protein 1, which is a chimeric protein that contains a catalytic domain related to that of astacin, but also contains segments that are clearly homologous with noncatalytic parts of the complement components C1r and C1s, which are in the chymotrypsin family.[33] Bone morphogenetic protein is placed in the family of astacin, and not that of chymotrypsin.

No entirely satisfactory method has been found for naming families of peptidases, or indeed other proteins. Typically, a term of the type "chymotrypsin family" is adopted by convention, chymotrypsin being a member for which the sequence was determined early, but there is no reason why "trypsin family" or "elastase family" could not be considered equally correct. In addition to providing the name of the family, the founder member has commonly also provided the system for the numbering of amino acid residues that is essential in later comparisons of molecular structures within the family. As an optional alternative to this way of naming families, Rawlings and Barrett[28] established a numbering system in which each family of peptidases was assigned a code constructed of a letter denoting the catalytic type (S, C, A, M, or U, for serine, cysteine, aspartic, metallo-, or unknown), followed by an arbitrarily assigned number. Such a code can be used for quick and unambigous reference to a particular family.

The system of families that is arrived at on the basis of primary structures as described above almost certainly contains several sets of families that in fact represent single evolutionary lines. That is to say, all of the

[32] M. O. Dayhoff, W. C. Barker, and L. T. Hunt, this series, Vol. 91, p. 524.
[33] N. D. Rawlings and A. J. Barrett, Biochem. J. 266, 622 (1990).

proteins in these sets of families have diverged from a single ancestral protein, but they have diverged so far that we can no longer prove their relationship by comparison of the primary structures, and have to place them in several separate families. The term "clan" was introduced by Rawlings and Barrett[28] to describe such a group of families that have a common ancestor. The indications of distant relationship come mainly from the linear order of catalytic-site residues, clusters of conserved amino acids around the catalytic residues, and the tertiary structures of the proteins. Sometimes the folds of the proteins in different families are so similar as to give a very persuasive indication of relationship at the clan level. As yet, however, there are no generally accepted, objective methods by which to decide whether these similarities truly reflect divergent relationships as opposed to convergence of structures under evolutionary pressures. Even when the similarities in tertiary structure are so strong as to leave little doubt that a clan is monophyletic, the component families recognizable from the primary structures continue to have their value, because the members are sufficiently close in structure to have a strong tendency to share other properties, too.

Chapters in this volume and in Volume 288 will reflect the results of classification of peptidases by family. These will summarize the distinctive characteristics of over 20 families of serine peptidases (see [2]), nearly as many families of cysteine peptidases (see [32]), about 30 families of metallopeptidases,[34] and also a number of families of aspartic peptidases and those of as yet unknown catalytic type.[35] Moreover, the chapters in this volume on individual enzymes are arranged so as to bring those dealing with members of a single family together.

Possible Integration of Three Approaches to Classification of Peptidases

Over 300 peptidases have been characterized with sufficient rigor to justify entries in *Enzyme Nomenclature*,[1] and there is clear evidence of the existence of many hundreds more. Confronted with the need to devise a good system for the classification of this large group of enzymes, it is attractive to consider the possibility of making use of all three of the methods that we have reviewed above. The system currently used for the peptidases in *Enzyme Nomenclature* depends primarily on the reactions catalyzed, as it does for all enzymes, but exceptionally, some groups of peptidases are subdivided on the basis of catalytic type, as we have seen. There is little scope in this system for recognizing families, because mem-

---

[34] N. D. Rawlings and A. J. Barrett, this series, Vol. 248, Chapter 13.
[35] N. D. Rawlings and A. J. Barrett, this series, Vol. 248, Chapter 7.

bers of these are commonly distributed among several sub-subsections in the first stage of classification (by reaction catalyzed). The literature abounds with sweeping statements about the properties of peptidases of serine, cysteine, or metallo-type that should properly be restricted to the enzymes of the chymotrypsin, papain, and thermolysin families, respectively. Such oversimplifications may be hard to eradicate so long as the family divisions cannot be presented in the standard scheme of classification.

An alternative system for the classification of peptidases can be envisaged in which the enzymes are grouped first by catalytic type, second by family, and finally by reaction catalyzed. This seems to have much to commend it, and may well merit experimental use. The arrangement of chapters in the present volume and in Volume 248, alluded to above, which is first by catalytic type of the peptidases and then by families, may be treated as an early experiment in the use of this approach.

Section I

Serine Peptidases

[2] Families of Serine Peptidases

By Neil D. Rawlings and Alan J. Barrett

Introduction

Proteolytic enzymes dependent on a serine residue for catalytic activity are widespread and very numerous. Serine peptidases are found in viruses (Table I), bacteria, and eukaryotes, and they include exopeptidases, endopeptidases, oligopeptidases, and omega peptidases.

By the criteria we use to distinguish families of peptidases (see [1] in this volume), over 20 families of serine peptidases are recognized (Table II). On the basis of three-dimensional structures, and some less direct evidence, most of these families can be grouped together into about six clans that may well have common ancestors. The structures that are known for members of four of the clans, chymotrypsin, subtilisin, carboxypeptidase C, and *Escherichia* D-Ala-D-Ala peptidase A, show them to be totally unrelated, however. We therefore envisage at least four separate evolutionary origins of serine peptidases, and suspect that there were considerably more.

There are similarities in the reaction mechanisms for several of the peptidases with different evolutionary origins. Thus, the peptidases of the chymotrypsin, subtilisin, and carboxypeptidase C clans have in common a "catalytic triad" of the three amino acids: serine (nucleophile), aspartate (electrophile), and histidine (base). The geometric orientations of these are closely similar between families, despite the fact that otherwise the protein folds are quite different. This striking example of convergent evolution has led researchers to look for the same catalytic residues in other serine peptidases. However, the catalytic mechanisms of clans SE* (*Escherichia* D-Ala-D-Ala peptidase A) and SF (repressor LexA) are already known to be very different, and there is little doubt that some of the other

---

* In reviewing the families of serine peptidases, we shall use the short codes for reference to the clans and families that we introduced previously (N. D. Rawlings and A. J. Barrett, *Biochem. J.* **290**, 205, 1993). Thus, all the codes for groups of serine peptidases start with "S" and are completed by a letter of the alphabet for a clan (giving SA, SB, etc.) or an arabic numeral for a family (giving S1, S2, etc.) (Table II).

Amino acid and nucleotide sequences are specified by database codes; most of these are from the Swiss-Prot Database (release 26), but a code given in parentheses is an EMBL database accession number. For some viral sequences, the code given is that of the viral polyprotein. For some viruses, numerous variants with only minor differences have been described, and only a single example of each has been included here.

TABLE I

FAMILIES OF PEPTIDASES FOUND IN VIRUSES

Type of virus	Peptidase family[a]
Double-stranded DNA	
Enveloped	
Baculovirus	C1
Herpes virus	S8, S21
Pararetrovirus	A2
Nonenveloped	
Adenovirus	C5
Syphovirus	S24, U7
Myovirus	U9
Single-stranded RNA	
Enveloped	
Coronavirus	S32, C16, U24
Flavivirus	S7
Pestivirus	S31
Retrovirus	A2
Togavirus	S3, S29, C9, C18
Nonenveloped	
Comovirus	C3
Nepovirus	C3
Picornavirus	C3, U31
Potyvirus	S30, C4, C6
Tymovirus	C21
Double-stranded RNA	
Unclassified	C7, C8

[a] The numbering of the families of peptidases is that of N. D. Rawlings and A. J. Barrett, *Biochem. J.* **290,** 205 (1993), as slightly revised in [2] and [32] of this volume, and [7] and [13] in Volume 248 of this series. All of the peptidases known to be encoded by viruses are endopeptidases, and it will be noted that there is no metallopeptidase among them.

families of serine peptidases also will prove to have distinctive mechanisms of action, without the "classic" Ser, His, Asp triad.

The arrangements of the catalytic residues in the linear sequences of members of the various families commonly reflect their relationships at the clan level (Table II).

The way in which certain glycine residues tended to be conserved in

TABLE II

CLANS AND FAMILIES OF SERINE PEPTIDASES[a]

Clan	Family	Representative enzyme	Known catalytic residues
SA	S1	Chymotrypsin	---------H---------D------------------S----------
"	S2	α-Lytic endopeptidase	--------------H-----D------------------S----------
"	S3	Sindbis virus core endopeptidase	---------H-------D----------------S-----------------
"	S5	Lysyl endopeptidase	------------H---------D--------------S--------------
"	S6	IgA-specific serine endopeptidase	--------------S-------------------------------------
"	S30	Tobacco etch virus 35 kDa endopeptidase	----------------------------------H-D---S-------------
"	S7	Yellow fever virus NS3 endopeptidase	-----HD----S---------------------------------------
"	S29	Hepatitis C virus NS3 endopeptidase	--------------S-------------------------------------
"	S31	Cattle diarrhea virus p80 endopeptidase	----------------S-----------------------------------
"	S32	Equine arteritis virus putative endopeptidase	--
SB	S8	Subtilisin	------D----H--------------------------S---------
SC	S9	Prolyl oligopeptidase	--S----D-H-
?	S10	Carboxypeptidase C	--------------------------------S-----------------D-----H-
?	S28	Lysosomal Pro-X carboxypeptidase	--
?	S15	*Lactococcus* X-Pro-peptidase	----------------------S-----------------------
SE	S11	*Escherichia* D-Ala-D-Ala peptidase A	------SK---
"	S12	*Streptomyces* R61 D-Ala-D-Ala peptidase	----------SK---------------------------------------
"	S13	*Actinomadura* R39 D-Ala-D-Ala peptidase	-----SK--
SF	S24	Repressor LexA	----------------------------------S--------K----------
"	S26	Leader peptidase	---------------S---------K-------------------------
"	S27	Eukaryote signal peptidase	--
SG	S14	Clp endopeptidase (subunit clpP)	----------------------------S-----H------------------
?	S16	Endopeptidase La	--S------
?	S25	Multicatalytic endopeptidase complex	--
-	S18	Omptin	--
-	S19	*Coccidioides* endopeptidase	--
-	S21	Assemblin	--------------H-----------S-----------------------
-	S17	*Bacteroides* extracellular endopeptidase	--
-	S23	*Escherichia* protease I	--

[a] The order in which the families are listed here is that in which they are to be found in the text. The arrangement of catalytic residues in the representative enzymes is illustrated diagrammatically by use of lines of normalized length to depict the polypeptide chains, although in reality the chains vary considerably in length, of course. For some of the families, catalytic residues remain to be identified.

the vicinity of the catalytic serine residue in the first serine peptidases to be sequenced, to form the motif Gly-Xaa-**Ser**-Yaa-Gly[1] led to an early expectation that this motif would be found in all serine peptidases. Though it is true that most of the families do show conserved glycine residues

[1] S. Brenner, *Nature* (*London*) **334**, 528 (1988).

near the essential serine, the exact positions of these are variable, as is shown in Fig. 1.

Chymotrypsin Clan (SA)

As is shown in Table II, we include 10 families in clan SA, and all the active members of these families are endopeptidases. For the families of chymotrypsin (S1), α-lytic endopeptidase (S2), Sindbis virus core endopeptidase (S3), and *Achromobacter* lysyl endopeptidase (S5), the three-dimensional structures are known to be similar. The families of yellow fever virus NS3 endopeptidase (S7) and tobacco etch virus 35-kDa endo-

Clan	Family	
SA	S1	· · G · SG · · · ·
	S2	· · G · SG · · · ·
	S3	· · G · SG · · · ·
	S5	· · G · SG · · · ·
	S6	· · G · SG · · · ·
	S30	· · G · SG · · · ·
	S7	· · G · SG · · · ·
	S29	· · G · SGG · · ·
	S31	· · G · SG · · · ·
SB	S8	· · G · S · · · · ·
SC	S9	· · G · S · GG · ·
	S10	· · G · S · · G · ·
	S15	· · G · S · · G · ·
SE	S11	· · · · S · · K · ·
	S12	· · · · S · · K · ·
	S13	· · · · S · · K · ·
SF	S24	· · G · S · · · · G
	S26	· · · · S · · · · ·
SG	S14	G · · · S · · · · ·
	S16	· · G · S · G · · ·
-	S21	· · · · S · · · · ·

FIG. 1. Glycine residues totally conserved in the vicinity of the catalytic serine and lysine residues in the known members of various families of serine peptidases. The codes for clans and families are as in Table II.

peptidase (S30) have the same order of catalytic residues as chymotrypsin. The *Neisseria* IgA-specific serine endopeptidase (S6), hepatitis C virus NS3 endopeptidase (S29), cattle diarrhea virus p80 endopeptidase (S31), and equine arteritis virus putative endopeptidase (S32) show somewhat similar sequences around the catalytic serine residues, so these families are included with a lower degree of confidence.

The order of catalytic residues in the polypeptide chain in clan SA is His/Asp/Ser, and Fig. 2 shows sequences around these catalytic residues for selected members of the clan.

Chymotrypsin Family (S1)

The members of the chymotrypsin family (Table III) are almost entirely confined to animals. The exceptions are trypsin-like enzymes from actinomycetes of the genera *Streptomyces* (TRYP_STRGR) and *Saccharopolyspora* (TRYP_SACER), as well as one recently sequenced from the fungus *Fusarium oxysporum* (EMBL : S63827). The members of this family are inherently secreted proteins. Each is synthesized with a signal peptide that targets it to the secretory pathway. The animal enzymes may be secreted directly (e.g., proenzymes of coagulation factors in the liver parenchymal cell), or packaged in vesicles for regulated secretion (e.g., chymotrypsinogen in the pancreas), or they may be retained in leukocyte granules (e.g., elastase, chymase, or granzymes in polymorphonuclear leukocytes, mast cells, or cytotoxic lymphocytes, respectively). Proteolytic activation of the proenzymes occurs extracellularly, or sometimes in storage organelles. Functions are occasionally intracellular, as in intracellular digestion of bacteria by neutrophils, but most commonly extracellular. Examples of extracellular functions are digestion of food proteins in the intestine by the enzymes from the pancreas, and coagulation, fibrinolysis, and complement activation by the enzyme systems in the plasma.

There are about 200 complete amino acid sequences known, more than for any other family of peptidases. Most enzymes are monomers, but granzyme A is a disulfide-bonded homodimer (see [4]), and tryptase is a homotetramer (see [6]).

The essential catalytic unit of these peptidases is a polypeptide chain of about 220 amino acid residues. However, many members of the family are mosaic proteins in which the molecule is extended N-terminally by addition of unrelated peptide segments (Fig. 3). The catalytic unit almost invariably forms the C-terminal domain: only acrosin and complement

	4	5	6	10	11	19	20
	012345678901234567901			234567890123		89012345678901	
	*			*		*	

Family S1

1	HFCGGSLINENWVVTAAHCGV	DITLLKLSTAAS	VSSCMGDSGGPLVC
2	HFCGGSLINDQWVVSAAHCYK	DIMLIKLSSPVK	KDSCQGDSGGPVVC
3	AICGGFLIREDFVLTAAHCEG	DIMLLKLKSKAK	RASFRGDSGGPLVC
4	HTCGGTLIRRNWVMTAAHCVS	DIALLRLAQSVT	RSGCQGDSGGPLHC
5	LLCGASLISDRWVLTAAHCLL	DIALMKLKKPVA	GDACEGDSGGPFVM
6	GRGGGALLGDRWILTAAHTLY	DIALLELENSVT	QDACQGDSGGVFAV
7	PWAGGALINEYWVLTAAHVVE	DIALVRLKDPVK	MDSCKGDSGGAFAV
8	HLCGGSLIGHQWVLTAAHCFD	DIALIKLQAPLN	KDACKGDSGGPLVC
9	HFCGGTLISPEWVLTAAHCLE	DIALLKLSSPAV	TDSCQGDSGGPLVC
10	YVCGGSLMSPCWVISATHCFI	DIALLKIRSKEG	TDSCQGDSGGPLVC
11	FLCGGFLISSCWILSAAHCFQ	DIALLQLKSDSS	HDACQGDSGGPLVC
12	GFCGGTILSEFYILTAAHCLY	DIAVLRLKTPIT	KDACQGDSGGPHVT
13	HLCGGSIIGNQWILTAAHCFY	DIALLKLETTVN	KDACKGDSGGPLSC
14	FQCGGIIVHRQWVLTAAHCIS	DLMLIRLTEPAD	KDTCVGDSGGPLMC
15	YLCGGVLIDPSWVITAAHCYS	DLMLLHLSEPAD	KDTCAGDSGGPLIC
16	VQCGGALVTNRHVITASHCVV	DIAILTLNDTVT	KDACQGDSGGPMML
17	HFCGATLIAPNFVMSAAHCVA	DIVILQLNGSAT	AGVCFGDSGSPLVC
18	SRCGGFLVREDFVLTAAHCWG	DIMLLQLSRRVR	KAAFKGDSGGPLLC
19	HFCGGSLIHPQWVLTAAHCVG	DIALLELEEPVN	RDSCQGDSGGPLVC
20	HLCGGSLLSGDWVLTAAHCFP	DIALVHLSSPLP	IDACQGDSGGPFVC
21	HLCGGVLVAEQWVLSAAHCLE	DLLLLQLSEKAT	RDSCKGDSGGPLVC
22	PWCGGSLLNRNTVLTAAHCVS	DLAILKLSTSIP	KDSCQGDSGGPIVD
23	MGCGGALYAQDIVLTAAHCVS	DWALIKLAQPIN	VDTCQGDSGGPMFR

Family S2

24	VGFSVTRGATKGFVTAGHCGT	DRAWVSLTSAQT	ACMGRGDSGGSWIT
25	LGFNVSVNGVAHALTAGHCTS	DYGIIRHSNPAA	VCAQPGDSGGSLFA
26	LGFNVRSGSTYYFLTAGHCTD	DYGIVRYTNTTI	VCAEPGDSGGPLYS
27	GSGVIIDADKGYVVTNNHVVD	DIALIQIQNPKN	AAINRGNSGGALVN
28	FIASGVVVGKDTLLTNKHVVD	DLAIVKFSPNEQ	LSTTGGNSGSPVFN
29	TSATGVLIGKNTVLTNRHIAK	DLALIRLKPDQN	GFTVPGNSGSGIFN
30	AAFNVTKGGARYFVTAGHCTN	DYGIVRYTDGSS	ACSAGGDSGGAHFA

Family S3

31	DVIGHALAMEGKVMKPLHVKG	DMEFAQLPVNMR	GVGGRGDSGRPIMD
32	KVTCYACLVGDKVMKPAHVKG	DLECAQIPVHMR	GAGKPGDSGRPIFD

Family S5

33	LVNNTANDRKMYFLTAHHCGM	DFTLLELNNAAN	GVTEPGSSGSPIYS

Family S30

34	ARVKRFEGSVQLFASVRHMYG	DLRIDNWQQETL	SKLTFGSSGLVLRQ

Family S6

35			NYGVLGDSGSPLFA

Family S7

36	IGACVYKE--GTFHTMWHVTR	DLISYGGGWKLE	LDFSPGTSGSPIVD

Family S29

37	TQSFLATCVNGVCWTVYHGAG	DQDLVGWQAPSG	VSYLKGSSGGPLLS

Family S31

38	LETGWAYTHQGGISSVDHVTA	DSGCPDGARCYV	LKNLKGWSGLFIFE

Family S32

39	TGSVWTRNNEVVVLTASHVVG	DFAEAVTTQSEL	AWTTSGDSGSAVVQ

component C2 are known to have C-terminal extensions beyond the peptidase domain, though coagulation factor X has a 16-residue C-terminal extension that may be removed during activation.[2]

Proteolytic cleavage at the N terminus of the catalytic domain of a proenzyme in the chymotrypsin family forms a new N-terminal amino acid residue with a hydrophobic side chain. The new terminal α-amino group forms a salt bridge with Asp-194, which leads to the assembly of the functional catalytic site.[3] The N-terminal residue is commonly Ile, but may be Leu, Val, or Met, and the salivary plasminogen activator from the vampire bat contains Ser in this position.[4]

The propeptide maintaining the proenzyme in its inactive state can be as small as two amino acid residues, as in cathepsin G[5] and granzyme B (see [5]), but many are much longer. Not uncommonly, these are disulfide bonded to the catalytic domain, so that they remain part of the active enzyme. These N-terminal extensions often contain one or more copies of several domains with different structures (Table IV). Domains such as "kringles" and "apples" are confined to this family of proteins, but others,

---

[2] K. Fujikawa, K. Titani, and E. W. Davie, *Proc. Natl. Acad. Sci. U.S.A.* **72**, 3359 (1975).
[3] R. Huber and W. Bode, *Acc. Chem. Res.* **11**, 114 (1978).
[4] J. Krätzschmar, B. Haendler, G. Langer, W. Boidol, P. Bringmann, A. Alagon, P. Donner, and W.-D. Schleuning, *Gene* **105**, 229 (1991).
[5] G. Salvesen, D. Farley, J. Shuman, A. Przybyla, C. Reilly, and J. Travis, *Biochemistry* **26**, 2289 (1987).

FIG. 2. Conservation of sequences around the catalytic triad residues in the chymotrypsin clan (SA). Residues are numbered according to bovine chymotrypsinogen. Asterisks indicate the catalytic and presumed catalytic residues. Residues identical to bovine chymotrypsin are shown in white on black. Key to sequences: 1, bovine chymotrypsin; 2, rat trypsin 1; 3, mouse granzyme A; 4, rat pancreatic elastase 1; 5, human thrombin; 6, human complement component $\overline{C1r}$; 7, human complement component $\overline{C1s}$; 8, human plasma kallikrein; 9, human plasmin; 10, human u-plasminogen activator; 11, human t-plasminogen activator; 12, human coagulation factor Xa; 13, human coagulation factor XIa; 14, human tissue kallikrein; 15, rat tonin; 16, limulus proclotting enzyme; 17, human leukocyte elastase; 18, human cathepsin G; 19, human tryptase; 20, human hepsin; 21, human complement component D; 22, *Fusarium* trypsin; 23, *Streptomyces* trypsin; 24, *Lysobacter* α-lytic endopeptidase; 25, streptogrisin A; 26, streptogrisin B; 27, *Escherichia* protease Do; 28, *Staphylococcus* glumatyl endopeptidase 1; 29, *Staphylococcus* epidermolytic factor A; 30, *Streptomyces* glutamyl endopeptidase II; 31, Sindbis virus core protein; 32, Semliki Forest virus core protein; 33, *Achromobacter* lysyl endopeptidase; 34, tobacco etch virus 35-kDa peptidase; 35, *Neisseria* IgA-specific serine endopeptidase; 36, dengue virus NS3 endopeptidase; 37, hepatitis C virus NS3 endopeptidase; 38, bovine diarrhea virus p80 endopeptidase; 39, equine arteritis virus putative proteinase. For sequences 37 and 38 only the catalytic serine residues have been identified with confidence. Catalytic residues are only presumed for sequence 39.

TABLE III
Peptidases of Chymotrypsin Clan (CA)[a]

Peptidase	EC	Database code
Family S1: Chymotrypsin		
Achelase (*Lonomia*)	-	ACH1_LONAC, ACH2_LONAC
Acrosin	3.4.21.10	ACRO_*
Ancrod	-	ANCR_AGKRH
Arginine esterase	3.4.21.35	ESTA_CANFA
Brachyurin	3.4.21.32	COGS_UCAPU
Calcium-dependent serine proteinase	-	CASP_MESAU
Cathepsin G	3.4.21.20	CATG_HUMAN, (M96801)
Cercarial elastase (*Schistosoma*)	-	CERC_SCHMA
Chymase (includes forms I and II)	3.4.21.39	MCP1_*, TRYM_CANFA, MCP2_*, MCP4_MOUSE, MCP5_MOUSE, (M69136)
Chymotrypsin (includes forms A, B, II and 2)	3.4.21.1	CTR2_*, CTRA_BOVIN, CTRB_*, CTR2_CANFA, (U03760)
Chymotrypsin 1 (*Penaeus*)	3.4.21.1	(X66415)
Chymotrypsin-like protease (*Haliotis*)	-	(X71438)
Clotting factor C (*Limulus*)	3.4.21.84	(D90271)
Clotting factor B (*Limulus*)	3.4.21.85	[b]
Clotting enzyme (*Tachypleus*)	3.4.21.86	PCE_TACTR
Coagulation factor VII	3.4.21.21	FA7_*
Coagulation factor IX	3.4.21.22	FA9_*
Coagulation factor X	3.4.21.6	FA10_*
Coagulation factor XI	3.4.21.27	FA11_HUMAN
Coagulation factor XII	3.4.21.38	FA12_HUMAN
Complement component C$\overline{1\,r}$	3.4.21.41	C1R_HUMAN
Complement component C$\overline{1\,s}$	3.4.21.42	C1S_HUMAN
Complement component C2	3.4.21.43	CO2_*
Complement factor B	3.4.21.47	CFAB_*
Complement factor D	3.4.21.46	CFAD_*
Complement factor I	3.4.21.45	CFAI_HUMAN
Crotalase	3.4.21.74	[c]
easter gene product (*Drosophila*)	-	EAST_DROME
Enteropeptidase	3.4.21.9	[d]
Epidermal growth factor-binding protein (includes forms 1, 2 and 3)	3.4.21.35	EGBA_MOUSE, EGBB_MOUSE, EGBC_MOUSE
Flavoboxin (habu snake)	-	FLVB_TRIFL
Gilatoxin	-	[e]
Granzyme A	3.4.21.78	GRAA_*, GRAX_MOUSE
Granzyme B	3.4.21.79	GRAB_*
Granzymes C, D, E, F, G and Y	-	GRAC_MOUSE, GRAD_MOUSE, GRAE_MOUSE, GRAF_MOUSE, GRAG_MOUSE, GRAH_HUMAN
Hepsin	-	HEPS_HUMAN
Hypodermin A	-	(L24914)
Hypodermin B	-	(L24915)
Hypodermin C	3.4.21.49	COGS_HYPLI
Leukocyte elastase	3.4.21.37	ELNE_HUMAN
Medullasin	-	ELNE_HUMAN
Myeloblastin	3.4.21.76	MELB_HUMAN, PTN3_HUMAN
Natural killer cell protease 1	-	NKP1_RAT
7S Nerve growth factor (includes α and γ chains)	3.4.21.35	NGFA_MOUSE, NGFG_MOUSE

TABLE III (*continued*)

Peptidase	EC	Database code
Pancreatic elastase I	3.4.21.36	EL1_*, (M27347)
Pancreatic elastase II (includes forms A and B)	3.4.21.71	EL2A_HUMAN, EL2B_HUMAN, EL2_*,
Pancreatic endopeptidase E (includes forms A and B)	3.4.21.70	EL3A_HUMAN, EL3B_HUMAN
Plasma kallikrein	3.4.21.34	KAL_*
Plasmin	3.4.21.7	PLMN_*, (M62832)
t-Plasminogen activator	3.4.21.68	UROT_*
u-Plasminogen activator	3.4.21.73	UROK_*
Salivary plasminogen activator (vampire bat)	3.4.21.68	UROT_DESRO
Protein C	3.4.21.69	PRTC_*
Proteinase RVV-V (Russell's viper) (includes forms α and γ)	-	RVVA_VIPRU, RVVG_VIPRU
Ra-reactive factor component P100	-	(D16492)
γ-Renin	3.4.21.54	RENG_MOUSE
Semenogelase	3.4.21.77	PROS_HUMAN
snake gene product (*Drosophila*)	-	SNAK_DROME
stubble gene product (*Drosophila*)	-	(L11451)
Serine protease (rat)	-	(L05175)
Serine protease (*Haematobia*)	-	(Z22567)
Serine proteases 1 and 2 (*Drosophila*)	-	SER1_DROME
Thrombin	3.4.21.5	THRB_*
Tissue kallikrein	3.4.21.35	KAG1_*, KAG2_*, KAG3_*, KAG5_MOUSE, KAG_PIG, KAGB_MOUSE, KAGP_RAT, KAGR_*
Tonin	3.4.21.35	TONI_RAT
Trypsin (includes forms I, II, III, IV, Va and Vb)	3.4.21.4	TRYP_*, TRY1_*, TRY2_*, TRY3_*, TRY4_RAT, TRYA_RAT, TRYB_RAT, (L04749), (L08428), (L15632), (L16805), (L16807), (L16808), (M77814), (M96372), (S63827), (X56744), (X70074), (X72781), (Z18889), (Z18890)
Trypsin-like enzyme (*Choristoneura*)	-	(L04749)
Tryptase (includes forms 1, 2 and 3)	3.4.21.59	TRYT_CANFA, TRYA_HUMAN, TRYB_HUMAN, (M33493), (M30038), MCP6_MOUSE, MCP7_MOUSE
Venombin A	3.4.21.74	BATX_BOTAT, PTCA_AGKCO
Vitellin-degrading endopeptidase (*Bombyx*)	-	(D16232)
Family S2: α-Lytic endopeptidase		
Epidermolytic toxins A and B (*Staphylococcus*)	-	ETA_STAAU, ETB_STAAU
Glutamyl endopeptidase (*Staphylococcus*)	3.4.21.19	STSP_STAAU
Glutamyl endopeptidase (*Bacillus*)	-	GSEP_BACLI
Glutamyl endopeptidase II (*Streptomyces*)	3.4.21.82	(D12470), *f*
α-Lytic endopeptidase	3.4.21.12	PRLA_*
"Metalloprotease" (*Bacillus subtilis*)	-	*g*
Protease Do	-	HTRA_*, (M24777), (M31119)
Protease I (*Rarobacter*)	-	(D10753)

TABLE III (continued)

Peptidase	EC	Database code
Streptogrisin A (*Streptomyces griseus*)	3.4.21.80	PRTA_STRGR
Streptogrisin B (*Streptomyces griseus*)	3.4.21.81	PRTB_STRGR
Family S3: Sindbis virus core endopeptidase		
Core endopeptidase	-	POLS_EEEV, POLS_RRVN, POLS_SFV, POLS_SINDV, POLS_WEEV
Family S5: Lysyl endopeptidase		
Lysyl endopeptidase (*Achromobacter*)	3.4.21.50	API_ACHLY
Family S6: IgA-specific serine endopeptidase		
IgA-specific serine endopeptidase	3.4.21.72	IGA_NEIGO, (X64357)
Family S30: Tobacco etch virus 35 kDa endopeptidase		
35 kDa endopeptidase	-	POLG_PPVD, POLG_PVYN, POLG_TEV, POLG_TVMV
Family S7: Yellow fever virus NS3 endopeptidase		
NS3 endopeptidase	-	POLG_DEN2J, POLG_JAEVJ, POLG_KUNJM, POLG_MVEV, POLG_TBEVS, POLG_WNV, POLG_YEFV1
Family S29: Hepatitis C virus NS3 endopeptidase		
NS3 endopeptidase	-	POLG_HCV1
Family S31: Cattle diarrhea virus p80 endopeptidase		
p80 endopeptidase	-	POLG_BVDVN, POLG_HCVA
Family S32: Equine arteritis virus putative endopeptidase		
Equine arteritis virus putative endopeptidase	-	RPOL_EAV

[a] EC is the enzyme nomenclature number (Nomenclature Committee of the International Union of Biochemistry and Molecular Biology, "Enzyme Nomenclature 1992," Academic Press, Orlando, Florida, 1992, and Supplement); a dash (–) indicates that no EC number has been assigned. Literature references to the individual proteins are generally to be found in the database entries for which the codes are given.

[b] T. Muta, T. Oda, and S. Iwanaga, *J. Biol. Chem.* **268**, 21384 (1993).

[c] H. Pirkle, F. S. Markland, I. Theodor, R. Baumgartner, S. S. Bajwa, and H. Kirakossian, *Biochem. Biophys. Res. Commun.* **99**, 715 (1981).

[d] A. Light and H. Janska, *J. Protein Chem.* **10**, 475 (1991).

[e] P. Utaisincharoen, S. P. Mackessy, R. A. Miller, and A. T. Tu, *J. Biol. Chem.* **268**, 21975 (1993).

[f] I. Svendsen, M. R. Jensen, and K. Breddam, *FEBS Lett.* **292**, 165 (1991).

[g] A. Sloma, C. F. Rudolph, G. A. J. Rufo, B. J. Sullivan, K. A. Theriault, D. Ally, and J. Pero, *J. Bacteriol.* **172**, 1024 (1990).

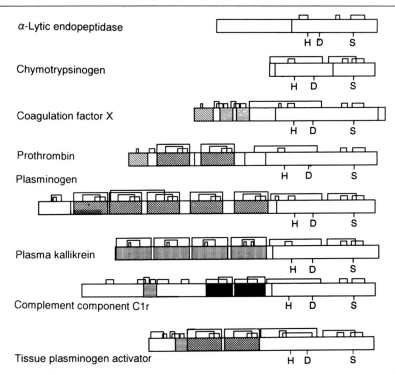

FIG. 3. Polypeptide chain structures of some members of the chymotrypsin family. Vertical lines represent positions of posttranslational cleavages, brackets indicate disulfide bonds, and H, D, and S mark the catalytic residues. Shaded segments correspond to epidermal growth factor-like (horizontal), Ca^{2+}-binding (diagonal), kringle (cross-hatched), apple (vertical), and sushi (black).

such as "sushi" and epidermal growth factor-like domains, are widely distributed among protein families. One domain is found not only in the chymotrypsin family (complement subcomponents C1r and C1s) but also in the astacin family of metallopeptidases.[6] The process by which so much "exon shuffling" has occurred in this family has been suggested to depend heavily on a conserved phase 1 intron N-terminal to the peptidase domain.[7]

A surprising feature of the genes for peptidases of the chymotrypsin family is the diversity of codons used for the active site serine residue. The six available codons for serine fall into two groups (TCA, TCG, TCT, TCC vs. AGC, AGT) such that interconversion between groups requires

[6] N. D. Rawlings and A. J. Barrett, this series, Vol. 248, Chapter 13.
[7] L. Patthy, *Semin. Thromb. Hemostasis* **16**, 245 (1990).

TABLE IV
OCCURRENCE OF ADDITIONAL DOMAINS IN MOSAIC PROTEINS
OF CHYMOTRYPSIN FAMILY

				Domain				
Protein	Kringle[a]	Sushi[b]	Apple[c]	Growth factor module[a]	Finger module[a]	Ca^{2+}-binding module[a]	C1r module[d]	Lectin (C-type) module[b]
Plasminogen	5	—	—	—	—	—	—	—
Apolipoprotein	37	—	—	—	—	—	—	—
Hepatocyte growth factor	4	—	—	—	—	—	—	—
Coagulation factor XII	1	1	—	2	1	—	—	—
Prothrombin	2	—	—	—	—	1	—	—
t-Plasminogen activator	2	—	—	1	1	—	—	—
u-Plasminogen activator	1	—	—	1	—	—	—	—
Protein C	—	—	—	2	—	1	—	—
Coagulation factor VII	—	—	—	2	—	1	—	—
Coagulation factor IX	—	—	—	2	—	1	—	—
Coagulation factor X	—	—	—	2	—	1	—	—
Complement factor B	—	3	—	—	—	—	—	—
Complement factor 2	—	3	—	—	—	—	—	—
Complement subcomponent C1r	—	2	—	1	—	—	1	—
Complement subcomponent C1s	—	2	—	1	—	—	1	—
Haptoglobin	—	1	—	—	—	—	—	—
Limulus clotting factor C	—	3	—	1	—	—	—	1
Plasma kallikrein	—	—	4	—	—	—	—	—
Coagulation factor XI	—	—	4	—	—	—	—	—

[a] L. Patthy, *Cell* (*Cambridge, Mass.*) **41,** 657 (1985).
[b] T. Muta, T. Miyata, Y. Misumi, F. Tokunaga, T. Nakamura, Y. Toh, Y. Ikehara, and S. Iwanaga, *J. Biol. Chem.* **266,** 6554 (1991).
[c] B. A. McMullen, K. Fujikawa, and E. W. Davie, *Biochemistry* **30,** 2050 (1991).
[d] L. Patthy, *Semin. Thromb. Hemostasis* **16,** 245 (1990).

two base changes. Thus, any single point mutation results in loss of the serine, and would doubtless result in loss of catalytic activity in a serine peptidase containing it. In line with expectation, only codons of the TCX group seem to be used for the catalytic Ser in the subtilisin family (S8). However, all six codons are used for active site serine residues in the chymotrypsin family; the most common are from the TCX set, but hepsin, plasmin, thrombin, coagulation factor IX, protein C, and complement components C1r and C1s have AGT or AGC codons. Brenner[1] pointed out that codons in both groups can be derived from cysteine codons in a single step, and suggested that separate evolutionary lines of serine peptidases may have diverged from an ancestral cysteine peptidase. However, our phylogenetic analysis suggests that the "AGX" members do not represent a monophyletic group, and that the change of codon usage has occurred independently several times. It is possible that the "AGX" members have evolved from "TCX" members via proteins without catalytic activity. Codons from both sets also occur in the catalytic serine in the α-lytic endopeptidase family (S2) of clan SA, as well as in two unrelated lines of serine peptidases, the families of prolyl oligopeptidase (S9) and carboxypeptidase C (S10).

The chymotrypsin family contains several proteins that lack peptidase activity. Among these, haptoglobin (HPT_RAT, HPT1_HUMAN, HPT2_HUMAN) has the active site His replaced by Lys, and the active site Ser by Ala, whereas the essential Asp residue is retained. In protein Z (PRTZ_*) and azurocidin (CAP7_*), both the active site Ser and His residues have been replaced. The complete triad is present in cattle procarboxypeptidase A subunit III (CAC3_BOVIN), and the inactivity of this has been attributed to lack of two N-terminal hydrophobic residues.[8] In rhesus monkey apolipoprotein (a) (APOA_MACMU) the active site Ser is replaced by Asn, but in the human protein (APOA_HUMAN) all three residues of the catalytic triad are present. The expected hydrophobic residues at positions 16 and 17 (chymotrypsinogen numbering) are also present in the human apolipoprotein, so that it has the appearance of a fully functional proenzyme. Conversion to an active peptidase would apparently require cleavage of a Ser+Ile bond, and proteolytic activity of apolipoprotein (a) has been reported.[9]

The fact that members of family S1 occur in both eukaryotes and prokaryotes has been the subject of considerable debate. The actinomycete trypsins have sequences too similar to eukaryote trypsins to be consistent with a divergence 3500 million years ago, which is when the common ancestor of prokaryotes and eukaryotes is generally thought to have lived. Hartley[10] attempted to explain this apparent anomaly in terms of a "horizontal" gene transfer from higher organisms to bacteria, whereas Young et al.[11] presented a dendrogram with bacterial sequences at the root, along with an estimate of divergence at about 1300 million years ago, and the implication that because this corresponded to the time that mitochondria were introduced into eukaryotes, a horizontal gene transfer had occurred.

[8] N. Venot, M. Sciaky, A. Puigserver, P. Desnuelle, and G. Laurent, *Eur. J. Biochem.* **157**, 91 (1986).

[9] T. M. Chulkova and V. V. Tertov, *FEBS Lett.* **336**, 327 (1993).

[10] B. S. Hartley, *Proc. R. Soc. London B* **205**, 443 (1979).

[11] C. L. Young, W. C. Barker, C. M. Tomaselli, and M. O. Dayhoff, *in* "Atlas of Protein Sequence and Structure" (M. O. Dayhoff, ed.), Vol. 5, Suppl. 3, p. 73. National Biomedical Research Foundation, Washington, D.C., 1978.

Our calculations confirm this date, and we suggest that the gene transfer is most likely to have been from the protomitochondrion to the eukaryote host.

α-Lytic Endopeptidase Family (S2)

Members of family S2 (listed in Table III) are known only from eubacteria, and they are secreted to act extracellularly. For example, α-lytic endopeptidase produced by the soil bacterium *Lysobacter* degrades cell walls of other soil bacteria, allowing *Lysobacter* to feed on them. Like the chymotrypsin family, family S2 contains endopeptidases that show specificity for P1 residues that are basic, hydrophobic, or alanine, but in addition, the α-lytic endopeptidase family includes several enzymes cleaving glutamyl bonds.

The endopeptidases of family S2 show only slight similarity in sequence to those of family S1, but James and co-workers[12] prepared alignments based on the tertiary structures of the enzymes, and concluded that the small microbial enzymes had common ancestry with the more sophisticated pancreatic endopeptidases.

The recently determined sequences of protease Do from *Escherichia coli* and *Salmonella typhimurium* show protease Do to be a distant homolog not only of α-lytic endopeptidase but also of the glutamyl endopeptidase from *Staphylococcus*. Significant RDF scores are obtained between *E. coli* protease Do and *Lysobacter* α-lytic endopeptidase (7.3) and between *E. coli* and *Salmonella* proteases Do and *Staphylococcus* epidermolytic toxin B (8.3 and 7.4, respectively).

Crystallographic structures have been determined for α-lytic endopeptidase,[13] streptogrisins A and B,[14,15] and *Streptomyces griseus* glutamyl endopeptidase (see [8]). The catalytic triad residues are included in the alignment of Fig. 2, which also shows our prediction for the catalytic triad of protease Do.

The mechanism of activation is not known for many peptidases in this family, but *Lysobacter* α-lytic endopeptidase is synthesized with a long propeptide, and the same apparently applies to *Streptomyces* glutamyl endopeptidase and *Rarobacter* protease I. Not all members of the family

---

[12] L. T. J. Delbaere, W. L. B. Hutcheon, M. N. G. James, and W. E. Thiessen, *Nature (London)* **257,** 758 (1975).
[13] L. T. J. Delbaere, G. D. Brayer, and M. N. G. James, *Nature (London)* **279,** 165 (1979).
[14] M. N. G. James, A. R. Sielecki, G. D. Brayer, L. T. J. Delbaere, and C.-A. Bauer, *J. Mol. Biol.* **144,** 43 (1980).
[15] M. Fujinaga, R. J. Read, A. Sielecki, W. Ardelt, M. Laskowski, Jr., and M. N. G. James, *Proc. Natl. Acad. Sci. U.S.A.* **79,** 4868 (1982).

retain Asp-194, which is crucial to activity in family S1 as described above. Although Ile-16 is conserved in α-lytic endopeptidase and the streptogrisins, it does not form an ion pair with Asp-194, and remains on the surface of the molecule.[12]

As in family S1, the active site serine residues in family S2 are generally encoded by TCX codons, but the *Bacillus* enzymes and streptogrisin A use the AGX codons.[16,17]

We report here the existence of two homologs of protease Do from *E. coli* (EMBL : M24777) and *Chlamydia trachomatis* (EMBL : 31119). In both cases, the protease Do-like sequences are interrupted by frameshifts, but these may reflect sequencing errors. The *E. coli* partial sequence is on the strand complementary to that containing the gene for malate dehydrogenase,[18] whereas the *Chlamydia* sequence is complementary to a gene that was proposed to be an open reading frame for a 59-kDa immunogenic protein.[19]

Sindbis Virus Core Endopeptidase Family (S3)

Togaviruses are single-stranded RNA viruses (Table I) that are vertebrate pathogens transmitted by arthropods, and examples are Semliki Forest virus and Sindbis virus. The genome encodes two polyproteins, p130 and p270. Polyprotein p270 contains a cysteine endopeptidase (see [32], family C8), whereas p130 contains a serine endopeptidase. Polyprotein p130 also contains structural proteins for the nucleocapsid core and for the glycoprotein spikes that protrude from the lipid bilayer surrounding the core and carry the recognition site for the host cell receptor. The nucleocapsid core protein forms the N terminus of the p130 polyprotein, and is the serine endopeptidase. The cleavage of the polypeptide occurs at a Trp✝Ser bond, and once the core protein is released, it retains no detected peptidase activity. Mutagenesis studies have identified a His/Asp/Ser catalytic triad in the core protein, and X-ray crystallography of the Sindbis virus core protein has shown a structure very similar to those of chymotrypsin and α-lytic endopeptidase.[20]

[16] S. Kakudo, N. Kikuchi, K. Kitadokoro, T. Fujiwara, E. Nakamura, H. Okamoto, M. Shin, M. Tamaki, H. Teraoka, H. Tsuzuki, and N. Yoshida, *J. Biol. Chem.* **267,** 23782 (1992).
[17] G. Henderson, P. Krygsman, C. J. Liu, C. C. Davey, and L. T. Malek, *J. Bacteriol.* **169,** 3778 (1987).
[18] R. F. Vogel, K. D. Entian, and D. Mecke, *Arch. Microbiol.* **149,** 36 (1987).
[19] S. Kahane, Y. Weinstein, and I. Sarov, *Gene* **90,** 61 (1990).
[20] H.-K. Choi, L. Tong, W. Minor, P. Dumas, U. Boege, M. G. Rossmann, and G. Wengler, *Nature (London)* **354,** 37 (1991).

The active site serine is encoded by AGX codons in both Sindbis virus[21] and Semliki Forest virus.[22]

Lysyl Endopeptidase Family (S5)

The recently determined tertiary structure of the lysyl endopeptidase from *Achromobacter lyticus* shows a clear similarity to that of trypsin (see [9]), showing that the enzyme is a member of the chymotrypsin clan. The active site residues are shown in Fig. 2. The salt bridge from the amino terminus to Asp-194 that is essential to the structure of the active site in the chymotrypsin family does not exist in lysyl endopeptidase, but a new disulfide bridge, Cys^6-Cys^{216}, may serve a similar function (see [9]).

A domain C-terminal to the potential active site residues of lysyl endopeptidase (residues 474–653) is homologous to a segment of *Vibrio* collagenase (family M9), which establishes the mosaic nature of both proteins. Both this domain and a long N-terminal peptide are removed during proteolytic activation of lysyl endopeptidase.[23]

IgA-specific Serine Endopeptidase Family (S6)

Family S6 contains similar enzymes from two gram-negative, pathogenic bacteria, *Neisseria gonorrhoeae* and *Haemophilus influenzae*. Both peptidases cleave the heavy chains of immunoglobulin A at certain prolyl bonds in the hinge region (see [10]). Unrelated metalloendopeptidases from organisms including *Streptococcus sanguis* also exhibit similar specificity.[24]

Inhibition characteristics identify the enzymes of *Neisseria* and *Haemophilus* as serine peptidases. They contain serine residues in sequences similar to that of chymotrypsin (Fig. 2), and mutation of these inactivates the enzymes.[25]

The precursor of the *Neisseria* enzyme is a mosaic protein of more than 1500 amino acid residues. The proprotein contains an N-terminal leader peptide that directs the protein to the periplasmic space and is removed by a leader peptidase. Following this, there is the peptidase domain, and then a C-terminal "helper" domain believed to create a pore

[21] C. M. Rice and J. H. Strauss, *Proc. Natl. Acad. Sci. U.S.A.* **78,** 2062 (1981).

[22] H. Garoff, A.-M. Frischauf, K. Simons, H. Lehrach, and H. Delius, *Proc. Natl. Acad. Sci. U.S.A.* **77,** 6376 (1980).

[23] T. Ohara, K. Makino, H. Shinagawa, A. Nakata, S. Norioka, and F. Sakiyama, *J. Biol. Chem.* **264,** 20625 (1989).

[24] A. G. Plaut, this series, Vol. 248, Chapter 38.

[25] W. W. Bachovchin, A. G. Plaut, G. R. Flentke, M. Lynch, and C. A. Kettner, *J. Biol. Chem.* **265,** 3738 (1990).

in the outer membrane of the bacterium to permit secretion of the active peptidase. The helper domain is cleaved and remains associated with the outer membrane.[26] The cleavage sites are not conserved between the species variants, and there are very few identical amino acids in the susceptible regions. The helper domain is homologous to domains found in other proteins: *E. coli* initiation factor 2 (IF2_ECOLI), an α-helical domain of unusual composition in *E. coli* tolA protein (TOLA_ECOLI), *Enterococcus faecium* P54 protein (P54_ENTFC), and rat plectin (PLEC_RAT).

Tobacco Etch Virus 35-kDa Endopeptidase Family (S30)

The tobacco etch virus is a potyvirus (Table I), the polyprotein of which contains three proteinases, two of which are cysteine proteinases (see [32], families C4 and C6). Most of the polyprotein processing is performed by the NIa cysteine proteinase, but one N-terminal cleavage (of a Tyr+Ser bond) is made by the 35-kDa serine endopeptidase.[27]

The residues His-214, Asp-223, and Ser-256 are conserved in the family, and have been shown to be essential for activity by site-directed mutagenesis.[28] The His/Asp/Ser order of catalytic residues, and the sequence surrounding the catalytic Ser residues (Fig. 2), are consistent with the inclusion of family S30 in the chymotrypsin clan of serine peptidases. His-214 and Asp-223 are much closer together than the corresponding residues in other peptidases of the clan, however (Table II).

Yellow Fever Virus NS3 Endopeptidase Family (S7)

Flaviviruses, which include dengue virus, yellow fever virus, and encephalitis viruses, are single-stranded RNA viruses (Table I). The RNA encodes a single polyprotein, which is processed by a viral endopeptidase as well as by cellular enzymes. Unlike the Sindbis virus core endopeptidase, the flavivirus endopeptidase is a nonstructural (NS) protein, and occurs internally in the polyprotein, not at the end. The flavivirus endopeptidase cleaves on the C-terminal side of paired basic amino acids, and excises all the nonstructural proteins from the polyprotein. (Excision of structural proteins, which are at the N terminus of the polyprotein, is believed to be mediated cotranslationally by a host cell peptidase associated with the endoplasmic reticulum.) Endopeptidases with a specificity for cleaving C-terminally to paired basic residues are also found in family S8.

[26] T. Klauser, J. Pohlner, and T. F. Meyer, *BioEssays* **15**, 799 (1993).
[27] J. Verchot, E. V. Koonin, and J. C. Carrington, *Virology* **185**, 527 (1991).
[28] J. Verchot, K. L. Herndon, and J. C. Carrington, *Virology* **190**, 298 (1992).

His-53, Asp-77, and Ser-138 have been identified as the members of a catalytic triad in the yellow fever virus NS3 protein by site-directed mutagenesis[29] (Fig. 2). The catalytic residues occur in the N-terminal half of protein NS3, the C-terminal half of which may be a helicase. It is notable that the serine endopeptidases of other enveloped, single-stranded RNA viruses, in families S29, S31, and S32, also occur as N-terminal domains of proteins, the C-terminal parts of which are helicases.

Hepatitis C Virus NS3 Endopeptidase Family (S29)

The hepatitis C virus is a togavirus (Table I). The viral RNA encodes a polyprotein for nonstructural proteins, among which NS3 has been identified as a serine peptidase. Site-specific mutagenesis has implicated Ser-159 in the catalytic mechanism, and other members of a His/Asp/Ser catalytic triad have been proposed[30] (Fig. 2).

Cattle Diarrhea Virus p80 Endopeptidase Family (S31)

Pestiviruses, which include cattle diarrhea virus and hog cholera virus, are closely related to flaviviruses (Table I). These are single-stranded RNA viruses, each of which encodes a large polyprotein. The p80 protein, which is approximately in the middle of the polyprotein, has been identified as the peptidase responsible for processing all nonstructural pestivirus proteins, and site-specific mutagenesis of Ser-311 (Ser-1842 in the polyprotein) prevented processing.[31]

On the basis of sequence similarities to other serine peptidases that we place in clan SA, Bazan and Fletterick[32] correctly predicted the catalytic nature of the p80 protein of pestiviruses, and the active site serine. They also predicted a catalytic triad including His-217 and Asp-254 and proposed that the peptidase domain of the p80 protein would possess a fold similar to that of chymotrypsin.

Equine Arteritis Virus Putative Endopeptidase (S32)

Equine arteritis virus, once thought to be a togavirus, is now regarded as more closely related to coronaviruses. The viral genome encodes several polyproteins, one of which includes a helicase and a polymerase,

---

[29] T. J. Chambers, R. C. Weir, A. Grakoui, D. W. McCourt, J. F. Bazan, R. J. Fletterick, and C. M. Rice, *Proc. Natl. Acad. Sci. U.S.A.* **87**, 8898 (1990).

[30] R. Bartenschlager, L. Ahlborn-Laake, J. Mous, and H. Jacobsen, *J. Virol.* **67**, 3835 (1993).

[31] M. Wiskerchen and M. S. Collett, *Virology* **184**, 341 (1991).

[32] J. F. Bazan and R. J. Fletterick, *Virology* **171**, 637 (1989).

and is assumed also to contain a serine peptidase with a catalytic triad of His-1103, Asp-1129, and Ser-1184.[33]

Clan SB: Subtilisin Family (S8)

The subtilisin family is the second largest family of serine peptidases so far identified, and is extremely widespread, members having been found in eubacteria, archaebacteria, eukaryotes, and viruses (Tables I and V). The great majority of the enzymes are endopeptidases, but there is also a tripeptidyl-peptidase. Crystallographic structures have been determined for several members of the family, and these show a catalytic triad composed of the same residues as in peptidases of the chymotrypsin clan. However, these occur in a different order in the sequence (Asp/His/Ser), and the three-dimensional structures of the molecules bear no resemblance to that of chymotrypsin, so that it is clear that subtilisin and chymotrypsin are not evolutionarily related.

The proprotein-processing endopeptidases kexin, furin, and related enzymes, so far known from yeasts and animals, form a distinct subfamily, which we shall term the kexin subfamily (see [11]–[13]). These preferentially cleave C-terminally to paired basic amino acids. A member of the kexin subfamily can be identified from the subtly different motifs around the active site residues. As can be seen in Fig. 4, the members of the kexin subfamily have Asp in place of Ser or Thr in position 139, Arg in place of His in position 172, and Ala in place of Met in position 324.

In subtilisins, the oxyanion hole is formed by the active site Ser and Asn-262 (numbering according to preprosubtilisin BPN'). Unusually, in the mammalian furin known as PC2, Asn-262 is replaced by Asp (see [11]), and Asp-119 for Asn.

The alignment of the sequences (made by the PILEUP program[34]) gives clues to the structural basis of the selectivity of the kexin subfamily for basic residues. Thus, acidic residues Glu-210 and Asp-273, which are found only in the kexins, occur in parts of the enzymes predicted to form the S2 and S1 binding pockets, respectively.

Some members of the subtilisin family have a requirement for thiol activation. These include endopeptidases R, T, and K from the yeast *Tritirachium,* and the cuticle-degrading endopeptidase from *Metarhizium,*

---

[33] J. A. Den Boon, E. J. Snijder, E. D. Chirnside, A. A. F. De Vries, M. C. Horzinek, and W. J. M. Spaan, *J. Virol.* **65**, 2910 (1991).
[34] Genetics Computer Group, *in* "Program Manual for the GCG Package" University of Wisconsin, Madison, 1991.

TABLE V
PEPTIDASES OF SUBTILISIN FAMILY (S8)[a]

Peptidase	EC	Database code
Family S8: Subtilisin		
Alkaline elastase (*Bacillus*)	-	ELYA_*
Alkaline endopeptidase (*Aspergillus*)	-	(M96758), (Z11580)
Alkaline endopeptidase (*Acremonium*)	-	ALP_CEPAC
Alkaline endopeptidase (*Trichoderma*)	-	ALP_TRIHA
Aqualysin I (*Thermus*)	-	AQL1_THEAQ
Bacillopeptidase F (*Bac. subtilis*)	-	SUBF_BACSU
C5a peptidase (*Streptococcus*)	-	SCPA_STRPY
Calcium-dependent extracellular endopeptidase A (*Vibrio*)	-	PROA_VIBAL
Calcium dependent endopeptidase (*Anabaena*)	-	PRCA_ANAVA
Cell-wall associated endopeptidase (*Lactococcus*) (includes forms PI, PII and PIII)	-	P1P_LACLA, P2P_*, P3P_LACLA, (X14130)
Cerevisin	3.4.21.48	PRTB_YEAST
Cuticle-degrading protease (*Metarhizium*)	-	CUDP_METAN
Endopeptidase K	3.4.21.64	PRTK_TRIAL
Endopeptidase R (*Tritirachium*)	-	PRTR_TRIAL
Endopeptidase T (*Tritirachium*)	-	PRTT_TRIAL
Epidermin processing protease (*Staphylococcus*)	-	EPIP_STAEP
Extracellular endopeptidase (*Serratia*)	-	PRTS_SERMA, PRTT_SERMA
Extracellular endopeptidase (*Xanthomonas*)	-	PROA_XANCP
Halolysin	-	HLY_HAL17
Intracellular serine endopeptidase (*Bacillus*)	-	ISP_*
Major intracellular endopeptidase (*Bacillus*)	-	ISP1_BACSU
Neutral endopeptidase (*Bacillus*)	-	NPRE_*
Nisin operon serine protease (*Lactococcus*)	-	(L11061)
Oryzin	3.4.21.63	AEP_YARLI, ORYZ_*
Serine endopeptidase (*Bac. subtilis*)	-	(PIR S11504)
Serotype-specific antigen precursor (*Pasteurella*)	-	SSA1_PASHA
Subtilisin	3.4.21.62	SUBT_*, SUBE_BACSU, SUBV_BACSU, (L24202)
Hypothetical subtilisin (*Saccharomyces*)	-	YCT5_YEAST, (D14063)
Subtilisin-like protease (*Dichelobacter*)	-	(L08175), (Z16080)
Subtilisin-like protease III (*Saccharomyces*)	-	YSP3_YEAST
Thermitase	3.4.21.66	THET_THEVU
Thermostable serine endopeptidase	-	(S50880)
Tripeptidyl-peptidase II	3.4.14.10	TPP2_HUMAN
Kexin	3.4.21.61	KEX2_YEAST, KEX1_KLULA, (L16238)
Furin	-	FURI_*, FUR1_*, FURL_DROME, FURS_DROME
Pituitary convertase (includes PC1, PC2, PC3, PC6, PACE4)	-	NEC1_*, NEC2_*, NEC3_MOUSE, NECA_HYDAT, NECB_HYDAT, PAC4_HUMAN, (D12619)
56 kDa Serine protease (catfish herpes virus)	-	VG47_HSVI1

[a] See Table III for explanation.

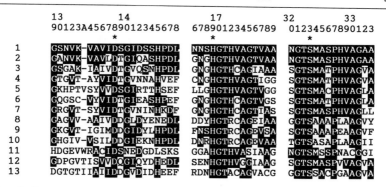

FIG. 4. Conservation of sequence around the catalytic triad residues in the subtilisin family. Residues are numbered according to that of preprosubtilisin BPN', and those identical to residues in subtilisin BPN' are shown in white on black. Asterisks indicate the catalytic triad residues. Key to sequences: 1, subtilisin BPN'; 2, subtilisin Carlsberg; 3, thermitase; 4, *Vibrio alginolyticus* calcium-dependent serine "exopeptidase" A; 5, *Yarrowia lipolytica* extracellular endopeptidase; 6, endopeptidase K; 7, cerevisin; 8, kexin; 9, human pituitary convertase PC2; 10, rat furin; 11, human tripeptidyl-peptidase II; 12, halolysin; 13, *Anabaena variabilis* calcium-dependent endopeptidase.

as well as the members of the kexin subfamily. This thiol dependence is attributable to Cys-173 near the active site histidine.[35]

It has been thought that none of the bacterial subtilisins contains cysteine residues,[35] but the subtilisin from *Bacillus smithii* is an exception to this (EMBL : L24202).

The sole viral member of family S8 is a 56-kDa proteinase from the herpes virus 1 that infects the channel catfish.[36] Herpes viruses are related to baculoviruses, both of which groups encode proteins homologous to mammalian endopeptidases (Table I). The catfish herpes virus is rather divergent, and its genome has a large insert containing some 65 extra open reading frames as compared to other α-herpes viruses. The gene encoding the putative peptidase is one of these, hence no homologs are found in other herpes viruses. This gene is presumably one "captured" from a host.

The subtilisin family contains some mosaic proteins. The secreted endopeptidases from *Vibrio alginolyticus* (known confusingly as "exopeptidase A"[37]) and *Xanthomonas campestris* contain homologous C-terminal domains similar to some found in metallopeptidases, including an aminopeptidase from *Vibrio proteolyticus* (family M28) and endopeptidases of the thermolysin family (M4). Homologs of subtilisin that are located in

[35] K.-D. Jany, G. Lederer, and B. Mayer, *FEBS Lett.* **199**, 139 (1986).
[36] A. J. Davison, *Virology* **186**, 9 (1992).
[37] S. M. Deane, F. T. Robb, S. M. Robb, and D. R. Woods, *Gene* **76**, 281 (1989).

the cell membrane of *Lactococcus* share a homologous domain with other bacterial membrane proteins such as the *Streptococcus* fibrinogen- and-immunoglobulin-binding protein (MRP4_STRPY).

Family S8 contains several enzymes with N- or C-terminal extensions that show no relationship to other proteins. The kexin subfamily includes endopeptidases with a variety of C-terminal extensions (see [13]). Tripeptidyl-peptidase II is a 135-kDa mammalian cytosolic enzyme active at neutral pH, which releases N-terminal tripeptides from polypeptides. The peptidase domain forms the N-terminal third of the protein, whereas the C-terminal two-thirds show no resemblance to any other protein.[38]

Carboxypeptidase Clan (SC)

Tertiary structures available for enzymes of family S10 (carboxypeptidase C) are of the "α/β-hydrolase fold" type, also seen for a wide variety of other enzymes including acetylcholinestase, lipases and *Xanthomonas* haloalkane dehalogenase. The sequence of *Neisseria* prolyl aminopeptidase (family S33) shows it to be homologous to the haloalkane dehalogenase, so we can conclude that families S10 and S33 have similar tertiary folds and linear arrangement of catalytic residues (Ser/Asp/His), and form a clan (SC).

The distinctive Ser/Asp/His arrangement of catalytic residues is also found in family S9 (prolyl oligopeptidase) (Table II), which raises the possibility that family S9 is also a member of clan SC.[39] The clan might additionally contain *Lactococcus* X-Pro-peptidase (family S15) and lysosomal Pro-X carboxypeptidase (family S28).[40] In support of this, there is limited conservation of sequence around the active site Ser residues (Fig. 5), but this is not strong enough to be conclusive.

As we shall see, the peptidases of family S9 differ biologically from those of families S10 and S28 in that the homologs of prolyl oligopeptidase are either cytosolic or integral membrane proteins, whereas the carboxypeptidases are secreted or lysosomal enzymes. Also, peptidases of families S9 and S15 do not seem to have proenzymes, whereas those of S10 and S28 do.

Prolyl Oligopeptidase Family (S9)

Family S9 contains peptidases with a very varied range of quite restricted specificities (see [14]–[17]).[41] Members are known from prokary-

[38] B. Tomkinson and A.-K. Jonsson, *Biochemistry* **30**, 168 (1991).
[39] L. Polgár, *FEBS Lett.* **311**, 281 (1992).
[40] F. Tan, P. W. Morris, R. A. Skidgel, and E. G. Erdös, *J. Biol. Chem.* **268**, 16631 (1993).
[41] N. D. Rawlings, L. Polgár, and A. J. Barrett, *Biochem. J.* **279**, 907 (1991).

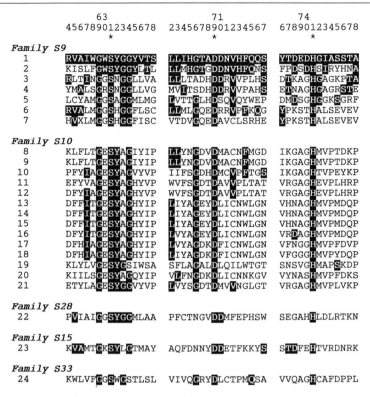

FIG. 5. Conservation of sequence around the catalytic triad residues in the families of prolyl oligopeptidase (S9), *Lactococcus* X-Pro-peptidase (S15), carboxypeptidase C (S10), lysosomal Pro-X carboxypeptidase (S28) and *Neisseria* prolyl aminopeptidase (S33) forming clan SC. Residues are numbered according to rat dipeptidyl-peptidase IV, and residues identical to those in that sequence are shown in white on black. Asterisks indicate the catalytic and presumed catalytic residues; those that have been identified with confidence are Ser, Asp, and His in families S9 and S10. Key to sequences: 1, rat dipeptidyl-peptidase IV; 2, yeast dipeptidyl-peptidase B; 3, pig prolyl oligopeptidase; 4, *Flavobacterium meningosepticum* prolyl oligopeptidase; 5, *Escherichia coli* oligopeptidase B; 6, pig acylaminoacyl-peptidase; 7, human DNF1552 protein (ACPH_HUMAN); 8, human lysosomal carboxypeptidase A; 9, mouse lysosomal carboxypeptidase A; 10, barley carboxypeptidase I; 11, barley carboxypeptidase II; 12, wheat carboxypeptidase II; 13, barley carboxypeptidase III; 14, wheat carboxypeptidase III; 15, rice carboxypeptidase III; 16, *Arabidopsis thaliana* serine carboxypeptidase; 17, yeast carboxypeptidase Y; 18, *Candida albicans* carboxypeptidase Y; 19, *Schizosaccharomyces pombe* carboxypeptidase Y; 20, yeast carboxypeptidase D; 21, *Naegleria fowleri* virulence-related protein; 22, human lysosomal Pro-X carboxypeptidase; 23, *Lactococcus lactis* X-Pro dipeptidyl-peptidase; 24, *Neisseria* prolyl aminopeptidase.

otic and eukaryotic organisms (Table VI). The family contains soluble as well as membrane-bound peptidases. The cytosolic enzymes include two oligopeptidases with different P1 specificities: prolyl oligopeptidase, which cleaves prolyl and some alanyl bonds (see [14]), and oligopeptidase B from eubacteria, which cleaves arginyl and lysyl bonds (see [15]).

Acylaminoacyl-peptidase (see [17]) is also a cytosolic enzyme that has been sequenced from mammals. The enzyme releases an N-acetyl or N-formyl amino acid from the N terminus of a peptide, and is thus an omega peptidase.

The active site Ser, Asp, and His residues have been identified in

TABLE VI

PEPTIDASES OF FAMILIES OF PROLYL OLIGOPEPTIDASE (S9), CARBOXYPEPTIDASE C (S10), LYSOSOMAL PRO-X CARBOXYPEPTIDASE (S28), AND *Lactococcus* X-PRO DIPEPTIDYL-PEPTIDASE (S15)[a]

Peptidase	EC	Database code
Family S9: Prolyl oligopeptidase		
Acylaminoacyl-peptidase	3.4.19.1	ACPH_*
Dipeptidyl-peptidase IV	3.4.14.5	DPP4_*
Dipeptidyl-peptidase IV-like protein	-	(M76429), (M96860)
Dipeptidyl aminopeptidase A (*Saccharomyces*)	-	(L21944)
Dipeptidyl aminopeptidase B (*Saccharomyces*)	-	DAP2_YEAST
Prolyl oligopeptidase	3.4.21.26	PPCE_*, PPCF_FLAME, (D14005), (M61966)
Protease II (*Escherichia coli*)	-	PTRB_ECOLI
Family S10: Carboxypeptidase C		
Carboxypeptidase C (forms I and III)	3.4.16.5	CBP1_HORVU, CBP3_*, (D10985), (D17586)
Carboxypeptidase D	3.4.16.6	KEX1_YEAST, CBP2_*
Carboxypeptidase Y (*Saccharomyces*)	3.4.16.1	CBPY_*, (D10199)
Carboxypeptidase Y-like protein (*Arabidopsis*)	-	CBPX_ARATH
Carboxypeptidase Y-like protein (rice)	-	(D17587)
Lysosomal carboxypeptidase A	3.4.16.5	PRTP_*
Serine-type carboxypeptidase (*Caenorhabditis*)	-	(M75784)
Serine-type carboxypeptidase (*Aedes*)	-	(M79452)
Virulence-related protein (*Naegleria*)	-	(M88397)
Family S28: Lysosomal Pro-X carboxypeptidase		
Lysosomal Pro-X carboxypeptidase	3.4.16.2	(L13977)
Family S15: *Lactococcus* X-Pro dipeptidyl peptidase		
X-Pro Dipeptidyl peptidase (*Lactococcus*)	3.4.14.5	DPP_*, (Z14230)

[a] See Table III for explanation.

family S9,[42] and the conservation of amino acids around them is shown in Fig. 5. In all known members of the family these residues are within about 130 residues of the C terminus, and the N-terminal parts of the molecules are more or less variable. The membrane-bound members of this family contain membrane-spanning domains near the N terminus.

Dipeptidyl-peptidase IV releases N-terminal dipeptides, preferentially cleaving prolyl bonds (see [16]). It is an integral membrane glycoprotein of lymphocytes and intestinal brush border, in particular, that exists as a homodimer or homotetramer. An enzyme with similar specificity from the membrane of the yeast vacuole, dipeptidyl peptidase B, is involved in processing the α-mating factor precursor.[43]

A notable feature of all members of this family is the lack of proteolytic processing, and the enzymes are apparently synthesized in active form.[44,45] Presumably, their restricted specificities prevent indiscriminate degradation of other proteins.

Like a number of other cytosolic oligopeptidases, including thimet oligopeptidase (family M3) and insulysin (M16), prolyl oligopeptidase and acylaminoacyl-peptidase are thiol dependent, but dipeptidyl-peptidase IV is not. This mix of thiol-dependent and thiol-independent enzymes is also seen in families M3 and M16. Unusually for a serine peptidase, the active site serine of prolyl oligopeptidase reacts with diazomethanes,[44] but that of oligopeptidase B reportedly does not do so (see [15]).

Two mammalian proteins, DPPX-S and DPPX-L, closely related to dipeptidyl-peptidase IV, have the catalytic Ser replaced by Asp. mRNAs for both sequences are abundant in the brain, and that for DPPX-S is also found in kidney, ovary, prostate, and testis.[46,47] To date, the corresponding proteins have not been isolated, but we should expect them to be inactive members of the family.

Carboxypeptidase C Family (S10)

The carboxypeptidase C family includes carboxypeptidases that are unusual among serine-dependent enzymes in that they are maximally active at acidic pH. The members of the family are shown in Table VI. The

[42] F. David, A.-M. Bernard, M. Pierres, and D. Marguet, *J. Biol. Chem.* **268**, 17247 (1993).
[43] C. J. Roberts, G. Pohlig, J. H. Rothman, and T. H. Stevens, *J. Cell Biol.* **108**, 1363 (1989).
[44] D. Rennex, B. A. Hemmings, J. Hofsteenge, and S. R. Stone, *Biochemistry* **30**, 2195 (1991).
[45] A. Kanatani, T. Masuda, T. Shimoda, F. Misoka, X. S. Lin, T. Yoshimoto, and D. Tsuru, *J. Biochem. (Tokyo)* **110**, 315 (1991).
[46] K. Wada, N. Yokotani, C. Hunter, K. Doi, R. J. Wenthold, and S. Shimasaki, *Proc. Natl. Acad. Sci. U.S.A.* **89**, 197 (1992).
[47] N. Yokotani, K. Doi, R. J. Wenthold, and K. Wada, *Hum. Mol. Genet.* **2**, 1037 (1993).

family appears to be restricted to eukaryotes, and sequences are known for enzymes of protozoa, fungi, plants, and animals. These carboxypeptidases have a lysosomal distribution in animals, and in plants and fungi are found in vacuoles. The fungal and plant enzymes can be divided into variants of carboxypeptidase C (EC 3.4.16.5), with a preference for a hydrophobic amino acid in P1', and those of carboxypeptidase D (EC 3.4.16.6), which release a C-terminal arginine or lysine. However, this may represent an oversimplification of the specificities (see [18]).

The mammalian enzyme is known as "protective protein" because it complexes β-galactosidase and neuroaminidase in the lysosome, protecting them from proteolytic degradation. Its absence therefore leads to functional deficiency in these enzymes, and the lysosomal storage disease, galactosialidosis. The protein expresses the acid carboxypeptidase characteristics of other carboxypeptidase C forms, and there is little doubt that it is the enzyme originally known as cathepsin A, and subsequently as lysosomal carboxypeptidase A.[48] Like other carboxypeptidases C, it also exhibits deamidase and esterase activity.[49,50]

The sequences around the catalytic Ser/Asp/His residues are shown in Fig. 5. In all the peptidases, the residue preceding the catalytic Ser is a Glu residue that is believed to be responsible for the acidic pH optimum of the enzymes (see [18]). The differences in substrate specificity between carboxypeptidases C and D have been attributed to two residues that are Glu in carboxypeptidase D but are both hydrophobic in character (Phe, Leu, or Met) in carboxypeptidase C (see [18]).

Yeast carboxypeptidase Y (one of the C-type enzymes) is inhibited by p-hydroxymercuribenzoate, presumably because of the presence of the free thiol of Cys-341 in the S1 binding pocket.[51]

Consistent with the lysosomal/vacuolar location of the serine carboxypeptidases, they are synthesized with N-terminal signal peptides and propeptides.

The crystal structures of several members of the family have been determined (see [18]). The fold of these proteins bears no relation to those of chymotrypsin, subtilisin, or D-Ala-D-Ala peptidases. However, the tertiary structures of serine carboxypeptidases do strongly suggest a distant relationship to a variety of nonpeptidase enzymes. The sequence relation-

[48] N. J. Galjart, H. Morreau, R. Willemsen, N. Gillemans, E. J. Bonten, and A. D'azzo, *J. Biol. Chem.* **266**, 14754 (1991).

[49] H. L. Jackman, F. Tan, H. Tamei, C. Beurling-Harbury, X.-Y. Li, R. A. Skidgel, and E. G. Erdös, *J. Biol. Chem.* **265**, 11265 (1990).

[50] K. Breddam, *Carlsberg Res. Commun.* **51**, 83 (1986).

[51] J. R. Winther and K. Breddam, *Carlsberg Res. Commun.* **52**, 263 (1987).

ship between human acetylcholinesterase and human protective protein, although distant, is statistically significant (RDF score = 6.75 SD units). In turn, cholinesterases are related to a wide range of enzymes and other proteins, including lipases and thyroglobulin, and form a distinct subfamily.[52,53] The differences in sequence between the two subfamilies are too great for us to be able to estimate when the divergence occurred.

The relationship of the cholinesterases to the serine carboxypeptidases invites reconsideration of the long-standing controversy over possible endopeptidase activity of the cholinesterases. At the time of writing, however, the evidence for endopeptidase activity expressed by the cholinesterases does not appear strong.[54–56]

The nonpeptidase enzymes that show no amino acid sequence similarity to carboxypeptidase C, but are thought to be distant relatives because of their similar tertiary structures, include cholinesterases, haloalkane dehalogenases, and carboxymethylenebutenolidases (dienelactone hydrolases) in the "α/β hydrolase fold" clan.[52,53]

Lysosomal Pro-X Carboxypeptidase Family (S28)

This family has only one known member, a carboxypeptidase specific for cleavage of prolyl bonds. Like members of the carboxypeptidase C family (S10), Pro-X carboxypeptidase has an acidic pH optimum, is lysosomal, and is synthesized with a signal peptide and propeptide.[40] The suggestion has been made that lysosomal Pro-X carboxypeptidase is evolutionarily related to both carboxypeptidase C and prolyl oligopeptidase.[40] Active site residues have not been biochemically identified for lysosomal Pro-X carboxypeptidase, but have been postulated to be Ser-134, Asp-333, and His-411 on the basis of similarities to the catalytic residues of carboxypeptidase C (Fig. 5).[40] The glutamate residue preceding the active site serine in carboxypeptidase C, and believed to be responsible for the acidic pH optimum of carboxypeptidase activity (see [18]), is replaced by Gly in lysosomal Pro-X carboxypeptidase, however (Fig. 5).

[52] D. L. Ollis, E. Cheah, M. Cygler, B. Dijkstra, F. Frolow, S. M. Franken, M. Harel, S. J. Remington, I. Silman, J. Schrag, J. L. Sussman, K. H. G. Verscheuren, and A. Goldman, *Protein Eng.* **5**, 197 (1992).

[53] M. Cygler, J. D. Schrag, J. L. Sussman, M. Harel, I. Silman, M. K. Gentry, and B. P. Doctor, *Protein Sci.* **2**, 366 (1993).

[54] M. De Serres, D. Sherman, W. Chestnut, B. M. Merrill, O. H. Viveros, and E. J. Diliberto, Jr., *Cell. Mol. Neurobiol.* **13**, 279 (1993).

[55] S. Michaelson and D. H. Small, *Brain Res.* **611**, 75 (1993).

[56] R. V. Rao and A. S. Balasubramanian, *J. Protein Chem.* **12**, 103 (1993).

Lactococcus X-Pro-Peptidase Family (S15)

Lactococcus lactis contains an enzyme known as "X-prolyl dipeptidyl aminopeptidase" with specificity similar to that of the mammalian dipeptidyl-peptidase IV, cleaving Xaa-Pro↓peptide bonds to release N-terminal dipeptides. An indication of relationship of the *Lactococcus* enzyme to others in families S9, S10, or S28 is similarity of sequence in the vicinity of the catalytic Ser residue (Fig. 5).[57] Like members of family S9, *Lactococcus* X-Pro-peptidase does not have a proenzyme,[58] and like dipeptidyl-peptidase IV specifically, the *Lactococcus* enzyme exists as a homodimer.[59] Proposals regarding the catalytic Asp and His residues are made in [15].

Neisseria Prolyl Aminopeptidase Family (S33)

Prolyl aminopeptidase is a 35 kDa peptidase from *Neisseria gonorrhoea* that selectively hydrolyses N-terminal Pro residues.[60] The amino acid sequence shows no relationship to that known for any other peptidase, but is homologous to those of *Pseudomonas* 2-hydroxymuconic semialdehyde hydrolase (XYLF_PSEPU), *Xanthobacter* haloalkane dehalogenase (HALO_XANAU), and human and rat epoxide hydrolase (HYEP_*). These enzymes, in turn, are known to be structurally related to acetylcholinesterase (ACES_*) and carboxypeptidase C (family S8). Most of the "α/β hydrolase fold" enzymes have Ser at the active site, and prolyl aminopeptidase also has Ser at this position (Fig. 5).

Serine-Type D-Ala-D-Ala Peptidases, Families S11, S12, and S13, Forming Clan SE

Both gram-positive and gram-negative bacterial cell walls are complex polymers of amino sugars and amino acids. Chains of alternating *N*-acetylglucosamine and *N*-acetylmuramic acid units are linked to one another by short peptides. In *E. coli* the structure of the link peptide is L-alanyl-D-isoglutamyl-L-*meso*-diaminopimelyl-D-alanine, but the nature of the third amino acid in the chain varies with the bacterial species. These chains are cross-linked, usually between the carboxyl group of D-alanine and the

---

[57] J.-F. Chich, M.-P. Chapot-Chartier, B. Ribadeau-Dumas, and J.-C. Gripon, *FEBS Lett.* **314,** 139 (1992).

[58] B. Mayo, J. Kok, K. Venema, W. Bockelmann, M. Teuber, H. Reinke, and G. Venema, *Appl. Environ. Microbiol.* **57,** 38 (1991).

[59] M. Nardi, M.-C. Chopin, A. Chopin, M.-M. Cals, and J.-C. Gripon, *Appl. Environ. Microbiol.* **57,** 45 (1991).

[60] N. H. Albertson and M. Koomey, *Mol. Microbiol.* **9,** 1203 (1993).

TABLE VII
PEPTIDASES OF FAMILIES OF D-Ala-D-Ala PEPTIDASES (CLAN SE)[a]

Peptidase	EC	Database code
Family S11: *Escherichia* D-Ala-D-Ala peptidase A		
Escherichia D-Ala-D-Ala peptidase A	3.4.16.4	DACA_ECOLI
D-Ala-D-Ala peptidase	3.4.16.4	DACA_*, DACC_ECOLI, (M37688), (M85047), (X68587)
Sporulation-specific penicillin-binding protein (*Bacillus*)	-	(M84227)
Family S12: *Streptomyces* R61 D-Ala-D-Ala peptidase		
Streptomyces R61 D-Ala-D-Ala peptidase	3.4.16.4	DAC_STRSP
D-Aminopeptidase (*Ochrobactrum*)	-	(M84523)
Family S13: *Actinomadura* R39 D-Ala-D-Ala peptidase		
Actinomadura R39 D-Ala-D-Ala peptidase	3.4.16.4	(X64790)
Penicillin-binding protein 4	3.4.16.4	PBP4_ECOLI

[a] See Table III for explanation.

free amino group of diaminopimelate. In the biosynthesis of the cell wall peptidoglycan the precursor has the four-residue structure above, but with an additional C-terminal D-alanine residue. The D-Ala-D-Ala transpeptidases and carboxypeptidases are involved in the metabolism of the cell wall components.[61] Some of these peptidases are serine enzymes, whereas others have zinc at the catalytic center.[62]

D-Ala-D-Ala peptidases are synthesized with leader peptides to target them to the cell membrane. The peptidases are retained in the membrane by a C-terminal membrane anchor, after removal of the leader peptide, except in *Streptomyces* K15 transpeptidase (family S11) and *E. coli* penicillin-binding protein 4 (family S13), in which a C-terminal propeptide is removed and where overexpression can lead to secretion of the enzymes into the medium.[62]

The members of the three families of serine-type D-Ala-D-Ala peptidases are listed in Table VII. These enzymes are known as low molecular mass penicillin-binding proteins, and probably are distantly related to the high molecular mass penicillin-binding proteins (at the clan level).

[61] J. T. Park, in "*Escherichia coli* and *Salmonella typhimurium*. Cellular and Molecular Biology" (F. C. Neidhart, ed.), Vol. 1, p. 663. American Society for Microbiology, Washington, D.C., 1987.

[62] J.-M. Ghuysen, *Annu. Rev. Microbiol.* **45**, 37 (1991).

The antibiotic action of penicillin is due to its binding to the high molecular mass penicillin-binding proteins and not the D-Ala-D-Ala peptidases (see [19]). Enzymes that are capable of degrading pencillins and related antibiotics are β-lactamases, among which the class C β-lactamases are homologous to D-Ala-D-Ala peptidases (S12), and the class A β-lactamases are more distantly related, as is revealed by the three-dimensional structures.[63] Some of the D-Ala-D-Ala peptidases perform a transpeptidation reaction in which the peptidoglycan monomer minus the C-terminal D-Ala is transferred to an exogenous acceptor; the *Streptomyces* K15 peptidase (EMBL : X59965) performs this reaction exclusively, and does not hydrolyse peptide bonds.[62]

Three-dimensional structures are known for *Streptomyces* R61 D-Ala-D-Ala peptidase[64] and*Citrobacter* class C β-lactamase[63] (both from family S12), and class A β-lactamases from *Streptomyces*[65] and *E. coli*[66] (not members of the peptidase families, but members of the clan). The catalytic mechanism elucidated for the class A β-lactamase of *E. coli* involves Ser-70, Lys-73, Ser-130, and Glu-166, Ser-70 acting as the nucleophile, and Glu-166 and Lys-73 being general bases (Fig. 6). The closely spaced Ser-70 and Lys-73 are completely conserved throughout the clan.

Family S11 contains only D-Ala-D-Ala peptidases and the strict transpeptidase of *Streptomyces* K15. Some of the enzymes in this family are partially inhibited by thiol-blocking reagents such as *p*-chloromercuribenzoate, and in the *Streptomyces* K15 peptidase there are two cysteine residues, one of which is close to the general base.[67]

Family S12 contains enzymes with the most diverse specificities. In addition to the *Streptomyces* R61 D-Ala-D-Ala peptidase there are class C β-lactamases (AMPC_*), a D-aminopeptidase from *Ochrobactrum,* a lipolytic esterase from *Pseudomonas* (EMBL : M68491), and proteins from *Bacteroides nodosus* (FMDH_BACNO, FMDD_BACNO). The *Ochrobactrum* enzyme is one of the rare aminopeptidases that are not metalloenzymes, and the only aminopeptidase known to be specific for D-amino acids.[68] The proteins from *B. nododus* may be involved in the assembly

[63] C. Oefner, A. D'Arcy, J. J. Daly, K. Gubernator, R. L. Charnas, I. Heinze, C. Hubschwerlen, and F. K. Winkler, *Nature (London)* **343**, 284 (1990).
[64] J. A. Kelly, J. R. Knox, P. C. Moews, G. J. Hite, J. B. Bartolone, H. Zhao, B. Joris, J.-M. Frère, and J.-M. Ghuysen, *J. Biol. Chem.* **260**, 6449 (1985).
[65] J. Lamotte-Brasseur, F. Jacob-Dubuisson, G. Dive, J.-M. Frère, and J.-M. Ghuysen, *Biochem. J.* **282**, 189 (1992).
[66] N. C. J. Strynadka, H. Adachi, S. E. Jensen, K. Johns, A. Sielecki, C. Betzel, K. Sutoh, and M. N. G. James, *Nature (London)* **359**, 700 (1992).
[67] P. Palomeque-Messia, S. Englebert, M. Leyh-Bouille, M. Nguyen-Distèche, C. Duez, S. Houba, O. Dideberg, J. Van Beeumen, and J.-M. Ghuysen, *Biochem. J.* **279**, 223 (1991).
[68] Y. Asano, Y. Kato, A. Yamada, and K. Kondo, *Biochemistry* **31**, 2316 (1992).

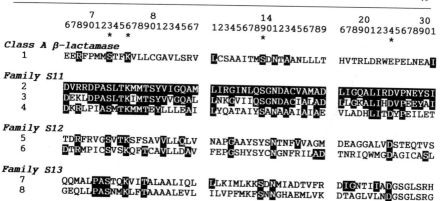

FIG. 6. Comparison of sequences in the vicinity of the catalytic residues of D-Ala-D-Ala carboxypeptidases from families S11, S12, and S13 (clan SE). Residues are numbered according to *Escherichia coli* D-Ala-D-Ala carboxypeptidase A. Residues identical to *E. coli* D-Ala-D-Ala carboxypeptidase A are shown in white on black. Key to sequences: 1, *Escherichia coli* class A β-lactamase; 2, *E. coli* D-Ala-D-Ala carboxypeptidase A; 3, *E. coli* D-Ala-D-Ala carboxypeptidase C; 4, *Bacillus subtilis* D-Ala-D-Ala carboxypeptidase A; 5, *Streptomyces* D-Ala-D-Ala carboxypeptidase; 6, *Ochrobactrum anthropi* D-aminopeptidase; 7, *E. coli* penicillin-binding protein 4; 8, *Actinomadura* R39 D-Ala-D-Ala carboxypeptidase. Asterisks indicate catalytic residues identified in *E. coli* class A β-lactamase.

of fimbriae, which are external appendages found on the surface of gram-negative bacteria and used for cell adhesion. The fimbrial subunits that are related to D-Ala-D-Ala peptidases are found only in some strains, and seem to represent a recent lateral transfer of genes. The function of these proteins is unknown, but the active site motif Ser-Xaa-Xaa-Lys is retained.

Family S13 contains D-Ala-D-Ala peptidases from *Actinomadura* and *E. coli*. The *E. coli* enzyme is also known as penicillin-binding protein 4, and has been reported to have endopeptidase as well as carboxypeptidase activity.[69] Although no crystal structures are available for this family, conservation of sequences around the catalytic residues suggests similar tertiary structures and very distant evolutionary relationship to the other members of the proposed clan.

Possible Ser/Lys Dyad Clan (SF) Comprising Families S24, S26, and S27

Although the majority of serine peptidases contain catalytic triads of serine, histidine, and aspartic acid, the D-Ala-D-Ala peptidases show that this is not the only way in which a serine-dependent catalytic site can be

[69] B. Korat, H. Mottl, and W. Keck, *Mol. Microbiol.* **5**, 675 (1991).

formed (Table II). A further example of a novel type of catalytic site is provided by the families of repressor LexA (S24), bacterial leader peptidase 1 (S26), and eukaryote signal peptidase (S27), which seem to comprise a clan (SF) of serine peptidases that use a Ser/base catalytic dyad.

More detailed descriptions of these families follow, but at this stage it may be useful to draw attention to the common features of the enzymes of the proposed clan. As shown in Fig. 7, there are similarities in the sequences around the Ser and Lys residues that are known to have catalytic activity in families S24 and S26, and the Ser and His residues that may be equivalent in the eukaryotic signal peptidases (S27).

The properties of the catalytic serine residues in clan SF are dissimilar from those of serine residues in the "catalytic triad" peptidases of clans SA, SB, and SC. Thus, they show little or no reactivity with diisopropyl fluorophosphate (DFP) and other reagents commonly used for serine peptidases, and the serine can be replaced by cysteine with partial retention of activity (see [20] and [21]).

Finally, the peptidases in all three families show a preference for cleavage of alanyl bonds. Thus Ala is the preferred P1 residue for both

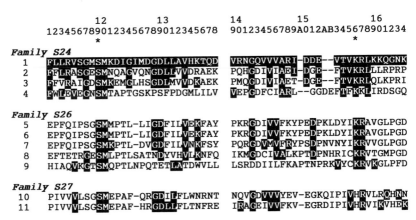

FIG. 7. Conserved residues in the vicinity of the catalytic residues in the families of repressor LexA (S24), leader peptidase (S26), and eukaryote signal peptidase (S27) forming clan SF. Residues are numbered according to *Escherichia coli* repressor LexA, and those identical in *E. coli* repressor LexA are shown in white on black. Asterisks indicate catalytic residues that have been identified in families S24 and S26, but are only presumed for family S27. Key to sequences: 1, *E. coli* repressor LexA; 2, *Salmonella typhimurium* ImpA protein; 3, *S. typhimurium* SamA protein; 4, bacteriophage lambda (λ) cl repressor; 5, *E. coli* leader peptidase 1; 6, *S. typhimurium* leader peptidase 1; 7, *Pseudomonas fluorescens* leader peptidase 1; 8, yeast mitochondrial inner membrane leader peptidase; 9, yeast microsomal signal peptidase 18-kDa subunit; 10, dog microsomal signal peptidase 21-kDa subunit; 11, dog microsomal signal peptidase 18-kDa subunit.

prokaryotic and eukaryotic leader/signal peptidases,[70] and is highly conserved in the cleavage sites of the LexA family (see [20]). However, by no means all alanyl bonds are cleaved, and the specificity sites are apparently extended ones in which residues other than P1 are of great importance. No cleavage of simple substrates such as alanyl nitroanilides has been reported, and for leader peptidase it was found that a pentapeptide was the smallest substrate detectably cleaved.[71] Mutations several residues away on either side of the cleavage site can greatly affect the rate of autolytic cleavage of LexA (see [20]), again consistent with an extended substrate-binding site.

Repressor LexA Family (S24)

The protein known as LexA represses about 20 genes of the SOS regulon that are involved in DNA repair in *E. coli*. After treatments of the bacterium that damage DNA or inhibit DNA replication, LexA is inactivated, leading to derepression of the genes for DNA repair (see [20]). LexA is known from other bacteria too, but more surprisingly, homologous repressors also occur in lambdoid phages[72] (Table VIII). Cleavage of the phage repressors results in phage induction.

In *E. coli*, the derepression of the DNA repair system is effected by the RecA protein, which acquires "coprotease" activity in the presence of single-stranded DNA, causing it to undergo an interaction with LexA that results in inactivation of LexA by proteolytic cleavage. The bond cleaved by the peptidase activity of LexA is the Ala^{84}+Gly bond, which disrupts the DNA-binding part of the molecule, inactivating it. At one time, it was thought that the peptidase activity that causes the derepression was mediated by a catalytic site on the RecA protein, but it is now clear that this is an autolytic reaction, in which the catalytic activity is that of LexA (see [20]). Isolated LexA autolyses slowly when incubated *in vitro* at alkaline pH, but the reaction triggered by RecA is much more rapid.

The molecules of LexA and related repressors consist of about 200 amino acid residues, of which the N-terminal 90 or so form the DNA-binding domain, for which crystallographic structures are available.[73] The C-terminal part of the molecules contains the residues that form the catalytic site for autolytic peptide bond cleavage. Of the amino acid residues

[70] G. von Heijne, *Nucleic Acids Res.* **14,** 4683 (1986).

[71] I. K. Dev, P. H. Ray, and P. Novak, *J. Biol. Chem.* **265,** 20069 (1990).

[72] R. T. Sauer, R. R. Yocum, R. F. Doolittle, M. Lewis, and C. O. Pabo, *Nature (London)* **298,** 447 (1982).

[73] D. H. Ohlendorf, W. F. Anderson, M. Lewis, C. O. Pabo, and B. W. Matthews, *J. Mol. Biol.* **169,** 757 (1983).

TABLE VIII
PEPTIDASES OF FAMILIES OF REPRESSOR LexA (S24), BACTERIAL LEADER
PEPTIDASE 1 (S26), AND EUKARYOTE SIGNAL PEPTIDASE (S27)[a]

Peptidase	EC	Database code
Family S24: LexA repressor		
Bacteriophage repressor protein (includes C1 and C2)	-	RPC1_LAMBD, RPC2_BPP22
ImpA protein (*Salmonella*)	-	IMPA_SALTY
LexA repressor	-	LEXA_*
MucA protein (*Salmonella*)	-	MUCA_SALTY
SamA protein (*Salmonella*)	-	SAMA_SALTY
SOS regulatory protein DinR (*Bacillus*)	-	DINR_BACSU
Umud protein (*Salmonella*)	-	UMUD_*
Family S26: Bacterial leader peptidase 1		
Bacterial leader peptidase 1	3.4.99.36	LEP_*, (X75604),(Z27458)
Mitochondrial inner membrane peptidase subunit 1	-	IMP1_YEAST
Mitochondrial inner membrane peptidase subunit 2	-	[b]
Family S27: Eukaryote signal peptidase		
Eukaryote signal peptidase	-	SC11_YEAST, SPC3_CANFA, SPC4_CANFA, (L11319)

[a] See Table III for explanation.
[b] J. Nunnari, T. D. Fox, and P. Walter, *Science* **262**, 1997 (1993).

that are conserved throughout the family, only Ser-119 (*E. coli* LexA numbering) and Lys-156 could not be replaced by Ala with retention of activity.[74] LexA had no significant reactivity with 1 mM DFP, but later work showed that Ser-90 is detectable labeled with 20 mM [^{3}H]DFP.[75]

The autolytic cleavage of LexA is an intramolecular reaction, and as such is difficult to study enzymologically, because the concentrations of substrate and enzyme cannot be altered independently. This difficulty has been overcome by the development of a system in which the N-terminal, substratelike part of the molecule and the C-terminal, catalytic domain are separated (see [20]).

Various other proteins are related to repressor LexA. The DinR protein from *Bacillus subtilis* is a repressor of the *dinR* and *dinC* genes and, similarly to the LexA protein, is inactivated during the SOS response on association with the RecA protein.[76] Other protein homologs lack the N-terminal DNA-binding region of the LexA protein, and are therefore

[74] S. N. Slilaty and J. W. Little, *Proc. Natl. Acad. Sci. U.S.A.* **84**, 3987 (1987).
[75] K. L. Roland and J. W. Little, *J. Biol. Chem.* **265**, 12828 (1990).
[76] W. H. Koch, D. G. Ennis, A. S. Levine, and R. Woodgate, *Mol. Gen. Genet.* **233**, 443 (1992).

not repressors. The UmuD, MucA, ImpA, and SamA proteins are involved in mutagenesis following ultraviolet light irradiation, and their cleavage is a prerequisite for mutagenesis.[76,77]

Bacterial Leader Peptidase 1 Family (S26)

Eubacteria contain at least three leader peptidases. Murein prelipoprotein peptidase (leader peptidase 2), which removes the leader peptide from one of the components of the bacterial outer membrane, may be an aspartic endopeptidase.[78] Type IV prepilin leader peptidase is a cysteine peptidase (see [32]). The serine-type leader peptidase has the more general function of removing the leader peptides from the other secreted proteins and proteins that are targeted to the periplasm and periplasmic membrane. It is thus most closely analogous to the eukaryote signal peptidase, but differs in being monomeric. The catalytic residues of E. coli leader peptidase 1 have been identified as a Ser/Lys dyad (Fig. 7) (see [21]).

Mitochondria possess three metallopeptidases that remove N-terminal targeting peptides, the mitochondrial processing peptidase (family M16) and the mitochondrial intermediate peptidase (M3), but the inner membrane leader peptidase is a serine enzyme. In fact, the enzyme is composed of two subunits, both homologous with the bacterial leader peptidase 1. The enzyme is responsible for removing leader peptides from both nuclear- and mitochondrial-encoded proteins destined for the inner mitochondrial space, and it is believed that both subunits are active, with distinct specificities.[79]

Eukaryote Signal Peptidase Family (S27)

The eukaryote microsomal signal peptidase complex, responsible for removing signal peptides from secretory proteins as they are transported into the lumen of the endoplasmic reticulum, is much more complex than its eubacterial and mitochondrial counterparts (see [22]). Varying numbers of subunits have so far been isolated; in yeast there are four, in chicken oviduct there are two, and in dog pancreas there are five. In each species, one of the subunits is glycosylated, but as far as is known, the glycosylated subunits are not directly involved in the peptidase activity. Of the other subunits so far sequenced, one from yeast, one from rat, and two from dog are homologous, and form family S27 (Table VIII). These subunits

[77] J. Hauser, A. S. Levine, D. G. Ennis, K. M. Chumakov, and R. Woodgate, J. Bacteriol. 174, 6844 (1992).
[78] K. Sankaran and H. C. Wu, this series, Vol. 248, Chapter 12.
[79] J. Nunnari, T. D. Fox, and P. Walter, Science 262, 1997 (1993).

(about 20 kDa) contain sequences reminiscent of those in which the catalytic residues of bacterial leader peptidase 1 are located (Fig. 7). The alignment suggests that the eukaryote signal peptidases depend on a Ser/His dyad for activity, in contrast to the Ser/Lys dyad of the eubacterial leader peptidases 1. The existence of two subunits in dog that may have catalytic activity is analogous to the situation in yeast mitochondrial inner membrane leader peptidase (family S26).

Indications of a Clan (SG) of ATP-Dependent Endopeptidases Embracing Families S14, S16, and S25

Although the serine endopeptidase components of the Clp endopeptidase (family S14), the multicatalytic endopeptidase complex (MEC; family S25), and endopeptidase La (S16) do not show similarities in sequence, all the peptidase domains are associated with ATP-binding domains that are homologous to each other, and in each case the interaction couples the hydrolysis of peptide bonds to that of ATP. In view of these similarities, it is natural to look closely for any indication that the proteins of these three families may have a single origin.

Clp and the subunits of MEC associate in approximately hexameric rings. These further associate with ATPase subunits, to produce complexes that appear to have similar topology in the electron microscope.[80] Immunological cross-reactivity between *E. coli* ClpP and subunits of MEC has been reported by two groups.[80,81] In addition, marked similarities in the chymotrypsin-like activities of Clp and MEC led Arribas and Castaño[80] to suggest that ClpP might be evolutionarily related to subunits expressing chymotrypsin-like activity in MEC.

In endopeptidase La the endopeptidase and ATPase components are parts of the single polypeptide chain, but there are other groups of enzymes in which catalytic and regulatory domains are fused in some members, and comprise separate subunits in others. Goldberg[82] has given a detailed account of the similarities and differences between endopeptidases La and Clp. In reviewing these indications of possible relationship between the three types of ATP-dependent endopeptidase, Rechsteiner *et al.*[83] provide an imaginative three-dimensional model to depict the way in which the structures may have evolved. However, there are no similarities around the catalytic residues in these enzymes. Moreover, though the

---

[80] J. Arribas and J. G. Castaño, *J. Biol. Chem.* **268,** 21165 (1993).

[81] K. Tanaka, T. Tamura, A. Kumatori, T. H. Kwak, C. H. Chung, and A. Ichihara, *Biochem. Biophys, Res. Commun.* **164,** 1253 (1989).

[82] A. L. Goldberg, *Eur. J. Biochem.* **203,** 9 (1992).

[83] M. Rechsteiner, L. Hoffman, and W. Dubiel, *J. Biol. Chem.* **268,** 6065 (1993).

TABLE IX
PEPTIDASES OF FAMILIES OF ClpP ENDOPEPTIDASE SUBUNIT (S14), ENDOPEPTIDASE
La (S16), AND MULTICATALYTIC ENDOPEPTIDASE COMPLEX (S25)[a]

Peptidase	EC	Database code
Family S14: ClpP endopeptidase subunit		
ATP-dependent endopeptidase (ClpP subunit)		
(*E. coli*)	-	CLPP_ECOLI, (L07793)
Chloroplast ATP-dependent endopeptidase	-	CLPP_MARPO, CLPP_TOBAC,
		CLPP_ORYSA, CLPP_WHEAT,
		CLPP_EPIVI
Family S16: Endopeptidase La		
Endopeptidase La	3.4.21.53	LON_*, (D12923), (D13204)
Endopeptidase La-like protein	-	(X74215), (X76040)
Family S25: Multicatalytic endopeptidase complex		
Multicatalytic endopeptidase complex subunits	3.4.99.46	PRC1_*, PRC2_*, PRC3_*,
		PRC4_*, PRC5_*, PRC7_*,
		PRC8_*, PRC9_*, PRCA_*,
		PRCB_*, PRCC_*, PRCD_*,
		PRCU_YEAST, PRCX_YEAST,
		PRCZ_YEAST, PR28_DROME,
		PR29_DROME, PR35_DROME,
		(D10754), (D10755), (D10757),
		(D10758), (D21799), (L17127),
		(L22212), (L22213), (M63641),
		(M64992), (U00790), (X57210),
		(X70304)
SCL1 suppressor protein	-	SCL1_YEAST

[a] See Table III for explanation.

order of catalytic residues is Ser/His in ClpP, this cannot be the case in endopeptidase La.

ClpP Endopeptidase Subunit Family (S14)

The members of families S14, S16, and S25 are listed in Table IX. Clp endopeptidase is a bacterial, ATP-dependent endopeptidase (see [23]). Historically, the name "clp" arose as an acronym for "caseinolytic protease"; the enzyme has also been termed endopeptidase Ti.[82] The Clp endopeptidase exists as a complex of two types of subunits (ClpP and ClpA), of composition $ClpP_{12}ClpA_6$. The ClpP subunit is a heat-shock protein synthesized under the control of σ^{32} factor.[84] Alone, ClpP has peptidase activity that is restricted to peptides of five residues or

[84] H. E. Kroh and L. D. Simon, *J. Bacteriol.* **172**, 6026 (1990).

less, so it could be described as an oligopeptidase, but full endopeptidase activity is expressed by the complex with ClpA (or ClpB or ClpX; see [23]). The additional subunits are homologous to the ATP-binding domain of the monomeric endopeptidase La, and will be considered below. The active site serine (Ser-111) and histidine (His-136) residues of ClpP have been identified for the *E. coli* enzyme, but no third member of a possible catalytic triad.

Proteins with sequences homologous to ClpP have been detected in chloroplast inner envelope membrane and soluble fraction,[85] and are encoded by the chloroplast genome.[86] The chloroplast protein has not been characterized, but ATP-dependent proteolysis has been detected in these organelles.[87] The chloroplast ClpP-like sequences have the active site serine and histidine residues conserved, but as yet no chloroplast ClpA component has been sequenced. We report here that the gene sequence of phosphoproteins of photosystem II from the chloroplast of the evening primrose (*Oenothera argillicola*)[88] includes a fragment of a homolog of ClpP at the 3' end of the complementary strand.

A surprising homolog of ClpP is the 5' untranslated region of potato leaf-roll luteovirus (EMBL : D00530). This virus is a single-stranded RNA virus most closely related to the picornaviruses. The genomic RNA sequence[89] may be truncated at the 5' end. The ClpP homolog retains a residue equivalent to the active site serine, but not the active site histidine, and so is almost certainly inactive. Presumably, the viral sequence was acquired by horizontal genetic transfer from the potato host.

It has been suggested[83] that ClpP is homologous to ubiquitin-peptidase 2 from yeast, a cysteine peptidase (see [32]). In the comparison with the *E. coli* ClpP, 19.6% identity can be detected over a region of 189 amino acids, but there is no statistically significant relationship (RDF score = 4.90). Moreover, the similarity is seen only with the *E. coli* ClpP and not with the chloroplast homologs, and the active site Ser of ClpP is replaced by Asp in the ubiquitin peptidase.

Endopeptidase La Family (S16)

Unlike the Clp endopeptidase, endopeptidase La (product of the *lon* gene of *E. coli*) is a single-chain mosaic protein, containing an ATP-binding domain and a peptidase domain (see [25]). The ATP-binding domain is

[85] T. Moore and K. Keegstra, *Plant Mol. Biol.* **21**, 525 (1993).
[86] J. C. Gray, S. M. Hird, and T. A. Dyer, *Plant Mol. Biol.* **15**, 947 (1990).
[87] X.-Q. Liu and A. T. Jagendorf, *FEBS Lett.* **166**, 248 (1984).
[88] K. Offermann-Steinhard and R. G. Herrmann, *Nucleic Acids Res.* **18**, 6452 (1990).
[89] M. A. Mayo, D. J. Robinson, C. A. Jolly, and L. Hyman, *J. Gen. Virol.* **70**, 1037 (1989).

homologous to the ClpA and ClpB subunits of the Clp endopeptidase, and a number of other ATP-binding proteins such as bacterial NIF-specific and XYLR regulatory proteins (NIFA_KLEPN, XYLR_PSEPU).

To date, only the catalytic serine residue (Ser-679) has been identified.[90] The sequences contain no conserved His C-terminal to this.

Homologs of endopeptidase La have been discovered in the mitochondria of eukaryotes (see [26]). These are from yeast and human cells.

Multicatalytic Endopeptidase Complex Family (S25)

The multicatalytic endopeptidase complex (MEC), also called the proteasome, is a 700-kDa endopeptidase containing about 24–28 similarly sized, homologous subunits. The complex occurs in the cytoplasm and nuclei of eukaryotic cells generally, as well as those of the archaebacterium *Thermoplasma acidophylum* (see [24]). In the electron microscope, MEC is seen as a cylinder, formed as a stack of four rings, each of about seven subunits.

In the archaebacterium, MEC is composed of subunits of only two kinds, α and β.[91] In eukaryotes, there are far more different subunits, but all are variants of the α or β type of structure, and are divided into A and B groups on this basis.

Three types of peptidase activity are simply recognized by use of synthetic substrates, and these can be described as trypsin-like (although cleavage is predominantly of Arg+ bonds), chymotrypsin-like (cleaving Leu+, Tyr+, and Phe+ bonds), and glutamyl peptidase (cleaving Glu+ bonds). The *Thermoplasma* enzyme shows only the chymotrypsin-like activity,[92] but the eukaryotic endopeptidase complex shows all three, and indeed at least five separate catalytic sites are distinguishable in careful work (see [24]). Whether the endopeptidase activity against β-casein is attributable to any of the sites active on synthetic substrates is not clear.[93] It has yet to be established which subunits, and which catalytic residues, are responsible for the various activities, but most of these are inhibited by 3,4-dichloroisocoumarin, and are probably those of atypical serine peptidase catalytic sites.

An alignment of the available sequences made with the PILEUP pro-

---

[90] A. Y. Amerik, V. K. Antonov, A. E. Gorbalenya, S. A. Kotova, T. V. Rotanova, and E. V. Shimbarevich, *FEBS Lett.* **287**, 211 (1991).

[91] P. Zwickl, A. Grziwa, G. Pühler, B. Dahlmann, F. Lottspeich, and W. Baumeister, *Biochemistry* **31**, 964 (1992).

[92] B. Dahlmann, L. Kuehn, A. Grziwa, P. Zwickl, and W. Baumeister, *Eur. J. Biochem.* **208**, 789 (1992).

[93] M. E. Pereira, T. Nguyen, B. J. Wagner, J. W. Margolis, B. Yu, and S. Wilk, *J. Biol. Chem.* **267**, 7949 (1992).

gram (with the standard defaults) shows very few residues completely conserved. The possibility must be borne in mind that the minimal catalytic unit comprises both an A and a B type of subunit, with each contributing some of the active site residues. Among the few residues that are completely conserved in the sequences (both A and B type) are Gly-80, Asp-84, and Gly-166 (numbered according to the *Thermoplasma* α subunit). The potential active site residues completely conserved in the A subunits are Ser-16, Tyr-26, Asp-84, and perhaps Arg-130 (although Arg has not yet been proved to be a catalytic residue in any peptidase). In the B subunit, Asp-84 is the only completely conserved potential catalytic residue.

There is some evidence for proteolytic processing in the rat B subunits, several of which have N-terminal threonine residues, implying a Gly+Thr cleavage at a site near the N terminus that is conserved in the great majority of B subunits.[94]

Despite some contrary reports,[95,96] the 26S ATP-dependent, ubiquitin conjugate-degrading enzyme complex is now generally believed to contain components of MEC as its catalytic unit. Additional components then confer the ATP dependence and the specificity for ubiquitin conjugates.[83]

The one additional subunit of the 26S complex (EMBL : L02426) that has been sequenced includes putative ATP-binding domains that show significant similarities to those in ClpA (RDF score 6.93) and endopeptidase La (RDF score 5.99).[97] The subunit also shows a relationship to human Tat-binding protein 1 (TBP1_HUMAN), yeast cell division control protein 48 (CD48_YEAST), and a yeast peroxisome biosynthesis protein (PAS1_YEAST), among others.

Other Serine Peptidase Families

The members of families S18, S19, S21, S17, and S23 are listed in Table X.

Omptin Family (S18)

The *E. coli* outer membrane endopeptidase omptin (the product of the *ompT* gene), previously known as protease VII, cleaves between paired basic amino acid residues (see [27]). Omptin is inhibited by serine peptidase inhibitors such as DFP, phenylmethylsulfonyl fluoride (PMSF), and

---

[94] K. S. Lilley, M. D. Davison, and A. J. Rivett, *FEBS Lett.* **262,** 327 (1990).
[95] A. Seelig, P.-M. Kloetzel, L. Kuehn, and B. Dahlmann, *Biochem. J.* **280,** 225 (1991).
[96] L. Kuehn, B. Dahlmann, and H. Reinauer, *Arch. Biochem. Biophys.* **295,** 55 (1992).
[97] W. Dubiel, K. Ferrell, G. Pratt, and M. Rechsteiner, *J. Biol. Chem.* **267,** 22699 (1992).

TABLE X
MEMBERS OF OTHER FAMILIES OF SERINE PEPTIDASES[a]

Peptidase	EC	Database code
Family S18: Omptin		
Omptin (*Escherichia coli*)	3.4.21.87	OMPT_ECOLI
Outer membrane protease (*Escherichia*)	-	(X74278)
Coagulase/fibrinolysin (*Yersinia*)	-	COLY_YERPE
Phosphoglycerate transport system activator		
(*Salmonella*)	-	PGTE_SALTY
Family S19: *Coccidioides* endopeptidase		
Chymotrypsin-like protease (*Coccidioides*)	-	(X63114)
Family S21: Assemblin, herpesvirus		
Assemblin	-	VP40_*, (M64627)
Family S17: *Bacteroides* endopeptidase		
Extracellular endopeptidase (*Bacteroides*)	-	PRTE_BACNO
Family S23: *Escherichia coli* protease I		
Protease I	-	TESA_ECOLI

[a] See Table III for explanation.

Tos-Phe-CH$_2$Cl, but to date the active site residues have not been identified.[98]

There are two homologs of omptin from other bacterial species. The E protein from *Salmonella typhimurium* is functionally analogous in that it is also located in the outer membrane and is capable of cleaving substrates similar to those cleaved by omptin. However, whereas omptin has a typical signal peptide for transport to the outer membrane, the N terminus of the protein E precursor is longer, with little similarity.[99]

A second homolog is known from the plague organism, *Yersinia pestis,* and is believed to have a role in the transmission of the disease. The enzyme expresses both coagulant and fibrinolytic activities, according to the temperature.[100] The coagulant activity causes the blood meal to clot in the gut of the flea, which responds to ejecting the gut contents back into the host bloodstream together with the plague organism. The coagulant activity does not seem to be mediated by prothrombin activation, so the enzyme may act directly on fibrinogen. Fibrinolytic activity, in contrast, *is* brought about by activation of host plasminogen, producing the

[98] K. Sugimura and T. Nishihara, *J. Bacteriol.* **170,** 5625 (1988).
[99] J. Grodberg and J. J. Dunn, *J. Bacteriol.* **171,** 2903 (1989).
[100] K. A. McDonough and S. Falkow, *Mol. Microbiol.* **3,** 767 (1989).

same N terminus as host plasminogen activator, cleaving an Arg+Val bond. The *Yersinia* enzyme also seems to be capable of preventing an inflammatory reaction by destroying complement component C3, a component important in the production of the chemoattractant C5a fragment.[101]

Coccidioides Endopeptidase Family (S19)

Coccidioides imitis is a soil fungus, and the causative agent of coccidioidomycosis. Infection results from the inhalation of air-borne spores. The endopeptidase, described as chymotrypsin-like, is associated with the cell wall during active growth of the pathogenic cells.[102] The peptidase has been shown to be collagenolytic, elastinolytic, and able to cleave human IgG and IgA. Activity is inhibited by Tos-Phe-CH$_2$Cl, chymostatin, and α_1-proteinase inhibitor.[103] The sequence shows no relation to any other protein, and the active site residues have not been determined.

Assemblin Family (S21)

Cytomegaloviruses, which include the herpes simplex and Epstein–Barr viruses, are double-stranded DNA viruses. The virus particles are assembled in a process similar to that of bacteriophages, in which the proteins of the head are built on a scaffold called the assembly protein. The prohead so formed becomes the head of the mature virus once the assembly protein has been degraded and the empty space filled with DNA.

Degradation of the assembly protein is an autolytic event. The assembly protein has the proteinase, assemblin, as its C-terminal domain. Assemblin is inhibited by serine peptidase inhibitors such as DFP and PMSF, and cleaves Ala+Ser or Ala+Val bonds to release a polypeptide from the C terminus of the assembly protein. Assemblin may also release itself from its precursor, which would entail cleavage at an Ala+Ser bond, and exceptionally an Ala+Asn bond in infectious laryngotracheitis virus.

In the assemblin of simian cytomegalovirus, catalytic residues have been identified by site-directed mutagenesis as His-47 and Ser-118.[104]

Bacteroides Endopeptidase Family (S17)

Bacteroides nodosus is the causal agent of ovine footrot, in which the hoof separates from the underlying epidermis. The presence of an

---

[101] O. A. Sodeinde, Y. V. B. K. Subrahmanyam, K. Stark, T. Quan, Y. Bao, and J. D. Goguen, *Science* **258**, 1004 (1992).

[102] G. T. Cole, S. Zhu, L. Hsu, D. Kruse, K. R. Seshan, and F. Wang, *Infect. Immun.* **60**, 416 (1991).

[103] L. Yuan and G. T. Cole, *Infect. Immun.* **55**, 1970 (1987).

[104] A. R. Welch, L. M. McNally, M. R. T. Hall, and W. Gibson, *J. Virol.* **67**, 7360 (1993).

endopeptidase with chymotrypsin-like specificity able to digest elastin, keratin, fibrinogen, and collagen implies that the disease is a proteolytic process. The enzyme has been identified as a divalent cation-dependent serine endopeptidase, but the sequence reported by Moses et al.[105] shows no resemblance to any other protein. A subtilisin homolog from this bacterium has been sequenced,[106] raising the possibility that the protein identified by immunological screening of an expression library by Moses et al. may not be the peptidase after all.

Escherichia coli Protease I Family (S23)

This is a one-member family containing a periplasmic enzyme from E. coli that cleaves Tyr+ and Phe+ bonds in ester substrates such as Z-Tyr-p-nitrophenyl ester. The enzyme had no amidase activity, and low activity on polypeptides.[107] Activity was inhibited by DFP, but no active site residues are known.

The published sequence[108] is identical to that of acyl-CoA thioesterase I, in which the active site serine was identified as Ser-10.[109] His-157 occurs within a Gly-Ile-His motif that is conserved in other serine thioesterases and may be part of the active site. The enzyme is synthesized with a 26-residue leader peptide and is predicted to be a periplasmic enzyme (although previously thought to be cytoplasmic). There have been doubts for some time that protease I is a peptidase.[110]

[105] E. K. Moses, J. I. Rood, W. K. Yong, and G. G. Riffkin, *Gene* **77,** 219 (1989).
[106] G. G. Lilley, D. J. Stewart, and A. A. Kortt, *Eur. J. Biochem.* **210,** 13 (1992).
[107] M. Pacaud, L. Sibilli, and G. Le Bras, *Eur. J. Biochem.* **69,** 141 (1976).
[108] S. Ichihara, Y. Matsubara, C. Kato, K. Akasaka, and S. Mizushima, *J. Bacteriol.* **175,** 1032 (1993).
[109] H. Cho and J. E. Cronan, Jr., *J. Biol. Chem.* **268,** 9238 (1993).
[110] J. D. Kowit, W.-N. Choy, S. P. Champe, and A. L. Goldberg, *J. Bacteriol.* **128,** 776 (1976).

[3] Myeloblastin: Leukocyte Proteinase 3

By John R. Hoidal, N. V. Rao, and Beulah Gray

Myeloblastin (EC 3.4.21.76) is a ~29,000-Da neutral serine endopeptidase produced during myeloid differentiation and stored in the azurophilic granules of polymorphonuclear leukocytes (PMNs), where it is present in amounts comparable to elastase and cathepsin G. Myeloblastin was initially described by Baggiolini *et al.*[1] in 1978 and designated proteinase 3 (PR-3) as a result of observations made after nondenaturing polyacrylamide gel electrophoresis indicated that it represented a third neutral serine proteinases in azurophilic granule extracts, distinct from elastase and cathepsin G. In the late 1980s investigators independently described a serine proteinase involved in the control of growth and differentiation of human leukemia cells, myeloblastin;[2] an enzyme in PMNs that degraded elastin *in vitro* and caused extensive tissue damage and emphysema, proteinase 3[3] or proteinase 4;[4] a 29,000-Da protein of the azurophilic granule that had potent microbicidal activity, AGP7[5] or p29b;[6] and a serine proteinase that was the target antigen of cytoplasmic-staining antineutrophil cytoplasmic autoantibodies (c-ANCA) circulating in Wegener's granulomatosis.[7] Subsequent characterization established that these proteins with seemingly distinct functions were actually identical; the enzyme is now referred to as myeloblastin or, more commonly, as proteinase 3. Given the potential importance of the enzyme in human health and disease, it has become of considerable practical importance to determine the physical and functional properties of myeloblastin.

Assay for Myeloblastin

Assays for myeloblastin are based on its preference for small aliphatic amino acids at the P1 site of susceptible peptide bonds. Though myeloblastin degrades a wide variety of extracellular matrix molecules, including elastin, fibronectin, vitronectin, collagen type IV, and the core protein of proteoglycans, the esterolytic or amidolytic activity of the enzyme against selected blocked amino acid or oligopeptide substrates has proved to be most useful in assessing enzyme activity. Rao *et al.*[8] studied the substrate specificity of myeloblastin with a series of commercially available chromogenic and fluorogenic substrates. They found that myeloblastin hydrolyzed

---

[1] M. Baggiolini, U. Bretz, B. Dewald, and M. E. Feigenson, *Agents Actions* **8**, 3 (1978).

[2] D Bories, M. C. Raynd, D. H. Solomon, Z. Darzynkiewicz, and Y. E. Cayre, *Cell* **59**, 959 (1989).

[3] R. C. Kao, N. G. Wehner, K. M. Skubitz, B. H. Gray, and J. R. Hoidal, *J. Clin. Invest.* **82**, 1963 (1988).

[4] K. Ohlsson, C. Linder, and M. Rosengren, *Biol. Chem. Hoppe-Seyler* **371**, 549 (1990).

[5] J. E. Gabay, R. W. Scott, D. Campanelli, J. Griffith, C. Wilde, M. N. Marra, M. Seeger, and C. F. Nathan, *Proc. Natl. Acad. Sci. U.S.A.* **86**, 5610 (1989).

[6] D. Campanelli, P. A. Detmers, C. F. Nathan, and J. E. Gabay, *J. Clin. Invest.* **85**, 904 (1990).

[7] J. Lüdemann, B. Utecht, and W. L. Gross, *J. Exp. Med.* **171**, 357 (1990).

[8] N. V. Rao, N. G. Wehner, B. C. Marshall, W. R. Gray, B. H. Gray, and J. R. Hoidal, *J. Biol. Chem.* **266**, 9540 (1991).

p-nitrophenyl esters of alanine or valine and was active against selected tri- or tetrapeptide substrates with alanine or valine at the P1 site.[8] Myeloblastin had an overlapping, but narrower, range of activity than elastase toward these substrates and lower catalytic efficiency (k_{cat}/K_m). To date, no substrate has been identified that is hydrolyzed by myeloblastin, but not human leukocyte elastase (HLE). This presents a practical problem when attempting to assess myeloblastin activity of biological fluids or solutions containing a mixture of the enzymes. One possible solution is to assess activity in the presence and absence of secretory leukoprotease inhibitor, because it efficiently inhibits elastase but not myeloblastin (*vide infra*).

Spectrophotometric Assay

t-butyloxy carbonyl p-nitrophenylester (Boc-Ala-OPhNO$_2$) is a nonspecific but sensitive substrate that is well suited to work with myeloblastin particularly in studies evaluating inhibitors.

Principle. Release of p-nitrophenol is measured by the increase in absorbance at 347.5 nm.

Reagents. Substrate: Boc-Ala-OPhNO$_2$ is dissolved in methanol as a 20 mM solution and the stock solution is stored at 4°. Buffer: 0.1 M sodium phosphate buffer (pH 7.5) containing 0.1% (v/v) Triton X-100.

Procedure. 100 μl of buffered enzyme (1 μg of pure enzyme or equivalent) is mixed with 400 μl of buffer and 475 μl of water in a thermostatted cuvette at 25°. The reaction is initiated by the addition of 25 μl of substrate solution and the increase in absorbance is monitored for 3 min. A blank without enzyme is used to determine the spontaneous hydrolysis of the substrate under the assay conditions.

Units of Activity. An enzyme unit is defined as liberation of 1 μmol of p-nitrophenol per minute based on its extinction coefficient 5500 M^{-1} cm^{-1}.

Active Site Titration

At present, there are no selective active site titrants available for myeloblastin. We use an indirect method to determine the active site molarity of myeloblastin.[8] Briefly, porcine pancreatic trypsin is titrated with p-nitrophenyl 4'-guanidinobenzoate.[9] The titrated trypsin is used as a primary standard in determining the active α_1-proteinase inhibitor concentration. The standardized α_1-proteinase inhibitor is used as a secondary standard against purified myeloblastin to determine the active enzyme concentration using Boc-Ala-OPhNO$_2$ as the substrate.

[9] T. Chase, Jr., and E. Shaw, *Biochemistry* **8**, 2212 (1969).

Purification

Preparation of Leukocyte Granule Extracts

Myeloblastin is purified from leukocyte concentrates prepared from pooled normal human blood or obtained by leukapheresis of patients with chronic myelogenous leukemia or of normal donors. Several laboratories are in the developmental stages of producing a recombinant form of the enzyme that may provide an alternative source in the future.

The following procedure has worked well in our laboratories.[3]

Isolation of Leukocytes. Leukocyte concentrates are subjected to red cell sedimentation using 6.0% dextran in physiologic saline. The supernatant is removed and centrifuged at 400 g for 10 min at 4°. To lyse contaminating red blood cells, pellets are resuspended in plasma and 0.87% (w/v) NH_4Cl for 4 min, recentrifuged, and then subjected to hypotonic shock. After centrifugation at 400 g for 10 min at 4°, leukocytes are washed and resuspended in Hanks' balanced salt solution (HBSS), counted, and a sample sedimented onto a glass slide, then stained with Wright–Giemsa stain for determination of differential cell populations. Of the isolated cells, 85–95% is polymorphonuclear leukocytes.

Isolation of Cytoplasmic Granules and Extraction of Granule Proteins. Approximately 5×10^8 leukocytes are suspended in 2 ml of 0.34 M sucrose and homogenized at 4° with a motor-driven homogenizer with Teflon pestle for two intervals of 3 min each. The homogenate is centrifuged at 200 g for 10 min at 4°, the supernatant removed, and centrifuged at 8700 g for 20 min at 4°. The pellet contains the mixed granule fraction and is extracted immediately or stored at −70°. For extraction, granule pellets from 3×10^{10} leukocytes are suspended in 12 ml of cold 0.01 M NH_4Cl and sonicated at 45 kHZ for three 30-sec intervals. The extract is centrifuged at 105,000 g for 1 hr at 4°, the supernatant fluid decanted, and used immediately for purification of proteinases by column chromatography.

Chromatographic Separation of Proteinases from Granule Extracts

Dye–Ligand Affinity Chromatography. Proteins in the granule extract from approximately 3×10^{10} PMNs are first separated by dye–ligand affinity with a 28×1.8-cm Matrex Gel Orange A (Amicon Corp., Danvers, MA) column. Approximately 150 mg protein is applied to the column and allowed to bind at room temperature for 1 hr. The column is washed with 100 ml of 0.08 M citrate–phosphate buffer (CPB), pH 5.6, followed by 150 ml of 0.1 M NaCl in CPB. Bound proteins are eluted with a gradient from 0.1 to 1.6 M NaCl in CPB. Samples of alternate 5-ml fractions are removed for determination of protein and esterase activity using α-naphthyl acetate.[10] NaCl content of the fractions is determined by measuring the conductivity. The fractions containing myeloblastin elute between 0.90 and 1.1 M NaCl, whereas the fractions containing elastase and cathepsin G elute between 0.30 and 0.54 M NaCl. The presence of esterase activity is used to locate the fractions containing proteinases.

Cation-Exchange Chromatography. A Bio-Rex 70 cation-exchange resin is converted with 0.5 M NaOH and washed thoroughly with 0.08 M CPB. A 7 × 1-cm column is poured and equilibrated to 0.05 M NaCl in CPB, pH 7. Fractions containing myeloblastin elute from the Matrex Gel Orange A column, are diluted to 0.05 M NaCl, and applied to the cation exchanger, which is subsequently washed with 0.05 M NaCl in CPB, pH 7.0. Bound proteins are eluted with a gradient from 0.05 to 1.0 M NaCl in CPB, pH 7.0. Protein, conductivity, and nonspecific esterase activity are determined on samples from alternate 5-ml fractions. Myeloblastin elutes as a single peak of esterase activity at approximately 0.11 M NaCl. The enzyme is stored at 4° and is stable for several months.

Comment. The recovery of myeloblastin from Matrex Gel Orange A varies considerably between batch preparations of the dye–ligand resin. Several batches of the Matrex Gel Orange A may need to be tested to identify one that efficiently binds myeloblastin.

Alternative Purification Schemes. Weislander *et al.*[11] have described a high-performance liquid chromatography/fast protein liquid chromatography (HPLC/FPLC)-based strategy for purification of myeloblastin. The scheme involves initial ion-exchange chromatography using DEAE–Sephacel, which removes contaminating anionic proteins but does not bind myeloblastin. This is followed by a second ion-exchange step on a cation exchanger (MonoS, Pharmacia, Piscataway, NJ), to which myeloblastin binds and can be eluted by a salt gradient. The final purification is by gel filtration on a TSK SW 3000 column. Information on yield with this strategy has not been published.

Gabay *et al.*[5] described a purification strategy in which granule extracts were initially fractionated by chromatography on a Bio-Sil TSK-125 (Bio-Rad, Richmond, CA) high-performance size-exclusion column equilibrated and developed in 50 mM glycine, pH 2.0, 100 mM NaCl. This was followed by reversed-phase HPLC using a Vydac wide-base C_4 column (Rainin Instruments, Emeryville, CA) equilibrated in 0.1% (v/v) trifluoroacetic acid and eluted in 0–100% acetonitrile in 0.1% trifluoroacetic acid.

[10] A. J. Barrett, *Anal. Biochem.* **47**, 280 (1972).

[11] J. Wieslander, N. Rasmussen, and P. Bygren, *APMIS* **97**(Suppl. 6), 42 (1989).

Finally, a solid-phase immunoaffinity purification procedure has been described by Lüdemann et al.[12] In this procedure an immunoadsorbent column is prepared with antibody precipitated from sera of patients with Wegener's granulomatosis (or from antisera raised against purified myeloblastin) coupled to CNBr-activated Sepharose. As reported, the one-step procedure results in pure enzyme from supernatants of leukocytes that have been triggered by phorbol myristate acetate to degranulate.

Properties

Myeloblastin is a basic protein and appears as three isoforms in polyacrylamide gel electrophoresis at neutral pH.[3] The molecular weight of myelobastin is close to 29,000 in SDS–gel electrophoresis and the isoelectric point is about 9.1.[3] The amino acid composition of myeloblastin is similar to that of HLE and cathepsin G, the greatest difference being the comparatively lower amount of arginine.[13] The arginine content of myeloblastin (11 residues/mole) is one-half that of HLE and approximately one-third that of cathepsin G. The NH_2-terminal amino acid sequence is homologous to HLE, cathepsin G, and human lymphocyte protease.[8] The first four N-terminal residues (Ile-1 to Gly-4) were identical to that of HLE, whereas the eight-residue stretch from Pro-9 to Ala-16 was identical to that of cathepsin G, murine granzymes A to F, rat mast cell proteases, and human lymphocyte proteases. Over the first 40 residues myeloblastin has approximately 60% identity to HLE, cathepsin G, and human lymphocyte proteases. When compared with the granzyme family of serine proteases, myeloblastin exhibited strong resemblance to granzyme B, and less to granzymes A and C–F and to rat mast cell proteases. Myeloblastin hydrolyzes hemoglobin and to a lesser extent azocasein. It also degrades structural proteins such as elastin, fibronectin, laminin, vitronectin, and collagen type IV. It has no or minimal activity against interstitial collagens types I and III, respectively.

Myeloblastin is inhibited by typical inhibitors of serine peptidases, such as phenylmethylsulfonyl fluoride and diisopropyl fluorophosphate, but not by cysteine, aspartic, or metalloproteinase group inhibitors.[3] Of the two HLE-specific chloromethylketone inhibitors tested, MeOSuc-Ala-Ala-Pro-Val-CH_2Cl and MeOSuc-Ala-Ala-Pro-Ala-CH_2Cl, only MeOSuc-Ala-Ala-Pro-Val-CH_2Cl inactivates myeloblastin ($k_{obs}/[I] = 10\,M^{-1}\,sec^{-1}$).[8] Thus, the rate of inhibition by chloromethanes paralleled myeloblastin activity on the respective chromogenic substrates.

[12] J. Lüdemann, B. Utecht, and W. L. Gross, J. Immunol. Methods 114, 167 (1988).
[13] N. V. Rao, N. G. Wehner, B. C. Marshall, A. B. Sturrock, T. P. Huecksteadt, G. V. Rao, B. H. Gray, and J. R. Hoidal, Ann. N.Y. Acad. Sci. 624, 60 (1991).

Human plasma proteinase inhibitors α_1-proteinase inhibitor and α_2-macroglobulin inactivate myeloblastin.[8] The inhibitor present in the mucous secretions and interstitial fluid, secretory leukoproteinase inhibitor, does not inhibit myeloblastin, but rather is degraded by myeloblastin with no loss of its serine proteinase inhibitory activity.[14] Elafin, another recently described inhibitor with homology to secretory leukoprotease inhibitor at the inhibitory site region, is an efficient inhibitor of myeloblastin.[15] Other protein inhibitors of myeloblastin include soybean trypsin inhibitor and an inhibitor from leech, eglin c. However, eglin c is a weak inhibitor of myeloblastin based on its k_a value for myeloblastin ($4.2 \times 10^4\ M^{-1}\ sec^{-1}$), a value 1000- and 100-fold lower than that for HLE ($1.3 \times 10^7\ M^{-1}\ sec^{-1}$) and cathepsin G ($2.0 \times 10^6\ M^{-1}\ sec^{-1}$), respectively. Aprotinin, an efficient inhibitor of HLE and cathepsin G, does not inhibit myeloblastin.

The gene for myeloblastin spans approximately 6.5 kilobase pairs and consists of five exons and four introns.[16] The genomic organization is similar to that of other serine proteinases expressed in hemopoetic cells. The gene has been localized to chromosome 19p13.3 in a cluster of these elastase-like genes including azurocidin and leukocyte elastase.[17,18] Gene expression is restricted to the promyelocytic/myelocytic stage of myelomonocytic development as an approximately 1.3-kb mRNA.[16,19] Myeloblastin RNA has been detected in a renal cancer cell line[20] and human umbilical vein endothelial cells.[21]

[14] N. V. Rao, B. C. Marshall, B. H. Gray, and J. R. Hoidal, *Am. J. Respir. Cell. Mol. Biol.* **8,** 612 (1993).

[15] J.-M. Sallenave and A. P. Ryle, *Biol. Chem. Hoppe-Seyler* **372,** 13 (1991).

[16] A. B. Sturrock, K. F. Franklin, G. Rao, B. C. Marshall, M. B. Rebentisch, R. S. Lemons, and J. R. Hoidal, *J. Biol. Chem.* **267,** 21193 (1992).

[17] A. B. Sturrock, R. Espinosa III, J. R. Hoidal, and M. M. LeBeau, *Cytogenet. Cell Genet.* **64,** 33 (1993).

[18] M. Zimmer, R. L. Medcalf, T. M. Fink, C. Mattmann, P. Lichter, and D. E. Jenne, *Proc. Natl. Acad. Sci. U.S.A.* **89,** 3215 (1992).

[19] D. Campanelli, M. Melchior, Y. Fu, M. Nakata, H. Shuman, C. Nathan, and J. E. Gabay, *J. Exp. Med.* **172,** 1709 (1990).

[20] W. J. Mayet, E. Herman, E. Csernak, A. Knath, T. Poralla, W. L. Gross, and K. H. Meyer zum Büschenfelde, *J. Immunol. Methods* **143,** 57 (1991).

[21] W. J. Mayet, E. Csernak, C. Szymkowiak, W. L. Gross, and K. H. Meyer zum Büschenfelde, *Blood* **82,** 1221 (1993).

[4] Granzyme A

By MARKUS M. SIMON and MICHAEL D. KRAMER

Introduction

Cytotoxic T lymphocytes (CTLs) and natural killer (NK) cells of human, mouse, and rat origin express a large number of homologous serine proteinases (for a comprehensive review, see Sitkovsky and Henkart[1]). The proteinases are termed granzymes, because they are stored within cytoplasmic granules, but they have many other synonyms.[2-4] To date, seven mouse, six rat, and four human lymphocyte-associated serine proteinases have been described.[4] Some members of the granzyme family are constitutively expressed by NK cells.[4,5] In T lymphocytes the granzymes appear only after previous activation[6-8] and are stored together with other molecules, including the lethal protein perforin, within cytoplasmic granules of effector cells.[1] On effector–target cell contact the granule content, including the granzymes, is delivered to the extracellular space.[1] The enzymatically active granzymes A and B are characterized by highly restricted and distinct proteolytic activities[9-13] and have been implicated

[1] M. V. Sitkovsky and P. A. Henkart (eds.), "Cytotoxic Cells. Recognition, Effector Function, Generation, and Methods." Birkhaeuser, Boston, Basel, and Berlin, 1993.

[2] P. Haddad, D. E. Jenne, O. Krähenbühl, and J. Tschopp, *in* "Cytotoxic Cells. Recognition, Effector Function, Generation, and Methods" (M. V. Sitkovsky and P. A. Henkart, eds.), p. 251. Birkhaeuser, Boston, Basel, and Berlin, 1993.

[3] M. M. Simon, K. Ebnet, and M. D. Kramer, *in* "Cytotoxic Cells. Recognition, Effector Function, Generation, and Methods" (M. V. Sitkovsky and P. A. Henkart, eds.), p. 278. Birkhaeuser, Boston, Basel, and Berlin, 1993.

[4] D. Hudig, G. R. Ewoldt, and S. L. Woodard, *Curr. Opin. Immunol.* **5,** 90 (1993).

[5] C. Manyak, G. P. Norton, C. G. Lobe, R. C. Bleackley, H. K. Gershenfeld, I. L. Weissman, V. Kumar, N. H. Sigel, and G. C. Koo, *J. Immunol.* **142,** 3707 (1989).

[6] J. A. Garcia-Sanz, G. Plaetinck, F. Velotti, D. Masson, J. Tschopp, H. R. Macdonald, and M. Nabholz, *EMBO J* **6,** 933 (1987).

[7] H.-G. Simon, U. Fruth, C. Eckerskorn, F. Lottspeich, M. D. Kramer, G. Nerz, and M. M. Simon, *Eur. J. Immunol.* **18,** 855 (1988).

[8] K. Ebnet, J. Chluba-deTapia, U. Hurtenbach, M. D. Kramer, and M. M. Simon, *Int. Immunol.* **3,** 9 (1991).

[9] M. D. Kramer, L. Binninger, V. Schirrmacher, H. Moll, M. Prester, G. Nerz, and M. M. Simon, *J. Immunol.* **136,** 4644 (1986).

[10] M. M. Simon, H. Hoschützky, U. Fruth, H.-G. Simon, and M. D. Kramer, *EMBO J.* **5,** 3267 (1986).

[11] U. Fruth, F. Sinigaglia, M. Schlesier, J. Kilgus, M. D. Kramer, and M. M. Simon, *Eur. J. Immunol.* **17,** 1625 (1987).

in a variety of T lymphocyte-mediated functions, such as target cell lysis, extravasation, induction of B cell proliferation, and control of virus replication.[3,10,14] Among all granzymes, granzyme A (from human or mouse) is unique in occurring as disulfide-linked homodimers with two catalytic sites; all other granzymes consist of a single polypeptide chain with one active site.[10,11,15-18] A serine proteinase(s) with some similarity to human and mouse granzyme A has been isolated from a rat NK tumor cell line.[4,19] It is the purpose of this chapter to describe the purification, structure, expression pattern, and possible functions of granzyme A (EC 3.4.21.36) from mouse and man.

Purification

Human and mouse granzyme A can be purified either from enriched cytoplasmic granules or from cell lysates of T lymphocyte populations[10,11,15,20-23] as summarized in the Appendix to this chapter.

Mouse Granzyme A

Cytoplasmic granules are isolated from long-term cultured murine CD8$^+$ cytotoxic T lymphocyte (CTL) lines by nitrogen cavitation and subsequent enrichment via Percoll density gradient centrifugation. After solubilization of the granule-associated proteins—at least seven molecules

[12] D. Masson, M. Zamai, and J. Tschopp, *FEBS Lett.* **208**, 84 (1986).

[13] S. Odake, C. M. Kam, L. Narasimhan, M. Poe, J. T. Blake, O. Krahenbuhl, J. Tschopp, J. C. Powers, *Biochemistry* **30**, 2217 (1991).

[14] M. D. Kramer and M. M. Simon, *Immunol. Today* **8**, 140 (1987).

[15] D. Masson and J. Tschopp, *Cell (Cambridge, Mass.)* **49**, 679 (1987).

[16] D. Masson and J. Tschopp, *Mol. Immunol.* **25**, 1283 (1988).

[17] D. Gurwitz, M. M. Simon, U. Fruth, and D. D. Cunningham, *Biochem. Biophys. Res. Commun.* **161**, 300 (1989).

[18] U. Fruth, C. Eckerskorn, F. Lottspeich, M. D. Kramer, M. Prester, and M. M. Simon, *FEBS Lett.* **237**, 45 (1988).

[19] L. Shi, C. M. Kam, J. C. Powers, R. Aebershold, and A. H. Greenberg, *J. Exp. Med.* **176**, 1521 (1993).

[20] O. Krähenbühl, C. Rey, D. Jenne, A. Lanzavecchia, P. Groscurth, and J. Tschopp, *J. Immunol.* **141**, 3471 (1988).

[21] A. Hameed, D. M. Lowrey, M. Lichtenheld, and E. R. Podack, *J. Immunol.* **141**, 3142 (1988).

[22] M. Poe, C. D. Bennett, W. E. Biddison, J. T. Blake, G. P. Norton, J. A. Rodkey, N. H. Sigal, R. V. Turner, J. K. Wu, and H. J. Zweerink, *J. Biol. Chem.* **263**, 13215 (1988).

[23] U. Vettel, G. Brunner, R. Bar-Shavit, I. Vlodavsky, and M. D. Kramer, *Eur. J. Immunol.* **23**, 279 (1993).

ranging from 27 to 70 kDa are detectable by SDS–polyacrylamide gel electrophoresis—the lytic perforin (66 kDa) is removed on gel filtration (TSK 400). The remainder of the granule proteins is pooled and loaded onto a cation-exchange column (Mono S). All major components are separable with a gradient of increasing NaCl concentrations and granzyme A elutes at $\approx$780 mM NaCl.[15]

Another method relies on the high binding affinity of mouse granzyme A to heparin. It applies Triton X-100 lysates of CD8[+] CTLs in 0.01 M Tris-HCl, 150 mM NaCl, pH 8.5. The lysates are loaded onto a heparin–Sepharose column, and molecules with low to moderate affinity to heparin are removed by washing with 500 mM NaCl. Granzyme A is eluted from the column at high NaCl concentration (1 M).[23]

Human Granzyme A

Cytoplasmic granules are isolated by Percoll density gradient centrifugation either from human CD8[+] T lymphocyte clones[20] or from human peripheral blood mononuclear cells previously propagated in recombinant interleukin 2 for 14–20 days.[21] Granzyme A is purified from solubilized granule material in a procedure similar to that described for mouse granzyme A.[15,20] Alternatively, granule proteins are solubilized in buffer containing benzamidine. Following centrifugation, the supernatant is separated by gel filtration (Sephacryl S-300). Fractions containing granzyme A activity (corresponding to 60–70 kDa) are dialyzed and loaded onto a cation-exchange column (Mono S). Applying a linear gradient of NaCl (0–1 M), granzyme A is eluted at $\approx$500 mM NaCl.[21]

The third method relies on the binding affinity of granzyme A to benzamidine. Supernatants of Triton X-100 lysates of long-term cultured CD8[+] cytotoxic T lymphocytes are adjusted to 1 mg/ml of heparin and loaded onto a benzamidine–Sepharose column. Proteins bound to the gel matrix with low affinity are removed on increasing the NaCl concentration. Granzyme A is eluted from the column with 1 M arginine in buffer without heparin.[18]

Structure of Mouse and Human Granzyme A

The most remarkable feature of mouse and human granzyme A is that they are disulfide-linked homodimers of 60 kDa (mouse[10,15]) and 50 kDa (human[11,20]), with both catalytic sites being active.[16,17] Under reducing conditions their molecular weights shift to 35,000 and 25,000, respectively.[10,11,15,20]

The mouse and human granzyme A-specific cDNAs encode active enzymes of 232 amino acids (mouse[24]) and 234 amino acids (human[25]). Alignment of the amino acid sequences to those of other well-classified members of the chymotrypsin (S1) family, such as rat mast cell protease (RMCP) I and II, elastase, or trypsin, reveals partial homology and the typical features of genuine serine proteinases: they express the amino acid residues His, Asp, and Ser, which form the active site of such enzymes[26] at equivalent positions. Adjacent residues are also highly conserved. In addition to the six cysteine residues expressed by all members of the granzyme family,[2,3] which are involved in intramolecular disulfide bonding, granzyme A contains three additional cysteine residues. Two of them form a fourth intramolecular disulfide bridge, and the third cysteine residue is responsible for the disulfide linkage of the homodimer.[2,3,24,25]

The granzyme A gene of the mouse consists of five exons and has striking similarities with that of other mouse granzymes.[27] The active enzyme is encoded by four exons. The nucleotide triplets coding for the catalytic triad (His, Asp, Ser) are on exons 2, 3, and 5, corresponding to the location in other granzymes. The fact that the promoter sequence of mouse granzyme A is distinct from that of other granzymes (B and C) indicates that regulation of its expression differs from that of other granzymes.[27] Protein and cDNA analyses of the human granzyme A gene suggest that its organization is similar to that of the mouse gene.[18,25]

The mRNAs of mouse and human granzyme A encode a typical signal peptide required for translocation of proteins into the endoplasmic reticulum.[24,25,28] Moreover, they encode a two-amino acid propeptide, indicating that granzyme A in both species is synthesized as an inactive precursor. Both are stored, however, in their active form within cytoplasmic granules. Most probably the propeptide is removed during transport of the enzymes into the cytoplasmic granules.[28]

Catalytic Activity

Substrate Specificity, Effect of pH, and Inhibitors

The enzyme activities of human and mouse granzyme A have been explored by hydrolysis of synthetic peptidenitroanilides (pNA),[10,11]

[24] H. K. Gershenfeld and I. L. Weissman, Science 232, 854 (1986).
[25] H. K. Gershenfeld, R. J. Hershberger, T. B. Shows, and I. L. Weissman, Proc. Natl. Acad. Sci. U.S.A. 85, 1184 (1988).
[26] J. Kraut, Annu. Rev. Biochem. 46, 331 (1977).
[27] K. Ebnet, M. D. Kramer, and M. M. Simon, Genomics 13, 502 (1992).
[28] D. E. Jenne and J. Tschopp, Immunol. Rev. 103, 53 (1988).

TABLE I
CHARACTERISTICS OF MOUSE AND HUMAN GRANZYME A

Parameter	Granzyme A[a]	
	Mouse	Human
M_r nonreduced/reduced	60/35	50/25
Substrates	D-Pro-Phe-Arg-pNA	Tos-Gly-Pro-Arg-pNA
	Pro-Phe-Arg-AMC	Gly-Pro-Arg-AMC
	Z-Lys-SBzl	Val-Pro-Arg-AMC
pH optimum	7.5–8.5 (D-Pro-Phe-Arg-pNA)	8 (Gly-Pro-Arg-AMC)
		11 (Tos-Gly-Pro-Arg-pNA)
Synthetic inhibitors	DFP	DFP
	PMSF	PMSF
	Benzamidine	Benzamidine
	D-Pro-Phe-Arg-CH$_2$Cl	MUGB
Natural inhibitors	Soybean trypsin inhibitor	Soybean trypsin inhibitor
	Aprotinin	Aprotinin
	Leupeptin	Leupeptin
Physiological inhibitors	α_2-Macroglobulin	α_2-Macroglobulin
	Antithrombin III	Antithrombin III
	C1 esterase inhibitor	α_1-Proteinase inhibitor
	Protease nexin-1	

[a] DFP, Diisopropyl fluorophosphate; PMSF, phenylmethylsulfonyl fluoride; MUGB, 4-methylumbelliferyl-p-guanidinobenzoate.

7-amino-4-methylcoumarins (AMC),[12,13] and thiobenzyl esters (SBzl).[13,29] Mouse and human granzyme A have trypsinlike activity in that they cleave best after Arg and Lys residues (Table I). In addition, they are more reactive than simple amino acid derivatives toward extended peptide substrates.[10] When tested on peptide nitroanilides the best substrate for mouse granzyme A is D-Pro-Phe-Arg-pNA,[10] and for human granzyme A Tos-Gly-Pro-Arg-pNA.[11] Other representative substrates for mouse granzyme A are Z-Lys-SBzl and Pro-Phe-Arg-AMC[12,13] and for human granzyme A Gly-Pro-Arg-AMC and Val-Pro-Arg-AMC.[13,20] Substrates composed of other amino acids at positions 2 and 3 adjacent to L-arginine are either not cleaved or are only marginally cleaved.[10,11] These findings are evidence for the highly restricted specificity of human and mouse granzyme A, and indicate that both enzymes do not exert merely degradative but rather protein-processing functions.[30,31]

[29] M. S. Pasternack and H. N. Eisen, Nature (London) 314, 743 (1985).
[30] H. Neurath and K. A. Walsh, Proc. Natl. Acad. Sci. U.S.A. 73, 3825 (1976).
[31] H. Neurath, Trends Biochem. Sci. 14, 268 (1989).

The pH optimum of mouse granzyme A for the cleavage of Pro-Phe-Arg substrates (peptide-pNA and peptide-AMC) is between 7.5 and 8.5.[10,12,13] The pH optimum of human granzyme A is ≈8 for the cleavage of Gly-Pro-Arg-AMC and ≈11 for the cleavage of Tos-Gly-Pro-Arg-pNA.[11,20]

Both mouse and human granzyme A react strongly with diisopropyl fluorophosphate, and other serine proteinase inhibitors, including phenyl-methylsulfonyl fluoride, benzamidine, aprotinin, leupeptin, and soybean trypsin inhibitor, and substituted isocoumarins.[10–13] The human granzyme A is also inhibited by the synthetic antiprotease 6-amidino-2-naphthyl-4-guanidinobenzoate.[22] The peptidyl chloromethane, D-Pro-Phe-Arg-CH$_2$Cl, is a potent inhibitor of mouse granzyme A.[32] Tos-Lys-CH$_2$Cl and Tos-Phe-CH$_2$Cl, good inhibitors of trypsin and chymotrypsin, respectively, have little or no effect on mouse and human granzyme A. In addition, neither of the two enzymes is inhibited by inhibitors of aspartic (pepstatin), metallo- (1,10-phenanthroline), or thiol- (4-hydroxymercuribenzoate) proteinases.[10,11] The enzyme activity of mouse and human granzyme A is not dependent on the presence of Ca^{2+} or Mg^{2+}, because the chelation of divalent cations by EDTA or EGTA has no inhibitory effect.[10,11,33] However, in the presence of Ca^{2+}, Zn^{2+}, or reducing agents such as dithiothreitol at 1 mM, the activity of mouse granzyme A is significantly reduced.[10]

The activity of mouse granzyme A is also susceptible to various vascular and extravascular protease inhibitors; its interaction with α_2-macroglobulin, an inhibitor for several serine proteinases, leads to formation of a complex, resulting in the inhibition of enzyme activity on large proteins such as fibronectin, but not on peptide substrates.[34] This indicates that granzyme A could still play a role in T lymphocyte-mediated processes *in vivo* by cleaving small proteins, even after its complexation with α_2-macroglobulin. Finally, antithrombin III and C1 esterase inhibitor, which are present in the vascular system,[16,17,34] as well as protease nexin-1, an inhibitor associated with extracellular matrices,[17] inhibit mouse granzyme A. In the presence of heparin, antithrombin III and protease nexin-1 at physiological concentrations form covalently linked complexes with both subunits of the granzyme A molecule.[16,17] This indicates an important function of vascular and extravascular inhibitors in protecting surrounding tissue in T lymphocyte-mediated processes.

[32] M. M. Simon, M. Prester, M. D. Kramer, and U. Fruth, *J. Cell. Biochem.* **40**, 1 (1989).
[33] J. D.-E. Young, L. G. Leong, C. C. Liu, A. Damiano, D. A. Wall, and Z. A. Cohn, *Cell* (*Cambridge, Mass.*) **47**, 183 (1986).
[34] M. M. Simon, T. Tran, U. Fruth, D. Gurwitz, and M. D. Kramer, *Biol. Chem. Hoppe-Seyler* **371**(Suppl), 81 (1990).

Expression and Localization of Granzyme A

The analysis of lymphoid and nonlymphoid cell populations and tissues shows that with few possible exceptions,[2] production of granzyme A is restricted to T lymphocytes, their thymic precursors, and NK cells.[2-4] The enzyme is not expressed in resting T lymphocytes but is induced on antigen stimulation *in vitro* and *in vivo*.[6-8,35] Granzyme A is stored together with other members of the granzyme family and perforin within cytoplasmic granules of effector cells.[36-38] On appropriate conjugation of cytolytic cells (T lymphocytes, NK cells) with target cells the enzyme is vectorially secreted into the extracellular space.[11,32,39]

Mouse granzyme A has been detected in various CD8[+] and CD4[+] T lymphocyte lines, NK cells, lymphokine-activated killer cells, and gamma/delta[+] (γ/δ^+) Thy 1[+] dendritic epidermal T lymphocyte lines.[2-4] The enzyme is also expressed in the majority of CD4[-]CD8[+] and a fraction of CD4[+]CD8[-] T lymphocytes previously sensitized *in vitro* by antigen or lectin.[40]

Mouse granzyme A is expressed *in vivo* in virus-specific CD8[+] T lymphocytes,[8,41] as well as in T lymphocytes infiltrating allografts,[42,43] or pancreatic islets of nonobese diabetic mice during the development of autoimmune diabetes.[44,45] The availability of monoclonal antibodies allowed the localization of granzyme A in cytoplasmic granules of CD8[+]CD4[-] T lymphocytes during viral infection,[46] as well as in T lymphocytes infiltrating

[35] D. Masson, P. Corthesy, M. Nabholz, and J. Tschopp, *EMBO J.* **4,** 2533 (1985).
[36] P. J. Peters, J. Borst, V. Oorschot, M. Fukuda, O. Krähenbühl, J. Tschopp, and H. J. Geuze, *J. Exp. Med.* **173,** 1099 (1991).
[37] D. Masson, M. Nabholz, C. Estrade, and J. Tschopp, *EMBO J.* **5,** 1595 (1986).
[38] U. Fruth, M. Prester, J. Golecki, H. Hengartner, H. G. Simon, M. D. Kramer, and M. M. Simon, *Eur. J. Immunol.* **176,** 613 (1987).
[39] H. Takayama, G. Trenn, W. Humphrey, J. A. Bluestone, P. A. Henkart, and M. V. Sitkovsky, *J. Immunol.* **138,** 566 (1987).
[40] U. Fruth, G. Nerz, M. Prester, H.-G. Simon, M. D. Kramer, and M. M. Simon, *Eur. J. Immunol.* **18,** 773 (1988).
[41] C. Müller, D. Kägi, T. Aebischer, B. Odermatt, W. Held, E. R. Podack, R. M. Zinkernagel, and H. Hengartner, *Eur. J. Immunol.* **19,** 1253 (1989).
[42] C. Müller, H. K. Gershenfeld, C. G. Lobe, C. Y. Okada, R. C. Bleackley, and I. L. Weissman, *J. Exp. Med.* **167,** 1124 (1988).
[43] C. Müller, H. K. Gershenfeld, C. G. Lobe, C. Y. Okada, R. C. Bleackley, and I. L. Weissman, *Transplant. Proc.* **20,** 251 (1988).
[44] W. Held, H. R. Macdonald, I. L. Weissman, M. W. Hess, and C. Müller, *Proc. Natl. Acad. Sci. U.S.A.* **87,** 2239 (1990).
[45] G. Griffiths and C. Mueller, *Immunol. Today* **12,** 415 (1991).
[46] M. D. Kramer, U. Fruth, H.-G. Simon, and M. M. Simon, *Eur. J. Immunol.* **19,** 151 (1989).

skin lesions of *Leishmania major*-infected mice.[47] Granzyme A was also found in cytoplasmic granules of natural killer cell-like "granulated metrial gland cells" of the murine placenta, which are thought to be involved in the immune surveillance at the maternal–fetal interphase.[48] Transcripts specific for mouse granzyme A have also been detected in a small population of CD4⁻CD8⁻ thymocytes.[49] It was speculated that the expression of the enzyme in the thymus is an early event in T lymphocyte differentiation.

Human granzyme A is expressed in activated T lymphocytes, including normal peripheral blood lymphocytes previously sensitized by interleukin 2 or mitogen (phytohemagglutinin), alloreactive CD8⁺ and CD4⁺ T lymphocyte clones, as well as NK cells and large granular lymphocytes.[11,50] It is also induced *in vitro* by lectin in both CD4⁻CD8⁺ and CD4⁺CD8⁻ T lymphocytes.[11]

In vivo, human granzyme A was found to be expressed in T lymphocytes associated with inflamed tissues from various dermatoses,[51] and rheumatoid arthritis[52] or with cellular infiltrates in the myocardium of cardiac transplants undergoing rejection.[45,53]

The findings that granzyme A is mainly expressed by *in vitro-* and *in vivo*-activated T lymphocytes but not by resting T lymphocytes and most normal tissues suggest the enzyme as a useful marker for the detection of cytolytic effector cells (T lymphocytes or NK cells) *in vivo.*[45]

Function of Granzyme A

The biological role of granzyme A has yet to be determined. Any hypothesis on the involvement of granzyme A in T or NK cell-mediated processes has to consider the fact that granzyme A is only operative under conditions of the extracellular milieu and is apparently inactive under acidic conditions existing in the intracytoplasmic granules.[2,3] Exocytosis of granzyme A into the extracellular space can be induced by engagement of cell surface receptors, which include the classical T lymphocyte recep-

---

[47] H. Moll, C. Müller, R. Gillitzer, H. Fuchs, M. Röllinghoff, M. M. Simon, and M. D. Kramer, *Infect. Immun.* **59**, 4701 (1991).

[48] L. M. Zheng, D. M. Ojcius, C. C. Liu, M. D. Kramer, M. M. Simon, E. L. Parr, and J. D. E. Young, *FASEB J.* **5**, 79 (1991).

[49] W. Held, H. R. Macdonald, and C. Müller, *Int. Immunol.* **2**, 57 (1990).

[50] J. Schmid and C. Weissmann, *J. Immunol.* **139**, 250 (1987).

[51] G. S. Wood, C. Müller, R. A. Warnke, and I. Weissman, *Am. J. Pathol.* **133**, 218 (1988).

[52] G. M. Griffiths, S. Alpert, E. Lambert, J. McGuire, and I. L. Weissman, *Proc. Natl. Acad. Sci. U.S.A.* **89**, 549 (1992).

[53] G. M. Griffiths, R. Namikawa, C. Müller, C. C. Liu, J. D.-E. Young, M. Billingham, and I. Weissman, *Eur. J. Immunol.* **21**, 687 (1991).

tor for antigen[11,39] and/or extracellular matrix receptors of the integrin family.[54] The regulated secretion and specific proteolytic activity of granzyme A suggest, therefore, specific functions for the enzyme. A number of studies have provided indirect evidence for its participation in T or NK cell-mediated activities such as target cell lysis, regulation of immune responses, migration of T lymphocytes to inflammatory foci, and control of viral replication.[2,10,32,55-57]

The most compelling evidence exists for the involvement of granzyme A in T or NK cell-mediated cytolysis. On conjugate formation of cytolytic T lymphocytes/NK cells, the cytolytic molecule perforin, as well as granzymes, including granzyme A, are released from the granules into the extracellular space. Granzyme A does not exert cytolytic activity on its own,[10] but seems to participate indirectly in target cell killing, most probably by processing proteins of the effector and/or target cell. This is supported by the finding that cytolytic activity of isolated cytoplasmic granules is inhibited by granzyme A-specific inhibitors[32,56] and that transfection of an alloreactive cytotoxic T lymphocyte clone with an antisense granzyme A vector reduces its cytolytic potential.[58] Furthermore, rat basophilic leukemia (RBL) cells transfected with genes for granzyme A *and* perforin showed both potent target cell killing *and* DNA breakdown in target cells. In contrast, RBL/perforin transfectants were only cytolytic, and did not induce DNA fragmentation. RBL/granzyme A transfectants showed neither of the two activities.[59,60] Together, these data indicate that granzyme A may be operative in various stages of T or NK cell-mediated target cell lysis.

Granzyme A may also play a role in the extravasation of T lymphocytes. In fact, it was shown that mouse granzyme A cleaves various components of the subendothelial extracellular matrix, such as proteoglycans,[57] collagen type IV,[61] laminin,[33] and fibronectin.[62] The enzyme may

[54] K. Takahashi, T. Nakamura, H. Adachi, H. Yagita, and K. Okumura, *Eur. J. Immunol.* **21,** 1559 (1991).

[55] H. G. Simon, U. Fruth, M. D. Kramer, and M. M. Simon, *FEBS Lett.* **223,** 352 (1987).

[56] M. M. Simon, U. Fruth, H. G. Simon, and M. D. Kramer, *Ann. Inst. Pasteur-Immunol.* (*Paris*) **17,** 309 (1987).

[57] M. M. Simon, H.-G. Simon, U. Fruth, J. Epplen, H. K. Müller-Hermelink, and M. D. Kramer, *Immunology* **60,** 219 (1987).

[58] A. Talento, M. Nguyen, S. Law, J. K. Wu, M. Poe, J. T. Blake, M. Patel Sigal, I. L. Weissman, R. C. Bleackley, E. R. Podack, M. L. Tykocinski, and G. C. Koo, *J. Immunol.* **149,** 4009 (1992).

[59] J. W. Shiver and P. A. Henkart, *Cell (Cambridge, Mass.)* **64,** 1175 (1991).

[60] J. W. Shiver, L. Su, and P. A. Henkart, *Cell (Cambridge, Mass.)* **71,** 315 (1992).

[61] M. M. Simon, M. D. Kramer, M. Prester, and S. Gay, *Immunology* **73,** 117 (1991).

[62] M. M. Simon, M. Prester, G. Nerz, M. D. Kramer, and U. Fruth, *Biol. Chem. Hoppe-Seyler* **369**(Suppl), 107 (1988).

therefore allow stimulated T lymphocyte and NK cells to penetrate vessel walls and to reach appropriate targets in tissues. This is further supported by the finding that exocytosis of granzyme A can be triggered by ligand binding to integrins, which mediate binding of the cells to extracellular matrix proteins.[54] Moreover, by activating the proenzyme of the plasminogen activator urokinase (pro-urokinase-type plasminogen activator), granzyme A can recruit the extracellular matrix-degrading proteolytic potential of plasmin.[63]

Granzyme A has also been suggested to be involved in the regulation of B cell growth. In fact, granzyme A, similar to other serine proteinases (trypsin and thrombin), induces proliferation of B cells in the absence of antigen *in vitro*[10] and it is possible that it controls—at least partially—B cell expansion *in vivo*.

A possible role for granzyme A in the control of viral infection by CD8[+] CTLs is suggested by the finding that proteins essential for viral replication (the enzyme reverse transcriptase) or for viral infectivity (the envelope protein gp70) from the Moloney murine leukemia retrovirus are cleaved by granzyme A.[3,55] One could speculate that during interaction of cytolytic T lymphocytes with virus-infected target cells the infectivity of mature virus particles is decreased by the proteolytic activity of exocytosed granzyme A.

The goal of this review has been to bring together pertinent information about the molecular characteristics of mouse and human granzyme A and to relate these properties to their biological function(s). The most intriguing findings are that only activated and not resting T lymphocytes express granzyme A, and that the enzyme is constitutively produced by NK cells. The biochemical properties of granzyme A as well as its storage in enzymatically inert form within secretory cytoplasmic granules suggest that the enzyme is operative only after its release into the extracellular space. Although granzyme A has been proposed to be involved in a multitude of physiological processes, including T lymphocyte differentiation in the thymus, extravasation, cytolysis, control of B lymphocyte responses, and virus replication, the evidences for these assumptions are only indirect so far. With the molecular genetic techniques available it should now be possible to inactivate granzyme A by homologous recombination and to generate a granzyme A-deficient mouse. The rapid progress toward this goal may help to reveal the precise function(s) of granzyme A in T or NK cell-mediated processes.

[63] G. Brunner, M. M. Simon, and M. D. Kramer, *FEBS Lett.* **260**, 141 (1990).

Appendix: Purification of Granzyme A

Purification of Mouse Granzyme A

Procedure 1: Isolation of Mouse Granzyme A from Cytoplasmic Granules

Suspension of 5×10^8 cells of the cytotoxic T cell line B6.1 in 12 ml of 100 mM KCl, 3.5 mM NaCl, 3.5 mM MgCl$_2$, 1 mM ATP, 1.25 mM EGTA, 10 mM PIPES, pH 6.8.[15]

Isolation of cytoplasmic granules by nitrogen cavitation and Percoll density gradient centrifugation.

Solubilization of granule proteins by adjusting the buffer to 1.5 M in NaCl and 10 mM in benzamidine.

Ultracentrifugation at 170,000 g for 2 hr at 4°. Supernatant used for further purification.

Removal of perforin by nonspecific retardation via passage over TSK G 4000 SWG gel matrix.

Dialysis of nonretarded proteins against 10 mM Bis–Tris, 50 mM NaCl, pH 6.0.

Loading of granule proteins onto cation-exchange column (Mono S) and elution with a linear gradient from 50 mM to 1 M NaCl.

Elution of granzyme A at $\approx$700 mM NaCl.

Procedure 2: Isolation of Mouse Granzyme A from Detergent Lysates of Cells

Preparation of cell lysates from 5×10^7–2×10^8 cells of the murine cytotoxic T cell line 1.3E6SN in 100 mM Tris-HCl, 150 mM NaCl, 0.1% Triton X-100, pH 8.5.[23]

Centrifugation of lysates at 10,000 g for 5 min at 4°.

Loading of lysates onto a heparin–Sepharose CL-6B column equilibrated with 100 mM Tris-HCl, 150 mM NaCl, pH 8.5.

Incubation for 1 hr at 4°.

Washing of column with 100 mM Tris-HCl, 150 mM NaCl, pH 8.5.

Elution of proteins with low and moderate affinity to heparin–Sepharose with 100 mM Tris-HCl, 500 mM NaCl, pH 8.5.

Elution of granzyme A with 10 mM Tris-HCl, 1 M NaCl, pH 8.5.

Purification of Human Granzyme A

Procedure 1: Isolation of Human Granzyme A from Cytoplasmic Granules

In vitro propagation of human peripheral blood mononuclear cells (PBMCs) for 14–20 days in the presence of 1000 U/ml recombinant human interleukin 2.[21]

Isolation of cytoplasmic granules from 2×10^9–2×10^{10} PBMCs by nitrogen cavitation and Percoll density gradient centrifugation.

Solubilization of granule contents with 1 M NaKHPO$_4$, pH 6.5, 10 mM benzamidine hydrochloride, 1 mM EDTA, 0.02% NaN$_3$.

Centrifugation for 30 min at 20,000 rpm in a Sorvall centrifuge. Pellet discarded and supernatant used for further purification.

Separation of supernatant by Sephacryl S-300 molecular sieve chromatography using 100 mM phosphate buffer, pH 6.5, 0.5 M NaCl, 1 mM benzamidine, 1 mM EDTA, 3 mM NaN$_3$ as running buffer.

Analysis of fractions for granzyme A activity.

Dialysis of pooled fractions with granzyme A activity (corresponding to 60–70 kDa) against 10 mM NaKHPO$_4$, pH 6.0, and loading of dialyzed fraction onto Mono S cation-exchange chromatography column.

Elution of bound proteins with a gradient of 0–2 M NaCl in 10 mM NaKHPO$_4$, pH 6.0.

Elution of granzyme A at $\approx$500 mM NaCl.

Procedure 2: Isolation of Human Granzyme A from Detergent Lysates of Cells

Preparation of detergent lysates of 1×10^9 cells of the human cytotoxic T cell line B34.C7 in 10 mM Tris-HCl, 0.1% Triton X-100, pH 8.0, at a density of 1×10^7/ml.[18]

Centrifugation at 10,000 g for 15 min.

Adjusting supernatant to 1 mg/ml of heparin.

Loading of supernatant onto a benzamidine–Sepharose column.

Elution of nonspecifically bound proteins with a linear gradient of 0–1 M NaCl in 2.5 column volumes of 10 mM Tris-HCl, pH 8.0, 1 mg/ml heparin.

Elution of granzyme A with 1 M arginine in 10 mM Tris-HCl, pH 8.0, in the absence of heparin.

[5] Granzyme B

By Manuel C. Peitsch and Juerg Tschopp

Introduction

Cell-mediated killing by cytotoxic T lymphocytes (CTLs) is an important immunologic defense mechanism against tumor cell proliferation, viral infection, and transplanted tissue. Activated CTLs have dense cytoplasmic granules containing a variety of proteins. On CTL–target cell contact, these cytoplasmic granules are exocytosed and their content is released into the intercellular space. Besides perforin, which is responsible for the lesions often associated with target cell lysis, several serine peptidases (granzymes) are found within these granules. Although mouse CTLs contain seven granzymes (A, B, C, D, E, F, and G),[1,2] human CTLs possess only granzymes A, B, and H[3–8]; rat CTLs contain at least granzymes A and B.[9] The cloning and sequencing of the cDNAs coding for these proteins unequivocally identified them as serine peptidases of the chymotrypsin family (family S1; see [1] in this volume). The immature granzymes are activated by the cleavage of a two-residue propeptide by the dipeptidyl-peptidase I (cathepsin C). It is the purpose of this chapter to describe the purification and assay methods for granzyme B (EC 3.4.21.79).

The highest levels of granzyme B[2,10,11] (also called CTLA-1,[12] CCPI,[13] HLP,[4] HSE26.1,[5] SECT,[6] CSPB,[14] or RNKP-1[9]) are found in the granules

[1] D. E. Jenne and J. Tschopp, *Curr. Top. Microbiol. Immunol.* **140**, 33 (1988).
[2] D. Masson and J. Tschopp, *Cell (Cambridge, Mass.)* **49**, 679 (1987).
[3] H. K. Gershenfeld, R. J. Hershberger, T. B. Shows, and I. L. Weissman, *Proc. Natl. Acad. Sci. U.S.A.* **85**, 1184 (1988).
[4] J. Schmid and C. Weissmann, *J. Immunol.* **139**, 250 (1987).
[5] J. A. Trapani, J. L. Klein, P. C. White, and B. Dupont, *Proc. Natl. Acad. Sci. U.S.A.* **85**, 6924 (1988).
[6] A. Caputo, D. Fahey, C. Lloyd, R. Vozab, E. McCairns, and P. B. Rowe, *J. Biol. Chem.* **263**, 6363 (1988).
[7] J. L. Klein, A. Selvakumar, J. A. Trapani, and B. Dupont, *Tissue Antigens* **35**, 220 (1990).
[8] A. Hameed, D. M. Lowrey, M. Lichtenheld, and E. R. Podack, *J. Immunol.* **141**, 3142 (1988).
[9] S. J. Zunino, R. C. Bleakley, J. Martinez, and D. Hudig, *J. Immunol.* **144**, 2001 (1990).
[10] D. Masson, M. Nabholz, C. Estrade, and J. Tschopp, *EMBO J.* **5**, 1595 (1986).
[11] M. Poe, J. T. Blake, D. A. Boulton, M. Gammon, N. H. Sigal, J. K. Wu, and H. J. Zweerink, *J. Biol. Chem.* **266**, 98 (1991).
[12] J. L. Crosby, R. C. Bleackley, and J. H. Nadeau, *Genomics* **6**, 252 (1990).
[13] C. G. Lobe, C. Havele, and R. C. Bleackley, *Proc. Natl. Acad. Sci. U.S.A.* **84**, 1448 (1986).
[14] J. L. Klein, T. B. Shows, B. Dupont, and J. A. Trapani, *Genomics* **5**, 110 (1989).

of CTLs. Lower levels of granzyme B are found in helper T cells[15,16] and in thymocytes.[17] Proposed physiological roles of granzyme B include the inhibition of colon cancer cell growth[18] and the induction of chromatin DNA degradation during T cell-mediated apoptosis.[19,20]

Assay Methods

Materials and Reagents

Spectrofluorimeter.
Spectrophotometer or plate reader.
Assay buffer: 100 mM HEPES, pH 7.0, containing 0.3 M NaCl, 1 mM EDTA, and 0.05% (v/v) Triton-X-100.[11,21]

Glutamyl-2-naphthylamidase (GluNHNap) Assay

Substrate solution A: 1 mM L-glutamyl-2-naphthylamide (L-Glu-βNA; Bachem, Bubendorf, Switzerland) in assay buffer; freshly prepared. Small aliquots of granzyme B-containing fractions are added to the substrate solution A. The activity is measured at 21° on a spectrofluorimeter with fluorescence excitation at 340 mm and fluorescence emission recorded at 415 nm (both with 5-nm bandwidth). The fluorescence increase is measured for 10–40 min.

Boc-Ala-Ala-Asp-SBzl Hydrolase assay

Substrate solution B: 1% Boc-Ala-Ala-Asp-SBzl stock (v/v) and 1% DTNB stock (v/v) in assay buffer; freshly prepared.

Boc-Ala-Ala-Asp-SBzl stock: 10 mM N-α-tert-butoxycarbonyl-L-ala-nyl-L-alanyl-L-aspartylthiobenzyl ester (custom synthesis by Bachem, Bu-bendorf, Switzerland) in dimethyl sulfoxide; freshly prepared.

[15] J. A. Garcia Sanz, H. R. MacDonald, D. E. Jenne, J. Tschopp, and M. Nabholz, *J. Immunol.* **145**, 3111 (1990).

[16] K. Ebnet, J. Chluba de Tapia, U. Hurtenbach, M. D. Kramer, and M. M. Simaon, *Int. Immunol.* **3**, 9 (1991).

[17] W. Held, H. R. MacDonald, and C. Müller, *Int. Immunol.* **2**, 57 (1990).

[18] T. J. Sayers, T. A. Wiltrout, R. Sowder, W. L. Munger, M. J. Smyth, and L. E. Henderson, *J. Immunol.* **148**, 292 (1992).

[19] L. Shi, R. P. Kraut, R. Aebersold, and A. Greenberg, *J. Exp. Med.* **175**, 553 (1992).

[20] L. Shi, C.-M. Kam, J. C. Powers, R. Aebersold, and A. Greenberg, *J. Exp. Med.* **176**, 1521 (1992).

[21] S. Odake, C.-M. Kam, L. Narasimhan, M. Poe, J. T. Blake, O. Krähenbuhl, J. Tschopp, and J. C. Powers, *Biochemistry* **30**, 2217 (1991).

DTNB stock: 11 mM dithiobis(2-nitrobenzoic acid) (Sigma, St. Louis, MO) in dimethyl sulfoxide; stored at −20°.

Small aliquots of fractions containing granzyme B are added to the substrate solution B at 21° and the rate of absorbance increase is monitored at 405 nm on a plate reader or spectrophotometer. Absorbance increases are converted to enzymtic rates using an extinction coefficient of 13,100 cm^{-1} M^{-1} at 405 nm and pH 7.0.

Purification

Human and mouse granzyme B are isolated from cytoplasmic granules of activated CTL lines grown in the presence of recombinant interleukin 2.[2,10,11,22]

Reagents

3M NaCl.
PHA: Phytohemagglutinin purified from red kidney beans (Burroughs Wellcome, Research Triangle Park, NC).
L-Glutamine.
Heat-inactivated human serum.
Heat-inactivated fetal calf serum.
RMPI 1640 culture medium.
Human peripheral blood leukocytes.[23]
2-Mercaptoethanol.
Concanavalin A supernatant.[23]
Human and mouse recombinant interleukin 2.
Percoll (Pharmacia, Uppsala, Sweden).
FPLC (fast protein liquid chromatography) system (Pharmacia, Uppsala, Sweden).
TSK 4000 column (LKB, Bromma, Sweden).
SDS–PAGE system.
FPLC Mono S cation-exchange column HR 5/5 (Pharmacia, Uppsala, Sweden).
Sephacryl S300 (Pharmacia, Uppsala, Sweden).
Relaxation buffer: 10 mM PIPES, pH 6.8, containing 100 mM KCl, 3.5 mM MgCl$_2$, 1.25 mM EGTA, and 1 mM ATP.

[22] O. Krähenbuhl, C. Rey, D. Jenne, A. Lanzavecchia, P. Groscurth, S. Carrel, and J. Tschopp, *J. Immunol.* **141**, 3471 (1988).
[23] J. E. Coligan, A. M. Kruisbeek, D. H. Marguiles, E. M. Shevan, and W. Strober, *in* "Current Protocols in Immunology." Green Publ. & Wiley(Interscience), New York, Chichester, Brisbane, Toronto, and Singapore, 1991.

TSK buffer: 10 mM Sodium acetate, pH 4.5, containing 1 M NaCl and 1 mM EDTA.
Buffer A: 10 mM Bis–Tris, pH 6.0, containing 50 mM NaCl.
Buffer B: 10 mM Bis–Tris, pH 6.0, containing 1 M NaCl.
Buffer C: 0.2 M Sodium acetate, pH 3.77.

Cells

The human CTL clone KV10,[22] or a similar clone, is grown in RPMI 1640 medium supplemented with 2 mM L-glutamine, 5×10^{-5} M 2-mercaptoethanol, 5% heat-inactivated human serum, and 100 U/ml recombinant human interleukin 2. Cells are restimulated every 3 to 5 weeks with 1 mg/ml PHA and 3000 rad irradiated allogenic peripheral blood leukocytes.[23] KV10 cells are harvested 2 to 4 weeks after restimulation.

The mouse CTL line B6.1,[10] or a similar line, is grown in RPMI 1640 medium supplemented with 2 mM L-glutamine, 5×10^{-5} M 2-mercaptoethanol, 5% fetal calf serum, and 12% concanavalin A supernatant or 20 U/ml recombinant mouse interleukin 2.

Preparation of Cytoplasmic Granules

Cytoplasmic granules are usually isolated by centrifugation in a Percoll density gradient and disrupted by high-salt buffers: 10^9 cells are sedimented at 270 g and are resuspended at a concentration of 5×10^7 cells/ml in ice-cold relaxation buffer. The cell suspension is pressurized to 30 bar under N_2 at 0° for 20 min and then lysed by sudden decompression. The homogenate is centrifuged at 270 g at 4°. The pellet is washed twice with relaxation buffer and the supernatants combined. Portions (5 ml each) of the supernatant are layered onto 20 ml of a linear gradient from 0 to 90% Percoll (density 1.00 to 1.18 g/ml) in relaxation buffer. The gradients are then centrifuged at 60,000 g for 45 min and 0.8-ml fractions are collected. The fractions containing the narrow band of highest turbidity at a density of 1.10 g/ml are combined.

Disruption of Cytoplasmic Granules

The purified granules are disrupted in 1.5 volumes of 3 M NaCl to separate the membranes from proteins. This suspension is then centrifuged for 30 min at 38,000 g, and the resulting clear supernatant (solubilized granule proteins) is kept for protein separation.

Protein Separation

Procedure 1. Up to 300 μg of the solubilized granule proteins (in 5 ml of 1.5 M NaCl) is loaded onto a TSK 4000 column (60 × 2.5 cm) connected

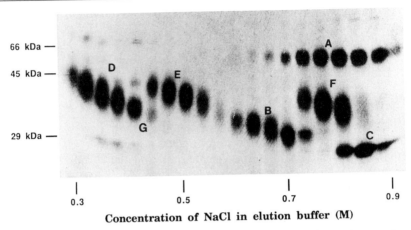

Fig. 1. Elution pattern of granzymes loaded on a Mono S cation-exchange column. Granule proteins of the CTL line B6.1, devoid of perforin, were loaded onto a Mono S column. Granzymes A through G were eluted by 30 ml of a linear gradient from 0.5 to 1 M NaCl; 50 μl of each collected 1-ml fraction was loaded on 10% SDS–PAGE. Following the electrophoreses, the gel was stained with Coomassie blue.

to an FPLC system. The column is equilibrated with TSK buffer and run at 5 ml/min in the same buffer. This column allows the separation of perforin from the other granule proteins.[24] The proteins with granzyme B and granzyme A activity (see [4] in this volume), devoid of perforin, are pooled and briefly dialyzed against buffer A (2 hr; longer dialysis results in precipitates). The dialysate is then loaded on a Mono S cation-exchange column equilibrated in buffer A. After washing the column with 20 ml of buffer A, the proteins are eluted with 30 ml of a linear gradient from 0 to 100% buffer B at a flow rate of 1 ml/min. Fractions of 1 ml are collected and 50 μl of each is analyzed by 10% SDS–PAGE (Fig. 1). The fractions with highest Boc-Ala-Ala-Asp-SBzl esterase activity contain granzyme B. The contamination with granzyme A can be assessed by the Boc-Leu-SBzl esterase assay (see [4] in this volume).

Procedure 2. The solubilized granule proteins (in 5 ml of 1.5 M NaCl) are loaded onto a Sephacryl S300 (2.6 × 90 cm) column equilibrated in TSK buffer. Fractions of 6 ml are collected and tested for their Boc-Ala-Ala-Asp-SBzl and Boc-Leu-SBzl esterase activity. Fractions with the highest Boc-Ala-Ala-Asp-SBzl and lowest Boc-Leu-SBzl esterase activity (see [4] in this volume) are pooled and further separated by cation-exchange chromatography on a Mono S column as described above. Because

[24] D. Masson and J. Tschopp, *J. Biol. Chem.* **260**, 9069 (1985).

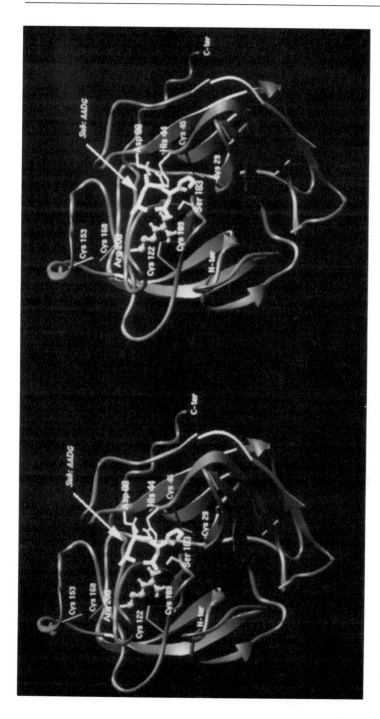

Fig. 2. Stereo ribbons representation of mouse granzyme B. The backbone of mouse granzyme B is represented as a gray ribbon, with arrows showing the orientation of the β strands. The gray sticks represent the disulfide bonds (the cysteines are numbered). The active site triad is represented by white sticks, and Arg-208 by a ball and sticks representation. The substrate Ala-Ala-Asp-Gly is shown in the active site of granzyme B. Note the proximity of the guanidino group of Arg-208 and the side chain of Asp in the substrate.

TABLE I
GRANZYME B INHIBITORS

Inhibitor	Substrate	Concentration	Inhibition (%)	Ref.
Lima bean trypsin inhibitor	GluNHNap	0.25 mg/ml	50 ± 3	11
Soybean trypsin inhibitor	GluNHNap	0.25 mg/ml	52 ± 14	11
Human α_1-protease inhibitor	GluNHNap	0.01 mg/ml	85 ± 13	11
Bovine aprotinin (Trasylol)	GluNHNap	0.25 mg/ml	40 ± 19	11
Human α_2-macroglobulin	GluNHNap	0.01 mg/ml	66 ± 2	11
Human antithrombin III	GluNHNap	5 g/ml	−3 ± 6	11
Phosphoramidon	GluNHNap	0.05 mg/ml	65 ± 5	11
Chymostatin	GluNHNap	0.05 mg/ml	42 ± 16	11
Antipain	GluNHNap	0.05 mg/ml	35 ± 18	11
N-α-Cbz-L-Phe-CH$_2$Cl	GluNHNap	0.05 mg/ml	16 ± 12	11
Elastatinal	GluNHNap	0.05 mg/ml	11 ± 13	11
Pepstatin	GluNHNap	0.03 mg/ml	8 ± 5	11
Leupeptin	GluNHNap	0.05 mg/ml	1 ± 36	11
Benzamidine	GluNHNap	0.10 mM	20 ± 14	11
			Inhibition $k_{obs}/[I]$ (M^{-1} sec^{-1})	
3,4-Dichloroisocoumarin	Boc-Ala-Ala-Asp-SBzl	8.2 μM	3700	21
4-Chloro-3-methoxyisocoumarin	Boc-Ala-Ala-Asp-SBzl	39 μM	21	21

granzyme B is unstable at pH 6.0, combine 0.2 ml buffer C with each fraction of interest, to readjust the pH to 4.5.

Substrates

Based on the experimentally determined three-dimensional (3D) structure of rat mast cell protease II, detailed molecular models were built for human[25] and mouse[26] granzyme B. The residues lining the S1 substrate-binding pocket are found in positions −6, +15 to +17, and +25 relative to the active serine in the aligned sequences,[27] and determine the nature of the side chain preceding the cleaved peptide bond of the substrate. In mouse granzyme B, these residues are Ala-177, Ser-198, Tyr-199, Gly-200, and Arg-208. The molecular model (Fig. 2) shows that the side chain of Arg-208 is oriented toward the active site, and partially fills the binding pocket. Granzyme B would thus preferentially cleave substrates with

[25] M. E. P. Murphy, J. Moult, R. C. Bleackley, H. Gershenfeld, I. L. Weissman, and M. N. G. James, *Proteins: Struct. Funct. Genet.* **4**, 190 (1988).
[26] M. C. Peitsch and J. Tschopp, unpublished (1989).
[27] J. Kraut, *Annu. Rev. Biochem.* **46**, 331 (1977).

negatively charged side chains, and should prefer aspartic over glutamic acid, because aspartic acid has a shorter side chain. Experimental evidence was provided by Poe *et al.*[11] and Odake *et al.*,[21] who demonstrated that granzyme B cleaves synthetic peptide substrates preferentially after aspartic acid and more weakly after a glutamic acid. Granzyme B also exhibits a weak esterase activity toward substrates containing a serine (Boc-Ala-Ala-Ser-SBzl[21]) or an asparagine (Boc-Ala-Ala-Asn-SBzl[21]) because these two residues can form hydrogen bonds with the Arg-208. Furthermore, granzyme B preparations cleave substrates with a methionine Boc-Ala-Ala-Met-SBzl[21]), but it is not clear whether this activity is due to granzyme B or another contaminating esterase in these preparations.

Inhibitors

Granzyme B is inhibited by a variety of inhibitors. A list of the known inhibitors and their relative efficiency is given in Table I.

Properties

pH Dependence

The Boc-Ala-Ala-Asp-SBzl esterase activity of granzyme B is effective in the range of pH 6.0 to 8.0 with a maximum at pH 7.5.

Ion Dependence

The Boc-Ala-Ala-Asp-SBzl esterase activity of granzyme B is enhanced by the presence of 50 mM CaCl$_2$, whereas the Boc-Ala-Ala-Met-SBzl esterase activity is highest in the presence of 0.1–1.0 M NaCl.

Stability

Granzyme B is stable at pH 4.5 for over 6 months at $-20°$.

[6] Tryptase: A Mast Cell Serine Protease

By Lawrence B. Schwartz

Introduction

Tryptase (EC 3.4.21.59) is the principal enzyme accounting for the trypsinlike activity first detected in human mast cells by histochemical techniques.[1,2] Mast cells and basophils are the major effector cells in allergic diseases. Tryptase, along with chymase, carboxypeptidase, and a cathepsin G-like protease, are the dominant protein components of secretory granules in human and rodent mast cells.[3] Such enzymes serve as selective markers that distinguish mast cells from other cell types, including basophils, and different mast cell subpopulations from one another. Whereas human and rodent mast cells are rich in serine proteases of the trypsin and chymotrypsin families, human basophils have relatively little esterase/protease activity. Because of the selective expression of these proteases in mast cells, one could hypothesize that these enzymes play an important role in the biology of mast cells, that such enzymes might serve as sensitive and specific markers of mast cells and of mast cell activation, and that regulation of the transcription of their corresponding genes is related to the regulation of the differentiation of this cell type. This article will concentrate on tryptase, particularly that derived from human mast cells, except where a different species of origin is specified.

Purification of Tryptase from Human Tissues

Tryptase was first purified to apparent homogeneity from dispersed and enriched lung mast cells in 1981,[1] and later from crude lung cell dispersions,[4] HMC-1 cells,[5] and from lung,[6] pituitary,[7] and skin[8] tissue.

[1] L. B. Schwartz, R. A. Lewis, and K. F. Austen, *J. Biol. Chem.* **256**, 11939 (1981).
[2] V. K. Hopsu and G. G. Glenner, *J. Cell Biol.* **17**, 503 (1963).
[3] L. B. Schwartz (ed.), "Neutral Proteases of Mast Cells." Karger, Basel, 1990.
[4] L. B. Schwartz and T. R. Bradford, *J. Biol. Chem.* **261**, 7372 (1986).
[5] J. H. Butterfield, D. A. Weiler, L. W. Hunt, S. R. Wynn, and P. C. Roche, *J. Leukocyte Biol.* **47**, 409 (1990).
[6] T. J. Smith, M. W. Hougland, and D. A. Johnson, *J. Biol. Chem.* **259**, 11046 (1984).
[7] J. A. Cromlish, N. G. Siedah, M. Marcinkiewicz, J. Hamelin, D. A. Johnson, and M. Chretein, *J. Biol. Chem.* **262**, 1363 (1987).
[8] I. T. Harvima, N. M. Schechter, R. J. Harvima, and J. E. Fräki, *Biochim. Biophys. Acta* **957**, 71 (1988).

The original technique involved sequential chromatography on Dowex 1-X2, DEAE–Sephadex, and heparin–agarose, but required dispersed mast cells of at least 20% purity. Subsequently, hydrophobic chromatography was included to permit purification from cruder sources. An immunoaffinity chromatography technique has been developed with a mouse monoclonal immunoglobulin G (IgG) recognizing the active form of tryptase.[9]

In our laboratory purification of tryptase by conventional chromatography proceeds as follows. Dispersed lung cells are suspended in 0.01 M 2-(N-morpholino)ethanesulfonic acid (MES), pH 6.5, and 1 M NaCl at 10 to 20 × 10^8 cells/ml, sonicated with a cell disrupter (model W-225R, Heat Systems-Ultrasonics, Inc., Plainview, NY) with microtip attachment at power 3 to 4 for 20 pulses times two at 50% pulse cycle, and centrifuged at 40,000 g for 40 min at 4°. Tryptase is released from mast cell granules by this procedure and is solubilized from endogenous heparin proteoglycan. The low pH and temperature and high salt concentration tend to stabilize tryptase in its active form.

Cell sonicates are next brought to 80% saturated ammonium sulfate at 4°, incubated for 60 min, and centrifuged at 40,000 g for 60 min at 4°. Pellets that contain all of the tryptase are resuspended in 45% ammonium sulfate, sonicated if necessary to encourage solubilization of tryptase, and centrifuged at 40,000 g for 45 min to remove insoluble debris. The supernatant, which should contain about 75% of the starting amount of tryptase activity, is applied directly to a decyl-agarose column equilibrated in 45% ammonium sulfate (25 to 50 × 10^6 cell equivalents/ml bed volume), and the column is washed with 3 column volumes of this solution. Tryptase is step-eluted with 25% ammonium sulfate and dialyzed against 0.01 M MES, pH 6.5, to at least 0.02 M NaCl equivalents as determined by conductivity (model CDM-3, Radiometer, Copenhagen). The yield on decyl-agarose typically approaches 90%.

The dialyzate is applied to a Sephadex A-25 (DEAE-Sephadex) column equilibrated with 0.01 M MES, pH 6.5, 2 mM CaCl$_2$ ($\leq$3 mg protein/ml bed volume). After washing with 3 column volumes of 0.01 M MES, pH 6.5, 0.05 M NaCl, tryptase is eluted with a 10-column volume linear gradient from 0.05 to 2.0 M NaCl. A single peak of tryptase activity is collected, the yield being about 75%, and applied directly to heparin–agarose ($\leq$3 mg/protein/ml bed volume), washed with 3 column volumes of 0.45 M NaCl, and eluted with a linear gradient of 0.45 to 2 M NaCl in 0.01 M MES, pH 6.5, made to 20% glycerol. The yield from heparin

[9] L. B. Schwartz, T. R. Bradford, D. C. Lee, and J. F. Chlebowski, *J. Immunol.* **144,** 2304 (1990).

chromatography typically approaches 100%. The overall yield of tryptase should be between 50 and 60% from the starting sonicate. All chromatographic steps are performed at 4°. Protein determinations are made by the bicinchoninic acid method.[10]

Methods for Enzymatic Assay

Tryptase activity can be measured by cleavage of small synthetic ester or peptide substrates with Arg or Lys in the C-terminal position. Cleavage of tosyl-L-arginine methyl ester (TAME) is determined by addition of tryptase to 1 ml of 0.04 M Tris-HCl, pH 8.1, containing 0.15 M NaCl and 1 mM TAME at either room temperature or at 37°. The molar extinction coefficient for the change in absorbance at 247 nm equals 540. The concentration of TAME can be confirmed by measuring the absorbance at 247 nm after complete cleavage of the substrate by excess trypsin. Purified tryptase typically exhibits a specific activity of 100 to 120 U/mg.

A variety of tripeptide derivatives of p-nitroanilide also are used to assay tryptase.[11] Interestingly, substrates with basic amino acids at the S1 and S2 positions appear to be among the preferred substrates for tryptase. The pH optima for peptide substrates tends to be slightly lower than for ester substrates. In our laboratory rates of hydrolysis of tosyl-L-Gly-Pro-Lys-p-nitroanilide are determined by addition of 2 to 10 μl of tryptase to 0.6 ml of 0.05 M HEPES, pH 7.6, 0.12 M NaCl, and 0.1 mM substrate at room temperature and monitoring the absorbance at 405 nm. The molar extinction coefficient for released p-nitroanilide is 8800 at this wavelength. The concentration of the tripeptide substrate can be checked by measuring the change in absorbance at 405 nm after complete cleavage with excess trypsin, or by measuring the absorbance at 316 nm where the molar extinction coefficient for uncleaved substrate is 12,700.

Physicochemistry

Tryptase is a tetrameric endoprotease of 134,000 Da with subunits of 31,000 to 34,000 Da, each with an active enzymatic site[1] and common antigenic sites.[12] A reduction in molecular mass of tryptase subunits of 2000 to 4000 after treatment with endoglycosidase indicates that carbohy-

[10] P. K. Smith, R. I. Krohn, G. T. Hermanson, A. K. Mallia, F. H. Gartner, M. D. Provenzano, E. K. Fujimoto, N. M. Goeke, B. J. Olson, and D. C. Klenk, *Anal. Biochem.* **150,** 76 (1985).
[11] T. Tanaka, B. J. McRae, K. Cho, R. Cook, J. E. Fraki, D. A. Johnson, and J. C. Powers, *J. Biol. Chem.* **258,** 13552 (1983).
[12] L. B. Schwartz, *J. Immunol.* **134,** 526 (1985).

drate is present on each subunit.[7] Within mast cell granules and after secretion, tryptase associates with proteoglycan, presumably heparin, by ionic interactions. Because NaCl concentrations of at least 0.7 M are needed to dissociate tryptase from heparin, the mechanism for dissociation to occur *in vivo*, if one exists at all, is not clearly defined. However, antithrombin III, when present in large excess *in vitro*, can dissociate up to one-third of the tryptase bound to heparin. Chymase and carboxypeptidase appear to residue in a complex with proteoglycan that is distinct and separable from the macromolecular tryptase–proteoglycan complex,[13] suggesting that tryptase is processed and packaged separately from these other two proteases.

Enzymatic Properties

Tryptase hydrolyzes proteins and peptides on the C-terminal side of the basic residues, lysine and arginine.[11] Tripeptide substrates with basic amino acid residues in the S1 and S2 positions are preferred by tryptase.[7] Such substrates also are cleaved by other trypsinlike enzymes and therefore are not specific. Although classified as a serine protease, the absence of inhibition by the classical inhibitors of serine esterases present in plasma, lung, and urine, as well as by lima bean, soybean, and ovomucoid trypsin inhibitors, clearly distinguishes tryptase from pancreatic trypsin and from most other serine esterases.[14] Tryptase is inhibited by small molecular weight substances such as leupeptin, diisopropyl fluorophosphate, and phenylmethylsulfonyl fluoride. Divalent cations such as calcium,[15] and benzamidine and its derivatives,[16,17] are competitive inhibitors of tryptase. Interestingly, histamine shifts the substrate dose–response curve to the right and to a sigmoidal pattern, suggesting cooperative behavior.[15]

Instead of the enzymatic activity of tryptase being regulated by the usual inhibitors of serine proteases, tryptase is uniquely stabilized in its active tetrameric form by heparin,[4] to which it is ionically bound under physiologic conditions (Fig. 1). When free in solution, tryptase subunits irreversibly dissociate from one another into inactive monomers, without

[13] S. M. Goldstein, J. Leong, L. B. Schwartz, and D. Cooke, *J. Immunol.* **148**, 2475 (1992).
[14] S. C. Alter, J. A. Kramps, A. Janoff, and L. B. Schwartz, *Arch. Biochem. Biophys.* **276**, 26 (1990).
[15] S. C. Alter and L. B. Schwartz, *Biochim. Biophys. Acta* **991**, 426 (1989).
[16] J. Stürzebecher, D. Prasa, and C. P. Sommerhoff, *Biol. Chem. Hoppe-Seyler* **373**, 1025 (1992).
[17] G. H. Caughey, W. W. Raymond, E. Bacci, R. J. Lombardy, and R. R. Tidwell, *J. Pharmacol. Exp. Ther.* **264**, 676 (1993).

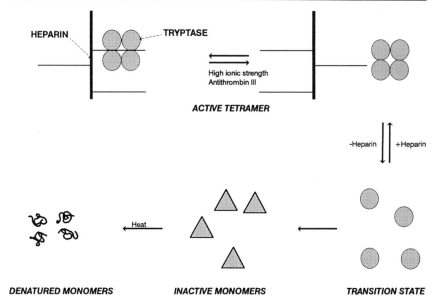

FIG. 1. Regulation of tryptase. Tryptase is stabilized in its active tetrameric form by binding to heparin proteoglycan. At high ionic strength or in the presence of excess antithrombin III, at least a portion of tryptase is solubilized from heparin. Free tryptase spontaneously dissociates into monomers that soon become irreversibly inactive in association with changes in conformation. Heating tryptase will denature the protein.

any evidence for autodegradation. Substantial conformational changes occur during this process, as evidenced by circular dichroic spectral shifts and by distinct epitopes being detected on the active tetramer and inactive monomers.[9] It also appears that inactivation proceeds as a multistep process, because limited recovery of lost enzyme activity reportedly can be detected at an early stage of the inactivation process.[18] Chondroitin sulfate E and polyglutamic acid stabilize tryptase somewhat less well than heparin, and chondroitin monosulfates are little better than buffer at stabilizing tryptase under physiologic conditions. Thus, negative charge density rather than carbohydrate structure is the primary determinant for stabilizing tryptase activity.[19] The physiologic means of dissociating tryptase from heparin or chondroitin sulfate E *in vivo* is not totally clear, but the capacity of divalent cations[15] and of heparin-binding proteins such as antithrombin III[14] to destabilize heparin-bound tryptase at physiologic concentrations *in vitro* may be relevant to the situation *in vivo*. *In vitro*,

[18] N. M. Schechter, G. Y. Eng, and D. R. McCaslin, *Biochemistry* **32**, 2617 (1993).

TABLE I
POTENTIAL BIOLOGICAL ACTIVITIES OF HUMAN TRYPTASE

Enzymatic event	Potential significance
Fibrinogenolysis	Prevents clotting at local sites of mast cell activation
Augmentation of histamine-mediated contractility of airway smooth muscle	Increases airway hyperreactivity to histamine and allergens
Degradation of vasoactive intestinal peptide	Decreases endogenous bronchodilatory activity in lung
Activation of prostromelysin	Facilitates activation of procollagenase and tissue remodeling
Degradation of fibronectin	Tissue remodeling
Stimulation of fibroblast proliferation	Enhances fibrogenesis or would repair

the relationship of heparin-stabilized tryptase activity to ionic strength is somewhat complex, because it reflects the destabilizing effect of dissociation of tryptase from heparin, the stabilizing effect of high ionic strength on free tetramer, and the direct effects of increasing ionic strength to raise the K_m and increase K_{cat} values.[7,19]

Potential Biological Activities

Even though the biologic role of tryptase has not been convincingly demonstrated, several activities of potential biologic interest have been examined in vitro (Table I). Tryptase rapidly cleaves and inactivates fibrinogen as a coagulable substrate for thrombin[20]; the lack of fibrin deposition and rapid resolution of urticaria/angioedema reactions may, in part, reflect this same activity in vivo. Tryptase activates latent collagenase derived from rheumatoid synovial cells, apparently by first activating prostromelysin (metalloproteinase III),[21] which in turn activates latent collagenase. Tryptase directly degrades fibronectin.[22] The increased numbers of mast cells found in rheumatoid synovium[23] and in inflammatory

[19] S. C. Alter, D. D. Metcalfe, T. R. Bradford, and L. B. Schwartz, *Biochem. J.* **248,** 821 (1987).

[20] L. B. Schwartz, T. R. Bradford, B. H. Littman, and B. U. Wintroub, *J. Immunol.* **135,** 2762 (1985).

[21] B. L. Gruber, M. J. Marchese, K. Suzuki, L. B. Schwartz, Y. Okada, H. Nagase, and N. S. Ramamurthy, *J. Clin. Invest.* **84,** 1657 (1989).

[22] J. Lohi, I. Harvima, and J. Keski-Oja, *J. Cell. Biochem.* **50,** 337 (1992).

[23] D. G. Malone, R. L. Wilder, A. M. Saavedra-Delgado, and D. D. Metcalfe, *Arthritis Rheum.* **30,** 130 (1987).

cutaneous lesions of scleroderma[24] together with the increased tissue turnover that occurs in each condition suggest a related role for tryptase. Tryptase does not appear to affect complement anaphylatoxin or bradykinin generation[25]; bradykinin, lysylbradykinin, C3a, and C5a, if generated *in vivo* during immediate hypersensitivity events, are not directly due to tryptase activity. Tryptase does degrade calcitonin gene-related peptide[26,27] and vasoactive intestinal peptide,[28] two neuropeptides that relax smooth muscle. If human tryptase destroys vasoactive intestinal peptide in lung, then bronchial tone might increase and increased airway hyperreactivity ensue. Tryptase also exhibits mitogenic activity on fibroblast lines *in vitro*,[29] suggesting one mechanism by which mast cells might participate in fibrogenesis in diseases such as scleroderma.

Molecular Biology

Two tryptase cDNA molecules (α and β) have been cloned from a human lung mast cell cDNA library and sequenced,[30,31] and three (1, 2, and 3) from a human skin mast cell library, as summarized in Table II.[32] The α-tryptase cDNA codes for a 27,682-Da, 245-amino acid catalytic protein with two putative carbohydrate-binding sites.[30,30a] The β-tryptase cDNA codes for a 245-amino acid catalytic protein that is 92% identical to α-tryptase and 98 to 99% identical to tryptases 1, 2, and 3. There are two N-linked carbohydrate-binding sites in the α, 1, and 3 types of tryptase, and one carbohydrate-binding site in the other tryptase sequences, suggesting an explanation for a portion of the electrophoretic heterogeneity of natural tryptase in SDS–polyacrylamide gels. The amino acid sequence for α-tryptase contains 24 amino acid differences and a single amino acid dele-

[24] R. A. Hawkins, H. N. Claman, R. A. Clark, and J. C. Steigerwald, *Ann. Intern. Med.* **102**, 182 (1985).

[25] L. B. Schwartz, *Monogr. Allergy* **27**, 90 (1990).

[26] A. F. Walls, S. D. Brain, A. Desai, P. J. Jose, E. Hawkings, M. K. Church, and T. J. Williams, *Biochem. Pharmacol.* **43**, 1243 (1992).

[27] E. K. Tam and G. H. Caughey, *Am. J. Respir. Cell Mol. Biol.* **3**, 27 (1990).

[28] G. H. Caughey, F. Leidig, N. F. Viro, and J. A. Nadel, *J. Pharmacol. Exp. Ther.* **244**, 133 (1988).

[29] T. Hartmann, S. J. Ruoss, W. W. Raymond, K. Seuwen, and G. H. Caughey, *Am. J. Physiol.* **262**, L528 (1992).

[30] J. S. Miller, E. H. Westin, and L. B. Schwartz, *J. Clin. Invest.* **84**, 1188 (1989).

[30a] R. Huang, M. Åbrink, A. E. Globl, G. Nilsson, M. Aveskogh, L. G. Larsson, K. Nilsson, and L. Hellman, *Scand. J. Immunol.* **38**, 359 (1993).

[31] J. S. Miller, G. Moxley, and L. B. Schwartz, *J. Clin. Invest.* **86**, 864 (1990).

[32] P. Vanderslice, S. M. Ballinger, E. K. Tam, S. M. Goldstein, C. S. Craik, and G. H. Caughey, *Proc. Natl. Acad. Sci. U.S.A.* **87**, 3811 (1990).

TABLE II
MOLECULAR CHARACTERISTICS OF HUMAN TRYPTASE

Property	α-Tryptase	β-Tryptase (tryptases 1, 2, 3, β)
Number of amino acids		
Catalytic regions	245	245
Prepro regions	30	30
Homologies of catalytic regions		
α to β		92% identical
β to 1, 2, 3		98 to 99% identical
Homologies of prepro regions		
α to β		87%
β to 1, 2, 3		100%
Carbohydrate-binding regions	2	1 to 2
Calculated peptide molecular weights		
Catalytic region	27,682	27,449–27,585
Prepro region	3048	3089
Chromosome localization	16	16

tion compared to the β-tryptase sequence. Interestingly, the amino acid region of β-tryptase corresponding to where the deletion occurs is identical to the sequence reported for dog tryptase.[33]

The prepropeptide portion is 30 amino acids long in all cases; these regions are identical in the β, 1, 2, and 3 forms of tryptase, but differ by 4 amino acids from the α-tryptase prepropeptide. The activation peptide for tryptase has not been clearly defined, but its termination in Gly is unusual. The propeptide sequences for pancreatic serine proteases terminate in a basic residue; most leukocyte and mast cell granule proteases terminate in an acidic residue. Neither the prepropeptide nor the propeptide portions of tryptase have been detected in enzyme purified from human lung mast cells, consistent with tryptase being sorted in an active form. The prepropeptide may be responsible for directing tryptase to secretory granules and/or the assembly of tryptase as an active tetramer, and most likely are removed before or shortly after tryptase enters the secretory granules.

By applying polymerase chain reaction (PCR) techniques to DNA taken from human : hamster somatic cell hybrids, at least two tryptase sequences were localized to the normal human haploid genome on chromosome 16.[31] One sequence corresponds to α-tryptase and tryptase III, the other to β-tryptase, tryptase I, and tryptase II. One tryptase gene has been sequenced that appears to match tryptase I cDNA. It contains six exons and five introns. The first intron falls between the 5'-noncoding

[33] P. Vanderslice, C. S. Craik, J. A. Nadel, and G. H. Caughey, Biochemistry 28, 4148 (1989).

and the ATG-methionine start site, which is rare, if not unique, among proteases. Positions of the four introns within the coding region are similar to those for trypsin and glandular kallikrein, but different from those for proteases involved in blood coagulation and for mast cell chymases. The 5' region has several consensus sites for regulating transcription, including TATA and CAAT boxes and a region showing homology to the enhancer elements for the rat chymase II gene[34] and for protease genes expressed in the pancreas. If the latter proves to be a mast cell-specific enhancer region, it would be of interest to examine its regulation during the differentiation of human mast cells.

Cellular Marker of Mast Cells during Development and in Tissues

Based on double-labeling techniques for tryptase by immunohistochemistry, and for mast cells by histochemistry, tryptase was found to be in all mast cells and only in mast cells in human skin, lung, and bowel. In contrast, chymase, carboxypeptidase, and a cathepsin G-like protease were detected together in a subset of mast cells called MC_{TC} cells.[35] Those mast cells with only tryptase were called MC_T cells. Substantial amounts of tryptase are present in MC_{TC} cells derived from foreskin (35 pg/cell) and in MC_T cells derived from lung (10 pg/cell), where it is located in secretory granules and is released in parallel with histamine during degranulation.[36] These levels appear to account for perhaps 20% of the entire protein in the mast cell. Small amounts have been measured in human basophils (0.04 pg/cell). The basophil leukemia cell line KU-812 expressed only tryptase 1 mRNA, a β-like sequence, but whether this corresponds to normal basophils is not known.[37] Other cell types found in peripheral blood, such as eosinophils, neutrophils, monocytes, and lymphocytes, have no detectable tryptase. Thus the enzyme is a discriminating marker of human mast cells.

Tryptase expression occurs at the time granule formation begins during the development of human mast cells.[38] Thus the enzyme serves as a useful differentiation marker for this cell type. Studies of Kit-ligand-mediated differentiation of mast cells from fetal liver cells[39] and from cord blood

[34] J. Sarid, P. N. Benfey, and P. Leder, *J. Biol. Chem.* **264,** 1022 (1989).
[35] A. A. Irani and L. B. Schwartz, *in* "Neutral Proteases of Mast Cells" (L. B. Schwartz, ed.), p. 146. Karger, Basel, 1990.
[36] L. B. Schwartz, R. A. Lewis, D. Seldin, and K. F. Austen, *J. Immunol.* **126,** 1290 (1981).
[37] T. Blom and L. Hellman, *Scand. J. Immunol.* **37,** 203 (1993).
[38] S. S. Craig, N. M. Schechter, and L. B. Schwartz, *Lab. Invest.* **60,** 147 (1989).
[39] A. A. Irani, G. Nilsson, U. Miettinen, S. C. Craig, L. K. Ashman, T. Ishizaka, K. M. Zsebo, and L. B. Schwartz, *Blood* **80,** 3009 (1992).

mononuclear cells[40] *in vitro* have utilized tryptase expression as a critical marker for identifying commitment to a mast cell lineage.

Clinical Utility as Soluble Marker of Mast Cell-Dependent Events

The two major barriers to the assessment of mast cell involvement in clinical events have been the localization of this cell type in tissue rather than in circulation and the lack of a marker, detectable in biologic fluids, with sufficient sensitivity and specificity to indicate activation of mast cells occurring in tissues.

Several mediators of mast cells have been considered as potentially useful indicators of mast cell or basophil activation. Certain neutral proteases have greater potential utility than do other mediators because of their abundant presence in and selective localization to mast cell secretory granules. A sensitive and specific immunoassay for tryptase was developed that distinguished tryptase from other trypsinlike enzymes and was used to quantify tryptase in complex biologic fluids[41,42] such as serum,[43] bronchoalveolar lavage fluid,[44] nasal lavage fluid,[45] skin chamber fluid,[46,47] and tears.[48] This has enabled the measurement of tryptase release *in vivo* and thereby a precise assessment of mast cell activation. In serum, tryptase levels in normal subjects are undetectable (<1 ng/ml), whereas elevated levels are detected in systemic mast cell disorders, such as anaphylaxis and mastocytosis. Thus, tryptase levels provide a more precise assessment of mast cell activation *in vivo* than previously available.

Tryptase quantities in serum are elevated in systemic anaphylaxis.[43] During bee sting-induced anaphylaxis circulating levels are maximal at 60

[40] H. Mitsui, T. Furitsu, A. M. Dvorak, A.-M. A. Irani, L. B. Schwartz, N. Inagaki, M. Takei, K. Ishizaka, K. M. Zsebo, S. Gillis, and T. Ishizaka, *Proc. Natl. Acad. Sci. U.S.A.* **90**, 735 (1993).

[41] S. Wenzel, A. M. Irani, J. M. Sanders, T. R. Bradford, and L. B. Schwartz, *J. Immunol. Methods* **86**, 139 (1986).

[42] I. Enander, P. Matsson, J. Nystrand, A.-S. Andersson, E. Eklund, T. R. Bradford, and L. B. Schwartz, *J. Immunol. Methods* **138**, 39 (1991).

[43] L. B. Schwartz, D. D. Metcalfe, J. S. Miller, H. Earl, and T. Sullivan, *N. Engl. J. Med.* **316**, 1622 (1987).

[44] S. E. Wenzel, A. A. Fowler III, and L. B. Schwartz, *Am. Rev. Respir. Dis.* **137**, 1002 (1988).

[45] M. Castells and L. B. Schwartz, *J. Allergy Clin. Immunol.* **82**, 348 (1988).

[46] M. Shalit, L. B. Schwartz, N. Golzar, C. vonAllman, M. Valenzano, P. Fleekop, P. C. Atkins, and B. Zweiman, *J. Immunol.* **141**, 821 (1988).

[47] L. B. Schwartz, P. C. Atkins, T. R. Bradford, P. Fleekop, M. Shalit, and B. Zweiman, *J. Allergy Clin. Immunol.* **80**, 850 (1987).

[48] S. I. Butrus, K. I. Ochsner, M. B. Abelson, and L. B. Schwartz, *Ophthalmology* **97**, 1678 (1990).

to 120 min, and then decline with a half-life of 1.5 to 2.5 hr.[49] The magnitude of the tryptase level correlates with the clinical severity as measured by the drop in mean arterial pressure.[50] Elevated levels of mast cell tryptase in postmortem sera reflect antemortem mast cell activation and suggest that tryptase levels also may be used as a practical indicator of fatal anaphylaxis.[51] Elevated levels of tryptase in postmortem sera of victims of sudden infant death syndrome (SIDS) has implicated mast cell activation in the pathogenesis of this disorder.[52]

Ongoing mast cell activation in asthma, detected by elevated levels of tryptase and PGD_2 in bronchoalveolar lavage fluid, higher spontaneous release of histamine by mast cells obtained from the bronchoalveolar lavage fluid of asthmatics, and ultrastructural analysis of mast cells in pulmonary tissue, appears to be a characteristic of the inflammation of asthma.[44,53-55] The early and late response to endobronchial challenge of atopic subjects without asthma at different doses of allergen showed findings analogous to those in the skin, with eosinophil and perhaps basophil recruitment and activation occurring during the late-phase response.[56] Mast cell activation appears to be involved in the immediate respiratory and vascular responses to oral aspirin challenge of sensitive subjects.[57] On the other hand, mast cell activation may not be a feature of exercise-induced asthma.[58]

[49] L. B. Schwartz, J. W. Yunginger, J. S. Miller, R. Bokhari, and D. Dull, *J. Clin. Invest.* **83,** 1551 (1989).

[50] P. G. Van der Linden, C. E. Hack, J. Poortman, Y. C. Vivié-Kipp, A. Struyvenberg, and J. K. Van der Zwan, *J. Allergy Clin. Immunol.* **90,** 110 (1992).

[51] J. W. Yunginger, D. R. Nelson, D. L. Squillace, R. T. Jones, K. E. Holley, B. A. Hyma, L. Biedrzycki, K. G. Sweeney, W. Q. Sturner, and L. B. Schwartz, *J. Forensic Sci.* **36,** 857 (1991).

[52] M. S. Platt, J. W. Yunginger, A. Sekula-Perlman, A. A. Irani, and L. B. Schwartz, *J. Allergy Clin. Immunol.,* in press (1993).

[53] M. C. Liu, E. R. Bleecker, L. M. Lichtenstein, A. Kagey-Sobotka, Y. Niv, T. L. McLemore, S. Permutt, D. Proud, and W. C. Hubbard, *Am. Rev. Respir. Dis.* **142,** 126 (1990).

[54] P. H. Howarth, J. Wilson, R. Djukanovic, S. Wilson, K. Britten, A. Walls, W. R. Roche, and S. T. Holgate, *Int. Arch. Allergy Appl. Immunol.* **94,** 266 (1991).

[55] N. N. Jarjour, W. J. Calhoun, L. B. Schwartz, and W. W. Busse, *Am. Rev. Resp. Dis.* **144,** 83 (1991).

[56] J. B. Sedgwick, W. J. Calhoun, G. J. Gleich, H. Kita, J. S. Abrams, L. B. Schwartz, B. Volovitz, M. Ben-Yaakov, and W. W. Busse, *Am. Rev. Respir. Dis.* **144,** 1274 (1991).

[57] J. V. Bosso, L. B. Schwartz, and D. D. Stevenson, *J. Allergy Clin. Immunol.* **88,** 830 (1991).

[58] D. H. Broide, S. Eisman, J. W. Ramsdell, P. Ferguson, L. B. Schwartz, and S. I. Wasserman, *Am. Rev. Respir. Dis.* **141,** 563 (1990).

Tryptase from Nonhuman Sources

Rat mast cell tryptase has been purified from peritoneal mast cells[59] and skin.[60] Only small amounts (approximately 0.5 pg/cell) have been measured in serosal mast cells, where it was purified in association with an inhibitor of the enzyme called trypstatin, which is related to inter-α-trypsin inhibitor. Unlike human tryptase, rat tryptase derived from peritoneal mast cells is inhibited by soybean trypsin inhibitor, α_1-antitrypsin, antipain, and aprotinin. In contrast, skin-derived tryptase was not found to be associated with an inhibitor, and was not inhibited by soybean trypsin inhibitor and α_1-antitrypsin, but like tryptase from peritoneal mast cells, it was inhibited by antipain and aprotinin. In neither case did tryptase bind to heparin in physiologic buffer, a property also deduced with subcellular fractions of rat peritoneal mast cells.[61] The size of rat mast cell tryptase has been reported to vary from 145,000 to 160,000 Da, with four equally sized glycosylated subunits of 30,000 to 40,000 Da.

The distribution of tryptase among mast cells in rats is distinct from that in humans. Polyclonal antibodies prepared against skin-derived tryptase were used to show that the enzyme was present only in a portion of the safranin-positive mast cells (connective tissue type).[62] For example, for peritoneal mast cells 95% were safranin positive and 20% of these were tryptase positive; mast cells in bowel mucosae were essentially all safranin negative and tryptase negative; lung mast cells were 53% safranin positive and 94% of these were tryptase positive; skin mast cells were 80% safranin positive and of these only 6% were tryptase positive. Thus, rat mucosal mast cells lack this enzyme, analogous to mouse cells, wherein, among mature mast cells, only those of the connective tissue type are positive for tryptase mRNA.

Murine mast cell tryptase is represented by two genes and the corresponding mRNA molecules.[63,64] Mouse mast cell protease (MMCP)-6 is expressed at the mRNA level in connective tissue mast cells, whereas MMCP-7 mRNA expression has only been detected on the immature IL-3-dependent bone marrow-derived mast cells. MMCP-7 has only five exons, unlike MMCP-7 and human tryptase, which have six. This results

[59] H. Kido, Y. Yokogoshi, and N. Katunuma, *J. Biol. Chem.* **263**, 18104 (1988).
[60] V. J. Braganza and W. H. Simmons, *Biochemistry* **30**, 4997 (1991).
[61] D. Lagunoff, A. Rickard, and C. Marquardt, *Arch. Biochem. Biophys.* **291**, 52 (1991).
[62] Z. Chen, A. A. Irani, T. R. Bradford, S. S. Craig, G. Newlands, H. Miller, T. Huff, W. H. Simmons, and L. B. Schwartz, *J. Histochem. Cytochem.* **41**, 961 (1993).
[63] D. S. Reynolds, R. L. Stevens, W. S. Lane, M. H. Carr, K. F. Austen, and W. E. Serafin, *Proc. Natl. Acad. Sci. U.S.A.* **87**, 3230 (1990).
[64] W. Chu, D. A. Johnson, and P. R. Musich, *FASEB J.* **4**, A2159 (1991).

from a point mutation at an acceptor splice site. MMCP-6 is located on chromosome 17, separate from the chymase genes, at least four of which are on chromosome 14, suggesting that gene clustering is not critical for coordinate expression.

Tryptase from canine mast cells[65] has been purified from dog mastocytoma cells and cloned.[66] Dog tryptase has a molecular mass of 132,000 to 144,000 Da and is a tetramer composed of two closely related subunits of 30,000 to 35,000 Da each.[65] Like human tryptase, dog tryptase is insensitive to inhibition by α_1-antitrypsin and plasma and is stabilized by heparin. Dog tryptase potentiates histamine-mediated contraction of guinea pig pulmonary smooth muscle[67] and proliferation of fibroblasts.[68]

Bovine tryptase has been purified from liver capsule.[69] Two subunits of 39,000 and 41,000 Da have been reported, with a molecular mass for the native of 360,000, suggesting it may not be a tetramer, unlike tryptase from rat, dog, and human sources. It is sensitive to inhibition by aprotinin and by high ionic strength, but not by soybean trypsin inhibitor and α_1-trypsin inhibitor. Like human and dog tryptase, bovine tryptase binds to heparin.

[65] G. H. Caughey, N. F. Viro, J. Ramachondran, S. C. Lazarus, D. B. Borson, and J. A. Nadel, *Arch. Biochem. Biophys.* **258**, 555 (1987).
[66] G. H. Caughey, E. H. Zerweck, and P. Vanderslice, *J. Biol. Chem.* **266**, 12956 (1991).
[67] K. Sekizawa, G. H. Caughey, S. C. Lazarus, W. M. Gold, and J. A. Nadel, *J. Clin. Invest.* **83**, 175 (1989).
[68] S. J. Ruoss, T. Hartmann, and G. H. Caughey, *J. Clin. Invest.* **88**, 493, (1991).
[69] L. Fiorucci, F. Erba, and F. Ascɔli, *Biol. Chem. Hoppe-Seyler* **373**, 483 (1992).

[7] Hepsin

By KOTOKU KURACHI, ADRIAN TORRES-ROSADO, and AKIHIKO TSUJI

Introduction

Proteases play critical roles in many physiological and pathological processes, such as protein catabolism, blood coagulation, cell growth and migration, tissue rearrangement and morphogenesis in development, inflammation, tumor growth, and metastasis.[1] Although a large number of proteases have been implicated in these processes, our knowledge on the biological role(s) and structure–function relationship of cell surface

[1] J. S. bond, *Biomed. Biochim. Acta* **50**, 775 (1991).

(plasma membrane-associated) serine proteases is greatly limited at the present time.[1]

In order to identify and isolate hitherto unidentified serine proteases, we have taken a molecular approach that is based on amino acid sequences highly conserved among serine proteases.[2] In this approach, a constructed human liver cDNA library was originally screened by an oligonucleotide probe, 5'-CC(G, A, T, or C)GC(G or A)CA(G or A)AACAT-3' (parentheses indicate mixed nucleotide sequences), which corresponds to an amino acid sequence, MFCAG, found approximately 15 amino acids prior to the active site serine residue of many serine endopeptidases of the chymotrypsin family. Thirty-one strongly hybridizing cDNA clones, obtained by screening a liver cDNA library consisting of about 14,000 recombinant cDNA colonies, were sorted by cross-hybridization into subgroups. Nucleotide sequencing analyses found that one of the cDNA subgroups codes for a novel serine protease, which was named hepsin. Subsequently, another human liver cDNA library constructed with λgt11 phage vector was screened with the cDNA as probe, identifying more hepsin cDNA clones, including full-length cDNA clones as well as ones containing a part of possible intron sequence. The frequency of the hepsin cDNA clone found was approximately 70 out of 960,000 recombinant clones when a HepG2 cell line cDNA library was used, suggesting that the hepsin gene is expressed at a moderately high level. The nucleotide sequence of hepsin has a single, long, open reading frame of 417 amino acid residues. Other proteins also identified in this screening approach include known proteases such as prothrombin (the largest subgroup identified), C1r, and factor IX, in addition to unknown proteins that are yet to be characterized.[3]

Predicted Amino Acid Sequence of Hepsin

The nucleotide sequence of the hepsin cDNA and the amino acid sequence predicted from the nucleotide sequence are shown in Fig. 1.[2] The cDNA is about 1.8 kb in length, which is in good agreement with its mRNA size (1.8–1.85 kb) estimated by RNA blot analysis.[4] The predicted amino acid sequence shows that hepsin is synthesized as a protein of 417 amino acid residues in length, starting with Met at amino acid (aa) +1. No typical signal peptide is contained, and instead, an internal hydrophobic

[2] S. P. Leytus, K. R. Loeb, F. S. Hagen, K. Kurachi, and E. W. Davie, *Biochemistry* **27**, 1067 (1988).
[3] S. P. Leytus, K. Kurachi, K. S. Sakariassen, and E. W. Davie, *Biochemistry* **25**, 4855 (1980).
[4] A. Tsuji, A. Torres-Rosado, T. Arai, M. M. Le Beau, R. S. Lemons, S.-H. Chou, and K. Kurachi, *J. Biol. Chem.* **266**, 16948 (1991).

sequence stretch (27 amino acid residues in length) is located at residues 18–44. A sequence, RIVGG, which is typical for the proteolytic activation site of many serine protease zymogens, is also contained at residues 62–66. If this sequence is cleaved between Arg and Ile, in an analogy to the activation mechanism of other serine protease zymogens, hepsin may be converted to an activated form, consisting of a light chain (the amino-terminal half, noncatalytic subunit) and a heavy chain (carboxyl-terminal half, protease catalytic subunit) linked by a disulfide bond. The light chain is composed of a short amino-terminal leading sequence (17 residues) followed by the 27-residue-long hydrophobic sequence and a link sequence (118 amino acid residues). The heavy chain (protease catalytic subunit) is composed of 255 amino acid residues (residues 163–417). The heavy chain contains all features of a typical, functional serine protease, including amino acid residues involved in a formation of the active site triad. The overall similarity of the heavy chain to chymotrypsin, trypsin, and the catalytic subunit of plasmin are 39, 43, and 42%, respectively. These results have indicated that hepsin is a novel serine protease that is likely membrane associated. The light chain of hepsin has no significant similarity to other known protein sequences, except the typical hydrophobic sequence stretch, which is similar to other general membrane-spanning sequences.

The hepsin molecule is predicted to have 19 cysteine residues, 10 in the catalytic subunit and 9 in the noncatalytic subunit. Based on the reasonable structural similarity to serine endopeptidases of known structure, 9 out of 10 Cys residues in the catalytic subunit may form four intrasubunit disulfide bonds (paired: 188/204, 291/359, 322/338, and 349/381) and an intersubunit disulfide bond involving Cys-277. This leaves Cys-372 with no analogous counterpartner in other serine proteases. This residue may be involved in linking hepsin to other ligands or proteins, including the second hepsin molecule. The latter possibility, however, is not supported by protein blotting analyses. Nine Cys residues predicted in the noncatalytic subunit may be involved in forming four intrasubunit disulfide bonds and one intersubunit disulfide bond involving Cys-153 and

---

FIG. 1. Nucleotide sequence of the cDNA coding for hepsin. The predicted amino acid sequence is shown above the nucleotide sequence. The inverted solid triangle indicates the location of the inserted intronlike sequence (not included here) found in some cDNA clones. The boxed amino acid sequence represents a potential transmembrane sequence. The arrow identifies an Arg-Ile peptide bond that may be cleaved in the proteolytic activation process. The active site His, Asp, and Ser residues are circled. The underlined nucleotide sequence is the site that hybridizes to the synthetic oligonucleotide probe. Reproduced from ref. 2 with permission.

Cys-277 in the catalytic subunit, as predicted from the similarity to other serine proteases.

Hepsin contains at least one potential attachment site for N-linked carbohydrates located at residue 112.

Antibody Preparation

A set of polyclonal antihepsin antibodies used in the following experiments is prepared by immunizing rabbits with synthetic peptides (P1, P2, P3, P4, and P5) conjugated with keyhole limpet hemocyanin.[4] These peptides correspond to residues 1–17 (P1), 246–257 (P2), 294–305 (P3), 360–372 (P4), and 398–417 (P5). Antibodies are affinity purified on columns containing the immobilized specific peptides. Among the antibodies raised, those against the catalytic subunit (P2, P3, P4, and P5) are very useful because of their good specificities. Among them, HAbP5, which is raised against P5 (the carboxyl-terminal region, EWIFQAIKTHSEASGMV-TQL), has become particularly useful for most of the experiments, including immunoflurescent staining and growth inhibition. Antibody (HAbP1) raised against P1 (the amino-terminal region at the cytosolic side) is not very specific, possibly due to sequence similarity of this region to other unidentified proteins. Antibodies designated HAbM are prepared against PM (the equimolar mixture of synthetic peptides P2, P3, and P4). Both HAbP5 and HAbPM can cross-react with rodent hepsin, suggesting highly conserved hepsin structures among different species. Preparation of polyclonal and monoclonal antibodies against purified hepsin is currently in progress.

Structure and Subcellular Distribution

Western blot analyses demonstrate the presence of hepsin in various mammalian cells, such as hepatoma cells [HepG2 cells and Alexander cells (PLC/PRF/5)], baby hamster kidney (BHK) cells, and others (Fig. 2).[4] In HepG2 cells, hepsin is primarily present in plasma membranes as molecular species of 51 kDa (major) and 28 kDa (minor), whereas in BHK cell membrane, only a 51-kDa species is present. Hepsin is undetectable in cytosol as well as in culture medium. The *in vitro* translational products of hepsin mRNA prepared from the cDNA with T7 promoter gave a molecular size of about 45 kDa, which is in good agreement with the apparent molecular size of cellular hepsin (51 kDa), assuming that some carbohydrate chains are attached to cellular hepsin. Carbohydrate chains are absent in the *in vitro* translation product. The minor molecular species (28 kDa) observed for hepsin may be a specific degradation product of hepsin. [^{3}H]DFP (diisopropyl fluorophosphate) labeling showed that a

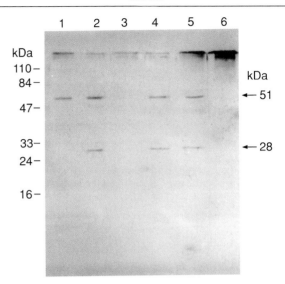

FIG. 2. Immunoblot analysis of HepG2 and BHK cells. Aliquots (7.5 μg) of proteins of cell subcomponents and media are loaded in each lane. Lane 1, BHK cell plasma membranes; lane 2, HepG2 cell plasma membranes; lane 3, HepG2 cytosol; land 4, HepG2 cell mitochondria fraction; lane 5, HepG2 cell nucleus fraction; lane 6, HepG2 cell media. Numbers on the left and right show the positions of size markers and protein bands, respectively. Reproduced from ref. 4 with permission.

hepsin fragment with an apparent size of 46 kDa, which is derived from the intact 51 kDa by limited proteolysis, has an active site serine residue.[5] Whether the difference in molecular mass between 51 and 46 kDa can be accounted for by a release of a possible activation peptide, or by other proteolytic modifications, is not known at the present time.

Hepsin is also present in other subcellular fractions, such as the mitochondria or nuclei of HepG2 cells (Fig. 2), indicating that hepsin may be involved in multiple cellular functions. The half-life of cellular hepsin is about 3.5 hr as estimated by pulse-labeling experiments with Alexander cells.[6]

Molecular Topology at Cell Surfaces

The 15-amino acid residue sequences in the immediate flanking regions of the hydrophobic sequence have three net positive charges for the amino-

[5] A. Torres-Rosado and K. Kurachi, unpublished data.
[6] A. Torres-Rosado, K. S. O'Shea, A. Tsuji, S.-H. Chou, and K. Kurachi, *Proc. Natl. Acad. Sci. U.S.A.* **90,** 7181 (1993).

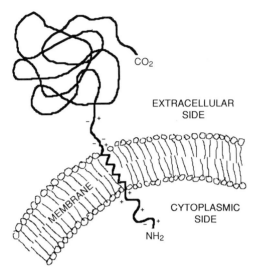

FIG. 3. The type II membrane topology of hepsin at the plasma membrane. The distribution of positively ($+$) and negatively ($-$) charged amino acid residues in the immediate flanking regions of the hydrophobic sequence is indicated; the extracellular portion of hepsin contains the protease catalytic module.

terminal side and one negative charge for the carboxyl-terminal side.[2] As shown in Fig. 3, this distribution of negative and positive charges in the regions supports a typical topology of type II membrane-spanning protein for hepsin.[4,7,8] Immunofluorescent staining analyses of HepG2 cells with HAbP5 and mild proteinase K digestion of cells, resulting in cleavage of the extracellular portion of hepsin while maintaining the cell viability, strongly support this molecular orientation of hepsin at the cell surface, which has its protease catalytic subunit (carboxyl-terminal half) sticking outward into the extracellular space and its amino-terminal short sequence (17 amino acid residues in length) located at the cytosolic side. This molecular orientation categorizes hepsin in a unique subgroup of serine proteases, which includes several others, such as a 67-kDa cell surface serine protease and guanidinobenzoatase.[1,4,6,9,10] These cell surface serine proteases, which are apparently different from hepsin in their molecular

[7] G. D. Parks and R. A. Lamb, *Cell (Cambridge, Mass.)* **64**, 777 (1991).
[8] A. Tsuji, A. Torres-Rosado, T. Arai, S.-H. Chou, and K. Kurachi, *Biomed. Biochim. Acta* **50**, 491 (1991).
[9] G. K. Scott, H. F. Seow, and C. A. Tse, *Biochim. Biophys. Acta* **1010**, 160 (1989).
[10] F. S. Steven and M. M. Griffin, *Biol. Chem. Hoppe-Seyler* **369**(Suppl.), 137 (1988).

weight, have been reported to have important roles in cell proliferation as well as other cellular functions.

Tissue Distribution

As shown in Fig. 4, when tested with various tissues obtained from a normal young adult baboon, hepsin is present in most tissues, but at a particularly high level in liver.[4] This ubiquitous expression pattern suggests that hepsin may play an essential role in cell functions. In culture, some cells, such as cancer cells (HepG2 and Alexander cells or breast cancer cells) and nerve cells (PC12), express hepsin at very high levels, whereas BHK cells and endothelial cells (both human umbilical cord or rat capillary) express hepsin at lower (but significant) and undetectable levels,

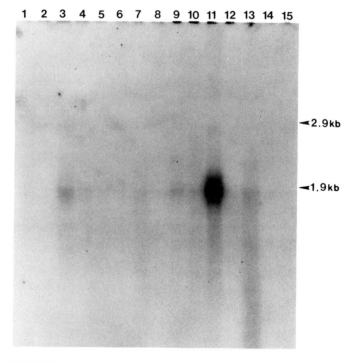

FIG. 4. RNA blot analysis of young adult baboon tissue. Each lane contained 20 μg of total RNAs isolated from a young adult baboon. Lanes 1–15 contain hypothalamus, small intestine, pancreas, testis, salivary gland, skeletal muscle, lung, adrenal gland, thyroid, pituitary gland, liver, spleen, kidney, brain, and thymus, respectively. The size and positions of RNAs are shown at the right. A hepsin cDNA (1.8 kb) was used as the radiolabeled probe in this experiment.

respectively. These indicate that hepsin plays a crucial role(s) in most, if not all, types of mammalian cells.

Substrate Specificity

Hepsin is purified as a mixture of activated and nonactivated forms by using an antihepsin antibody column (HAbP5) from plasma membranes that are solubilized by 0.25% (v/v) Nonidet P-40 (NP-40).[6] Hepsin has a substrate specificity typical of trypsin-type serine proteases and can hydrolyze synthetic substrates, including N-benzoyl-Leu-Ser-Arg-pNA · HCl, N-benzoyl-Ile-Glu-Phe-Ser-Arg-pNA · HCl, and N-benzoyl-Phe-Val-Arg-pNA · HCl. This is consistent with its amino acid sequence typical of a serine protease and the observed [³H]DFP incorporation into its derivative. The purified nonactivated hepsin can be progressively activated by a catalytic amount of trypsin (1 : 200 trypsin : hepsin preparation, w/w). The activity of activated hepsin toward N-benzoyl-Leu-Ser-Phe-pNA · HCl is about 1.2% that of trypsin in 100 mM Tris-HCl, pH 7.5, at 37°. HAbP5 substantially suppresses the activation of hepsin and also inhibits the amidolytic activity toward the same synthetic substrate by about 70%. Precise substrate specificity, activation kinetics, and the underlying mechanism, as well as natural substrate(s) of hepsin, are yet to be determined.

Localization and Cell Cycle-Dependent Expression of Hepsin Gene

The human hepsin gene is located on chromosome 19 as determined by synteny test of the hepsin gene and human chromosomes in rodent–human hybrid cell clones.[4] The majority of hybridizing grains with radiolabeled hepsin cDNA probe are localized to bands q11–13.2, with the largest number of grains accumulated at q13.1. Southern blot analysis does not show that hepsin has any pseudogene(s). The hepsin gene, however, seems to belong to a new family of genes encoding membrane-associated serine proteases, which may only weakly cross-hybridize with the hepsin cDNA probe.

The hepsin gene expression is cell cycle-dependent as tested with synchronously growing Alexander cells (Fig. 5).[5] The hepsin gene expression is induced in the middle of S phase and its high expression continues through G_2 phase, followed by an abrupt decrease in M phase and only marginal expression through G_1 phase. Because the half-life of hepsin is estimated to be about 3.5 hr, this hepsin gene regulation may result in a maximal level of cellular hepsin during the middle of G_2 phase through

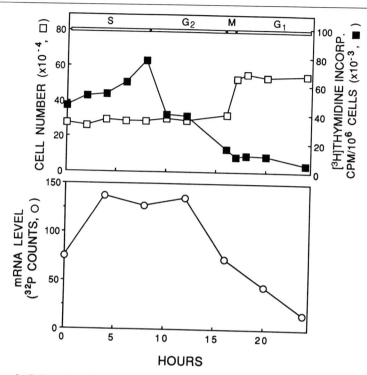

FIG. 5. Cell cycle analysis of hepsin expression. Cell cycle analysis was carried out according to Thommes et al. [P. Thommes, T. Reiter, and R. Knippers, *Biochemistry* **25,** 1308 (1986)] with modification. *Upper:* One full cycle of Alexander cell division is shown with cell cycle phases. Cells with adjusted cell numbers were synchronized for growth by treating with 7 mM hydroxyurea added to culture medium [Dulbecco's modified Eagle's medium (DMEM) + 10% (v/v) fetal calf serum (FCS)] for 24 hr. The medium is then switched to the normal medium without hydroxyurea, but 1 μCi/ml [³H]thymidine. New DNA synthesis, as shown by [³H]thymidine incorporation, and cell numbers were determined at various time points. *Lower:* Expression levels of hepsin at various cell cycle stages. A set of growth-synchronized cells as described for the upper panel (no [³H]thymidine added) was used to prepare total RNAs at the various time points. Aliquots (50 μg) of RNA were analyzed in RNA blot analysis for hepsin RNA levels by using hepsin cDNA as a ³²P-labeled prove. Radioactivities (cpm) of the hepsin bands on the blot were quantitated by Betascope 603 Blot Analyzer (Betagen, Waltham, MA).

M phase, which is then followed by a much reduced level throughout G₁ phase and most of the S phase.

Biological Role

The precise biological role(s) of hepsin and the underlying mechanism are not known. Hepsin may be involved in membrane protein catabolism, processing, and/or activation of other membrane-associated or extracellular proteins and peptides, or may serve as a receptor for a specific ligand(s). Its protease activity may or may not be essential for these potential biological functions. Hepsin also may possibly exert its biological activity (or activities) by molecular properties other than its protease activity.

The first evidence to indicate the crucial biological role of hepsin in cell growth was obtained by using affinity-purified antihepsin antibody (HAbP5).[6] HAbP5 added to the culture medium suppresses the growth of hepatoma cells (HepG2 and Alexander cells). Alexander cells are then used in a series of experiments as a model cell line. On day 4, wells with the antibody added have a cell number only 11% that of wells without antibody added. This suppression of cell growth accompanies no noticeable cell morphology changes or decrease in hepsin production, suggesting the importance of the presence of free hepsin molecule on the cell surface. The cell growth-suppressive effect disappears on removing the antibody from the culture medium. Other control antibodies (nonspecific rabbit IgG or antiangiogenin antibodies) do not show any such negative effects on cells, supporting the specificity of the suppressive effect of antihepsin antibody.

The importance of hepsin for cell growth is more clearly demonstrated by a series of experiments employing hepsin-specific antisense oligonucleotides.[6] Successful utilization of antisense oligonucleotides for delineating biological functions of various gene products is well documented.[9] A series of single-stranded antisense and sense oligonucleotides used in this experiment are listed in Table I. Both phosphodiester oligonucleotides and their phosphorothioate derivatives, a much stabler form than phosphodiester oligonucleotides, can be used for the same purpose in these experiments. The optimal concentration of oligonucleotides for observing the specific effects on cell growth varies significantly from one cell type to another, probably depending on the expression level of hepsin. For Alexander cells and HepG2 cells, the optimal concentration is 12.5–15 μM. Nonspecific effects are observed at 20 μM and more clearly at higher concentrations. Antisense oligonucleotides, AS-pd-oligo237 (phosphodiester regular oligonucleotide) and AS-pt-oligo237 (phosphorothioate derivative), which were designed to the first ATG site area, show gross inhibitory effects leading to a total arrest of cell growth after 3–4 days of treatment (Fig. 6). The lag of the negative effects of antisense oligonucleotides is in good agreement with the estimated half-life of hepsin (~3.5 hr). On treatment with these oligonucleotides, gross changes in morphology (flattening and enlargement like a pancake) are induced. The arrest of cell growth and change in cell morphology are reversible by removing anti-

TABLE I
SINGLE-STRANDED OLIGONUCLEOTIDES SYNTHESIZED AT
VARIOUS SITES OF HEPSIN cDNA SEQUENCE[a]

Oligonucleotide	Sequence
AS-pt-oligo237	5'-GGCAGTGACATGGCGCAGAAG-3'
RAS-pt-oligo237	5'-CCCGTCGTATCGATCCGTTCC-3'
SS-pt-oligo237	5'-CTTCTGCGCCATGTCACTGCC-3'
AS-pt-oligo952	5'-GCCCCCGTGGTAGACCACAG-3'
RAS-pt-oligo952	5'-CTCAGAGCACGCTCAGCGGC-3'
SS-pt-oligo952	5'-TGTGGTCTACCACGGGGGCT-3'
AS-pt-oligo1720	5'-GGCACCTAGACAGGAGTCCC-3'
RAS-pt-oligo1720	5'-CGAGTCAGACCGGGCCTAAC-3'
SS-pt-oligo1720	5'-GGGACTCCTGTCTAGGTGCC-3'
AS-FIX-oligo	5'-GCTATGTAACATTTTCGAT-3'

[a] AS, Antisense; RAS, randomized, SS, sense; pt, phosphorothioate; oligo, oligonucleotide; FIX, factor IX.

sense oligonucleotides from media. However, a prolonged treatment of
cells with the antisense oligonucleotides (longer than about 1 week) results
in a gradual loss of cell viability. Various control oligonucleotides, including sense strand oligonucleotides (SS-pd-oligo237 or SS-pt-oligo237), oligonucleotides with a randomized sequence of antisense oligonucleotides
(RAS-pd-oligo237 or RAS-pt-oligo237), or an unrelated antisense oligonucleotide that is designed to the factor IX sequence (AS-FIX-oligo), show
no significant effects on cell growth or cell morphology, indicating the
specific effects of antisense oligonucleotides. It is also noted that on antisense oligonucleotide treatment, some subcellular structures such as mitochondria and RER (rough endoplasmic reticulum) are substantially dis-
turbed.[5] These inhibitory effects of antisense oligonucleotides on cell
growth are due to the specific inhibition of hepsin biosynthesis, as shown
by immunofluorescent staining of cells treated with antisense oligonucleotides (see ref. 6). Cells treated with various control oligonucleotides do
not show this effect. Hepsin level in Alexander cells on exposure to
antisense oligonucleotides is rapidly lowered to 29% after 1 day and to
8% after 2 days.[6] The antisense oligonucleotide RAS-pt-oligo952, which
is designed to a much downstream sequence still in the coding region,
also showed a specific, but weaker, inhibitory effect on cell growth (about
60% inhibition). Antisense oligonucleotides designed to the general 3'-
untranslated region away from the known functional significance showed
no significant inhibitory effects. The weaker or noninhibitory effects of
antisense oligonucleotides designed to the downstream regions are in good

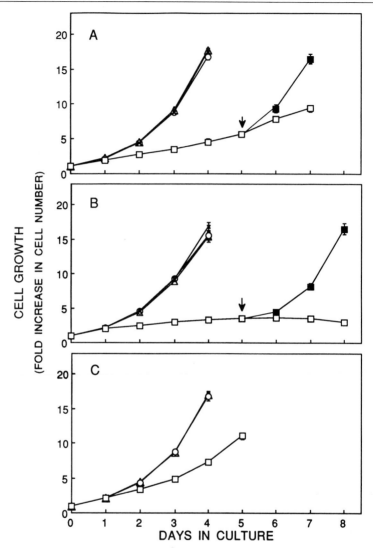

Fig. 6. Effects of antihepsin antibodies and antisense oligos on the growth of Alexander cells. (A) Cells cultured in medium to which water (△), nonspecific rabbit IgG (○), antiangiogenin IgG (×), or antihepsin (□) was added daily. Arrow indicates change from an antihepsin antibody-containing medium (□) to an antihepsin-free medium (■). Bars indicate standard error ranges for days 1–5 and range of values for days 6 and 7. (B) Cells cultured in medium to which water (△), RAS-pt-oligo237 (○), SS-pt-oligo237 (×), or AS-pt-oligo237 (□) was added. Bars indicate standard error ranges. Arrow indicates a change from an oligonucleotide-containing (□) to an oligonucleotide-free (■) medium. (C) Cells cultured in medium to which water (△), RAS-pt-oligo952 (○), or AS-pt-oligo952 (□) was added. Mean values ($n = 5$) were plotted. All other conditions are the same as in B. Reproduced from ref. 6 with permission.

agreement with the general inhibitory effects of this type of antisense oligonucleotides for biosynthesis of various gene products.[11] These antisense oligonucleotides are considered to exert their inhibitory effects by passively interfering with mRNA binding to ribosomes.

These observations with antihepsin antibodies and hepsin-specific antisense oligonucleotides demonstrate that hepsin plays essential roles in cell growth as well as in maintenance of the normal cell morphology. This is the very first time to demonstrate that a cell surface serine protease has such a crucial cellular role.

As a protease, hepsin may be involved in the degradation of extracellular proteins to create spaces for their growth, which requires cell shape change and migration. If hepsin plays such a role in cell growth, it may be produced at a significant level, particularly in actively growing and rearranging tissues. This hypothesis is proved to be true by examining hepsin expression in the developing mouse embryo.[6] In the early stage of development (tested as early as day 10 of development), hepsin is expressed at a relatively low level throughout the embryo. By day 13, hepsin is sparsely deposited in the forming surface ectoderm and colonic epithelium, increasing significantly over the next several days both in intensity and in number of tissues stained, such as skin, liver, and muscle. These observations demonstrate that hepsin plays an important role in cell growth and tissue rearrangement, particularly at highly circumscribed developmental stages.

The protease activity of hepsin is very probably required in these processes to create space for cell migration and process extension through an extracellular matrix and the cell-filled milieu. Some cell surface proteases have also been shown to function as receptors for viruses.[12–14] Whether hepsin also assumes such a role is yet to be determined.

Concluding Remarks

Our studies on hepsin have clearly demonstrated for the first time, at the molecular level, its importance in cell growth and maintenance of the normal cell morphology. The underlying mechanism(s) for these biological activities, in addition to other yet unidentified function(s) of hepsin, need

[11] J. Goodchild, in "Oligodeoxynucleotides: Antisense Inhibitors of Gene Expression" (J. S. Cohen, ed.), p. 53. CRC Press, Boca Raton, Florida, 1989.

[12] H. Kido, A. Fukutomi, and N. Katunuma, Biomed. Biochim. Acta 50, 781 (1991).

[13] B. Delmas, J. Gelfi, R. L'Haridon, L. K. Vogel, H. Sjostrom, O. Noren, and H. Laude, Nature (London) 357, 417 (1992).

[14] C. L. Yeager, R. A. Ashmun, R. K. Williams, C. B. Cardellichio, L. H. Shapiro, A. T. Look, and K. V. Holmes, Nature (London) 357, 420 (1992).

to be established in the near future. The knowledge obtained regarding hepsin will not only greatly help us in better understanding other cell surface serine proteases with the similar topology, but also will aid in developing a potentially new approach(es) for controlling malignant cell growth.

[8] Glutamyl Endopeptidases

By JENS J. BIRKTOFT and KLAUS BREDDAM

Serine endopeptidases can be divided into two major groups based on the importance of the S1 binding site[1] for substrate cleavage. Some enzymes, e.g., the subtilisins,[2,3] do not exhibit a particularly strong requirement with respect to the nature of the amino acid residue at the P1 position. Rather amino acids at positions away from the scissile bond are equally important for cleavage site selectivity. These serine proteases, by virtue of being relatively nonspecific, are highly suitable for general degradation purposes but usually are not particularly convenient for specific polypeptide cleavage.

The other major group of serine endopeptidases is characterized by being fairly selective for a specific amino acid or for types of amino acids at the P1 position. Traditionally this classification has resulted in a division into three groups based on their primary substrate preference: the trypsin-like, which cleaves after positively charged residues; the chymotrypsin-like, which cleaves after large hydrophobic residues; and the elastase-like, which cleaves after small hydrophobic residues.[4,5] For these enzymes the P1 residue nearly exclusively dictates the site of peptide bond cleavage. Residues at other positions only affect the rate of cleavage but not the primary specificity. However the subsite interactions can in some instances limit the permissible amino acids around the specific P1 residue to a unique set or sets of sequences. Examples of this are observed with

---

[1] The binding site notation is that of I. Schechter and B. Berger, *Biochem. Biophys. Res. Commun.* **27,** 157 (1967). Accordingly, amino acid residues in the substrate are referred to as P1, P2, . . . , Pi in the amino-terminal direction away from the scissile bond and P1', P2', . . . , Pi' in the carboxy-terminal direction away from the scissile bond. Enzyme subsites are denoted S in correspondence with the substrate.

[2] H. Philipp and M. L. Bender, *Mol. Cell. Biochem.* **51,** 5 (1983).

[3] I. Svendsen, *Carlsberg Res. Commun.* **41,** 237 (1976).

[4] B. S. Hartley, *Philos. Trans. R. Soc. London B* **257,** 77 (1970).

[5] R. M. Stroud, *Sci. Am.*, 456 (1974); J. Kraut, *Ann. Rev. Biochem.* **46,** 331 (1977).

endopeptidases participating in blood clotting, fibrinolysis, and hormone processing.

This initial three-group classification of the serine endopeptidases has subsequently been expanded with proteases for which the requirement of the P1 residue is limited to a specific amino acid. Examples of this are the lysine-specific enzyme from *Achromobacter*,[6] the arginine-specific clostripain from *Clostridium*,[7] and the postproline-specific enzyme from *Flavobacterium*.[8] Another large and expanding group of endopeptidases is that containing enzymes specific for glutamic and aspartic acid residues. Such enzymes have been isolated from a variety of sources, including *Staphylococcus aureus*,[9] *Actinomyces species*,[10] *Streptomyces thermovulgaris*,[11] *Streptomyces griseus*,[12,13] and *Bacillus licheniformis*.[14] Several of these glutamic/aspartic acid-specific proteases are widely employed for the fragmentation of polypeptides prior to amino acid sequence determinations and as catalysts for the synthesis of peptide bonds.[15,16] Common to most of these enzymes is a general preference for cleavage of peptide bonds following glutamic acids, compared to those following aspartic acids. The serine endopeptidase from *S. aureus*, strain V8 (GluV8) was until 1987 the only well-characterized representative of the Glu/Asp serine endopeptidases.[9,17,18] Since then similar enzymes have been isolated and two of these, one from *S. griseus* (GluSGP)[12,13] and the other from *B. licheniformis* (GluBL),[14] can easily be purified in large quantities from commercially obtainable fermentation extracts. The GluV8 protease can be obtained in pure form from several commercial sources, and these will suffice when only minor quantities are required. Large-scale protein digestions and peptide syntheses necessitate a less expensive enzyme source, warranting a description of the purification of GluSGP and GluBL.

[6] T. Masaki, K. Nakamura, M. Isono, and M. Soejima, *Agric. Biol. Chem.* **42**, 1443 (1978).
[7] A. M. Gilles, J. M. Imhoff, and B. Keil, *J. Biol. Chem.* **254**, 1462 (1979).
[8] T. Yoshimoto, R. Walter, and D. Tsuru, *J. Biol. Chem.* **225**, 4786 (1980).
[9] G. R. Drapeau, Y. Boily, and J. Houmard, *J. Biol. Chem.* **247**, 6720 (1972).
[10] O. V. Moslova, G. N. Rudenskaya, V. M. Stepanov, O. M. Khodova, and I. A. Tsaplina, *Biokhimiya* **52**, 414 (1987).
[11] N. V. Khaidarova, G. N. Rudenskaya, L. P. Revina, V. M. Stepanov, and N. S. Egorov, *Biokhimiya* **54**, 46 (1989).
[12] N. Yoshida, S. Tsuruyama, K. Nagata, K. Hirayama, K. Noda, and S. Makisumi, *J. Biochem. (Tokyo)* **104**, 451 (1988).
[13] I. Svendsen, M. R. Jensen, and K. Breddam, *FEBS Lett.* **292**, 165 (1991).
[14] I. Svendsen and K. Breddam, *Eur. J. Biochem.* **204**, 165 (1992).
[15] V. de Filippis and A. Fontana, *Int. J. Pept. Protein Res.* **35**, 219 (1990).
[16] R. Seetharam and A. S. Acharya, *J. Cell. Biochem.* **30**, 87 (1986).
[17] G. R. Drapeau and J. Houmard, *Proc. Natl. Acad. Sci.* **69**, 3506 (1972).
[18] G. R. Drapeau, this series, Vol. 47, p. 189.

These two enzymes are also characterized by their high catalytic potential compared to GluV8.[19,20]

Several other proteases with specificity for glutamic and/or aspartic acids that have important medically relevant biological roles have been identified. Epidermolytic toxins A and B from *S. aureus* that induce staphylococcal scalded skin syndrome in newborns[21] are serine proteases homologous to GluV8, the glutamic acid-specific protease also from *S. aureus*, strain V8.[9,17,18] The aspartic acid-specific interleukin 1β converting enzyme, present in monocytes, activates interleukin 1β, which serves an important role in the pathogenesis of inflammatory diseases.[22] Granzyme B found in cytotoxic T lymphocytes, where it functions in the defense against tumor cell proliferation and viral infection, is likewise an aspartic acid-specific protease.[23] Additionally, a number of viral proteases have been demonstrated to cleave after glutamic acid and glutamine residues. These viral proteases include those found in hepatitis A, human polio virus, and human rhinovirus (common cold), where they serve a principal role in the processing of viral polyproteins into functional gene products.[24]

A previous description of this group of proteases in Vol. 47, this series was limited to brief account of GluV8[18]; in this review we will describe some of the properties of GluV8, GluSGP, and GluBL, as well as provide a brief description of the recently determined crystal structure of GluSGP.[25]

Enzyme Sources and Purification

Among the three proteases with acid specificity, the one from *S. aureus*, GluV8, can be obtained in highly purified forms from a number of commercial sources, including ICN ImmunoBiochemicals (Lisle, IL), Miles Laboratories (Elkhart, IN), and Worthington Biochemicals (Freehold, NJ). The proteases from *S. griseus*, GluSGP, and *B. licheniformis*,

---

[19] K. Breddam and M. Meldal, *Eur. J. Biochem.* **206**, 103 (1992).
[20] K. Nagata, N. Yoshida, F. Ogata, M. Araki, and K. Noda, *J. Biochem. (Tokyo)* **110**, 859 (1991).
[21] S. J. Dancer, R. Garrett, J. Saldanha, H. Jhoti, and R. Evans, *FEBS Lett.* **268**, 129 (1990).
[22] N. A. Thornberry, H. G. Bull, J. R. Calaycay, K. T. Chapman, A. D. Howard, M. J. Kostura, D. K. Miller, S. M. Molineaux, J. R. Weidner, J. Aunins, K. O. Elliston, J. M. Ayala, F. J. Casano, J. Chin, G. J. F. Ding, L. A. Egger, E. P. Gaffney, G. Limjuco, O. C. Palyha, S. M. Raju, A. M. A. Rolando, J. A. Schmidt, and M. J. Tocci, *Nature (London)* **356**, 768 (1992).
[23] S. Odake, C. M. Kam, L. Narasimhan, M. Poe, J. T. Blake, O. Krahenbuhl, J. Tschopp, and J. C. Powers, *Biochemistry* **30**, 2217 (1991).
[24] J. Wellink and A. van Kammen, *Arch. Virol.* **98**, 1 (1988).
[25] V. L. Nienaber, K. Breddam, and J. J. Birktoft, *Biochemistry* **32**, 3456 (1993).

GluBL, can be isolated in large amounts from commercial extracts, and their purification will be described here.

Materials

Sepharose 4B and bacitracin can be obtained from Pharmacia (Sweden) and Sigma (St. Louis, MO), respectively, and the preparation of bacitracin–Sepharose is described in Ref. 26. Fractogel TSK CM 650 (M) is a product of Merck (Germany). Reagents and solvents are of analytical grade obtainable from commercial sources, including Applied Biosystems (Foster City, CA) and Rathburn (Walkerburn, UK).

Purification of Glutamic Acid-specific Endopeptidase from
  Bacillus licheniformis

To 100 ml Alcalase (Novo Nordisk Industries, Bagsværd, Denmark) add 615 ml water and adjust to pH 6.2 by addition of 1 M NaOH.[14] This solution is applied to a 5 × 25-cm Fractogel TSK CM 650 (M) column, equilibrated with 0.01 M NaH$_2$PO$_4$, pH 6.2. The column is washed with approximately 1600 ml of the equilibration buffer and eluted with a linear salt gradient from 0 to 0.35 M in the same buffer (total 5 liters). The enzyme elutes at approximately 0.25 M NaCl. The fractions containing the activity toward N-carbobenzoxy-Glu-p-nitroanilide are pooled, the pH is adjusted to 8.2 by addition of 1 M NaOH, and is then applied to a 57.5-cm bacitracin–Sepharose column, equilibrated with 20 mM Bicine, 2 mM CaCl$_2$, pH 8.2. The enzyme is eluted with a gradient from 0 to 1 M NaCl in the same buffer. The enzyme elutes at approximately 0.9 M NaCl. The fractions containing activity toward N-carbobenzoxy-Glu-p-nitroanilide are dialyzed against 0.01 M NaH$_2$PO$_4$, pH 6.2, and are then again applied to the Fractogel TSK CM 650 (M) column, equilibrated, and eluted in the same way. Fractions containing activity are pooled and applied to the bacitracin–Sepharose column, equilibrated, and eluted as described above. The eluted fractions with activity are concentrated by ultrafiltration utilizing a membrane with a molecular weight cutoff at 20,000. The concentrated enzyme is dialyzed against 10 mM 2-(N-morpholino)ethanesulfonic acid (MES), 0.1 M NaCl, pH 6.0, and then frozen at −18°.

Purification of Glutamic Acid-specific Endopeptidase from
  Streptomyces griseus

Dissolve 200 g of pronase (Actinase, Kaken Seiyaku, Tokyo, Japan), a culture filtrate of *S. griseus*, in 2700 ml water and adjust to pH 5.25 by

[26] V. M. Stepanov and G. N. Rudenskaya, *J. Appl. Biochem.* **5**, 420 (1983).

addition of 10% acetic acid.[13] This sample is applied to a 10 × 17-cm CM-52 cellulose column equilibrated with 10 mM sodium acetate, pH 5.25. The column is washed with 1.5 liters of buffer until A_{280} is below 0.4, and the enzyme is then eluted with a salt gradient from 0 to 0.2 M NaCl in the acetate buffer (2 × 7 liters). The fractions with activity toward N-carbobenzyoxy-Glu-p-nitroanilide are pooled, concentrated, and diafiltrated toward 10 mM MES, pH 6.0. The concentrate is applied to a 2.6 × 19-cm ε-[N-(aminocaproyl)p-aminobenzyl]succinyl–Sepharose column[27] equilibrated with 10 mM MES, pH 6.0. The column is washed with 110 ml of buffer and the enzyme is eluted with a salt gradient from 0 to 0.35 M NaCl in the MES buffer (2 × 600 ml). The fractions with activity toward N-carbobenzoxy-Glu-p-nitroanilide are concentrated and diafiltrated against 50 mM HEPES, pH 7.5. To remove residual activity toward N-acetyl-Arg-p-nitroanilide, due to a trypsinlike enzyme, this sample is applied to a 1.6 × 5-cm arginine–Sepharose 4B (Pharmacia) column equilibrated and then washed with 50 mM HEPES, pH 7.5. The breakthrough is concentrated, diafiltrated against water, and frozen at −18°.

Enzyme Activity Assays

Substrates for Activity Assays

A large number of glutamic acid-containing peptides can be employed in assaying the enzymatic activities of glutamyl endopeptidases. The substrates listed in this section are among those that have been found to be most useful. The preparation of the assay substrates o-aminobenzoyl-(anthraniloyl)-Ala-Ala-Glu-Val-Tyr(NO₂)-Asp-OH (and related anthraniloyl- and 3-nitrotyrosine-based substrates) is described in Ref. 28; benzyloxycarbonyl-Leu-Leu-Glu-β-naphthylamide and carbobenzyloxycarbonyl-Phe-Leu-Glu-p-nitroanilide can be obtained from Sigma (St. Louis, MO) and Boehringer Mannheim, respectively. The preparation of additional glutamic acid-containing substrates, including O-phenyl esters and nitroanilides, has been described.[29] Substrates suitable for monitoring the activities of contaminating trypsin- and chymotrypsin-like enzymes such as N-acetyl-Arg-p-nitroanilide and N-acetyl-Phe-p-nitroanilide can be obtained from Bachem (Switzerland), but other substrates can also be employed.

[27] K. Breddam, S. B. Sørensen, and M. Ottesen, *Carlsberg Res. Commun.* **48,** 217 (1983).
[28] M. Meldal and K. Breddam, *Anal. Biochem.* **195,** 141 (1991).
[29] J. Houmard, *J. Pept. Protein Res.* **8,** 199 (1976).

Activity Assays

Depending on the substrate employed, different techniques are used to monitor enzyme activities: (a) nitroanilide substrate hydrolysis is determined spectrophotometrically at 410 nm; (b) substrates containing the anthraniloyl and 3-nitrotyrosine groups are assayed by monitoring the fluorescence emission at 420 nm on excitation at 320 nm;[28] (c) the release of β-naphthylamide is followed fluorometrically.[30] A standard assay mixture typically contains 965 μl 50 mM Bicine, pH 8.25, plus 25 μl of 5–50 mM substrate dissolved in water or methanol. To these solutions 10–50 μl enzyme solution is then added to start the reaction. When using GluBL, 2–5 mM CaCl$_2$ should be included in the assay mixture. As discussed in the next section the optimal pH is dependent on the specific enzymes used, varying between pH 7.3 and 9.0, and also on the substrate employed.[19]

pH Dependence of Glu Proteases

The pH dependence for GluV8 was established in a detailed study of the hydrolysis of Abz-Ala-Phe-Ala-Phe-Glu-Val-Phe-Tyr(NO$_2$)-Asp-OH and showed a bell-shaped profile with an optimum at pH 7.2.[31] The GluV8 activity is dependent on two ionizable groups with apparent pK_a values of 5.8 and 8.4. Previously, investigations with hemoglobin as substrate displayed two pH optima, at pH 4.0 and pH 7.8,[9] which was attributed to an effect of pH on the hemoglobin substrate as well as on the GluV8 protease. When the substrates *tert*-butyloxycarbonyl-Ala-Leu-Leu-Asp-*p*-nitroanilide and *tert*-butyloxycarbonyl-Ala-Ala-Leu-Glu-*p*-nitroanilide were employed, GluV8 displayed pH optima at pH 7 and pH 8, respectively.[20] Taken together these studies suggest that a pH of about 7.5–8.0 should be employed in GluV8 digestions. GluBL and GluSGP both display bell-shaped pH optimum curves between pH 7 and 10, with optimum pH values of 8.0 and 9.0, respectively.[14,32,33] For studies of peptide substrate cleavage the following buffer systems are recommended:[19] (a) GluBL—0.05 M Bicine, 2 mM CaCl$_2$, pH 8.0; (b) GluSGP—0.05 M cyclohexylaminoethane sulfonic acid (CHES), pH 9.0; (c) GluV8—0.05 M HEPES, pH 7.3.

[30] H. Djaballah and A. J. Rivett, *Biochemistry* **31**, 4133 (1992).
[31] S. B. Sørensen, T. L. Sørensen, and K. Breddam, *FEBS Lett.* **294**, 195 (1991).
[32] C. Dammann, S. B. Mortensen, P. Budtz, and S. Eriksen, *Int.* Patent WO 91/13553 (PTC/DK91/00069) (1991).
[33] K. Breddam, unpublished (1993).

Inhibition of Glu/Asp Proteases

GluBL, GluV8, and GluSGP are all inhibited by diisopropyl fluorophosphate but not by phenylmethylsulfonyl fluoride.[9,12,32] GluSGP is inhibited by a modified form of the third domain of turkey ovomucoid, Glu[18]-SSTKKY3,[34] and by an inhibitor from the seeds of bitter gourd.[20] In contrast, neither of these inhibitors has any activity toward GluV8.[20] Human α_1-proteinase inhibitor inhibits GluSGP and is inactivated by GluV8.[20] Chloromethane analogs of glutamic and aspartic acid-containing substrates have been shown to inhibit GluV8 and GluSGP.[35]

Effect of Small Molecules/Ligands on Glu/Asp Proteases

Earlier reports[17,36] have indicated that GluV8 protease is inhibited by monovalent anions when carbobenzyoxy-Glu-O-phenyl ester is used as substrate. This inhibition by chloride, acetate, and nitrate ions was confirmed using the larger peptide, o-aminobenzoyl-(anthraniloyl)-Ala-Phe-Ala-Phe-Glu-Val-Phe-Tyr(NO$_2$)-Asp-OH as substrate[31] and increasing concentrations of anions. Presumably, the inhibition is partially due to ionic and/or hydrogen bond interactions being important for the interaction between enzyme and substrate, and such interactions are adversely affected by increasing ionic strength. However, in addition it was found that phosphate and sulfate had an activating effect on the enzyme that may be ascribed to the tetrahedral configuration of these anions as compared to the planar configuration of the inhibitory acetate, borate, nitrate, and bicarbonate ions. The nature of this apparent specific interaction of phosphate and sulfate with the enzyme is unclear.

The influence of anions on the hydrolysis by GluV8 of longer substrates such as o-aminobenzoyl-(anthraniloyl)-Ala-Phe-Ala-Phe-Xaa-Val-Phe-Tyr(NO$_2$)-Asp-OH (Xaa = Glu or Asp) was studied in different buffer systems. The results revealed that the substrate with Xaa = Glu is hydrolyzed 3000–5000 times faster than that with Xaa = Asp. Furthermore, the nature of the buffer has only little influence on this ratio. Extending these investigations to even larger substrates, the initial cleavage of the B chain of oxidized insulin (Glu[13]-Ala[14]) in phosphate and bicarbonate buffer, respectively, was followed by high-performance liquid chromatography (HPLC) at low enzyme concentration (0.001 mg/ml). It was found that

[34] T. Komiyama, T. L. Bigler, N. Yoshida, K. Noda, and M. Laskowski, Jr., *J. Biol. Chem.* **266,** 10727 (1991).
[35] J. C. Powers, personal communication.
[36] J. Houmard, *Eur. J. Biochem.* **68,** 621 (1976).

the rate of cleavage was approximately 10 times faster in phosphate (0.030 mM/min) as compared to bicarbonate (0.0028 mM/min).

To investigate further the selectivity of the GluV8 protease for Glu-Xaa bonds, performic acid-oxidized insulin and glucagon were both digested in phosphate and bicarbonate buffer at high and low enzyme concentrations. Insulin contains no Asp-Xaa bonds but two Glu-Xaa bonds in both the A and B chains, Glu^{4A}-Gln^{5A}, Glu^{17A}-Asn^{18A}, Glu^{13B}-Ala^{14B}, and Glu^{21B}-Arg^{22B}. Glucagon contains no Glu-Xaa bonds, but the following Asp-X bonds: Asp^9-Tyr^{10}, Asp^{15}-Ser^{16}, and Asp^{21}-Phe^{22}. In phosphate, at low enzyme concentration (0.001 mg/ml), the Glu^{13B}-Ala^{14B} bond was cleaved very quickly whereas Glu^{17A}-Asn^{18A} and Glu^{21B}-Arg^{22B} were cleaved at a much lower rate. Glu^{4A}-Gln^{5A} and all peptide bonds involving aspartic acid were not cleaved. In bicarbonate, at low enzyme concentration, the same glutamic acid peptide bonds were cleaved but at lower rates.

At high enzyme concentration (0.2 mg/ml), all the peptide bonds involving glutamic acid were cleaved almost instantaneously except Glu^{4A}-Gln^{5A}, which was cleaved slowly. Compared to the Glu-Xaa bonds, the Asp-Xaa bonds were cleaved at much lower rates. The following decreasing order was observed in phosphate buffer: Asp^{15}-Ser^{16} > Asp^{21}-Phe^{22} > Asp^9-Tyr^{10}. In bicarbonate the rates were even lower, but the pattern was the same. Thus, the Asp-Xaa bonds are also cleaved in ammonium bicarbonate, although at lower rates than in sodium phosphate, just as is the case for the Glu-Xaa bonds. Consequently, the preference for glutamic acid over aspartic acid is not significantly affected by the nature of the buffer as previously suggested.[17,18] The beneficial effects previously attributed to bicarbonate are probably due to the extensive inhibition of the enzyme by bicarbonate that reduces its activity toward the preferred Glu-Xaa bonds.

The activity of GluBL is adversely affected by high ionic strength and beneficially affected by the inclusion of calcium ions in the assay medium. It is partially inhibited by EDTA. Though no enzymatic role can be attributed to calcium in GluBL activity, it likely serves in a structurally stabilizing function. No effect of calcium ions on the activities of GluSGP and GluV8 has been demonstrated.

Digestion of Proteins by GluV8 and Related Endopeptidases

Since its discovery[17] GluV8 has been employed extensively in the digestion of polypeptides and proteins. However, in consideration of the conclusions drawn from the studies described in the previous section, the procedures for the widespread use of the *S. aureus* strain V8 protease for generating peptides from large proteins for amino acid sequence studies should be modified. Cleavage in ammonium bicarbonate at a high enzyme : substrate ratio (1 : 30, w/w) should be abandoned in favor of a much lower enzyme concentration in phosphate buffer. Similar precautions should be used when employing GluBL and GluSGP. A representative description of the digestion of porcine glucagon and insulin can be found in Ref. 31. Similar conditions apply for peptide and protein digestion by the related GluSGP and GluBL proteases, with the exception that GluBL requires the addition of calcium ions.

Primary and Secondary Subsite Specificity

All three Glu-specific proteases display a pronounced preference for Glu-Xaa bonds versus Asp-Xaa bonds, as already mentioned. As measured by k_{cat}/K_m, GluBL prefers glutamic acid at P1 over aspartic acid by a factor of 1000, whereas the corresponding Glu/Asp preference factors for GluSGP and GluV8 are 105 and 1240, respectively.[19]

For most endopeptidases the subsite interactions have a noticeable influence on the efficiency of substrate hydrolysis. In the case of GluV8, GluSGP, and GluBL the effects of placing Ala, Val, Phe, Ser, Asp, Arg, and Pro at substrate positions P4–P2' have been described.[19] Common to all three enzymes is a preference for aspartic acid at P4, whereas alanine and valine are preferred at P3. At P2 GluSGP has a preference for proline and valine, whereas GluBL and GluV8 are most effective with phenylalanine at this position. Proline is in general disfavored at P3, P1', and P2' and aspartic acid is disfavored at P1'. Overall, these observations are in general agreement with the results of a similar study of shorter peptides using GluV8 and GluSGP.[20] It has previously been observed that Glu-Glu, Glu-Asp, and Glu-Pro peptide bonds were not hydrolyzed by GluV8.[37] As is the case with most other endopeptidases, peptide bond cleavage is not observed at positions located within two or three residues of the C- and N-terminal ends.[37]

Amino Acid Sequence

The amino acid sequence of mature GluV8 consists of 268 amino acids.[38] A notable feature is the precedence of a unique 12-fold repeat of a tripeptide—Pro-Asn/Asp-Asn (one tripeptide has a Glu in the third position)—in the last part of the sequence. The role of this sequence motif is unknown, although a role in stabilization of the prepro form of the

[37] B. M. Austen and E. L. Smith, *Biochem. Biophys. Res. Commun.* **72**, 411 (1976).
[38] C. Carmona and G. L. Gray, *Nucleic Acids Res.* **15**, 6757 (1987).

enzyme has been suggested.[38] The isolated GluBL consists of one peptide chain of 222 amino acid residues and has a calculated molecular mass of 23,589 Da.[14] The amino acid sequence of GluBL shows significant homology with the Glu, Asp-specific enzymes previously isolated from *S. aureus*,[18,38] *S. thermovulgaris*,[10] and *Actinomyces* species,[11] suggesting a common evolutionary origin and common three-dimensional structure for these members of the V8-type proteases. It should be noted that until now only enzymes with specificity toward acidic amino acids have been shown to be homologous to this group of proteases. Furthermore, no three-dimensional structure has been determined for any of these V8-type proteases. There is little if any evidence for homology, at least at the amino acid sequence level, between this group of serine proteases and members of other protease groups, such as the trypsin or subtilisin families.

The amino acid sequence of GluSGP[13] contains 187 amino acids and displays a high level of sequence identity to *S. griseus* protease A and B, both having chymotrypsin-like specificity, as well as to the elastase-like α-lytic protease from *Myxobacter* 495. GluSGP also shows similarities to members of the trypsin family, but the extent of sequence identity is much lower. Because these proteases contain about 30 fewer amino acids than the pancreatic enzymes, they have been classified into a subgroup of pancreatic serine proteases, the small bacterial serine proteases.[39] (family S2; see [2] in this volume). The three-dimensional structure of GluSP has been determined[25] and will be briefly described here.

Crystallographic Studies of GluSGP

Crystallization

Using GluSGP isolated according to ref. 13, high-quality crystals can be prepared as follows. After incubation overnight with a 10-fold molar excess of *tert*-butyloxycarbonyl-Ala-Ala-Pro-Glu-*p*-nitroanilide (or another suitable substrate or inhibitor) at 4° in 20 m*M* HEPES, pH 7.5, the solution is brought to a final protein concentration of 33 mg/ml in 0.2 *M* MgCl$_2$, 0.1 *M* HEPES. Crystals are grown by vapor diffusion against 50% saturated sodium citrate, pH 7.5, at 20° using the sitting drop method.[40] Over a period of 36–48 hr they reach a size of ~0.5 mm. The crystals belong to space group *C2*, having cell dimensions of *a* = 77.26 Å, *b* =

---

[39] M. N. G. James, L. T. J. Delbaere, and G. D. Brayer, *Can. J. Biochem.* **56**, 396 (1978).
[40] C. W. Carter (ed.) *Methods (San Diego)* **1** (1990).

36.29 Å, c = 51.22 Å, β = 101.8°. Crystals have been generated that diffract to better than 0.9 Å resolution.

Three-Dimensional Structure of GluSGP

Using molecular replacement techniques, the structure of GluSGP, complexed with the *tert*-butyloxycarbonyl-Ala-Ala-Pro-Glu tetrapeptide ligand, was determined and refined to an R-factor of 17% at 1.5 Å resolution. The overall fold of GluSGP closely resembles that observed in the pancreatic-type serine proteases.[41] Like trypsin and related enzymes it is composed of two β-barrel cylindrical structures and a C-terminal α-helix. The first β-barrel contains His-57 and Asp-102 of the catalytic triad; the second β-barrel contains Ser-195 of the catalytic triad and the S1 substrate specificity pocket. The structure is most similar to those of *S. griseus* proteases A,[42] and B,[43] and α-lytic protease,[44] and somewhat less similar to those of the pancreatic serine proteases.[5,41] The major difference is that loops, located on the protein surface connecting β-strands, have been shortened or eliminated. The overall mode of the interactions between GluSGP and the Ala-Ala-Pro-Glu tetrapeptide ligand is similar to that observed in the related trypsin family of serine proteases and is shown in a schematic form in Fig. 1. The peptide substrate forms a short antiparallel β-sheet structure that is stabilized via hydrogen bonds between the amido nitrogen of Glu-P1 and the carbonyl oxygen of Ser-214, between the amido nitrogen of residue P3 and the carbonyl oxygen of Ser-216, and between the carbonyl oxygen of residue P3 and the amido nitrogen of Ser-216. One oxygen of the free α-carboxyl group is bound in the oxyanion hole, forming hydrogen bonds with the amido nitrogens of Ser-195 and Gly-193. One of the two ε-oxygens of the side-chain carboxyl group of Glu-P1, Oε1, points toward the back side of the substrate-binding pocket, forming hydrogen bonds with His^{213}-Nε2 and Ser^{192}-Oγ, while Oε2 points toward solvent and is also hydrogen bonded to Ser^{216}-Oγ and with solvent molecules. The overall geometry of the GluSGP S1 site is very similar to that observed in the other small bacterial proteases. *Streptomyces griseus* proteases A and B both have an alanine at 192, a threonine at 213, and a glycine at 216. In α-lytic protease, methionines are at positions 192 and 213, while a glycine is at position 216. Thus, none of the residues implicated

---

[41] J. J. Birktoft and D. M. Blow, *J. Mol. Biol.* **68,** 187 (1972).

[42] A. R. Sielecki, W. A. Hendrickson, C. G. Broughton, L. T. J. Delbaere, G. D. Brayer, and M. N. G. James, *J. Mol. Biol.* **134,** 781 (1979).

[43] R. J. Read, M. Fujinaga, A. R. Sielecki, and M. N. G. James, *Biochemistry* **22,** 4420 (1983).

[44] M. Fujinaga, L. T. J. Delbaere, G. D. Brayer, and M. N. G. James, *J. Mol. Biol.* **184,** 479 (1985).

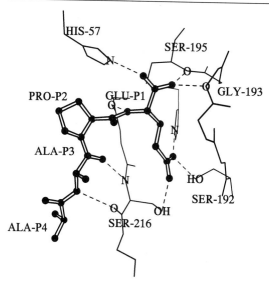

Fig. 1. Schematic representation of the active site of GluSGP, showing residues His-57, Ser-195, His-213, Ser-192, and Ser-216 and the main chain of residues 192–195 and 214–216. The bound Ala-Ala-Pro-Glu is shown as thick bonds and solid circles. Residues are labeled with sequence number and residue type. Hydrogen bonds between the peptide and GluSGP are represented by dashed lines between the ligand and protein atoms. The latter are denoted with N or O (OH). The side chain of Glu-P1 is hydrogen bonded at the back side of the substrate-binding pocket with His-213, which is shown in a thinner outline, and with Ser^{192}-Oγ and with Ser^{216}-Oγ. Other interactions are discussed in the text.

in the binding specificity of GluSGP are present in these other related enzymes. However, though the overall fold of GluSGP closely resembles that observed in the small bacterial proteases and pancreatic-type serine proteases, stabilization of the negatively charged substrate, when bound to this protein, appears to involve a more extensive part of the protease than previously observed. The substrate carboxylate is bound to a histidine side chain, His-213, which provides the primary electrostatic compensation for the negative charge on the substrate, and to two serine hydroxyls, Ser-192 and Ser-216. GluSGP displays maximum activity at pH 8.3, and assuming normal pK_a values the glutamate side chain and His-213 will be negatively charged and neutral, respectively, at this pH. In order for His-213 to carry a positive charge at the pH for optimal enzymatic activity, its pK_a will have to be raised by at least two units. An alternative mechanism for substrate charge compensation can be suggested that involves a novel histidine triad, His-213, His-199, and His-228, not observed in any other serine protease. The C-terminal α-helix, ubiquitous to all pancreatic-type proteases, is directly linked to this histidine triad and may also play a role in substrate stabilization.

As mentioned, no three-dimensional structures of other proteases with specificity toward acidic amino acids have been reported. However, some suggestions as to the nature of the interactions in the primary substrate-binding site have been made based on comparisons of amino acid sequences near the primary substrate-binding site in trypsin family proteases with those of distantly related viral and bacterial proteases.[25,45] These considerations and comparison with a recently reported structure of a chymotrypsin-like protease from Sindbis virus core protein[46] tentatively suggest that a putative general substrate-binding scheme for proteases with specificity toward glutamic acid may involve a histidine residue, homologous to His-213, and a hydroxyl function at sequence position 192.

[45] J. F. Bazan and R. J. Fletterick, *Virology* **1**, 311 (1990).
[46] L. Tong, G. Wengler, and M. G. Rossmann, *J. Mol. Biol.* **146**, 337 (1993).

[9] Lysyl Endopeptidase of *Achromobacter lyticus*

By Fumio Sakiyama *and* Takeharu Masaki

Achromobacter lysyl endopeptidase (EC 3.4.21.50), formerly known as *Achromobacter* protease I (API), is one of the extracellular proteases synthesized by *Achromobacter lyticus* M497-1, a gram-negative bacterium that lyses both gram-positive and gram-negative bacteria. The protease is specific for the cleavage of lysyl bonds, including the lysylproline bond.[1,2] In addition to lysine specificity, the protease has distinctive properties, such as a high peptidase activity, a wide pH optimum, and stability against denaturants. These favorable properties as a lysine-specific protease make it useful as a tool for peptide fragmentation in protein sequence analysis.

Assay Methods

Depending on the purpose of assay, four methods can be used to assay the lysyl endopeptidase. For proteolytic activity, the method of Hagihara

[1] T. Masaki, K. Nakamura, M. Isono, and M. Soejima, *Agric. Biol. Chem.* **42**, 1443 (1978).
[2] T. Masaki, M. Tanabe, K. Nakamura, and M. Soejima, *Biochim. Biophys. Acta* **660**, 44 (1981).

et al.[3] using 0.7% heat-denatured casein in 50 m*M* Tris-HCl buffer (pH 9.0) at 40° is used. The hydrolysis of Bz-Lys-*p*-nitroanilide (Bz-Lys-*p*NA) can be followed at 405 nm by the method of Tuppy *et al.*[4] using 0.25 m*M* Bz-Lys-*p*NA in 0.17 *M* 2-amino-2-methyl-1,3-propanediol buffer (pH 9.0). Esterolytic activity with Tos-Lys-OMe in 80 m*M* Tris-HCl (pH 8.0) is determined at 247 nm by the method of Schwert and Takenaka.[5]

A more sensitive assay of peptidase activity is established by modifying the method of Zimmerman *et al.*[6] described for chymotrypsin to use as substrate Boc-Val-Leu-Lys-AMC [AMC, 7-(4-methyl)coumarylamide]. A typical procedure is described below.

Assay Solutions

(a) Substrate solution. Boc-Val-Leu-Lys-AMC is dissolved at 50 µ*M* in 0.2 *M* Tris-HCl (pH 9.0) containing 1% (v/v) dimethylformamide.
(b) Enzyme solution. API is dissolved at 2 n*M* in 0.2 *M* Tris-HCl (pH 9.0) and cooled to 0° until used. Enzyme concentration is determined for a 2 µ*M* solution based on the molar absorbance of 52,000 at 280 nm.

Substrate solution (1.95 ml) is placed in a cuvette equipped with a small magnetic stirrer and thermostatted with circulating water at 37° on a Hitachi F4000 spectrofluorometer. After keeping the enzyme solution at this temperature for a few minutes, 50 µl is quickly added to the substrate solution. Immediately, the fluorescence of the liberated 7-amino-4-methyl-coumarin is automatically recorded at 440 nm (slit, 5 nm). The excitation wavelength and slit width are 380 and 20 nm, respectively. The rate of hydrolysis is calculated from the slope of the linear increase in fluorescence.

Purification Procedure

Growth of Bacteria. The microorganism is inoculated into 400 ml of nutrient broth [1% sucrose/1% polypeptone/0.5% milk casein/0.01% KH_2PO_4/0.01% K_2HPO_4/0.01% $MgSO_4 \cdot 7H_2O$ in deionized water (pH 7.2)] and cultivated at 28° for 100 hr on a shaker (135 revolutions per min). The cells are removed by centrifugation and the supernatant is collected.

[3] B. Hagihara, H. Matsubara, M. Nakai, and K. Okunuki, *J. Biochem.* (*Tokyo*) **45**, 185 (1958).
[4] H. Tuppy, U. Wiesbauer, and E. Winterberger, *Z. Physiol. Chem.* **329**, 278 (1962).
[5] G. W. Schwert and Y. Takenaka, *Biochim. Biophys. Acta* **16**, 570 (1955).
[6] M. Zimmerman, E. Yurewicz, and G. Patel, *Anal. Biochem.* **70**, 258 (1976).

Purification. The following procedure is an improved modification of the procedure previously described.[2] All steps are carried out at 4° unless otherwise mentioned.

Step 1. Benzalkonium chloride treatment and acetone fractionation. To the supernatant (20.6 liters) obtained above, solid benzalkonium chloride is slowly added with stirring to 0.04% (w/v) concentration, kept overnight, and centrifuged. The supernatant is then mixed with 3 volumes of acetone (precooled at −15°) with stirring and kept at −15° for 12 hr, followed by centrifugation. The precipitate is collected, washed twice with cold acetone, and dried under reduced pressure (acetone powder). Almost all the lysyl endopeptidase activity in the culture broth is recovered in the precipitate.

Step 2. CM-cellulofine treatment. The acetone powder (43.8 g) is suspended in 10 mM Tris-HCl, pH 8.0 (Buffer A) (400 ml), allowed to stand overnight, and centrifuged (23,600 *g*, 10 min). Precipitates are washed with a small amount of Buffer A and centrifuged in the same way. To the combined supernatants (750 ml), CM-cellulofine (Seikagaku) (900 g) previously equilibrated with Buffer A is slowly added under stirring. After 40 min, the CM-cellulofine is collected by filtration with a glass filter (26G-3) and washed with Buffer A. At this step, α- and β-lytic proteases are adsorbed on CM-cellulofine and a brownish crude solution of lysyl endopeptidase (1.55 liters) is obtained.

Step 3. DEAE-cellulofine treatment. DEAE-cellulofine A-200 (Seikagaku) (700 g) previously equilibrated with Buffer A is added to the crude preparation, gently stirred for 2 hr, and collected by filtration on a glass filter (26G-3) at ambient temperature. The filtrate (2.5 liters) is concentrated by ultrafiltration (Amicon 2000, YM10 membrane) to 510 ml, which is dialyzed against 2 mM Tris-HCl, pH 8.0 (Buffer B) and concentrated with the ultrafiltration membrane to 470 ml. Part of the colored material and most of the contaminating proteins are adsorbed on DEAE-cellulofine.

Step 4. AH-Sepharose 4B chromatography. The concentrated enzyme solution (235 ml) is applied to AH-Sepharose 4B (4.0 × 20 cm) equilibrated with Buffer B. After washing with Buffer B (1.4 liters) to remove *A. lyticus* protease II, lysyl endopeptidase is eluted with a linear gradient to 1.1 M NaCl (1 liter) in Buffer B (1 liter). The active protease fractions (2.36–3.28 liters) eluted with 0.5 M NaCl are collected, dialyzed against Buffer B, and concentrated. The dialyzed enzyme solution (2.5 liters total) is concentrated by an Amicon 2000 (Danvers, MA) ultrafiltration membrane to 48 ml.

Step 5. Isoelectric focusing. The concentrated enzyme solution (48 ml) is divided into two parts that are processed separately by the isoelectric

TABLE I
PURIFICATION OF *Achromobacter* LYSYL ENDOPEPTIDASE

Step	Volume (ml)	Total protein ($A_{280\,nm}$)	Total activity[a] (U)	Specific activity ($U/A_{280\,nm}$)	Recovery (%)
Acetone powder	750	13,300	1940	0.146	100
CM-cellulofine	1550	10,500	1800	0.171	92.8
DEAE-cellulofine	470	4020	1670	0.415	86.1
AH-Sepharose 4B	48	1320	1710	1.30	88.1
Ampholine (pH 6–8)	12	469	1220	2.60	62.9
Sephadex G-50	6.9	357	973	2.73	50.2

[a] Lysyl endopeptidase activity was assayed with Bz-Lys-pNA at pH 9.0.

focusing (carrier ampholyte, pH 6–8). The active fraction (270–345 ml) corresponding to pI 6.9 is collected, dialyzed thoroughly against Buffer B, and concentrated to 12 ml as mentioned earlier. After gel filtration on Sephadex G-50 (2.0 × 94 cm) using the same buffer, the enzyme fractions are collected, concentrated to 6.9 ml, and stored at −15°. The purified enzyme is stable for over 1 year under these conditions.

The results of purification are summarized in Table I.

Enzymatic Properties

The optimum pH values for amidolytic, esterolytic, and caseinolytic activities of API are 9.0–9.5, 7.8–8.2, and 8.5–10.7, respectively. The enzyme is stable in a pH range of 4.0–10.0 and retains all activities up to 40°. Tos-Lys-CH$_2$Cl and Z-Leu-Leu-lysinal[7] are strong competitive irreversible and reversible inhibitors, respectively. DFP, PMSF, n-alkyl-amine,[8,9] Zn^{2+},[10] alkali metal ions, and ammonium ion inhibit the activity (Table II). However, trishydroxymethylaminomethane does not inhibit the activity and is used as a component of buffer solutions for digestion.

API hydrolyzes lysyl and S-aminoethyl peptide bonds at the same rate.[11] The arginyl bond cannot be cleaved, but the ornithyl bond can be

[7] T. Masaki, T. Tanaka, S. Tsunasawa, F. Sakiyama, and M. Soejima, *Biosci. Biotech. Biochem.* **56**, 1604 (1992).

[8] T. Masaki, T. Fujihashi, K. Nakamura, and M. Soejima, *Biochim. Biophys. Acta* **660**, 51 (1981).

[9] T. Masaki, T. Fujihashi, and M. Soejima, *Nippon Nogeikagaku Kaishi* **58**, 865 (1984).

[10] M. Soejima and T. Masaki, *Tanpakushitsu Kakusan Koso* (*Proteins, Nucleic Acids and Enzymes*) **29**, 1532 (1984).

[11] Y. Kawata, F. Sakiyama, and H. Tamaoki, *Eur. J. Biochem.* **176**, 683 (1988).

TABLE II
REVERSIBLE INHIBITORS OF LYSYL ENDOPEPTIDASE

Inhibitors	K_i (mM)	Ref.
ZnCl$_2$	0.58 (0.4)[a]	10
BaCl$_2$	3.4 (1.6)	2
NH$_4$Cl	28.0 (2.5)	2
CsCl	40.0 (5.5)	2
KCl	45.0 (5.2)	2
NaCl	86.0 (15.0)	2
LiCl	90.0 (24.0)	2
Bz-Lysine	0.38 (0.3)	9
Bz-Lysinal	0.3 (0.7)	7
Z-Val-lysinal	0.0065 (0.12)[b]	7
Z-Pro-lysinal	0.04 (0.08)[b]	7
Z-Leu-Leu-lysinal	0.04 (0.04)[b]	7
Methylamine	4.6 (2.7)	8
Ethylamine	1.0 (0.7)	8
n-Propylamine	0.25 (0.32)	8
n-Butylamine	0.05 (0.03)	8
n-Amylamine	0.03 (0.04)	8
n-Hexylamine	0.35	8
Isobutylamine	0.45	9
Cyclohexylamine	0.15	9
Lysine	10	8
6-Amino-1-hexanoic acid	2.2	9
Hydrazine	0.4	9

[a] K_i values were estimated with Tos-Lys-OMe in 0.1 M Tris-HCl buffer, pH 8, at 30°. Values in parentheses were similarly estimated with Bz-Lys-pNA.
[b] Noncompetitive inhibition.

cleaved slowly (Table III). API shows an order of magnitude higher lysyl endopeptidase activity compared to trypsin. Accordingly, in the digestion of peptide chains in protein sequence analysis, a molar ratio of enzyme to substrate of 200–400 to 1 has been used in 0.1 M Tris-HCl buffer, pH 9, for 6 hr at 37°.[12] Under these conditions, all lysyl peptide bonds are usually cleaved completely. An exception is the lysylproline bond, which is hydrolyzed less efficiently. The hydrolysis rate is also affected by the nature of adjacent amino acids, as expected, in view of the existence of three binding subsites.[13] For instance, the hydrolysis of the lysyl bond is

[12] S. Tsunasawa, T. Masaki, M. Hirose, M. Soejima, and F. Sakiyama, J. Biol. Chem. 264, 3832 (1989).
[13] F. Sakiyama, M. Suzuki, A. Yamamoto, S. Aimoto, S. Norioka, T. Masaki, and M. Soejima, J. Protein Chem. 9, 297 (1990).

TABLE III
KINETIC PARAMETERS FOR HYDROLYSIS OF VARIOUS SUBSTRATES

Substrates[a]	K_m (mM)	k_{cat} (sec^{-1})	k_{cat}/K_m (mM^{-1}, sec^{-1})	Assay pH at 30°	Ref.
Bz-Lys-NH$_2$	0.32	1.54	4.81	8.5	1
Bz-Lys-pNA	0.07	0.86	12.3	9.5	1
Lys-pNA	0.08	0.14	1.75	9.2	1
Bz-Lys-OMe	0.091	225.3	2480	8.0	8
Tos-Lys-OMe	0.1	570.0	5700	8.0	8
Tos-Arg-OMe	35.7	5.04	0.14	8.0	8
Bz-Orn-OMe	1.0	10.4	10.4	8.0	8
Boc-Lys-AMC	41.0	3.02	0.07	9.0	—
Boc-Ala-Lys-AMC	3.11	32.9	10.6	9.0	—
Boc-Ala-Ala-Lys-AMC	3.09	49.0	15.2	9.0	—
Bovine trypsinogen[b]	0.0019	0.103	54.2	8.0	21

[a] OMe, Methyl ester; AMC, 7-amino-4-methylcoumarin; Orn, ornithine.
[b] Activation to trypsin at 25°.

slow when one or two basic amino acid residues precede it,[14,15] or an acidic residue is adjacent. A lysyl bond at the N terminus is hydrolyzed very slowly. For both protein and polypeptide substrates, very pure lysyl endopeptidase is highly specific for the lysyl peptide bond, and cleavage of other bonds is rare or nonexistent. Hydrolysis at Arg-Ser,[16] Arg-Ala,[17–19] Gly-Ala,[11] and Phe-Lys[20] has been reported, although the latter two cases are quite exceptional. API of sequenching grade is commercially available as lysyl endopeptidase (Wako Pure Chemical Industries, Ltd., Chuoku, Osaka, Japan; fax: +81-6-222-1203).

API retains most of its activity in the presence of 20% organic solvents such as methanol, ethanol, and 2-propanol at 37°, but tends to be inactivated in the presence of acetonitrile under the same conditions.[22] The

[14] S. Tsunasawa, S. Sugihara, T. Masaki, F. Sakiyama, Y. Takeda, T. Miwatani, and K. Narita, *J. Biochem. (Tokyo)* **101**, 111 (1987).
[15] Y. Yoshida, S. Wakabayashi, H. Matsubara, T. Hashimoto, and K. Tagawa, *FEBS Lett.* **170**, 138 (1984).
[16] T. Yonetsu, K. Higuchi, S. Tsunasawa, S. Takagi, F. Sakiyama, and T. Takeda, *FEBS Lett.* **203**, 149 (1986).
[17] S. Isemura, E. Saitoh, and K. Sanada, *J. Biochem. (Tokyo)* **96**, 489 (1984).
[18] Y. Kitagawa, S. Tsunasawa, N. Tanaka, Y. Katsube, F. Sakiyama, and K. Asada, *J. Biochem. (Tokyo)* **99**, 1289 (1986).
[19] K. Okamura, T. Miyata, S. Iwanaga, K. Takamiya, and M. Nishimura, *J. Biochem. (Tokyo)* **101**, 957 (1987).
[20] T. Miyata, M. Hiranaga, M. Umezu, and S. Iwanaga, *J. Biol. Chem.* **259**, 8924 (1984).
[21] T. Masaki and M. Soejima, *Agric. Biol. Chem.* **49**, 1867 (1985).
[22] K. G. Welinder, *Anal. Biochem.* **174**, 54 (1988).

peptidase activity is unaffected by 4 M urea or 0.1% (w/v) SDS in 40 mM Tris-HCl, pH 8, for 20 min at 30°. In the presence of 0.25 M guanidine hydrochloride, API loses half of its activity. However, in certain cases, it is possible to digest proteins even in the presence of 4 M guanidine hydrochloride [enzyme/substrate, 1/100 (w/w), in 50 mM Tris-HCl, pH 9.0, for 4 hr at 37°][23] or 8 M urea.[24] Heating for 30 min at 60° or dissolution into concentrated formic acid inactivates the enzyme completely. Treatment of an API-containing sample in concentrated formic acid for 5 min at room temperature is recommended to suppress autolysis during manipulation prior to SDS–PAGE.

API catalyzes the formation of a lysyl peptide bond between N^α-acylated lysine and an appropriate amino component. By this method, threonine ester was coupled to the C terminus of the B chain of porcine insulin in place of the normal alanine.[25,26]

Semisynthesis of Human Insulin. To zinc-free porcine insulin (100 mg) dissolved in 0.1 M NH$_4$HCO$_3$ (20 ml), add API (0.32 mg). After a 90-hr incubation at 37°, the reaction mixture is lyophilized. More than 95% of the insulin is converted to des-alanine insulin. The lyophilizate is dissolved to make 10 mM des-alanine insulin and 7 μM API with threonine *tert*-butyl ester (195 mg, 0.5 M), in ethanol (0.5 ml) and dimethylformamide (0.5 ml), and is kept at 37° for 20 hr. The yield of lysyl bond synthesis is 83%. The reaction mixture is passed through Sephadex G-50 (4.2 × 30 cm) in 0.5 M acetic acid at 4°. The fraction corresponding to human insulin ester is collected, lyophilized, loaded on DEAE-Sephadex A-25 (2.2 × 22 cm) equilibrated with 0.01 M Tris-HCl buffer (pH 7.4) containing 7 M urea and is chromatographed in a gradient of 0–0.3 M NaCl at 4°. The first eluting peak (elution starts at around 0.1 M NaCl) is dialyzed against 0.5 M acetic acid at 4° and is lyophilized (56 mg). Deprotection is carried out with trifluoroacetic acid in the presence of anisole for 30 min at room temperature, yielding 52 mg of human insulin.

Chemical Properties

API is stable at pH 4–10 and can be stored for several years at $-20°$ in 1 mM to 0.1 M Tris-HCl buffer at pH 6–10 at enzyme concentrations higher than 0.1 mg/ml. $A^{1\%}_{280\,nm}$ is 18.8 (pH 8)[2]; M_r 27,728; pI 6.9.

API consists of a single chain of 268 amino acid residues, and three

[23] H. Hayashi, T. Hayashi, and Y. Hanaoka, *Eur. J. Biochem.* **205**, 105 (1992).
[24] T. Katoh and F. Morita, *J. Biol. Chem.* **268**, 2380 (1993).
[25] K. Morihara, T. Oka, H. Tsuzuki, Y. Tochino, and T. Kanaya, *Biochem. Biophys. Res. Commun.* **92**, 396 (1980).
[26] K. Morihara, Y. Ueno, and K. Sakina, *Biochem. J.* **240**, 803 (1986).

a

```
-205                          -200                                          -190
ATG AAA CGC ATT TGT GGT TCC CTG CTG TTG CTC GGT TTG TCG ATC AGC GCC GCG CTC GCC
Met Lys Arg Ile Cys Gly Ser Leu Leu Leu Leu Gly Leu Ser Ile Ser Ala Ala Leu Ala

GCC CCG GCC TCG CGC CCC GCG GCG TTC GAT TAC GCC AAT CTT TCC AGC GTC GAC AAG GTC
Ala Pro Ala Ser Arg Pro Ala Ala Phe Asp Tyr Ala Asn Leu Ser Ser Val Asp Lys Val

GCC TTG CGC ACC ATG CCG GCG GTC GAC GTG GCC AAG GCC AAG GCC GAA GAT TTG CAG CGC
Ala Leu Arg Thr Met Pro Ala Val Asp Val Ala Lys Ala Lys Ala Glu Asp Leu Gln Arg

GAC AAG CGC GGC GAC ATC CCG CGC TTC GCC CTG GCG ATC GAC GTG GAC ATG ACC CCT CAG
Asp Lys Arg Gly Asp Ile Pro Arg Phe Ala Leu Ala Ile Asp Val Asp Met Thr Pro Gln

AAT TCC GGC GCG TGG GAA TAC ACC GCC GAC GGC CAG TTC GCC GTA TGG CGC CAG CGC GTT
Asn Ser Gly Ala Trp Glu Tyr Thr Ala Asp Gly Gln Phe Ala Val Trp Arg Gln Arg Val

CGT TCG GAG AAG GCG CTG TCA CTG AAC TTC GGT TTC ACC GAC TAC TAC ATG CCC GCC GGC
Arg Ser Glu Lys Ala Leu Ser Leu Asn Phe Gly Phe Thr Asp Tyr Tyr Met Pro Ala Gly

GGC CGC CTG CTG GTA TAT CCG GCG ACT CAG GCG CCG GCC GGC GAT CGC GGC TTG ATC AGC
Gly Arg Leu Leu Val Tyr Pro Ala Thr Gln Ala Pro Ala Gly Asp Arg Gly Leu Ile Ser

CAG TAC GAC GCC AGC AAC AAC AAC TCG GCG CGC CAA CTG TGG ACG GCG GTG GTG CCG GGC
Gln Tyr Asp Ala Ser Asn Asn Asn Ser Ala Arg Gln Leu Trp Thr Ala Val Val Pro Gly

GCC GAA GCG GTG ATC GAA GCG GTG ATC CCG CGC GAC AAG GTC GGC GAG TTC AAG CTG CGC
Ala Glu Ala Val Ile Glu Ala Val Ile Pro Arg Asp Lys Val Gly Glu Phe Lys Leu Arg

CTG ACC AAG GTC AAC CAC GAC TAC GTC GGT TTC GGC CCG CTC GCG CGC CGC CTG GCC GCT
Leu Thr Lys Val Asn His Asp Tyr Val Gly Phe Gly Pro Leu Ala Arg Arg Leu Ala Ala
                    -1
GCG TCC GGC GAG AAG
Ala Ser Gly Glu Lys
```

FIG. 1. The amino acid sequences of prepropeptide (a), mature API (b), and the C-terminal extension (c). These three portions are linked in this order in preproAPI. Disulfide linkages exist at Cys[6]–Cys[216], Cys[12]–Cys[80], and Cys[36]–Cys[58] in mature API.

disulfide bonds are formed at Cys[6]–Cys[216], Cys[12]–Cys[80], and Cys[36]–Cys[58] (Fig. 1). His-57, Asp-113, and Ser-194 compose the catalytic triad and Asp-225 has been identified as the primary specificity determinant for lysine.[27] His-210, Gly-211, and Gly-212 are perhaps involved in substrate-binding subsites because subsite mapping analysis using synthetic peptide substrates revealed the presence of three sites toward the N terminus from the scissile lysyl bond. A positively charged group corresponding to the amino-terminal Ile-16 (chymotrypsinogen numbering) in bovine trypsin is absent and the acidic amino acid residue corresponding to Asp-194 is serine, so the formation of an ion pair between these residues is impossible in API. Presumably, the Cys[6]–Cys[216] linkage stabilizing the tertiary structure serves the role of this interaction, and gives stability at alkaline pH. The Cys[6]–Cys[216] bridge has not been found in members of the chymotrypsin (S1) family listed by Rawlings and Barrett.[27a] When

[27] S. Norioka and F. Sakiyama, in "Methods in Protein Sequence Analysis" (K. Imahori and F. Sakiyama, eds.), p. 101. Plenum, New York and London, 1993.
[27a] N. D. Rawlings and A. J. Barrett, Biochem. J. 290, 205 (1993).

b

```
1                                    10                                              20
GGC GTG TCG GGT TCG TGC AAC ATC GAC GTG GTC TGC CCC GAA GGC GAC GGC CGC CGC GAC
Gly Val Ser Gly Ser Cys Asn Ile Asp Val Val Cys Pro Glu Gly Asp Gly Arg Arg Asp
                                              30                                    40
ATC ATC CGC GCG GTC GGT GCG TAC TCG AAG AGC GGC ACG CTG GCC TGT ACC GGT TCG CTG
Ile Ile Arg Ala Val Gly Ala Tyr Ser Lys Ser Gly Thr Leu Ala Cys Thr Gly Ser Leu
                                              50                                    60
GTC AAC AAC ACC GCC AAC GAC CGC AAG ATG TAC TTC CTG ACC GCG CAC CAC TGC GGC ATG
Val Asn Asn Thr Ala Asn Asp Arg Lys Met Tyr Phe Leu Thr Ala His His Cys Gly Met
                                              70                                    80
GGC ACG GCC TCG ACC GCG GCG TCG ATC GTG GTG TAC TGG AAC TAT CAG AAC TCG ACC TGC
Gly Thr Ala Ser Thr Ala Ala Ser Ile Val Val Tyr Trp Asn Tyr Gln Asn Ser Thr Cys
                                              90                                    100
CGC GCG CCC AAC ACG CCG GCC AGC GGC GCC AAC GGC GAC GGC TCG ATG AGC CAG ACC CAG
Arg Ala Pro Asn Thr Pro Ala Ser Gly Ala Asn Gly Asp Gly Ser Met Ser Gln Thr Gln
                                              110                                   120
TCG GGT TCG ACG GTC AAG GCG ACC TAC GCC ACC TCC GAC TTC ACC CTG CTC GAG TTG AAC
Ser Gly Ser Thr Val Lys Ala Thr Tyr Ala Thr Ser Asp Phe Thr Leu Leu Glu Leu Asn
                                              130                                   140
AAT GCG GCC AAC CCC GCG TTC AAC CTG TTC TGG GCC GGT TGG GAC CGT CGC GAC CAG AAC
Asn Ala Ala Asn Pro Ala Phe Asn Leu Phe Trp Ala Gly Trp Asp Arg Arg Asp Gln Asn
                                              150                                   160
TAT CCC GGC GCG ATC GCC ATC CAC CAT CCC AAC GTC GCC GAG AAG CGC ATC AGC AAC TCC
Tyr Pro Gly Ala Ile Ala Ile His His Pro Asn Val Ala Glu Lys Arg Ile Ser Asn Ser
                                              170                                   180
ACC AGC CCG ACC TCG TTC GTG GCC TGG GGC GGC GGC GCC GGC ACC ACG CAT TTG AAC GTG
Thr Ser Pro Thr Ser Phe Val Ala Trp Gly Gly Gly Ala Gly Thr Thr His Leu Asn Val
                                              190                                   200
CAG TGG CAG CCC TCG GGC GGC GTG ACC GAG CCG GGT TCG TCG GGT TCG CCG ATC TAC AGC
Gln Trp Gln Pro Ser Gly Gly Val Thr Glu Pro Gly Ser Ser Gly Ser Pro Ile Tyr Ser
                                              210                                   220
CCG GAA AAG CGC GTG CTC GGC GAG CTG CAC GGC GGC CCG TCG AGC TGC AGC GCC ACC GGC
Pro Glu Lys Arg Val Leu Gly Glu Leu His Gly Gly Pro Ser Ser Cys Ser Ala Thr Gly
                                              230                                   240
ACC AAC CGC AGC GAC CAG TAC GGC CGC GTG TTC ACC TCG TGG ACC GGC GGC GGC GCC GCG
Thr Asn Arg Ser Asp Gln Tyr Gly Arg Val Phe Thr Ser Trp Thr Gly Gly Gly Ala Ala
                                              250                                   260
GCC TCG CGC CTG AGC GAT TGG CTC GAT CCG GCC AGC ACC GGC GCG CAG TTC ATC GAC GGC
Ala Ser Arg Leu Ser Asp Trp Leu Asp Pro Ala Ser Thr Gly Ala Gln Phe Ile Asp Gly
                                              268
CTG GAT TCG GGC GGC GGC ACG CCG
Leu Asp Ser Gly Gly Gly Thr Pro
```

FIG. 1. (*continued*)

compared with bovine trypsin, API holds an extension composed of several amino acid residues at the N terminus and a long extension (about 26 residues) at the C terminus.

API Gene, Prepro-API, and Activation Mechanism

API is synthesized as a big precursor protein (prepro-API) composed of a 20-residue signal peptide, a 185-residue propeptide, a 268-residue mature protease, and a 180-residue hydrophilic C-terminal peptide, which are arranged in this order from the initiator methionine.[28] The API gene

[28] T. Ohara, K. Makino, H. Shinagawa, A. Nakata, S. Norioka, and F. Sakiyama, *J. Biol. Chem.* **264**, 20625 (1989).

C
```
    270                                    280
AAC ACT CCG CCG GTG GCG AAC TTC ACC TCC ACC ACC AGC GGC CTG ACC GCG ACC TTC ACC
Asn Thr Pro Pro Val Ala Asn Phe Thr Ser Thr Thr Ser Gly Leu Thr Ala Thr Phe Thr

GAC AGC TCC ACC GAC AGC GAC GGT TCG ATC GCC TCG CGT AGC TGG AAC TTC GGC GAC GGC
Asp Ser Ser Thr Asp Ser Asp Gly Ser Ile Ala Ser Arg Ser Trp Asn Phe Gly Asp Gly

AGC ACC TCG ACC GCG ACC AAC CCG AGC AAG ACC TAC GCC GCG GCG GGC ACC TAC ACC GTC
Ser Thr Ser Thr Ala Thr Asn Pro Ser Lys Thr Tyr Ala Ala Ala Gly Thr Tyr Thr Val

ACC CTG ACG GTC ACC GAC AAC GGC GGC GCC ACC AAC ACC AAG ACC GGT TCG GTC ACC GTG
Thr Leu Thr Val Thr Asp Asn Gly Gly Ala Thr Asn Thr Lys Thr Gly Ser Val Thr Val

TCC GGC GGC CCG GGT GCG CAG ACC TAC ACC AAC GAC ACC GAT GTG GCG ATC CCG GAC AAC
Ser Gly Gly Pro Gly Ala Gln Thr Tyr Thr Asn Asp Thr Asp Val Ala Ile Pro Asp Asn

GCG ACG GTC GAA AGC CCG ATC ACC GTG TCC GGC CGC ACC GGC AAC GGC TCG GCG ACC ACG
Ala Thr Val Glu Ser Pro Ile Thr Val Ser Gly Arg Thr Gly Asn Gly Ser Ala Thr Thr

CCG ATC CAG GTG ACG ATC TAC CAC ACC TAC AAG AGC GAT CTG AAG GTG GAC CTG GTC GCG
Pro Ile Gln Val Thr Ile Tyr His Thr Tyr Lys Ser Asp Leu Lys Val Asp Leu Val Ala

CCG GAC GGC ACC GTC TAC AAC CTG CAC AAC CGC ACC GGC GGC AGC GCG CAC AAC ATC ATC
Pro Asp Gly Thr Val Tyr Asn Leu His Asn Arg Thr Gly Gly Ser Ala His Asn Ile Ile

CAG ACC TTC ACC AAG GAC CTG TCG AGC GAA GCG GCT CAA CGG GCA CCT GGA AGC TGC GGG
Gln Thr Phe Thr Lys Asp Leu Ser Ser Glu Ala Ala Gln Arg Ala Pro Gly Ser Cys Gly
449
TGA
```

FIG. 1. (*continued*)

consists of 1959 bp rich in G + C, amounting to 70% (Fig. 1). The propeptide portion is supposed to be necessary for correct folding of pro-API, but the C-terminal extension is required for neither folding nor enzyme activity. It would appear that this threonine/serine-rich extension might serve for the penetration of active API through the outer membrane in *A. lyticus* cells. A gene quite similar to that of API has been cloned from *Lysobacter enzymogenes* (ATCC 29487, Rockville, MD). The nucleotide sequences of mature protease portions in this and API genes were totally identical.[29]

The cloned gene of prepro-API is expressed in *Escherichia coli*; pro-API is secreted to the periplasm and activated autocatalytically. No secretion to the outside of *E. coli* cells takes place. This mode of expression of the API gene is seen with or without the C-terminal extension of the molecule.

Isolation of API Synthesized by Escherichia coli. The *E. coli* cells transformed by an appropriate expression vector bearing the API gene are incubated overnight in L-broth (1 liter) and collected by centrifugation.

[29] T. Ohara, A. Yamamoto, H. Shinagawa, A. Nakata, S. Norioka, and F. Sakiyama, *Abst. 5th Protein Soc. Symp.*, 120 (1991).

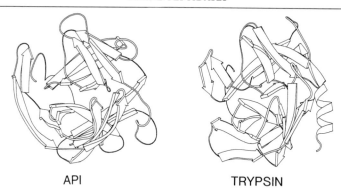

API   TRYPSIN

FIG. 2. Tertiary structure of *Achromobacter* lysyl endopeptidase compared to that of trypsin.

The cell pellet is suspended in 10 mM Tris-HCl, pH 7.5, containing 30 mM NaCl (100 ml), centrifuged for 10 min, and washed with 33 mM Tris-HCl, pH 7.5 (100 ml). The washed cells thus obtained are suspended in 33 mM Tris-HCl, pH 7.5, containing 40% sucrose and 0.1 mM EDTA (100 ml), incubated at 37° for 10 min and centrifuged at 10,000 rpm for 10 min. The precipitates are suspended in 0.5 mM MgCl$_2$ (100 ml) and vigorously shaken for 10 min under cooling with ice. The supernatant containing the periplasmic proteins is collected by centrifugation as described above, followed by dialysis against 10 mM Tris-HCl, pH 9.0 (Buffer C) overnight at 4°. The dialyzate was passed through a DEAE-cellulose column (3.6 × 50 cm) equilibrated with Buffer C and the unadsorbed fraction containing API activity is pooled. The pooled fraction is then adsorbed on a column (1.6 × 25 cm) of chicken ovomucoid-Sepharose 4B (8.5 mg or 0.3 mmol of ligand per ml wet gel) and eluted with 10 mM ammonium acetate, pH 3.0. Care is needed to immediately neutralize the effluent with a dilute ammonium hydroxide solution, followed by gel filtration on a TSK gel G2000 SW column in 0.2 M ammonium acetate, pH 7. Active fractions are collected and stored at −80°.

Tertiary Structure

The crystal structure of API has been solved (Fig. 2).[30] The main chain folds as a two-β-barrel structure similar to that in bovine trypsin. His-57 and Asp-113, residing in a barrel, and Ser-194, residing in another barrel, constitute the catalytic triad. As anticipated, there is no ion pair between

---

[30] Y. Kitagawa, Y. Katsube, K. Sasi, Y. Matsuura, S. Norioka, and F. Sakiyama, *Abst. XVIth Intern. Congr. General Assembly, IUC, August, 1993, Beijing, China.*

Gly-1 and Ser-193 and instead the Cys^6–Cys^{216} bond is formed to connect the regions around the above residues. A relatively open, shallow pocket is formed for the binding of the P1 lysine side chain and Asp-225 lies inside this pocket. Unlike the substrate-binding subsites of trypsin, which are formed by a set of Ser-214, Trp-215, and Gly-216, API does not have the former two amino acid residues but has His-210 and Gly-211 at subsites S1 and S2, respectively. The indole ring of Trp-169 lies near the pocket in API. As a result, the orientation of this aromatic ring is different from that of Trp-215, and these two rings are disposed mutually perpendicular. Apparent features, except the position of the essential negative charge in the S1 pocket and the side chains of binding subsites S1 and S2, have a close resemblance to those in the active site of trypsin.

[10] IgA-Specific Prolyl Endopeptidases: Serine Type

By Andrew G. Plaut and William W. Bachovchin

Introduction

IgA proteinases are a group of endopeptidases produced by medically important bacteria in the genera *Streptococcus, Neisseria, Haemophilus, Ureaplasma, Clostridium, Capnocytophaga,* and *Bacteroides* (Table I). The enzymes are secreted to the extracellular environment, and all have pronounced substrate specificity for human IgA_1 immunoglobulins, one of the two IgA isotypes that are the dominant form of antibody in human secretions. In IgA_1 each proteinase attacks a single, specific peptide bond in the heavy (α) polypeptide chain hinge region, resulting in formation of hydrolysis products consisting of the intact antigen-binding Fab and the Fc region of these antibody proteins (Fig. 1).

The IgA proteinases have differences in catalytic mechanism, with serine type, metallo-, and cysteine peptidases all represented. This chapter deals with the IgA-specific serine endopeptidase, IUBMB classification EC 3.4.21.72, and peptidase evolutionary family S6 in the scheme of Rawlings and Barrett[1] (see also [1] in this volume). The IgA-specific metalloproteinases are discussed elsewhere in this series.

[1] N. Rawlings and A. J. Barrett, *Biochem. J.* **290**, 205 (1993).

TABLE I
INFECTIONS CAUSED BY BACTERIA SYNTHESIZING IgA₁ PROTEINASE

Bacterial pathogen	Disease produced[a]	Comments on IgA proteinase
Haemophilus influenzae	Otitis media, sinusitis, bronchitis, epiglottitis, pneumonia, meningitis	At least two cleavage types. Serogroup: type 1 in serotypes a, b, d, f; type 2 in c and e
Haemophilus aegyptius	Conjunctivitis	—
Neisseria gonorrhoeae	Gonorrhea, invasive gonococcemia	Two cleavage types; type 1 common in invasive strains
Neisseria meningitidis	Meningitis, meningococcemia	Two cleavage types; type 1 common in epidemic and invasive strains
Streptococcus sanguis; Streptococcus oralis	Dental plaque and caries; bacterial endocarditis, sepsis	Metalloproteinases
Streptococcus pneumoniae	Disease spectrum similar to that of *Haemophilus influenzae*	Presumptive metalloproteinase
Ureaplasma ureolyticum	Genital infection, urethritis, prostatitis	Presumptive serine proteinase; may remain cell associated (not secreted)
Bacteroides melaninogenicus; other *Bacteroides* species	Destructive periodontal disease	Inhibited by thiol proteinase inhibitors; activity requires reducing conditions
Capnocytophaga species	Destructive periodontal disease	Inhibited by thiol proteinase inhibitors; activity requires reducing conditions
Clostridium ramosum	Soft tissue infection	Protease cleaves both IgA₁ and IgA₂

[a] Most IgA proteinase-positive bacteria can also colonize mucosal surfaces without causing symptoms.

Bacteria-Producing IgA Proteases

Bacteria known to synthesize IgA proteinases and the human diseases they cause are listed in Table I. The enzymes are synthesized *in vivo,* having been found in human mucosal secretions colonized or infected by these bacteria, in suspensions of dental plaque, and in spinal fluid of patients with bacterial meningitidis. The biological consequences of IgA hydrolysis are discussed in the summary at the end of this chapter.

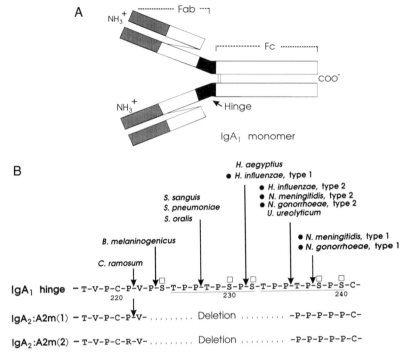

FIG. 1. (A) Diagram of monomeric human IgA$_1$ protein showing variable regions (light shading) and constant regions (nonshaded and hinge) of the heavy and light polypeptide chains. Fab and Fc fragments are produced by IgA$_1$ proteinase cleavage at the heavy chain hinge. (B) Primary sequence of the IgA$_1$ hinge (dark shading in A) showing peptide bonds cleaved by each IgA$_1$ proteinase. Black dots identify the serine-type enzymes. The duplicated octapeptide in IgA$_1$ is underlined, and open squares at serine residues represent O-linked glycosylation sites. Both IgA$_2$ allotypes have the hinge deletion that confers resistant to all but *C. ramosum* proteinase. Amino acid numbers are those of Y. Tsuzukida, C. C. Wang, and F. W. Putnam, *Proc. Natl. Acad. Sci. U.S.A.* **76,** 1104 (1979).

Synthesis and Secretion of Serine-Type IgA Proteinases

The IgA proteinases of the gram-negative bacteria *Neisseria gonorrhoeae, Neisseria meningitidis,* and *Haemophilus influenzae* are encoded by a single chromosomal *iga* gene. There is no known gene regulation, and the enzyme is constitutively expressed *in vitro*. Nonpathogenic species within the genera *Haemophilus* and *Neisseria* do not synthesize IgA proteinases,[2] and DNA from such strains fails to hybridize with *iga* gene

[2] M. H. Mulks and A. G. Plaut, *N. Engl. J. Med.* **299,** 973 (1978).

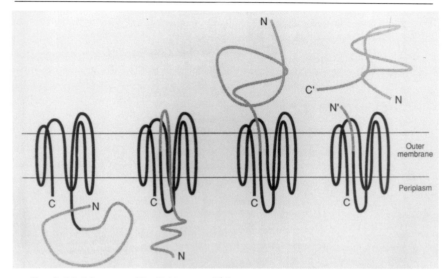

FIG. 2. Model proposed by Pohlner *et al.*[4] for translocation of IgA proteinase across the the *N. gonorrhoeae* outer membrane (see text for details). Translocation is depicted (left to right) as pore formation by an integral outer membrane carboxy-proximal helper (black line), translocation of enzyme domain (lighter line) from periplasm through the pore, autocatalytic proteolysis of the precursor, and enzyme release. N and C are amino and carboxy termini, respectively. Reproduced from A. P. Pugsley, *Microbiol. Rev.* **57**, 50 (1993), with permission.

probes; thus they are truly *iga* negative, not merely harboring an allelic counterpart that is silent or defective.[3]

Enzyme secretion to the culture medium takes place throughout the logarithmic phase of bacterial growth. The secretion pathway for IgA proteinases in *N. gonorrhoeae* and *H. influenzae* has been defined (Fig. 2). Extracellular secretion requires that the enzyme traverse the inner membrane, periplasmic space, and outer membrane. The pathway was originally identified in *N. gonorrhoeae* by Pohlner *et al.*[4], and has subsequently been shown to be essentially the same for *H. influenzae.*[5] All information needed for enzyme activity and secretion in gram-negative bacteria is embodied in a single polypeptide chain. The gonococcal *iga* gene encodes a 169-kDa protein precursor of 1532 amino acids arranged into four functional domains: an amino-terminal signal sequence, the IgA proteinase enzyme, a strongly polar and helical α region, and a large

[3] J. M. Koomey and S. Falkow, *Infect. Immun.* **43**, 101 (1984).
[4] J. Pohlner, R. Halter, K. Beyreuther, and T. F. Meyer, *Nature (London)* **325**, 458 (1987).
[5] F. Grundy, A. G. Plaut, and A. Wright, *J. Bacteriol.* **169**, 4442 (1987).

carboxy-terminal domain. The signal peptide directs passage of the precursor through the bacterial inner membrane and is removed proteolytically. The amphipathic carboxy-terminal domain is essential to direct the enzyme domain to the extracellular environment; it forms a porelike structure in the bacterial outer membrane through which the α domain and enzyme pass to the medium. During secretion the proteinase acquires an active conformation, and a final secretion step involves autocatalytic proteolysis of the precursor at several peptide bonds in the α domain to yield the 105-kDa mature, extracellular enzyme (this step identifies the precursor as an additional substrate). The α domain, like the IgA_1 hinge region, is proline rich, and in the *N. gonorrhoeae* proteinase precursor it has three autoproteolytic cleavage sites (+ is the cleaved bond): -PVKPAP+SPAA-, -IVVAPP+SPQA-, and -ILPRPP+APVF-.[4] Because of these multiple sites the released enzyme has length heterogeneity; the longest form, 121 kDa, is secondarily processed to the 109- and 106-kDa mature forms. No additional function for the 45-kDa carboxy-terminal domain persisting in the cell outer membrane has been established. Further details of the membrane targetting and secretion of the proteinase in *Neisseria* have recently been reported.[5a]

Secretion by this pathway also occurs when the *iga* genes of *Neisseria* or *H. influenzae* are expressed in *Escherichia coli*. Recombinant strains have shown the critical importance of the carboxy-terminal region for secretion; *E. coli* transformed with *N. gonorrhoeae* or *H. influenzae iga* DNA that is truncated at the 3' end yields full enzyme activity that accumulates in the periplasmic space, and is not directed extracellularly. The reader is referred to a review of secretion in gram-negative bacteria that places the IgA protease secretion pathway into a more general context.[6]

Specificity

IgA proteinases cleave both serum and secretory human IgA_1 proteins.[7] The polymeric immunoglobulin receptor (formerly called secretory component) covalently bound to mucosal dimeric IgA does not interfere with cleavage. Neither of the human IgA_2 allotypes A2m(1) or A2m(2) is susceptible to cleavage because of the large hinge region deletion in this isotype (Fig. 1B). The single exception is the IgA proteinase of *Clostridium ramo-*

[5a] T. Klauser, J. Kramer, K. Otzelberger, J. Pohlner, and T. F. Meyer, *J. Mol. Biol.* **234,** 579 (1993).

[6] A. Pugsley, *Microbiol. Rev.* **57,** 50 (1993).

[7] M. Kilian, J. Mestecky, R. Kulhavy, M. Tomana, and W. T. Butler, *J. Immunol.* **124,** 2596 (1980).

sum, which cleaves a -P-V- peptide bond (amino acids 221–222) at the amino-proximal extent of the hinge region in both IgA_1 and IgA_2 A2m(1) proteins.[8] IgA_1 and IgA_2 proteins are isotypic, found in all normal individuals. Although secretory IgA contains both isotypes, the relative amounts of each vary at various mucosal sites; in the upper respiratory tract, one of the major sites at which IgA proteinase-positive bacteria colonize, 90% of the IgA-producing B cells synthesize IgA_1. The biological role of the two isotypes has been reviewed.[9]

Figure 1B shows the human IgA_1 hinge region peptide bonds cleaved; the enzymes invariably cleave prolyl bonds, and each attacks one peptide bond in the hinge despite the tandemly duplicated octapeptide sequence. Type 1 and type 2 cleavage assignment refers to the peptide bond cleaved. In both *H. influenzae* and *N. gonorrhoeae* two independent cleavage types have been identified, each bacterial isolate yielding only one type, and specificity remains unchanged despite repeated subculture, and when *iga* genes are cloned into *E. coli*. The structural basis for cleavage type specificity in the *H. influenzae* IgA proteinase is discussed in the section on structure, below. In *H. influenzae*, proteinase cleavage type correlates with capsular serotype, types a, b, d, and f typically producing type 1 enzyme, and types c and e, type 2. Among clinical isolates of *N. gonorrhoeae* and *N. meningitidis*, type 1 cleavage correlates with invasive disease, as discussed below.

Searches for additional natural substrates other than human, gorilla, and chimpanzee IgA have not been successful (earlier reports that rhesus monkey IgA is a substrate have not been confirmed). The IgA_1 heavy chain devoid of light chains is cleaved, and several synthetic peptides and proteins based on the IgA_1 hinge structure have been reported to be susceptible, but none is in common use for assay. Our experience has been that these are cleaved slowly, if at all. Glycosylation of the IgA_1 hinge region has not been entirely excluded as relevant in substrate recognition. Five hinge region serines bear O-linked sugars (*N*-acetylgalactosamine alone on Ser-224, galactosyl-β-(1→3)-*N*-acetylgalactosamine on the four other serines residues; see Fig. 1A). In certain proteins these have terminal sialic acid residues. Though several investigators have shown that neuraminidase pretreatment of IgA_1 does not materially change its proteinase susceptibility, a contribution of the remaining oligosaccharide to substrate recognition has not been rigorously excluded.

[8] Y. Fujiyama, K. Kobayashi, S. Senda, Y. Benno, T. Bamba, and S. Hosoda, *J. Immunol.* **134**, 573 (1985).
[9] J. Mestecky and M. Kilian, *Monogr. Allergy* **19**, 277 (1986).

Assay of IgA Proteinases

Substrate Purification

IgA proteinase activity is determined using human IgA_1 substrate, the preferred source being the monoclonal paraproteins in plasma of patients with multiple myeloma. These proteins are often at very high concentration and are easy to purify. IgA_1 in normal human plasma[10] is also suitable, but this IgA_1 is polyclonal, and thus yields Fab fragments having some size and charge heterogeneity.

For kinetic studies, any IgG contamination of the substrate must be avoided because IgG contains inhibiting antibodies (see below). A two-step final purification[11] of IgA_1 is therefore recommended:

a. Passage over Affi-Gel protein A (Bio-Rad, Richmond, CA). A small transfer pipette containing the absorbent held in place with glass wool is equilibrated with 0.05 M Tris-HCl, pH 7.5, and aliquots of 10–15 mg IgA_1 are applied at room temperature. Fractions containing IgA but free of IgG by Ouchterlony double-diffusion analysis using commerical antihuman immunoglobulin antisera are pooled and concentrated by ultrafiltration through an Amicon PM10 membrane (Amicon Corp., Lexington, MA).

b. The IgA concentrates are applied to a 1.0 × 20.0-cm, 15-ml Bio-Gel P-300 column, 100–200 mesh (Bio-Rad) prepared in 0.05 Tris-HCl, pH 7.5. After applying the IgA the column is eluted at 8.0 cm/hr linear flow. Fractions containing IgA and free of IgG by Ouchterlony analysis are concentrated as described above, and protein levels are determined using a standard protein assay.

Enzyme Assay

Qualitative assay of IgA proteinases is useful in screening wild-type and recombinant bacterial isolates and for monitoring fractions during enzyme purification. A method using IgA_1-rich human serum was described in an earlier volume.[12] At present, our laboratory does qualitative assay by brief incubation of test samples with ^{125}I-labeled IgA_1; digests are analyzed by autoradiographs of SDS–PAGE, as described in detail below. IgA products of unlabeled substrate can also be analyzed by West-

---

[10] L. M. Loomes, W. W. Stewart, R. L. Mazengera, B. W. Senior, and M. A. Kerr, *J. Immunol. Methods* **141**, 209 (1991).
[11] W. W. Bachovchin, A. G. Plaut, G. R. Flentke, M. Lynch, and C. A. Kettner, *J. Biol. Chem.* **265**, 3738 (1990).
[12] A. G. Plaut, this series, Vol. 165, p. 117.

ern blots using commercial antihuman IgA antisera. Because proteinase-positive bacteria, e.g., *Streptococcus pneumoniae,* may also elaborate carbohydrases, IgA fragments may be enzymatically deglycosylated as well as proteolyzed, thereby yielding fragments having lower than expected M_r on SDS–PAGE.[13,14] For enzyme screening among numerous recombinant lytic plaques or bacterial colonies on agar plates, an ^{125}I-labeled IgA$_1$ overlay method has been developed.[15]

Quantitative assay with sufficient precision for kinetic and inhibitor studies involves cleavage of iodinated IgA$_1$, followed by quantitation of the products separated by SDS–PAGE.[16] The stock substrate solution contains 2 mg/ml purified IgA$_1$ labeled with [125]I (2% labeled) in 0.05 M Tris-HCl buffer, pH 7.5, containing 0.5% (w/v) bovine serum albumin (BSA). For assay, 25 μl of substrate, 25 μl of the Tris-HCl/BSA buffer, and 25 μl of IgA proteinase solution are incubated in a water bath at 37°; 10-μl aliquots are removed at 10-min intervals over a 40-min period to test for IgA$_1$ hydrolysis. The aliquot is prepared for electrophoresis by adding to 100 μl sample buffer consisting of 12.5% (w/v) glycerol, 1.2% sodium dodecyl sulfate, 1.2% 2-mercaptoethanol, and 0.001% bromphenol blue; the mixture is then boiled for 5 min. Electrophoresis is on 9% polyacrylamide gels, which are then stained with 0.05% Coomassie brilliant blue, dried, and autoradiographed at −70° using Kodak (Rochester, NY) XAR5 film with an intensifying screen. The autoradiograph is used as a template to cut the dried gel into segments containing the residual, uncleaved IgA α chain (approximately 50 kDa) and the approximately 22-kDa Fd region of the α chain, that portion in the Fab fragment (Fig. 1A). These segments are counted on a gamma counter, and background counts from control digests are subtracted from each value. The percentage of the heavy chain cleaved is calculated by the formula: (cpm Fd × 100)/ (cpm Fd + cpm heavy chain). Note: We use the Fd band for quantitation because this region in IgA$_1$ often becomes more heavily iodinated.

Purification of IgA Proteinases

Purification of native serine-type IgA proteinases is from spent bacterial culture media; a method for the *N. gonorrhoeae* enzyme was provided in an earlier volume of this series[12]; alternative approaches utilizing phenyl-Sepharose (Pharmacia LKB Biotechnology, Piscataway, NJ) and mono-

[13] E. V. G. Frandsen, J. Reinholdt, and M. Kilian, *Infect. Immun.* **55,** 631 (1987).
[14] J. Reinholdt, M. Tomana, S. B. Mortensen, and M. Kilian, *Infect. Immun.* **58,** 1186 (1990).
[15] J. V. Gilbert and A. G. Plaut, *J. Immunol. Methods* **57,** 247 (1983).
[16] J. V. Gilbert, A. G. Plaut, B. Longmaid, and M. E. Lamm, *Mol. Immunol.* **20,** 1039 (1983).

clonal antibodies have also been published.[17] *Haemophilus influenzae* enzyme is purified from cells grown in brain heart infusion medium supplemented with 10 μg/ml β-NAD and hemin (Sigma, St. Louis, MO), V and X growth factors essential for *H. influenzae* growth in laboratory media.

Although extracellular secretion and the large size of IgA proteinases are advantages in purification, protein levels in spent media are low. Enzymes can also be purified from recombinant *iga*$^+$ bacteria, with the important caveat that even active enzyme from a foreign environment may be structurally incomplete relative to the native form.[18]

Storage

Loss of activity of purified and crude enzymes stored in assay and purification buffers occurs slowly (from weeks to months) at 4°; the frozen enzymes are stable for at least a year.

Structure

A complete nucleotide and deduced amino acid sequence for the type 2 IgA proteinase of *N. gonorrhoeae*[4] and the type 1 proteinase of *H. influenzae* serotype b[19] have been reported. Both recombinant proteinases were secreted to the extracellular medium. The amino acid sequences have distinct, localized differences but are about 50% homologous, confirming earlier DNA hybridization studies of the two *iga* genes in these distantly related genera.[3] The amino acid sequences of these two enzymes have no homology to the metallo-type IgA proteinase produced by *Streptococcus sanguis*.

Neisseria gonorrhoeae strain MS 11 *iga* gene, 4596 bp, specifies a single polypeptide chain having 1532 amino acids with a calculated molecular mass of 169 kDa, and consisting of four distinct domains, as discussed earlier. The proteinase domain contains the sequence -G-V-L-G-D-S-G-S-P-L-F-A-, which is very similar to the conserved sequence surrounding the active site serine in the chymotrypsin–trypsin family of serine endopeptidases. The precursor contains only two cysteine residues, both in the proteinase domain, and separated by 10 amino acids.

The *H. influenzae* serotype b strain HK 368 *iga* gene has an open reading frame of 4646 nucleotides that encodes a protein of 1541 amino

[17] M. S. Blake and C. Eastby, *J. Immunol. Methods* **144**, 215 (1991).
[18] J. V. Gilbert, A. G. Plaut, and A. Wright, *Infect. Immun.* **59**, 7 (1991).
[19] K. Poulsen, J. Brandt, P. Hjorth, H. C. Thogersen, and M. Kilian, *Infect. Immun.* **57**, 3097 (1989).

acids having a calculated molecular mass of 169 kDa. Like the gonococcal enzyme, this value is far higher than the 100 kDa estimated for the mature proteinase; the difference is again attributable to autoproteolytic removal of the carboxy-proximal helper segment during secretion through the outer membrane (see section on synthesis and secretion, above). The *H. influenzae* enzyme domain has an -A-L-G-D-S-G-S-P-L-F-V- sequence containing the catalytic serine, and it also contains two cysteines that flank an identical stretch of 10 amino acids as in the gonococcal enzyme. When the entire *H. influenzae* sequence is optimally aligned with that of *N. gonorrhoeae* proteinase there is high similarity in a stretch from residues 16 to 47 near their amino termini. This segment includes the proposed signal peptide cleavage site, but its importance in translocation, proteinase activity, or specificity has not been determined. A strikingly dissimilar region of unknown function in the two enzymes is at residues 980–1240, a hydrophilic stretch near the amino-terminal end of the helper domain.[19]

Heteroduplex analysis of the *iga* genes of *H. influenzae* encoding type 1 and 2 proteinase specificity show extensive areas of sequence similarity.[5] There is, however, an 0.8-kb region of low similarity starting 0.2 kb from the start of the genes, a region that encodes a part of the enzymes involving their cleavage specificity determinant (CSD). Grundy *et al.* examined the basis for the cleavage type differences between type 1 and type 2 *H. influenzae* proteinases by preparing a series of recombinant *iga* genes encoding type 1–type 2 hybrid enzymes.[20] Assay of these hybrids localized the CSD to a stretch of 123 amino acids near the amino terminus of the enzyme proteins, corresponding to the area of low similarity shown by heteroduplex analysis. A second region of dissimilarity between the type 1 and 2 *H. influenzae* genes is a deletion–substitution loop located 1.2 kb from the carboxy-terminal end of the DNA insert. This segment, which is 0.9 kb in the type 1 and 0.2 kb in the type 2 *iga* genes, is within the region encoding the helper domain of the enzyme precursor.

Catalytic Mechanism

Assignment of the *Haemophilus* and *Neisseria* IgA proteinases to the class of serine-type enzymes is based on several lines of evidence. The deduced sequences of both type 2 *Neisseria* and type 1 *H. influenzae* enzymes contain the aforementioned, identical nine-amino acid segment that is highly similar to other enzymes in the chymotrypsin/trypsin family of serine endopeptidases. The sequence -G-D-S-G-S-P-L- of the IgA proteinases is closely matched by the -G-D-S-G-G-P-L- of the Gly[193]-Leu[199]

---

[20] F. J. Grundy, A. G. Plaut, and A. Wright, *Infect. Immun.* **58,** 320 (1990).

TABLE II
INHIBITION OF IgA PROTEINASES BY PEPTIDE PROLYLBORONIC ACIDS[a]

| | K_i (nM) | | | |
| | N. gonorrhoeae | | H. influenzae | |
Inhibitors	Type 1	Type 2	Type 1	Type 2
Ac-Ala-Pro-boro-Pro-OH	16	4	1300	13
Boc-Ala-Pro-boro-Pro-OH	63	35	5900	30
H-Ala-Pro-boro-Pro-OH	62,000	85,000	c	47,000
Boc-Ala-Pro-boro-Val-OH	b	>100,000	b	b
MeOSuc-Ala-Ala-Pro-boro-Pro-OH	52	28	7000	19
MeOSuc-Ala-Ala-Pro-boro-Val-OH	b	>100,000	>100,000	25,000

[a] Adapted from W. W. Bachovchin, A. G. Plaut, G. R. Flentke, M. Lynch, and C. A. Kettner, *J. Biol. Chem.* **265**, 3738 (1990).
[b] Not determined.
[c] No inhibition at 10^{-3} M or less.

in the chymotrypsin numbering scheme, and the Ser (for Gly) in the IgA proteinase is also found in other serine proteinases, e.g., the mouse cytotoxic T cell proteinase and the glutamyl endopeptidases of *Staphylococcus aureus*.[21] In addition, site-specific oligonucleotide-directed mutagenesis of the putative active site serine in *iga* genes encoding either type 1 or type 2 proteinases inactivates the enzyme encoded by *E. coli* transformants. And third, the *H. influenzae* and *N. gonorrhoeae* proteinases are inactivated by 4×10^{-4} M diisopropyl fluorophosphate (DFP)[11]; on this basis the proteinase of *Ureaplasma ureolyticum* has also been identified as serine type because it is inhibited by both DFP and by 100 μM 3,4-dichloroisocoumarin.[22] The susceptibility of the *Neisseria* and *Haemophilus* proteases to inhibition by peptide prolylboronic acids is discussed below.

Synthetic Peptide Inhibitors

A series of peptide analogs of the human IgA$_1$ hinge substrate region has been synthesized with α-aminoboronic acid analogs of proline (boro-Pro) as the carboxy-terminal residue.[11] These are potent inhibitors of the serine-type IgA proteinases (Table II) but are ineffective against the metalloproteinsae-type IgA proteinases. These compounds are the first

[21] S. Brenner, *Nature (London)* **334**, 528 (1988).
[22] R. K. Spooner, W. C. Russell, and D. Thirkell, *Infect. Immun.* **60**, 2544 (1992).

low molecular weight compounds with high affinity for the enzyme active site; earlier IgA hinge peptide analogs without the boro-Pro group were found to inhibit only weakly, and were not useful as substrates. The high affinity suggests that these inhibitors are binding to the active site as transition-state analogs, providing evidence that these are serine proteinases, because peptide boronic acids are much less effective against other classes.

Modification of the peptide portion of successful peptide prolylboronic acids provides information about the enzyme active site and specificity requirements. Tri- and tetrapeptides with blocked amino termini are inhibitors of both types 1 and 2 neisserial and type 2 $H.$ $influenzae$ proteinase (K_i 4–63 nM for tripeptides and 19–52 nM for tetrapeptides); inhibition of type 1 $H.$ $influenzae$ enzyme was in the micromolar range. Boro-Pro did not inhibit, presumably because it has a free amino group that cannot be accommodated in the active site of these endopeptidases. Both type 1 and 2 enzymes exhibit strong preference for proline at P1, because substitution of either Ala or Val at this position results in loss of affinity (Table II).

The finding that peptide boronic acids typically inhibit both the type 1 and 2 proteinases raises the question as to why these enzymes do not cleave more than one peptide bond in the hinge. It may be that the conformation of the inhibiting peptides is more flexible than the more rigid conformation of the IgA$_1$ hinge peptide. Alternatively, there is some evidence that IgA proteinases bind to segments of IgA$_1$ outside the hinge region. None of these peptides has yet been clinically tested, so their capacity to prevent or modify infectious illness is unknown.

Antibody Inhibition of IgA Proteinases

Antibodies inhibiting the serine-type IgA proteinases are present in most normal human sera and secretions. Serum titers rise substantially after systemic infections with, or colonization by, IgA proteinase-producing bacteria.[16,23] Serum antibody levels can be conveniently measured by ELISA (enzyme-linked immunosorbent assay),[23] but direct assay is needed to quantify the levels of inhibitory antibody. To measure inhibition we preincubate the test enzyme and antibody source at 37° for 30 min, and then measure activity in the standard quantitative assay (see assay section). We express the inhibitory power of antibodies in serum or secre-

---

[23] G. F. Brooks, C. J. Lammel, M. S. Blake, B. Kusecek, and M. Achtman, $J.$ $Infect.$ $Dis.$ **166,** 1316 (1992).

tions as the IC_{50} titer, the dilution of the antibody source that causes a 50% reduction in IgA_1 cleavage. In serum the antibodies are mainly of the IgG isotype; inhibitor coelutes with IgG on column chromatography, and purified human polyclonal IgG inhibits both serine-type IgA proteinases. Plasma and serum of severely hypogammaglobulinemic patients having extremely low or zero levels of the major antibody isotypes IgG, IgA, and IgM do not inhibit IgA proteinases. The inhibitory antibodies, of high titer in patients who have recovered from invasive infections, often show great specificity for the enzyme cleavage type of the infecting microorganism, suggesting that antibody targets are epitopes in or near the active sites, or in the cleavage site determinant of the enzyme protein.[24] Experimental animals immunized with either *Haemophilus* or *Neisseria* serine-type IgA proteinases develop inhibiting antibodies that help classify the proteinases, as discussed below.

In human milk the inhibiting antibodies are secretory IgA, more specifically, secretory IgA_1 in the few milk samples in which this has been tested.[25] Thus a subpopulation of the polyclonal secretory IgA_1 antibodies in milk inhibit, while all milk secretory IgA_1 antibodies are subject to IgA proteinase hydrolysis. Milk secretory IgA_1 inhibits through an antibody mechanism and not as a competitive substrate,[16] and inhibition has been localized to the antigen-binding Fab region of these proteins.[25] Human milk supporting growth of 10^8 *H. influenzae*/ml contains sufficient antibody to precipitate and inhibit all mature IgA proteinase produced, and to protect bystander secretory IgA_1 from hydrolysis. Because autocatalysis is a critical step in proteinase release from the bacterial outer membrane, milk inhibition has the potential to block processing. But recent experiments following size, activity, and distribution of the precursor and mature enzyme during growth of *H. influenzae* in milk indicate that processing was blocked to only a minor extent. It was interesting to note that milk IgA aggregated wild-type *iga*$^+$ but not *iga*$^-$ mutant *H. influenzae* cells during growth in milk; aggregation may favor bacterial colonization at mucosal surfaces.[26] These *in vitro* studies do not reproduce host–bacteria events on mucosal surfaces so they cannot determine if milk antiproteinase antibodies are protective for the nursing infant. But they do show that IgA proteinase inhibition preserves the structure of secretory IgA_1 antibodies regardless of antibody specificity, and this may be of value in mucosal defense.

[24] A. G. Devenyi, A. G. Plaut, F. J. Grundy, and A. Wright, *Mol. Immunol.* in press.
[25] A. G. Plaut, J. Qiu, F. Grundy, and A. Wright, *J. Infect. Dis.* **166**, 43 (1992).
[26] E. R. Moxon and R. Wilson, *Rev. Infect. Dis.* **13**(Suppl. 6), S518 (1991).

Epidemiologic Investigation of IgA Proteinases

Diversity among IgA proteinases in strains collected over long periods or during epidemics has been studied to better understand dynamics of microbial populations, and to determine if IgA proteinases may be vaccine candidates. For example, Poulsen and colleagues[19] classified 60 *H. influenzae* type b IgA proteinases into four groups (I–IV) by restriction analysis of their *iga* genes, and raised rabbit antisera to purified proteinases from each group. Enzymes in groups I, II, and III were inhibited by antibodies raised to any one of these groups, but anti-I, -II, and -III sera could not inhibit group IV enzymes (these represent only 2% of *H. influenzae* type b isolates). In a related study of *H. influenzae* type b strains collected from many countries over a 40-year period, 88% had the I or II *iga* genotype. These strains were otherwise highly diverse, having many different outer membrane protein and lipopolysaccharide phenotypes, and 25 of the 32 known multilocus genotypes.[27] Kilian and Thomsen had already shown a close correlation between IgA proteinase antigenic type and capsular serotype in *H. influenzae*.[28]

The epidemiology and molecular biology of *N. meningitidis* IgA proteinases have also been examined using strains collected during years 1917–1988 in 19 countries. Among 58 isolates from epidemics of invasive meningococcal disease all but one produced type 1 IgA proteinase; this strongly contrasted with equally distributed type 1 and 2 cleavage patterns among 72 nonepidemic disease or control strains. The explanation for the association of type 1 enzyme cleavage type with severe, invasive epidemic disease is unknown, but type 1 IgA proteinase is also characteristic of invasive isolates of *N. gonorrhoeae*.[29] Restriction fragment length polymorphism suggests the occurrence of frequent interstrain recombination among *N. meningitidis* strains *in vivo*.[30]

Role of IgA Proteinases in Infectious Process

Although the role of IgA proteinases in the infectious process is not yet defined, locally produced IgA is the principal antibody isotype in human milk and in secretions that bathe the mucosal surfaces of the oral cavity, upper respiratory tract, intestine, colon, genital tract, and the

[27] J. M. Musser, D. M. Granoff, P. E. Pattison, and R. K. Selander, *Proc. Natl. Acad. Sci. U.S.A.* **82**, 5078 (1985).
[28] M. Kilian and B. Thomsen, *Infect. Immun.* **42**, 126 (1983).
[29] M. H. Mulks and J. S. Knapp, *Infect. Immun.* **55**, 931 (1987).
[30] H. Lomholt, K. Poulsen, D. A. Caugant, and M. Kilian, *Proc. Natl. Acad. Sci. U.S.A.* **89**, 2120 (1992).

conjunctiva.[31,32] Because bacteria colonize or infect these sites as their first encounter with the human host, IgA hydrolysis may allow these microbes to circumvent immunity, or possibly even recruit antibodies or their fragments as a step in the infectious process. IgA proteinases may not only establish a relatively localized immunodeficiency,[33] but IgA_1 cleavage products may foster microbial immune evasion, because Fab fragments that retain antigenic binding ability[34] can still bind their antigenic targets on the microbial surface. This may not only obstruct binding of intact antibodies, but Fab coating may promote bacterial adherence to host tissues by increasing microbial surface hydrophobicity.[35-37] Finally, IgA proteinases could cleave other host proteins not yet identified, or these large enzyme proteins may have other, possibly nonenzymatic, properties that contribute to microbial adhesiveness or other pathogenic mechanisms. In this regard, Provence and Curtiss[38] have recently cloned and characterized an *E. coli* hemagglutinating protein (designated Tsh) in strains that cause respiratory disease in poultry. Tsh, for which a proteolytic activity has not been described, has homology to the serine-type IgA proteinases, including the region surrounding the active site serine.

Acknowledgment

Supported by NIH DE 09677 and the GRASP Digestive Disease Research Center, NIH DK-34928.

[31] M. H. Mulks, in "Bacterial Enzymes and Virulence" (I. A. Holder, ed.), p. 82, CRC Press, Boca Raton, Florida, 1985.
[32] M. Kerr, *Biochem. J.* **271,** 285 (1990).
[33] C. H. Sorensen and M. Kilian, *Acta Pathol. Microbiol. Immunol. Scand.* (*Sect. C*) **92,** 85 (1984).
[34] B. Mansa and M. Kilian, *Infect. Immun.* **52,** 171 (1986).
[35] G. Hajishengallis, E. Nikolova, and M. W. Russell, *Infect. Immun.* **60,** 5057 (1992).
[36] T. Ahl and J. Reinholdt, *Infect. Immun.* **59,** 563 (1991).
[37] M. W. Russell, J. Reinholdt, and M. Kilian, *Eur. J. Immunol.* **19,** 2243 (1989).
[38] D. L. Provence and R. Curtiss, III, *Infect. Immun.* **62,** 1369 (1994).

[11] Biochemical and Genetic Methods for Analyzing Specificity and Activity of a Precursor-Processing Enzyme: Yeast Kex2 Protease, Kexin

By Charles Brenner, Alison Bevan, and Robert S. Fuller

Introduction

The Ca^{2+}-dependent, transmembrane Kex2 protease of the yeast *Saccharomyces cerevisiae* (kexin, EC 3.4.21.61) was discovered through analysis of strains bearing mutations in the *KEX2* gene that blocked posttranslational processing of the precursors of two secreted peptides, the α-mating pheromone and M_1 killer toxin.[1–3] Kex2 protease has emerged as the prototype of a family of eukaryotic enzymes that cleave polypeptide precursors, including prohormones, neuropeptide precursors, and the precursors of a variety of secreted and integral membrane proteins, in late compartments of the secretory pathway.[4] Most such cleavages occur at the carboxyl side of pairs of basic residues, especially -Lys-Arg- and -Arg-Arg-, although cleavages also occur at monobasic and polybasic sites. Additional members of the Kex2 family are discussed in [12] and [13] in this volume.[5,6]

Kex2 protease and its metazoan homologs form a distinct branch of the subtilisin family of serine peptidases, previously characterized members of which were degradative enzymes with relatively little substrate discrimination.[7–9] As a prototype of a family of enzymes involved in biosynthetic proteolysis, Kex2 is being investigated to elucidate the kinetic and structural properties that provide the remarkable specificity that these enzymes display in processing polypeptide precursors.

Several factors, divisible into two broad categories, can contribute to the specificity of a processing enzyme within the cell. First, intracellular

[1] M. J. Leibowitz and R. B. Wickner, *Proc. Natl. Acad. Sci. U.S.A.* **89,** 73 (1976).
[2] D. Julius, L. Blair, A. Brake, G. Sprague, and J. Thorner, *Cell (Cambridge, Mass.)* **32,** 839 (1983).
[3] R. S. Fuller, R. E. Sterne, and J. Thorner, *Annu. Rev. Physiol.* **50,** 345 (1988).
[4] D. F. Steiner, S. P. Smeekens, S. Ohagi, and S. J. Chan, *J. Biol. Chem.* **267,** 23,435 (1992).
[5] K. Nakayama, this volume [12].
[6] M. Chretien and N. G. Seidah, this volume [13].
[7] R. S. Fuller, A. J. Brake, and J. Thorner, *Science* **246,** 482 (1989).
[8] R. J. Siezen, W. M. deVos, J. A. M. Leunissen, and B. W. Dijkstra, *Protein Eng.* **4,** 719 (1991).
[9] P. Gluschankof and R. S. Fuller, *EMBO J.* **13,** 2280 (1994).

targeting and cocompartmentalization of enzymes and substrates can limit the range of molecules exposed to an enzyme. Conditions within the processing compartment may further modulate the activity or specificity of the enzyme and the conformation or accessibility of potential substrates. Although these "cellular" factors will not be the subject of this chapter, it is worthy of note that Kex2 protease is localized quite selectively to a late compartment of the Golgi complex in yeast, most likely equivalent to the mammalian trans-Golgi network.[10,11] Retention of Kex2 protease in this compartment requires both its single transmembrane domain and a "retention signal" in the C-terminal cytosolic tail sequence of the enzyme that resembles the tyrosine-internalization signals found in mammalian cell surface receptors.[7,12] Mutation of the retention signal lowers the steady-state level of Kex2 protease in the pro-α-factor processing compartment, resulting in decreased processing efficiency.[12] Clearly for Kex2 and undoubtedly for other processing proteases as well, correct localization within the secretory pathway is critical for biochemical function within the cell.

The second and more obvious set of factors affecting specificity governs the direct physical interaction between the enzyme and substrate. These factors determine the degree of fidelity in substrate discrimination that the enzyme can achieve, ensuring efficient cleavage of correct bonds in intended substrates and not of incorrect bonds in substrates or nonsubstrates. At a first level, the enzyme recognizes substrate residues near the cleaved bond. Such "primary" specificity is ideally probed using small, conformationally unconstrained peptide substrates and inhibitors. However, the physiological substrates of Kex2 and other precursor processing enzymes are polypeptides that presumptively possess both secondary and tertiary structure. The degree to which higher order structural features influence processing specificity is still an open question. One straightforward way to assess the influence of higher order structure is to compare the kinetics of cleavage of a site in a native precursor with the cleavage of a conformationally unconstrained peptide having the corresponding sequence.

The information specifying localization of Kex2 is found in C-terminal sequences that are exclusive of the catalytic domain on which analysis of enzymatic specificity focuses.[7,9,12,13] Thus a genetically engineered, C-terminally truncated form of Kex2 protease has provided a convenient

[10] K. Redding, C. Holcomb, and R. S. Fuller, *J. Cell Biol.* **113,** 527 (1991).
[11] C. A. Wilcox and R. S. Fuller, *J. Cell Biol.* **115,** 297 (1991).
[12] C. A. Wilcox, K. Redding, R. Wright, and R. S. Fuller, *Mol. Biol. Cell* **3,** 1353 (1992).
[13] R. S. Fuller, A. J. Brake, and J. Thorner, *Proc. Natl. Acad. Sci. U.S.A.* **86,** 1434 (1989).

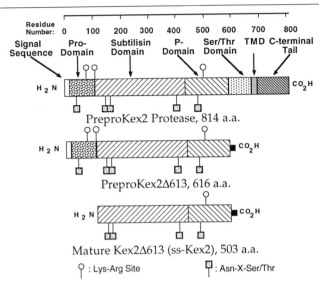

FIG. 1. Schematic structures of prepro-Kex2 protease, prepro-Kex2Δ613, and mature ss-Kex2. Features of the sequence of prepro-Kex2 protease, indicated by arrows, are described in the text. TMD, Transmembrane domain. Prepro-Kex2Δ613 ends with Kex2 residue Glu-613 plus three linker-encoded residues (Arg-Asn-Arg). See P. Gluschankof and R. S. Fuller, *EMBO J.* **13**, 2280 (1994).

model for studies of enzymatic activity and specificity (Fig. 1).[14] Methods of purification, active site titration, and characterization of this secreted, soluble form of Kex2 protease (ss-Kex2) are described in the first three sections of this chapter. We have taken several approaches to studying the enzymatic specificity of Kex2 protease. The uses of peptidyl methylcoumarylamide (MCA) substrates[14] and peptidyl chloromethanes[15] as probes of Kex2 specificity have been described. Such peptides permit analysis of the interactions with substrate residues N-terminal to the cleaved bond; that is, in the P1, P2, etc., positions. Internally quenched fluorogenic peptide substrates permit sensitive and quantitative analysis of the effects of substitutions of residues on both sides of the cleaved bond.[16] Use of such substrates based on cleavage sites in pro-α-factor will be described elsewhere.[17] Certain general conclusions can be drawn from these studies. First, wild-type Kex2 protease exhibits a high degree of

[14] C. Brenner and R. S. Fuller, *Proc. Natl. Acad. Sci. U.S.A.* **89**, 922 (1992).
[15] H. Angliker, P. Wikstrom, E. Shaw, C. Brenner, and R. S. Fuller, *Biochem. J.* **293**, 75 (1993).
[16] E. D. Matayoshi, G. T. Wang, G. A. Krafft, and J. Erickson, *Science* **247**, 954 (1990).
[17] C. Brenner, G. Wang, G. Krafft, and R. S. Fuller, unpublished data (1993).

primary specificity, displaying its highest k_{cat}/K_M of $\sim10^7$ M^{-1} sec^{-1} for the substrate acetyl-Pro-Met-Tyr-Lys-Arg-MCA, which is based on three of the four Kex2 cleavage sites in pro-α-factor. Kex2 shows a strict requirement for an Arg residue at P1. With Arg at P1, nearly all substrates exhibit a k_{cat} in the range of 20 to 40 sec^{-1}, whereas Lys at P1 reduces k_{cat} >100-fold and k_{cat}/K_M up to 3000-fold. Substitutions at P2 principally affect K_M, with Lys being preferred to Arg by $\sim$4-fold.

As alluded to previously, analysis of cleavage of small peptides may tell only part of the specificity story, because higher order structure in an authentic precursor may affect its interaction with a processing enzyme. The analysis of cleavage of pro-α-factor by Kex2 protease, both in a reconstituted system and *in vivo*, will be discussed in the fourth and fifth sections of this chapter. The yeast genetic system affords additional tools for the study of Kex2 activity and specificity. Because Kex2 protease catalyzes a required step in synthesis of α-factor, the mating pheromone secreted by haploid cells of the α mating type, *kex2* mutants are sterile.[1,2] This phenotype is easily scored either qualitatively or quantitatively on petri plates by the simple appearance or frequency of appearance of prototrophic **a**/α diploids from the mating of auxotrophic **a** and α haploids. These bioassays can be used to monitor the effects of mutations in pro-α-factor on Kex2 cleavage, or alternatively, to screen for mutations in the enzyme that affect its activity or specificity. An example of this second approach is provided in the final section on "one-step site-directed mutagenesis" of the Kex2 oxyanion hole.[18]

Methods

Engineering and Purification of Secreted, Soluble Kex2 Protease

Kex2 protease contains a domain of $\sim$155 amino acids, carboxyl to the domain with 30% identity to subtilisin, termed the P domain, which is conserved in metazoan Kex2 homologs and is required for biosynthesis of active Kex2 protease.[9,19] Truncation of the Ser/Thr-rich domain, transmembrane domain, and cytosolic tail carboxyl to the P domain (see Fig. 1) results in secretion of the ss-Kex2 enzyme into culture medium.[9,14] Both wild-type and active, C-terminally truncated forms of prepro-Kex2 undergo intramolecular, autoproteolytic removal of the N-terminal prodo-

[18] C. Brenner, A. Bevan, and R. S. Fuller, *Curr. Biol.* **3**, 498 (1993).
[19] R. S. Fuller, C. Brenner, P. Gluschankof, and C. A. Wilcox, *in* "Methods in Protein Sequence Analysis" (H. Jörnvall, J.-O. Hoog, and A.-M. Gustavsson, eds.), p. 205. Birkhaeuser, Basel, 1991.

main by cleavage after Lys^{108}-Arg^{109} shortly after translocation into the endoplasmic reticulum.[9,11] In the Golgi, two N-terminal dipeptides (Leu^{110}-Pro^{111} and Val^{112}-Pro^{113}) are removed by the STE13-encoded type IV dipeptidyl-peptidase.[14]

For purposes of purification, the truncated KEX2 gene, KEX2Δ613, is expressed by a powerful constitutive promoter of the TDH3 gene (glyceraldehyde-3-phosphate dehydrogenase), on the high copy number yeast episomal plasmid pG5 containing the URA3 gene as a selectable marker.[14] We have found that buffering growth medium to pH 7.2 and inclusion of 0.5% casamino acids (Difco, Detroit, MI) leads to consistently high yields. Growth of ss-Kex2-secreting cells in unbuffered medium leads to extremely low yields of enzyme due to the instability of the enzyme in medium acidified by yeast fermentation. Curiously, growth of ss-Kex2-producing cells to stationary phase in 10-ml quantities in 18 × 150-mm culture tubes, but not in bulk in shake flasks, leads to very high levels of accumulation of active enzyme in culture medium, up to ~30 mg/liter when using yeast strain ABY01. Preparations as large as several liters have been prepared by growing hundreds of 10-ml cultures. The following methods are adapted from Brenner and Fuller.[14]

Preparaton of ss-Kex2

Day 1. The lithium acetate transformation method[20] is used to introduce ss-Kex2-overproducing plasmid pG5-KEX2Δ613[14] into strain ABY01 (MATa ura3 trp1 ade2 his3). Transformants are selected by plating at 30° on synthetic complete medium lacking uracil (SDC-Ura).[21]

Day 2. A pump is used to deliver 4 liters of 1040 medium (below) in 10-ml volumes to 18-mm × 150-mm culture tubes, which are then capped and autoclaved.

Day 3. ~10 Ura+ transformants are inoculated into 2-ml overnight cultures of SDC-Ura medium and grown overnight to saturation at 30° in a Rollordrum (New Brunswick Scientific, Edison, NJ).

Day 4. 1 ml of each saturated SDC-Ura culture is used to inoculate a 10-ml volume of 1040 medium for overnight growth at 30°.

Day 5. 2 μl of a 1 : 10 dilution of culture medium of the 1040 overnights are assayed for Kex2 activity (below). Cultures exhibiting ≥300–400 U/μl of medium are used to inoculate the remaining 10-ml volumes of 1040 medium at a 1 : 20 dilution. Cultures are grown for 24 hr and cell-

---

[20] R. H. Schiestl and R. D. Gietz, *Curr. Genet.* **16,** 339 (1989).

[21] M. D. Rose, F. Winston, and P. Hieter, "Methods in Yeast Genetics." Cold Spring Harbor Laboratory, Cold Spring Harbor, New York, 1990.

free medium is harvested from pooled cultures by centrifugation. Enzyme accumulates to greater than 50% of the total protein in the medium. medium is harvested from pooled cultures by centrifugation. Enzyme accumulates to greater than 50% of the total protein in the medium.

Day 6. Subsequent steps are carried out at 4°. Cell-free medium is diluted with 3 volumes of 40 mM Bis–Tris base and mixed with 20 ml Fast Flow Q-Sepharose (Pharmacia, Piscataway, NJ) per liter of culture medium. The resin is collected in a column, washed with one column volume of 40 mM Bis–Tris-Cl, pH 7.0, 2 mM CaCl$_2$, and eluted with a linear gradient (3 column volumes) of 0–0.5 M NaCl in the same buffer. The active peak is pooled and pressure dialyzed against NBC buffer (100 mM NaCl, 40 mM Bis–Tris-Cl, pH 7.0, 10 mM CaCl$_2$) using a YM30 membrane (Amicon, Danvers, MA). Because of the high expression of ss-Kex2 obtained with strain ABY01, the enzyme is ≥97% pure at this stage. An additional chromatography step on Mono Q (Pharmacia) as used in the original purification method may afford a higher degree of purity.[14]

1040 Medium. Per liter: 1.7 g yeast nitrogen base without amino acids or (NH$_4$)$_2$SO$_4$ (Difco), 1.32 g (NH$_4$)$_2$SO$_4$, 5 g NH$_4$Cl, 5 g vitamin assay casamino acids (Difco), 8.37 g [bis(2-hydroxyethyl)imino]tris(hydroxy-methyl)methane (Bis–Tris) base, 20 g D-glucose, 0.12 g L-tryptophan, and 0.24 g adenine hydrochloride. The pH should be 7.2.[14]

Standard Kex2 Assay. The standard assay[14] is performed in 50 μl of 200 mM Bis–Tris-HCl (pH 7.0), 1 mM CaCl$_2$, 0.01% (v/v) Triton X-100, 0.5% (v/v) dimethyl sulfoxide, and 100 μM *tert*-butoxycarbonyl-Gln-Arg-Arg-MCA (Peninsula Labs, Burlingame, CA). Enzyme (≤75 U) is added, reactions are incubated at 37° for 4 min, and are terminated on ice with 0.95 ml 125 mM ZnSO$_4$. 7-Amino-4-methylcoumarin (AMC) released is measured fluorimetrically relative to a standard (λ_{ex} 385 nm; λ_{min} 465 nm). One unit of Kex2 activity equals 1 pmol AMC released per min.[13]

Quantitation of Protein and Active Sites

For values of k_{cat} to be meaningful, the concentration of active enzyme must be known.[14] Further, the fraction of a purified enzyme preparation that is active is a measure of how native it is and how authentically it represents the cellular activity. The numerator of such a fraction is determined by active site titration. The denominator can be determined by measurement of total protein concentration or by amino acid analysis as described below.

Determining Molar Protein Concentration by Amino Acid Analysis

1. Approximately 2 μg of highly purified protein diluted into 1 ml of water is precipitated by the addition of 0.1 ml of 72% (w/v) trichloroacetic acid in a clean glass test tube. The dried pellet is resuspended in 0.1 ml

of 6 *N* HCl and protein is hydrolyzed during a 24-hr incubation at 110° *in vacuo*.

2. The hydrolyzate is resuspended with 0.5 nmol norleucine as an internal standard and loaded on a Beckman 7300 amino acid analyzer (Beckman Instruments, Palo Alto, CA) for ninhydrin quantitation of amino acid residues.

3. The picomole yields of 14 amino acids (naturally occurring amino acids except Gly, Met, Cys, Trp, Asn, and Gln) are converted to molarity by dividing by the initial volume. Molar yields are plotted against predicted occurrences of amino acids determined from the amino acid sequence of the protein. Occurrences of Asp and Glu are taken as the sum of Asp and Asn and the sum of Glu and Gln, respectively. The slope of the best-fit line through the origin equals the molar concentration of the polypeptide. We have found this method to give good agreement with values determined from measurement of total protein, e.g., by the BCA assay (Pierce, Rockford, IL).

Initial Burst Titration of Active Sites. Serine peptidase active sites may be determined by detection of a leaving group or product in the first turnover if a rate-limiting step occurs later in the reaction cycle than acylation.[21a] For instance, in the hydrolysis of *p*-nitrophenyl ethyl carbonate, chymotrypsin exhibited biphasic kinetics. One molar equivalent of *p*-nitrophenol was released rapidly, followed by a slow, steady-state accumulation of the alcohol leaving group.[22] Initial burst titrations have been performed with ester substrates that acylate rapidly but are hydrolyzed slowly. Surprisingly, burst kinetics have been observed with the best available amide substrate of Kex2, Ac-Pro-Met-Tyr-Lys-Arg-MCA. The kinetic and mechanistic basis of this reaction will be described elsewhere. The following method is adapted from Brenner and Fuller.[14]

1. A quenched-flow mixer is fitted with an enzyme syringe containing at least 10 pmol of ss-Kex2 protease per reaction loop volume and a substrate syringe containing 40 μM Ac-Pro-Met-Tyr-Lys-Arg-MCA; 0.75 *M* citric acid pH 3.0 is used as a quench.

2. Time points of 4, 8, 12, 16, 20, 100, 200, 300, 400, and 500 msec are taken in triplicate. Zero time points are created manually by adding enzyme to the quench solution before substrate is added. From 0 to 50 pmol of AMC is added to the zero time points to create a standard curve of product.

3. With the RQF mixer (KinTek, University Park, PA), longer time points contain more quenching solution. Prior to reading the fluorescence

[21a] C. G. Knight, this series.
[22] B. S. Hartley and B. A. Kilby, *Biochem. J.* **56**, 288 (1954).

of samples, time points are adjusted with the appropriate volume of quench to render them uniform.

4. Product fluorescence is plotted against time. A best-fit line of the data from later time points ($t \geq 20$ msec) is used to calculate the magnitude of the initial burst, i.e., the y intercept.

Active-site-directed irreversible inhibitors such as peptidyl chloromethanes offer an alternative approach to active site titration and examination of specificity.[15] Increasing concentrations of Pro-Nvl-Tyr-Lys-Arg-CH_2Cl and Phe-Ala-Lys-Arg-CH_2Cl were incubated with a known quantity of ss-Kex2 protease for 2 hr. Inhibited samples were then added to saturating substrate and assayed briefly. When activity remaining was plotted against the ratio of nominal inhibitor concentration to enzyme, 1.5 to 2 equivalents of the peptidyl chloromethanes were required to inactivate the enzyme. Because the enzyme concentration had been determined by protein assay, amino acid analysis, and initial burst assay, this titration was taken to adjust the true concentration of the inhibitors.[15] However, this assay, the full protocol of which is provided in the original literature, adds to the arsenal of techniques that may be used to assess the purity and the concentration of active ss-Kex2 protease.

Assays of Stability and Activity as Function of pH and Buffer System

Measurements of pH dependence of enzyme activity are usually intended to probe the pK_a values of functional groups. Titratable groups include not only catalytic groups in the enzyme but also moieties in both the enzyme and substrate that affect substrate binding. Unfortunately, variations in pH and the chemical nature of the buffers used can also have unanticipated effects on protein stability. "Activity" measured in different buffers not only reflects intrinsic activity at a given pH but also an inactivation rate specific for the buffer, pH, and temperature. To avoid this pitfall, it is advisable to determine the effects of buffers on enzyme stability before investing substantial effort in pH titration of enzyme activity.

Determination of Half-Times for Loss of Activity

1. Reaction premixes (0.4 ml) are assembled on ice with various buffers at 200 mM, plus 0.01% Triton X-100, 1 mM $CaCl_2$, and 1000 U of ss-Kex2 protease. Two 40-μl controls are withdrawn from each premix and mixed with 0.8 ml of 125 mM $ZnSO_4$ plus 10 μl of 500 μM Boc-Gln-Arg-Arg-MCA; 100 pmol of AMC is added to one sample from each buffer.

2. The premixes are incubated at 37°. At intervals ranging from 1 to 300 min, 40 μl of each premix are removed to 10 μl of 500 μM Boc-Gln-

Arg-Arg-MCA and incubated at 37° for 1 min to assay enzyme activity remaining. Reactions are terminated with 0.8 ml of 125 mM ZnSO$_4$. 3. AMC fluorescence of reactions is compared to that of nonincubated samples in corresponding buffers. At pH 7, in Bis–Tris, the first incubated samples should have liberated 125 pmol of AMC, corresponding to 2.5% of the substrate. Activity is plotted as a function of time of preincubation at 37° and fitted to a first-order decay curve, $y = \text{limit}[1 - \exp(-kt)]$. The half-time for loss of activity is $(\ln 2)/k$.

As can be seen from the measured half-times for loss of activity (Table I), measurements of pH dependence of enzyme activity must proceed with extreme caution. For the purposes of studying enzyme chemistry, buffer inactivation rates are uninteresting on their own. They are, however, essential to measure to optimize the standard assay and to evaluate the limitations of performing assays in different buffers. In the case of Kex2 protease, the pH dependencies of initial rates have been measured using brief end point incubations[14] and the pH dependence of k_{cat}/K_m was measured using continuous monitoring of product formation at low concentrations of substrate.[18] Once the effects of buffer and pH on enzyme stability have been controlled, Kex2 exhibits a relatively simple titration curve with a pK_a of ~5.7, indicative of titration of a single ionizable group, most likely the catalytic His-213.[14]

Cleavage of Purified α-Factor Precursor

Prepro-α-factor consists of an N-terminal signal peptide and hydrophilic leader, or "pro," segment (about 64 residues) followed by four

TABLE I
HALF-TIMES FOR LOSS OF INITIAL VELOCITY AT 37°

	Half-time (min) for buffer					
pH	Sodium acetate	Na-MES	Bis–Tris-Cl	Na-HEPES	Na-Bicine	Na-CHES
5.0	36					
5.5		5				
6.0		18				
6.5		17				
7.0			250	6		
7.5				22		
8.0				5		
8.5					44	
9.0					17	
9.5						2

copies, in tandem, of α-factor separated by spacers that contain sites for proteolytic maturation: the pair of basic residues, -Lys-Arg-, followed by two or three Glu-Ala or Asp-Ala dipeptides. The first Kex2 site (-Ser-Leu-Asp-Lys-Arg + Glu-Ala-) abuts the pro segment and thus differs from the remaining three sites (-Pro-Met-Tyr-Lys-Arg + Glu-Ala-), which occur between α-factor repeats. Reactions with these and related sequences have been investigated in the context of peptidyl-MCA substrates,[14] internally quenched fluorogenic peptide substrates,[17] and peptidyl chloromethanes.[15] However, to test to what degree presentation of these sites as part of a folded polypeptide affects their cleavage by Kex2 protease, pro-α-factor has been purified from recombinant bacterial cells and assayed with purified Kex2 protease.[23] Heterologous expression of pro-α-factor was required because, in wild-type yeast, the precursor is cleaved by Kex2 protease, and in *kex2* mutant cells, it becomes heterogeneously glycosylated. Thermal and chaotrope denaturation curves measured by circular dichroism and intrinsic tryptophan fluorescence indicate that pro-α-factor possesses substantial secondary and probably tertiary structure.[23]

Pro-α-factor processing presents a complex kinetic problem because cleavage at the four sites generates five products and as many as nine possible proteolytic intermediates. Using a combination of SDS–PAGE, Edman degradation, and amino acid analysis, it has been possible to demonstrate that ss-Kex2 cleaves pro-α-factor efficiently and only at the expected sites. At a concentration of 6 μM, α-factor precursor is cleaved to 50% completion in ~80 min and to apparent completion in 12 hr by 3 nM ss-Kex2 at 30° without further degradation of either the pro segment or the α-factor repeats.[23] The first and most abundant intermediates lack the C-terminal one or two repeats, suggesting that the C-terminal two cleavage sites are the most accessible in the folded pro-α-factor structure. An inhibition assay has been used to attempt to quantify the interaction of ss-Kex2 and purified pro-α-factor. Added pro-α-factor inhibits hydrolysis of peptidyl-MCA substrates competitively. However, the apparent K_i for pro-α-factor exhibited a striking dependence on pro-α-factor concentration. The lowest $K_{i,app}$, ~2 μM, was observed with 1 μM pro-α-factor, with the measured "constant" increasing at higher precursor concentrations. This may be the result of pro-α-factor aggregation in this concentration range, evidence for which has come from the behavior of pro-α-factor in chromatography and equilibrium analytical ultracentrifugation.[24] These results strongly underscore the importance of understanding the effects of precursor aggregation, thought to occur in the case of many prohor-

---

[23] A. Bevan, C. Brenner, and R. S. Fuller, unpublished data (1993).
[24] A. Bevan, T. Holzman, and R. S. Fuller, unpublished data (1993).

mones and neuropeptide precursors in the trans-Golgi network, on cleavage by processing enzymes.

Genetic Analysis of Substrate Specificity

To complement the biochemical analysis of substrate recognition, we have developed a genetic system for analyzing and manipulating the specificity of Kex2 protease.[25] Because Kex2 cleavage of pro-α-factor is required for yeast α cells to mate with **a** cells, we have been able to use yeast mating assays to measure the effects of mutations in a pro-α-factor cleavage site on Kex2 recognition. The yeast mating system provides sensitive bioassays for monitoring cleavage of pro-α-factor by Kex2 protease. The "halo" assay permits direct assessment of α-factor secretion as demonstrated by a zone of growth inhibition in a lawn of sensitive *MATa* cells.[2,26] The quantitative mating assay, although a less direct measurement of α-factor secretion, has enormous sensitivity, and a dynamic range of $\geq 10^6$.[27] Mating can also be scored in a qualitative fashion by a replica-plating assay (see next section), which can be used to screen for mutations in Kex2 that alter the specificity of the enzyme and permit cleavage of mutant forms of pro-α-factor.

The yeast strain constructed for this system, CB012,[25] is a *MATα* haploid, so that mating depends on production of α-factor. CB012 bears disruptions of the *KEX2* structural gene and the *MFα1* and *MFα2* genes, which encode the major and minor forms of prepro-α-factor. Therefore, the tester strain is sterile unless it carries two plasmids, one with a functional *KEX2* gene and the other with a functional α-factor precursor gene. Auxotrophic markers allow selection and maintenance of plasmids: *trp1* for the *KEX2* plasmid and *ura3* for the α-factor precursor plasmid. Additional auxotrophic markers permit the selection, on minimal yeast plates, of prototrophic diploids formed by mating of CB012 with a *MATa* tester strain.

For controlled expression and ease of mutagenesis, the *MFα1* gene has been placed under the control of the galactose-inducible, glucose-repressible *GAL1* promoter on the single-copy centromere plasmid pBM258.[10] The precursor gene was further modified by deleting three α-factor repeats, fusing the first Lys-Arg + Glu-Ala-Glu-Ala processing site to the fourth repeat of mature α-factor. Mating of CB012 containing this plasmid depends entirely on cleavage of the single, remaining site, and

[25] C. Brenner, A. Bevan, and R. S. Fuller, unpublished data (1993).

[26] D. L. Julius, L. Blair, A. Brake, G. Sprague, and J. Thorner, *Cell* (*Cambridge, Mass.*) **32**, 839 (1983).

[27] L. H. Hartwell, *J. Cell Biol.* **85**, 811 (1980).

the effects of amino acid substitutions on both sides of this cleavage site can be monitored by qualitative or quantitative mating assays. Unique restriction sites upstream and downstream from the Kex2 cleavage site permit alteration of the cleavage site by cassette mutagenesis.[28] A number of substitutions have been made this way for the P2 Lys residue.[25] For these substitutions, the following order of cleavage preference was established: Lys-Arg = 1.0 ≫ (Thr-Arg > Pro-Arg > Ile-Arg > Asn-Arg) = 10^{-2}–10^{-4} ≫ (Phe-Arg, Leu-Arg) ≤ 4 × 10^{-6}, where the number indicates the quantitative mating efficiency relative to the normal Lys-Arg site. Mating by strains containing the Thr-Arg, Pro-Arg, Ile-Arg, and Asn-Arg substitutions, although feeble, depended on the presence of the *KEX2* plasmid. Phe-Arg and Leu-Arg substitutions resulted in mating at levels as low as with negative controls (no *KEX2* plasmid, no *MFα1* plasmid, and insertion of a translational termination codon in the P2 position), suggesting that these molecules were not cleaved at all. Strains containing the *MFα1-100* plasmid produced smaller α-factor halos compared to the wild-type *MATα* strain, as expected from the reduced mating efficiency. Halo assays failed to detect any α-factor secretion by strains containing *any* of the mutant precursor genes, suggesting that α-factor production was much less than 1% of wild type,[29] and dramatically demonstrating the greater sensitivity of the mating assay. Mutations in the *KEX2* gene that act as second-site suppressors of these α-factor precursor mutations are being sought by random mutagenesis.

One-Step Site-Directed Mutagenesis

Though the structural basis for substrate specificity can be approached with random mutagenesis and genetic selection, other aspects of Kex2 activity can be investigated using site-directed mutagenesis.[18] We have developed a rapid new method of site-directed mutagenesis that relies on the high efficiency of homologous recombination in mitotic yeast cells combined with the rapidity of the polymerase chain reaction (PCR).[30] We have applied this method, "one-step site-directed mutagenesis," to assess the effects of amino acid substitutions for the Kex2 oxyanion hole Asn-314 on the biosynthesis and activity of the enzyme.[18] The strategy is depicted in Fig. 2. Mutations are obtained in a yeast expression plasmid simply by transforming yeast with a linearized plasmid along with a PCR product primed with a mutagenic primer. In the mutagenesis of Asn-314,

---

[28] J. A. Wells, M. Vasser, and D. B. Powers, *Gene* **34**, 315 (1985).
[29] J. Kurjan, *Mol. Cell. Biol.* **5**, 787 (1985).
[30] R. K. Saiki, D. H. Gelfand, S. Stoffel, S. J. Scharf, R. Higuchi, G. T. Horn, K. B. Mullis, and H. A. Erlich, *Science* **239**, 487 (1988).

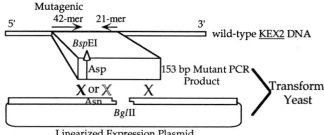

Fig. 2. One-step site-directed mutagenesis strategy. A 153-bp fragment encoding an amino acid substitution at codon 314 and silent substitutions at codon 312, which introduce a site for BspEI, was generated by PCR using the indicated primers. This fragment was cotransformed with a BglII digested yeast expression plasmid pG5-KEX2Δ613. Ligation of the linear plasmid or recombination on both sides of the BglII site is required for the plasmid to be replicated in yeast. On the left side of the BglII site, recombination to the left of the substitutions (distal), indicated by the **X**, generates a mutated plasmid and recombination to the right of the substitutions (proximal), indicated by X̶, generates wild-type plasmid.

transformants containing a plasmid with the desired substitutions of Ala and Asp for the oxyanion hole Asn-314 in Kex2 were identified by three methods: (i) assaying secreted Kex2 proteolytic activity, (ii) detecting a new restriction site in the plasmid-borne *KEX2* gene after PCR amplification, and (iii) performing qualitative mating assays.[18] Advantages of the method are that it requires no subcloning or passaging of constructs through *Escherichia coli* and that mutant proteins are immediately expressed in a yeast host strain and can therefore be purified and characterized without delay.

One-Step Site-Directed Mutagenesis of Kex2 Oxyanion Hole. As shown in Fig. 2, a 153-bp mutant PCR product that begins 5′ to Asn-314 and ends 3′ to *Bgl*II is generated from a wild-type *KEX2* template. The 5′ PCR primers are mutagenic 42-mers that introduce a silent *Bsp*EI site at codon 312 and change codon 314 from Asn to Ala or Asp at primer positions 25 to 32. This allows efficient priming by the 3′ ends of the primers and permits recombination distal (i.e., 5′) to the altered sites. The 3′ primer, a 21-mer, contains only wild-type sequences. To obtain recombinant plasmids, yeast strains containing a wild-type or a mutant chromosomal *KEX2* gene are transformed with the PCR product and pG5-KEX2Δ613 plasmid DNA digested at *Bgl*II site 87 nucleotides 3′ to the Asn-314 codon. It is probably important to utilize a site for linearization in the target plasmid that is as close as possible to the site of mutation in order to increase the frequency of recovering mutant recombinants. Because 80% of the PCR product 5′ to the *Bgl*II site is 3′ to codon 314, most recombinational transformants are expected to regenerate the wild-type plasmid. Consistent with this expectation, 15% of the transformants contain the desired mutations.[18]

Detection of One-Step Site-Directed Mutations by PCR

1. Analytical-scale PCR (25-μl reactions) contain 25 mM Na MES, pH 6.5, 50 mM KCl, 3 mM MgCl$_2$, 250 μM each deoxynucleoside triphosphate, 1 μM each primer, and 100 μg/ml bovine serum albumin (BSA); 1 μl of saturated yeast culture is added as a source of template DNA. Premixes are heated 7 min at 94°, 5 U Taq polymerase (Perkin-Elmer, Norwalk, CT) are added, and then reactions are performed in 30 cycles of 30 sec at 55°, 60 sec at 70°, and 30 sec at 94°. The primers employed are the nonmutagenic 3' primer used to create the initial PCR product and a 21-mer primer corresponding to the nonmutagenic 5' half of the original mutagenic 42-mers.

2. PCR products are digested with BspEI (New England Biolabs, Beverly, MA) and electrophoresed on a 2.5% agarose gel to determine whether the resulting PCR products contain the introduced mutations. As shown in Fig. 3, positive colonies obtained in strain ABY01 with an intact $KEX2$ gene have a doublet after digestion. The upper band corresponds to the unaltered chromosomal $KEX2$ gene. The lower band corresponds to the mutated gene, now rendered sensitive to BspEI. DNA sequence analysis proves that only the desired mutations have been introduced.[18]

Mating Assay for Presence of Site-Directed Mutations

1. A $MAT\alpha kex2\Delta$::$TRP1$ $leu2$ $trp1$ $ade2$ $his3$ $ura3$ strain, CB017, is transformed with linearized pG5-$KEX2\Delta613$ plus Asn-314-Ala PCR product. Because of the low activity of these mutants, the site-directed mutant transformants are not expected to mate but the background of regenerated pG5-$KEX2\Delta613$ transformants are expected to mate.

2. CB017 transformants are grown on SDC-Ura plates, replica plated onto a lawn of ~5 × 10^5 DC14 cells ($MATa$ $his1$) on a YPD plate (rich,

153 bp
126 bp

Fig. 3. PCR assay for presence of site-directed mutations. $KEX2$ DNA was amplified from 16 yeast transformants as described in the text, digested with BspEI, electrophoresed, and stained with ethidium bromide. Six lanes have doublets, indicating the presence of site-directed mutations.

nonselective medium[21]), incubated for 12 hr, and then replica plated onto an SD plate (minimal selective medium[21]), on which only diploids formed from mating of CB017 and DC14 cells can grow. A representative selective plate, photographed after 48 hr of incubation at 30°, is shown in Fig. 4.

Perspectives

Through the identification of the Kex2 protease as the authentic physiological pro-α-factor processing enzyme, yeast genetics and molecular biology have played a seminal role in the identification of the specific processing proteases likely responsible for maturation of a wide range of secretory precursors.[1–4] As a system, yeast continues to provide advantages in the study of kexin, the Kex2 protease. Rapid methods of site-directed and randomized mutagenesis combined with sensitive biological assays for enzyme activity and substrate specificity allow facile identification of mutant enzymes of significant interest, which can be overproduced and purified within a few days in quantities sufficient for both biochemical and structural analysis. In addition, the capability of purifying large amounts

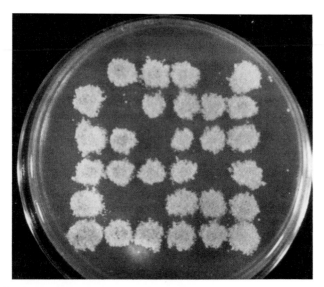

Fig. 4. Mating assay for presence of site-directed mutations. The CB017 transformant at the upper left contained plasmid pG5 (vector without *KEX2*), the next transformant to the right contained pG5-*KEX2Δ613*, and the other 34 were chosen at random from transformants of *Bgl*II-cleaved pG5-*KEX2Δ613* with the Asn[314]-Ala PCR product. Six out of the 34 transformants tested did not mate, indicating the presence of site-directed mutations.

of modified pro-α-factor produced through bacterial expression will permit quantitative biochemical analysis of the interactions between enzyme and substrate variants identified by *in vivo* assays. As a result, not only can Kex2 serve as a powerful model for the family of precursor processing proteases, but it also can stand alone as a powerful protein engineering system for studying catalysis and enzyme–substrate interactions.

Acknowledgments

This work was supported by NIH Grant GM39697 and a Lucille P. Markey Scholar award to R.S.F. A.B. and C.B. were graduate fellows of the Medical Scientist Training Program (GM 2T326M07365) and Cancer Biology Program (NCI 5T32CA09302).

[12] Purification of Recombinant Soluble Forms of Furin Produced in Chinese Hamster Ovary Cells

By Kazuhisa Nakayama

Introduction

The maturation of biologically active peptides and proteins often involves the limited proteolysis of larger precursors. In endocrine and neuronal cells, precursors of bioactive peptides are cleaved at paired basic amino acids within the regulated pathway of secretion.[1,2] By contrast, in nonneuroendocrine cells various secretory and membrane proteins are produced from precursors through cleavage at sites often marked by the consensus sequence, Arg-Xaa-(Lys or Arg)-Arg(RXK/RR), within the constitutive secretory pathway.[3] The regulated pathway is characteristic of endocrine and neuronal cells, and serves to store the peptides in secretory granules and to release them in response to external stimuli.[4] On the other hand, the constitutive pathway is present in all types of cells, and serves to release molecules continuously without storage.[4]

Research on processing endopeptidases has advanced with investigation of kexin, the Kex2 protease of the yeast *Saccharomyces cerevisiae*. It is a Ca^{2+}-dependent serine protease with a subtilisin-like catalytic domain and is involved in processing of pro-α-factor and pro-killer toxin at

[1] K. Docherty and D. F. Steiner, *Annu. Rev. Physiol.* **44**, 625 (1982).
[2] Y. P. Loh, M. J. Brownstein, and H. Gainer, *Annu. Rev. Neurosci.* **7**, 189 (1984).
[3] M. Hosaka, M. Nagahama, W.-S. Kim, T. Watanabe, K. Hatsuzawa, J. Ikemizu, K. Murakami, and K. Nakayama, *J. Biol. Chem.* **266**, 12127 (1991).
[4] T. L. Burgess and R. B. Kelly, *Annu. Rev. Cell Biol.* **3**, 243 (1987).

paired basic residues.[5,5a] Initial efforts to isolate mammalian proteases related to Kex2 were without success. However, the availability of the Kex2 sequence led to the serendipitous discovery of homology between the Kex2 protein and furin (EC 3.4.21.75), a product of the human *fur* gene.[6] It was originally detected as an open reading frame immediately upstream of the c-*fes/fps* protooncogene [hence, c-*fes/fps* upstream region (*fur*)].[7] Later, using polymerase chain reaction and cross-hybridization, cDNAs for five additional furin/Kex2-related proteases (PC2, PC1/3, PC4, PACE4, and PC6) were identified.[8-13] Furin mRNA is detectable in all tissues and cell lines examined so far.[14,15] By coexpression experiments in cultured cells using gene transfer techniques, furin has been shown to cleave a wide variety of precursor proteins at RXK/RR sites within the constitutive secretory pathway.[3,8-13] These include precursors for growth factors, serum proteases, receptors, viral envelope glycoproteins, and bacterial toxins.

Although transfection studies have been extremely useful to determine the relative degrees of specificity of each endopeptidase, they are often complicated by the presence of endogenous cellular endopeptidase. For example, cleavages attributed to the transfected protease may also occur indirectly through the activation of the endogenous enzyme. Therefore, *in vitro* experiments using pure endopeptidases are required. However, purification of furin to homogeneity from tissues and cells was not successful, because furin is a Golgi membrane-bound protease and is expressed at a very low level in cells. To overcome these problems, we[16] and Molloy *et al.*[17] prepared a construct of furin that lacked the putative transmem-

[5] G. Thomas, B. A. Thorne, and D. E. Hruby, *Annu. Rev. Physiol.* **50**, 323 (1988).

[5a] R. S. Fuller, C. Brenner, and A. Bevan, this series, Vol. 244.

[6] R. S. Fuller, A. J. Brake, and J. Thorner, *Science* **246**, 482 (1989).

[7] A. J. M. Roebroek, J. A. Schalken, J. A. M. Leunissen, C. Onnekink, H. P. J. Bloemers, and W. J. M. Van de Ven, *EMBO J.* **5**, 2197 (1986).

[8] P. J. Barr, *Cell (Cambridge, Mass.)* **66**, 1 (1991).

[9] I. Lindberg, *Mol. Endocrinol.* **5**, 1361 (1991).

[10] N. G. Seidah and M. Chrétien, *Trends Endocrinol. Metab.* **3**, 133 (1992).

[11] A. Rehemtulla and R. J. Kaufman, *Curr. Opin. Biotechnol.* **3**, 560 (1992).

[12] D. F. Steiner, S. P. Smeekens, S. Ohagi, and S. J. Chan, *J. Biol. Chem.* **267**, 23435 (1992).

[13] S. P. Smeekens, *Bio/Technology* **11**, 182 (1993).

[14] J. A. Schalken, A. J. M. Roebroek, P. P. C. A. Oomen, S. S. Wagenaar, F. M. J. Debruyne, H. P. J. Bloemers, and W. J. M. Van de Ven, *J. Clin. Invest.* **80**, 1545 (1987).

[15] K. Hatsuzawa, M. Hosaka, T. Nakagawa, M. Nagase, A. Shoda, K. Murakami, and K. Nakayama, *J. Biol. Chem.* **265**, 22075 (1990).

[16] K. Hatsuzawa, M. Nagahama, S. Takahashi, K. Takada, K. Murakami, and K. Nakayama, *J. Biol. Chem.* **267**, 16094 (1992).

[17] S. S. Molloy, P. A. Bresnahan, S. H. Leppla, K. R. Klimpel, and G. Thomas, *J. Biol. Chem.* **267**, 16369 (1992).

brane and cytoplasmic domains, and introduced this into mammalian cells. These cells secreted truncated forms of furin retaining catalytic activity. Our methods for obtaining the soluble form of furin (furin Δ704) are described below. Also described are transfection experiments using, as host, human colon carcinoma LoVo cells, which lack endogenous processing activity toward RXK/RR sites because they have a mutation of furin.[18]

Expression of Truncated Form of Furin in Chinese Hamster Ovary Cells and Preparation of Conditioned Medium

To produce large amounts of a secretory form of furin, the cDNA for a furin protein truncated up to the transmembrane domain from the COOH terminus (furin Δ704) is expressed under the control of the simian virus 40 (SV40) early promoter in Chinese hamster ovary (CHO) cells. The expression of furin in CHO cells is then amplified using methotrexate, and clonal cell lines are screened for secretion of furin. The amplification and clonal selection are important because the level of furin activity in the conditioned medium of the amplified, selected cells is much (>100-fold) higher than that of cell pools without amplification.

Plasmid Construction

Mouse furin cDNA,[15] which was subcloned into the *Eco*RI site of the pBluescript-II vector (Stratagene, La Jolla, CA), was digested with *Pst*I, blunt-ended with T4 DNA polymerase, ligated with *Xba*I linkers (catalog No. 1083, New England Biolabs, Beverly, MA) to introduce an in-frame stop codon, and digested with *Xho*I and *Xba*I. The ~2.2-kb *Xho*I–*Xba*I fragment was subcloned into the pSVDAS vector,[19] which is a derivative of the pSVD vector[20] containing the multiple cloning sites from the pBluescript-II vector (from *Xho*I to *Sac*II sites) downstream of the SV40 early promoter and which possesses the mouse dihydrofolate reductase (*dhfr*) gene as a selection marker. The resulting expression plasmid was designated as pSVD/FurΔ704 (schematically shown in Fig. 1) and was deduced to encode a furin protein terminated at the amino acid residue 704 before the transmembrane domain.

[18] S. Takahashi, K. Kasai, K. Hatsuzawa, N. Kitamura, Y. Misumi, Y. Ikehara, K. Murakami, and K. Nakayama, *Biochem. Biophys. Res. Commun.* **195,** 1019 (1993).
[19] K. Hatsuzawa, K. Murakami, and K. Nakayama, *J. Biochem.* (*Tokyo*) **111,** 296 (1992).
[20] K. Hatsuzawa, W.-S. Kim, K. Murakami, and K. Nakayama, *J. Biochem.* (*Tokyo*) **107,** 854 (1992).

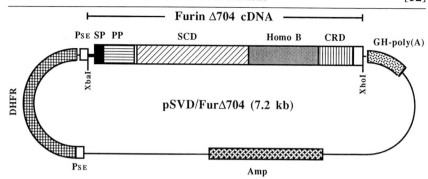

FIG. 1. Schematic representation of the structure of the expression plasmid for truncated furin, pSVD/FurΔ704. SP, Signal peptide; PP, propeptide; SCD, subtilisin-like catalytic domain; Homo B, homo B domain; CRD, Cys-rich domain; GH-poly(A), DNA fragment of bovine growth hormone gene containing a poly(A) addition site; Amp, β-lactamase gene; PSE, SV40 early promoter; DHFR, mouse *dhfr* gene.

Establishment of CHO Cell Line Overproducing Truncated Form of Furin

CHO/DXB11 cells (a *dhfr*⁻ strain)[21] are transfected with the pSVD/FurΔ704 using a CellPhect transfection kit (Pharmacia-LKB Biotechnology, Uppsala, Sweden), and the stable transfectants are selected under thymidine-free conditions (in Dulbecco's modified Eagle's medium supplemented with 10% dialyzed fetal calf serum and nonessential amino acids). Clones of cells are then adapted stepwise to growth in the presence of a *dhfr* inhibitor, methotrexate (Sigma Chemical Co., St. Louis, MO). The conditioned medium from each clone is assayed for furin activity as described below. A cell line that shows the highest furin activity is designated as CHO/Δ704 and used for purification of furin.

Preparation of Conditioned Medium

CHO/Δ704 cells at ~70% confluence on a Cell Factory-10 plate (Nunc, Roskilde, Denmark) are cultured in 2 liters of complete serum-free medium (S-clone SF-O2, Sanko Junyaku Co., Tokyo, Japan) for 48 hr. The conditioned medium is centrifuged at 3,000 *g* for 10 min at 4° to remove cell debris and insolubles, and stored at −80°C until use.

Assay for Furin

Assay for furin is based on the coexpression of furin and a substrate precursor in cultured cells, in which furin can cleave various precursor

[21] G. Urlaub and L. A. Chasin, *Proc. Natl. Acad. Sci. U.S.A.* **77**, 4216 (1980).

proteins after sequences fitting the RXK/RR motif. Our preliminary study revealed that furin was also capable of cleaving short fluorogenic peptides with a sequence fitting the RXK/RR motif.[18] Therefore, we routinely use a commercially available fluorogenic peptide, *tert*-butoxycarbonyl (Boc)-Arg-Val-Arg-Arg-4-methylcoumarin-7-amide (MCA) or <Glu-Arg-Thr-Lys-Arg-MCA (Peptide Institute, Inc., Osaka, Japan), as a substrate for furin.

A furin sample is incubated with 20 nmol of Boc-Arg-Val-Arg-Arg-MCA in a total volume of 250 μl of 100 mM 2-(N-morpholino)ethanesulfonic acid (MES)–NaOH buffer, pH 7.0, containing 1 mM CaCl$_2$ and 5 mg/ml bovine serum albumin (BSA) at 37° for 1–4 hr. The reaction is stopped by the addition of 3 ml of 5 mM EDTA, and the amount of 7-amino-4-methylcoumarin (AMC) released from the substrate is measured in a fluorescence spectrophotometer with excitation at 380 nm and emission of 460 nm. One unit of activity is defined as the amount of furin that can release 1 nmol of AMC from the substrate under the above conditions in 1 min.

Purification of Furin from Conditioned Medium of CHO/Δ704 Cells

Using a three-step procedure, we have successfully purified furin to homogeneity from the conditioned medium of CHO/Δ704 cells. The procedure includes ammonium sulfate precipitation, batchwise fractionation on Blue-Toyopearl, and DEAE-Toyopearl chromatography.

All purification steps are performed at 0–4°, and all buffers used contain 1 mM CaCl$_2$.

Step 1: Ammonium Sulfate Precipitation

1. To 500 ml of the conditioned medium of CHO/Δ704 cells, slowly add ammonium sulfate while stirring to give a final concentration of 40% saturation and leave for at least 1 hr.
2. Pellet precipitated proteins by centrifugation at 10,000 g for 30 min. Discard the pellet.
3. To the supernatant, slowly add ammonium sulfate again while stirring to give a final concentration of 60% saturation and leave for at least 1 hr.
4. Pellet precipitated proteins by centrifugation at 10,000 g for 30 min. Discard the supernatant.
5. Solubilize precipitated proteins in 10 ml of 10 mM MES–NaOH buffer, pH 6.0, and dialyze overnight against the same buffer. Change the dialysis buffer several times.

6. Centrifuge the dialyzed solution at 10,000 g for 10 min to remove insolubles.

Step 2: Blue-Toyopearl Batchwise Fractionation

1. To the dialyzed solution from Step 1, add 4 ml (wet volume) of AF-Blue-Toyopearl 650MH (Tosoh, Tokyo, Japan) equilibrated with 10 mM MES–NaOH buffer, pH 6.0, and gently mix for at least 30 min.
2. Centrifuge the suspension at 3000 g for 2 min. Discard the supernatant.
3. Suspend the pellet in 20 ml of 10 mM MES–NaOH buffer, pH 6.5, and gently mix for at least 30 min.
4. Centrifuge the suspension at 3000 g for 2 min. Discard the supernatant.
5. Resuspend the pellet in 15 ml of 10 mM MES–NaOH buffer, pH 7.0, containing 300 mM NaCl and gently mix.
6. Centrifuge the suspension at 3000 g for 2 min.
7. Dialyze the supernatant overnight against 10 mM MES–NaOH buffer, pH 7.0.
8. Centrifuge the dialyzed solution at 10,000 g for 10 min to remove insolubles.

Step 3: DEAE-Toyopearl Chromatography

1. Apply the dialyzed solution from Step 2 to a column (0.9 × 20 cm) of DEAE-Toyopearl 650S (Tosoh) equilibrated with 10 mM MES–NaOH buffer, pH 7.0.
2. Wash the column with at least 30 ml of the same buffer containing 50 mM NaCl.
3. Elute bound proteins with an 80-ml linear gradient of 50–150 mM NaCl in the same buffer. Collect fractions of 1.5 ml at a flow rate of 2 ml/min.

Comments on the Purification

Table I shows the summary of the purification. On the final DEAE-Toyopearl chromatography, furin activity is eluted as two peaks, pools I and II, at NaCl concentrations of ~90 and ~120 mM, respectively (Fig. 2A). The overall purification from 500 ml of the conditioned medium of CHO/Δ704 cells is typically ~100-fold with yields of ~260 and ~50 μg for pools I and II, respectively. Pools I and II contain a mixture of 83- and 81-kDa forms and a 96-kDa form, respectively, of furin as visualized

TABLE I
PURIFICATION OF FURIN FROM CONDITIONED MEDIUM OF CHO/Δ704 CELLS

Step	Total protein (mg)	Total activity (units)	Specific activity (units/mg)	Purification (-fold)	Yield (%)
Conditioned medium	282	475	1.68	1	100
(NH₄)₂SO₄ fractionation	5.30	253	47.7	28.4	53.3
Blue-Toyopearl	0.95	144	152	90.5	30.3
DEAE-Toyopearl					
Pool I	0.26	46.4	178	106	9.77
Pool II	0.050	7.64	153	91.1	1.61

by SDS–PAGE (Fig. 2B). Immunoblot analyses have suggested that the 96-kDa form is mature furin Δ704, and that the 83- and 81-kDa forms are produced from the 96-kDa form through differential COOH-terminal processing.[16] However, the two preparations show essentially the same enzymatic properties: specific activity (150–180 U/mg protein), pH profile (pH optimum, around 7.0), substrate specificity, and inhibitor profile.[16]

It should be noted that furin is very unstable at low protein concentrations; the final preparations stored at −80° lose activity over time, with only 5–20% of the activity remaining after 2 weeks. For the sake of convenience, we routinely store the final preparations with 5 mg/ml BSA

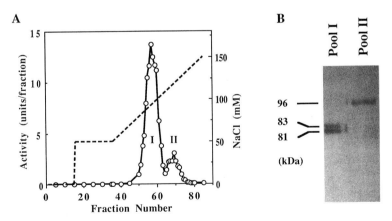

FIG. 2. Purifiction of truncated furin from the conditioned medium of CHO/Δ704 cells. (A) Elution profile of furin activity from a DEAE-Toyopearl 650S column. (B) SDS–polyacrylamide (7.5%, w/v) gel electrophoresis of the pool I (200 ng protein) and pool II (100 ng protein) preparations (silver staining).

at −20°, or with 5 mg/ml BSA and 0.02% NaN$_3$ at 4°; under these conditions, furin is stable with >95% of the activity remaining after 2 months.

Concluding Remarks

Thus far, we have shown by *in vitro* experiments using the above furin preparations that furin is able to cleave a variety of precursor proteins at RXK/RR sites. These include proalbumin,[22] complement pro-C3,[22] the Fo glycoprotein of Newcastle disease virus,[23] the proreceptor of hepatocyte growth factor (HGF),[24] and diphtheria toxin.[25] Molloy *et al.*[17] have shown that truncated human furin partially purified in a manner similar to ours cleaves the protective antigen of anthrax toxin.

Although transfection experiments have also been useful to determine the substrate specificities of furin and other processing endoproteases, they may often be complicated by the presence of endogenous endopeptidase, such as furin. However, this problem has been overcome by the identification of a processing-deficient cell line, human colon carcinoma LoVo cells.[26,27] We have confirmed that the processing deficiency is caused by a mutation of furin.[18] By transfection experiments using LoVo cells as the host, we have shown that furin can cleave HGF proreceptor[24] and diphtheria toxin.[25] Moehring *et al.*[28] have reported that a mutant strain, named RPE.40, of CHO cells shows properties similar to those of LoVo cells, although it remains to be elucidated whether furin is also impaired in RPE.40 cells. Therefore, the *in vitro* experiments in combination with transfection experiments using the processing-deficient cell lines will be useful to identify substrates of furin.

It is also of note that the envelope glycoproteins of many viruses, including human immunodeficiency virus, are or may be matured through cleavage at RXK/RR sites catalyzed by furin. Because the cleavage is a

[22] K. Oda, Y. Misumi, Y. Ikehara, S. O. Brennan, K. Hatsuzawa, and K. Nakayama, *Biochem. Biophys. Res. Commun.* **189**, 1353 (1992).

[23] B. Gotoh, Y. Ohnishi, N. M. Inocencio, E. Esaki, K. Nakayama, P. J. Barr, G. Thomas, and Y. Nagai, *J. Virol.* **66**, 6391 (1992).

[24] M. Komada, K. Hatsuzawa, S. Shibamoto, F. Ito, K. Nakayama, and N. Kitamura, *FEBS Lett.* **328**, 25 (1993).

[25] M. Tsuneoka, K. Nakayama, K. Hatsuzawa, M. Komada, N. Kitamura, and E. Mekada, *J. Biol. Chem.* **268**, 26461 (1993).

[26] B. Drewinko, M. M. Romsdahl, L. Y. Young, M. J. Ahearn, and J. M. Trujillo, *Cancer Res.* **36**, 467 (1976).

[27] A. Mondino, S. Giordano, and P. M. Comoglio, *Mol. Cell. Biol.* **11**, 6084 (1991).

[28] J. M. Moehring, N. M. Inocencio, B. J. Robertson, and T. J. Moehring, *J. Biol. Chem.* **268**, 2590 (1993).

prerequisite for fusion of the viral envelope and host cell membrane, furin can be one of the targets of antiviral agents. Therefore, the *in vitro* furin assay described here will become a powerful tool for screening of new agents of this kind.

[13] Pro-Protein Convertases of Subtilisin/Kexin Family

By Nabil G. Seidah and Michel Chrétien

In 1967, Chrétien and Li[1] and Steiner *et al.*[2] advanced the hypothesis that prohormones exist and need to be cleaved postranslationally at pairs of basic amino acids in order to release bioactive end products. Since then, it has become apparent that the limited proteolysis of precursors at specific pairs of basic residues and/or at single basic amino acids is a widespread mechanism by which the cell expresses a repertoire of biologically active proteins and peptides.[3] The molecular identification of the tissue-specific and developmentally regulated proteinases responsible for such conversions has finally been achieved by a remarkable series of events, which are described in more detail in Refs. 3 and 4 and Chapters 11 and 12 of this volume.

Mammalian Family of Pro-protein and Prohormone Convertases

Using the concept of the conservation of sequence around active site residues in a family of serine proteinases, six different mammalian convertases were identified.[4] These include the ubiquitously expressed furin[5] and PACE4,[6] the enzyme PC5[7] (also called PC6),[8] which is expressed in some endocrine and some nonendocrine cells, and the neural and endocrine

[1] M. Chrétien and C. H. Li, *Can. J. Biochem.* **45,** 1163 (1967).

[2] D. F. Steiner, D. Cunningham, L. Spiegelman, and B. Aten, *Science* **157,** 697 (1967).

[3] N. G. Seidah and M. Chrétien, *Trends, Endocrinol. Metab.* **3,** 133 (1992).

[4] N. G. Seidah, R. Day, M. Marcinkiewicz, S. Benjannet, and M. Chrétien, *Enzyme* **45,** 271 (1991).

[5] A. J. M. Roebroek, J. A. Schalken, J. A. M. Leunissen, C. Onnekink, H. P. J. Gloemers, and W. J. M. Van de Ven, *Embo. J.* **5,** 2197 (1986).

[6] M. C. Kiefer, J. E. Tucker, R. Joh, K. E. Landsberg, D. Saltman, and P. J. Barr, *DNA Cell Biol.* **10,** 757 (1991).

[7] J. Lusson, D. Vieau, J. Hamelin, R. Day, M. Chrétien, and N. G. Seidah. *Proc. Natl. Acad. Sci. U.S.A.* **90,** 6691 (1993).

[8] T. Nakagawa, M. Hosaka, S. Torij, T. Watanabe, K. Murakami, and K. Nakayama, *J. Biochem. (Tokyo)* **113,** 132 (1993).

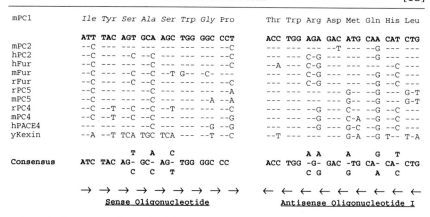

FIG. 1. Consensus oligonucleotide sequences chosen for the polymerase chain reaction amplification of mRNAs coding for the prohormone convertases.

convertases PC1[9,10] (also called PC3)[11] and PC2,[9,12] and finally the enzyme PC4,[13,14] which is expressed primarily in testicular germ cells, including pachetene spermatocytes and round spermatids, and also in ovaries.[14]

The oligonucleotides used for the discovery of the enzymes PC1, PC2, PC4, and PC5 were based on the conservation of sequence around the active sites of subtilisin-like enzymes, including furin (Fig. 1). We have chosen a set of two degenerate oligonucleotides, one preceding the catalytically important Asn (sense oligo) and the other following the active site Ser (antisense oligo). We found that this pair of oligonucleotides, when combined with PCR amplification, allowed us to amplify a cDNA fragment of about 450 bp for any of the six convertases known. These conserved *IleTyrSerAlaSerTrpGlyPro* and *ThrTrpArgAspMetGlnHisLeu* sequences, which did not include the active sites Asn and Ser, are usually separated by a relatively constant number of amino acids in all the subtilisin-like proteinases studied. These criteria were quite useful in eliminating false positives on DNA sequencing the subcloned 450-bp polymerase chain

[9] N. G. Seidah, L. Gaspar, P. Mion, M. Marcinkiewicz, M. Mbikay, and M. Chrétien, *DNA* **9**, 415 (1990).

[10] N. G. Seidah, M. Marcinkiewicz, S. Benjannet, L. Gaspar, G. Beaubien, M. G. Mattei, C. Lazure, M. Mbikay, and M. Chrétien, *Mol. Endocrinol.* **5**, 111 (1991).

[11] S. P. Smeekens, A. S. Avruch, J. LaMendola, S. J. Chan, and D. F. Steiner, *Proc. Natl. Acad. Sci. U.S.A.* **88**, 340 (1991).

[12] S. P. Smeekens and D. F. Steiner, *J. Biol. Chem.* **265**, 2997 (1990).

[13] K. Nakayama, W.-S. Kim, S. Torij, M. Hosaka, T. Nakagawa, J. Ikemizu, T. Baba, and K. Murakami, *J. Biol. Chem.* **267**, 5897 (1992).

[14] N. G. Seidah, R. Day, J. Hamelin, A. Gaspar, M. W. Collard, and M. Chrétien, *Mol. Endocrinol.* **6**, 1559 (1992).

reaction (PCR)-amplified fragments, as one could easily eliminate these unrelated structures if one or both of these sites was absent in the sequenced fragment.

Comparative Protein Structures of Mammalian Paired Basic
    Residue Convertases

The comparative architectural features of the six mammalian subtilisin-like pro-protein convertases known so far as those of the subtilisin BPN' and the yeast kexin are presented in Fig. 2, showing that the sizes of the eukaryotic members vary from 637 to 969 amino acids. It is seen that all

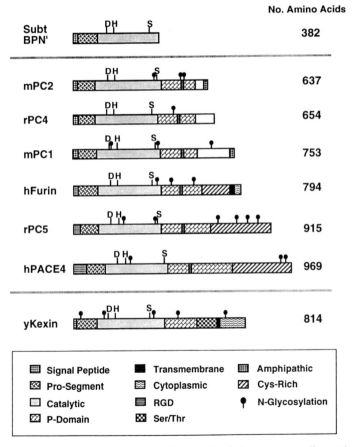

FIG. 2. Comparative structural elements found in the six known mammalian prohormone convertases, and those found in bacterial subtilisin (subt) BPN' and yeast(y) kexin.

the members contain a signal peptide, and hence will enter the secretory pathway, which includes the endoplasmic reticulum (ER) and Golgi apparatus. This alignment suggests that by analogy to prosubtilisin,[15] each convertase's unique prosegment might have to be excised in order to generate an active proteinase.[3,4] Accordingly, the rate of removal of the N-terminal prosegment conceivably could represent a mechanism by which the cell would control the rate of pro-protein processing.[16] Recent studies with the convertases PC1 and PC2[17] demonstrated that the excision of the N-terminal prosegment of PC1 and PC2 is much more efficient for pro-PC1 than it is for pro-PC2. Furthermore, whereas pro-PC1 is rapidly cleaved into PC1 within the ER, pro-PC2 slowly exits from the ER and is processed into PC2 within the trans-Golgi network (TGN). The consequence of this different temporal activation of PC1 and PC2 is that in cells that express both enzymes, PC1 will cleave precursors before PC2, leading to an ordered cleavage mechanism. This hypothesis is in agreement with the observation that within the cells of the pituitary pars intermedia, proopiomelanocortin (POMC) is first processed into β-LPH and then into β-endorphin,[18] which are peptide products requiring the concerted action of PC1 and PC2.[19] Similar to prosubtilisin and prokexin, the activation of profurin was shown to be autocalytic, but the intracellular site has not yet been well defined.[20] No information is yet available for similar studies on the rate of activation of pro-PACE4, pro-PC4, and pro-PC5. Following the prosegment, all the members contain a catalytic domain comprising the active sites Asp, His, and Ser found at positions equivalent to those in subtilisins and kexin. The catalytic region of each convertase represents the segment exhibiting the highest protein sequence identity between the convertases (about 50–60%). PC2 seems to be the most distant member of the mammalian kexinlike convertases, as it shows the least sequence identity to other members of the family. In addition, PC2 is the only convertase[9,12] that shows the presence of an Asp residue in place of the catalytically important Asn residue found in all other subtilisin-like proteinases.

[15] J. A. Wells, E. Ferrari, D J. Henner, D. A. Estell, and E. Y. Chen, *Nucleic Acids Res.* **18**, 7911 (1983).
[16] S. Benjannet, T. Reudelhuber, C. Mercure, N. Rondeau, M. Chrétien, and N. G. Seidah, *J. Biol. Chem.* **267**, 11417 (1992).
[17] S. Benjannet, N. Rondeau, L. Paquet, A. Boudreault, C. Lazure, M. Chrétien, and N. G. Seidah, *Biochem. J.* **294**, 735 (1993).
[18] P. Crine, F. Gossard, N. G. Seidah, L. Blanchettte, M. Lis, and M. Chrétien, *Proc. Nat. Acad. Sci. U.S.A.* **76**, 5085 (1979).
[19] S. Benjannet, N. Rondeau, R. Day, M. Chrétien, and N. G. Seidah, *Proc. Nat. Acad. Sci. U.S.A.* **88**, 3564 (1991).
[20] R. Leduc, S. S. Molloy, B. A. Thomas, and G. Thomas, *J. Biol. Chem.* **267**, 14304 (1992).

```
            Arg Arg Gly Asp Leu Ala Ile
rFurin      CGC CGT GGT GAC CTG GCT ATC
hPace4      CGC CGA GGA GAC CTC CAG ATC
rPC5        CGC CGC GGG GAC CTG GAG ATC
rPC1        CGT AGA GGA GAC CTT CAT GTC
rPC2        AGG AGA GGA GAC CTG AAC ATC
rPC4        CGC CGC GGG GAC CTG GAC ATC

            T A T   G       T GCT G
Consensus   CGC CGA GGA GAC CTG CAG ATC
RGD         G   C   T         C A C

            ← ← ← ← ← ← ← ←
            Antisense Oligonucleotide II
```

FIG. 3. Consensus RGD oligonucleotide sequence used for the polymerase chain reaction amplification of rat (r) PC5 and PACE4.

Primary sequence analysis of all the mammalian subtilisin-like convertases[4] revealed the presence of a hydrophilic Arg*ArgGlyAsp*Leu sequence absent from kexin and subtilisin BPN'. The tripeptide ArgGlyAsp (RGD) sequence is found in a number of extracellular matrix proteins and is a key recognition sequence by cell surface integrins implicated in cell adhesion.[21] Although the function of this sequence in the convertases is not yet defined, it may be important for the folding of the molecule within the ER, or it could be involved in the intracellular sorting of the convertases via their interaction with integrins during their migration in the endoplasmic reticulum and/or in the Golgi apparatus. In fact, the alignment of the nucleotide sequences around this RGD sequence, revealed a conserved sequence that formed the basis for the synthesis of a consensus RGD antisense oligonucleotide (Fig. 3; oligonucleotide II), which allowed us to extend the sequence information of PC5 during our original cloning of this cDNA.[7] This consensus sequence was also used to clone a fragment of rat PACE4 useful for tissue distribution studies by *in situ* hybridizations in the rat (in preparation). As shown in Fig. 2, this RGD sequence is found within a segment called P- (or Homo B-) domain, just following the catalytic domain. This region has been reported to be important for the folding of the proteinase within the ER and the elaboration of the full cellular enzymatic activity of kexin[22] and furin.[23] Therefore, cleavage of the convertases at the ArgArg↓GlyAsp sequence may provide a mechanism by which the activity of the convertases may be controlled. Preliminary results in our laboratories suggest that in PC1,[17] but not in PC2, such cleavage does indeed occur late along the secretory pathway (not shown).

[21] E. Ruoslahti, *Annu. Rev. Biochem.* **57,** 375 (1988).
[22] C. A. Wilcox and R. Fuller, *J. Cell. Biol.* **115,** 297 (1991).
[23] K. Hatsuzawa, K. Murakami, and K. Nakayama, *J. Biochem.* (*Tokyo*) **111,** 296 (1992).

Following the P-domain, each enzyme exhibits a unique C-terminal sequence. Furin, PACE4, and PC5[7] all contain a segment rich in Cys residues, which are arranged in a repetitive consensus motif Cys XX Cys XXX Cys X_{5-7} Cys XX Cys X_{8-13} Cys XXX Cys. This Cys-rich domain may control the cellular localization and/or stability of PACE4 and PC5. Only furin contains a transmembrane domain, apparently responsible for its retention within the Golgi apparatus. Although the structures of PC1 and PC2 do not predict the presence of a transmembrane domain, their C-terminal segments code for an amphipathic structure that could interact with membranes in a pH-dependent manner.[9,10] PC4 is the first mammalian convertase in which at least three mRNAs coding for different C-terminal sequences have been reported, possibly resulting from differential splicing.[14,24]

It is interesting to note that all six convertases contain between one and six N-glycosylation sites, and, in contrast to kexin, none appears in the prosegment. Recent studies demonstrated that whereas the N-glycosyl moieties of PC1 are of the complex type, those of PC2 are of the high-mannose type.[17] Furthermore, our data show that under conditions wherein POMC levels are not affected, prevention of N-glycosylation by prior treatment of the cells with tunicamycin causes a dramatic destruction of PC1 and PC2 by resident hydrolases in the ER.[17] Therefore, N-glycosylation plays an important role in the folding and the ability of these convertases to exit from the ER.

Other posttranslational modifications that were studied include Tyr sulfation and phosphorylation. The data showed that PC1 and PC2 are Tyr sulfated possibly at Tyr-694 and Tyr-171, respectively.[17] Interestingly, the very conserved sequence around the Tyr-sulfation site found in mouse PC2 was also found in PC2 of the mollusc *Aplysia californica*.[25] However, furin is not sulfated and neither PC1, PC2, nor furin are phosphorylated.[25a]

Single and Paired Basic Amino Acid Cleavage Specificity of PC1 and PC2

The cleavage specificity of PC1 and PC2 was studied either by coexpression of these convertases with precursor substrates and defining the processed products or *in vitro* with a number of peptide substrates. As

[24] M. Mbikay, M.-L. Raffin-Sanson, H. Tadros, F. Sirois, N. G. Seidah, and M. Chrétien, *Genomics* **20**, 231 (1994).
[25] T. Ouimet, A. Mammarbachi, T. Cloutier, N. G. Seidah, and V. Castellucci, *FEBS Lett.* **330**, 343 (1993).
[25a] S. Benjannet, M. Chrétien, and N. G. Seidah, (in preparation) (1994).

shown in Table I, the precursors studied were mouse[19,26] and porcine[27] POMC, human proenkephalin,[28] rat prodynorphin,[29] human and mouse prorenin 2,[16] rat prosomatostatin,[30] human proinsulin,[31] and rat pro-luteinizing hormone releasing hormone (LHRH).[32] In all cases it was shown that both enzymes selectively cleaved these precursors at distinct pairs of basic residues. Furthermore, data with prodynorphin demonstrated that both convertases are also capable of cleaving at a single Arg residue, also known to be processed *in vivo*.[29] From the available data, it is clear that only PC1, but not PC2 or furin,[33] can cleave substrates with an aliphatic residue (Leu, Val, Ile) at the C terminus of the cleavage site. For example, this was observed in the case of human renin, where a Leu follows the cleaved LysArg Pair (Table I). Furthermore, it is also evident that the presence of a basic residue (most often Arg) at positions -4, -6, or -8 from the cleavage site favors processing at the basic residue at position -1 by either PC1 or PC2 (Table I), as was also observed for furin.[33] These conclusions were also largely confirmed for PC1, as purification of either mouse or human PC1 allowed the study of its specificity *in vitro* with a number of peptides, including a proparathyroid hormone peptide substrate.[34] It was found that PC1 is a Ca^{2+}-dependent enzyme that exhibits a pH optimum of about pH 5.5–6.0. Furthermore, this work reported the synthesis of two tetrapeptide semicarbazone inhibitors of general structure Arg-X-LysArg-semicarbazone. Although these peptides are not selective for PC1, they can inhibit its activity at the low micromolar range.[34] In general, then, the available cleavage specificity data demonstrate that both PC1 and PC2 cleave precursors both at specific single and pairs of basic residues and that the four combinations, Lys-Arg↓, Arg-Arg↓, Arg-Lys↓,

[26] L. Thomas, R. Leduc, B. A. Thorne, S. P. Smeekens, D. F. Steiner, and G. Thomas, *Proc. Natl. Acad. Sci. U.S.A.* **99**, 5297 (1991).
[27] N. G. Seidah, H. Fournier, G. Boileau, S. Benjannet, N. Rondeau, and M. Chrétien, *FEBS Lett.* **310**, 235 (1992).
[28] M. B. Breslin, I. Lindberg, S. Benjannet, J. P. Mathis, C. Lazure, and N. G. Seidah, *J. Biol. Chem.* **268**, 27084 (1993).
[29] R. Day, A. C. Dupuy, H. Akil, M. Chrétien, and N. G. Seidah, "Program of the 22nd Annual Meeting of the Society of Neurosciences," p. 194. Anaheim, California (Abstract 91-8) (1992).
[30] A. Galanopoulou, G. Kent, S. N. Rabbani, N. G. Seidah, and Y. C. Patel, *J. Biol. Chem.* **268**, 6041 (1993).
[31] S. P. Smeekens, A. G. Montag, G. Thomas, C. Albigez-Rizo, R. Carrol, M. Benig, L. A. Phillips, S. Martin, S. Ohagi, P. Gardner, H. H. Swift, and D. F. Steiner, *Proc. Natl. Acad. Sci. U.S.A.* **89**, 8822 (1992).
[32] W. C. Wetsel, L. Thomas, J. S. Hayflick, Z. Liposits, N. G. Seidah, and Gary Thomas, (in preparation) (1994).
[33] T. Watanabe, K. Murakami, and K. Nakayama, *FEBS Lett.* **320**, 215 (1993).
[34] F. Jean, A. Basak, N. Rondeau, S. Benjannet, G. Hendy, N. G. Seidah, M. Chrétien, and C. Lazure, *Biochem. J.* **292**, 891 (1993).

TABLE I
CLEAVAGE SPECIFICITY OF PC1 AND PC2

Pro-hormones	PC1	PC2
mPOMC	ProSerProArgGluGlyLysLysArg → SerTyr	ProSerProArgGluGlyLysLysArg → SerTyr
	AlaPheProLeuGluPheLysArg → GluLeu	AlaPheProLeuGluPheLysArg → GluLeu
	SerAsnProProLysLysArg → TyrGly	SerAsnProProLysLysArg → TyrGly
	GlyLysProValGlyLysLysArg - ArgPro	GlyLysProValGlyLysLysArg → ArgPro
		AlaGlySerAlaAlaGlnArgArg → AlaGlu
	GlnProLeuThrGluAsnProArg - LysTyr	GlnProLeuThrGluAsnProArg →↑ LysTyr
	IleIleLysAsnAlaHisLysLys - GlyGln	IleIleLysAsnAlaHisLysLys → GlyGln
hPro-Renin	GluTrpSerGlnMetProLysArg → LeuThr	GluTrpSerGlnMetProLysArg - LeuThr
mPro-Renin2	GluTrpAspValPheThrLysArg → SerSer	
hPro-Enkephalin	ArgTyrGlyGlyPheMetLysLys → AspAla	ArgTyrGlyGlyPheMetLysLys - AspAla
	GlyPheMetArgGlyLeuLysArg → SerPro	GlyPheMetArgGlyLeuLysArg → SerPro
	TrpTrpMetAspTyrGlnLysArg → TyrGly	TrpTrpMetAspTyrGlnLysArg → TyrGly
	ArgTyrGlyGlyPheLeuLysArg → PheAla	ArgTyrGlyGlyPheLeuLysArg → PheAla
	AsnGluGluValSerLysArg → TyrGly	AsnGluGluValSerLysArg → TyrGly
	GluValProGluMetGluLysArg → TyrGly	GluValProGluMetGluLysArg → TyrGly
rPro-Dynorphin	ArgGlnPheLysValValThrArg → SerGln	ArgGlnPheLysValValThrArg → SerGln
	AspLeuArgGlysGlnAlaLysArg → TyrGly	AspLeuArgGlysGlnAlaLysArg → TyrGly
	GlyHisGluAspLeuTyrLysArg → TyrGly	GlyHisGluAspLeuTyrLysArg → TyrGly
	PheLeuArgLysTyrProLysArg - SerSer	PheLeuArgLysTyrProLysArg → SerSer
	LeuLysTrpAspAsnGlnLysArg - TyrGly	LeuLysTrpAspAsnGlnLysArg → TyrGly
rPro-Somatostatin	AlaMetAlaProArgGluArgLys → AlaGly	AlaMetAlaProArgGluArgLys - AlaGly
hPro-Insulin	PheTyrThrProLysThrArgArg → GluAla	PheTyrThrProLysThrArgArg - GluAla
	LeuGluGlySerLeuGlnLysArg → GlyIle	LeuGluGlySerLeuGlnLysArg → GlyIle
rPro-LHRH	GlyLeuArgProGlyGlyLysArg - AspAla	GlyLeuArgProGlyGlyLysArg → AspAla

and Lys-Lys+ are possible cleavage sites for these convertases. Therefore, these results are consistent with the proposed hypothesis concerning the physiological roles of PC1 and PC2 as distinct pro-protein convertases acting alone or together to produce a set of tissue-specific maturation products both in the brain and in peripheral tissues.[1,6,8] So far, little is known about the cleavage specificity of PC4 and PC5 except that PC4 also cleaves fluorogenic peptide substrates at pairs of basic residues (N. G. Seidah, A. Bazak, and M. Chrétien, unpublished results, 1994).

Tissue Distribution of PC1, PC2, PC4, and PC5

From the data presented above, it became clear that PC1 and PC2 are more adapted to process precursors that exit from the cells via the regulated secretory route, i.e., precursors that are routed toward secretory granules. On the other hand, furin seems to be a better processing enzyme for precursors that negotiate the constitutive secretion route and that are normally synthesized in cells devoid of secretory granules. It was therefore important to define the tissue distribution of these convertases and to correlate the observed data with those obtained from substrate cleavage specificity data.

Northern and *in situ* hybridization analyses demonstrated a tissue and cellular specificity of expression of PC1 and PC2, which were found at variable relative proportions only within endocrine and neuroendocrine cells.[9,10] *In situ* hybridization results demonstrated a distinct localization of the PC1 and PC2 transcripts in mouse[9,10] and rat[35,36] pituitary and brain. By *in situ* hybridization, we examined the specific distribution of PC1 and PC2 in the anterior lobe corticotrophs. Double-labeling colocalization studies indicated the presence of PC1 in POMC-containing cells, whereas very little PC2 mRNA could be detected in adult corticotrophs.[35] Ontogeny studies in the mouse pituitary not only confirmed both at the mRNA and protein levels that PC1 predominates over PC2 in mature corticotrophs, but also pointed out the prevalence of PC2 over PC1 within the first 2 weeks after birth.[37] This is one example emphasizing the plasticity of the endocrine and neuroendocrine system in which the ratio of PC1 and PC2 can vary under different physiological stimuli. The unique localization patterns of PC1 and PC2 indicate the differential and possible cooperative roles of these convertases in the brain, and may have some significance

[35] R. Day, M. K.-H. Schäfer, S. J. Watson, M. Chrétien, and N. G. Seidah, *Mol. Endocrinol.* **6**, 485 (1992).

[36] M. K-H. Schäfer, R. Day, W. E. Cullinan, M. Chrétien, N. G. Seidah, and S. J. Watson, *J. Neurosci.* **13**, 1258 (1993).

[37] M. Marcinkiewicz, R. Day, N. G. Seidah, and M. Chrétien, *Proc. Natl. Acad. Sci. U.S.A.* **90**, 4922 (1993).

in terms of our understanding of the "tissue-specific posttranslational processing" of pro-proteins. By contrast to PC1 and PC2, the reported distribution of furin suggests a rather ubiquitous expression in most tissues[36,38,39] and cell lines[5,35] analyzed, befitting its proposed function as a convertase of constitutively secreted precursors. However, we have reported that furin, not PC1 or PC2, is the processing enzyme of the regulated precursor pro-7B2.[40] This is one example in which the role of furin also includes the processing of some precursors, of which the products are concentrated within dense core secretory granules and are secreted via the regulated pathway.

The recent discovery of PC4[13,14] and PC5[7,8] led to extensive studies of their tissue distribution and gene regulation both by Northern blot analysis and by in situ hybridization. Tissue distribution studies demonstrated that the expression of PC4 is restricted to germ cells of the testis within round spermatids and pachytene spermatocytes, but not in the elongating spermatids, and no detectable levels are observed in Leydig cells, peritubular cells, and Sertoli cells stimulated with cAMP. In the female, we also observed the expression of PC4 only within the ovary and that this expression is up-regulated on activation of ovulation. To establish the site of PC4 mRNA synthesis in the rat testis, tissue sections were analyzed by in situ hybridization. The data showed that PC4 mRNA was detected in all the tubules, although at different levels of expression. The tubular distribution pattern of PC4 is consistent with germ cells as the primary site of expression of PC4, and we also noted that the cellular expression pattern of PC4 matches that of proenkephalin in the spermatocytes and round spermatids. From the ontogeny studies in the rat testis, it was deduced that PC4 mRNA could not be detected from postnatal days 1 to 19 and that the expression could only be weakly detected at day 22. Thereafter, we observed increasing levels of expression up to postnatal day 60.[14]

PC5 is widely distributed over many tissues, with a high abundance in the gut, adrenal glands, ovaries, and lungs. Within the gut, the duodenum, jejunum, and ileum are the richest sources and PC5 transcripts were localized mainly within the epithelial cells of the small intestine. We also detected rat PC5 mostly in the anterior lobe of the pituitary and in the thyroid. In general, within the central nervous system the levels of PC5 are lower than those observed for PC1 and PC2.[7] Within the reproductive

[38] K. Hatsuzawa, M. Hosaka, T. Nakagawa, M. Nagase, A. Shoda, K. Murakami, and K. Nakayama, J. Biol. Chem. 265, 22075 (1991).

[39] R. Day. M. K-H. Schäfer, W. E. Cullinan, S. J. Watson, M. Chrétien, and N. G. Seidah, Neurosci. Lett. 149, 27 (1993).

[40] L. Paquet, F. Bergeron, A. Boudreault, N. G. Seidah, M. Chrétien, M. Mbikay, and C. Lazure, J. Biol. Chem., in press (1994).

organs the levels of PC5 are more elevated in the female (ovaries) than in the male (testis). From the other tissues examined, we deduced that PC5 is also expressed mainly in the adrenal cortex, oviduct, heart atria and ventricles, and esophagus, with very low levels in seminal vesicles, kidney, thymus, spleen, muscle, submaxillary gland, and pancreas.

We have also examined the distribution of PC5 within a number of cell lines, including AtT-20, GH3, βTC3, BSC40, and Y1 cells. In general, we find low levels of PC5 in most of these cells, except for the Y1 cells, where important amounts of PC5 mRNA are detected, especially after stimulation with adrenocorticotropic hormone (ACTH), suggesting a regulation by cAMP. Within the central nervous system, *in situ* hybridization of rat brain tissues demonstrated a unique distribution of PC5 as compared to PC1, PC2, and furin, with an abundance of PC5 within the CA3 region of the hippocampus, with lower levels in the dentate gyrus, and important amounts within the amygdaloid nucleus.[7]

The major mRNA form of rat PC5 observed migrated with an apparent molecular size of 3.8 kb. Minor forms of approximate sizes 6.5 and 7.5 kb were also observed, the ratio of which varies in a tissue-specific manner. For example, in the gut we observed mainly the 7.5-kb form and in the lung the 6.5-kb form, whereas both forms are equally expressed in the adrenal gland.[7] It was reported[41] that the 7.5-kb form codes for a C-terminally extended membrane-bound form of PC5 (called PC6-B) in which the Cys-rich domain contains 22 copies (instead of 5 in PC5/PC6) of the 40-amino acid Cys-rich motif mentioned in the previous section on comparative protein structure. It was predicted[41] that the PC5/PC6-B form is composed of about 1877 amino acids, representing the longest mammalian convertase known. It is therefore also possible that other forms of PC5 will be described, which probably arise from differential splicing. In our laboratory, we also identified a new form of PC5, in which the exon coding for the catalytically important Asn residue is skipped (D. Vieau, S. Benjannet, M. Chrétien, and N. G. Seidah, (in preparation) (1994)). The catalytic activity of such differentially spliced forms has not yet been demonstrated.

Genes Coding for PC1, PC2, Furin, and PC4

The genes expressing PC1, PC2, furin,[10,42,43] and PC4[14] are on human chromosomes 5, 20, 15, and 21, respectively. In contrast, the gene coding for PACE4 was found within 5 megabases from the furin gene on human

[41] T. Nakagawa, K. Murakami, and K. Nakayama, *FEBS Lett.* **327**, 165 (1993).
[42] N. G. Seidah, M. G. Mattei, L. Gaspar, S. Benjannet, M. Mbikay, and M. Chrétien, *Genomics* **11**, 103 (1991).
[43] N. G. Copeland, D. J. Gilbert, M. Chrétien, N. G. Seidah, and N. A. Jenkins, *Genomics* **13**, 1356 (1992).

chromosome 15.[6] The chromosomal localization of PC5 is not yet determined. The estimated size of the mouse *PC1* gene of 42 kb[44] is intermediate between that of human *fur* (10 kb)[45] or mouse *PC4* (9.5 kb)[24] and that of either human *PC2* (>135 kb),[46] or mouse *PC2* (>140 kb; M. L. Raffin-Sanson, N. G. Seidah, M. Chrétien, and M. Mbikay, unpublished results). Although the *hfur* (15 introns), *mPC4* (14 introns), and *mPC1* (14 introns) genes carry 3 to 4 more introns as compared to *hPC2* (11 introns), the latter gene is 4- to 15-fold longer than the former genes. The main difference lies in the size of some of the introns found in *hPC2,* five of which are bigger than 13 kb.[46]

Within the coding sequence up to the RGD domain, the intron/exon organization of the *PC1, PC2, fur,* and *PC4* genes is very similar. This conservation is most remarkable in the region of high sequence homology, which constitutes the N-terminal two-thirds of the molecules, reinforcing the view that these genes either evolved from a common ancestral sequence or one from another by deletions, insertions, and other mutations around essential functional domains. Structural variations among these genes are clustered at the 3' end and may be associated with distinctive appendages that confer particular properties to each enzyme, such as its intracellular localization and/or its catabolism. Thus, the last exon of the furin gene that encodes the Cys-rich as well as the transmembrane-spanning domain characteristic of this enzyme exhibits no significant homology with the corresponding exons of the other genes. The 3' divergence is even observed between *PC1* and *PC2* genes, both of which are expressed in the neuroendocrine system.

Analysis of the distinct proximal promoter segments of the *PC1, PC2, PC4,* and *fur* genes demonstrated that all lack a functional TATA box, and primer extension analysis revealed the presence of multiple transcription start sites for each of these genes. A similar observation was also made for the gene coding for carboxypeptidase E,[47] suggesting that possibly all the enzymes involved in the posttranslational cleavage and processing of pro-proteins may be TATA-less genes that do not exhibit a strict control on the transcription initiation site. This added degree of flexibility may be important for the expression of these genes in widely diverse types of tissues and cells.

[44] N. Ftouhi, R. Day, M. Mbikay, M. Chrétien, and N. G. Seidah, *DNA Cell Biol.* **13,** 395 (1994).
[45] A. M. W. Van den Ouweland, J. J. M. Van Groningen, A. J. M. Roebroek, C. Onnekink, and W. J. M. Van de Ven, *Nucleic Acids Res.* **17,** 7101 (1989).
[46] S. Ohagi, J. LaMendola, M. M. LeBeau, R. Espinosa, J. Takeda, S. P. Smeekens, S. J. Chan, and D. F. Steiner, *Proc. Natl. Acad. Sci. U.S.A.* **89,** 4977 (1992).
[47] Y.-K. Jung, C. J. Kunczt, R. K. Pearson, J. E. Dixon, and L. D. Fricker, *Mol. Endocrinol.* **5,** 1257 (1991).

From the deduced PC4[14] and PC5/PC6[41] cDNA and gene[24] structures, it was predicted that by differential splicing multiple forms of these enzymes could be generated. In the case of furin, PC1, and PC2, no differential splicing has yet been observed. However, from our recent data on the gene structure of PC1,[44] it also became evident that the 3- and 5-kb mRNA forms of PC1 that arise by alternate choices of polyadenylation sites are differentially regulated by dopamine. These results reinforce the notion that the multiple forms of the mRNAs and/or proteins observed for the convertases increase the diversity of these enzymes and allow for a more flexible regulation of expression.

Conclusions

Over 25 years elapsed since the discovery that prohormones are activated at pairs of basic residues. The enzymes involved in the posttranslational processing of protein and hormonal precursors were only recently identified and molecularly characterized. The cDNA cloning of the mRNA coding for the six known convertases—PC1/PC3, PC2, furin, PACE4, PC4, and PC5/PC6—as well as the carboxypeptidase E and amidation enzyme PAM, opened new avenues in our detailed understanding of the mechanism of action and specificity of these enzymes. Even though other PC-like enzymes may yet be discovered, the general principles that govern the zymogen activation, proteolytic specificity, gene expression, and regulation and cellular localization of the convertases are being worked out in detail in a number of laboratories. Aside from prohormones and proenzymes, other potential substrates for the PC-like enzymes are growth factors, some of their receptors, and viral envelope glycoproteins. The unequivocal identification of the cognate processing enzyme(s) for each of these precursors is a challenge that still has to be met. Specific inhibition of an endogenous convertase (i.e., with antisense RNA) in a cell line that produces a given processible precursor can help in this determination. From the perspective of evolution, it appears to be a conservative path for an organism to preserve the same cleavage motifs in its large panoply of precursors for signaling peptides, so that a limited family of proteases placed at key positions within the cell could ensure their processing. The basic structure of these enzymes is preserved while diversity is created by subtle variations in signal sequence requirements, tissue distribution, and subcellular localization. Given the role of hormones and growth factors in uncontrolled cell proliferation, if their precursors are proved to be natural substrates for the convertases, then one can envision inhibition of the production or activity of these enzymes as a possible means to arrest this proliferation. A probable advantage of targeting these enzymes for inactivation lies in the fact that a variety of substrates can be adversely affected at the same time, thus amplifying the antiproliferative effect. This is an avenue for investigation that the discovery of the convertases has opened. It might well be possible that the posttranslational proteolysis of progrowth factors and viral envelope pro-proteins by convertases will become as important and as general in cancerology and virology as it has turned out to be in the study of neuronal and endocrine peptidergic systems.

[14] Prolyl Oligopeptidases

By László Polgár

Introduction

Prolyl oligopeptidase (EC 3.4.21.26; also called prolyl endopeptidase, post-proline cleaving enzyme) degrades a variety of proline-containing peptides by cleaving the peptide bond at the carboxy side of proline residues. It was discovered by Walter et al.[1] in human uterus as an oxytocin-degrading enzyme. Prolyl oligopeptidase is a cytosolic serine peptidase that may be involved in the metabolism of peptide hormones and neuropeptides.[2–4] It is a representative of a new family of serine peptidases,[5,6] not related to the long-known trypsin and subtilisin families.

Assay

Principle

Hydrolysis of serine proteases proceeds through the formation of an acyl-enzyme intermediate according to Eq. (1), where E, S, ES, EA, P_1, P_2, K_S, k_2, and k_3 are the free enzyme, the substrate, the enzyme–substrate complex, the acyl-enzyme, the amine product, the acyl product, the dissociation constant of the ES complex, the first-order acylation constant,

[1] R. Walter, H. Shlank, J. D. Glass, I. L. Schwartz, and T. D. Kerényi, *Science* **173**, 827 (1971).
[2] R. Walter, W. H. Simmons, and T. Yoshimoto, *Mol. Cell. Biochem.* **30**, 111 (1980).
[3] S. Wilk, *Life Sci.* **33**, 2149 (1983).
[4] R. Mentlein, *FEBS Lett.* **234**, 251 (1988).
[5] D. Rennex, B. A. Hemmings, J. Hofsteenge, and S. R. Stone, *Biochemistry* **30**, 2195 (1991).
[6] N. D. Rawlings, L. Polgár, and A. J. Barrett, *Biochem. J.* **279**, 907 (1991).

and the deacylation constant, respectively.[7,8] Equation (2) defines the second-order acylation constant (k), which is equal to k_{cat}/K_m.[7,8]

$$E + S \underset{+\,P_1}{\overset{K_S}{\rightleftharpoons}} ES \xrightarrow{k_2} EA \xrightarrow{k_3} E + P_2 \tag{1}$$

$$E + S \xrightarrow{k} EA + P_1 \tag{2}$$

The activity of prolyl oligopeptidase can be measured spectrofluoro-metrically with Z-Gly-Pro-Nap (Z, benzyloxycarbonyl; Nap, 2-naphthyl-amine) substrate. Because fluorescence measurements give arbitrary values, the enzyme is best characterized by the second-order acylation rate constant. This specificity rate constant can be determined under first-order conditions, i.e., at substrate concentrations below the K_m value. The first-order rate constant can be calculated from the complete progress curve, which is, in contrast to the initial rate, not affected by quenching problems. Furthermore, the determination of first-order rates requires less substrate as compared with initial rate measurements. The second-order rate constant is obtained by dividing the first-order constant by the total enzyme concentration present in the reaction mixture. The rate constant so determined is more precise than the k_{cat}/K_m obtained from Line-weaver–Burk plots.

Procedure

The hydrolysis of Z-Gly-Pro-Nap (Bachem Inc.) is continuously monitored at 25° in 50 mM HEPES buffer, pH 8.0, containing 0.5 M NaCl, 1 mM EDTA, and 1 mM dithioerythritol.[9] The excitation and emission wavelengths are 340 and 410 nm, respectively.[10] The excitation and emission slit widths are 1.5 and 5 nm, respectively, using a Jasco FP 777 spectrofluorometer equipped with a jacketed cell holder. The substrate stock solution (3 mg/ml in acetonitrile) is diluted 240-fold with water before use, a 200 μl of diluted solution is added to the reaction mixture of 3.0 ml final volume (1.93 μM substrate). Under this condition the acetonitrile concentration is sufficiently low (0.026%) so that it does not interfere with the activity of prolyl oligopeptidase. The enzyme concentration in the cell is 2–4 nM, which gives a half-life $(t_{1/2})$ of 1–2 min for the

[7] L. Polgár, *New Compr. Biochem.* **16**, 159 (1987).
[8] L. Polgár, "Mechanisms of Protease Action," p. 87. CRC Press, Boca Raton, Florida, 1989.
[9] L. Polgár, *Eur. J. Biochem.* **197**, 441 (1991).
[10] H. Knisatschek, H. Kleinkauf, and K. Bauer, *FEBS Lett.* **111**, 157 (1980).

first-order decrease in the substrate concentration (see section on cis–trans isomeric specificity). In such diluted solutions many proteins tend to denature. This is not the case with prolyl oligopeptidase, which gives the same second-order rate constant at enzyme concentrations two times higher and lower. Addition of 0.01% serum albumin into the reaction mixture, which is known to stabilize proteins, also does not affect the rate constant.

Liberation of naphthylamine from the substrate can also be measured spectrophotometrically.[11] However, this method is less sensitive by two orders of magnitude and is applicable only at higher concentrations of substrate and organic solvent.

The enzyme concentration can be calculated by using a molecular mass of 80,751 Da (from pig brain cDNA)[5] and 78,705 Da (*Flavobacterium meningosepticum*).[12] The latter may have a mature periplasmic form of 76,784 Da.[13] The A_{280} is 1.4 for the 1 mg/ml solution of the *Flavobacterium* enzyme.[14] A somewhat higher value, 1.64, is obtained for the pig enzyme when calculated from its amino acid sequence.

Other Substrates

The 7-(4-methylcoumaryl)amide of Z-Gly-Pro (Z-Gly-Pro-MCA) is also a good substrate of prolyl oligopeptidase.[15] It is also a fluorescent substrate, more sensitive than the naphthylamine derivative, but more expensive, and displays a somewhat lower rate constant.[16] An even lower rate constant is obtained with Z-Gly-Pro-4-nitroanilide.[16] An advantage of this substrate is that its hydrolysis can be monitored spectrophotometrically at 410 nm, but this method is less sensitive than the fluorescence measurements.

When the succinyl group is substituted for the benzyloxycarbonyl group, the substrates become much more soluble, and do not require organic solvent even at high concentration. Therefore, such substrates are often used for kinetic measurements,[17,18] but these compounds are

[11] F. Willenbrock and K. Brocklehurst, *Biochem. J.* **222**, 805 (1984).
[12] T. Yoshimoto, A. Kanatani, T. Shimoda, T. Inaoka, T. Kokubo, and D. Tsuru, *J. Biochem. (Tokyo)* **110**, 873 (1991).
[13] S. Chevallier, P. Goeltz, P. Thibault, D. Banville, and J. Gagnon, *J. Biol. Chem.* **267**, 8192 (1992).
[14] T. Yoshimoto, M. Ando, K. Ohta, K. Kawahara, and D. Tsuru, *Agric. Biol. Chem.* **46**, 2157 (1982).
[15] T. Yoshimoto, K. Ogita, R. Walter, M. Koida, and D. Tsuru, *Biochim. Biophys. Acta* **569**, 184 (1979).
[16] L. Polgár, *Biochem. J.* **283**, 647 (1992).
[17] J. Heins, P. Welker, C. Schönlein, I. Born, B. Hartrodt, K. Neubert, D. Tsuru, and A. Barth, *Biochim. Biophys. Acta* **954**, 161 (1988).
[18] A. Moriyama, M. Nakanishi, O. Takenaka, and M. Sasaki, *Biochim. Biophys. Acta* **956**, 151 (1988).

poor substrates, their specificity rate constants being lower by two orders of magnitude than those of the benzyloxycarbonyl derivatives.[19]

Enzyme Purification

Enzyme Extraction

Fresh or frozen pig muscle (750 g), free of fat and connective tissues, is homogenized in cold 20 mM phosphate buffer, pH 6.5, containing 1 mM EDTA and 5 mM Na$_2$SO$_3$ (1.5 liters).[9,19] The homogenization is carried out in appropriate portions in a kitchen-type blender for 3 × 10 sec, allowing a 10-sec cooling period between runs. The temperature of the mixture remains below 4°, even though the procedure is performed at room temperature. The homogenate is kept at room temperature for an additional 15 min and is stirred with a glass rod every 2–3 min. It is then centrifuged on a Beckman J2-21 centrifuge at 3000 g (JA 10 rotor), at 2° for 25 min. The supernatant is filtered through a cotton layer in a glass funnel to remove some frozen fat particles.

Acetone Precipitation

The supernatant (1.6 liters) is treated in an ice bath with doubly-distilled acetone precooled to −18°. To 1 liter of supernatant, 470 ml of organic solvent is added, resulting in a final concentration of 32%. The solvent is allowed to flow slowly from a separating funnel into the supernatant (15 min) while stirring and keeping the temperature below 4° with an ice bath. The precipitate is removed by centrifugation in closed polypropylene tubes at 3000 g, at −2° for 20 min. The precipitate, about 5% by volume, is discarded, and the supernatant is treated with ammonium sulfate.

Ammonium Sulfate Precipitation

To the supernatant (2.2 liters) containing the enzyme in a solution of 32% acetone, solid ammonium sulfate is added (235 g/liter) for 30 min. The mixture is mechanically stirred in an ice bath during the addition of the ammonium sulfate, and for a further 15 min thereafter. A negligible amount of crystals may not dissolve during that time. The mixture is centrifuged at 3000 g, at 0° for 20 min. The precipitated proteins accumulate at the interphase of water and acetone. The liquid phases are completely decanted from the protein cake, which is extracted with 100 ml 20 mM phosphate buffer, pH 6.5, containing 1 mM EDTA and 5 mM Na$_2$SO$_3$.

[19] L. Polgár, *Biochemistry* **31,** 7729 (1992).

After centrifugation at 10,000 g (JA 14 rotor), at 4° for 20 min, the extraction is repeated with 50 ml of buffer.

Gel Filtration

The combined supernatants (170 ml) are loaded onto a Sephadex G-50 column (8 × 50 cm) equilibrated with 20 mM phosphate buffer, pH 6.5, containing 1 mM EDTA and 5 mM Na$_2$SO$_3$. Along with the ammonium sulfate and acetone some colored compounds are also separated from the prolyl oligopeptidase during desalting.

Chromatography on DEAE-Cellulose

The combined fractions (350 ml) from the previous step are applied to a Whatman (Clifton, NJ) DE-32 cellulose column (2 × 16 cm) that has previously been washed with 20 mM phosphate buffer, pH 6.5, containing 1 mM EDTA and 1 mM dithioerythritol (standard buffer). The flow rate is 150 ml/hr. The column is washed with about 100 ml of standard buffer, while the absorbance at 280 nm diminishes to a value close to zero. The enzyme is eluted with a linear NaCl gradient (0–0.25 M) in standard buffer (350 plus 350 ml). Fractions of 8 ml are collected at a flow rate of 24 ml/hr. The fractions that contain the active enzyme are combined (40–48 ml), concentrated with an Amicon (Danvers, MA) PM30 membrane, diluted about 10-fold with 20 mM phosphate buffer, pH 7.2, containing 1 mM EDTA and 1 mM dithioerythritol, and concentrated again.

Blue-Sepharose Chromatography

Two preparations, corresponding to 2 × 750 g muscle, are combined (1–2 ml) and loaded on a Dyematrex Blue A column (Pierce Chemical Co., Rockford, IL 1 × 13 cm), which has previously been washed with 20 mM phosphate buffer, pH 7.2, containing 1 mM EDTA and 1 mM dithioerythritol. The chromatography gives a higher yield of prolyl oligopeptidase at pH 7.2 than at the pH of the standard buffer (6.5). The enzyme is eluted with the equilibration buffer at a flow rate of 24 ml/hr. Its activity is found in the breakthrough protein peak, representing 5–10% of the total absorbance applied to the column. A total of 80–85% of the protein is eluted with 1.5 M NaCl in the equilibration buffer as an inactive peak. The rest is eluted wth 8 M urea when the column is regenerated. The enzyme is concentrated by ultrafiltration and stored in the presence of 40% (v/v) ethylene glycol at −18° until the final purification step (FPLC, fast protein liquid chromatography).

FPLC Method

Prior to chromatography on the Mono Q column, the concentration of ethylene glycol in the enzyme solution is reduced by a 10-fold dilution with the standard buffer. The solution is concentrated with an Amicon PM30 membrane and applied to a Mono Q HR 5/5 column (Pharmacia, Piscataway, NJ) at room temperature. In one run 1–2 mg of protein is used. The enzyme is eluted with an NaCl gradient, using the standard buffer (solvent A) and 0.5 M NaCl in the standard buffer (solvent B). The gradient, from 18 to 35% solvent B, is developed with a flow rate of 0.5 ml/min during 22 min. The active peak elutes at about 24% solvent B, and contains the enzyme at approximately 1 mg/ml concentration. The largest contamination, which travels slightly before prolyl oligopeptidase during sodium dodecyl sulfate (SDS)–polyacrylamide gel electrophoresis, is eluted from the Mono Q column by washing with 100% solvent B. A total of 2.0–2.5 mg of prolyl oligopeptidase is obtained from 1.5 kg of muscle. The enzyme is stored in the presence of 40% ethylene glycol at $-18°$.

Comments on Procedure

The method of preparation has been reproduced many times in the author's laboratory. The only critical step is the acetone treatment, which may unfavorably affect the yield of the enzyme. Preferably freshly distilled solvent is used, and the temperature of the enzyme solution should be carefully controlled during and after the treatment. The liquid phases from the protein cake obtained after the ammonium sulfate precipitation should be removed as completely as possible. The extract of the cake should immediately be gel filtered in order to remove the remaining acetone. Indeed, the activity in the extract before gel filtration is always lower than that found after the removal of the acetone. Uncertain data related to this extract are not included in Table I, which summarizes the purification procedure. The total amount of activity can be expressed in arbitrary units,[9] or by multiplication of the first-order rate constant with the total volume of the enzyme. The activity measured in the supernatant of homogenate is taken as 100%.

Other Sources

Prolyl oligopeptidase is widely distributed in various tissues.[3] It has been isolated from several organs (kidney, brain, liver, muscle) of different mammalian species (rat, rabbit, lamb, cow, pig),[3] as well as from *F.*

TABLE I
PURIFICATION OF PROLYL OLIGOPEPTIDASE[a]

Purification step	Volume (ml)	Total A_{280}	Total activity (%)	Rate constant ($\text{m}M^{-1}\ \text{sec}^{-1}$)
Homogenate	3200	95,000	100	0.36
Gel filtration	650	8900	78	2.5
DEAE chromatography	100	280	67	58
Blue-Sepharose + concentrate	1.6	11	54	1200
FPLC	15	3.4^b	38	4700

[a] From 1.5 kg of pig muscle.
[b] The total yield corresponds to 3.4/1.64 = 2.07 mg protein.

meningosepticum,[13,20] mushroom,[21,22] and carrot.[23] With the exception of the *Flavobacterium* enzyme (pI 9.6), prolyl oligopeptidases are slightly acidic proteins (pI 5).

Effects of Ionic Strength

In contrast to a number of extensively studied serine endopeptidases, such as chymotrypsin and subtilisin, prolyl oligopeptidase displays a strong dependence on ionic strength of the reaction mixture. The increase of the second-order rate constant is approximately linear up to 0.3–0.5 M NaCl, and the activity levels off at higher salt concentrations.[9] KCl and NaNO$_3$ exert similar effects. The rate increase depends on the pH and the nature of the substrate. In the presence of 0.5 M NaCl, the enhancement is about 2.5-fold with Z-Gly-Pro-Nap at pH 8.0.

Effects of Organic Solvents

Uncharged amino acid or peptide derivatives are often sparingly soluble in water, and, therefore, their hydrolysis is usually carried out in the presence of organic solvents. Thus, the activity of prolyl oligopeptidase

[20] T. Yoshimoto, R. Walter, and D. Tsuru, *J. Biol. Chem.* **255,** 4786 (1980).
[21] T. Yoshimoto, A. K. M. A. Sattar, W. Hirose, and D. Tsuru, *J. Biochem.* (*Tokyo*) **104,** 622 (1988).
[22] A. K. M. A. Sattar, N. Yamamoto, T. Yoshimoto, and D. Tsuru, *J. Biochem.* (*Tokyo*) **107,** 256 (1990).
[23] T. Yoshimoto, A. K. M. A. Sattar, W. Hirose, and D. Tsuru, *Biochim. Biophys. Acta* **916,** 29 (1987).

TABLE II
EFFECTS OF SOLVENTS ON ACTIVITY OF PROLYL OLIGOPEPTIDASE[a]

Solvent %	Activity (%)		
	Acetonitrile	Dimethylformamide	Dioxane
0[b]	100	100	100
0.1	91	78	95
0.2	86	62	93
0.5	70	41	88
1.0	51	21	72
5.0	—	—	20
10.0	—	—	6

[a] Measured with Z-Gly-Pro-Nap substrate in 50 mM HEPES buffer, pH 8.0, containing 0.5 M NaCl, 1 mM EDTA, and 1 mM dithioerythritol.
[b] Calculated by extrapolation (see text).

is sometimes measured at dioxane concentrations as high as 15–20%.[24] Prolyl oligopeptidase, however, is remarkably sensitive to organic solvents, as seen from Table II, which shows that even 1% acetonitrile inhibits the enzyme by about 50%. The effect is greater with dimethylformamide, but less with dioxane. However, in the presence of 10% dioxane the remaining activity is only 6%. Although dioxane shows less inhibitory effects at low concentrations compared with acetonitrile, the latter may be preferred as solvent because impurities in commercial dioxane often impair the enzyme. The detrimental effect of dioxane has also been noted in a study of the prolyl oligopeptidase from *F. meningosepticum*.[17] A plot of the rate constants against the acetonitrile concentrations and extrapolation of the data to zero solvent concentration show that the inhibition under the standard assay conditions (0.028% acetonitrile) is less than 3%.[19]

Cis–Trans Isomeric Specificity

It is known that the peptide bond at the N-terminal side of a proline residue exists as a mixture of cis and trans isomers.[8] For example, nuclear magnetic resonance (NMR) studies have demonstrated that the tetrapeptide Gly-Gly-Pro-Ala contains 20% cis isomer in D_2O at 25°.[25] It has also been shown that prolyl oligopeptidase isolated from *F. meningosepticum* hydrolyzes only the substrate when the Gly-Pro bond adopts the trans

[24] T. Yoshimoto, M. Fischl, R. C. Orlowski, and R. Walter, *J. Biol. Chem.* **253,** 3708 (1978).
[25] C. Grathwohl and K. Wütrich, *Biopolymers* **20,** 2623 (1981).

form.[26,27] We have confirmed this result with the pig muscle enzyme under the standard assay conditions. Specifically, when the substrate hydrolysis has been performed at high enzyme concentration (>0.1 μM), two well-separated kinetic phases can be observed. The first phase of the reaction is fast and dependent on the enzyme concentration; the second phase is relatively slow and independent of enzyme concentration. The semilogarithmic plot of the substrate concentration against time has indicated that both the fast and the slow reactions can be treated as first-order processes. Apparently, the fast phase represents the direct hydrolysis of the substrate possessing the Gly-Pro bond in the trans conformation, whereas the slow phase is determined by the cis–trans isomerization of those substrate molecules that initially have the Gly-Pro bond in cis form. This implies that the cis form must isomerize to the trans form before cleavage can occur.

The first-order rate constant of isomerization of Z-Gly-Pro-Nap was found to be 0.053 sec^{-1} ($t_{1/2} = 13$ sec), a value comparable to the data obtained for the related substrate Z-Gly-Pro-MCA (0.030–0.038 sec^{-1} at $23.2°$).[26] Extrapolation of the slow-phase data points of the semilogarithmic plot to zero time gave 24% cis isomer content for Z-Gly-Pro-Nap substrate. A similar value was obtained with Z-Gly-Pro-MCA.[26]

The effects of cis–trans isomerization should always be considered in kinetic studies. To obtain reliable rate constants, the first-order enzyme reaction must be much slower ($t_{1/2} > 1$ min) than the cis–trans isomerization, as proposed in the assay section.

Specificity

Prolyl oligopeptidase preferentially cleaves the peptide bond that contains a proline residue in the P1 position. It also cleaves at alanine residues,[3,18,20,27–29] as well as at the N-methyl derivatives of alanine and glycine,[29] but the rate of hydrolysis is usually much slower in these cases.

For hydrolysis to occur, it is essential that the S2 subsite of prolyl oligopeptidase be occupied by an amino acid of L configuration.[2,17] The interaction between the P2 residue of the substrate and the S2 subsite of the enzyme involves the formation of a hydrogen bond, as demonstrated by using a thiono substrate Z-Gly[CS-NH]Pro-Nap.[30] Specifically, hydrolysis by prolyl oligopeptidase is not detectable with the thiono substrate, and this indicates that the specificity rate constant is at least five orders

[26] L.-N. Lin and J. F. Brandts, *Biochemistry* **22**, 4480 (1983).
[27] G. Fischer, J. Heins, and A. Barth, *Biochim. Biophys. Acta* **742**, 452 (1963).
[28] A. Moriyama and M. Sasaki, *J. Biochem. (Tokyo)* **94**, 1387 (1983).
[29] K. Nomura, *FEBS Lett.* **209**, 235 (1986).
[30] L. Polgár, *FEBS Lett.* **322**, 227 (1993).

of magnitude lower with the thiono than with the corresponding oxo substrate.

Prolyl oligopeptidase is assumed to have five binding subsites from S3 to S2'.[2] However, elongation of the classic dipeptide Z-Gly-Pro-Nap substrate with 1–3 residues decreases the specificity rate constants, a major difference from the finding with the well-characterized serine endopeptidases, chymotrypsin and subtilisin.[19] Insertion of a charged residue into the substrate, such as lysine or aspartic acid, considerably affects the rate constants, which are higher with the positively charged substrates and lower with the peptides bearing a negative charge.[19] This can be interpreted in terms of a negatively charged active site, which exerts electrostatic attraction or repulsion toward charged substrates.

Prolyl oligopeptidase does not hydrolyze large proteins even if they are denatured.[2,3] For example, the hemoglobin β-chain, and its two CNBr peptides with M_r of 10,200 and 6400, respectively, are not hydrolyzed, whereas their tryptic peptides with M_r of less than 3000 can be cleaved under the same conditions.[18]

Inhibitors

Like most serine peptidases, prolyl oligopeptidase can readily be inhibited with diisopropyl fluorophosphate, but phenylmethylsulfonyl fluoride, another potent inhibitor of chymotrypsin and subtilisin, does not prevent the catalytic activity.[2] The enzyme is also not inhibited by 3,4-dichloroisocoumarin,[30a] or by pancreatic or soybean trypsin inhibitor, or by the ovomucoid from chicken egg white.[2] It is, however, inhibited by chloromethyl ketones, such as Z-Gly-pro-CH$_2$Cl.[2,3] Chloromethyl ketones are known to alkylate the catalytic histidine residues of serine peptidases.[7,8]

Aldehyde analogs of peptide substrates are potent inhibitors of serine peptidases. The aldehyde interacts with the active site serine to form a tetrahedral transition-state analog.[7,8] Such a peptide aldehyde, Z-Pro-prolinal, proved to be an effective inhibitor of prolyl oligopeptidase with a K_i of 14 nM, which was three orders of magnitude lower than the K_i of the corresponding acid or alcohol.[3,31] It was also found that introduction of a sulfur atom into the pyrrolidine ring effectively increased the inhibitory activity. Thus, Z-thioprolylthiazolidine and Z-thioprolylthioprolinal showed K_i values of 0.36 and 0.01 nM, respectively.[32] The K_i for Z-Pro-prolinal was 3.7 nM in this study.

[30a] S. Kalwant and A. G. Porter, *Biochem. J.* **276**, 237 (1991).
[31] S. Wilk and M. Orlowski, *J. Neurochem.* **41**, 69 (1983).
[32] D. Tsuru, T. Yoshimoto, N. Koriyama, and S. Furukawa, *J. Biochem.* (*Tokyo*) **104**, 580 (1988).

A special case of inhibition of prolyl oligopeptidase concerns its reactions with thiol reagents. It appears that one of the cysteine residues of the enzyme is close enough to the active site so that a large reagent attached to this thiol group excludes the substrate from the catalytic entity, whereas a small one only exerts an incomplete steric hindrance. Thus, p-chloromercuribenzoate prevents the catalysis almost completely and instantaneously; N-ethylmaleimide causes about 85% inhibition; and the small iodoacetamide inhibits the activity only 45–50%.[9] It should be emphasized, however, that the prolyl olgiopeptidases isolated from *F. meningosepticum*[20] and a mushroom (*Lyophyllum cinerascens*) enzyme[21] are not inhibited with thiol reagents, apparently because they do not have cysteine residues at or near the active sites.

Mechanism of Action

Proline, as an imino rather than an amino acid, does not possess the free hydrogen atom that is essential to the catalysis by chymotrypsin and subtilisin for forming an S1P1 hydrogen bond with a backbone carbonyl oxygen.[7,8] Indeed, these serine peptidases do not hydrolyze peptides at prolyl bonds. This suggests that the active site structure of prolyl oligopeptidase may display a unique feature and somewhat altered mechanism compared with the well-characterized serine peptidases.

pH Dependence Studies

The formation and hydrolysis of the intermediate acyl enzyme in serine peptidase catalysis are promoted by a histidine residue, which operates as a general base/acid catalyst and exhibits a pK_a of about 7.[7,8] The ionization of this residue governs the pH dependence of the catalysis, which conforms to a simple dissociation curve. (Chymotrypsin reactions exhibit bell-shaped curves because the enzyme is inactivated with increasing pH due to the dissociation of a proton from the α-ammonium group of Ile-16 forming a salt-bridge with Asp-194.[7,8]) The acylation of prolyl oligopeptidase with Z-Gly-Pro-Nap is more complicated. It exhibits a doubly sigmoidal curve, which tends to decrease above pH 9. This implies a mechanism involving two enzyme forms of different activities, which interconvert with changing pH, as indicated by Eq. (3), where X is an unknown ionizing group whose dissociation perturbs the pK_a of the imidazole group of the catalytic histidine residue.[9]

$$EH(ImH)XH \rightleftharpoons EH(Im)XH \rightleftharpoons EH(Im)X \rightleftharpoons E(Im)X \qquad (3)$$
$$\text{inactive} \qquad\quad \text{active} \qquad\quad \text{active} \qquad\quad \text{inactive}$$

The pH–rate profiles are markedly dependent on the substrate structure and ionic strength.[19] The pH dependencies of the rate constants with longer neutral substrates, such as Z-Ala-Ala-Gln-Gly-Pro-Nap, exhibit roughly bell-shaped curves. The reactions with charged substrates markedly deviate from the bell-shaped pH–rate profile and clearly demonstrate the existence of two active enzyme forms as indicated by Eq. (3). Thus, the physiologically competent high-pH form prefers positively charged substrates [Z-Lys-Pro-2-(4-methoxy)naphthylamide, Z-Ala-Lys-Gln-Gly-Pro-Nap], whereas the low-pH form reacts faster with the negatively charged substrate (Z-Asp-Gly-Pro-Nap). The results can be interpreted in terms of a negatively charged active site that exists at high pH and exerts electrostatic attraction or repulsion toward charged substrates. These electrostatic effects, as expected, are influenced by ionic strength in a pH-dependent way.[19]

Deuterium Isotope Effects

The two active forms of prolyl oligopeptidase discussed in the preceding paragraph have different mechanistic features, as demonstrated by kinetic deuterium isotope measurements.[9] Such studies on chymotrypsin and other serine peptidases have shown that both the acylation and deacylation reactions proceed slower in heavy water than in water by a factor of 2–3, indicating that general acid/base catalysis by the active site histidine is rate limiting in both steps.[7,8] However, in the hydrolysis of Z-Gly-Pro-Nap by prolyl oligopeptidase, only the low-pH form of the enzyme displays a significant kinetic deuterium isotope effect; the physiologically competent high-pH form has virtually no kinetic isotope effect. These results have suggested that a general base–acid-catalyzed acylation step is partially rate-limiting in the lower pH range and that an isotopically silent step, probably a conformational change preceding acylation, dominates the reaction in the physiological pH range.[9] This result has been confirmed by a study showing that acylations of prolyl oligopeptidase with the nitrophenyl ester and several amide derivatives of Z-Gly-Pro proceed at similar rates.[16] In contrast, in the chymotrypsin and subtilisin catalyses, wherein the chemical step, rather than a conformational change, is rate limiting, the activated nitrophenyl ester is hydrolyzed several orders of magnitude faster than the corresponding amide substrates.

Because the hydrolysis of the specific Z-Gly-Pro-Nap is very fast, the rate-limiting conformational change in the free enzyme must also be a rapid process. Consequently, the conformational change cannot be rate limiting with a slow substrate if the enzyme isomerization is independent of the substrate binding. In this case the chemical step, i.e., the general base-catalyzed acyl-enzyme formation, should be rate limiting. On the other hand, if the physical step remains rate limiting with the slow substrate, general base catalysis may not be observed kinetically. This mechanism is illustrated by Eq. (4), where E' is the catalytically active form of E, and the transformation of ES into E'S is the rate-limiting process.

$$E + S \rightleftharpoons ES \rightleftharpoons E'S \rightarrow E'A + P_1 \qquad (4)$$

The mechanism represented by Eq. (4) has been confirmed by using the very poor thiono substrate, Z-Gly-Pro[CS-NH]Nap.[30] The lack of deuterium isotope effects with this compound supports the view that a substrate-induced conformational change is the rate-limiting step for both the specific and nonspecific substrates.

Structural Aspects

Prolyl oligopeptidase is a single-chain protein having a molecular mass about three times that of chymotrypsin or subtilisin. This large molecule contains a peptidase domain at the C-terminal region of the polypeptide chain. The active site serine[5] and histidine[33] were identified as Ser-554 and His-680, respectively, in the pig brain enzyme. This order of the catalytic residues is the reverse of that found with the trypsin and subtilisin amino acid sequences, but corresponds to some lipase sequences. A structural relationship between lipases and peptidases of the prolyl oligopeptidase family is supported by the observation of short regions of similarity around the catalytic residues.[34] Comparison of the sequences of lipases and oligopeptidases suggests that the third member of the catalytic triad is Asp-642,[34] although Asp-529 as another candidate has also been considered.[35,36]

The peptidase domain of prolyl oligopeptidase has not yet been separated from the rest of the molecule. Limited proteolysis by trypsin cleaved the Lys[196]-Ser[197] bond, resulting in an enhanced catalytic activity of the low-pH form of the enzyme.[37] The two fragments of the nicked molecule did not separate during size exclusion chromatography under nondenaturing conditions. The noncatalytic portion distinguishes prolyl oligopeptidase from simple digestive enzymes and strengthens the idea that it functions as a regulatory enzyme.

[33] S. R. Stone, D. Rennex, P. Wikstrom, E. Shaw, and J. Hofsteenge, *Biochem. J.* **276**, 837 (1991).
[34] L. Polgár, *FEBS Lett.* **311**, 281 (1992).
[35] A. J. Barrett and N. D. Rawlings, *Biol. Chem. Hoppe-Seyler* **373**, 353 (1992).
[36] N. D. Rawlings and A. J. Barrett, *Biochem. J.* **290**, 205 (1993).
[37] L. Polgár and A. Patthy, *Biochemistry* **31**, 10796 (1992).

[15] Oligopeptidase B: Protease II from *Escherichia coli*

By DAISUKE TSURU and TADASHI YOSHIMOTO

In 1975, a trypsinlike enzyme was isolated from cells of *Escherichia coli* by Pacaud and Richaud[1] and was designated *E. coli* protease II. It catalyzes hydrolysis of synthetic substrates and oligopeptides exclusively at the carboxyl side of basic amino acid residues, but is only slightly active toward high molecular mass peptides. The enzyme is inhibited by diisopropyl fluorophosphate (DFP), tosyl-L-lysyl chloromethane (TLCK), and antipain, but is insensitive to protein inhibitors of trypsin from soybean, lima bean, egg white, and pancreas.[1,2] This enzyme has now been called oligopeptidase B (EC 3.4.21.83). Similar enzymes have also been found in plant seeds,[3] *Trypanosoma brucei*,[4] and *Rhodococcus erythropolis*,[5] but their physicochemical properties are somewhat different from each other.* These oligopeptidases B seem to be distributed widely in nature, although their physiological role has remained ambiguous.

This chapter mainly deals with oligopeptidase B from *E. coli* (*E. coli* protease II), with special relevance to the structure–activity relationships, in comparison with other related enzymes.

Assay Methods

Method A: Spectrophotometric Assay

Reagents

2 m*M* Benzoyl-L-arginine-β-naphthylamide (Bz-Arg-βNA) dissolved in 20 m*M* Tris-HCl buffer, pH 8.0, containing 40% dioxane, kept at 4°.

[1] M. Pacaud and C. Richaud, *J. Biol. Chem.* **250**, 7771 (1975).
[2] A. Kanatani, T. Masuda, T. Shimoda, F. Misoka, L.-S. Xu, T. Yoshimoto, and D. Tsuru, *J. Biochem.* (*Tokyo*) **110**, 315 (1991).
[3] M. Nishikata, *J. Biochem.* (*Toyko*) **95**, 1169 (1984).
[4] M. J. Kornblatt, G. W. N. Mpimbaza, and J. D. Lonsdale-Eccles, *Arch. Biochem. Biophys.* **293**, 25 (1992).
[5] J. D. Shannon, J. S. Bond, and S. G. Bradley, *Arch. Biochem. Biophys.* **219**, 80 (1982).
* Kato *et al.* have reported *E. coli* protease In, a trypsinlike enzyme, which is insensitive to DFP and inert toward benzoyl-L-arginine-*p*-nitroanilide but is strongly inhibited by aprotinin: M. Kato, T. Irisawa, M. Ohtani, and M. Muramatsu, *Eur. J. Biochem.* **210**, 1007 (1992).

0.1% Fast Garnet GBC solution dissolved in 1 M acetate buffer, pH 4.0, containing 10% Triton X-100: prepare just before use.[2]

To the mixture of 800 μl of 20 mM Tris-HCl buffer, pH 8.0, and 100 μl of 2 mM Bz-Arg-βNA solution, 100 μl of enzyme sample is added at 37°. After 10 min incubation, the reaction is stopped by addition of 500 μl of 0.1% Fast Garnet GBC solution, and the solution is kept at 37° for 5 min. The absorbance of the reaction mixture is measured at 550 nm. One unit of activity is defined as the amount of enzyme that releases 1 μmol of β-naphthylamine per min under the conditions. Bz-Arg-p-nitroanilide and other chromogenic substrates are also used for the activity assay of the oligopeptidase B and related enzymes.[1-5]

Method B: Fluorometric Assay for Rhodococcus erythropolis Oligopeptidase B[5]

Reagents

A stock solution of substrate[6]: 10 mM Bz-Arg-7-amino-4-methylcoumarin (Bz-Arg-NHMec) in dimethyl sulfoxide, kept at 4°: dilute with 0.1% Brij 35 solution just before use.[5]

Stop solution: 0.1 M chloroacetate solution containing 30 mM sodium acetate and 70 mM acetic acid, pH 4.0.

30 mM Tris-HCl buffer, pH 7.1.

The reaction mixture contains 0.7 ml of 30 mM Tris-HCl buffer, 0.3 M NaCl, pH 7.1, 0.25 ml of 20 μM Bz-Arg-NHMec in 0.1% Brij 35, and 0.5 ml of enzyme sample. After incubation for 5 to 10 min at 37°, the reaction is stopped with 1 ml of the stop solution. The fluorescence is measured in a spectrophotofluorometer with an excitation wavelength of 345 nm and emission wavelength of 440 nm. One unit of activity is defined as that quantity releasing 1 nmol of 7-amino-4-methylcoumarin per min under the conditions described.

Screening, Cloning, and Expression of Oligopeptidase B Gene from *Escherichia coli*

Chromosomal DNA of *E. coli*[2] HB101 prepared by the method of Saito-Miura[7] is digested with *Pst*I, inserted into the *Pst*I site of plasmid pBR322, and used to transform *E. coli* DH1 by Hanahan's method.[8] From about 500 transformants grown in LB broth on 96-well assay plates, one

[6] A. J. Barrett, *Biochem. J.* **187**, 909 (1980).
[7] H. Saito and K. Miura, *Biochim. Biophys. Acta* **72**, 1513 (1963).
[8] D. Hanahan, *J. Mol. Biol.* **166**, 557 (1983).

strain that shows a significant activity of the enzyme is selected. The recombinant plasmid (pPROII-1) in this transformant is extracted, and the chromosomal DNA fragment (7 kbp) inserted is further digested with EcoRV and subcloned into a plasmid, pUC19, following the protocols of Sambrook et al.[9] Escherichia coli JM83 transformed with the recombinant plasmid, pPROII-12, which contains an EcoRV–EcoRV fragment (2.4 kbp) inserted into the EcoRV site of pUC19, is found to show about 90-fold higher enzyme activity than the host. High accumulation of the enzyme protein in the transformant harboring pPROII-12 is confirmed by SDS–PAGE. When the fragment is reversely inserted (pPROII-13), the enzyme productivity markedly decreases (Fig. 1).

Purification and Enzymatic Properties of Expressed Enzyme[2]

Purification

Cultivation of Transformant and Extraction of Enzyme. The transformant of E. coli JM83 harboring pPROII-12 is grown in 10 liters of N broth [1% meat extract, 1% polypeptone, and 0.5% NaCl, containing ampicillin (25 μg/ml)] at 30° using a jar fermenter. After 18 hr, cells are collected by centrifugation and washed with 20 mM Tris-HCl buffer, pH 7.0. The washed cells (40 g wet weight) are suspended in the above buffer and disrupted for 10 min in the cold with glass beads in a Dyno-Mill. After removal of glass beads by decantation, cell debris is centrifuged off and the supernatant is used for enzyme purification.

Ammonium Sulfate Fractionation and Ion-Exchange Chromatography. Proteins in the supernatant are precipitated by 35% saturation of ammonium sulfate. The precipitate is dissolved in a small volume of the above buffer and then subjected to gel filtration on a Sephadex G-25 column. The active fractions are applied to a DEAE-Toyopearl column (2.5 × 30 cm) equilibrated with the same buffer. The adsorbed enzyme is eluted with an increasing linear gradient system consisting of 1.2 liters of 20 mM Tris-HCl buffer, pH 7.0 (the mixing chamber), and an equal volume of the same buffer containing 0.7 M NaCl (the reservoir). The enzyme is eluted at 0.24–0.26 M NaCl concentration. The active fractions are combined and the enzyme is precipitated by 80% saturation of ammonium sulfate, dissolved in a small volume of 20 mM Tris-HCl buffer, pH 7.0, and desalted by gel filtration on a column of Sephadex G-25. After repeated DEAE-Toyopearl chromatog-

[9] J. Sambrook, F. E. Fritsch, and T. Maniatis, *in* "Molecular Cloning: A Laboratory Manual." Cold Spring Harbor Laboratory, Cold Spring Harbor, New York, 1989.

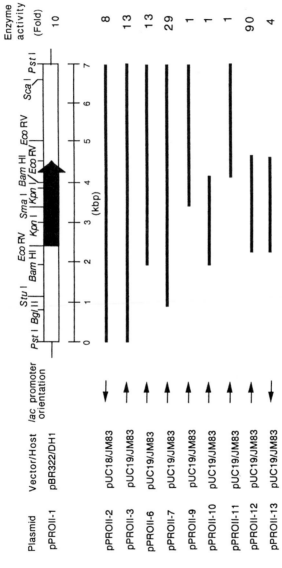

FIG. 1. Restriction endonuclease map of the 7-kbp chromosomal DNA fragment of *E. coli* HB101 and the oligopeptidase B activities of the transformants harboring plasmids. The box indicates the DNA fragment. The enzyme activities were assayed by method A, and expressed as values relative to that of the host, *E. coli* JM83. The position and direction of the enzyme gene are indicated by the large arrow in the box. The small arrows show the orientation of *lac* promoter. From A. Kanatani, T. Masuda, T. Shimoda, F. Misoka, T. Yoshimoto, and D. Tsuru, *J. Biochem. (Tokyo)* **110**, 315 (1991).

raphy, the enzyme is precipitated with ammonium sulfate and dissolved in a minimum volume of the buffer.

Gel Filtration on Sephadex G-150. The enzyme solution is applied to a column (3.0 × 120 cm) of Sephadex G-150 equilibrated with the above buffer containing 100 mM KCl. The active fractions are combined, dialyzed against 5 mM Tris-HCl buffer, pH 7.0, for 2 days, and concentrated. The enzyme solution is kept at −20° until use or lyophilized.

The final preparation gives a single protein band on SDS–PAGE, and its specific activity is 2.0 U/mg, as checked by Method A, the recovery being about 17%. The results are summarized in Table I.

Some Enzymatic and Physicochemical Properties

Table II shows some enzymatic and physicochemical properties of the oligopeptidase B from *E. coli*, compared with those of similar enzymes. The *E. coli* enzyme has a molecular weight of 81,000, which is in good agreement with that calculated from the amino acid sequence deduced from the nucleotide sequence of the enzyme gene, as shown later. All of the enzymes in Table II cleave several N-blocked, chromogenic, and fluorogenic substrates at the carboxyl side of basic amino acid residues at pH 7 to 8, and are also active toward biologically active peptides composed of less than 30 amino acid residues. They are, however, only slightly active toward large polypeptides, unlike trypsin (Table III). They are inhibited by DFP, TLCK, and antipain, but are not affected by protein inhibitors of trypsin from animals and plants, with a few exceptions (Table IV).

TABLE I
PURIFICATION OF *Escherichia coli* OLIGOPEPTIDASE B[a]

Procedure	Protein (mg)	Total activity (units)	Specific activity (units/mg)	Yield (%)	Purification (−fold)
Cell-free extract	53,820	1204	0.022	100.0	1
Chromatography on					
1st DEAE-Toyopearl	1364	722	0.53	60.0	23.8
2nd DEAE-Toyopearl	384	476	1.20	39.5	55.6
Sephadex G-150	101	202	2.00	16.8	89.7

[a] A. Kanatani, T. Masuda, T. Shimoda, F. Misoka, L.-S. Xu, T. Yoshimoto, and D. Tsuru, *J. Biochem. (Tokyo)* **110,** 315 (1991).

TABLE II
ENZYMATIC AND PHYSICOCHEMICAL PROPERTIES OF *Escherichia coli* Oligopeptidase B
and Similar Enzymes from Other Species

		Enzyme source		
Property	*E. coli*[a]	*Rhodococcus erythropolis*[b]	*Trypanosoma brucei*[c]	Soybean seed[d]
Optimum pH	8.0	7.0–7.2	8.2	8.5
Stable pH	5.5–9.5	7–9	n.d.	n.d.
Thermal stability[e]	45°	37°	n.d.	n.d.
Molecular weight	81,800	82,000–90,000	80,000	59,000
p*I*	5.2	n.d.	5.1	n.d.
Subunit	Monomer	Monomer	Monomer	Monomer

[a] A. Kanatani, T. Masuda, T. Shimoda, F. Misoka, L.-S. Xu, T. Yoshimoto, and D. Tsuru, *J. Biochem. (Tokyo)* **110**, 315 (1991).
[b] J. D. Shannon, J. S. Bond, and S. G. Bradley, *Arch. Biochem. Biophys.* **219**, 80 (1982).
[c] M. J. Kornblatt, G. W. N. Mpimbaza, and J. D. Lonsdale-Eccles, *Arch. Biochem. Biophys.* **293**, 25 (1992).
[d] M. Nishikata, *J. Biochem. (Tokyo)* **95**, 1169 (1984).
[e] Of the initial activity, 50% remained after incubation for 30 min at pH 8.0.
[f] n.d., Not determined.

Nucleotide Sequence of Oligopeptidase B Gene of *Escherichia coli* and Deduced Amino Acid Sequence of Enzyme

The recombinant plasmid, pPR0II-12, was isolated by the alkaline lysis procedure,[9] precipitated with polyethylene glycol, and the nucleotide sequence of a 2.4 kbp DNA fragment was determined by the method of Hattori–Sakaki,[10] except that the Klenow fragment was replaced by Sequenase. Figure 2 shows the entire nucleotide sequence of a 2.4 kbp DNA fragment and the deduced amino acid sequence of the enzyme. Within this sequence, there was an open reading frame consisting of 2121 nucleotide bp beginning at an ATG initiation codon and encoding 707 amino acid residues. Ten bases upstream from the proposed initiation codon was an AGAAAG sequence, which seems to be a ribosome-binding site (called SD). Twenty-five bases upstream from the start codon was the six-base sequence, TAAGAT, which is the so-called −10 region, or Pribnow box. Furthermore, the six-base TTGCAT sequence around 55 bases upstream from the start codon seems to be the −35 region. The amino-terminal sequence of the enzyme was established to be Met-Leu-

[10] M. Hattori and Y. Sakaki, *Anal. Biochem.* **152**, 232 (1986).

TABLE III
SUBSTRATE SPECIFICITY OF OLIGOPEPTIDASES B[a]

| | E. coli[b,c] | | Soybean seed[d] | | T. brucei[e] | | R. erythropolis[f] |
Substrate	K_m (μM)	V_{max}	K_m (μM)	V_{max} (nmol/sec/U)	K_m (μM)	k_{cat} (sec^{-1})	V_{max} (nmol/min/mg)
Bz-Arg-βNA	250	0.024[g]	—	—	—	—	—
Bz-Lys-βNA	920	0.044[g]	—	—	—	—	—
Bz-Arg-pNA	500	23.6[h]	11.0	20.5	9.89	108.9	233
Bz-Arg-NHMec	—	—	2.2	15.7	3.12	145.2	—
Bz-Arg-OEt	480	157.6[h]	77.0	25.1	—	—	—
Tos-Arg-OMe	—	—	160.0	11.0	—	—	—
Tos-Lys-OMe	—	—	120.0	4.6	—	—	—
Z-Arg-Arg-NHMec	—	—	—	—	0.09	72.9	—
Z-Phe-Arg-NHMec	—	—	—	—	0.73	83.9	624
Z-Arg-NHMec	—	—	—	—	—	—	242

Proteolytic activity toward polypeptides (activity ratio to trypsin)

	E. coli[b,c]		Soybean seed[d]		T. brucei[e]		R. erythropolis[f]
Hemoglobin	$7.7 \times 10^{-4,b}$		—		—		3.2×10^{-2}
Casein	$1.2 \times 10^{-2,c}$		—		—		5.6×10^{-2}
Azocasein			—		—		2.4×10^{-4}
Oxidized insulin B	Active		Inactive		Inactive		
Plasminogen	—		—				—

[a] Bz, Benzoyl; Z, benzyloxycarbonyl; Tos, tosyl; NHMec, 7-amino-4-methylcoumarin; OEt, ethyl ester; OMe, methyl ester; pNA, p-nitroanilide; βNA, β-naphthylamide; —, not examined.
[b] A. Kanatani, T. Masuda, T. Shimoda, F. Misoka, S-L. Xu, T. Yoshimoto, and D. Tsuru, J. Biochem. (Tokyo) 110, 315 (1991).
[c] M. Pacaud and C. Richaud, J. Biol. Chem. 250, 7771 (1975).
[d] M. Nishikata, J. Biochem. (Tokyo) 95, 1169 (1984).
[e] M. J. Kornblatt, G. W. N. Mpimbaza, and J. D. Lonsdale-Eccles, Arch. Biochem. Biophys. 293, 25 (1992).
[f] J. D. Shannon, J. S. Bond, and S. G. Bradley, Arch. Biochem. Biophys. 219, 80 (1982).
[g] k_{cat} (sec^{-1}).
[h] V_{max} (μmol/min/mg).

TABLE IV

EFFECT OF SERINE PEPTIDASE INHIBITORS ON OLIGOPEPTIDASES B

	Enzyme source			
Inhibitor	$E. coli^{a,b}$	Soybean seed[c]	$T. brucei^d$	$R. erythropolis^e$
DFP[f]	+[g]	+	+	+
TLCK	+	+	+	+
Antipain	+	+	+	+
Leupeptin	+	+	+	+
Z-Leu-Lys-CHN$_2$	n.e.	n.e	+	n.e.
4-Aminobenzamidine	n.e.	n.e	n.e.	+
PMSF	–	–	–	–
Trypsin inhibitors				
Aprotinin	–	Partially	Partially	–
Lima bean	–	–	n.e.	–
Soybean	–	–	n.e.	–
Ovomucoid	–	–	–	+
α_1-Antitrypsin	n.e.	n.e.	n.e.	–

[a] A. Kanatani, T. Masuda, T. Shimoda, F. Misoka, S.-L. Xu, T. Yoshimoto, and D. Tsuru, *J. Biochem.* (*Tokyo*) **110**, 315 (1991).
[b] M. Pacaud and C. Richaud, *J. Biol. Chem.* **250**, 7771 (1975).
[c] M. Nishikata, *J. Biochem.* (*Tokyo*) **95**, 1169 (1984).
[d] M. J. Kornblatt, G. W. N. Mpimbaza, and J. D. Lonsdale-Eccles, *Arch. Biochem. Biophys.* **293**, 25 (1992).
[e] J. D. Shannon, J. S. Bond, and S. G. Bradley, *Arch. Biochem. Biophys.* **219**, 80 (1982).
[f] DFP, Diisopropyl fluorophosphate; TLCK, tosyl-L-lysylchloromethane; PMSF, phenyl-methylsulfonyl fluoride.
[g] +, Inhibited; –, not inhibited; n.e., not examined.

Pro-Lys-Ala-Ala-Arg-Ile- by protein sequencing, indicating that the mature protein-encoding nucleotide sequence just starts at the ATG initiation codon of the open reading frame. In addition, amino acid sequences of the CNBr fragments and amino acid composition of the enzyme were coincidental with those deduced from the nucleotide sequence of the gene. Thus, the oligopeptidase B from *E. coli* was concluded to be composed of 707 amino acid residues with a calculated molecular weight of 81,858 and with Met as its amino terminus.[2]

Homology Survey and Some Consideration on Active Site Structure

Active Site Labeling

Escherichia coli oligopeptidase B, 1 mg, was incubated with a 100-fold molar excess of [³H]DFP at 37° in 1 ml of 20 mM Tris-HCl buffer, pH 8.0. After 1 hr, unlabeled DFP is added to obtain a final concentration of 1 mM, and the mixture is incubated for an additional 1 hr before extensive dialysis against distilled water and is then lyophilized.[2] The tritium-labeled enzyme is incubated with 1 mg of CNBr in 1 ml of 70% formic acid under nitrogen gas in the dark for 24 hr at room temperature, diluted with water, and lyophilized. The resultant fragments are separated and purified by reversed-phase HPLC[11]: longer peptides, which have no radioactivity, are absorbed on a Vydac C_4 column (4.6 × 250 mm) equilibrated with 0.075% trifluoroacetic acid (TFA) at 25°, and the breakthrough fraction is applied to a Vydac C_{18} column (4.6 × 250 mm) equilibrated with the same solution and the absorbed peptides are eluted with an increasing linear gradient of acetonitrile : 2-propanol (3 : 1) containing 0.06% TFA (reservoir) at a flow rate of 1.0 ml/min. Elution of peptides is monitored by measuring absorbance at 214 nm.

The amino acid composition of the radioactive peptide is Ser (0.8), Gly (4.0), Ala (1.0), and homoserine (1.1), and the amino acid sequence is established to be Gly-Gly-Ser-Ala-Gly-Gly-Met, which corresponds to residues 530–536, where Ser-532 is labeled by tritium. These results clearly indicate that E. coli oligopeptidase B is a serine-type peptidase. The active site sequence conforms to a motif, Gly-X-Ser-X-Gly, that is also found in trypsin and α-chymotrypsin, in the chymotrypsin family.

Homology Survey

Homology analysis indicated that the amino acid sequence of E. coli oligopeptidase B was quite different from those of the chymotrypsin and subtilisin families, except around the active site Ser residues. Surprisingly, however, the E. coli enzyme was found to be significantly similar to prolyl endopeptidases (now called prolyl oligopeptidases: EC 3.4.21.26[12]) from porcine brain,[13] *Flavobacterium meningosepticum*,[14,15] and *Aeromonas hydrophila*.[16] The amino acid sequence of E. coli oligopeptidase B is 24–25% identical to those of these three proline-specific peptidases. In particular, strong similarity was observed in the regions around the putative active sites of these four enzymes.[16] These findings led to the recogni-

[11] R. Kobayashi, T. Yoshimoto, and D. Tsuru, *Agric. Biol. Chem.* **53**, 2737 (1989).

[12] N. D. Rawlings, L. Polgar, and A. J. Barrett, *Biochem. J.* **279**, 907 (1991).

[13] D. Rennex, B. A. Hemmings, J. Hofsteenge, and S. R. Stone, *Biochemistry* **30**, 2195 (1991).

[14] T. Yoshimoto, A. Kanatani, T. Shimoda, T. Inaoka, T. Kokubo, and D. Tsuru, *J. Biochem.* (*Tokyo*) **110**, 873 (1991).

[15] S. Chevallier, P. Goeltz, P. Thibault, D. Banville, and J. Gagnon, *J. Biol. Chem.* **267**, 8192 (1992).

[16] A. Kanatani, T. Yoshimoto, A. Kitazono, T. Kokubo, and D. Tsuru, *J. Biochem.* (*Tokyo*) **113**, 790 (1993).

```
         -109  ATCCGCGCCATTACCTGTCATCATCTAAGCAATGACTCCCCTGTTTCGC  -61
 -60  TTGCATCCCCGGTGAGTTTTGCCACCCTTATAAGATGTTTCAACCAGAAAGAACAATAAC  -1
      -35                 -10              SD
  1  ATGCTACCAAAAGCCGCCCGCATTCCCCACGCCATGACGCTTCATGGCGATACGCGCATC  60
  1  M  L  P  K  A  A  R  I  P  H  A  M  T  L  H  G  D  T  R  I   20
 61  GATAATTACTACTGGCTGCGGGACGATACGCGTTCTCAGCCAGAAGTCCTGGACTACCTG  120
 21  D  N  Y  Y  W  L  R  D  D  T  R  S  Q  P  E  V  L  D  Y  L   40
121  CAACAAGAAAATAGTTACGGTCATCGGGTGATGGCCTCACAACAAGCCTTGCAGGATCGC  180
 41  Q  Q  E  N  S  Y  G  H  R  V  M  A  S  Q  Q  A  L  Q  D  R   60
181  ATCTTAAAGGAAATCATCGACCGCATTCCGCAACGAGAAGTTTCTGCGCCCTACATCAAA  240
 61  I  L  K  E  I  I  D  R  I  P  Q  R  E  V  S  A  P  Y  I  K   80
241  AATGGCTACCGCTATCGGCATATTTATGAACCAGGCTGTGAATATGCTATCTACCAGCGT  300
 81  N  G  Y  R  Y  R  H  I  Y  E  P  G  C  E  Y  A  I  Y  Q  R   100
301  CAATCGGCATTCAGTGAAGAGTGGGATGAGTGGGAAACATTGCTCGATGCCAATAAGCGC  360
101  Q  S  A  F  S  E  E  W  D  E  W  E  T  L  L  D  A  N  K  R   120
361  GCAGCTCATAGTGAGTTTTATTCGATGGGCGGAATGGCGATTACGCCCGATAACACCATT  420
121  A  A  H  S  E  F  Y  S  M  G  G  M  A  I  T  P  D  N  T  I   140
421  ATGGCGCTGGCAGAAGATTTTCTTTCCCGACGCCAGTACGGCATTCGTTTTCGTAATCTG  480
141  M  A  L  A  E  D  F  L  S  R  R  Q  Y  G  I  R  F  R  N  L   160
481  GAAACTGGTAACTGGTACCCGGAACTGCTGGATAACGTTGAACCCAGCTTTGTCTGGGCA  540
161  E  T  G  N  W  Y  P  E  L  L  D  N  V  E  P  S  F  V  W  A   180
541  AATGACTCCTGGATTTTCTACTATGTTCGCAAGCATCCGGTGACGCTGCTGCCTTATCAG  600
181  N  D  S  W  I  F  Y  Y  V  R  K  H  P  V  T  L  L  P  Y  Q   200
601  GTCTGGCGTCACGCCATCGGTACGCCAGCATCGCAAGATAAACTGATCTACGAAGAAAAA  660
201  V  W  R  H  A  I  G  T  P  A  S  Q  D  K  L  I  Y  E  E  K   220
661  GACGATACCTATTACGTCAGCCTGCATAAAACGACCTCGAAGCACTATGTAGTCATTCAT  720
221  D  D  T  Y  Y  V  S  L  H  K  T  T  S  K  H  Y  V  V  I  H   240
721  TTGGCCAGCGCCACCACCAGTGAAGTTCGCCTGCTGGACGCGGAAATGGCCGATGCCGAG  780
241  L  A  S  A  T  T  S  E  V  R  L  L  D  A  E  M  A  D  A  E   260
781  CCGTTTGTTTTTCTGCCGCGCCGCAAAGATCACGAATACAGCCTTGATCACTACCAGCAT  840
261  P  F  V  F  L  P  R  R  K  D  H  E  Y  S  L  D  H  Y  Q  H   280
841  CGTTTTTATCTGCGTTCCAACCGCCACGGCAAAAACTTTGGCTTATACCGTACCCGTATG  900
281  R  F  Y  L  R  S  N  R  H  G  K  N  F  G  L  Y  R  T  R  M   300
901  CGTGATGAGCAACAGTGGGAAGAGTTAATTCCGCCACGCGAAAACATCATGCTGGAAGGG  960
301  R  D  E  Q  Q  W  E  E  L  I  P  P  R  E  N  I  M  L  E  G   320
961  TTTACGCTGTTTACCGACTGGCTGGTGGTTGAAGAGCGTCAGCGCGGGTTAACCAGTTTG  1020
321  F  T  L  F  T  D  W  L  V  V  E  E  R  Q  R  G  L  T  S  L   340
1021  CGCCAAATTAACCGCAAGACCCGGGAAGTCATTGGTATTGCCTTTGATGATCCGGCCTAT  1080
341  R  Q  I  N  R  K  T  R  E  V  I  G  I  A  F  D  D  P  A  Y   360
```

FIG. 2. Nucleotide sequence of the *E. coli* oligopeptidase B gene and the deduced amino acid sequence of the enzyme. The nucleotide sequence is numbered from the top of the initiation codon of the open reading frame. Numbering of the amino acid sequence starts at the amino terminus of the enzyme. The putative −35 and −10 sequences and potential SD sequence are overlined. Amino acid sequences of the amino-terminal region and CNBr fragments, determined by Edman degradation, are specified below the amino acid sequence by underlines. From A. Kanatani, T. Masuda, T. Shimoda, F. Misoka, T. Yoshimoto, and D. Tsuru, *J. Biochem. (Tokyo)* **110**, 315 (1991).

```
1081  GTGACCTGGATTGCCTACAATCCAGAACCTGAAACCGCGCGATTGCGTTATGGTTATTCT  1140
361   V  T  W  I  A  Y  N  P  E  P  E  T  A  R  L  R  Y  G  Y  S   380

1141  TCCATGACTACACCAGACACTTTGTTTGAACTGGATATGGATACCGGTGAGCGTCGTGTA  1200
381   S  M  T  T  P  D  T  L  F  E  L  D  M  D  T  G  E  R  R  V   400

1201  TTAAAACAAACGGAAGTTCCTGGTTTTTATGCGGCGAATTACCGCAGTGAACACCTGTGG  1260
401   L  K  Q  T  E  V  P  G  F  Y  A  A  N  Y  R  S  E  H  L  W   420

1261  ATAGTCGCCCGTGATGGCGTCGAAGTTCCGGTTTCGTTGGTCTACCATCGCAAACATTTT  1320
421   I  V  A  R  D  G  V  E  V  P  V  S  L  V  Y  H  R  K  H  F   440

1321  CGCAAAGGACACAACCCGTTGCTGGTGTATGGCTATGGTTCTTACGGCGCAAGTATTGAT  1380
441   R  K  G  H  N  P  L  L  V  Y  G  Y  G  S  Y  G  A  S  I  D   460

1381  GCCGATTTCAGTTTTAGCCGCTTGAGTTTGTTAGATCGTGGCTTTGTCTACGCCATTGTC  1440
461   A  D  F  S  F  S  R  L  S  L  L  D  R  G  F  V  Y  A  I  V   480

1441  CATGTTCGCGGCGGTGGTGAGCTGGGGCAACAATGGTACGAAGACGGAAAATTTCTGAAG  1500
481   H  V  R  G  G  G  E  L  G  Q  Q  W  Y  E  D  G  K  F  L  K   500

1501  AAGAAAAATACGTTTAATGATTATCTTGATGCCTGCGATGCATTGTTAAAAACTGGGCTAT  1560
501   K  K  N  T  F  N  D  Y  L  D  A  C  D  A  L  L  K  L  G  Y   520

1561  GGCTCTCCTTCGCTTTGTTATGCGATGGGCGGGAGTGCGGGGGGCATGTTGATGGGCGTT  1620
521   G  S  P  S  L  C  Y  A  M  G  G  S  A  G  G  M  L  M  G  V   540

1621  GCAATTAATCAACGCCCGGAATTATTCCACGGCGTTATCGCCCAGGTACCGTTTGTTGAT  1680
541   A  I  N  Q  R  P  E  L  F  H  G  V  I  A  Q  V  P  F  V  D   560

1681  GTTGTAACAACGATGCTTGATGAATCAATTCCTCTTACCACTGGTGAGTTTGAAGAGTGG  1740
561   V  V  T  T  M  L  D  E  S  I  P  L  T  T  G  E  F  E  E  W   580

1741  GGTAACCCGCAGGATCCGCAATATTACGAGTACATGAAAAGCTACAGCCCGTATGACAAC  1800
581   G  N  P  Q  D  P  Q  Y  Y  E  Y  M  K  S  Y  S  P  Y  D  N   600

1801  GTCACCGCACAGGCTTATCCGCATTTACTGGTAACGACCGGTTTGCACGATTCTCAGGTG  1860
601   V  T  A  Q  A  Y  P  H  L  L  V  T  T  G  L  H  D  S  Q  V   620

1861  CAATATTGGGAACCGGCAAAATGGGTCGCTAAATTGCGCGAGCTGAAAACCGATGACCAT  1920
621   Q  Y  W  E  P  A  K  W  V  A  K  L  R  E  L  K  T  D  D  H   640

1921  CTTTTATTGCTCTGTACCGACATGGACTCAGGCCATGGCGGTAAATCTGGTCGCTTTAAA  1980
641   L  L  L  L  C  T  D  M  D  S  G  H  G  G  K  S  G  R  F  K   660

1981  TCGTACGAAGGCGTAGCGATGGAATATGCTTTTCTGGTCGCGCTGGCGCAGGGAACATTA  2040
661   S  Y  E  G  V  A  M  E  Y  A  F  L  V  A  L  A  Q  G  T  L   680

2041  CCCCTACGCCTGCGGACTAAGTATTTTCCAGATAATGTTTCAGTGTTAAACGCAGCTCCG  2100
681   P  L  R  L  R  T  K  Y  F  P  D  N  V  S  V  L  N  A  A  P   700

2101  GGCTCATGCTGTCCAGGTTATTAAATAACCAGCGCAGATAGCCCGGATCGCGTTCGCCAA  2160
701   G  S  C  C  P  G  Y  *  707

2161  CGTCGGAAACCGCTTTGCCACGGTATTTGCCAAAGGTGAAGGTCGTCATCAACGACGGAC  2220
2221  GTCCGGTGAT  2230
```

FIG. 2. (*continued*)

tion of a new family of serine peptidases related to prolyl oligopeptidase.[2,12,14,17]

In 1991, Barrett and co-workers[12] pointed out that there is a significant
sequence homology among porcine brain prolyl oligopeptidase,[13] dipepti-

[17] A. J. Barrett and N. D. Rawlings, *Biol. Chem. Hoppe-Seyler* **373**, 353 (1992).

```
X-PDP   VDKAPYR--FTHGW-TYSLH-DYFLTRGFASIYVA----GVGTRSSDGFQTSGDYQQIYSHTAV
RDPIV   SKKYPLLIDVYAGF--CSQKADAAF-RLHWATYLASTEHIIVASFDG--RGSGYQGDKIHHAI
HDPIV   SKKYPLLLDVYAGF--CSQKADAAF-RLHWATYLASTEHIIVASFDG--RGSGYQGDKIHHAI
DP-B    SDHYPVFFFAYGGF--HSQQVVKTF-SVGFHEVVASQLHAIVVVVDG--RGTGFKGQDFRSLV
F-PEP   DGKHPTI-LYSYGGFHISLQPAFSVVHAIWHEH------GG--IYAVPHIRGGGEYGKKWHDAG
A-PEP   DGSHPTI-LYGYGGFDVSLTPSFSVSVAHWLDL------GG--IYAVAHLRGGGEYGQAWHLAG
P-PEP   DGSHPAF-LYGYGGFHISITPHYSVSRLIFVRHH-----GG--VLAVAHIRGGGEYGETWHKGG
E-PII   KGHHPLL-VYGYGSYGASIDADFSFSRLSLLDR------GF-VYAIVHVRGGGELGQQWYHDG
PACYL   KTQVPHVVHPHGGF--HSSFVTAWHLFFAHLCKH-----GFAVLLV-HYRGSTGFGQDSILSL
RACYL   KTQVPHVVHPHGGF--HSSFVTAWHLFFAHLCKH-----GFAVLLV-HYRGSTGFGQDSILSL
3P21    RPKCPWWSCF-TGA--HSSFVTAWHLFFAHLCKH-----GFAVLLV-HYRGSTGFGQDSILSL
                          ?                 *
X-PDP   ----------IDWLHGRA--//  //---WAHGKVAHTGKSYLGTHAYGAATTGVSGLHVILAHAG
RDPIV   HKRLGTLHVHDHIHAARQFLKHG--FVDSKRVAIWGWSYSGGYVTSHVLGSGSG-VFKCGIA-V
HDPIV   HRRLGTLHVHDHIHAARQFSKHG--FVDHKRVAIWGWSYGGGYVTSHVLGSGSG-VFKCGIA-V
DP-B    RDRLGDYHARDQISAAS--LYGSLTFVDPQKISLFGWSYGGYLTLKTLHKDGGRHFKYGHS-V
F-PEP   TKHQKKHVFHDFIAAGHY-LQKHGYTSKHYH-ALSGRSHGGLLVGATHTHRFD-LAKVAFPGV
A-PEP   TRHHKQHVFDDFIAAHSY-LKAHGYTRTDRL-AIRGGSHGGLLVGAVHTQRFD-LHRVACQAV
P-PEP   ILAHKQHCFDDFQCAAHY-LIKHGYTSFKRL-TIHGGSHGGLLVATCAHQRFD-LFGCVIAQV
E-PII   KFLKKKHTFHDYLDACDA-LLKLGYGGFSLC-YAHGGSAGGHLHGVAIHQRFS-LFHGVIAQV
PACYL   PGHVGHQDVKDVQFAVHQVLQHEHFDAGRV---ALHGGSHGGFLSCHLIGQYPH-TYSACVVRH
RACYL   PGHVGHQDVKDVQFAVHQVLQHEHFDARRV---ALHGGSHGGFLSCHLIGQYPH-TYSACIARH
3P21    PGHVGHQDVKDVQFAVHQVLQHEHFDASHV---LHGGSHGGFISCHLIGQYPH-TYRACL--R
                        ?                    *
X-PDP   I----------SSWYHYYRHHGLVRSPGG-FPGHDLDVLAALT-YSRHLD--//  //--LGKT
RDPIV   APV--------SRW-HYYDSVYTHRYHGLPTPHDHLDHYRHSTVHSRAHH----------FKQ-
HDPIV   APV--------SRW-HYYDSVYTHRYHGLPTPHDHLDHYRHSTVHSRAHH----------FKQ-
DP-B    APV--------TDW-RFYDSVYTHRYH--HTPQHHFDGYVSSVHHVTAL----------AQA-
F-PEP   GVLD----HLRYHKFTAGAGWA-YDYGTAHDSHHFH--------YLKSYSPVHHVK----AGTC
A-PEP   GVLD----HLRYHKFTAGAGWA-YDYGTSADSHAHFD--------YLKGYSPLHSVR----AGVS
P-PEP   GVHD----HLKFHKYTIGHAHY-TDYGCS-DSKQHFH--------WLIKYSPLHHVKLFRADDIQ
E-PII   PFVDVVTTHL-DHSIPLTTGHF-HHWGHPQDF-QYYH--------YHKSYSPYDHVT----AQA-
PACYL   PVIHIA-SHHGSTDIFD---WCHVHAGFSYSHDHHFHLDSVWAAHLDK-SPIKYIP----QV
RACYL   PVIHIA-SHHGSTDIFD---WCHVHTGFFYSHHSCLPDLHVWHHHLDK-SPIKYIP----QV
3P21    TRDHHA-SHLGSTDILTGAWW----RLASFSSDCLPDLSVWAHHLDK-SPIRYSS------GH
                   ?                                   *
X-PDP   AVSFAQFDHHYDDHTFKKYSKDFHVFKKDLFHH--//  //--LYSTDFSHTVRDHRKVTTHI
RDPIV   -VHYLLIHGTADDHVHFQQSAQIS---KALVDAGVD----FQAHHYTDHDHGIASSTAHQHIY
HDPIV   -VHYLLIHGTADDHVHFQQSAQIS---KALVDVGVD----FQAHHYTDHDHGIASSTAHQHIY
DP-B    -HRFLLHHGTGDDHVHFQHSLKFL---DLLDLHGVH----HYDVHVFPDSDHSIRYHHAHVIVF
F-PEP   YPSTHVITSDHDDRVVPAHSFKFGSHLQ----AKQSCKHFILIRIHTHAGHGAGRSTSQVVAX
A-PEP   YPSTLVTTADHDDRVVPAHSFKFAATLQ----ADDAGFHPQLIRIHTHAGHGAGTFVAKLIHQ
P-PEP   YPSHLLLTADHDDRVVPLHSLKFIATLQYIVGRSRKQHHFPLIHVDTHAGHGAGKFTAKVIRH
E-PII   YPHLLVTTGLHDSQVQYWHPAKWVAKLR-----HLKTDDHLLLLCTDHDSGHG-GKSGRFKSYH
PACYL   KTPVLLHLGQHDDRVPFKQGHHYRVLKARHVP------VRLLLY--PKSTHAL--SHVHVHSD
RACYL   KTPVLLHLGQHDDRRVPFKQGHHYYHRALKARHHVP---VRLLLY--PKSHHAL--SHVHAHSD
3P21    DTTV-TDVGQHD-AVCLSR-HHYTSSRP-HCA------VRLLLY--PKSTHAL--SHVHVHSA
```

FIG. 3. Alignment of the C-terminal domains of enzymes belonging to the prolyl oligopeptidase family.[a] The asterisks indicate the active site serine and histidine residues, and the question marks indicate the putative active site aspartic acid residues. X-PDP, X-prolyl dipeptidyl aminopeptidase from *Lactococcus lactis*[b,c]; RDPIV, dipetidyl-peptidase IV from rat[d]; HDPIV, human dipeptidyl-peptidase IV[e]; DP-B, dipeptidyl aminopeptidase B from yeast[f]; F-PEP, A-PEP, and P-PEP, prolyl endopeptidases (prolyl oligopeptidases) from *Flavobacterium meningosepticum*,[g] *Aeromonas hydrophila*,[a] and porcine brain,[h,i] respectively; E-PII, *E. coli* oligopeptidase B (*E. coli* protease II)[j]; PACYL and RACYL, acylamino acid-releasing ezymes from pig[k] and rat[l], respectively; 3P21, human protein 3p21.[m] Key to references: (*a*) A. Kanatani, T. Yoshimoto, A. Kitazono, T. Kokubo, and D. Tsuru, *J. Biochem. (Tokyo)* 113, 789 (1993); (*b*) B. Mayo, J. Kok, K. Venema, W. Bockelmann, M. Teuber, H. Reinke, and G. Venema, *Appl. Environ. Microbiol.* 57, 38 (1991); (*c*) M. Nardi, M. Chopin, A. Chopin, M. Clas, and Gripon, *Appl. Environ. Microbiol.* 57, 45 (1991); (*d*) S. Ogata, Y. Misumi, and Y. Ikehara, *J. Biol. Chem.* 264, 3596 (1989); (*e*) D. Darmoul, M. Lasaca, L. Baricault, D. Marguet, C. Sapin, P. Trotot, A. Barbat, and G. Trugnan, *J. Biol. Chem.* 267, 4824 (1992); (*f*) C. J. Roberts, G. Pholing, J. H. Rothman, and T. H. Stevens, *J. Cell Biol.* 108, 1363 (1989); (*g*) T. Yoshimoto, A. Kanatani, T. Shimoda, T. Inaoka, T. Kokubo, and D. Tsuru, *J. Biochem. (Tokyo)* 110, 873 (1991); (*h*) D. Rennex, B. A. Hemmings, J. Hofsteenge, and S. R. Stone, *Biochemistry* 30, 2195 (1991); (*i*) S. R. Stone, D. Rennex, P. Wikstrom, E. Shaw, and J. Hofsteenge, *Biochem. J.* 276, 837 (1991); (*j*) A. Kanatani,

dyl peptidase IV from rat liver,[18] yeast dipeptidyl aminopeptidase B,[19] acylamino acid-releasing enzyme from rat liver,[20] and human protein 3p21,[21] especially in their C-terminal domains, and proposed that Ser-554 and His-680 are essential catalytic residues in porcine brain prolyl oligopeptidase. Figure 3 is an alignment of parts of the sequences of several proteins that have similarities to *E. coli* oligopeptidase B. The serine residues marked by an asterisk were shown to be DFP reactive in prolyl oligopeptidases from *F. meningosepticum*[14,15,22] and porcine brain,[13] *E. coli* oligopeptidase B,[2] rat liver dipeptidyl peptidase IV,[23] and X-prolyl dipeptidyl aminopeptidase from *Lactococcus lactis*.[24] Their surrounding regions have a common sequence, Gly-X-Ser-X-Gly-Gly, except in the *L. lactis* enzyme,which has the same substrate specificity as dipeptidyl-peptidase IV. Likewise, histidine residues marked by an asterisk in Fig. 3 were proved to be in the catalytic triad of prolyl oligopeptidase from *Flavobacterium*[22] and acylamino-releasing enzyme from human erythro-cytes.[25] In addition, there are two aspartic acid residues (marked by ? in Fig. 3) that are conserved in all members of this group. Replacements of these two aspartic acids in *Flavobacterium* enzyme by serine, individually, led to almost complete losses of the activity.[22] Thus, it seems impossible at present to decide which aspartic acid is the one involved in the catalytic triad.

Of great interest is the question of whether these enzymes are only slightly active on high molecular mass substrates. The reason for this

[18] S. Ogata, Y. Misumi, and Y. Ikenaka, *J. Biol. Chem.* **264**, 3596 (1989).
[19] C. J. Roberts, G. Pohlig, J. H. Rothman, and T. H. Stevens, *J. Cell Biol.* **108**, 1363 (1989).
[20] K. Kobayashi and J. A. Smith, *J. Biol. Chem.* **262**, 11435 (1987).
[21] S. L. Naylor, A. Marshall, C. Hensel, P. F. Martinez, B. Holley, and A. Y. Sakaguchi, *Genomics* **4**, 355 (1989).
[22] T. Okuma, T. Inaoka, H. Tokame, T. Kikuchi, Y. Abe, K. Takimoto, T. Yoshimoto, T. Kokubo, and D. Tsuru, *43th Symp. Protein Struct.,* 65 (1992).
[23] S. Ogata, Y. Misumi, E. Tsuji, N. Takami, K. Oda, and Y. Ikenaka, *Biochemistry* **31**, 2582 (1992).
[24] J.-F. Chich, M.-P. Chapot-Chartier, B. Ribadeau-Dumas, and J.-C. Gripon, *FEBS Lett.* **314**, 139 (1992).
[25] A. Scalon, W. M. Jones, D. Barra, M. Pospischil, S. Sassa, A. Popowicz, L. R. Manning, O. Schneewind, and J. M. Manning, *J. Biol. Chem.* **267**, 3811 (1992).

T. Masuda, T. Shimoda, F. Misoka, S.-L. Xu, T. Yoshimoto, and D. Tsuru, *J. Biochem.* (*Tokyo*) **110**, 315 (1991); (*k*) M. Mitta, K. Asada, Y. Uchimura, F. Kimizuka, I. Kato, F. Sakiyama, and S. Tsunasawa, *J. Biochem.* (*Tokyo*) **106**, 548 (1989); (*l*) K. Kobayashi, L. W. Lin, J. E. Yeadon, L. B. Klickstein, and J. A. Smith, *J. Biol. Chem.* **264**, 8892 (1989); (*m*) S. L. Naylor, A. Marshall, C. Hensel, P. F. Martinez, B. Holley, and A. Y. Sakaguchi, *Genomics* **4**, 355 (1989).

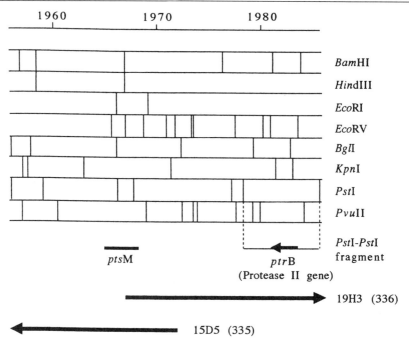

FIG. 4. Location of the *E. coli* oligopeptidase B gene (*ptrB*) [A. Kanatani, T. Yoshimoto, H. Nagai, K. Ito, and D. Tsuru, *J. Bacteriol.* **174,** 7881 (1992)] based on the revised map of Kohara *et al.* [Y. Kohara, K. Ariyama, and K. Isono, *Cell (Cambridge, Mass.)* **50,** 495 (1987); Y. Kohara, personal communication]. The arrows indicate orientation of gene. Map units are indicated in kilobase pairs. *ptsM*, phosphotransferase system gene.

specificity of oligopeptidases is unknown, but we suggest that the large N-terminal domains, where the active site amino acid residues are not located, regulate the binding of high molecular weight substrates.

In conclusion, it was shown that the *E. coli* oligopeptidase B belongs to the prolyl oligopeptidase family, and that the catalytic triad is constructed by Ser-532, His-652, and Asp-507 (or Asp-617).

Localization of Oligopeptidase B Gene (*ptrB*) on Physical Map of *Escherichia coli* Chromosome

The position of the *ptrB* gene on the *E. coli* chromosome[26] was determined by hybridization to the Kohara library.[27,28] With the *Eco*RV–

---

[26] A. Kanatani, T. Yoshimoto, H. Nagai, K. Ito, and D. Tsuru, *J. Bacteriol.* **174,** 7881 (1992).
[27] Y. Kohara, K. Akiyama, and K. Isono, *Cell (Cambridge, Mass.)* **50,** 495 (1987).
[28] Y. Kohara, personal communication.

*Eco*RV fragment of pPR0II-12 as a probe, only one clone 19H3 (phage 336) hybridized with the probe. Comparison of the *ptrB* sequence with the restriction map of Kohara *et al.*[27,28] placed the gene at 1981 to 1984 kbp on the *E. coli* physical map (Fig. 4). The location of the *ptrB* gene on the *Eco*RV–*Eco*RV fragment of 19H3 was also confirmed by the nucleotide sequencing.

[16] Dipeptidyl-peptidase IV from Rat Liver

By Yukio Ikehara, Shigenori Ogata, and Yoshio Misumi

Dipeptidyl-peptidase IV, or dipeptidyl aminopeptidase IV (DPPIV) (EC 3.4.14.5), is a serine protease that cleaves N-terminal dipeptides from oligo- and polypeptides with a penultimate prolyl residue.[1–3] DPPIV is a membrane-bound glycoprotein localized on the cell surface (ectoenzyme), in contrast to other DPPs localized in the lysosome (DPPI and DPPII) and in the cytoplasm (DPPIII). In polarized epithelial cells such as those in the liver, small intestine, and kidney proximal tubules, DPPIV is localized in the apical domain of the plasma membrane.[4–6] Although widely distributed in a variety of tissues, the enzyme has been purified to homogeneity from kidney,[7,8] small intestine,[9] submaxillary gland,[10] and liver.[11,12] The purified enzyme is found to be a dimer, comprising two identical subunits of 110–130 kDa that are variable depending on species and tissues, possibly due to the extent of glycosylation.

[1] V. Hopsu-Havu and G. G. Glenner, *Histochemie* **7,** 197 (1966).
[2] J. K. McDonald and C. Schwabe, *in* "Proteinases in Mammalian Cells and Tissues" (A. J. Barrett, ed.), p. 371. North-Holland Publ., Amsterdam 1976.
[3] A. J. Kenny, *in* "Proteinases in Mammalian Cells and Tissues" (A. J. Barrett, ed.), p. 417. North-Holland Publ., Amsterdam, 1976.
[4] K. M. Fukasawa, K. Fukasawa, N. Sahara, M. Harada, Y. Kondo, and I. Nagatsu, *J. Histochem. Cytochem.* **29,** 337 (1981).
[5] J. R. Bartles, L. T. Braiterman, and A. Hubbard, *J. Biol. Chem.* **260,** 12792 (1986).
[6] S. Hartel, R. Grossrau, C. Hanski, and W. Reutter, *Histochem. J.* **89,** 151 (1988).
[7] A. J. Kenny, A. G. Booth, S. G. George, J. Ingram, D. Kershaw, E. J. Wood, and A. R. Young, *Biochem. J.* **157,** 169 (1976).
[8] T. Yoshimoto and R. Walter, *Biochim. Biophys. Acta* **485,** 391 (1977).
[9] B. Svensson, M. Danielsen, M. Staun, L. Jeppesen, O. Noren, and H. Sjöström, *Eur. J. Biochem.* **90,** 489 (1978).
[10] K. Kojima, T. Hama, T. Kato, and T. Nagatsu, *J. Chromatogr.* **189,** 233 (1980).
[11] J. Elovson, *J. Biol. Chem.* **255,** 5807 (1980).
[12] K. M. Fukasawa, K. Fukasawa, B. Y. Hiraoka, and M. Harada, *Biochim. Biophys. Acta* **657,** 179 (1981).

Assay Methods

Direct Photometric Method

Principle. The release of *p*-nitroaniline from dipeptidyl-*p*-nitroanilides is photometrically determined at 385 nm after the indicated incubation period.[13] Gly-Pro-*p*-nitroanilide tosylate is used as a routine substrate, but any Xaa-Pro-*p*-nitroanilide can be used for the assay. This method is useful for following the progress of purification and for laboratories without a recording spectrophotometer or fluorometer.

Reagents

Glycine–NaOH buffer: 0.3 *M*, pH 8.7.
Substrate: 3 m*M* Gly-Pro-*p*-nitroanilide. Gly-Pro-*p*-nitroanilide tosylate (Peptide Institute, Inc., Osaka, Japan) is dissolved in 2% (v/v) Triton X-100 (1.4 mg of the substrate/ml), stored at 4° (not frozen), and should be used within 4 days.
Standard *p*-nitroaniline solution: 0.3 m*M* in 2% (v/v) methanol.
Acetate buffer: 1.0 *M*, pH 4.2.

Procedure. A reaction mixture (1.0 ml) contains 0.25 ml of 0.3 *M* glycine–NaOH buffer (pH 8.7), 0.5 ml of the substrate (1.5 µmol), 0.2 ml of H$_2$O, and 0.05 ml of an appropriately diluted enzyme solution. A blank tube contains the same assay mixture without the enzyme solution. The mixtures are incubated at 37° for 30 min. The reaction is stopped by adding 3.0 ml of 1 *M* acetate buffer (pH 4.2). The mixtures are centrifuged at 3000 *g* for 5 min at room temperature, if necessary. The absorbance of the sample is read at 385 nm and corrected by subtraction of the blank. The quantity of *p*-nitroaniline released is determined from the absorbance of the sample relative to that of a standard solution prepared by substituting 0.3 m*M p*-nitroaniline (0.15 µmol) for substrate in the assay mixture. One unit of activity is defined as the amount of enzyme that produces 1 µmol of *p*-nitroaniline per minute. The specific activity is expressed in units per milligram of protein.

Colorimetric Method

Principle. Diazotization of *p*-nitroaniline formed and coupling with *N*-(1-naphthyl)ethylenediamine develop a color that is measured at 548 nm.[13] This procedure is about 10-fold more sensitive than the direct photometric method described above.

---

[13] T. Nagatsu, M. Hino, H. Fuyamada, T. Hayakawa, S. Sakakibara, Y. Nakagawa, and T. Takemoto, *Anal. Biochem.* **74**, 466 (1976).

Reagents

Glycine–NaOH buffer: 0.3 M, pH 8.7.
Substrate: 3 mM Gly-Pro-p-nitroanilide tosylate.
Standard p-nitroaniline solution: 0.15 mM in 2% methanol.
5% (v/v) Perchloric acid.
0.2% (w/v) Sodium nitrite.
0.5% (w/v) Ammonium sulfamate.
0.05% (v/v) N-(1-Naphthyl)ethylendiamine in 95% (v/v) ethanol.

Procedure. A reaction mixture (0.2 ml) contains 50 μl of 0.3 M gly-cine–NaOH buffer (pH 8.7), 100 μl of 3 mM Gly-Pro-p-nitroanilide, 30 μl of H$_2$O, and 20 μl of an enzyme solution. A blank tube contains the same assay mixture except for the enzyme solution. The sample and blank tubes are incubated at 37° for 30 min, and the reaction is stopped by adding 0.8 ml of 5% perchloric acid. The same volume (20 μl) of the enzyme solution is added to the blank tube. The mixtures are centrifuged at 3000 g for 10 min, and 0.5 ml of each supernatant is removed. A standard solution, which contains 100 μl of 0.15 mM p-nitroaniline (15 nmol), 50 μl of the buffer, and 50 μl of H$_2$O, is treated as above. To all the tubes, 0.5 ml of 0.2% sodium nitrite is added, and the tubes are kept at 4° for 10 min. Freshly prepared 0.5% ammonium sulfamate (0.5 ml) is then added. After 1.0 ml of 0.05% N-(1-naphthyl)ethylenediamine in 95% etha-nol is added, all the tubes are incubated at 37° for 30 min in the dark. The absorbance of the sample is read at 548 nm after that of the blank is adjusted to 0. The quantity of p-nitroaniline enzymatically formed is calculated from the absorbance of the sample relative to that of the standard p-nitroaniline solution (15 nmol). In this case the value thus obtained should be multiplied by a factor of 2, because one-half of the reaction mixture has been used for color development.

Other Methods

Spectrophotometric Method using Gly-Pro-p-Nitroanilide as Sub-strate. The assay mixture (3.0 ml) contains 0.75 ml of 0.3 M glycine–NaOH (pH 8.7), 1.5 ml of 3 mM Gly-Pro-p-nitroanilide, and 0.6 ml of H$_2$O.[13] The temperature of the assay mixture is equilibrated to 37°, and the reaction is started by adding 0.15 ml of enzyme. The blank mixture is prepared by substituting the enzyme solution with H$_2$O. The reaction mixtures in the cuvette with a 1-cm light path are maintained at 37°. The absorbance of the sample is read at 385 nm by an automatic recording spectrophotometer that has been zeroed against the blank. The absorbance of a standard solution containing 0.45 μmol of p-nitroaniline is also recorded for calcula-tion of the quantity of the substrate hydrolyzed.

Fluorometric Method Using Gly-Pro-2-Naphthylamide as Substrate. Free 2-naphthylamine is intensely fluorescent whereas dipeptidyl-2-naphthylamides are only slightly fluorescent.[7,14] This fact permits continuous monitoring of the progress of hydrolysis by means of a recording fluorometer or spectrofluorometer. The assay mixture contains 1.0 ml of 0.3 M Tris-HCl (pH 8.0), 1.0 ml of 6.0 mM Gly-Pro-2-naphthylamide, and 0.8 ml of H_2O. After the assay mixture is warmed to 37°, the reaction is started by adding 0.2 ml of enzyme, followed by recording the rate of change in fluorescence at 410 nm relative to the 2-naphthylamine standard (0.01 mM). The excitation wavelength is 340 nm.

Purification Procedure

Two forms of DPPIV, papain-cleaved soluble form and Triton X-100-solubilized membrane form, are purified from plasma membranes of rat liver. Plasma membranes are prepared by the method of Ray[15] with a slight modification.[16] The specific activity of DPPIV in the isolated membranes is increased by about 25-fold as compared with that of homogenates. The plasma membranes are stored at $-20°$ until use.

Purification of the Soluble Form

Unless otherwise indicated, the following procedures are conducted at about 4°.[17]

Step 1. Papain Treatment. Plasma membranes (506 mg) are suspended in 30 ml of 20 mM Tris-HCl (pH 7.5) containing 5 mM L-cysteine. Papain (31.3 U/mg protein, type III from Sigma, St. Louis, MO) is added to the membrane suspension (final concentration of papain, 1 mg/ml), and the mixture is stirred at 37° for 3 hr. The mixture, to which $MgCl_2$ is added (to 1 mM) to inhibit papain activity, is then centrifuged at 105,000 g for 1 hr. About 90% of total DPPIV activity in the membranes is recovered in the supernatant.

Step 2. Ammonium Sulfate Precipitation. Saturated ammonium sulfate solution is added dropwise to the supernatant with continuous mixing to 60% saturation, followed by centrifugation of the mixture at 15,000 g for 20 min. The supernatant is then adjusted to 90% saturation of ammonium sulfate, and the mixture is again centrifuged as above. The precipitates

[14] R. D. C. Macnair and A. J. Kenny, *Biochem. J.* **179**, 379 (1979).
[15] T. K. Ray, *Biochim. Biophys. Acta* **196**, 1 (1970).
[16] Y. Ikehara, K. Takahashi, K. Mansho, S. Eto, and K. Kato, *Biochim. Biophys. Acta* **470**, 202 (1977).
[17] S. Ogata, Y. Misumi, and Y. Ikehara, *J. Biol. Chem.* **264**, 3596 (1989).

thus obtained with ammonium sulfate between 60 and 90% saturation are dissolved in 3 ml of 20 mM Tris-HCl (pH 7.5) containing 0.2 M NaCl and dialyzed overnight against 2 liters of the same solution.

Step 3. Gel Filtration. The sample is subjected to gel filtration through a Sephacryl S-300 column (2.5 × 100 cm, Pharmacia, Piscataway, NJ) equilibrated with 20 mM Tris-HCl (pH 7.5)/0.2 M NaCl. The column is eluted with the same buffer at a flow rate of 6 ml/hr, and 2-ml fractions are collected. DPPIV appears as a single peak at an elution position with M_r 220,000. Four fractions with DPPIV activity are combined and the sample is adjusted to contain 0.5 M NaCl in the above buffer.

Step 4. Wheat-Germ Agglutinin (WGA)-Sepharose Chromatography. The sample obtained by gel filtration is applied onto a WGA-Sepharose column (2 × 15 cm, Pharmacia) equilibrated with 20 mM Tris-HCl (pH 7.5)/0.5 M NaCl. The column is washed with 200 ml of the same solution, and adsorbed proteins including DPPIV are eluted with 0.2 M N-acetylglucosamine (GlcNAc) in the same buffer (Fig. 1). Fractions with DPPIV activity are combined and concentrated to about 1 ml in an ultrafiltration cell with an XM50 membrane (Amicon, Denvers, MA).

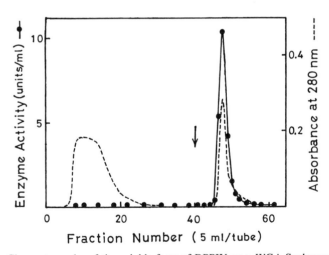

FIG. 1. Chromatography of the soluble form of DPPIV on a WGA-Sepharose column. An enzyme fraction obtained by Sephacryl S-300 chromatography was subjected to chromatography through a WGA-Sepharose column (2 × 15 cm) that had been equilibrated with 20 mM Tris-HCl (pH 7.5) containing 0.5 M NaCl. After the column was washed with 200 ml of the same solution, adsorbed proteins including DPPIV were eluted with 0.2 M GlcNAc in the same solution (indicated by an arrow). Protein concentration was measured by absorbance at 280 nm (----) and DPPIV activity was determined with Gly-Pro-p-nitroanilide as substrate (—●—).

Step 5. Polyacrylamide Gel Electrophoresis. Aliquots (70 µl/gel) of the sample are subjected to electrophoresis on disc gels (7.5% acrylamide, 1 × 13 cm) at pH 8.6 according to Davis.[18] Immediately after electrophoresis, the gels are stained for DPPIV activity at 37° for 5 min in 0.2 M Tris-HCl (pH 7.8) containing 0.5 mM Gly-Pro-2-naphthylamide and Fast Garnet GBC (1.25 mg/ml).[8] Stained areas of the gels are cut out. Segments obtained from 15 gels are packed into a 10-cm syringe and squeezed through a 21-gauge needle. The gel homogenates in 20 ml of 20 mM Tris-HCl (pH 7.5) are stirred for 2–3 hr and then centrifuged at 30,000 g for 30 min. The supernatant is concentrated to 1 ml in the ultrafiltration cell as above. The enzyme thus purified is found to be a single protein with 103 kDa when analyzed by SDS–polyacrylamide gel electrophoresis.[17]

Purification of the Membrane Form

Step 1. Extraction with Triton X-100. Plasma membranes (903 mg) are suspended in 60 ml of 20 mM Tris-HCl (pH 7.5) and adjusted to contain 0.5% Triton X-100. The mixture is stirred for 30 min and then centrifuged at 105,000 g for 1 hr. Of DPPIV activity in the mixture, 97% is recovered in the supernatant.[17]

Step 2. Chromatography on an Affi-Gel Blue Column. The sample is adjusted to contain 0.2 M NaCl in the solubilizing buffer and applied to an Affi-Gel Blue column (3 × 30 cm, Bio-Rad, Richmond, CA) which has been equilibrated with 20 mM Tris-HCl (pH 7.5)/0.2 M NaCl/0.5% Triton X-100. The column is subjected to stepwise elutions with 0.2 M NaCl (600 ml), 0.6 M NaCl (700 ml), 1.0 M NaCl (900 ml), and 1.5 M NaCl (600 ml) in the above buffer. About 85% of the enzyme activity applied is eluted with 1.0 M NaCl, as shown in Fig. 2. Fractions 152 to 190 of the activity peak are pooled and used for the next step.

Step 3. Chromatography on WGA-Sepharose Column. The sample is directly applied to the WGA-Sepharose column (2 × 15 cm) equilibrated with 20 mM Tris-HCl (pH 7.5)/0.5 M NaCl/0.5% Triton X-100. The column is washed with 200 ml of the equilibrating buffer and then with 200 ml of 20 mM Tris-HCl/0.1 M NaCl/0.1% Triton X-100. DPPIV adsorbed to the column is eluted with 0.2 M GlcNAc in the latter buffer. Fractions of the active peak are pooled and concentrated over the XM50 membrane in the ultrafiltration cell. In this case, concentrating the protein is accompanied by an increase in concentration of Triton X-100, which is, however,

[18] B. J. Davis, *Ann. N.Y. Acad. Sci.* **121,** 404 (1964).

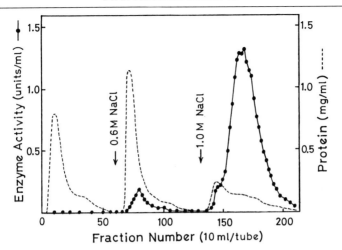

FIG. 2. Chromatography of the membrane form of DPPIV on an Affi-Gel Blue column. An enzyme fraction obtained by extraction of plasma membranes with 0.5% Triton X-100 was applied to an Affi-Gel Blue column (3 × 30 cm) that had been equilibrated with 20 mM Tris-HCl (pH 7.5) containing 0.2 M NaCl and 0.1% Triton X-100. Elution was carried out by stepwise increase of NaCl (0.2, 0.6, and 1.0 M) in the same buffer containing 0.1% Triton X-100. Protein concentration, ----; DPPIV activity, —●—.

considerably lowered by repeating a dilution of the condensed sample with 20 mM Tris-HCl (pH 7.5). The sample is finally concentrated to 1.5 ml.

Step 4. Polyacrylamide Gel Electrophoresis. Polyacrylamide gel electrophoresis and extraction of DPPIV from gels are carried out in the presence of 0.1% Triton X-100. All the other conditions are the same as those described for the soluble DPPIV. The purified membrane form, when analyzed by SDS–polyacrylamide gel electrophoresis, is found to be a single protein of 109 kDa, slightly larger than the soluble form.[17]

The purification procedures of the soluble and membrane forms of DPPIV are summarized in Table I.

Properties

Stability

DPPIV is a relatively stable enzyme in most respects.[19] The enzyme is known to survive autolysis for up to 24 hr at pH 4 and 37°, with a recovery of more than 50% of the activity.[7] The purified enzyme, however,

[19] T. Yoshimoto, M. Fischl, R. C. Orlowski, and R. Walter, *J. Biol. Chem.* **253**, 3708 (1978).

TABLE I

PURIFICATION OF SOLUBLE AND MEMBRANE FORMS OF RAT LIVER DPPIV

Step	Total protein (mg)	Total activity (units)	Specific activity (units/mg)	Purification factor	Yield (%)
Soluble form					
Plasma membranes[a]	506	170.0	0.34	1	100
Papain solubilized	184	156.2	0.85	2.5	91.9
Ammonium sulfate	33.8	144.7	4.28	12.6	85.1
Sephacryl S-300	5.6	120.6	21.5	63.2	70.9
WGA-Sepharose	2.4	112.0	46.7	137.4	65.9
Electrophoresis	1.5	99.4	66.3	195.0	58.5
Membrane form					
Plasma membranes[b]	903	343	0.38	1	100
Triton X-100 extract	275	333	1.21	3.2	97.1
Affi-Gel Blue	43.6	254.8	5.84	15.4	74.3
WGA-Sepharose	6.9	242.2	35.1	92.4	70.6
Electrophoresis	2.6	160.2	61.6	162.1	46.7

[a] Prepared from 60 rat livers (wet weight, 780 g).
[b] Prepared from 110 rat livers (wet weight, 1430 g).

becomes unstable below pH 5.0, but remains stable in neutral and alkaline solutions. No significant loss of the activity is caused by incubation at 45° and pH 7.5 for 30 min. It was reported that the purified enzyme was remarkably stable in 8 M urea, but was inactivated by a 4-hr exposure to 12 M urea.[20]

Activators

DPPIV requires neither metals nor any other cofactors for its activity.[2]

Inhibitors

The enzyme is very sensitive to diisopropyl fluorophosphate (DFP) but much less sensitive to other serine enzyme inhibitors such as phenylmethylsulfonyl fluoride (PMSF) and diethyl 4-nitrophenylphosphate (E600).[7] Most of the usual sulfhydryl reagents have little or no effect, nor do chelating agents.

[20] A. Barth, H. Schulz, and K. Neubert, *Acta Biol. Med. Ger.* **32,** 157 (1974).

TABLE II
SUBSTRATE SPECIFICITY OF PURIFIED RAT
LIVER DPPIV[a]

Substrate	Relative rate (%)
Gly-Pro-pNA	100
Lys-Pro-pNA	72.4
Arg-Pro-pNA	65.7
Glu-Pro-pNA	56.2
Ala-Ala-pNA	8.5
Gly-Ala-pNA	4.3
Gly-Hyp-pNA	11.8
Gly-Leu-pNA	0
Ala-pNA	0

[a] Relative rates of hydrolysis were determined photometrically using dipeptide p-nitroanilide (pNA) derivatives at 1.5 mM in 75 mM glycine–NaOH buffer (pH 8.7). Hyp, Hydroxyproline.

Substrate Specificity

The substrate specificity of DPPIV has been studied by many investigators using the enzymes purified from various sources in combination with various substrates.[2,7,10,19–24] Table II shows the results obtained with DPPIV purified from rat liver. The conclusions obtained by these specificity studies are summarized as follows. (1) DPPIV exhibits a strong preference for substrates having a penultimate prolyl residue, which can be replaced only by alanine[20,22] and hydroxyproline[10] with much lower rates of hydrolysis. (2) When the substrates have the penultimate prolyl residue, the identity of the N-terminal residue is not important for the enzyme activity. (3) No action is detected on NH$_2$-blocked derivatives such as benzyloxycarbonyl and *tert*-butyloxycarbonyl peptides,[2,22] indicating that the N-terminal residue must have a free amino group. (4) Peptides containing proline[23] or hydroxyproline[24] at the third position (X-Pro-Pro- or X-Pro-Hyp-) cannot serve as a substrates for the enzyme.

[21] H. C. Krutzsch and J. J. Pisano, *Biochim. Biophys. Acta* **576,** 280 (1979).
[22] V. K. Hopsu-Havu, P. Rintola, and G. G. Glenner, *Acta Chem. Scand.* **22,** 299 (1968).
[23] J. K. McDonald, B. B. Zeitman, and S. Ellis, *Nature (London)* **225,** 1048 (1970).
[24] H. Oya, M. Harada, and T. Nagatsu, *Arch. Oral. Biol.* **19,** 489 (1974).

Structural Features

Primary Structure

The entire amino acid sequence of DPPIV for rat,[17] human,[25,26] and mouse[27] has been predicted by molecular cloning and sequencing of the respective cDNAs (Fig. 3). The subunit of rat DPPIV contains 767-amino acid residues with a calculated size of 88,107 Da. The predicted N-terminal sequence with a characteristic sequence for the membrane translocation signal is completely identical to that chemically determined for the purified membrane form of DPPIV.[17] In addition, the N-terminal sequence of the papain-solubilized form is identified in the predicted sequence starting at the 35th position from the N terminus (Fig. 3). Thus, it is evident that the signal peptide of DPPIV is not cleaved off during biosynthesis but functions as the membrane-anchoring domain, as demonstrated for other ectoenzymes such as aminopeptidases A and N, sucrase-isomaltase, and γ-glutamyl transpeptidase (γ-glutamyltransferase). The presence of eight potential N-linked glycosylation sites in the molecule accounts for the difference in molecular mass between the predicted polypeptide (88 kDa) and the purified glycoprotein (109 kDa).

The predicted amino acid sequence of rat DPPIV exhibits 84.9 and 91.2% identity to that of the human and mouse enzyme, respectively. The C-terminal sequences of about 250 residues in these DPPIVs are more than 95% identical, and also have a significant similarity to those of other serine peptidases that belong to the prolyl oligopeptidase family.[28,29]

Active Sites

The active site serine of rat DPPIV has been identified by chemical analysis of [³H]DFP-labeled DPPIV and confirmed by site-directed mutagenesis/expression analysis.[30] The purified enzyme was labeled with the active-site -directed reagent [³H]DFP[7] and digested with lysyl endopeptidase. High-performance liquid chromatography of the digested peptides

[25] Y. Misumi, Y. Hayashi, F. Arakawa, and Y. Ikehara, *Biochim. Biophys Acta* **1131**, 333 (1992).

[26] D. Darmoul, M. Lacasa, L. Baricault, D. Marguet, C. Sapin, P. Trotot, A. Barbat, and G. Trugnan, *J. Biol. Chem.* **267**, 4824 (1992).

[27] D. Marguet, A. M. Bernard, I. Vivier, D. Darmoul, P. Naquet, and M. Pierres, *J. Biol. Chem.* **267**, 2200 (1992).

[28] N. D. Rawlings, L. Polgar, and A. J. Barrett, *Biochem. J.* **279**, 907 (1991).

[29] N. D. Rawlings and A. J. Barrett, *Biochem. J.* **290**, 205 (1993).

[30] S. Ogata, Y. Misumi, E. Tsuji, N. Takami, K. Oda, and Y. Ikehara, *Biochemistry* **31**, 2582 (1992).

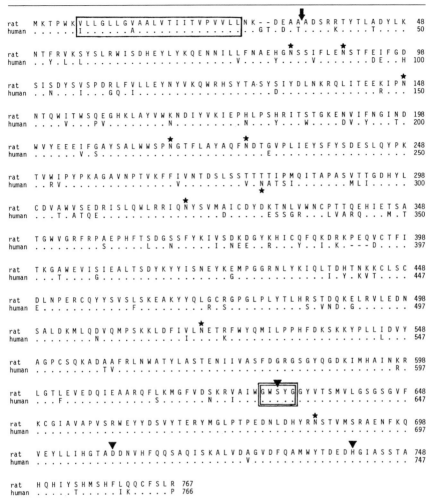

FIG. 3. Primary structures of rat and human DPPIV predicted by the cDNA sequences. Amino acid residues of human DPPIV that are identical with those of the rat enzyme are indicated by a dot. Dashes indicate gaps introduced into the sequences so that they could be aligned. The boxed N-terminal sequence represents a transmembrane domain that also functions as an uncleavable signal peptide during the biosynthesis of DPPIV. An arrow indicates the papain cleavage site for release of the soluble form. Asparagine residues with stars indicate potential N-glycosylation sites. The boxed sequence G-W-S-Y-G corresponds to the consensus active site sequence G-X-S-X-G proposed for serine proteinases. The catalytic triad residues serine, aspartic acid, and histidine are indicated by arrowheads.

yielded a single ^{3}H-labeled peptide, which was analyzed for amino acid sequence and radioactivity distribution. A comparison of the determined sequence with the predicted primary structure of DPPIV revealed that the ^{3}H]DFP was bound to Ser-631 within the sequence Gly-Trp-Ser-Tyr-Gly (positions 629–633), which corresponds to the consensus Gly-X-Ser-X-Gly motif proposed for serine proteases.[31] The DPPIV cDNA was modified by site-directed mutagenesis, followed by expression of the mutagenized cDNAs in COS-1 cells. The complete loss of the enzyme activity was caused by any single substitution of Gly-629, Ser-631, or Gly-633, indicating that the sequence Gly-X-Ser-X-Gly (positions 629–633) is essential for the expression of the DPPIV activity (Fig. 3).

A comparison of the DPPIV sequence (C-terminal 250 residues) with those of other serine peptidases of the prolyl oligopeptidase family suggested candidate residues, aspartic acid and histidine, which may form part of a catalytic triad.[28] Site-directed mutagenesis and expression of the cDNA in COS-1 cells have demonstrated that Asp-709 and His-741 are the other essential residues, possibly required for the catalytic triad of rat DPPIV together with Ser-631 (Fig. 3). This was also confirmed by the same techniques for the human (Asp-708 and His-740)[25] and mouse DPPIV (Asp-702 and His-734).[32] It is of interest to note that the sequential order Ser-631-Asp-709-His-741 of the putative catalytic triad of DPPIV is distinct from that of the classical serine proteases, including the chymotrypsin family (His-Asp-Ser) and the subtilisin family (Asp-His-Ser).

Mutation

Watanabe *et al.*[33] reported the defect of DPPIV in a substrain of Fischer-344 rats. The molecular basis for the enzyme deficiency was examined in the subsequent studies.[34,35] Cloning and sequencing of DPPIV cDNAs revealed a point mutation (G → A at nucleotide 1897) in the cDNA from the enzyme-defective rat, which leads to substitution of Gly-633 → Arg in the active site sequence Gly-Trp-Ser-Tyr-Gly (Fig. 3). Pulse-chase experiments with primary-cultured rat hepatocytes showed that the mutant DPPIV, although being synthesized as a precursor with the same molecular mass (103 kDa) as the wild type, was rapidly degraded within the endoplas-

[31] S. Brenner, *Nature (London)* **334**, 528 (1988).
[32] F. David, A.-M. Bernard, M. Peirres, and D. Margue, *J. Biol. Chem.* **268**, 17247 (1993).
[33] Y. Watanabe, T. Kojima, and Y. Fujimoto, *Experientia* **43**, 400 (1987).
[34] E. Tsuji, Y. Misumi, T. Fujiwara, N. Takami, S. Ogata, and Y. Ikehara, *Biochemistry* **31**, 11921 (1992).
[35] T. Fujiwara, E. Tsuji, Y. Misumi, N. Takami, and Y. Ikehara, *Biochem. Biophys. Res. Commun.* **185**, 776 (1992).

mic reticulum without being processed into the mature form (109 kDa), resulting in no expression of DPPIV on the cell surface. Site-directed mutagenesis and expression of the cDNA confirmed that the rapid degradation of DPPIV in the endoplasmic reticulum is caused by a single substitution not only of Gly-633 but also of Gly-629 to any other residue.[34,35] Thus, it is evident that the consensus active site motif is essential for the expression of DPPIV on the cell surface as well as for its catalytic activity.

[17] Acylaminoacyl-peptidase

By WANDA M. JONES, ANDREA SCALONI, and JAMES M. MANNING

Acylaminoacyl-peptidase (EC 3.4.19.1) catalyzes the removal of a blocked amino acid from a blocked peptide as described in the following equation:

$$X\text{-}aa_1 - aa_a \cdots aa_n \rightarrow X\text{-}aa_1 + aa_2 \cdots aa_n$$

The enzyme is also referred to by the names acylpeptide hydrolase,[1-3] acylamino acid-releasing enzyme[4,5] and acylaminoacyl-peptide hydrolase.[6] The products of the reaction are an acyl amino acid and a peptide with a free N terminus shortened by one amino acid. The enzyme acts on a variety of substrates, including peptides with different N-terminal acyl groups, i.e., acetyl, chloroacetyl, formyl, and carbamyl groups.[7] The optimum length of the blocked peptide substrate is 2–3 amino acids, but larger peptide substrates are also cleaved at slower rates.[8] For instance, the blocked 13-residue peptide, α-melanocyte-stimulating hormone (α-MSH), is a substrate.[7] On the other hand, N-terminally blocked proteins are not substrates for the enzyme.[9] Acylaminoacyl-peptidase could con-

[1] W. Gade and J. L. Brown, *J. Biol. Chem.* **253**, 5012 (1978).
[2] W. M. Jones and J. M. Manning, *Biochem. Biophys. Res. Commun.* **126**, 933 (1985).
[3] K. Kobayashi, L.-W. Lin, J. E. Yeadon, L. B. Klickstein, and J. A. Smith, *J. Biol. Chem.* **264**, 8892 (1989).
[4] S. Tsunasawa, K. Narita, and K. Ogata, *J. Biochem. (Tokyo)* **77**, 89 (1975).
[5] M. Mitta, K. Asada, Y. Uchimura, F. Kimizuka, I. Kato, F. Sakiyama, and S. Tsunasawa, *J. Biochem. (Tokyo)* **106**, 548 (1989).
[6] G. Radhakrishna and F. Wold, *J. Biol. Chem.* **264**, 11076 (1989).
[7] W. M. Jones, L. R. Manning, and J. M. Manning, *Biochem. Biophys. Res. Commun.* **139**, 244 (1986).
[8] W. M. Jones, A. Scaloni, F. Bossa, A. M. Popowicz, O. Schneewind, and J. M. Manning, *Proc. Natl. Acad. Sci. U.S.A.* **88**, 2194 (1991).
[9] T. C. Farries, A. Harris, A. D. Auffret, and A. Aitken, *Eur. J. Biochem.* **196**, 679 (1991).

ceivably act on nascent peptides during biosynthesis and also hydrolyze bioactive peptides, but its physiological function is not known with certainty.

Purification Procedure

Assay of Enzyme Activity

For purification of the enzyme, activity is most easily determined with acetylalanine p-nitroanilide as substrate. Because it is fast, sensitive, and easy to perform, this assay is used for monitoring fractions collected during column chromatography. The assay solution contains 4 mM substrate in 100 mM Bis–Tris, pH 7.4, and between 0.05 and 2 μg of enzyme in a final volume of 1.0 ml at 37°. The rate of formation of product, p-nitroaniline, is followed at 405 nm as a function of time. Specific activity is defined as micromoles of product formed per minute per unit of absorbance at 280 nm.

For kinetic studies, the hydrolysis of acetylated peptide substrates is followed with the fluorescamine (Sigma, St. Louis, MO) reagent to monitor the appearance of new amino groups. Incubations are performed at 37° and aliquots are removed periodically and placed into new culture tubes (16 × 100 mm). While vortexing, 0.5 ml of 0.04% fluorescamine in high-performance liquid chromatography (HPLC)-grade acetone is mixed in and, after at least 5 min, 1.5 ml of H$_2$O is added. The relative fluorescence intensity is read and a standard graph of the expected product (alanine in the case of acetyldialanine as substrate or dialanine in the case of acetyltrialanine as substrate) is employed to convert the fluorescence value into nanomoles of product.

Preparation of Red Cell Hemolysate

The enzyme is prepared from human red cells. Outdated, packed red blood cells (10 units; about 5 liters of whole blood) are suspended in 2 volumes of 0.9% (isotonic) sodium chloride. After centrifugation at 1000 g for 10 min at 4°, the supernatant is discarded and the packed cells are washed with 2 volumes of isotonic saline. After gentle mixing, the cells are again centrifuged at 1000 g for 10 min. This supernatant is also discarded. The cells are lysed by addition of 2 volumes of water containing 1 mM dithiothreitol (DTT) and then subjected to a freeze–thaw step in dry ice/acetone. The hemolysate is centrifuged for 30 min at 16,000 g. With care not to include particulate cell matter, most of the supernatant is removed and saved. An equal volume of water containing 1 mM DTT is added to the contents of the centrifuge bottle and, after careful mixing, the solution is centrifuged at 16,000 g for 30 min. This supernatant is removed and added to that saved previously.

Bulk DE-52 Chromatography

To remove hemoglobin, the hemolysate is applied to a large column (15 × 70 cm) containing a 15 × 15-cm bed of DEAE-52 (Whatman, Clifton, NJ) equilibrated in 50 mM Bis–Tris, pH 6.5, containing 1 mM DTT and 1 mM EDTA (Buffer A). To facilitate the large volume of hemolysate, a large column (70 × 15 cm) with a wide-bore stopcock outlet is used. After application of lysate, the resin is washed with Buffer A containing 0.15 M NaCl until the resin is no longer red and the effluent A_{280} value is in the range of 0.1–0.3. The enzyme is eluted in Buffer A containing 0.5 M NaCl. Fractions (~200 ml each) are collected in 250-ml centrifuge bottles. Each fraction is assayed for enzyme activity as described above. The active fractions are pooled and extensively dialyzed against 20 mM Bis–Tris, pH 6.5, 1 mM DTT, 1 mM EDTA.

Second DE-52 Chromatography

The bulk DE-52 pool described above is applied directly without concentrating to a DE-52 column (4 × 30 cm) equilibrated in Buffer A. The column is washed with 3 liters of Buffer A containing 0.15 M NaCl but no fractions are collected. A linear gradient consisting of 900 ml each of Buffer A containing 0.15 M NaCl and Buffer A containing 0.5 M NaCl is then applied. The A_{280} and activity of the collected fractions are determined and the fractions containing active enzyme are pooled and concentrated to a volume of ~15 ml.

Sephacryl S-300 Chromatography

The protein from the second DE-52 chromatography is applied in two steps to a Sephacryl S-300 column (3 × 65 cm), which is equilibrated and eluted in Buffer A; 3-ml fractions are collected. After determination of A_{280} and activity, fractions are divided into three pools on the basis of specific activity ranging from 100–400.

Toyopearl Chromatography

A column of Toyopearl DEAE 650 M (2.5 × 10 cm) is packed, washed with Buffer A containing 0.5 M NaCl, and then equilibrated in Buffer A alone. The partially purified enzyme from the Sephacryl column (30–40 A_{280} units) is applied to this column, which is eluted with a linear gradient consisting of 100 ml each of Buffer A containing 0.15 M NaCl and Buffer

A containing 0.5 M NaCl. Fractions (1.5 ml) are collected and after determination of A_{280} and enzyme activity, the specific activity of each fraction is calculated and the appropriate fractions pooled. Fractions near the peak containing the pure enzyme showed a single band in the Beckman Paragon nondenaturing gel system and have specific activities near 400. Fractions of specific activity 100–300 are less pure and are rechromatographed. This chromatography is repeated until all of the concentrated Sephacyl S-300 pool has been chromatographed. Fractions of specific activity lower than 100 are discarded.

Activity with Different Substrates

N-Terminal blocked peptides of 2–20 amino acids are substrates, with the optimal length being 2–3 amino acids.[8] The enzyme exhibits a wide pH optimum depending on the substrate used.[2] For example, with acetyl glutamate-p-nitroanilide, the pH optimum is around 6, but for acetylalanine p-nitroanilide, the pH optimum is around 8.4. For most acetylated peptide substrates, the pH optimum is between 7.3 and 7.6. Esters such as p-nitrophenyl propionate are substrates and even large esters such as α-naphthyl butyrate are hydrolyzed by the enzyme.

Enzyme Properties

The enzyme is composed of four identical subunits, each of 732 amino acids, for a total molecular weight of about 300,000.[1-4] The N-terminal amino acid of each subunit is blocked. Its isoionic point by isoelectric focusing has been determined to be 4. The enzyme maintains full activity indefinitely at $-80°$ in the presence of 0.1 mM dithiothreitol.

The purified enzyme is specifically cleaved by trypsin at only one position, i.e., the peptide bond between amino acids 193 and 194. After such cleavage there is no effect on enzyme activity.

Catalytic Mechanism

The enzyme is inhibited by several types of reagents, including diisopropyl fluorophosphate (DFP), p-hydroxymercuribenzoate,[10] diethyl pyrocarbonate, and some heavy metals, such as Hg^{2+}, Zn^{2+}, and Cd^{2+}.[1] The inhibition by DFP is due to modification at Ser-597, and acetylleucine chloromethyl ketone inactivates the enzyme by reaction at the active site His-707.[10] These two sites constitute two parts of the catalytic triad present

---

[10] A. Scaloni, W. M. Jones, D. Barra, M. Pospischil, S. Sassa, A. Popowicz, L. R. Manning, O. Schneewind, and J. M. Manning, *J. Biol. Chem.* **267**, 3811 (1992).

in the enzyme.[10,11] Studies have been initiated on the structure of the enzyme,[12] which has the features of α/β proteins, i.e., the catalytic histidine residues are located close to the C-terminal end. Although the enzyme is present in practically all normal tissues, as first shown by Narita and colleagues,[4] it is absent from cultured cell lines of small-cell lung carcinoma.[13]

Acknowledgments

The authors are grateful to Adelaide Acquaviva for her expert assistance with the manuscript.

[11] N. D. Rawlings, L. Polgar, and A. J. Barrett, *Biochem. J.* **279**, 907 (1991).
[12] M. Feese, A. Scaloni, W. M. Jones, J. M. Manning, and S. J. Remington, *J. Mol. Biol.* **223**, 546 (1993).
[13] A. Scaloni, W. Jones, M. Pospischil, S. Sassa, O. Schneewind, A. M. Popowicz, F. Bossa, S. L. Graziano, and J. M. Manning, *J. Lab. Clin. Med.* **120**, 546 (1992).

[18] Carboxypeptidases C and D

By S. JAMES REMINGTON and KLAUS BREDDAM

Introduction

Carboxypeptidases C and D, members of the serine carboxypeptidase family, are enzymes that specifically release amino acids from the C termini of peptides and proteins. Originally these enzymes were termed acid carboxypeptidases because the optimum for their hydrolysis of peptide bonds was in the acidic pH range.[1] However, this group of enzymes was renamed because the original name suggested that they utilized the mechanism of the acid proteases, and no evidence for this has been presented. On the contrary, all these enzymes are inhibited by diisopropyl fluorophosphate,[1] a specific inhibitor of the serine proteases. This implies the existence of a Ser/His/Asp catalytic triad similar to the one found in the serine endopeptidases. In the case of carboxypeptidase Y (CPD-Y) from *Saccharomyces cerevisiae* chemical modifications and site-directed mutagenesis have identified Ser-146 and His-397 in this capacity[1,2] and

[1] K. Breddam, *Carlsberg Res. Commun.* **51**, 83 (1986).
[2] L. M. Bech and K. Breddam, *Carlsberg Res. Commun.* **54**, 165 (1989).

the three-dimensional structure of CPD-Y[3] has confirmed Asp-338 as the third element in the catalytic triad.

Distribution and Function of Serine Carboxypeptidases

Serine carboxypeptidases are widely distributed among fungi, higher plants, and animals.[1] In all organisms wherein they appear they may have several functions. As vacuolar enzymes they participate in the general turnover of proteins and as extracellular enzymes they cleave off amino acids needed for nutrition from external peptides or proteins. More specific functions of serine carboxypeptidases have also been described, e.g., in proteolytic processing,[4,5] in regulation of peptide hormone levels, and in complexes with β-galactosidase and neuraminidase as a "protective protein"[6-11] that is required for assembly of the functional particle.

Assay and Isolation of Serine Carboxypeptidases

Carboxypeptidase activity is most conveniently determined spectro-photometrically using N-blocked dipeptides as substrates. The choice of blocking group determines the wavelength required to monitor the cleavage of the peptide bond. With the benzyloxycarbonyl (Z) and benzoyl (Bz) groups a rather short wavelength in the ultraviolet range is required, whereas with the furylacryloyl (FA) group wavelengths around 340 nm must be used. The fact that the FA substrates may be employed outside the range where proteins absorb is very convenient, in particular during purification of enzymes where large amounts of ultraviolet absorbing mate-

[3] J. Endrizzi, K. Breddam, and S. J. Remington, *Biochemistry*, in press.
[4] L. Latchinian-Sadek and D. Y. Thomas, *J. Biol. Chem.* **268**, 534 (1993).
[5] A. Dmochowska, D. Dignard, D. Henning, D. Y. Thomas, and H. Bussey, *Cell (Cambridge, Mass.)* **50**, 573 (1987).
[6] H. L. Jackman, F. Tan, H. Tamei, C. Beurling-Harbury, X.-Y. Li, R. A. Skidgel, and E. G. Erdös, *J. Biol. Chem.* **265**, 11265 (1990).
[7] C. E. Odya and E. G. Erdös, this series, Vol. 80, p. 460.
[8] H. L. Jackman, P. W. Morris, P. A. Deddish, R. A. Skidgel, and E. G. Erdös, *J. Biol. Chem.* **267**, 2872 (1992).
[9] J. Tranchemontagne, L. Michaud, and M. Potier, *Biochem. Biophys. Res. Commun.* **168**, 22 (1990).
[10] N. J. Galjart, N. Gillemans, D. Meijer, and A. d'Azzo, *J. Biol. Chem.* **265**, 4678 (1990).
[11] N. J. Galjart, N. Gillemans, A. Harris, G. T. J. van der Horst, F. W. Verheijen, H. Galjaard, and A. d'Azzo, *Cell (Cambridge, Mass.)* **54**, 755 (1988).

rial is present. The following general assay may be employed for a serine carboxypeptidase:

965 μl 0.05 M Acetic acid, 1 mM EDTA, pH 4.5

25 μl 8 mM FA-Xaa-Yaa dissolved in methanol

10 μl Enzyme solution

The cleavage of the Xaa-Yaa peptide bond can be monitored as a decrease in the absorption at 337 nm. The choice of Xaa-Yaa depends on the specificity of the enzyme. With carboxypeptidases C and D (see below) Phe-Phe and Ala-Lys, respectively, are good choices because these substrates are commercially available from various sources.

Serine carboxypeptidases have been isolated by various techniques but affinity chromatographic techniques using DL-benzylsuccinic acid as ligand and aminocaproic acid or glycyl-L-tyrosine as spacer have gained widespread use.[12-17] The adsorption of enzyme to the resin is accomplished at pH 4–5. With crude extracts typically 0.1–1 mg of enzyme may be loaded per ml of resin but with preparations of pure enzyme the loadings may be increased dramatically. In many cases extensive washing of the column with high concentrations of sodium chloride is possible without eluting the enzyme, but there are exceptions.[12,14,17] Elution of the enzyme from the affinity column is accomplished either by shifting the pH to 7, where the enzyme does not bind to the benzylsuccinic acid ligand,[13] or by addition of benzylsuccinic acid to the buffer.[14-17]

Biosynthesis of Serine Carboxypeptidases

The biosynthesis of a few serine carboxypeptidases has been investigated. CPD-Y from *S. cerevisiae* is synthesized as a preproenzyme containing a signal peptide of 20 amino acid residues that directs the precursor across the endoplasmic reticulum membrane. In the endoplasmic reticulum the signal peptide is cleaved off and folding takes place. A four-amino acid segment of the propeptide serves as signal for vacuolar targeting.[18] The propeptide renders the enzyme inactive and facilitates correct folding of the enzyme after denaturation, suggesting that it functions as an intramolecular chaperone during folding of the enzyme.[19] On arrival of the proen-

[12] F. D. Degan, B. Ribadeau-Dumas, and K. Breddam, *Appl. Environ. Microbiol.* **58**, 2144 (1992).

[13] J. T. Johansen, K. Breddam, and M. Ottesen, *Carlsberg Res. Commun.* **41**, 1 (1976).

[14] K. Breddam, S. B. Sørensen, and M. Ottesen, *Carlsberg Res. Commun.* **48**, 217 (1983).

[15] K. Breddam, S. B. Sørensen, and M. Ottesen, *Carlsberg Res. Commun.* **50**, 199 (1985).

[16] K. Breddam and S. B. Sørensen, *Carlsberg Res. Commun.* **52**, 275 (1987).

[17] K. Breddam, *Carlsberg Res. Commun.* **53**, 309 (1988).

[18] L. A. Valls, J. R. Winther, and T. H. Stevens, *J. Cell. Biol.* **111**, 361 (1990).

[19] J. R. Winther and P. Sørensen, *Proc. Natl. Acad Aci. U.S.A.* **88**, 9330 (1991).

zyme in the vacuole it is activated by degradation of the propeptide in a process involving the aspartic endopeptidase saccharopepsin (EC 3.4.23.25, also known as proteinase A).[20] In the endoplasmic reticulum and in the Golgi apparatus, CPD-Y is glycosylated (N linked) at four positions, but the carbohydrate side chains appear to influence only the rate of intracellular transport and not the vacuolar targeting or the activity of the enzyme.[21]

The information concerning the biosynthesis of serine carboxypeptidases from other organisms is much more limited. However, the cloning of carboxypeptidase I from germinated barley (CPD-MI) has provided an important piece of information. This enzyme is synthesized as a single-chain enzyme[22] but is later converted into a two-chain enzyme by excising a linker peptide in the middle of the structural gene.[23] However, it is not known whether this processing is required for activity. The structure of the genes for carboxypeptidase III from wheat (CPD-WIII) and rice has been determined[24,25] and the production of both enzymes is stimulated by the plant hormone gibberellic acid.

Enzymatic Properties and Substrate Preferences of Serine Carboxypeptidases

Among serine endopeptidases k_{cat} increases from 0 at acidic pH, dependent on the deprotonation of the essential His with a pK_a value around 7.[26] With CPD-Y, the most extensively studied serine carboxypeptidase, k_{cat} for the hydrolysis of peptides also increases with pH but is dependent on a pK_a value around 4.5,[27,28] which is unusually low for a His. Furthermore, k_{cat} does not approach 0 at acidic pH, with the result that it only increases by a factor of 4.5 over the pH range 3.7–7. High k_{cat} values at low pH values are also found with other serine carboxypeptidases[29] but the mechanistic features responsible for this have not been identified.

With CPD-Y as well as other serine carboxypeptidases, K_m for the hydrolysis of peptide substrates increases with pH.[1] A study using site-

[20] G. Ammerer, C. P. Hunter, J. H. Rothman, G. C. Saari, L. A. Valls, and T. Stevens, *Mol. Cell. Biol.* **6,** 2490 (1986).

[21] J. R. Winther, T. H. Stevens, and M. C. Kielland-Brandt, *Eur. J. Biochem.* **197,** 681 (1991).

[22] N. P. Doan and G. B. Fincher, *J. Biol. Chem.* **263,** 11106 (1988).

[23] S. B. Sørensen, K. Breddam, and I. Svendsen, *Carlsberg Res. Commun.* **51,** 475 (1986).

[24] D. C. Baulcombe, R. F. Barker, and M. G. Jarvis, *J. Biol. Chem.* **262,** 13726 (1987).

[25] K. Washio and K. Ishikawa, *Plant Mol. Biol.* **19,** 631 (1992).

[26] A. Fersht, "Enzyme Structure and Mechanism," 2nd Ed., Freeman, New York, 1985.

[27] U. Mortensen, S. J. Remington, and K. Breddam, *Biochemistry* **33,** 508 (1994).

[28] M. Fukuda and S. Kunugi, *Eur. J. Biochem.* **149,** 657 (1985).

[29] K. Breddam, *Carlsberg Res. Commun.* **50,** 309 (1985).

TABLE I
SERINE CARBOXYPEPTIDASES DIVIDED INTO SUBCLASSES CARBOXYPEPTIDASE C AND D

Carboxypeptidase C	Lys/Phe	P1[f]	Carboxypeptidase D	Lys/Phe	P1[f]
CPD-Y[a]	0.11	H	CPD-MII/CPD-WIII[b]	12	H/B
CPD-MIII[b]	0.0001	H	CPD-AII[c]	1300	H/B
CPD-MI[b]	0.001	H/B	KEX-1[a,g]	n.d.	n.d.
CPD-AI[c]	0.06	H/B	CPD-S1[e]	2	B
CPD-Pro[d]	n.d.	Pro			

[a] From *Saccharomyces cerevisiae*.[1,14]
[b] From barley/wheat.[14–16,29,32]
[c] From *Aspergillus niger*.[12]
[d] Human.[7]
[e] From *Penicillium janthinellum* (see text).[17]
[f] P1 preference: H, hydrophobic amino acid residues; B, basic amino acid residues; n.d., not determined.
[g] *Saccharomyces cerevisiae* *kex1* gene product.

directed mutagenesis on CPD-Y has shown that this is due to ionization of Glu-145 with a pK_a of 4.3,[27] resulting in charge repulsion between the carboxylate group of the substrate and that of Glu-145 (see later). This probably explains the acidic pH profile for peptide hydrolysis with other serine carboxypeptidases as well, considering that Glu-145 is situated in a part of the amino acid sequence that is conserved among serine carboxypeptidases.[30,31]

The substrate preference of a number of serine carboxypeptidases with respect to the P1 and P1' positions has been determined and compared.[12] The enzymes may conveniently be divided into two groups according to their P1' substrate preferences. The enzymes with preference for hydrophobic and basic P1' amino acid residues have been given the name carboxypeptidase C (EC 3.4.16.5) and carboxypeptidase D (EC 3.4.16.6), respectively. The serine carboxypeptidases for which substrate preference data are available may be distributed according to the k_{cat}/K_m ratio Z-Ala-Lys/Z-Ala-Phe (see Table I). This ratio is very low with carboxypeptidases I (CPD-MI) and III (CPD-MIII) from germinated barley, and these enzymes are therefore true members of the carboxypeptidase C group. With carboxypeptidase I from *Aspergillus niger* (CPD-AI) and CPD-Y, the ratios were significantly higher, reflecting that these enzymes hydrolyze peptides with basic C-terminal amino acid residues at appreciable rates.

[30] S. B. Sørensen, I. Svendsen, and K. Breddam, *Carlsberg Res. Commun.* **54**, 193 (1989).
[31] I. Svendsen, T. Hofman, J. Endrizzi, S. J. Remington, and K. Breddam, *FEBS Letters* **333**, 39 (1993).

Nevertheless, they do belong to the "C" group. Carboxypeptidase II from *A. niger* (CPD-AII) is by far the most specific of the "D" enzymes, but carboxypeptidase II from barley (CPD-MII) and wheat (CPD-WII)[32] also belongs to this group. The classification of the serine carboxypeptidase involved in α-factor processing (yeast *kex1* gene product) and the mammalian proline-specific enzyme (CPD-Pro, human prolyl carboxypeptidase) is based on an incomplete/nonsystematic characterization of substrate preference. With carboxypeptidase I from *Penicillium janthinellum* (CPD-SI) the Lys/Phe ratio is 1.8, and hence this enzyme cannot be classified as belonging to either group.

The preference of these enzymes with respect to the P1 position is also indicated in Table I. It is seen that a preference for Lys/Arg or hydrophobic amino acid residues in the P1' is not necessarily linked to the same preference in the P1 position. Both subclasses share the low preference for Gly, His, Pro, Asp, Asn, Glu, Gln, Ser, and Thr in both the P1 and the P1' positions.

Serine carboxypeptidases do not require a free C-terminal carboxylate group on the substrate for activity. All serine carboxypeptidases described so far hydrolyze alkyl esters of N-blocked amino acids and peptides at very high rates. The K_m for the hydrolysis of peptide esters does not increase with pH as observed with unblocked peptide substrates. Thus, the dissociation of Glu-145 does not influence the hydrolysis of ester substrates. In addition, serine carboxypeptidases act on peptide amides, releasing either ammonia (amidase activity) or amino acid amides (peptidyl amino acid amide hydrolase activity). The ratio of these two competing reactions is dependent on the C-terminal sequence of the peptide amide substrate, but it may be shifted in favor of amidase activity by chemical modification of Met-398, situated in the S1' binding site, with bulky groups.[1] The influence of pH on K_m for the hydrolysis of peptide amides has not been determined but it is probably similar to that found with ester substrates, because k_{cat}/K_m for the two types of substrates is identical.[1] The accommodation of an amino acid amide in the S1' binding site is probably aided by a beneficial hydrogen bond between the C-terminal NH_2 group of the substrate and the deprotonated form of Glu-145.[27]

Primary Structure of Serine Carboxypeptidases

Serine carboxypeptidases are glycoproteins with subunit molecular weights of 45,000 to 75,000. A large part of this difference is due to different extents of glycosylation. Thus, CPD-MIII[30] and CPD-SI[31] are glycosylated

---

[32] K. Breddam, S. B. Sørensen, and I. Svendsen, *Carlsberg Res. Commun.* **52**, 297 (1987).

only at a single position as compared to four and five positions in CPD-Y[1] and CPD-WII,[32] respectively. However, the size of the protein part also differs: most of the enzymes contain 411 to 433 amino acid residues, but the molecular masses of CPD-AI and CPD-AII have been found to be significantly larger, consistent with an additional 60 amino acid residues.[12]

The fact that some serine carboxypeptidases are proteolytically processed to give a two-chain enzyme while others are not gives rise to two different types of structures, both found within each of the two subclasses, carboxypeptidases C and D. The serine carboxypeptidases that have been isolated from fungi contain only a single polypeptide chain whereas those isolated from higher plants and animals in most cases contain two peptide chains linked by disulfide bridges.[1] However, a number of the described enzymes aggregate, forming dimers or oligomers under nondenaturing conditions. The available amino acid sequences of serine carboxypeptidases show that single-chain enzymes are about 30% identical[31] whereas the degree of identity drops to about 25% when the double-chain enzymes are included in the comparison. Enzymes isolated from related organisms and with the same function exhibit a much higher degree of identity, in some cases exceeding 90%. CPD-Y[33] and the corresponding enzyme from *Candida albicans*[34] represent such an example, and CPD-WIII from wheat,[24] barley,[30] and rice[25] represents another.

Three-Dimensional Structures of Serine Carboxypeptidases

The three-dimensional structure of a member from each of the carboxypeptidase C and D groups has been determined. The homodimeric serine carboxypeptidase II from wheat (CPD-WII) was the first example to be analyzed,[35] and an atomic model refined at 2.2 Å resolution has been discussed in detail.[36] A model for the monomeric yeast vacuolar enzyme (CPD-Y) has been refined against data collected to 2.7 Å resolution.[3] With about 25% overall sequence identity, the enzymes differ substantially in substrate specificity (hence the classification). CPD-WII hydrolyzes bonds with basic P1′ residues most efficiently whereas CPD-Y has preference for large hydrophobic side chains. It is therefore of interest to compare the two structures with this difference in mind.

[33] L. A. Valls, C. P. Hunter, J. H. Rothman, and T. H. Stevens, *Cell* (*Cambridge, Mass.*) **48**, 887 (1987).

[34] M. Mukhtar, D. A. Logan, and N. F. Käufer, *Gene* **121**, 173 (1992).

[35] D.-I. Liao and S. J. Remington, *J. Biol. Chem.* **265**, 6528 (1990).

[36] D.-I. Liao, K. Breddam, R. M. Sweet, T. Bullock, and S. J. Remington, *Biochemistry* **31**, 9796 (1992).

Overall Fold

The overall fold of the polypeptide backbone is very similar in the two molecules and is rather tyical for an enzyme. Surrounding either side of a central 10-strand β sheet (the N terminus contributes a short and irregular eleventh strand) are 15 α helices. Many of these helices form connections between the six parallel strands that form the center of the sheet and the active site cleft; however, residues 180–311 (the numbering for the wheat enzyme was chosen to correspond to that of the yeast enzyme after sequence alignment) comprise a largely α-helical insertion into the basic fold (Fig. 1a and b), which surrounds the active site and forms part of the cavity. For this reason, the overall fold has been termed $\alpha + \beta$.[35] In the dimeric wheat enzyme, the active sites are independent and there is no evidence that dimeric structure is important for activity.

The remarkable discovery has been made that the serine carboxypeptidases are members of a much larger family of enzymes that hydrolyze unrelated substrates.[37] These are dienelactone hydrolase for *Pseudomonas* sp. B13,[38] several triacylglycerol lipases (e.g., from *Geotrichum candidum*[39] and *Rhizomucor miehei*[40]), acetylcholinesterase from *Torpedo californica*,[41] and haloalkane dehalogenase from *Xanthobacter autotrophicus*.[42] The structures of these enzymes have been determined and there had been no hint that they would have related folds, with the exception that there is identifiable sequence homology between acetylcholinesterase and the *G. candidum* lipase. Members of the family range in size from 25 to 60 kDa, some are glycosylated and others are not, and some are oligomers and others monomers; however, there is no question that the basic fold of these enzymes is the same. Each enzyme contains a catalytic triad nucleophile–acid–histidine in order of amino acid sequence, where the nucleophile is either Ser, Asp, or Cys and the acid is either Asp or Glu. The basic fold of the family is shown schematically in Fig. 2 and consists (usually) of eight strands of β sheet connected by six α helices. The catalytic triads are invariably located on highly conserved secondary structures associated with this fold, and other elements of secondary structure

[37] D. L. Ollis, E. Cheah, M. Cygler, B. Dykstra, F. Frolow, S. Fraken, M. Harel, S. J. Remington, I. Silman, J. Schrag, J. Sussman, and A. Goldman, *Protein Eng.* **5**, 197 (1992).

[38] D. Pathak, K. L. Ngai, and D. Ollis, *J. Mol. Biol.* **204**, 435 (1988).

[39] J. D. Schrag, Y. Li, S. Wu, and M. Cygler, *Nature (London)* **351**, 761 (1991).

[40] A. M. Brzozowski, U. Derewenda, Z. S. Derewenda, G. G. Dodson, D. M. Lawson, J. P. Turkenburg, F. Bjorkling, B. Huge-Jensen, S. A. Patkar, and L. Thim, *Nature (London)* **351**, 491 (1991).

[41] J. L. Sussman, M. Harel, F. Frolow, C. Oefner, A. Goldman, L. Toker, and I. Silman, *Science* **253**, 872 (1991).

[42] S. M. Franken, H J. Rozeboom, K. H. Kalk, and B. W. Dijkstra, *EMBO J.* **10**, 1297 (1991).

a

b

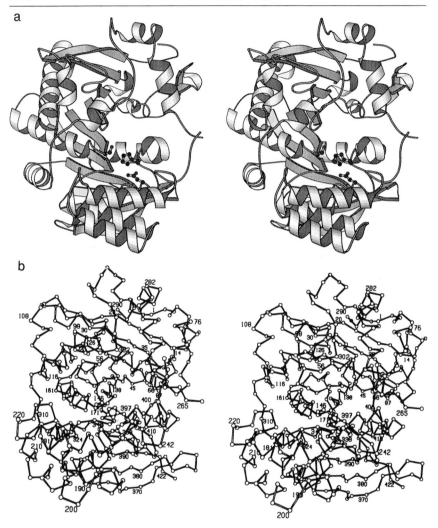

FIG. 1. (a) Stereo ribbon diagram of wheat serine carboxypeptidase II (CPD-WII; RIBBONS program by P. Kraulis). The catalytic residues Asp-338, His-397, and Ser-146 are shown as ball-and-stick models. The point of view is into the active site. (b) Stereo α-carbon trace, with selected residue numbering, from the same point of view.

can be regarded as insertions into the overall fold, termed the α/β hydro-lase fold.[37] Remarkably, the only feature other than the basic fold common to all of these enzymes is the histidine of the catalytic triad. This is therefore the first instance in which evolution has been observed to pre-

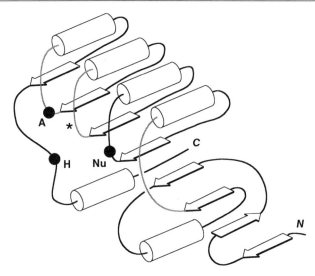

FIG. 2. Schematic diagram of the topology of the α/β hydrolase fold family of enzymes. Nu represents the catalytic nucleophile (Ser, Asp, or Cys), A represents the acid (Glu or Asp), and H represents the conserved histidine. The lightly shaded loops are positions in the fold that accommodate α-helical insertions of up to 200 residues in length. The asterisk shows the point at which residues 180–311 are inserted into the fold in serine carboxypeptidases to form part of the active site cavity and possibly also form part of the basis for substrate specificity.

serve the number and position of the elements of a catalytic triad, but not their identities.

Catalytic Residues

The active sites of carboxypeptidases C and D contain a catalytic triad Asp-338/His-397/Ser-146 similar in geometry but not identical to the geometries of the trypsin and subtilisin families of enzymes[36] (Fig. 3). The chief difference between the catalytic triads of the serine carboxypeptidases and the serine endopeptidases lies in the fact that the carboxylate of Asp-338 and the imidazole of His-397 are not coplanar, and the geometry is far from optical for transfer of a proton.[43] In addition, the relationships of the respective aspartates to the serines are different, so that overall, the three residues comprising each catalytic triad cannot be superimposed.

However, the backbone amides of residues 147 and 53 (the "oxyanion hole") are positioned and oriented in such a way as to be ideally suited to stabilize the negative charge on the tetrahedral intermediate formed in

[43] D. M. Blow, J. J. Birktoft, and B. S. Hartley, *Nature* (*London*) **221**, 337 (1969).

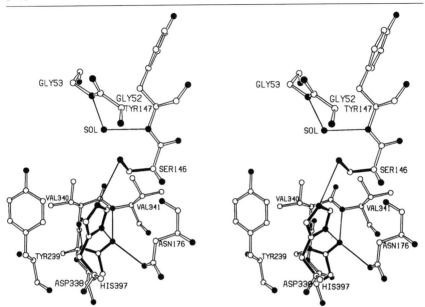

FIG. 3. Active site arrangement of CPD-WII. Members of the catalytic triad are drawn with filled bonds, and hydrogen bonds are indicated with thin lines. A presumed water molecule (labeled SOL) occupies the oxyanion hole.

the reaction. The positioning of the serine relative to the two amides is essentially identical in all three families of serine proteinases.[36] This reinforces the concept that this arrangement is a key aspect of the active site. The stereo diagram in Fig. 3 shows this arrangement in some detail, with the enzyme (CPD-WII) in the resting state, and the oxyanion hole occupied by a presumed water molecule.

In the proximity of His-397 of CPD-Y is found an unblocked Cys. Mutational replacement of this amino acid residue with Ser reduces the activity to approximately 10%.[44] Thus, Cys-341 is not an essential amino acid residue in spite of the fact that it is conserved among single-chain serine carboxypeptidases.[5,30,31] It appears to be involved in correctly positioning the side chain of His-397 and also to shield Asp-338 from solvent, a role played also by Val-341 in the wheat enzyme.

Carboxylate-Binding Site

A prominent feature of a carboxypeptidase must be the ability to recognize and bind the C-terminal carboxylate group of peptide substrates.

[44] J. R. Winther and K. Breddam, *Carlsberg Res. Commun.* **52**, 265 (1987).

The three-dimensional structures of CPD-Y and CPD-WII have suggested that a hydrogen bonding network consisting of Trp-49, Asn-51, Glu-145, and Glu-65 may function in this capacity.[3,35,36] In the complex of CPD-WII with arginine the carboxylate group interacts with Asn-51 and Glu-145,[36] suggesting that these two amino acid residues together function in binding the C-terminal carboxylate group of peptide substrates. This interaction is shown in Fig. 4, where hydrogen bonds from the carboxylate of arginine (filled bonds) to Glu-145, Asn-51, and the amide nitrogen of Gly-52 are shown as thin lines.

Also shown in Fig. 4 is a hydrogen bond between the carboxylates of Glu-145 and Glu-65, a very unusual feature of the active sites of CPD-Y and CPD-WII and conserved among known sequences. In order for this bond to form, one or both of these residues must be protonated. In order for Glu-145 to form a hydrogen bond with the substrate carboxylate, one or both of these must also be protonated. The side chain of Asn-51 is oriented by Trp-49 in such a way that it can donate a proton only in a hydrogen bond, which suggests that the C-terminal carboxylate binds in the charged form to the enzyme, and that Glu-145 must initially be protonated on binding. Thus, it was speculated that the increase of K_m for peptide

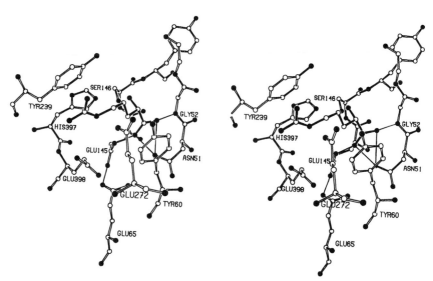

Fig. 4. Binding of the amino acid arginine (filled bonds) at the P1' site of CPD-WII. The hydrogen bonds from the carboxylate of arginine to residues Asn-51, Gly-52, and Glu-145 are shown as thin lines. In addition, the unusual interaction of Glu-65 and Glu-145 is also shown as a thin line. Glu-272 and Glu-398 interact electrostatically, but do not directly form a salt bridge, with the guanidinium moiety of arginine, and Tyr-239 and Tyr-60 accommodate alkyl or aromatic groups.

substrates above pH 5 would be due to deprotonation of Glu-145, and consequent unfavorable repulsive interactions with substrate.[36]

The mutations Trp-49 → Phe, Asn-51 → Ala, Glu-65 → Ala, and Glu-145 → Ala have only little effect on the esterase activity of the enzyme, consistent with the absence of a hydrogen bond acceptor in the P1' position of such substrates.[27] In contrast, the peptidase activity is reduced by all the mutations, in particular when Asn-51 and Glu-145 are replaced. The results are consistent with Trp-49 and Glu-65 serving to orient Asn-51 and Glu-145 correctly by forming hydrogen bonds with these residues. However, it appears that strong interactions are formed only in the transition state because the combined removal of Asn-51 and Glu-145 reduces k_{cat} about 100-fold and leaves K_m practically unchanged.[27]

Basis for Substrate Specificity

CPD-WII, a carboxypeptidase D, most effectively hydrolyzes bonds with basic P1' residues (lysine and arginine), although C-terminal hydrophobic residues are also good substrates. On the other hand, CPD-Y, a carboxypeptidase C, most effectively hydrolyzes bonds with large hydrophobic P1' side chains and has very little activity toward charged side chains. At least part of this difference can be accounted for in a relatively straightforward way. Inspection of Fig. 4 shows that the alkyl portion of the side chain of bound arginine is fully extended and stacks between two aromatic groups, Tyr-239 and Tyr-60, whereas the charged guanidinium moiety interacts with Glu-398 and Glu-272. Shorter substrate side chains such as Phe could be accommodated in the same fasion between Tyr-239 and Tyr-60, and would not necessarily have unfavorable interactions with the two glutamic acid residues. Therefore, from a structural point of view it is not surprising that both types of amino acids bind tightly to the enzyme.

In CPD-Y, glutamic acid residues 398 and 272 are replaced with methionine and leucine, respectively (Fig. 5) so that the pocket is entirely hydrophobic. Structural rearrangements also cause the binding pocket to open up somewhat relative to CPD-WII. The side chain of the C-terminal residue of peptide substrates would be accommodated by Tyr-256, Met-398, and Leu-272. The hydrophobicity of the pocket appears to account for the fact that CPD-Y has rather low activity toward any type of charged residue and the participation of Met-398 is consistent with changes in specificity observed on chemical modification of this residue.[1]

The active site of serine carboxypeptidases is rather deep, and the cavity could accommodate several amino acids N-terminal to the scissile bond.[36] No studies of peptides binding to this region have been reported as of this writing, but structural studies of CPD-Y modified by p-chloro-

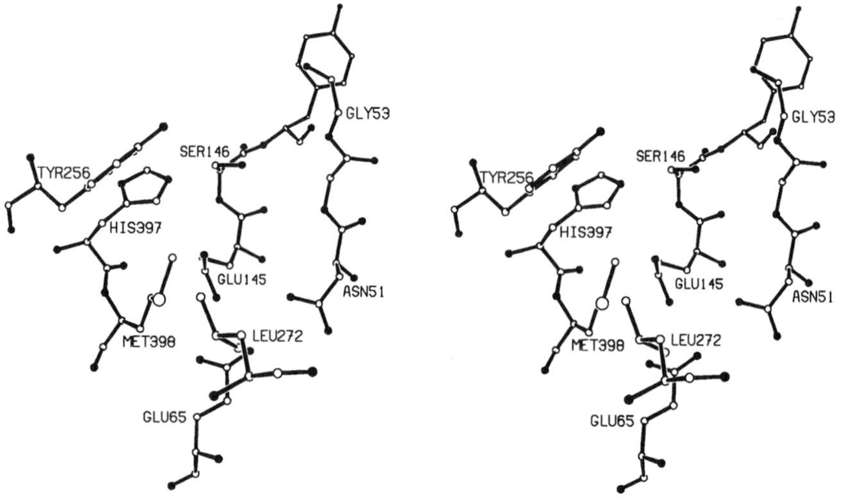

FIG. 5. P1' portion of the active site of CPD-Y from a point of view similar to that of Fig. 4. Glu-272 and Glu-398 of CPD-WII are replaced by leucine and methionine, respectively, to form a more hydrophobic cavity and consequently cause a change in substrate specificity.

mercuribenzoic acid (which alters substrate specificity at P1) have suggested that the pocket that accommodates the P1 side chain consists of Cys-341, Tyr-147, Leu-178, Trp-312, and Ile-340, which is also almost entirely hydrophobic.[3] Furthermore, adjacent to this site in CPD-Y are found a number of polar-to-hydrophobic substitutions relative to CPD-WII, forming an unusual exposed hydrophobic patch not found in CPD-WII.[3] Thus, a general theme of polar-to-hydrophobic substitution correlates with a change in substrate specificity from basic to hydrophobic.

Disulfide Bridges

CPD-WII has three disulfide bridges (56–303, 210–222, 246–268); CPD-Y has five (56–298, 193–207, 217–240, 224–233, and 262–268). One of these, that involving Cys-56, is in the active site and is undoubtedly important in stabilizing the conformation of residues that contact substrate, e.g., the loop 51–56 that forms part of the oxyanion hole and the carboxylate-binding site. Among the others, 210–222 and 217–240 are spatially conserved as are 246–268 and 262–268, although the latter is translated several angstroms relative to CPD-WII. Two new bridges appear in CPD-Y (193–207 and 224–233), but both are located adjacent to the active site. Only the disulfide involving Cys-56 belongs to the "hydrolase fold" portion of these enzymes, but it seems to be unique to the serine carboxypeptidase members of the family. However, all other disulfides are located in the region of the sequence 180–311, which is the largest insertion into the basic fold. This nonuniform distribution of disulfides suggests that they have an important function in anchoring together residues 180–311, which differ substantially between CPD-Y and CPD-WII (see next section) and may be important in substrate recognition.

Evolution of Serine Carboxypeptidases

Although overall sequence identity is only about 25% between CPD-Y and CPD-WII, it rises substantially when the most structurally conserved parts of the moelcules are compared. Overlays of α-carbons show that a 296-residue "core" superimposes with a root mean square deviation of 0.9 Å.[3] Of these, 96 are identical in sequence, leading to 33% sequence identity. Surprisingly, there is no sequence similarily in residues 180–311 in the two enzymes, the part of the molecule that is the largest insertion into the α/β hydrolase fold. The three-dimensional structures of these segments are surprisingly different as well. Figure 6 shows an overlay of these segments, based on the conserved core. Two of the helices are considerably longer in CPD-Y (open bonds) than in their counterparts in CPD-WII, and they are translated by up to a helix diameter. Because this

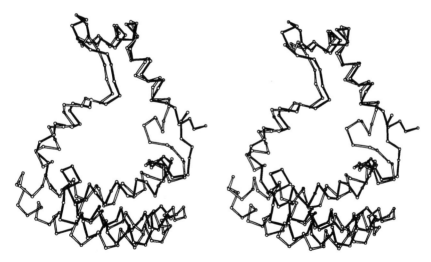

FIG. 6. Structural overlay and α-carbon trace of the "insertion domains," residues 180–311 of CPD-Y (open bonds) and CPD-WII (filled bonds). The overlay was based on an alignment of conserved residues in the cores of the moelcules. There is no sequence homology in these two domains, and large structural rearrangements have taken place during the course of evolution.

segment forms a large part of the active site cavity it may be important for substrate recognition. Circumstantial evidence from comparison of exposed surface residues in the two molecules supports this idea. In CPD-WII, there are a number of negatively charged residues that are all replaced by hydrophobic groups in CPD-Y, leading to the unusual exposed hydrophobic patch discussed previously.

Thus, it would appear that these large structural differences are of functional importance, and that parts of the molecule have evolved much more rapidly than the conserved core region. It is interesting that this rapid evolution seems to be restricted to a region regarded as an insertion into the basic fold. In fact, the hydrolase fold enzymes almost always have an insertion at this point in the structure, which is indicated by the asterisk in Fig. 2. It may be that the hydrolase fold portion of these molecules is most important for proper folding (as well as providing the framework for the catalytic groups), and the insertions are relatively unimportant for folding and are therefore free to change rapidly in the course of evolution.

Applications of Serine Carboxypeptidases

Determination of C-Terminal Sequences

The ability of carboxypeptidases to release amino acids sequentially from the C terminus of peptide chains has frequently been used in the elucidation of primary structures of peptides and proteins. In recent years the use of serine carboxypeptidases and, in particular, CPD-Y has become widespread.[45,46] CPD-Y exhibits a preference for peptides containing hydrophobic amino acid residues (except proline) but hydrophilic and charged amino acids are also released, albeit with very low rates. Although this permits digestion of most peptides with accessible C termini, the sequence analysis is complicated whenever a slowly cleaved peptide bond is succeeded by one or several faster cleaved bonds so that it appears that several amino acids are released as a block. In such cases the application of serine carboxypeptidases with significantly different substrate preferences is advantageous.[45] In particular, a combination of a member of each of the two subclasses carboxypeptidase C and D may be useful.[45] Because serine carboxypeptidases exhibit the highest k_{cat}/K_m for the hydrolysis of peptides at pH 4–5.5[1] it is convenient initially to carry out the digestion using the following procedure with CPD-Y: (1) Peptide is dissolved in 50

---

[45] K. Breddam and M. Ottesen, *Carlsberg Res. Commun.* **52**, 55 (1987).
[46] R. Hayashi, this series, Vol. 47, p. 84 (1977).

μl 0.05 M acetic acid, 1 mM EDTA at a concentration of 0.5 mM; (2) 50 μl 0.05 M sodium acetate, 1 mM EDTA is added, resulting in a pH of 4.5 to 5.0. (3) A 25-μl aliquot is removed and acidified with 4 μl 0.6 M HCl (zero time). (4) Reaction is started by addition of 2 μl CPD-Y (1 mg/ml). (5) At 3, 15, and 60 min 25-μl aliquots are removed and acidified (as described above). (6) The aliquots are applied directly to the amino acid analyzer.

Many peptides are soluble in 0.05 M acetic acid; if not, solubilization may often be accomplished by addition of up to 6 M urea. After dilution the urea concentration will be 3 M, which both CPD-Y and CPD-MII tolerate.[45] Sodium dodecyl sulfate may also be employed for solubilization provided that the final concentration does not exceed 0.1%. The pH of the buffer should in this case be raised to 6.5 because CPD-Y and CPD-MII are not stable in this medium at acidic pH.[45] After the first attempt it is usually necessary to adjust the time course of the reaction by changing the concentration of enzyme or the times the aliquots are removed from the reaction mixture.

Synthesis of Peptide Bonds

In serine carboxypeptidase-catalyzed hydrolysis of amide and ester bonds, water functions as nucleophile. When such reactions are allowed to proceed in aqueous solution with another nucleophile added, e.g., amines, alcohols,[1] amino acids, and various α-carboxylate-blocked derivatives,[1,17] a partitioning of the acyl-enzyme intermediate may take place to yield a hydrolysis product and, e.g., an aminolysis product. The enzyme is not restricted to the natural amino acids and their α-carboxylate-blocked derivatives; other amino acid derivatives, e.g., containing nitro groups and derivatives of benzylamine,[1] may also be accepted. It should be noted that the highest fractions of aminolysis are obtained with amino acid amides and amino acid methyl esters and not with unblocked amino acids, which otherwise would have been thought to be ideal nucleophiles considering the carboxypeptidase nature of the enzyme. It has been demonstrated that the nucleophile forms a complex with the enzyme prior to the nucleophilic attack.[1,47]

When a peptide ester is used as a substrate in a serine carboxypeptidase-catalyzed transacylation reaction, using an amino acid or amino acid derivative as nucleophile, the peptide chain is elongated by one amino acid residue.[1] Thus, serine carboxypeptidases may be used for stepwise synthesis of peptides, and the fact that the reactions may be carried out

[47] U. Christensen, H. B. Drøhse, and L. Mølgaard, *Eur. J. Biochem.* **210,** 467 (1992).

without protection groups under mild reaction conditions represents some advantages. However, the yields are typically lower than in corresponding chemical reactions.

When a peptide is used as a substrate, a transpeptidation reaction takes place where the C-terminal amino acid residue is replaced with the added nucleophile.[1] Transpeptidation reactions have been demonstrated with numerous proteases but serine carboxypeptidases are particularly suited because the site of reaction is restricted to the C-terminal peptide bond. Reactions of this type have been shown to proceed in high yield (>80%) with amino acid amides and o-nitrobenzylamine as nucleophiles.[17,29,48–50] These reactions can be applied to amidation of peptides[51,52] and incorporation of radioactive labels and other groups of various importance.[53] As an example, the following conditions were used for amidation of GRF(1–28)-Ala-OH, a fragment of growth hormone-releasing factor (GRF)[52]: 2 mM GRF(1–28)-Ala-OH, 1.5 M H-Arg-NH$_2$, 5 mM EDTA, pH 8.0, CPD-Y (0.002 mg/ml). After 150 min 97% of the substrate had been converted and the yield of the transpeptidation product GRF(1–28)-Arg-NH$_2$ was 87%. These reaction conditions may serve as a guideline but the amount of enzyme required is strongly dependent on the C-terminal sequence of the substrate, and the concentration of nucleophile required to saturate the enzyme is dependent on its nature. Furthermore, with some serine carboxypeptidases an influence of the C-terminal amino acid residue of the substrate is observed; with CPD-Y C-terminal Ala is the optimal choice of leaving group.[1]

[48] K. Breddam, F. Widmer, and J. T. Johansen, *Carlsberg Res. Commun.* **46**, 121 (1981).
[49] D. B. Henriksen, K. Breddam, and O. Buchardt, *Int. J. Pept. Protein Res.* **41**, 169 (1980).
[50] K. Breddam, F. Widmer, and J. T. Johansen, *Carlsberg Res. Commun.* **45**, 237 (1980).
[51] D. B. Henriksen, M. Rolland, M. H. Jakobsen, O. Buchardt, and K. Breddam, *Pept. Res.* **5**, 321 (1992).
[52] K. Breddam, F. Widmer, and M. Meldal, *Int. J. Pept. Protein Res.* **37**, 153 (1991).
[53] A. Schwarz, C. Wandrey, E. A. Bayer, and M. Wilchek, this series, Vol. 184, p. 160 (1990).

[19] Serine-Type D-Ala-D-Ala Peptidases and Penicillin-Binding Proteins

By BENOÎT GRANIER, MARC JAMIN, MAGGY ADAM, MORENO GALLENI,
BERNARD LAKAYE, WILLY ZORZI, JACQUELINE GRANDCHAMPS,
JEAN-MARC WILKIN, CLAUDINE FRAIPONT, BERNARD JORIS,
COLETTE DUEZ, MARTINE NGUYEN-DISTÈCHE, JACQUES COYETTE,
MÉLINA LEYH-BOUILLE, JEAN DUSART, LÉON CHRISTIAENS,
JEAN-MARIE FRÈRE, and JEAN-MARIE GHUYSEN

Substrates and Catalyzed Reactions

DD-Peptidase Activity

The catalyzed reaction is a double transfer of the R-L-aminoacyl(aa)-D-alanyl moiety of R-L-aa-D-alanyl-D-alanine carbonyl donors to the active site serine, with formation of a serine ester-linked acyl (R-L-aa-D-alanyl) enzyme and, from this, to an exogenous acceptor. Here E is the enzyme; D, the carbonyl donor; E · D, the Michaelis complex; E–D*, the acyl-enzyme; HY, the exogenous acceptor; P1, the leaving group of the enzyme acylation step; K, the dissociation constant, and k_{+2} and k_{+3} are the first-order rate constants. Reaction (1) occurs:

$$E + D \underset{}{\overset{K}{\rightleftharpoons}} \underset{\text{P1}}{E \cdot D} \overset{k_{+2}}{\longrightarrow} E-D^* \underset{\text{HY}}{\overset{k_{+3}}{\longrightarrow}} E + P2 \qquad (1)$$

Here P1 is D-alanine and P2 depends on the final acceptor. When HY is D-alanine, H_2O, or a suitably structured amino compound, the carbonyl donor is regenerated, hydrolyzed (carboxypeptidase activity), or transpeptidated (transpeptidase activity), respectively; $k_{cat} = k_{+2}k_{+3}/(k_{+2} + k_{+3})$; $K_m = Kk_{+3}/(k_{+2} + k_{+3})$; and $k_{cat}/K_m = k_{+2}/K$, i.e., the second-order rate constant for enzyme acylation.

Prototypic carbonyl donors are diacetyl (Ac_2)-L-Lys-D-Ala-D-Ala and N^{α}-acetyl-L-Lys-D-Ala-D-Ala.

Penicillin-Binding Activity

Penicillin is a suicide substrate. Because of the endocyclic nature of the scissile β-lactam amide bond, the leaving group of the enzyme acylation step remains part of the acyl enzyme. The first part only of the transfer cycle is achieved, leading to a long-lived, serine ester-linked acyl(penicilloyl)-enzyme, and the enzyme behaves as a penicillin-binding protein (PBP). Reaction (1) becomes

$$E + D \overset{K}{\rightleftharpoons} E \cdot D \overset{k_{+2}}{\to} E\text{-}D^* \overset{k_{+3}}{\to} E + P \quad \text{(with low } k+3 \text{ value)} \quad (2)$$

The inertness of the penicilloyl enzyme is not absolute. Hydrolytic breakdown causes release of penicilloate. Rupture of the C-5–C-6 bond causes release of the thiazolidine moiety and formation of an acylic acyl-enzyme that is hydrolytically and aminolytically labile. Enzyme recovery is always a very slow process. Under conditions where acyl-enzyme breakdown is negligible (half-life = $0.69/k_{+3}$), the β-lactam concentration and the time of incubation required to immobilize 99% of the enzyme as acyl-enzyme at the steady state of the reaction are related to k_{+2}/K by ([D] $t)_{0.99} = 4.6K/k_{+2}$.

Prototypic β-lactam carbonyl donors are benzylpenicillin, [^{14}C]- and [^{3}H]benzylpenicillin, ^{125}I-labeled hydroxybenzylpenicillin, and 5'-fluo-resceylglycyl-6-aminopenicillanate (5'-Flu-Gly-6APA) (Fig. 1).

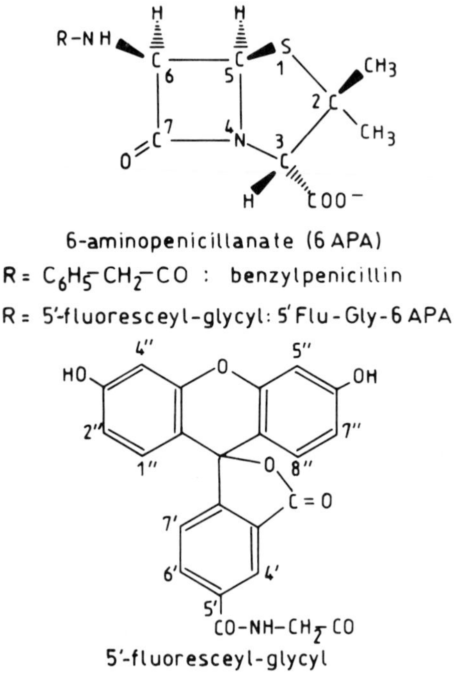

6-aminopenicillanate (6 APA)

R = C_6H_5-CH_2-CO : benzylpenicillin

R = 5'-fluoresceyl-glycyl: 5'Flu-Gly-6 APA

5'-fluoresceyl-glycyl

FIG. 1. Structure of benzylpenicillin and 5'-fluoresceyl-glycyl-6-amino-penicillanate (5'-Flu-Gly-6APA).

Esterase–Thiolesterase Activity

Reaction (1) applies. Prototypic carbonyl donors are ester S1e, C_6H_5-CONH-CH_2-COO-CH(CH_2-C_6H_5)-COO$^-$ (racemic mixture); thiolester S2a, C_6H_5-CONH-CH_2-COS-CH_2-COO$^-$; and thiolester S2d, C_6H_5-CONH-CH(CH_3)-COS-CH_2-COO$^-$ (D-isomer).

Low and High Molecular Weight Penicillin-Binding Proteins

All bacteria possess an assortment of low and high molecular weight membrane-bound PBPs [reaction (2)]. The PBPs belong to the penicilloyl serine transferase superfamily. Invariably, the active site serine residue belongs to the conserved tetrad SXXK.[1–3]

The low molecular weight PBPs are single catalytic entities. The bulk of the protein is on the outer face of the plasma membrane and bears a carboxy-terminal extension, the end of which serves as membrane anchor. The low molecular weight PBPs probably help control the extent of wall peptidoglycan cross-linking throughout the life cycle of the cells. The low molecular weight PBPs so far identified perform DD-peptidase, esterase, and thiolesterase activities.

The high molecular weight PBPs involved in wall peptidoglycan assembly and cell morphogenesis are multimodule proteins. They are membrane anchored at the amino end of the polypeptide chain whereas the bulk of the protein is on the outer face of the membrane and consists of an N-terminal module, fused to a C-terminal, penicillin-binding module. Those so far identified catalyze acyl transfer reactions on esters and thiolesters, but they lack activity on D-alanyl-D-alanine-terminated peptides.

High molecular weight PBPs are also involved in the β-lactam-induced derepression of β-lactamase synthesis in *Bacillus licheniformis* and *Staphylococcus aureus* (BlaR) and in derepression of low-affinity PBP2' synthesis in *S. aureus* (MecR). The penicillin-binding sensor is exposed on the outer face of the membrane and is fused to the carboxy end of a multi-transmembrane segment signal transducer. The BlaR sensor of *B. licheniformis* lacks DD-peptidase, esterase, and thiolesterase activity and is not discussed further in this article.

---

[1] J. M. Ghuysen, *Annu. Rev. Microbiol.* **45**, 37 (1991).

[2] J. M. Frère, M. Nguyen-Distèche, J. Coyette, and B. Joris, *in* "The Chemistry of β-Lactams" (M. I. Page, ed.), p. 148. Chapman & Hall, Glasgow, 1992.

[3] J. M. Ghuysen and R. Hakenbeck (eds.), *New Compr. Biochem.* **27**, (1994).

Bacterial Strains, Growth Conditions, and Enzyme Production

The low molecular weight PBPs/DD-peptidases described below are peculiar in that they are obtained as soluble proteins directly from the producing strains. The high molecular weight PBPs described below are truncated, water-soluble derivatives of the corresponding, membrane-bound proteins. The encoding genes have been cloned and sequenced except for that of the membrane-bound PBP3 of *Enterococcus hirae* [American Type Culture Collection (ATCC) 9790; Rockville, MD].

DD-Peptidase/Low Molecular Weight PBP of Streptomyces K15

This enzyme (the mature protein is 262 amino acid residues)[4] is a strict transpeptidase. Hydrolysis of Ac_2-L-Lys-D-Ala-D-Ala is negligible because the released D-alanine competes successfully with 55.5 M H_2O, performs aminolysis of the acyl-enzyme, and regenerates the original tripeptide. At millimolar concentrations, glycylglycine (which is structurally related to the peptidoglycan interpeptide bridge) overcomes the acceptor activities of D-alanine and water, leading to the quantitative conversion of the tripeptide into Ac_2-L-Lys-D-Ala-Gly-Gly. The *Streptomyces* K15 enzyme is membrane bound, but it lacks membrane anchors. Amplified expression in *Streptomyces lividans* TK24 harboring pDML225 results in the export of an appreciable proportion of the synthesized enzyme in the culture medium.[5]

Streptomyces lividans TK24/pDML225 is grown at 28° for 48 hr in a 20-liter Biolafitte fermentor containing 10 liters of TSB2-thiostrepton medium [trypsin soy broth (Gibco-BRL, Gaithersburg, MD), 30 g; $(NH_4)_2SO_4$, 2 g; thiostrepton, 50 mg; water, 1 liter] with an agitation speed of 250 rev/min and an airflow of 5 liters/min. The culture fluid contains the soluble DD-peptidase. Yield is 2 mg of enzyme per liter of culture.

DD-Peptidase/Low Molecular Weight PBP of Streptomyces R61

This enzyme (the mature protein is 349 amino acid residues)[6] is mainly a carboxypeptidase, but it also catalyzes transpeptidation reactions on specific donor–acceptor substrates and under proper experimental condi-

---

[4] P. Palomeque-Messia, S. Englebert, M. Leyh-Bouille, M. Nguyen-Distèche, C. Duez, S. Houba, O. Dideberg, J. Van Beeumen, and J. M. Ghuysen, *Biochem. J.* 279, 223 (1991).

[5] P. Palomeque-Messia, V. Quittre, M. Leyh-Bouille, M. Nguyen-Distèche, C. J. L. Gershater, I. K. Dacey, J. Dusart, J. Van Beeumen, and J. M. Ghuysen, *Biochem. J.* 288, 87 (1992).

[6] C. Duez, C. Piron-Fraipont, B. Joris, J. Dusart, M. S. Urdea, J. A. Martial, J. M. Frère, and J. M. Ghuysen, *Eur. J. Biochem.* 162, 509 (1987).

tions. The wild-type strain produces the enzyme as a secretory protein. Overexpression is achieved in *S. lividans* TK24 harboring pDML115.[7]

Streptomyces lividans TK24/pDML115 is grown at 28° for 72 hr in a 1-liter Erlenmeyer flask containing 500 ml of TSB1-thiostrepton medium [trypsin soy broth (Gibco-BRL), 30 g; thiostrepton, 25 mg; water, 1 liter] with orbital agitation (250 rev/min). The culture fluid contains the soluble DD-peptidase. Yield is 30 to 40 mg of enzyme per liter of culture.

DD-Peptidase/Low Molecular Weight PBP of Actinomadura R39

This enzyme (489 amino acid residues)[8] is also a secreted carboxypeptidase/transpeptidase. Its specificity profile as a DD-peptidase differs from that of the *Streptomyces* R61 enzyme and it has a much higher penicillin affinity.

Actinomadura R39 is grown at 28° for 48 hr in a 20-liter Biolafitte fermentor containing 14 liters of TAU medium [starch (Merck), 10 g; urea (Merck), 2 g; universal peptone (Merck), 5 g; yeast extract (Difco), 10 g; K_2HPO_4, 2 g; $MgSO_4 \cdot 6H_2O$, 0.6 g; trace element solution, 1 ml; 100 mM Tris-HCl (pH 8.0), 1 liter] with an agitation speed of 250 rev/min and an airflow of 28 liters/min. The trace element solution contains 1 g each of $ZnSO_4 \cdot 7H_2O$, $FeSO_4 \cdot 7H_2O$, $MnCl_2 \cdot 4H_2O$, and $CaCl_2$ for 1 liter of water. The culture fluid contains the soluble DD-peptidase. Yield is 10 mg of enzyme per liter.

Escherichia coli PBP3p

The high molecular weight PBP3 (588 amino acid residues)[9] is involved in cell septation and is a lethal target of penicillin in *E. coli*. About 20 copies of PBP3 occur per bacterial cell. A truncated form (G57-V577) of the PBP3, i.e., PBP3p, is overexpressed in the periplasm *E. coli* HB101 or RR1 harboring pDML232 (carrying the modified *ftsI* gene) and pDML230 (carrying the chaperone SecB-encoding gene).[10] Regulation of the synthesis is at the transcriptional level and is isopropyl-β-D-thiogalactopyranoside (IPTG) dependent. Overexpression above a certain level causes conversion of a large proportion of the enzyme into inactive aggregates during the purification steps.

---

[7] A. M. Hadonou, M. Jamin, M. Adam, B. Joris, J. Dusart, J. M. Ghuysen, and J. M. Frère, *Biochem. J.* **282**, 495 (1992).

[8] B. Granier, C. Duez, S. Lepage, S. Englebert, J. Dusart, O. Dideberg, J. Van Beeumen, J. M. Frère, and J. M. Ghuysen, *Biochem. J.* **282**, 781 (1992).

[9] M. Nakamura, I. N. Maruyama, M. Soma, J. I. Kato, H. Suzuki, and Y. Hirota, *Mol. Gen. Genet.* **191**, 1 (1983).

[10] C. Fraipont, M. Adam, M. Nguyen-Distèche, W. Keck, J. Van Beeumen, J. A. Ayala, B. Granier, H. Hara, and J. M. Ghuysen, *Biochem. J.* **298**, 189 (1994).

Escherichia coli HB101 (or RR1)/pDML232/pDML230 is grown at 37° in a 20-liter Biolafitte fermentor containing 15 liters of LB/kanamycin/ tetracycline medium [Bacto-tryptone (Difco), 12 g; Bacto-yeast extract, 5 g; NaCl, 10 g; kanamycin, 25 mg; tetracycline, 12.5 mg; water, 1 liter] with an agitation speed of 250 rev/min and an airflow of 5 liters/min. When the culture reaches an absorbance of 0.8 at 550 nm, 2 mM IPTG is added to induce PBP3p synthesis and the culture continues for 3 hr. The cells are collected by centrifugation and spheroplasted at 20° in 1150 ml of 30 mM Tris-HCl (pH 8.0)/27% sucrose/5 mM EDTA/600 mg of lysozyme. The spheroplast suspension is supplemented with 15 mM CaCl$_2$ and 0.5 M NaCl and centrifuged at 30,000 g for 15 min. The supernatant, i.e., the periplasmic fraction, contains the truncated PBP3p. Yield is about 2 mg of enzyme per liter of culture.

Streptococcus pneumoniae PBP2x*

The high molecular weight PBP2x (750 amino acid residues)[11] is a major target of the β-lactam antibiotics in *Streptococcus pneumoniae*. PBP2x is the first PBP to be altered in its amino acid sequence in cefotaxime-resistant laboratory mutants and transformation with DNA segments encoding low-affinity variants of PBP2x from clinical isolates gives rise to transformants with increased penicillin resistance. A truncated form of PBP2x lacking the 30 hydrophobic amino acid residues at positions 19–48, i.e., PBP2x*, is overexpressed in *E. coli* DH5α harboring pCG31.[12]

Escherichia coli DH5α/pCG31 is grown at 37° in a 20-liter Biolafitte fermentor containing 15 liters of LB-chloramphenicol medium [Bacto-tryptone (Difco), 10 g; Bacto-yeast extract, 5 g; NaCl, 10 g; chloramphenicol, 0.5 g; water, 1 liter] with an agitation speed of 200 rev/min and an airflow of 15 liters/min. When the culture reaches an absorbance of 1.2 at 550 nm, the cells are collected by centrifugation, suspended in 25 ml of 10 mM Tris-HCl (pH 8.0), and disrupted with a French press (1000 psi). After centrifugation at 20,000 g for 60 min, the supernatant, which contains PBP2x*, is dialyzed at 4° twice for 2 hr against 10 mM Tris-HCl (pH 8.0). Yield is 6–7 mg of enzyme per liter of culture.

Enterococcus hirae ATCC 9790 t-PBP3

The 78-kDa PBP3 of *E. hirae* ATCC 9790 is involved in cell septation and is the specific target of cefotaxime in this organism.[13] It can be con-

[11] G. Laible, R. Hakenbeck, M. A. Sicard, B. Joris, and J. M. Ghuysen, *Mol. Microbiol.* **3,** 1337 (1989).

[12] G. Laible, W. Keck, R. Luiz, H. Mottl, J. M. Frère, M. Jamin, and R. Hakenbeck, *Eur. J. Biochem.* **207,** 943 (1992).

[13] G. Piras, A. El Kharroubi, J. Van Beeumen, E. Coeme, J. Coyette, and J. M. Ghuysen, *J. Bacteriol.* **172,** 6856 (1990).

verted into a water-soluble 58-kDa t-PBP3 by trypsin digestion of the isolated membranes.

Enterococcus hirae ATCC 9790 is grown at 37° in a 100-liter fermentor containing 95 liters of SB medium [yeast extract, 10 g; peptone, 10 g; $NaH_2PO_4 \cdot H_2O$, 16.45 g; $Na_2HPO_4 \cdot 2H_2O$, 31.85 g; KH_2PO_4, 0.42 g; K_2HPO_4, 0.305 g; glucose, 20 g (sterilized separately as a 20% aqueous solution); H_2O, 1 liter, pH 6.5 at 20°] with an agitation speed of 150 rev/min and an airflow of 100 liters/min. When the culture reaches an absorbance of 5.0 at 550 nm (late exponential phase), the cells are collected and suspended in 2800 ml of 5 mM sodium phosphate (pH 7.0)/1 mM $MgCl_2$. The suspension is divided into seven samples and each sample is supplemented with 40 mg of lysozyme, 800 μg of DNase, 400 μg of RNase, and 4 mg of *Streptomyces globisporus* muramidase I (Sigma, St. Louis, MO), resulting in cell lysis. The membranes are collected by centrifugation at 40,000 g for 30 min and washed several times in 5 mM sodium phosphate (pH 7.0). Yield is 0.16 mg of membrane-bound PBP3 per liter of culture.

Enzyme Purification

Unless otherwise specified, the purification steps are routinely carried out at 4°.

Streptomyces K15 Enzyme

Step 1. The culture supernatant (10 liters, 11.7 g of total proteins, 20 mg of enzyme) is clarified by filtration through a 0.2-μm-pore size poly(vinylidene difluoride) membrane (Millipore, Bedford, MA), concentrated to 1 liter with a Flowgen Ultrasette Tangential Flow ultrafiltration cell and supplemented with an equal volume of cold ($-20°$) acetone. The mixture is maintained at $-20°$ for 30 min, resulting in the precipitation of the enzyme. The pellet, collected by centrifugation at 7500 g for 30 min, is dissovled in 100 ml of 25 mM Tris-HCl (pH 8.0)/0.2 mM dithiothreitol/ 0.5 M NaCl.

Step 2. The solution is dialyzed against the Tris–dithiothreitol buffer without salt, resulting in the precipitation of the enzyme. The pellet collected by centrifugation is dissolved in 10 ml of 25 mM Tris-HCl (pH 7.2)/0.2 mM dithiothreitol/0.4 M NaCl. The solution is divided into five aliquots, each of which is applied to a 35-ml Q-Sepharose column equilibrated against the same buffer. The enzyme is eluted in the void volume while the contaminating nucleic acids are absorbed onto the ion exchanger.

Final yield is 75% and purity is noted below. The preparation is stored at $-20°$ in the presence of a 0.4 M NaCl and at a protein concentration of 1 mg per ml to avoid precipitation.

Filtration of the enzyme preparation on Sephadex G-75 in 25 mM Tris-HCl (pH 8.2)/0.2 mM dithiothreitol/0.5 M NaCl yields a single protein fraction, the elution volume of which is compatible with a M_r value of 27,000 for the protein. SDS–PAGE, however, reveals the presence of a 52,000 M_r PBP in addition to the expected 27,000 M_r PBP. The two proteins each bind penicillin in a 1 : 1 molar ratio, react with the anti-*Streptomyces* K15 enzyme antiserum, and exhibit the same amino-terminal sequence. The 52-kDa PBP completely disappears when, prior to the SDS–PAGE, the preparation is treated with 7 M guanidinium chloride in 0.1 M Tris-HCl, pH 8.5, at 100° for 3 min, and is dialyzed successively against 8 M urea and 0.1% SDS, suggesting that the 52-kDa PBP might be due to incomplete denaturation of the protein.

Streptomyces R61 Enzyme

Step 1. The culture supernatant (1 liter; 300 mg of total proteins; ≈30–40 mg of enzyme) is stirred at 20° with 100 g of Amberlite CG-50 (Serva, Heidelberg, Germany) and the suspension is adjusted to pH 4.0 with glacial acetic acid. The exchanger is collected by filtration, suspended in 50 mM Tris-HCl (pH 8.0; 2 ml of buffer per gram of wet exchanger), and the pH is adjusted to 8.0 with concentrated NH_3. After filtration, the exchanger is treated as above with a second batch of buffer. The combined eluates are dialyzed against 10 mM Tris-HCl (pH 8.0)/50 μM EDTA and concentrated fivefold with an Amicon (Danvers, MA) ultrafiltration unit.

Step 2. The solution is loaded on a 2.5 × 20-cm Q-Sepharose fast flow column equilibrated against 10 mM Tris-HCl (pH 8.0)/50 μM EDTA. After washing, the enzyme is eluted with a linear NaCl gradient from 0 to 0.3 M over 800 ml at a flow rate of 8 ml/min. The active fractions (about 0.12 M NaCl) are concentrated to 20 ml.

Step 3. The solution is filtered through a 4 × 85-cm Sephadex G-75 column equilibrated against 10 mM Tris-HCl (pH 8.0)/50 μM EDTA. The active fractions are pooled, dialyzed against 10 mM Tris-HCl (pH 7.0)/50 μM EDTA, and concentrated to 30 ml.

Step 4. The solution is injected onto a prepacked Hiload 26/10 Q-Sepharose Fast Flow column (Pharmacia, Piscataway, NJ) conditioned with 10 mM Tris-HCl (pH 7.0)/50 μM EDTA. Treatment with a linear NaCl gradient from 0 to 0.2 M over 1100 ml at a flow rate of 5 ml/min yields two active fractions. One of them is eluted at about 73 mM NaCl and the other at about 85 mM NaCl. Each fraction is dialyzed against 10 mM Tris-HCl (pH 8.0)/50 μM EDTA and concentrated to 5 mg of enzyme per milliliter.

Final yield is ≈50% (taking the two fractions into account) and purity is noted below. The preparation is stored at −20°.

The two enzyme isoforms each behave as homogeneous proteins by nondenaturing PAGE, SDS–PAGE, and gel electrofocusing. They each bind penicillin in a 1 : 1 molar ratio with comparable constant values. They might result from different processings at the N or C terminus of the protein. Such a microheterogeneity is not observed when the enzyme is produced by the wild-type *Streptomyces* R61.

Actinomadura R39 Enzyme

Step 1. The culture supernatant (12 liters, 25 g of total proteins, 120 mg of enzyme) is stirred for 15 min at 20° with 240 g of DEAE-cellulose previously equilibrated against 50 mM Tris-HCl (pH 8.0). The ion exchanger is collected by filtration, packed in a 4 × 10-cm column, and washed with 2.5 liters of 100 mM Tris-HCl (pH 8.0)/0.15 M NaCl. The adsorbed enzyme is eluted with a curvilinear NaCl gradient in 100 mM Tris-HCl (pH 8.0; mixing chamber at constant volume, 250 ml of 0.15 M NaCl; added solution, 0.5 M NaCl; flow rate, 2 ml/min). The active fractions (0.35 M NaCl) are concentrated to 25 ml with an Amicon ultrafiltration unit.

Step 2. The solution is filtered through a 4 × 85-cm Sephadex G-100 column in 100 mM Tris-HCl (pH 8.0). The active fractions are pooled (140 ml) and loaded on a 2.6 × 10-cm Q-Sepharose Fast Flow column in the Tris buffer. After washing with 1 liter of 0.35 M NaCl in buffer, the enzyme is eluted with a linear NaCl gradient from 0.35 to 0.45 M over 700 ml at a flow rate of 5 ml/min. The active fractions are pooled, dialyzed against the Tris buffer, and concentrated to 2 mg of enzyme per milliliter.

Final yield is 74% and purity is higher than 95%. The preparation is stored at −20°.

Step 1. The periplasmic fraction (1.15 liter; 1.5 g of total proteins, 30 mg of enzyme) is dialyzed against 10 mM Tris-HCl (pH 8.0)/10% (v/v) glycerol, resulting in the precipitation of the enzyme. The pellet is collected by centrifugation and dissolved in 30 ml of Tris-glycerol/0.5 M NaCl. The solution is dialyzed against Tris–glycerol/0.2 M NaCl, under which conditions 70–80% of PBP3p remains in solution.

Step 2. The solution is applied to a 100-ml column of Procion blue MX4GD (mix 1591)-Fractogel (Merck) equilibrated against Tris–glycerol/0.2 M NaCl. The adsorbed PBP3 is eluted by a linear NaCl gradient from 0.15 to 1 M over 375 ml at a flow rate of 4 ml/min. The active fractions (0.5 M NaCl) are pooled and concentrated.

Final yield is 30% and purity is 80%. The preparation is stored at 20° in the presence of 0.5 M NaCl and at a protein concentration of 0.5 mg per ml to avoid precipitation.

Streptococcus pneumoniae PBP2x*

Step 1. The dialyzed cell extract (3.9 g of total proteins; 98 mg of PBP2x*; in 25 ml 10 mM Tris-HCl, pH 8.0) is loaded on a 2.6 × 14-cm Q-Sepharose Fast Flow column equilibrated against the Tris buffer. The adsorbed PBP2x* is eluted with a linear NaCl gradient from 0 to 1 M over 320 ml at a flow rate of 4 ml/min. The active fractions (0.5 M NaCl) are pooled and dialyzed for 2 hr against 10 mM sodium acetate adjusted to pH 5.0 with 12 N HCl.

Step 2. The solution is divided into five 20-ml samples, each of which is adsorbed on a Mono S HR5/5 Pharmacia column equilibrated against the sodium acetate buffer. The adsorbed PBP2x* is eluted by a linear NaCl gradient from 0 to 1 M over 20 ml at a flow rate of 1 ml/min. Three active fractions are collected at 0.25, 0.32, and 0.61 M NaCl, respectively, and are dialyzed against 10 mM sodium phosphate (pH 7.0). The three enzyme isoforms each are homogeneous by SDS–PAGE and bind penicillin in a 1 : 1 molar ratio with similar constant values.

Final yield is 68% (taking the three fractions into account) and purity is at least 95%. The preparations (2.5 mg of enzyme per ml) are stored at −20° in the presence of 10% glycerol.

Enterococcus hirae t-PBP3

Step 1. A membrane suspension [1 g of total proteins; 2.5 mg of membrane-bound PBP3 in 65 ml of 50 mM Tris-HCl (pH 7.5)/0.5% Triton X-100] is incubated in the presence of 10 mg of trypsin type XI for 30 min at 37°. After centrifugation at 40,000 g for 60 min, the supernatant (containing the 58-kDa t-PBP3) is loaded on a 2.6 × 40-cm Q-Sepharose Fast Flow column equilibrated against 20 mM Bis–Tris (pH 6.3)/0.015% (v/v) Triton X-100. The adsorbed t-PBP3 is eluted with a linear NaCl gradient from 0 to 1 M over 1500 ml, at a flow rate of 5 ml/min. The active fractions (0.2 M NaCl) are pooled and concentrated to 10 ml with an Amicon ultrafiltration unit, and the solution is dialyzed against 10 mM sodium phosphate (pH 7.0).

Step 2. Samples (5 ml each) are filtered through a 1-ml Mono Q HR5/ 5 column equilibrated against the phosphate buffer. The adsorbed t-PBP3 is eluted with a linear NaCl gradient from 0 to 1 M over 31 ml at a flow rate of 0.6 ml/min. The active fractions (0.25 M NaCl) are dialyzed against the phosphate buffer.

Yield is 35% and purity is 7.5%. The preparation contains no PBP other than the 58-kDa t-PBP3. It is stored at $-20°$ at a final concentration of 1.9 mg total proteins per milliliter.

Enzyme Assays and Kinetic Parameters

Assays are performed in 10 mM sodium phosphate (pH 7.0) for the *Streptomyces* R61 enzyme, *S. pneumoniae* PBP2x*, and *E. hirae* t-PBP3, in 10 mM phosphate (pH 7.0) or 50 mM Tris-HCl (pH 7.5)/4 mM MgCl$_2$ for the *Actinomadura* R39, in 10 mM phosphate (pH 7.0)/0.25 M NaCl for the *E. coli* PBP3p, and in 25 mM Tris-HCl (pH 7.2)/0.2 mM dithiothreitol/0.4 M NaCl for the *Streptomyces* K15 enzyme.

Table I gives the K_m, k_{cat}, and k_{cat}/K_m (i.e., k_{+2}/K) values for the peptide, ester, and thiolester carbonyl donors [reaction (1)]. Table II gives the k_{+2}/K and k_{+3} values for benzylpenicillin and 5'-Flu-Gly-6APA [reaction (2)].

Peptides

The enzymatic estimation of D-alanine has been modified, resulting in a twofold increased sensitivity.[14]

i. *o*-Dianisidine is replaced by *o*-dianisidine dihydrochloride (10 mg per ml of water); the pyrophosphate buffer is replaced by 100 mM Tris-HCl, pH 8.0; and the methanol–water mixture (which is added just before the spectrophotometric measurements) is replaced by a mixture of sulfuric acid/methanol/water (60 : 50 : 40; v/v/v). The readings are recorded at 535 nm.

ii. Alternatively, *o*-dianisidine is replaced by 2,2'-azinodi(3-ethylbenzthiazoline sulfonate) (ABTS) and the experimental conditions are modified as follows.

Reagents

Citrate–phosphate, pH 5.3: citric acid, 8.5 g; Na$_2$HPO$_4$, 20.5 g; water, 1 liter.

FAD (monosodium; Boehringer Mannheim, Germany): 0.1 mg per ml of 100 mM Tris-HCl, pH 8.0.

D-Amino-acid oxidase (Boehringer): a suspension of 5 mg per ml of ammonium sulfate solution, as supplied.

ABTS (Boehringer): 5 mg per ml of water.

---

[14] J.-M. Frère, M. Leyh-Bouille, J. M. Ghuysen, M. Nieto, and H. R. Perkins, this series, Vol. 45, p. 610.

TABLE I
KINETIC PROPERTIES OF ACYCLIC CARBONYL DONORS

Enzyme	Ac$_2$-L-Lys-D-Ala-D-Ala			α-Ac-L-Lys-D-Ala-D-Ala			Ester S1e			Thiol ester S2a			Thiol ester S2d		
	k_{cat} (sec^{-1})	K_m (mM)	k_{+2}/K (M^{-1} sec^{-1})	k_{cat} (sec^{-1})	K_m (mM)	k_{+2}/K (M^{-1} sec^{-1})	k_{cat} (sec^{-1})	K_m (mM)	k_{+2}/K (M^{-1} sec^{-1})	k_{cat} (sec^{-1})	K_m (mM)	k_{+2}/K (M^{-1} sec^{-1})	k_{cat} (sec^{-1})	K_m (mM)	k_{+2}/K (M^{-1} sec^{-1})
Streptomyces K15	0.31	6.2	50 (a)	0.45	9	48 (b)	ND	ND	<0.2	ND	ND	8	0.11 / 0.34	1.5 / 4.5	75 / 75 (c)
Streptomyces R61	55	14	4000	0.25	15	17	4.6	0.9	5200	5	0.05	100,000	70	0.1	700,000
Actinomadura R39	18	0.8	22,500	32	0.2	160,000	0.33	0.05	6600	0.9	0.08	11,500	6	0.015	400,000
E. coli PBP3p	No activity			No activity			No activity			No activity			0.25	3	80
S. pneumoniae PBP2x*	No activity			No activity			0.33	2.8	100	0.47	0.8	600	30	5.6	5000
E. hirae t-PBP3	No activity			No activity			No activity			—	—	<20	6	1.8	3200

[a] From reaction (1). The reactions are carried out at 37° in buffer or in buffer containing 2 mM Gly-Gly (a), 10 mM Gly-L-Ala (b), and 1.5 mM Gly-Gly (c). ND, Not determined.

TABLE II
KINETIC PROPERTIES OF β-LACTAM CARBONYL DONORS[a]

Enzyme	Benzylpenicillin		5'-Flu-Gly-6APA	
	k_{+2}/K $(M^{-1} \sec^{-1})$	k_{+3} $(\sec^{-1})$	k_{+2}/K $(M^{-1} \sec^{-1})$	k_{+3} $(\sec^{-1})$
eptomyces K15	135	1.1×10^{-4}	25	ND
eptomyces R61	18,000	1.1×10^{-4}	6300	5×10^{-5}
tinomadura R39	300,000	3×10^{-6}	70,000	8×10^{-6}
coli PBP3p	4000	4×10^{-5}	25	ND
pneumoniae PBP2x*	58,000 ($\pm$5000)	5.5 ($\pm$0.5) $\times 10^{-5}$	1600	4×10^{-5}
hirae t-PBP3	>100,000	ND	600	ND

[a] From reaction (2). The reactions are carried out in buffer at 37°. ND, Not determined.

Peroxidase (Boehringer): 50 μg per ml of water.

Enzyme–coenzyme mixture (freshly prepared). FAD: D-amino-acid oxidase, 30 : 1 (v/v).

H_2O_2 assay mixture (freshly prepared). Citrate–phosphate, pH 5.3 : ABTS : peroxidase, 100 : 1 : 1 (v/v/v).

Assay. Samples containing D-alanine (30 μl) and the enzyme–coenzyme mixture (60 μl) are incubated together for 5 min at 37°. The H_2O_2 assay mixture (400 μl) is added and the absorbance is read at 415 nm. Note that the D-amino-acid oxidase must be completely catalase-free.

Sensitivity. D-Alanine (10 nmol in a final mixture of 490 μl) gives an absorbance of 0.6–0.7.

Esters

The values of the kinetic parameters derive from complete time courses obtained by monitoring the absorbance increase at 254 nm ($\Delta\varepsilon = 500$ M^{-1} cm^{-1}).[15,16]

Thiol esters

The time-dependent decrease of absorbance is monitored at 250 nm ($\Delta\varepsilon = -2200$ M^{-1} cm^{-1}).[15,16] Low K_m values (<500 μM) are derived from complete time courses. In the other cases, the initial rates are determined

[15] M. Adam, C. Damblon, B. Plaitin, L. Christiaens, and J. M. Frère, *Biochem. J.* **270,** 525 (1990).

[16] M. Adam, C. Damblon, M. Jamin, W. Zorzi, V. Dusart, M. Galleni, A. El Kharroubi, G. Piras, B. G. Spratt, W. Keck, J. Coyette, J. M. Ghuysen, M. Nguyen-Distèche, and J. M. Frère, *Biochem. J.* **271,** 601 (1991).

and the k_{cat} and K_m values are obtained by fitting the results on a hyperbola with a nonlinear regression (Enzfitter, Elsevier Biosoft, Cambridge, UK) program or with the help of the Hanes linearization of the Henri–Michaelis equation.

Radioactive and Nonradioactive β-Lactams

Determination of the k_{+3} value of acyl-enzyme breakdown rests on measurements of the recovery of the enzyme activity as a function of time and/or the release of the acyl-enzyme from preformed radioactively labeled acyl-enzyme.[17,18] Knowing the k_{+3} value, determination of the k_{+2}/K value rests on measurements of the pseudo-first-order rate constants of acyl-enzyme formation at varying concentrations of the β-lactam. Direct measurement of acyl-enzyme formation requires the use of a radioactive β-lactam. If the β-lactam is not available in a radioactively labeled form, an indirect procedure can be used in which the enzyme left free after a first incubation with varying concentrations of the nonradioactive β-lactam is fully acylated in a second step, by a radioactive β-lactam. It is important that precise experimental conditions are observed to avoid misleading results.[17,18]

The procedure described below[2] allows the k_{+2}/K value for a nonradioactive β-lactam 1 to be determined in a one-step competition with a radioactive β-lactam 2 for which the k_{+2}/K value is known. With $[D_1]$ and $[D_2] \gg E_0$, the nonradioactive β-lactam 1 decreases the maximal extent of acylation of the protein by the radioactive β-lactam 2, and the k_{+2}/K value for the β-lactam 1 can be computed by measuring the amounts of radioactive acyl enzyme formed with the β-lactam 2 in the absence $([E - D_2^*]_0)$ and in the presence $([E - D_2^*])$ of the β-lactam 1. With $[E - D_1^*] = [E - D_2^*]_0 - [E - D_2^*]$, the use of the following equation is justified if the acylation reactions are completed within a time much shorter than the half-lives of the acyl-enzymes, so that the k_{+3} steps can be neglected.

$$\left(\frac{k_{+2}}{k}\right)_1 = \left(\frac{k_{+2}}{K}\right)_2 \frac{[E - D_1^*]}{[E - D_2^*]} \frac{[D_2]}{[D_1]}$$

The reactions can be made extremely fast by using high concentrations of the β-lactams 1 and 2, because $[E - D_1^*]/[E - D_2^*]$ depends only on $[D_1]/[D_2]$. Utilization of an internal standard is necessary to avoid errors due to losses on deposition of the samples on the electrophoresis gels.

[17] B. Joris and J. M. Frère, *CRC Crit. Rev. Microbiol.* **11**, 299 (1985).
[18] J. M. Ghuysen, J. M. Frère, M. Leyh-Bouille, M. Nguyen-Distèche, and J. Coyette, *Biochem. J.* **235**, 159 (1986).

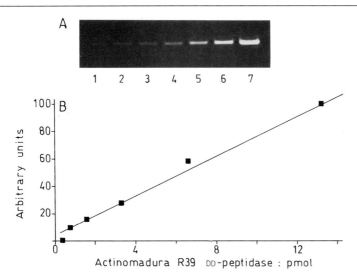

FIG. 2. Detection of the *Actinomadura* R39 DD-peptidase labelled with 5′-Flu-Gly-6APA, by SDS–PAGE and densitometry measurements. (A) Photograph of the gel illuminated with a UV lamp (λ_{max}, 312 nm). Enzyme amounts: 0.22, 0.44, 0.88, 1.65, 3.3, 6.6 and 13.2 pmol in lanes 1–6, respectively. (B) Calibration curve obtained after densitometry of (A). Reprinted with permission from Ref. 19.

Fluorescent β-Lactams

Fluorescein-coupled penicillins replace the radioactive β-lactams.[19] 5′-Flu-Gly-6APA (Fig. 1) serves to illustrate the technique.

i. Quantification by densitometry. The *Actinomadura* R39 enzyme (3 μM) is incubated with 5′-Flu-Gly-6APA (7 μM) in 50 mM Tris-HCl, pH 7.5, for 30 min at 37°, under which conditions the enzyme is fully acylated. Solutions containing varying concentrations of the acyl-enzyme, from 10 to 600 nM, in Tris buffer/2% (v/v) glycerol/0.2% (w/v) SDS/0.2% (v/v) 2-mercaptoethanol/0.02% (w/v) bromophenol blue are heated at 100° for 1 min. Samples (20 μl) are submitted to 12% SDS–PAGE using a 9.0 × 7.0 × 0.075-cm slab gel, for 45 min at 200 V (15 mA). Detection is made under UV light (λ_{max} = 312 nm) and quantification is made with the help of a two-dimensional densitometer (Cybertech CS-1, Dalton, Waalwijk, The Netherlands). Amounts as low as 0.44 pmol of enzyme can be detected with the naked eye and there is a good linear correlation between the fluorescence intensities and the amounts of enzyme (Fig. 2).

[19] M. Galleni, B. Lakaye, S. Lepage, M. Jamin, I. Thamm, B. Joris, and J. M. Frère, *Biochem. J.* **291,** 19 (1993).

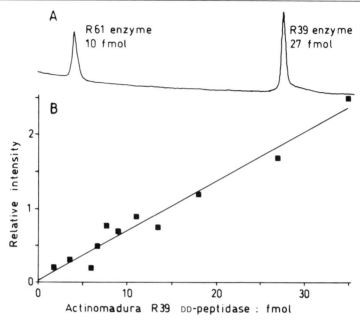

FIG. 3. Identification of *Actinomadura* R39 DD-peptidase and *Streptomyces* R61 DD-peptidase labeled with 5'-Flu-Gly-6APA, by SDS–PAGE using the ALF DNA sequencer. (A) Recording of the fluorescent tracing given by 27 fmol of the R39 enzyme and 10 fmol of the R61 enzyme. (B) Calibration curve showing the ratios of the areas under the peaks corresponding to the R39 and R61 enzymes (relative intensity) versus the amounts of the R39 enzyme. Reprinted with permission from Ref. 19.

ii. Quantification using the ALF automatic DNA sequencer. After reaction with 5'-Flu-Gly-6APA, solutions are made as above, containing varying concentrations, from 0.2 to 3.5 nM, of the acylated *Actinomadura* R39 enzyme and a fixed concentration, 1 nM, of the acylated *Streptomyces* R61 enzyme. The solutions are heated at 100° for 1 min. Samples (10 μl) are submitted to 12% SDS–PAGE using a 18 × 30 × 0.05-cm slab gel and an ALF DNA sequencer.[20] Detection of the fluorescent proteins is performed with a laser beam (λ_{ex}, 494 nm; λ_{em}, 512 nm) and the areas under the fluorescent bands are determined by triangulation. Amounts as low as 2 fmol of enzyme can be detected and satisfactory quantification is obtained down to 7 fmol (Fig. 3).

iii. Comparison with radioactive penicillin. Table II gives the k_{+2}/K and k_{+3} values for the interactions between benzylpenicillin and 5'-Flu-

[20] W. Ansorge, B. Sproat, J. Stegemann, C. Schawger, and M. Zenke, *Nucleic Acids Res.* **15**, 4593 (1987).

TABLE III
DETECTION OF *Streptomyces* R61 DD-PEPTIDASE/PBP WITH RADIOACTIVE AND
FLUORESCENT PENICILLINS[a]

Enzyme concentration (pmol)	[14C]Benzylpenicillin (55 mCi/mmol)	[3H]Benzylpenicillin (5000 mCi/mmol)	125I-Labeled hydroxybenzylpenicillin (2000 mCi/mmol)	5'-Flu-Gly-6APA
13.2	15 hr	15 hr	<6 hr	2 hr (A)
3.3	2 days	1 day	<6 hr	2 hr (A)
0.82	6 days	6 days	12 hr	6 hr (B)
0.22	10 days	7 days	1 day	6 hr (B)
0.05	>10 days	10 days	>1 day	6 hr (B)
0.01	>10 days	>10 days	>1 day	6 hr (B)

[a] Detection with 5'-Flu-Gly-6APA is made by densitometry (A) or with the ALF DNA sequencer (B).

Gly-6APA, respectively, and the enzymes and PBPs studied. Table III compares the total experience times required to detect varying amounts of the *Streptomyces* R61 DD-peptidase/PBP using [14C]benzylpenicillin, [3H]benzylpenicillin, 125I-labeled hydroxybenzylpenicillin, and 5'-Flu-Gly-6APA as labeling agents.

iv. Detection of the membrane-bound PBPs of freezed-thawed cells of *E. coli* HB101. Cells (from a 2-ml sample of a culture at an absorbance of 1.0 at 550 nm) are suspended in 50 μl of 50 mM sodium phosphate, pH 7.0, submitted to three cycles of freezing and thawing, and treated with 5'-Flu-Gly-6APA at a 10 μM final concentration, for 30 min at 30°. After centrifugation, the pellet is dissolved in 50 μl of denaturing buffer/2-mer-

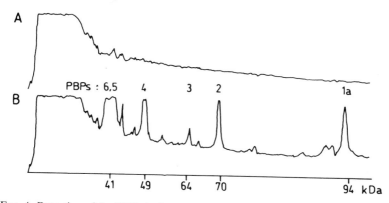

FIG. 4. Detection of the PBPs in freezed and thawed cells of *Escherichia coli* labeled with 5'-Flu-Gly-6APA, by SDS–PAGE using the ALF DNA sequencer. (A) Samples are pretreated with 200 μM benzylpenicillin before labelling with 10 μM 5'-Flu-Gly-6APA. (B) Samples are treated directly with 5'-Flu-Gly-6APA. Reprinted with permission from Ref. 19.

captoethanol and the suspension is heated at 100° for 1 min. Samples (10 μl) are analyzed as above using the ALF DNA sequencer. The same experiment is repeated on freezed-thawed cells preincubated with 200 μM nonradioactive benzylpenicillin, under which conditions all the PBPs are saturated. Figure 4 shows that multiple proteins in E. coli interact competitively with 5'-Flu-Gly-6APA and benzylpenicillin. The proposed protein numbering is based on the known molecular masses of the E. coli PBPs.

v. Specificity profile. Depending on the structure of the fluorescent β-lactam used, the rate of acylation of a given PBP may vary greatly. Thus, for example, 5'-Flu-Gly-6APA, 5'-fluoresceylampicillin, and 6'-fluoresceylampicillin acylate the E. coli PBP3p with k_{+2}/K values of 25 M^{-1} sec^{-1} (Table II), 13,000 M^{-1} sec^{-1}, and 2000 M^{-1} sec^{-1}, respectively.

Acknowledgments

This work was supported in part by the Belgian program on Interuniversity Poles of Attraction initiated by the Belgian State, Prime Minister's Office, Science Policy Programming (PAI No. 19), an Action Concertée with the Belgian Government (Convention 89/94-130), the Fonds de la Recherche Scientifique Médicale (Contract No. 3.4531.92), and a Convention Tripartite between the Région Wallonne, SmithKline Beecham, U.K., and the University of Liège.
Note added in proof: Biotinylated β-lactams can also be used for the analysis of Penicillin Binding Proteins (M. Dargis and F. Malouin, Antimicrob. Ag. Chemother. **38**, 973 (1994).

[20] Cleavage of LexA Repressor

By John W. Little, Baek Kim, Kenneth L. Roland, Margaret H. Smith, Lih-Ling Lin, and Steve N. Slilaty

Introduction

The SOS regulatory system of Escherichia coli controls the response of the cell to treatments that damage DNA or inhibit DNA replication.[1,2] In normally growing cells, about 20 SOS genes are turned off by the LexA repressor. On inducing treatments, LexA undergoes specific proteolytic cleavage; cleavage inactivates LexA and leads to derepression of the SOS genes. This specific cleavage reaction is therefore of biological interest, because it controls the state of the SOS regulatory system.

[1] J. W. Little and D. W. Mount, Cell (Cambridge, Mass.) **29**, 11 (1982).
[2] G. C. Walker, Microbiol. Rev. **48**, 60 (1984).

Cleavage is also of interest from a mechanistic point of view, because the mechanism here is unusual in several respects.[3,4] First, Lex A cleavage is catalyzed *in vivo*, and at neutral pH *in vitro*, by interaction with another regulatory protein, RecA, which is activated *in vivo* by inducing treatments. Second, cleavage proceeds *in vitro* at high pH in a spontaneous reaction termed autodigestion.[5] Many lines of evidence indicate a close mechanistic relationship between autodigestion and RecA-mediated cleavage. Accordingly, we have proposed[3] that activated RecA acts as a coprotease, stimulating autodigestion rather than acting directly as a protease. In this view, RecA acts as a catalyst, in that it stimulates a specific reaction, but the chemistry of cleavage is carried out by groups in LexA, rather than in RecA.

Mechanism of LexA Cleavage and Role of RecA

This view of LexA cleavage raises two major questions: What is the mechanism of LexA cleavage? How does RecA stimulate this reaction? In addressing these questions we have been hampered by two limitations. First, no structural information is available for the part of LexA involved in cleavage. Second autodigestion is an intramolecular reaction,[6] precluding the use of many tools available to enzymologists and protein chemists. Accordingly, we have resorted primarily to a combination of genetic and biochemical analysis of cleavage. This work involves the isolation of mutations that decrease or increase the rate of cleavage, and characterization of cleavage of these mutant proteins as a function of pH and of other reaction conditions. The use of mutant proteins with an increased rate of cleavage has recently allowed the development of an intermolecular cleavage reaction in which one molecule of LexA acts as an enzyme to attack other molecules,[7] offering the possibility of more detailed analysis of mechanism and kinetics.

Organization of LexA

LexA is 22.7 kDa in size,[8,9] and is organized into two structural and functional domains separated by a hinge region (Fig. 1). This protein is

[3] J. W. Little, *Biochimie* **73**, 411 (1991).
[4] J. W. Little, *J. Bacteriol.* **175**, 4943 (1993).
[5] J. W. Little, *Proc. Natl. Acad. Sci. U.S.A.* **81**, 1375 (1984).
[6] S. N. Slilaty, J. A. Rupley, and J. W. Little, *Biochemistry* **25**, 6866 (1986).
[7] B. Kim and J. W. Little, *Cell (Cambridge, Mass.)* **73**, 1165 (1993).
[8] B. E. Markham, J. W. Little, and D. W. Mount, *Nucleic Acids Res.* **9**, 4149 (1981).
[9] T. Horii, T. Ogawa, and H. Ogawa, *Cell (Cambridge, Mass.)* **23**, 689 (1981).

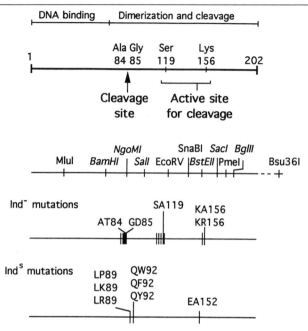

FIG. 1. Schematic depiction of LexA protein. Domain organization of the protein is shown at the top; locations of the cleavage site and active site are indicated. Below are indicated the positions of restriction sites used in recombinant DNA manipulations; naturally occurring sites are shown in roman font, and restriction sites introduced by silent changes are shown in italics. The *Bsu*36I site lies 320 bp beyond the end of *lexA*. All the sites shown are unique in the plasmid (pSHEP15, derived from pJWL184) that we currently use for genetic screens and selections. The *Bam*HI, *Sal*I, *Sna*BI, *Sac*I, and *Bsu*36I sites (and probably the *Pme*I site) are unique in the pET11a overproducing derivatives. At the bottom are shown the locations of *lexA* mutations affecting specific cleavage: Ind⁻ mutants (for noninducible) show little or no cleavage; Ind^S mutants (for superinducible) show increased cleavage rates. Several specific mutations are given. The convention for naming mutations is the wild-type amino acid (one-letter code), followed by the mutant amino acid and the amino acid residue in LexA. Adapted from J. W. Little, *J. Bacteriol.* **175**, 4943 (1993).

homologous to the better characterized phage λ CI repressor, which also undergoes specific cleavage. The LexA N-terminal region is the DNA-binding domain; its C-terminal domain contains the contacts for dimerization, and all the elements necessary for specific cleavage. The cleavage site, Ala⁸⁴-Gly⁸⁵, lies in the hinge region.[10] A tryptic fragment comprising residues 68–202 undergoes both autodigestion and RecA-mediated cleavage at normal rates.[5,6]

[10] T. Horii, T. Ogawa, T. Nakatani, T. Hase, H. Matsubara, and H. Ogawa, *Cell* (*Cambridge, Mass.*) **27**, 515 (1981).

LexA contains three types of sites involved in cleavage: the cleavage site, analogous to the substrate in an enzyme-catalyzed reaction; the active site, which contains a catalytic center and a binding pocket that binds the substrate and positions it optimally with respect to the catalytic center; and a RecA-binding site, where RecA binds to play its role in catalyzing cleavage. Genetic and biochemical analyses have shown the location of the cleavage site and the probable location of the active site; these sites are conserved among a large family of cleavable proteins.[11,12] By contrast, the location of the RecA-binding site in LexA is not yet known; some evidence[3,4] suggests that this site is not conserved among cleavable proteins.

Genetic Analysis of LexA Cleavage

Two types of *lexA* mutations affect cleavage (Fig. 1). Ind⁻ mutations reduce or eliminate specific cleavage; most of these were isolated by an *in vivo* screen for defects in RecA-dependent cleavage.[11] Most such mutant proteins are about equally defective in both cleavage reactions; curiously, no mutant proteins with specific defects in RecA-mediated cleavage, expected to be ones that no longer bind RecA, were found, in contrast to the existence of such mutant proteins in λ repressor.[13] Ind⁻ mutations lie in three regions: near the cleavage site, and at or near two residues, Ser-119 and Lys-156, that are crucial for the chemistry of cleavage (see next section).

Ind[S] mutations increase the rate of cleavage. To date, we have not had a systematic way to isolate Ind[S] mutations, and have isolated them by chance, by a screen of *amber* mutations with various *amber* suppressors, or by an inefficient screen for defects in repressor function. Known mutations[7,14,15] lie near the cleavage site or near Lys-156. They increase the rates of cleavage by large amounts; the most effective mutation, QW92, increases the rate of cleavage about 300-fold at neutral pH values.[16] QW92 and EA152 appear to act independently of one another.[17] The properties of mutant proteins at position 92 are consistent with a conformational moel (see below), which also can account for the role of RecA in stimulating cleavage.

[11] L. L. Lin and J. W. Little, *J. Bacteriol.* **170,** 2163 (1988).
[12] L. L. Lin and J. W. Little, *J. Mol. Biol.* **210,** 439 (1989).
[13] F. S. Gimble and R. T. Sauer, *J. Mol. Biol.* **192,** 39 (1986).
[14] S. N. Slilaty and H. K. Vu, *Protein Eng.* **4,** 919 (1991).
[15] M. H. Smith, M. M. Cavenagh, and J. W. Little, *Proc. Natl. Acad. Sci. U.S.A.* **88,** 7356 (1991).
[16] K. L. Roland, M. H. Smith, J. A. Rupley, and J. W. Little, *J. Mol. Biol.* **228,** 395 (1992).
[17] B. Kim and J. W. Little, unpublished observations (1992).

Mechanism of LexA Cleavage

Several lines of evidence suggest that the hydroxyl side chain of a particular serine residue, Ser-119, is the nucleophile that attacks the peptide bond. This residue is completely conserved among all cleavable proteins.[18,19] Changing it to Ala completely blocks cleavage; a change to Cys gives a protein that can cleave at a slower rate, but reactions always contain a residual unreactive fraction.[20] Ser-119 reacts specifically but at a slow rate with diisopropyl fluorophosphate; modification blocks cleavage.[21] From these data, we infer that LexA cleavage proceeds by a mechanism that shares some similarity with well-characterized serine endopeptidases.

However, LexA cleavage differs from classical serine proteases in that it does not appear to involve a Ser-His-Asp catalytic triad. Instead, cleavage appears to require a deprotonated lysine residue, Lys-156 in LexA. This residue is also completely conserved.[18,19] The pH rate profile of autodigestion[6] is consistent with the need to deprotonate a residue with a pK_a of about 10. Changing Lys-156 to Ala completely blocks cleavage.[20] In the KR156 variant, in which Lys is replaced with Arg, the pH rate profile is shifted to a higher pH value,[12] although it does not appear to reach a maximum value, because nonspecific hydrolysis occurs at the extreme pH values needed to titrate Arg. Although the role of uncharged Lys is not known, we have speculated[20] that it accepts the proton leaving the serine hydroxyl and provides a proton to the departing amino group during formation of an acyl-enzyme intermediate. A similar mechanism has received strong support from high-resolution structural work on β-lactamase-catalyzed hydrolysis.[22] In addition, a serine–lysine dyad mechanism has been proposed for leader peptidase[23-25] (see also [2] and [21] in this volume).

Although LexA cleavage may proceed by a mechanism resembling that of β-lactamase and leader peptidase, the biological setting of LexA differs from that of these proteins, in that LexA cleavage is constrained

[18] R. T. Sauer, R. R. Yocum, R. F. Doolittle, M. Lewis, and C. O. Pabo, *Nature (London)* **298**, 447 (1982).

[19] J. R. Battista, T. Ohta, T. Nohmi, W. Sun, and G. C. Walker, *Proc. Natl. Acad. Sci. U.S.A.* **87**, 7190 (1990).

[20] S. N. Slilaty and J. W. Little, *Proc. Natl. Acad. Sci. U.S.A.* **84**, 3987 (1987).

[21] K. L. Roland and J. W. Little, *J. Biol. Chem.* **265**, 12828 (1990).

[22] N. C. J. Strynadka, H. Adachi, S. E. Jensen, K. Johns, A. Sielecki, C. Betzel, K. Sutoh, and M. N. G. James, *Nature (London)* **359**, 700 (1992).

[23] M. T. Black, *J. Bacteriol.* **175**, 4957 (1993).

[24] W. R. Tschantz and R. Dalbey, this volume [21].

[25] M. O. Lively, A. L. Newsome, and M. Nusier, this volume [22].

to be an extremely slow reaction unless the protein interacts with activated RecA. By contrast, β-lactamase and leader peptidase are presumed to be fully active under physiological conditions. Accordingly, it will be of interest to learn whether the detailed mechanisms for these reactions are the same as that for LexA cleavage.

Role of RecA

LexA is apparently designed to cleave at a very slow rate under normal growth conditions, but to be capable of large rate increases on interaction with RecA. In principle, RecA might have this effect by contributing groups directly to catalysis, or indirectly by modulating the reactivity of LexA. We believe it unlikely that RecA contributes directly, because the RecA : LexA interaction is probably not conserved among cleavable proteins.[3,4]

We have proposed that the role of activated RecA in promoting autodigestion is that RecA reduces the pK_a of Lys-156 so that it is deprotonated even at neutral pH.[3,4,12,16] The pH rate profile for RecA-mediated cleavage of wild-type LexA[12] is practically flat between pH 6 and 11; by contrast, with a KR156 mutant protein substrate the rate is 1% the wild-type rate at pH 7, but rises with pH to be about 40% the wild-type rate at pH 9.5.

Conformational Model for LexA Cleavage

The properties of three Ind^S mutant proteins at position 92 provide a clue for how RecA might reduce the pK_a of Lys-156. These mutant proteins also display a reduction in the apparent pK_a of cleavage.[16] The pH rate profiles for these mutant proteins are consistent with a conformational model for LexA cleavage (Fig. 2). According to this model, LexA exists in two forms: An L* form, which is competent for cleavage, and an L form, which is not. Lys-156 can be protonated both in L* and in L, but the pK_a values differ, being about 5–6 for L* and the normal value of 10 for L. That is, in the L* form the local environment around Lys-156 favors deprotonation, perhaps by a high density of positive charge. The equilibrium constant K_{conf}, defined by [L*]/[L], is low in the wild type, but is increased by the mutant proteins and especially by interaction with RecA. At high values of K_{conf}, the equilibrium is driven toward L* even at neutral pH. In this view, the relationship between RecA-mediated cleavage and autodigestion is very close, and the role of RecA is indirect,

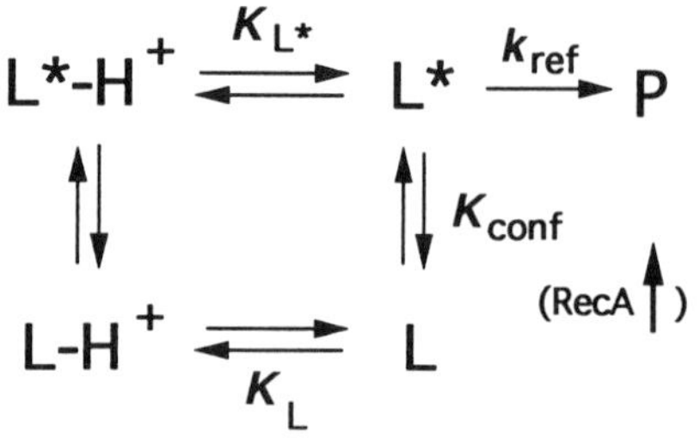

$$L^*\text{-}H^+ \underset{}{\overset{K_{L^*}}{\rightleftharpoons}} L^* \xrightarrow{k_{ref}} P$$

$$\Big\updownarrow \qquad\qquad \Big\updownarrow K_{conf}$$

$$(\text{RecA} \Big\uparrow)$$

$$L\text{-}H^+ \underset{K_L}{\rightleftharpoons} L$$

FIG. 2. Model for LexA cleavage reaction. According to this model, only the L^* form of LexA can autodigest. K_{conf} ($= [L^*]/[L]$) is an equilibrium constant between the L^* and forms; activated RecA is thought to stimulate cleavage by raising greatly the value of K_{conf}. K_L and K_{L^*} represent equilibrium constants for titration of Lys-156 in the L and L^* forms of LexA, respectively; according to the model pK_L and pK_{L^*} are 10 and 5–6, respectively; k_{ref} is the rate constant for cleavage of L^* (see text for details). Modified with permission from K. L. Roland, M. H. Smith, J. A. Rupley, and J. W. Little, *J. Mol. Biol.* **228**, 395 (1992).

perhaps being limited to making a specific protein–protein contact. This model has not yet been tested by physical means.

Conversion of LexA Cleavage into a Bimolecular Enzymatic Reaction

Availability of the Ind[S] mutant proteins has allowed the development of an intermolecular LexA cleavage reaction.[7] In this reaction, the purified C-terminal domain of LexA acts as an enzyme to cleave other molecules of LexA. The substrate for this reaction can either be an intact molecule unable to cleave itself, such as SA119, or a truncated molecule containing the N-terminal domain and a portion of the hinge extending 14 residues beyond the cleavage site. The rate of this trans-cleavage reaction is dictated by presence of Ind[S] mutations in the substrate and the enzyme. QW92 or LP89 in the substrate increases the rate of cleavage; these mutations have no effect on the quality of the enzyme. EA152 in the enzyme increases the rate of cleavage, but has no effect when present in an intact substrate. To date, this reaction, even with QW92 substrate and EA152 enzyme, appears to be below the K_m at a substrate concentration of 4 mM, so that the interaction is very weak. We believe it likely that this weak interaction is necessary in the biological context of the SOS system to prevent intramolecular LexA cleavage from proceeding in an uncontrolled manner. Current efforts are devoted to isolation of substrates that bind more tightly to the enzyme, because these should be useful in further analysis of the mechanism of LexA cleavage.

Purification of LexA and Derivatives

Overproduction of LexA

We use high-copy-number plasmids carrying various alleles of *lexA* fused to strong controllable promoters. Fusions to the *lac* and *tac* promoters make abundant LexA protein,[5,12,26] and are used for genetic screens and selections,[11,15,20,27] because expression of *lexA* is uncoupled from LexA function (LexA autoregulates its own promoter[26,28-30]), and LexA levels can readily be varied by addition of various levels of isopropyl-β-D-thiogalactoside (IPTG). However, the basal level of LexA from the *tac* promoter fusion is relatively high, and derivatives carrying *lexA* (Ind⁻) alleles make the cells sick,[11] presumably due to deleterious effects of high LexA levels, conferring selective pressure for loss of function. Accordingly, we do not generally use the *tac* promoter fusion except in studies of the Ind^S mutant proteins.

We now use[16,31] a fusion to a T7 promoter in the pET11a vector[32] for protein purification. Strains carrying a T7P :: *lexA* fusion make upward of 30–50% of the cell protein as LexA on induction of T7 RNA polymerase with IPTG. These overproducing strains carry the plasmid pLysS (this plasmid encodes T7 lysozyme, a protein that titrates out low levels of T7 RNA polymerase[33]), because these pET11a derivatives make the cells sick in the absence of pLysS. These strains cannot be frozen for long-term storage because they lyse on thawing, presumably due to the action of T7 lysozyme. Primary isolates of T7p :: *lexA* fusion plasmids are maintained in strains lacking T7 RNA polymerase to avoid selective pressure.

Modification of lexA Gene for Recombinant DNA Purposes

We use a modified *lexA* gene containing up to six new restriction sites (Fig. 1). These sites are unique in our *lacP* :: *lexA* fusion plasmids, such as the pJWL184 plasmid[16] (but several are not unique in the pET11a derivatives); these silent changes[34] were introduced by site-directed muta-

[26] J. W. Little, D. W. Mount, and C. R. Yanisch-Perron, *Proc. Natl. Acad. Sci. U.S.A.* **78,** 4199 (1981).
[27] J. W. Little and S. A. Hill, *Proc. Natl. Acad. Sci. U.S.A.* **82,** 2301 (1985).
[28] J. W. Little and J. E. Harper, *Proc. Natl. Acad. Sci. U.S.A.* **76,** 6147 (1979).
[29] R. Brent and M. Ptashne, *Proc. Natl. Acad. Sci. U.S.A.* **77,** 1932 (1980).
[30] R. Brent and M. Ptashne, *Proc. Natl. Acad. Sci. U.S.A.* **78,** 4204 (1981).
[31] J. W. Little, unpublished observations (1984, 1991).
[32] F. W. Studier, A. H. Rosenberg, J. J. Dunn, and J. W. Dubendorff, this series, Vol. 185, p. 60.
[33] F. W. Studier, *J. Mol. Biol.* **219,** 37 (1991).
[34] J. W. Little and D. W. Mount, *Gene* **32,** 67 (1984).

genesis. These sites facilitate separation of mutations, creation of multiple-mutant combinations, transfer of alleles to pET11a derivatives, and other recombinant DNA manipulations.

Purification of LexA

The following purification procedure[5] is largely adapted from early procedures for purification of RecA protein.[35] Several additional optional steps are described below, but these are generally omitted unless highly purified material is desired for purposes such as crystallography. For many purposes, columns can be simplified by using step elutions,[12] yielding material that is physically pure enough for analysis of cleavage. These shortcuts are particularly helpful when it is desired to analyze numerous mutant proteins; an experienced worker can purify four to six mutant proteins in 3 days. For most purposes, we purify protein from 1–2 liters of induced cells. For large-scale preparations, cells are grown in a 15-liter fermenter (Virtis Co., Gardiner, NY); modifications to the procedure for such preparations are given in parentheses, prefaced by the notation "15 l:"; comments are given after each step.

Growth of Cells. Derivatives of strain BL21(λDE3) of Studier *et al.*[32] carrying pLysS and a T7 :: *lexA* fusion plasmid with the desired allele of *lexA* are used. An overnight culture of the desired host strain is grown in λ broth (1% (w/v) Bacto-tryptone, 0.5% (w/v) NaCl, 1 μg/ml thiamin) plus 0.01% ampicillin. Cells are diluted about 100-fold into L broth (1% Bacto-tryptone (Difco, Detroit, MI), 0.5% Difco yeast extract, 1% NaCl, plus 0.12 g NaOH/liter) + 40 μg/ml ampicillin, followed by growth at 37°. Growth is monitored with a Klett meter (green filter); at a Klett value of 80 (~2–3 × 10⁸ cells/ml), IPTG is added to 0.5 m*M*, and aeration is continued until cell growth ceases; this generally occurs at a turbidity about twice that when IPTG is added. Cultures are then chilled rapidly by immersion of the flask in an ice–water slurry (15l: cultures are passed through a coil of ½-inch stainless-steel tubing immersed in an ice–water slurry) and are harvested by centrifugation. If the cells do not contain pLysS, cell pellets are frozen at −20° and can be stored for at least 1 month. If cells contain pLysS, they are resuspended in buffer A [50 m*M* Tris-HCl (pH 8.0), 0.5 m*M* EDTA, 10% (w/v) sucrose, 1 m*M* dithiothreitol (DTT), 200 m*M* NaCl] at 0.1–0.15 g cell pellet/ml suspension, and frozen at −20° in aliquots of a size suitable for centrifugation in the next step.

COMMENTS. (a) The above procedure can also be used to grow cells containing plasmids with *lacP* or *tacP* fusions, because these are induced

[35] N. L. Craig and J. W. Roberts, *Nature (London)* **283**, 26 (1980).

by IPTG. Such strains do not need to contain pLysS. We generally use an *E. coli* K12 host lacking chromosomal *lexA*, such as strain JL1436, for this purpose.[12] Strains with *lacP* :: *lexA* fusions continue growing after induction to high densities and are harvested 3–4 hr after IPTG addition. (b) If the fusion carries an Ind[S] allele of *lexA*, the protein is relatively unstable, and the protein level reaches a steady state, after which further growth does not provide more protein. Growth is halted after a period of time that is determined in a preliminary trial; for example, cells making QY92 protein, which has an *in vivo* half-life of about 1 hr, are chilled 1 hr after addition of IPTG.[16] (c) If the cells contain pLysS, it is crucial to suspend them before freezing, because they lyse on thawing.

Cell Lysis. If the cells do not contain pLysS, the following steps are employed.[36] Cell pellets are thawed on ice and suspended in buffer A as above. Aliquots are transferred to thin-walled beakers. Lysozyme (Sigma, St. Louis, MO), 5% solution in buffer A, is added to 250 μg/ml; suspensions are quick-frozen at $-70°$, swirled in a 20° bath until thawed, incubated 45 min at 0°, gently shaken 1 min at 37°, and chilled rapidly to 0° in an ice–water slurry. Lysis occurs during this heat pulse. All subsequent steps are carried out at 0–5°. The resulting highly viscous material is poured into centrifuge tubes, which are centrifuged 30 min at 30,000 g (15 l: 45 min at 27,000 rpm in the Beckman SW27 rotor). The DNA and cell debris form a pellet; the resulting nonviscous supernatant fluid and a small amount of fairly viscous material overlying the pellet are pooled (Fraction I).

COMMENTS. (a) If the cells contain pLysS, the cell suspensions are thawed on ice, resulting in lysis; lysates are centrifuged directly. (b) Volumes of cell suspensions are adjusted before cell lysis occurs, because lysates are so viscous that it is hard to control the volume when pouring. (c) LexA does not contain any cysteines[8,9] and does not require DTT for stability; DTT was added in early versions of this procedure and has been retained because omitting it might alter the behavior of other proteins. (d) Higher centrifugal forces give a firmer pellet and the recovery of LexA is correspondingly increased, because less of the supernatant is highly viscous.

Polymin P Treatment and Ammonium Sulfate Precipitation. To Fraction I is added 10% polymin P (polyethyleneimine, from Sigma), adjusted to pH 7.9 with HCl, to 0.35% final concentration; polymin P is added slowly with stirring over the course of 10–15 min at 0°. After stirring 10 min more, the suspensions are centrifuged 10 min at 15,000 g and the

---

[36] W. Wickner, D. Brutlag, R. Schekman, and A. Kornberg, *Proc. Natl. Acad. Sci. U.S.A.* **69**, 965 (1972).

supernatant fluid is saved. Then 0.4 g ammonium sulfate per milliliter of supernatant fluid is added slowly with stirring; after stirring 20 min more at 0°, suspensions are centrifuged 20 min at 20,000 g. Pellets are suspended in 0.5–1× the original volume of wash buffer (40 g ammonium sulfate per 100 ml buffer A) and recentrifuged. Pellets are then dissolved in a small volume of Buffer B [20 mM potassium phosphate (pH 7.0), 0.1 mM EDTA, 10% glycerol] + 1 mM DTT + 500 mM NaCl and dialyzed against the same buffer overnight with one change to give fraction II.

Phosphocellulose Column. Dialyzed Fraction II is diluted 2.5-fold with buffer B + 1 mM DTT to a final NaCl concentration of 200 mM, and applied to a phosphocellulose column [Whatman P-11 (Clifton, NJ), generally about 3–6 ml bed volume per liter starting culture] equilibrated with the same buffer; after washing with two column volumes buffer B + 200 mM NaCl (DTT is omitted here and in subsequent steps), protein is eluted with a linear gradient (3–5 column volumes total volume) in buffer B from 200 to 800 mM NaCl. LexA elutes at about 400 mM NaCl. Its position is identified by monitoring the OD_{254} of the effluent, followed by SDS–PAGE (13% gel); if the preparation contains hen egg white lysozyme, this protein elutes just after LexA. Pooled LexA-containing fractions are Fraction III.

COMMENTS. (a) Because few proteins bind to phosphocellulose, this column affords an excellent purification, giving material >90% pure. (b) Many mutant LexA proteins affecting cleavage do not bind to the column in 200 mM NaCl and must be applied at 100–125 mM NaCl; this must be determined by experiment. Wild-type and many mutant LexA proteins elute at the same NaCl concentration; certain mutant proteins elute at somewhat different NaCl concentrations. (c) For many purposes, a step elution with buffer B + 600 mM NaCl suffices.

Hydroxyapatite Chromatography. Fraction III is applied to a hydroxyapatite column [Bio-Rad (Richmond, CA) Bio-Gel HTP, about 2–3 ml bed volume per liter starting culture] equilibrated with buffer C (0.1 mM EDTA, 10% glycerol) + 50 mM potassium phosphate, pH 7.0. Following washing with the same buffer, LexA is eluted with a linear gradient (3–5 column volumes total volume) from 50 to 400 mM potassium phosphate, pH 7.0, in buffer C. LexA elutes at about 150 mM potassium phosphate and is dialyzed extensively against buffer D [10 mM PIPES–NaOH (pH 7.0), 0.1 mM EDTA, 10% (v/v) glycerol, 200 mM NaCl] to give Fraction IV.

COMMENTS. (a) For many purposes, a step elution with buffer C + 150 mM potassium phosphate, pH 7.0, suffices. (b) Hen egg white lysozyme, if present, passes through this column. (c) Fraction IV is >99% pure, as judged by visual estimates of stained SDS–PAGE gels. (d) Yields from the T7 system are typically 10–20 mg per gram of cell paste.

Optional Steps. Two additional steps can be incorporated when it is desired to obtain highly purified material for purposes such as crystallography.

PRELIMINARY PHOSPHOCELLULOSE COLUMN. Fraction II can be passed directly through a small column of phosphocellulose equilibrated with buffer B + 1 mM DTT + 500 mM NaCl. This column is intended to remove residual traces of polymin P.[37]

AFFI-GEL 501 COLUMN. Because LexA contains no cysteines, contaminants with Cys residues can be removed by adsorption to a Hg affinity column.[30] Fraction III is applied to a small column (0.5 ml bed volume per liter starting culture) of Affi-Gel 501 (Bio-Rad) equilibrated with buffer B + 500 mM NaCl, followed by washing with the same buffer. Apparently LexA binds to the column matrix, because it elutes gradually after the pass-through. Elution of LexA is monitored by OD_{254}, and washing ceases when this value drops to baseline. The location of LexA is verified by SDS–PAGE. Fractions containing LexA are pooled (Fraction IIIa) and further purified on hydroxyapatite as above. This column removes a large number of high molecular weight contaminants present in minute amounts. Some UV-absorbing material leaches from the column, but it passes through the hydroxyapatite column.

Storage of LexA. LexA appears to be a relatively stable protein. It can be maintained for long periods of time at 5° in buffer D; wild-type LexA slowly autodigests under these conditions,[5,26] but Ind⁻ mutant proteins do not. We routinely freeze aliquots and store them at −70°.

Purified LexA has one undesirable property. At high protein concentrations, under certain conditions (low salt concentrations and at pH values of about 6–7) the protein forms a sticky precipitate that cannot be redissolved.[31] Its calculated isoelectric point is about 6.5, and we surmise that precipitation results from limited solubility at or near the pI. To avoid this, we maintain NaCl at 200 mM and keep the pH at 7 or above. This precipitation is less of a problem at low protein concentrations, but experiments at low salt and neutral pH should be monitored for precipitation. This property of LexA appears to be localized to the C-terminal fragment, because preparations of this fragment (see below) also have this property, whereas N-terminal fragment preparations do not.[31]

Isolation of Radiolabeled LexA

Radiolabeled LexA has been prepared by several methods. Labeled protein can be prepared *in vivo*[6] by growing cells in minimal medium, then

[37] M. M. Cox, K. McEntee, and I. R. Lehman, *J. Biol. Chem.* **256,** 4676 (1981).

inducing cultures of cells containing overproducing plasmids, followed by addition of [^{35}S]methionine to 50 μCi/ml. Because the hosts used are methionine prototrophs, the label is added without carrier, and is taken up within about 1 min, so that this method is a pulse-to-exhaustion.[38] After 5 min, cultures are harvested and the protein is isolated by methods described above. If material of high specific activity is desired, labeling is done soon after IPTG induction (at 2 min with *tac* promoter fusions) to minimize dilution of labeled material with unlabeled material made before the pulse. The specific activity of the labeled protein can be estimated to within about 30% by comparisons with known amounts of intact LexA in stained gels. The specific activity of labeled proteins can then be adjusted by addition of known amounts of unlabeled protein. This approach involves the assumption that the chemical properties of the labeled and unlabeled preparations are the same. When the rates of autodigestion of such mixtures are assessed by assays A and B (see below), they are the same; because assay A examines all the molecules, whereas assay B examines only the labeled ones, this concordance is consistent with the assumption. This method of labeling is applicable to any mutant protein except those that cleave so rapidly that they are too unstable to accumulate to substantial levels *in vivo*.

LexA protein and mutant derivatives can also be labeled in a coupled *in vitro* transcription–translation system programmed with overproducing plasmids[16]; we have used operon fusions with the *lac* and *tac* promoters, and it should be possible to use fusions with the T7 promoter and addition of T7 RNA polymerase. Following labeling, LexA is radiochemically purified by batch adsorption to phosphocellulose and elution by high salt (see above), affording radiochemically pure material. Again, the specific activity of these preparations can be adjusted by the addition of unlabeled protein. Mutant proteins that autodigest extremely rapidly are contaminated after purification with cleavage products; for example, preparations of QW92 protein are typically 20–40% digested by the time the preparation is frozen.

Purification of LexA Cleavage Products

Purification of wild-type LexA autodigestion fragments is facilitated by the fact that the N-terminal fragment binds to phosphocellulose but not hydroxyapatite, whereas the C-terminal fragment binds to hydroxyapatite but not phosphocellulose. Purification of C-terminal fragments of Ind^S mutant proteins with an increased rate of cleavage is more complicated,

[38] J. W. Little, *J. Mol. Biol.* **167**, 791 (1983).

because these proteins autodigest *in vivo*.[7,15] The QY92 and EA152 proteins, which have an *in vivo* half-life of about 1 hr, can be purified in intact form by the above procedure if one works rapidly and does step elutions from the columns; more rapidly digesting proteins are present primarily as autodigestion products in lysates.[7]

Autodigestion. Purified wild-type LexA is autodigested at 37° in a buffer containing 50 mM CAPS–NaOH (pH 10) for 60 min or in 50 mM Tris-HCl (pH 9) for 4 hr. It is unnecessary to achieve complete autodigestion.

Purification of N-Terminal Fragment. A digest is dialyzed against buffer B + 200 mM NaCl, then diluted with buffer B to 100 mM NaCl, and applied to a phosphocellulose column equilibrated with buffer B + 100 mM NaCl.[39] After washing, the N-terminal fragment is eluted with a linear gradient in buffer B from 100 to 1000 mM NaCl; the N-terminal fragment elutes at about 250 mM NaCl. Because this fragment contains no tyrosines or tryptophans, it absorbs poorly at 254 nm; the fragment is detected by SDS–PAGE (15% gel). Peak fractions are pooled and passed through a hydroxyapatite column (equilibrated with buffer C + 50 mM NaCl). The pass-through is dialyzed against buffer D. This fragment comprises the DNA-binding domain, and is not involved in cleavage.

Purification of C-Terminal Fragment. Digests are dialyzed against buffer B + 500 mM NaCl, diluted to 175 mM NaCl (by error, this dilution step was not noted in Kim and Little[7]), and passed through a phosphocellulose column equilibrated with the same buffer. The pass-through is then purified on hydroxyapatite as described above for intact LexA, and dialyzed against buffer D.

Purification of C-Terminal Fragment from in Vivo Autodigestion of Ind^S Mutant Proteins. Crude extracts from induced cells are made as above, followed by polymin P precipitation, ammonium sulfate precipitation, and phosphocellulose chromatography as above; the C-terminal fragment passes through the column.[7] The flow-through is purified by hydroxyapatite chromatography as above. This material is considerably less pure than intact LexA. It could probably be purified greatly by further Affi-Gel 501 chromatography.

Purification of Truncated Substrates for Trans Cleavage. Substrates consist of amino acids 1–98 of LexA, except that the C-terminal Tyr residue is changed to Lys; these proteins are synthesized from T7p fusions of the type described above.[7] To purify these substrates, extracts are prepared as above; after polymin P treatment, ammonium sulfate precipitation, and dialysis, protein is diluted to 100 mM NaCl and purified on phosphocellulose (equilibrated with buffer B + 100 mM NaCl) as described

[39] B. Kim and J. W. Little, *Science* **255**, 203 (1992).

above. Substrates elute at about 175 mM NaCl, and are passed through a hydroxyapatite column, then dialyzed against buffer D. Concentrations are estimated by visual comparison with known amounts of N-terminal fragment in Coomassie blue-stained gels (see below).

Analysis of LexA Cleavage

In Vivo Assay of Cleavage Rates

In vivo rates of cleavage can be monitored by pulse-chase or pulse-to-exhaustion methods (as described above). Briefly, cells are grown in minimal medium to a Klett value of about 60 (about 2×10^8 cells/ml); cleavage rate is determined by pulse-labeling with [^{35}S]methionine, followed by sampling at intervals. Samples are quick-frozen; after thawing, extracts are made by sonication and LexA and its cleavage products are recovered by immunoprecipitation. Immunoprecipitates are analyzed by SDS–PAGE (15% gel) followed by fluorography. Estimates of half-lives are made by visual estimation. This method has been described in detail.[38] It can be used to follow cleavage rates after DNA damage and has been particularly useful in studying the return of induced cells to the normal growth state.

In Vitro Assays of LexA Cleavage

We describe conditions for three types of LexA cleavage: Autodigestion, RecA-mediated cleavage, and the intermolecular trans-cleavage reaction. In each case, cleavage reactions can be carried out under a variety of conditions; reaction mixtures are then analyzed by SDS–PAGE, using one of two gel systems.

Assay of Autodigestion. Standard reaction conditions are incubation at 37° in 50 mM CAPS–NaOH (pH 10.0). Incubations contain residual components of LexA storage buffer, but these appear to have no effect on the rate of the reaction; this reaction is also relatively insensitive to ionic strength and does not require metal ions.[6] At intervals, aliquots are removed and mixed with an equal volume of cracking buffer [4% (w/v) sodium dodecyl sulfate (SDS), 0.125 M Tris-HCl (pH 6.8), 10% (v/v) glycerol, 10% 2-mercaptoethanol, 0.05% (w/v) bromphenol blue]. Samples are then analyzed by SDS–PAGE (15% gels) using the Laemmli system.[40]

Assay of RecA-Mediated Cleavage. This reaction requires more exacting conditions.[12,16,35] It is carried out in two stages. First, RecA is activated

---

[40] U. K. Laemmli, *Nature (London)* **227**, 680 (1970).

at high concentrations by incubation with single-stranded DNA and a nucleotide cofactor; this leads to formation of a RecA filament polymerized on the DNA.[41-43] Second, activated RecA is added to prewarmed incubation mixtures containing all the other components, and incubation is carried out. RecA (U.S. Biochemicals, Cleveland, OH) is activated by incubation at 25 μM RecA in a buffer containing 20 mM Tris-HCl (pH 7.4), 2 mM MgCl$_2$, 1 mM DTT, 1 mM adenosine 5'-(γ-thio)triphosphate (ATPγS), 32 μg/ml heat-denatured calf thymus DNA at 0° for $\geq$20 min. RecA forms a polymeric filament on the DNA; this ternary complex is particularly stable in the presence of ATPγS. If the RecA concentration is changed in this incubation, the concentration of single-stranded DNA is changed proportionally. Reaction mixtures contain 20 mM Tris-HCl (pH 7.4), 5 mM MgCl$_2$, 2 mM DTT, 1 mM ATPγS, and LexA, typically at 3-5 μM, and are prewarmed to 37°; after a zero-time point is taken, reaction is initiated by addition of activated RecA, typically to 1 μM final concentration, followed by incubation at 37°. Aliquots are taken and analyzed as described for autodigestion.

Assay of Trans Cleavage. Three types of trans-cleavage reaction have been described,[7] which vary as to the nature of the enzyme and the substrate. In the type I reaction, the enzyme is the C-terminal fragment and the substrate is a truncated protein comprising residues 1-98 of LexA; in the type II reaction, the enzyme is the C-terminal fragment and the substrate is an intact LexA molecule that cannot autodigest but has a normal cleavage site (such as the SA119 protein); in the type III reaction, the enzyme is an intact form of LexA that cannot autodigest due to defects in the cleavage site (such as GD85 protein) and the substrate is either of the two forms just described. Reaction mixtures contain 20 mM CAPS-NaOH (pH 10.0), 200 mM NaCl; typical protein concentrations are 1-2 μM enzyme and 50 μM substrate. Aliquots are treated as above, or alternatively with 0.2 volumes of 6× stop buffer[44] [10% SDS, 36% glycerol, 0.5 M DTT, 0.175 M Tris-HCl (pH 6.8), 0.012% bromphenol blue], and analyzed in a 15% polyacrylamide gel containing 6 M urea, 0.1 M sodium phosphate (ph 7.2), 0.1% SDS with a 3.5% stacking gel in the same buffer; running buffer is 0.1 mM sodium phosphate (pH 7.3), 0.1% SDS. This gel system affords better separation of the truncated substrate and the N-terminal fragment than the Laemmli gel system; however, the

[41] M. M. Cox and I. R. Lehman, *Annu. Rev. Biochem.* **56,** 229 (1987).
[42] A. K. Eggleston and S. C. Kowalczykowski, *Biochimie* **73,** 163 (1991).
[43] R. M. Story, I. T. Weber, and T. A. Steitz, *Nature (London)* **355,** 318 (1992).
[44] F. M. Ausubel, R. Brent, R. E. Kingston, D. D. Moore, J. G. Seidman, J. A. Smith, and K. Struhl, "Current Protocols in Molecular Biology," Wiley (Interscience), New York, 1987.

Laemmli system gives sharper bands and is faster and easier, and would be preferable when intact substrates are used.

Analysis of Cleavage Rates

Two different assays are used, depending on the desired degree of precision.

Assay A: Visual Estimation. Gels are stained with a solution containing 25% (v/v) 2-propanol, 10% (v/v) acetic acid, 0.025% (w/v) Coomassie blue R-250 and are destained in 7% (v/v) acetic acid.[45] The fraction of cleaved material is estimated by eye. One useful landmark is the point at which the remaining intact protein and the C-terminal fragment stain with equal intensity. The C-terminal fragment of LexA stains far more intensely than the N-terminal fragment,[7] but somewhat less intensely than a comparable molar amount of the intact protein. Accordingly, when the bands of intact protein and C-terminal fragment are equally intense, roughly 60–70% of the protein has been cleaved. This judgment is easy to make and is reproducible, so that it is often used to compare relative rates between different samples. Such comparisons are based on the assumption that cleavage is a first-order reaction under the conditions being examined. This assumption has been verified[6] under a wide range of conditions for wild-type LexA using assay B. Assay A is of limited precision; values estimated in this way are correct to perhaps ±30%. It can be made somewhat more precise with the use of standards, but for many purposes this degree of precision suffices.

Assay B: Radioactive Counting. Reaction mixtures containing radiolabeled substrate (see above) are sampled as above, mixed with roughly 1 μg partially autodigested unlabeled LexA, and analyzed by gel electrophoresis (Laemmli system) as above. Bands of intact LexA and cleavage fragments are located by staining as above, excised with a razor blade, placed in scintillation vials containing 0.5 ml of 90% NCS tissue solubilizer (Amersham, Arlington Heights, IL), and incubated at 42° for 2–3 hr. Glacial acetic acid (15 μl) is added, followed by 5 ml ACS scintillation fluid (Amersham). Radioactivity is determined by scintillation counting,[6] and the extent of repressor cleavage is calculated as the ratio of the total number of counts in the two cleavage bands to the total radioactivity in these bands plus the intact band. Although this assay is considerably more tedious and requires labeled substrate, its precision is in the range of 5–10%. It is important to use highly purified SDS in the gel and gel buffers; impure SDS leads to broad bands of the N-terminal fragment.

[45] J. W. Little, *BioTechniques* **11**, 759 (1991).

With the increasing availability of direct counting devices such as the Phosphorimager (Molecular Dynamics, Sunnyvale, CA), this method could be simplified by the use of such an instrument.

Considerations in Design of Cleavage Experiments and Analysis of Data

Although autodigestion and RecA-mediated cleavage are probably closely related at the mechanistic level, in some ways it is difficult to compare these two reactions, due to the complexity of the RecA protein and its interaction with LexA.

Autodigestion. This reaction is intramolecular; it displays first-order kinetics and its rate can be described in terms of a rate constant or a half-life. The rate constant is independent of protein concentration[6] over the range of 1 nM to 10 μM. Accordingly, it is not necessary to know the protein concentration to measure this constant. This feature of the auto-digestion reaction expands the range of possible measurements. For instance, we were able to measure rate constants for rapidly cleaving Ind[S] mutant proteins, which could not be isolated from cells, by synthesizing them *in vitro* (see above). Although their molar concentration was not known, assay B showed that cleavage was first order and rate constants could be measured.

The intramolecular nature of the reaction limits the degree to which changes in the rate constant, conferred either by mutation or by changes in conditions, can be interpreted. As noted in the introduction, this reaction presumably involves binding of the cleavage site to the binding pocket of the active site. We believe it likely that this binding is not tight, for several reasons. First, the protein is designed to cleave slowly under physiological conditions, even though both reactants are present in the same molecule; tight binding might allow rapid reaction. Second, the conformational model (Fig. 2) suggests that a structural change is necessary to allow cleavage; the L-to-L* transition may simply represent binding of the cleavage site to the active site. Finally, the interaction in the intermolecular reaction is exceedingly weak ($K_m > 4$ mM). Accordingly, an observed change in rate constant might result from a change either in catalytic efficiency or in binding of cleavage site to active site. For example, the apparent titration of a basic group in the protein, inferred from the pH rate profile, might in principle affect either of these parameters. Availability of the trans-cleavage reaction[7] should permit these effects to be dissected, provided that we can work above the K_m.

RecA-Mediated Cleavage. This reaction is far more complex. This reaction can be treated as an enzyme-catalyzed reaction, with RecA as the enzyme and LexA as the substrate, even though groups in LexA carry out the actual catalysis. RecA can mediate cleavage of many LexA molecules, so it clearly acts as a catalyst. However, analysis of kinetics in this system is difficult for several reasons. First, activated RecA is a complex protein filament. Some evidence[46] suggests that LexA contacts two adjacent monomers in a deep groove of this helical filament, implying that the active species is not simply a single RecA molecule. Second, RecA is rather a weak catalyst, so that relatively high levels (≥ 0.1 μM) of activated RecA are needed.[47] Third, the requirement for high RecA levels complicates kinetic analysis; preliminary evidence[47] suggests that the K_m for LexA as substrate is in the range of 0.5 μM, similar to the enzyme levels required for catalysis, so that the levels of enzyme and substrate are comparable. Fourth, below 0.1 μM RecA the rate of this reaction is less than proportional to the RecA concentration.[47] Fifth, in some studies[16] the rate of cleavage as a function of LexA concentration does not show simple Michaelis–Menten behavior, continuing to increase at high protein concentrations, so it is unclear whether the RecA is being saturated. Accordingly, it is uncertain whether the measured rate reflects k_{cat}. In addition, it is not possible to measure this rate for rapidly cleaving Ind^S mutant proteins in the range of 1 μM, because these proteins cannot be isolated in quantity. Consequently, measurements with material radiolabeled *in vitro* are done at low protein concentrations and reflect k_{cat}/K_m values; one cannot interpret differences in such values unambiguously.

Comparing Different Cleavage Data. We believe it likely that comparisons among different autodigestion reactions are valid, within the limits discussed above. Comparisons among different RecA-mediated cleavage reactions have some validity, but differences cannot readily be ascribed to particular kinetic constants. The difficulty of interpreting RecA-mediated cleavage data makes it hard to compare these data directly with autodigestion data. For instance, the conformational model (Fig. 2) predicts that the maximum rate of autodigestion for Ind^S mutant proteins that increase k_{conf} should equal that for RecA-mediated cleavage for wild-type LexA. We have discussed at length[16] the uncertainties involved in assessing this prediction.

Acknowledgments

We are grateful to Julie Mustard and Roy Parker for comments on the manuscript. This work has been supported by grants GM24178 from the National Institutes of Health and DMB9004455 from the National Science Foundation.

[46] X. Yu and E. H. Egelman, *J. Mol. Biol.* **231**, 29 (1993).
[47] L. L. Lin, Ph.D. Dissertation, University of Arizona, Tucson (1988).

[21] Bacterial Leader Peptidase 1

By WILLIAM R. TSCHANTZ and ROSS E. DALBEY

Introduction

Proteins that are secreted out of the cell or translocated across lipid bilayers are usually made with N-terminal extension peptides, termed leader (or signal) sequences. These leader peptides have a basic amino-terminal region, a central hydrophobic domain of nine or more apolar residues, a helix-destabilizing residue (proline or glycine) at -5 or -6, and a carboxyl-terminal region containing small apolar amino acids at -1 and -3 relative to the cleavage site. After translocation across the membrane, the signal peptides are removed from exported proteins by a specific leader (or signal) peptidase. Signal (leader) peptidases have been purified in *Escherichia coli*,[1,2] yeast,[3] chicken,[4] and dog[5] and their genes have been identified in other prokaryotic organisms.[6]

In bacteria, there are two separate leader peptidases that process proteins. Leader peptidase 1 cleaves the majority of the preproteins destined to the cell surface, whereas leader peptidase 2 only processes preproteins that have been first modified with a lipid molecule at the $+1$ position. The *E. coli* leader peptidase 1 is an integral membrane protein of 323 amino acid residues. In β-octylglucoside, the purified protein has been shown to cleave the precursors of M13 procoat (bacterial inner membrane), the precursor to the leucine-binding protein, pre-β-lactamase, premaltose-binding protein (periplasmic space), pro-OmpA, and pre-λ phage receptor (outer membrane).[7] Its physiological role is to release exported proteins from the membrane by removing the leader sequence.[8]

Leader peptidase 1 has received increasing attention, and studies to determine its biochemical properties and its catalytic mechanism are underway in several laboratories. The present article describes the purification of leader peptidase, its substrate specificity, sequence homology of

[1] C. Zwizinski and W. Wickner, *J. Biol. Chem.* **255**, 7973 (1980).
[2] P. B. Wolfe, P. Silver, and W. Wickner, *J. Biol. Chem.* **257**, 7898 (1982).
[3] J. T. YaDeau and G. Blobel, *J. Biol. Chem.* **264**, 2928 (1989).
[4] R. K. Baker and M. O. Lively, *Biochemistry* **26**, 8561 (1987).
[5] G. Greensburg, G. S. Shelness, and G. Blobel, *J. Biol. Chem.* **264**, 15762 (1989).
[6] R. E. Dalbey and G. von Heijne, *Trends Biochem. Sci.* **17**, 474 (1992).
[7] C. Watts, R. Zimmermann, and W. Wickner, *J. Biol. Chem.* **258**, 2809 (1983).
[8] R. E. Dalbey and W. Wickner, *J. Biol. Chem.* **260**, 15925 (1985).

the leader peptidase 1 genes, and techniques to probe the structure and function of this protein.

Assay Methods

Leader peptidase can be assayed using a radiolabeled preprotein substrate or a peptide substrate. In the preprotein assay, leader peptidase is incubated with limiting amounts of ^{35}S-labeled procoat, the precursor to the major coat protein of the M13 phage. Cleavage of procoat is monitored by the production of ^{35}S-labeled coat protein, which is separated by polyacrylamide gel electrophoresis, and the bands are visualized by fluorography. The peptide assay involves incubating leader peptidase with a synthetic nonapeptide (for amino acid sequence, see below) corresponding to -7 to $+2$ of the precursor maltose-binding protein and separating the products by HPLC.[9] Although the peptide assay is not as sensitive as the preprotein assay, the kinetic parameters, k_{cat} and K_m, can be measured.

^{35}S-Labeled Procoat Assay

Procoat is radiolabeled using a cell-free system with an *E. coli* extract containing the pT712 plasmid encoding the procoat gene.[10] Each reaction (typically 30 μl) contains DNA, T7 RNA polymerase, *trans*-[^{35}S]methionine (30 μCi), Triton X-100 (0.4%), and S150-2.[11] The synthesis is stopped after 30 min by diluting into 0.2 ml of ice-cold Tris buffer (50 mM Tris, 0.1% Triton X-100, pH 8.0). In a standard assay, a 2-μl aliquot of leader peptidase (at various concentrations) is incubated with a 10-μl aliquot of the *in vitro*-synthesized ^{35}S-labeled procoat for 30 min at 37°. The amount of cleavage can be determined by separating the procoat and coat protein on a 23% SDS–polyacrylamide gel (acrylamide : bisacrylamide, 30 : 0.8).[12] One unit of leader peptidase activity is defined as the amount of enzyme that is needed to process ^{35}S-labeled procoat to produce equal amounts of ^{35}S-labeled procoat and ^{35}S-labeled coat bands on fluorographs of the gels.[1]

Peptide Assay

The synthetic peptide Phe-Ser-Ala-Ser-Ala-Leu-Ala-Lys-Ile coo$^-$ is the substrate that we use routinely and it is cleaved by leader peptidase between the -Ala-Lys- bond.[9] In our hands, k_{cat} and K_m are 76 hr^{-1} and

[9] I. K. Dev, P. H. Ray, and P. Novak, *J. Biol. Chem.* **265**, 20069 (1990).
[10] L. M. Shen, J. I. Lee, S. Cheng, H. Jutte, A. Kuhn, and R. E. Dalbey, *Biochemistry* **30**, 11775 (1991).
[11] K. Yamane, S. Khihara, and S. Mizushima, *J. Biol. Chem.* **262**, 2358 (1987).
[12] J. D. Boeke, M. Russel, and P. Model, *J. Mol. Biol.* **144**, 103 (1980).

0.33 mM, respectively. In a typical assay, 35 μl of 1 mg/ml of leader peptidase in 10% (v/v) polybuffer (PBE 94, Pharmacia Fine Chemicals, Piscataway, NJ) containing 1% (w/v) β-octylglucoside is added to 12.5 μl of 1.5 mg/ml of peptide substrate in phosphate buffer (83 mM sodium phosphate, pH 7.7) and incubated for 90 min at 37°. 47.5 μl of ice-cold 0.1% (v/v) trifluoroacetic acid (TFA) is added to quench the reaction. Just prior to loading the sample onto HPLC, the detergent is removed by centrifugation for 5 min in a microfuge. The reactant and product peptides are separated using a reversed-phase analytical C_{18} column (Vydac protein and peptide column; peptides are detected at a wavelength of 218 nm), utilizing an H_2O/acetonitrile gradient. The gradient at a flow rate of 1 ml/min is as follows, where solvent A is 0.1% TFA and solvent B is 0.1% TFA in acetonitrile: 0 to 5 min, 97%A : 3%B; 5 to 15 min, 60%A : 40%B; 15 to 20 min, 60%A : 40%B; 20 to 25 min, 97%A : 3%B.

Purification

Leader peptidase is purified by membrane isolation, Triton X-100 extraction, DEAE chromatography, and chromatofocusing. Except for the overproducing strain and the membrane isolation step, the procedure is very similar to the one described previously.[13] The strain MC1061/pRD8 overproduces leader peptidase 200- to 400-fold. The pRD8 vector was constructed by inserting the leader peptidase gene (*lepB*) into the pING vector,[8] a pBR322-derived plasmid that carries the arabinose regulatory elements and the *araB* promoter.[14] The cloning strategy is described in Fig. 1.

Ten liters of MC1061 cells transformed with pRD8 are grown at 37° in M9 media containing 0.5% (w/v) fructose, 0.2% (w/v) casamino acids, and 100 μg/ml ampicillin. When the OD at 600 nm reaches 0.4, arabinose (0.2% (w/v) final concentration) is added to induce synthesis of leader peptidase and the cells are grown for an additional 4 hr. Cells are harvested by centrifugation at 6400 g for 10 min at 4°, resuspended in an equal weight of 50 mM Tris-HCl (pH 7.5) containing 10% (w/v) sucrose, frozen by quickly pipetting the cell suspension into liquid nitrogen, and then kept at −80°.

The cell nuggets (110 g) are added at room temperature to 250 ml of 10 mM Tris-HCl (pH 8.5) containing 20% (w/v) sucrose and 5 mM EDTA. After thawing, lysozyme (0.15 mg/ml, final concentration) and DNase I (0.015 mg/ml final concentration) are added to the cell mixture. The cells

[13] P. B. Wolfe, W. Wickner, and J. M. Goodman, *J. Biol. Chem.* **258**, 12073 (1983).
[14] S. Johnston, J. H. Lee, and D. S. Ray, *Gene* **34**, 137 (1985).

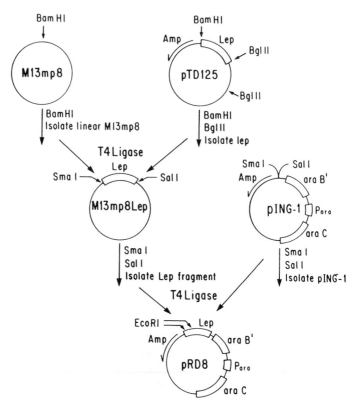

FIG. 1. Cloning strategy for generating pRD8. The leader peptidase gene (*lepB*) located on pTD125 was cloned into M13mp8 to form the vector M13mp8Lep and then transferred to pING-1, a pBR322-derived plasmid that contains the regulatory elements (*araC*) and the promoter of L-ribulokinase (*araB*). The resulting plasmid is called pRD8. Modified from Dalbey and Wickner.[8]

are lysed by freeze-thawing in a dry ice–ethanol bath. Magnesium acetate (final concentration 5 mM) is added to the cell lysate while stirring in a room-temperature water bath. The membrane fragments are then collected by centrifugation at 40,000 g for 30 min at 4°, resuspended in 200 ml of ice-cold 10 mM triethanolamine hydrochloride (pH 7.5) containing glycerol, and then subjected to centrifugation again (40,000 g, 30 min, 4°).

The membrane fragments are resuspended on ice in 80 ml of TEA buffer (10 mM triethanolamine hydrochloride, 10% (v/v) glycerol, 1% (v/v) Triton X-100, pH 7.5) by homogenizing five times with a 50-ml Potter–Elvehjem tissue grinder followed by 5 treatments with a 5-ml grinder. After the Triton X-100 extract is centrifuged at 40,000 g for 30 min at 4°, the supernatant is collected, with care not to transfer the white, opaque loose pellet, which contains the nonextracted membrane fragments.

The Triton X-100 extract (80 ml) is loaded on a column (16 × 5 cm) of DEAE-cellulose (Whatman DE-52 equilibrated in buffer A, consisting of 10 mM triethanolamine hydrochloride, pH 7.5, 5 mM magnesium chloride, 10% (v/v) glycerol, and 1% (v/v) Triton X-100). The column is washed with buffer A at a flow rate of 4 ml per min and 100 fractions (20 ml each) are collected. Leader peptidase elutes from the column in a very broad peak. A 12% SDS–PAGE gel is used to determine which fractions contain the protein. After pooling the desired fractions, they are dialyzed overnight against two volumes (9 liters each) of buffer B (25 mM imidazole acetate, 1% Triton X-100, pH 7.4).

The dialyzed protein fractions are pumped onto a 4-ml column of Polybuffer exchanger (PBE 94, Pharmacia Fine Chemicals), which has been equilibrated at 4° in buffer B. After loading the sample, the column is washed with 8 ml of buffer B. Leader peptidase is eluted from the column at a pH of approximately 6.9 with 150 ml of 10% (v/v) solution of Polybuffer 74 (Pharmacia Fine Chemicals), which contains 1% (w/v) β-octylglucoside. Analysis of the enzyme fractions by SDS–PAGE after various stages of the purification (Fig. 2) shows the enrichment of the leader peptidase protein.

Comments on Purification and Assay Methods

In a typical purification, approximately 40 mg of pure leader peptidase can be obtained from 10 liters of bacterial cells. The bacterial cells can be grown, induced, harvested, and frozen as a solid paste at −75° for an indefinite period of time with little loss in yield of purified leader peptidase. The chromatography steps should be done rapidly because during the purification a partially inactive 32-kDa leader peptidase fragment is generated by autocatalytic cleavage between residues 40 and 41.[15] The chromatofocusing step of the purification procedure not only partially separates the wild-type leader peptidase from the 32-kDa fragment but also serves to concentrate the leader peptidase protein. However, it is difficult to obtain pure leader peptidase free of the 32-kDa fragment. At −20°, leader peptidase is stable for many months, with the slow generation of the autocatalytic fragment after long periods of storage.

The procoat assay is suitable for measuring the activity of the wild-type leader peptidase but falls short in its use with mutant proteins because

[15] T. L. Talarico, I. K. Dev, P. J. Bassford, Jr., and P. H. Ray, *Biochem. Biophys. Res. Commun.* **181**, 650 (1991).

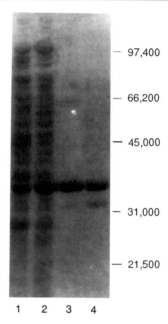

FIG. 2. Purification of leader peptidase. Each lane corresponds to 5 μg of total protein loaded from each step in the purification. The proteins were resolved on a 12% SDS–PAGE gel and were stained with Coomassie blue: cell lysate (lane 1), Triton X-100 extract (lane 2), DEAE pool (lane 3), and chromatofocusing pool (lane 4).

many that score inactive with this assay are active *in vivo*.[16–18] Another disadvantage is that this assay cannot be used to determine the k_{cat} and K_m constants because only a vanishingly small amount of substrate is used; it is very difficult to obtain chemical amounts of the wild-type procoat.

The advantage of the peptide substrate assay is that the kinetic parameters can be measured. However, this assay requires a large amount of enzyme compared to the procoat assay because of the low affinity and low turnover number with the peptide substrate.

Physical and Chemical Properties

Leader peptidase is composed of a single 36-kDa polypeptide with an isoelectric point of 6.9.[13] It is an essential protein.[8,19] The amino acid

[16] N. Bilgin, J. I. Lee, H. Zhu, R. Dalbey, and G. von Heijne, *EMBO J.* **9,** 2717 (1990).
[17] M. Sung and R. E. Dalbey, *J. Biol. Chem.* **267,** 13154 (1992).
[18] W. R. Tschantz, M. Sung, V. M. Delgado-Partin, and R. E. Dalbey, *J. Biol. Chem.* **268,** 27349 (1993).
[19] T. Date, *J. Bacteriol.* **154,** 76 (1983).

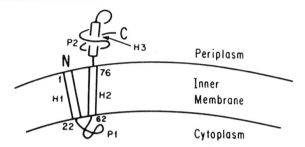

FIG. 3. The membrane topology of leader peptidase in the plasma membrane of *E. coli.* Apolar domains are represented by rectangles and polar regions are depicted by a line. H1, H2, and H3 are the hydrophobic stretches 1, 2, and 3, respectively; P1 and P2 are polar domains 1 and 2, respectively.

sequence of leader peptidase was deduced from the sequencing of its gene and indicates that the protein is synthesized without a cleaved leader sequence.[13] Interestingly, the amino terminus of the protein is blocked,[2] most likely by acetylation.[20] The significance of this modification is not known. The membrane orientation of leader peptidase has been determined by proteolysis[13,21] and gene fusion methods.[22] It is anchored to the membrane by two transmembrane segments (Fig. 3) and is oriented with a short polar domain facing the cytoplasm and a large carboxyl-terminal domain exposed to the periplasmic space.

Substrate Specificity and Inhibition

The substrate specificity of leader peptidase requires small, apolar amino acids at positions -1 and -3 and sometimes a helix-breaking residue at -5 or -6 relative to the cleavage site. This was first proposed based on the analysis of the primary sequence of a large number of signal peptides.[23,24] Site-directed mutagenesis has been used to test whether the -1 and -3 positions of the precursor to the M13 coat protein are important

[20] D. W. Kuo, H. K. Chan, C. J. Wilson, P. R. Griffin, H. Williams, and W. B. Knight, *Arch. Biochem. Biophys.* **303**, 274 (1993).
[21] K. E. Moore and S. Miura, *J. Biol. Chem.* **262**, 8806 (1987).
[22] J. L. San Millan, D. Boyd, R. Dalbey, W. Wickner, and J. Beckwith, *J. Bacteriol.* **171**, 5536 (1989).
[23] G. von Heijne, *Eur. J. Biochem.* **133**, 17 (1983).
[24] D. Perlman and H. O. Halvorson, *J. Mol. Biol.* **167**, 391 (1983).

for leader peptidase processing.[10] The results are summarized in Fig. 4A. In line with the predictions based on the above statistical approach,[23] almost any residue can be substituted in the +1, −2, −4, and −5 regions of this preprotein, confirming that these positions are not determinants for substrate binding. In contrast, only small residues at −1 and small and some larger aliphatic residues at the −3 position can be tolerated for leader peptidase processing, whereas other substitutions at these positions

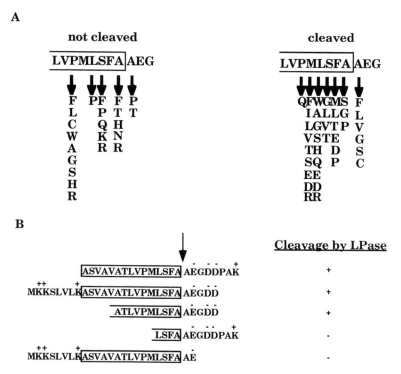

Fig. 4. Requirements for substrate processing by leader peptidase. The amino acid sequence of M13 procoat protein is shown, depicting the leader sequence and the mature coat domain. (A) Cleaved and uncleaved procoat mutants, indicating their substitutions in the +1 to −6 region. (B) Amino acid sequence of procoat peptide fragments. A, Ala; C, Cys; D, Asp; E, Glu; F, Phe; G, Gly; H, His; I, Ile; K, Lys; L, Leu; M, Met; N, Asn; P, Pro; Q, Gln; R, Arg; S, Ser; T, Thr; V, Val; W, Trp; Y, Tyr.

block processing. Similar results at the -1 and -3 position were found studying the maltose-binding protein precursor.[25] In addition, the helix breaker proline of procoat at the -6 position is critical for *in vivo* cleavage of procoat.[10] However, the -6 position is of secondary importance to that of -1 and -3 because most of the -6 mutant proteins, but not the -1 and -3 proteins, can be cleaved with 10^5 times the normal levels of leader peptidase.

To determine the minimum requirements for leader peptidase cleavage, peptide substrates have been used. Dierstein and Wickner[26] tested specific peptides derived from the precursor of M13 coat protein for cleavage (Fig. 4B). Strikingly, they found that a 16-residue peptide that spans the leader peptidase cleavage site is correctly cleaved despite the fact that most of the preprotein was deleted, including the basic amino terminus, the hydrophobic core of the leader sequence, as well as most of the mature region of the protein. In a more recent study,[9] the cleavage and kinetic properties of synthetic peptides corresponding to the cleavage site of the precusor to the maltose-binding protein were analyzed. In this study, the nonapeptide Phe-Ser-Ala-Ser-Ala-Ser-Ala-Leu-Ala+Lys-Ile-CONH$_2$ (-7 to $+2$) is cleaved between the alanyl and lysyl residue with a reported k_{cat} of 119 hr^{-1} and a K_m of 1 mM. Interestingly, the minimum substrate sequence for leader peptidase is the pentapeptide Ala-Leu-Ala-Lys-Ile.[9] However, these studies suggest that leader peptidase may recognize a specific conformation of the preprotein that is not fulfilled by peptide substrates, because the best peptide substrates have very poor catalytic parameters.

The enzymatic activity of leader peptidase has been found to be insensitive to serine-, cysteine-, aspartic acid-, and metallo-specific protease inhibitors (see Table I). This suggests that leader peptidase is an unusual protease. The activity is sensitive to high concentrations of sodium chloride (>160 mM), Mg^{2+} (>1 mM), and dinitrophenol.[27]

Sequence Homology of Type 1 Leader Peptidases with Subunits of Eukaryotic Signal Peptidases

The genes coding for leader peptidase 1 have been identified in *E. coli*,[19] *Salmonella typhimurium*,[28] *Pseudomonas fluorescens*,[29] and *Bacil-*

[25] J. D. Fikes, G. A. Barkocy-Gallagher, D. G. Klapper, and P. J. Bassford, Jr., *J. Biol. Chem.* **265**, 3417 (1990).

[26] R. Dierstein and W. Wickner, *EMBO J.* **5**, 427 (1986).

[27] C. Zwizinski, T. Date, and W. Wickner, *J. Biol. Chem.* **256**, 3593 (1981).

[28] J. M. van Dijl, R. van der Bergh, T. Reversma, H. Smith, S. Bron, and G. Venema, *Mol. Gen. Genet.* **223**, 233 (1990).

[29] M. T. Black, J. G. R. Munn, and A. E. Allsop, *Biochem. J.* **282**, 539 (1992).

TABLE I

INSENSITIVITY TO INHIBITORS[a]

Protease inhibitors[b]	Class	Ref.
o-Phenanthroline	Metallo	c
EDTA	Metallo	c
Phosphoramidon	Metallo	d
2,6-Pyridine dicarboxylic acid	Metallo	c
Bestatin	Metallo	e
TPCK	Serine	c
TLCK	Serine	c
PMSF	Serine	c
APMSF	Serine	e
ZPCK	Serine	e
Dichloroisocoumarin	Serine	e
Elastinal	Serine	d
Aprotinin	Serine	e
Chymostatin	Serine/cysteine	e
Leupeptin	Serine/cysteine	e
Antipain dihydrochloride	Serine/cysteine	e
Iodoacetamide	Cysteine	d, e
NEM	Cysteine	c, f
E-64	Cysteine	d, e
1,2-Epoxy-3-(p-nitrophenoxy)propane	Aspartic acid	d
Pepstatin	Aspartic acid	d
Diazoacetyl-DL-norleucine methyl ester	Aspartic acid	d

[a] Catalytic activity of leader peptidase is insensitive to protease inhibitors against the four well-defined protease groups.

[b] Abbreviations: TPCK, L-1-p-tosylamino-2-phenylethyl chloromethyl ketone; TLCK, 1-chloro-3-tosylamido-7-amino-2-heptanone hydrochloride; PMSF, phenylmethylsulfonyl fluoride; APMSF, aminophenylmethyl sulfonyl fluoride; ZPCK, N-carbobenzyloxy-L-phenylalanyl chloromethyl ketone; NEM, N-ethylmaleimide; E-64, trans-epoxysuccinyl-L-leucylamido(4-guanidino) butane.

[c] Zwizinski et al.[26]

[d] M. T. Black, J. G. R. Munn, and A. E. Allsop, Biochem. J. **282,** 539 (1992).

[e] Kuo et al.[20]

[f] Tschantz et al.[18]

lus subtilis.[30] Based on the alignment of the amino acid sequences of these proteins,[30] it is clear that there is sequence homology (Fig. 5). The degree of homology between the E. coli leader peptidase and the S. typhimurium, P. fluorescens, or B. subtilis protein is 93, 50, and 31%, respectively. The mitochondrial protease 1 from Saccharomyces cerevisiae has 28%

[30] J. M. van Dijl, A. de Jong, J. Vehmaanpera, G. Venema, and S. Bron, EMBO J. **11,** 2819 (1992).

	88 * 103	128 * 153	272 282
Ec	SGSMMPTLLIGDFILV	GD-IVVFKYPEDPKLDYIKRAVGLPGD	GDNRDNSADSR
St	SGSMMPTLLIGDFILV	GD-IVVFKYPEDPKLDYIKRAVGLPGD	GDNRDNSADER
Pf	SGSMKPTLDVGDFILV	GD-VMVFRYPSDPNVNYIKRVVGLPGD	GDNRDNSNDSR
Bs	GDSMYPTLHNRERVFV	GD-IVVL---NGDDVHYVKRIGGLPGD	GDNRRNSMDSR
Imp1	GESMLPTLSATNDYVH	GDCIVALK-PTDPNHRICKRVTGMPGD	GDNLSHSLDSR
Sec11	SGSMEPAFQRGDILFL	GD-VVVYE-VEGKQIPIVHRVLRQHNN	GDNNA-GNDIS
Spc18	SGSMEPAFHRGDLLFL	GE-IVVFR-IEGREIPIVHRVLKIHEK	GDNNA--VDDR
Spc21	SGSMEPAFHRGDLLFL	GE-IVVFK-VEGRDIPIVHRVIKVHEK	GDNNE--VDDR
Consensus	sgSM-Ptl--gd-###	Gd-ivvfk--------#vkRv#-#pgd	GDN-----D-R

FIG. 5. Conserved regions in signal peptidases. In the consensus motifs, an upper case letter indicates strictly conserved residues, a lower case letter indicates conservative substitutions, and a number sign (#) indicates conservation of hydrophobic residues. We also depict by an asterisk (*) the serine and lysine/histidine residues that may be involved in the catalytic mechanism. Ec, *E. coli;* St, *S. typhimurium;* Pf, *P. fluorescens;* Bs, *B. subtilis.* From Dalbey and von Heijne.[6]

sequence homology to the *B. subtilis* enzyme. There is also weak sequence homology between bacterial leader peptidase 1 and the eukaryotic yeast Sec11 and canine Spc18 and Spc21 subunits of the ER signal peptidases, suggesting that the signal peptidase in prokaryotes and eukaryotes may have evolved from a common ancestor.[6,30]

Structure and Function Relationships

Site-directed mutagenesis has been used[17,18] to try to classify leader peptidase into the four protease groups (serine, cysteine, metallo-, and aspartic acid proteases). Initially, the approach involved mutating residues known to be important in other proteases and determining whether these changes abolished activity. Each of the cysteine, histidine, serine, and aspartic acid residues in leader peptidase was mutated to alanine (except cysteine, which was changed to serine). We mutated histidine also because it is required as a general base in the serine and cysteine protease group and is needed as well in the metalloprotease family to coordinate the metal ion.

The mutant proteins were evaluated for their activity both *in vivo* and *in vitro,* and for their stability. The *in vitro* assay is based on a cell extract of MC1061 transformed with the pING plasmids. ^{35}S-Labeled procoat is used as the substrate. In the *in vivo* assay, processing is measured in the IT41 strain, which has a temperature-sensitive leader peptidase coded by the chromosomal *lepB* gene.[31] Figures 6 and 7 demonstrate these assays with various mutants. Leader peptidase S90A, where the serine-90 is

[31] T. Inada, D. L. Court, K. Ito, and Y. Nakamura, *J. Bacteriol.* **171,** 585 (1989).

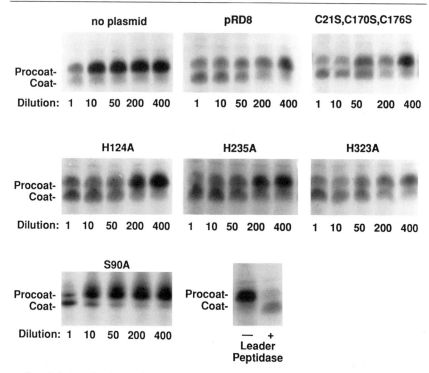

Fig. 6. *In vitro* leader peptidase processing of the M13 procoat protein. Cultures expressing the wild-type or mutant leader peptidase proteins were tested for leader peptidase activity by measuring the posttranslational processing of the M13 procoat at various dilutions of the cell extract. For procoat synthesis, a transcription/translation system was used. Briefly, undiluted or diluted (1:10, 1:50, 1:200, or 1:400) extracts were incubated for 30 min at 37° with *in vitro*-synthesized ^{35}S-labeled procoat. ^{35}S-Labeled procoat was separated from the processed coat protein by using a 23% SDS–polyacrylamide gel. In the bottom right panel, *in vitro*-synthesized procoat (10 μl) was incubated for 30 min at 37° with 2 μl of 1 mg/ml leader peptidase. From Sung and Dalbey.[17]

substituted with alanine, is inactive in both a detergent extract and in the intact cell. A pulse-chase experiment is used to determine the stability of leader peptidase within the cell. Figure 8 shows that leader peptidase D99A mutant is very unstable, with a half-life of less than 3 min.

In Vitro Leader Peptidase Processing

In MC1061, induction of the *lepB* gene on the pRD8 plasmid causes a substantial overproduction of leader peptidase (Fig. 6). Extracts of MC1061/pRD8 diluted 200-fold process procoat to coat at roughly the

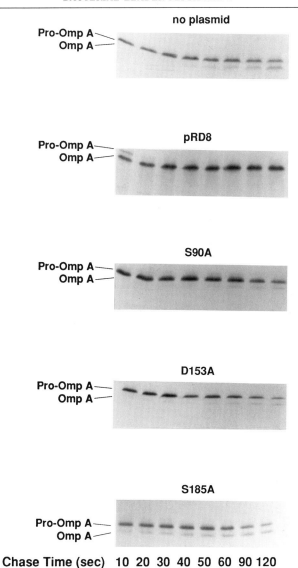

Chase Time (sec) 10 20 30 40 50 60 90 120

FIG. 7. *In vivo* processing of pro-OmpA at the nonpermissive temperature in IT41 bearing plasmids encoding wild-type or mutant leader peptidase molecules. IT41 without plasmid or with plasmids encoding wild-type leader peptidase (pRD8), leader peptidase S90A, D153A, or S185 was grown at 30° to the mid-log phase. Cultures were shifted to 42° for 1 hr and then induced with arabinose (0.2% final concentration). Pro-OmpA and OmpA were immunoprecipitated and separated on a 12% SDS–polyacrylamide gel with a 5% stacking gel. From Sung and Dalbey.[17]

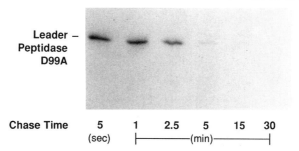

<table>
<tr><td>Leader –
Peptidase
D99A</td></tr>
</table>

Chase Time   5   1   2.5   5   15   30
          (sec)   ├————————(min)————————┤

FIG. 8. Destabilization of leader peptidase by the substitution of an alanine residue for an aspartic acid at position 99. Cells were pulsed for 30 sec and chased for the indicated times. Each sample was immunoprecipitated with antiserum to leader peptidase and then analyzed by SDS–polyacrylamide gel electrophoresis.

same extent as undiluted extracts of MC1061 without plasmid. MC1061 cells bearing a plasmid coding for wild-type or mutant leader peptidase are grown in M9 medium containing 0.5% (w/v) fructose and the 19 amino acids (lacking methionine) to an optical density at 600 nm of 0.2. After the addition of arabinose to a final concentration of 0.2% (w/v) to induce synthesis of leader peptidase, the cells are grown for 2 hr, followed by centrifugation for 1 min to concentrate the cells. The cells are then resuspended in 0.3 ml of lysis buffer containing 20% (w/v) sucrose, 10 mM Tris-HCl (pH 8.0), 10 mM EDTA, 1% (v/v) Triton X-100, lysozyme (1 mg/ml), deoxyribonuclease (5 μg/ml), ribonuclease (1 μg/ml), and phenylmethylsulfonyl fluoride (5 mM). Incubation is carried out for 30 min at room temperature, at which time the cell extracts are diluted (1 : 10, etc.) or used directly and incubated with ^{35}S-labeled procoat at 37° for 30 min. Procoat and coat proteins are resolved on a 23% polyacrylamide gel as described.[12]

In Vivo Leader Peptidase Processing

This assay takes advantage of IT41, which has a temperature-sensitive leader peptidase coded by the chromosomal *lepB* gene.[31] At the nonpermissive temperature (42°), the precursor to the outer membrane protein A (pro-OmpA) is processed slowly to OmpA with a $t_{1/2} > 90$ sec (Fig. 7). In contrast, processing is rapid (<10 sec) at 42° when IT41 is transformed with pRD8, which encodes the wild-type leader peptidase. IT41 with or without a plasmid is grown at 30° in M9 medium containing 0.5% fructose and the 19 amino acids except methionine. After reaching the early log phase, cultures are shifted to 42° for 1 hr to inactivate the temperature-sensitive leader peptidase. Leader peptidase is then induced by the addition of arabinose (0.2% final concentration) to the medium. Cells are labeled with 250 μCi of [^{35}S]methionine for 15 sec and chased with nonradioactive methionine (to a final concentration of 500 μg/ml). At the indicated times, samples are removed and treated with ice-cold 20% (w/v) trichloroacetic acid. Precipitates are collected after 30 min by centrifugation in a microfuge (5 min, 4°). After aspirating the supernatant away, pellet is washed with 1 ml of chilled acetone at 4° to remove traces of trichloroacetic acid and the pellet is collected at 4° by centrifugation for 5 min and the supernatant is again aspirated away. The pellet is dried at 95° for 5 min, at which time 100 μl of 10 mM Tris-HCl (pH 8.0) containing 2% (w/v) SDS is added to the tube. Immunoprecipitation is performed as described[2] using OmpA antiserum and the immunoprecipitates are adsorbed on Pansorbin cells (Calbiochem) and washed. Pro-OmpA and OmpA are separated on a 12% (w/v) SDS–polyacrylamide gel (acrylamide : bisacrylamide, 30 : 0.8) with a 5% (w/v) stacking gel using a discontinuous buffer system. The gels are then fixed and subjected to fluorography.[32]

Pulse-Chase Stability Assay

The stability of leader peptidase mutants is ascertained using a pulse-chase experiment (Fig. 8). MC1061 cells (1.5 ml) bearing the pING plasmid, which codes for the leader peptidase mutant, are grown in M9 medium containing 0.5% (w/v) fructose and the 19 amino acids (without methionine). After the cells reach the mid-log phase, arabinose is added to induce synthesis of leader peptidase (0.2% (w/v) final concentration). Cells (0.2 ml) are pulse labeled with 200 μCi of *trans*-[^{35}S]methionine for 30 sec and then incubated with nonradioactive methionine (500 μg/ml) for 5 sec and 1, 2.5, 5, 15, and 30 min. At each chase time, 0.2 ml of the labeled cells is added to an equal volume of chilled 20% (w/v) trichloroacetic acid to stop the reaction. Each sample is then analyzed by immunoprecipitation using leader peptidase antiserum, as described in the *in vivo* assay.

What have we learned about the structure and function of the *E. coli* leader peptidase using site-directed mutagenesis? First, cysteine and histidine residues are not required for activity of leader peptidase,[17] again consistent with leader peptidase not being a member of the four "standard" protease groups. Second, aspartic acid-99 is important for the stability of the protein and the serine-90 is important for catalysis.[17] We have been able to convert the serine-90 residue into a cysteine residue without significantly affecting the activity of the protein.[18] Interestingly, this thiol leader peptidase can be inactivated with *N*-ethylmaleimide under condi-

[32] K. Ito, T. Date, and W. Wickner, *J. Biol. Chem.* **255**, 2123 (1980).

tions that lead to little loss in activity in the wild-type protein.[18] We also showed that, of the three conserved basic residues (arginine-127, lysine-145, and arginine-146), only lysine-145 seems to be important for activity; its substitution with alanine, asparagine, and histidine leads to an inactive enzyme, based on the above assays.

We would like to stress that caution should be exercised in the interpretation of site-directed mutations. For example, we initially concluded incorrectly that aspartic acid-153 is important for catalysis because the D153 mutant was inactive. However, further studies showed that aspartic acid at position 153 could be replaced with asparagine or glutamic acid with little effect on activity, showing that the specific amino acid that is substituted at position 153 can influence the results that are obtained.[18]

All the available data so far suggest that serine-90 and lysine-145, which are conserved in prokaryotes, may play a direct role in catalysis.[17,18,29,33] Our working model is that the leader peptidase in prokaryotes is an unusual serine protease that catalyzes cleavage of preprotein substrates using a serine/lysine dyad, instead of the classical serine/histidine/aspartic acid triad.[34] Such a novel catalytic mechanism has been reported for the catalytic mechanism of autodigestion of the LexA repressor.[35] Interestingly, although the serine-90 is conserved as well in the eukaryotic Sec11 and Spc18 and Spc21 subunits of the endoplasmic reticulum signal peptidase,[6,30] the lysine-145 has been replaced by a histidine residue. This suggests that the eukaryotic signal peptidase may carry out catalysis utilizing a more conventional proteolytic mechanism. Further work will be required to verify the serine/lysine dyad hypothesis for the *E. coli* leader peptidase.

Conclusion

Leader peptidases are a group of intracellular proteases that play an important role in protein export and function to remove amino-terminal signal sequences that target exported proteins within the cell. Although the available data indicate that it does not belong to the four "standard" protease families—serine, cysteine, aspartic acid, or metallo-type proteases, structure–function studies suggest that leader peptidase carries out catalysis using a serine/lysine catalytic dyad. Laboratory studies are now underway to isolate active soluble fragments of leader peptidase that can

[33] M. T. Black, *J. Bacteriol.* **175,** 4957 (1993).
[34] H. Neurath, *in* "Proteolytic Enzymes: A Practical Approach" (R. J. Beynon and J. S. Bond, eds.), p. 1. IRL Press, Oxford and Washington, D.C., 1989.
[35] L. L. Lin and J. W. Little, *J. Bacteriol.* **170,** 2163 (1988).

be crystallized and used to obtain a high-resolution structure, revealing for the first time the catalytic residues and the substrate-binding pocket of this unusual protease.

Acknowledgments

We wish to thank Meesook Sung, who contributed to this work of leader peptidase, and the National Science Foundation, for Grant DCB-9020759 in support of this work.

[22] Eukaryote Microsomal Signal Peptidases

By MARK O. LIVELY, ANN L. NEWSOME, and MOHAMAD NUSIER

Introduction

Microsomal signal peptidase is the first proteolytic enzyme that acts on secretory proteins, membrane proteins, and certain intracellular proteins that pass through the secretory pathway of the endoplasmic reticulum (ER) as they progress toward their mature, functional structures properly localized at their sites of action.[1] Most proteins that are targeted to the ER are initially synthesized as precursors with 15- to 30-residue amino terminal extensions, known as signal peptides,[2] that are removed proteolytically once the growing protein is engaged at the site of translocation. Signal peptidase catalyzes the removal of these N-terminal extensions. The enzyme was first detected in preparations of rough ER microsomes from mouse myeloma cells, which were found to catalyze the cleavage of a precursor immunoglobulin light chain to its mature molecular weight.[3] The enzyme has since been detected in ER preparations from a wide range of eukaryotic cells. Signal peptidase has been purified from canine pancreas[4] and chicken oviduct[5,6] and has been partially purified from yeast microsomes.[7] It is a complex of integral membrane proteins that is localized largely on the lumenal side of the ER[8] at, or near, sites of

[1] T. A. Rapoport, *Science* **258**, 931 (1992).
[2] G. von Heijne, *J. Membr. Biol.* **115**, 195 (1990).
[3] C. Milstein, G. G. Brownlee, T. M. Harrison, and M. B. Mathews, *Nature (London) New Biol.* **239**, 117 (1972).
[4] E. A. Evans, R. Gilmore, and G. Blobel, *Proc. Natl. Acad. Sci. U.S.A.* **83**, 581 (1986).
[5] R. K. Baker and M. O. Lively, *Biochemistry* **26**, 8561 (1987).
[6] R. K. Baker and M. O. Lively, *J. Cell. Biochem.* **32**, 193 (1986).
[7] J. T. YaDeau, C. Klein, and G. Blobel, *Proc. Natl. Acad. Sci. U.S.A.* **88**, 517 (1991).
[8] G. S. Shelness, L. Lin, and C. V. Nicchitta, *J. Biol. Chem.* **268**, 5201 (1993).

translocation of nascent proteins. The bacterial signal (leader) peptidases are single-chain membrane proteins localized on the periplasmic side of the cytoplasmic membrane and are described in detail in [21] in this volume.[9,9a,10] Purification of chicken microsomal signal peptidase from oviducts of laying hens is described here.

Preparation of Buffers

Each step of chromatography is conducted in the presence of detergent to maintain the solubility of the enzyme and phosphatidylcholine is added in later steps to stabilize the enzyme. Glycerol is also included in chromatography buffers for stabilization of enzyme activity. All steps are performed at 4° unless otherwise noted. Dithiothreitol (DTT) is added as a solid to the buffers immediately before use. All chromatography buffers are filtered through 0.45-μm filters and stored at 4°.

Reagents

Stock phosphatidylcholine (PC), 25 mg/ml: egg L-α-phosphatidylcholine (Avanti Polar Lipids, Inc., Alabaster, AL) is obtained as a 20 mg/ml solution in $CHCl_3$. 50 ml of the PC in $CHCl_3$ is reduced to dryness under vacuum in a rotary evaporator to remove the $CHCl_3$. The residual PC is suspended in 40 ml H_2O, sonicated briefly in a bath sonicator, and frozen in 2-ml aliquots that are stored at $-20°$.

5× Assay buffer: 250 mM triethanolamine (TEA) acetate, pH 8.0, 250 mM KCH_3CO_2, 5 mM EDTA, 500 mM sucrose, 2.5 mg/ml PC (added using the 25 mg/ml stock PC), 50 mM DTT.

1× SDS-PAGE sample buffer: 60 mM Tris-HCl, pH 6.7, 4% (w/v) SDS, 10% (w/v) glycerol, 0.01% (w/v) bromphenol blue.

Low salt buffer (LSB): 50 mM TEA-HCl, pH 7.5, 50 mM KCl, 5 mM $MgCl_2$, 5 mM DTT.

Homogenization buffer: LSB containing 0.25 M sucrose.

Carbonate extraction buffer: 0.1 M Na_2CO_3, pH 11.5.

Stock NP-40: 10% (w/v) Nonidet P-40 (NP-40). Other related detergents in the Triton X-100 family can be substituted. Igepal CO630sp (Rhone-Poulenc, Cranbury, NJ) is a very similar detergent that gives good results.

DEAE chromatography buffer: 10 mM TEA-HCl, pH 8.6, at 4°, 5 mM $MgCl_2$, 10% (v/v) glycerol, 1% (w/v) NP-40 (or Igepal CO630).

[9] H. C. Wu, this volume [46].
[9a] Sankaran and H. C. Wu, this series.
[10] W. R. Tschantz and R. E. Dalbey, this volume [21].

CM buffer: 10 mM NaCH$_3$CO$_2$, pH 5.6, 10% (v/v) glycerol, 1% (w/v) NP-40 (or Igepal CO630).
CM buffer + PC: add 4 ml 25 mg/ml stock PC suspension to 200 ml CM buffer.
CM buffer + PC + 0.4 M NaCl: add 2 ml 25 mg/ml stock PC suspension and 2.34 g NaCl to 100 ml CM buffer.
HA column buffer: 0.5 mM sodium phosphate buffer, pH 6.8, 10% (v/v) glycerol, 0.5% (w/v) NP-40.
HA baseline buffer: 0.5 mM sodium phosphate buffer, pH 6.8, 10% (v/v) glycerol, 0.5% (w/v) 3-[(3-cholamidopropyl)dimethylammonio]-1-propane sulfonate (CHAPS).
Con A buffer: 10 mM Tris-HCl, pH 8.0, 250 mM NaCl, 1 mM CaCl$_2$, 1 mM MgCl$_2$, 10% (v/v) glycerol, 5 mM CHAPS, 0.08 mg/ml PC.
HA elution buffer: 0.1 M sodium phosphate, pH 8.0, in Con A buffer.

Assays

Signal peptidase is a highly selective enzyme whose only known substrates are secretory precursor proteins containing an N-terminal signal peptide or certain synthetic signal peptides with sequences based on known signal peptides.[11] Cleavage activity can be measured in intact microsomal vesicles using a cotranslational assay in which a cell-free protein synthesis system is actively translating a secretory protein mRNA in the presence of isolated ER microsomes.[12] The microsomes contain the necessary components for signal recognition, targeting to the ER, translocation across the lipid bilayer, and cleavage by signal peptidase so the newly synthesized precursor protein is translocated into the microsomal vesicles where the signal peptide is removed by proteolysis. Cleavage of the precursor protein is detected as a decrease in molecular weight of the substrate by analysis of the reaction mixture by polyacrylamide gel electrophoresis in the presence of sodium dodecyl sulfate (SDS–PAGE). Because the amount of protein made by cell-free protein synthesis is in the femtomole range, synthesis is performed in the presence of radiolabeled amino acids, usually [35S]Met or [35S]Cys, and the newly synthesized proteins are detected by autoradiography of the dried polyacrylamide gel. This method is only useful for signal peptidase assays using intact microsomes because conditions introduced by the signal peptidase purification fractions inhibit cell-free protein synthesis reactions.

[11] M. P. Caulfield, L. T. Duong, R. K. Baker, M. Rosenblatt, and M. O. Lively, *J. Biol. Chem.* **264**, 15813 (1989).
[12] G. Scheele, this series, Vol. 96, p. 94.

A posttranslational assay is used for measurement of solubilized signal peptidase activity.[13] The approach is basically the same as for the cotranslational assay except that the ^{35}S-labeled precursor protein is first synthesized by cell-free protein synthesis in the absence of microsomal membranes. Aliquots of the cell-free protein synthesis reaction mixtures containing the full-length secretory precursor protein are incubated with aliquots of the detergent-solubilized enzyme, then analyzed by SDS–PAGE and autoradiography. Many, but not all, precursor proteins retain conformations that permit recognition and cleavage by signal peptidase. The best precursor proteins to use for this purpose have molecular masses less than 40 kDa because it is frequently difficult to resolve the precursor from the cleaved product by SDS–PAGE when larger proteins are used as substrates. Additionally, some large proteins are not cleaved posttranslationally, presumably because they fold in ways that block cleavage by the peptidase. Two commonly used precursor protein substrates for studies of purified microsomal signal peptidase are preprolactin[13] and human preplacental lactogen (preHPL).[14] These mRNAs were selected initially because they are abundant and readily isolated from their source tissues: preprolactin from pituitaries and preHPL from human placenta delivered at term. Each encodes a precursor protein of approximately 25 kDa that is readily resolved from the cleaved protein product when analyzed by SDS–PAGE. Large amounts of mRNA can now be obtained by transcription of cloned cDNA *in vitro,* so any cDNA encoding a secretory precursor protein can be used to provide a signal peptidase substrate, although not all full-length precursor proteins are good substrates for the enzyme *in vitro.*[13] The assay described here uses preHPL as the substrate.

Preparation of preHPL Substrate

A cDNA encoding preHPL was excised from plasmid phPL815 (kindly provided by Dr. G. F. Saunders)[15] and subcloned into the *Bam*HI/*Sal*I cloning site of the vector pGEM-3Zf(−) (Promega Corp., Madison, WI) in the orientation for transcription by T7 RNA polymerase. mRNA is prepared by cell-free transcription using the protocol of the commercial supplier (Ambion, Inc., Austin, TX). This mRNA encoding preHPL is translated in a rabbit reticulocyte lysate protein synthesis system[16] in the

[13] R. C. Jackson, this series, Vol. 96, p. 784.
[14] A. W. Strauss, I. Zimmermann, B. Boime, B. Ashe, R. A. Mumford, and A. W. Alberts, *Proc. Natl. Acad. Sci. U.S.A.* **76,** 4225 (1979).
[15] H. A. Barrera-Saldaña, D. L. Robberson, and G. F. Saunders, *J. Biol. Chem.* **257,** 12399 (1982).
[16] R. J. Jackson and T. Hunt, this series, Vol. 96, p. 50.

presence of [^{35}S]Met according to the protocol provided by the supplier (Promega) to yield a translation reaction mixture containing [^{35}S]Met–preHPL. The wheat germ cell-free protein synthesis system,[17] also available commercially, is useful for precursor proteins with molecular masses less than 20 kDa. The wheat germ lysate does not contain the large amounts of globin present in the reticulocyte lysate that interfere with electrophoresis of proteins in the range of 15 kDa. Translation reaction mixtures are used in signal peptidase assays without additional treatment.

Proteolysis Assay

The final conditions for assay of signal peptidase are 50 mM TEA–acetate, pH 8.0, 50 mM KCH$_3$CO$_2$, 1 mM EDTA, 100 mM sucrose, 0.5 mg/ml egg L-α-phosphatidylcholine, 10 mM DTT. A typical assay contains 4 μl solubilized signal peptidase, 4 μl 5× assay buffer, 8 μl H$_2$O, and 4 μl translation reaction mixture containing [^{35}S]Met–preHPL. From 5 to 20 mM DTT is required for maximum activity. Reaction mixtures are incubated at 28° for 60 min and the reaction is stopped by addition of an equal volume of ice-cold 20% (w/v) trichloroacetic acid. Precipitated proteins are collected by centrifugation for 3 min at 14,000 g_{max}, washed twice with 50 μl 90% (v/v) ice-cold acetone, then dissolved in 20 μl 1 × SDS–PAGE sample buffer and heated at 100° for 3 min. The reaction products are separated by SDS–PAGE in a 12.5% (w/v) acrylamide gel[18] and detected by autoradiography of the dried gel using Kodak (Rochester, NY) X-Omat AR film.

Conversion of precursor to product is quantified by scanning densitometry of the autoradiography film.[13] Conversion of preHPL to HPL is typically in the range of 50 to 70% with good preparations of active enzyme. The inability routinely to obtain cleavage of 100% of substrate molecules appears to be dependent on some unknown property of the substrate, perhaps the conformation or state of aggregation of the precursor, which makes a portion of the molecules inaccessible to the enzyme. Consistent with this hypothesis, preincubation of the substrate with SDS can increase the extent of cleavage of some precursor proteins such as preprolactin.[4] The initial rate of cleavage is approximately linear for 10 min then slows as the percentage processing approaches a maximum value. The maximum conversion of precursor to product reached depends on the concentration of enzyme added. This unusual kinetic behavior may be due to the overall slow rate of reaction, the low concentration of substrate, and slow inactivation of signal peptidase during the incubation period. Oviduct signal pepti-

[17] A. H. Erickson and G. Blobel, this series, Vol. 96, p. 38.
[18] U. K. Laemmli, *Nature (London)* **227**, 680 (1970).

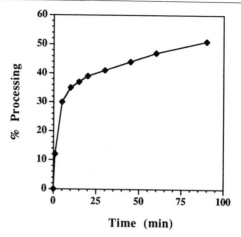

FIG. 1. Time course for signal peptidase cleavage of preHPL. Partially purified oviduct signal peptidase (CM pool) was incubated with [^{35}S]Met–preHPL prepared by cell-free protein synthesis in a rabbit reticulocyte lysate protein synthesis system. Aliquots of the reaction mixture were separated by SDS–PAGE, detected by autoradiography, and quantified by laser densitometry.

dase loses approximately 33% of its activity per hour at 28° under the conditions described here. Incubations are normally continued for 60–90 min to ensure maximal conversion of preHPL to product (Fig. 1).

Purification

Isolation of microsomal signal peptidase begins with the preparation of rough ER microsomes from an active secretory tissue such as pancreas[4] or chicken oviduct.[5,6] Rough microsomes are treated with conditions that remove soluble proteins associated with the microsomes before the membranes are solubilized. Solubilization is accomplished using mild detergents such as Triton X-100, Nonidet P-40, octylglucoside, deoxycholate, or Nikkol. The detergent concentration used must be high enough to totally dissolve the lipid bilayer, because the enzyme is an integral membrane protein[19] and requires detergent to remain soluble. Signal peptidase also requires a phospholipid environment for activity.[20] In the case of the purification of hen oviduct signal peptidase[5,6] described here, the solubilized enzyme is subjected to several chromatography steps, including

[19] M. O. Lively and K. A. Walsh, *J. Biol. Chem.* **258**, 9488 (1983).
[20] R. C. Jackson and W. R. White, *J. Biol. Chem.* **256**, 2545 (1981).

cation- and anion-exchange, hydroxylapatite, and concanavalin A–Sepharose affinity chromatography.

Isolation of Rough Microsomes

Oviducts are obtained from mature, laying hens supplied by local farmers. In a typical large-scale preparation, oviducts from 50 hens are processed to a total microsome fraction in a single day. The birds are anesthetized by intramuscular injection of a mixture containing 50 mg/ml ketamine and 50 mg/ml xylazine at a dosage of 30 mg/kg for each drug. After allowing 20 min for the drugs to take effect, the birds are sacrificed by decapitation. The oviduct is exposed by removing the abdominal contents.

The oviduct is a long tube that extends from the ovary (where numerous yellow yolks should be visible in actively laying hens) to the cloaca, the combined intestinal, urogenital orifice. The tube is divided into five distinct regions.[21] The magnum is the second region as one follows the tube from the ovary toward the cloaca, the direction of egg movement. The magnum is recognized by the change in tissue color to a dull pinkish-white. Yolks found in this region of the oviduct will be surrounded by egg white proteins but will not yet be enclosed in a membrane. The diameter of the oviduct and the thickness of the oviduct wall increase in this region. It is the longest single portion of the oviduct and its diameter narrows significantly at its end, which is recognized by a thin, translucent band approximately 2 mm wide that marks the beginning of the isthmus region.[21] The magnums are excised from the total oviduct and trimmed to remove attached ligaments and blood vessels. The magnum tissue is then cut into small slices using scissors. The tissue slices are weighed and suspended at a concentration of approximately 150 g wet tissue per liter cold homogenization buffer. Fifty laying hens normally yield 1100–1400 g magnum tissue.

The magnum tissue is homogenized at 4° using a Polytron probe homogenizer (Brinkman Instruments Inc., Westbury, NY) for 5–10 min until the suspension is homogeneous. The homogenate is filtered through two or three layers of cheesecloth, dispensed into 500-ml centrifuge bottles, and centrifuged for 10 min at 8000 rpm (10,000 g_{max}) in a DuPont GS3 rotor. The combined supernatants are then centrifuged at 4° for 45 min at 18,000 rpm (48,550 g_{max}) in a Beckman type 19 large-capacity, fixed-angle rotor. These supernatants are discarded and the total microsome pellets are combined in 50-ml conical screw-top tubes that are flash frozen in liquid nitrogen then stored at −70°. A typical preparation yields 150–200 g pelleted hen oviduct microsomes. This preparation is the total microsome

---

[21] R. N. C. Aitken, in "Physiology and Biochemistry of the Domestic Fowl" (D. J. Bell and B. M. Freeman, eds.), p. 1237. Academic Press, London and New York, 1971.

fraction that is significantly enriched with rough ER microsomes because of the extensive network of ER in the tubular gland cells. At the scale of this preparation method, density gradient centrifugation of this large amount of membranes to isolate rough ER is not considered practical or necessary. Cotranslationally active oviduct microsomes can be obtained by gradient centrifugation of small quantities of this preparation as described for dog pancreatic microsomes.[22]

Stripping and Solubilization of Microsomes

Soluble proteins present in the microsome preparation can be removed by treatment of the microsomes with ice-cold carbonate, conditions that dissociate soluble and peripheral membrane proteins and release proteins enclosed within the vesicles.[23] Integral membrane proteins, including signal peptidase, remain bound to the lipid bilayers of the microsomes, which are recovered by sedimentation. This treatment effectively removes as much as 80% of the protein associated with crude microsomes without apparent loss of signal peptidase activity.[5]

Approximately 100 g of crude hen oviduct microsomes is thawed and homogenized in 400 ml of homogenization buffer using a glass Dounce homogenizer (Wheaton Scientific, Millville, NJ). The $A_{280 \text{ nm}}$ of the suspension is adjusted to approximately 50 by dilution with homogenization buffer. The absorbance of the microsome suspension is conveniently determined by dissolving 10 μl of the homogenized suspension in 1ml of 5% SDS and measuring the $A_{280 \text{ nm}}$.

The carbonate stripping is performed in batches that can be accommodated by a single centrifugation in a Beckman type 19 fixed-angle rotor (235 ml/bottle). An 85-ml aliquot of the membrane suspension is poured into 1200 ml ice-cold carbonate extraction buffer and stirred at 4° for 30 min. The stripped lipid bilayers containing integral membrane proteins are collected by centrifugation at 4° in the type 19 rotor at 18,000 rpm (48,550 g_{max}) for 60 min. At 30 min before the end of the centrifugation run, a second 85-ml aliquot of the microsome suspension is added to 1200 ml ice-cold carbonate extraction buffer and stirring is begun. Following centrifugation, the supernatant is discarded and the combined pellets are gently rinsed with ice-cold homogenization buffer to neutralize the pH.

For solubilization, the stripped microsomes are resuspended in enough ice-cold LSB to obtain an approximate $A_{280 \text{ nm}}$ of 30. Signal peptidase is solubilized by adding one volume of 10% (w/v) NP-40 (Nonidet P-40) to three volumes of resuspended, stripped microsomes to achieve a final concentration of 2.5% (w/v) NP-40. This mixture is homogenized by a

[22] P. Walter and G. Blobel, this series, Vol. 96, p. 84.
[23] Y. Fujiki, A. L. Hubbard, S. Fowler, and P. B. Lazarow, *J. Cell Biol.* **93,** 97 (1982).

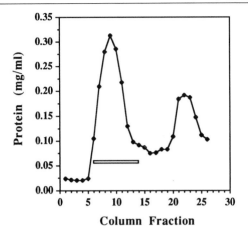

FIG. 2. Diethylaminoethyl cellulose chromatography of solubilized oviduct signal peptidase. Fractions (20 ml) containing signal peptidase were pooled, as indicated by the bar. Protein concentration was determined as described.[24]

few strokes with a Dounce homogenizer then allowed to stand at room temperature for 10 min. The solubilization mixture is then centrifuged at 15° for 3 hr at 28,000 rpm (141,000 g_{max}) in a Beckman SW 28 rotor. The supernatant is collected and dialyzed against three 2-liter changes of DEAE chromatography buffer. This preparation is stable indefinitely when frozen at −70°.

DEAE-Cellulose Chromatography

A 5× 30-cm column is packed with diethylaminoethyl cellulose (Whatman DE-52; Clifton, NJ) equilibrated in DEAE column buffer at 4°. The column is washed with DEAE buffer until the pH and conductivity of the eluted buffer are unchanged. The dialyzed, solubilized stripped microsomal fraction (typically 70 ml) is applied to the column at a flow rate of approximately 30 ml/hr and 20-ml fractions are collected. Once the entire sample has been applied, elution is continued with DEAE column buffer and 40 fractions are collected. Because of the high absorbance at 280 nm due to the NP-40 in the buffer, it is not possible to monitor the progress of chromatography using an absorbance monitor. All fractions are analyzed for protein content using a modification of the Lowry method.[24] The protein concentration of each DEAE fraction is determined using 100-μl aliquots. Signal peptidase activity is determined using 4-μl aliquots in the preHPL assay. Figure 2 shows a typical elution profile and the bar indicates

[24] G. L. Peterson, *Anal. Biochem.* **83**, 346 (1977).

the pooled fractions containing signal peptidase. Under these conditions, approximately 90% of the protein remains bound to the DEAE column and signal peptidase elutes in the unretained fractions. The pooled DEAE fractions are dialyzed overnight against three 2-liter changes of carboxymethyl (CM) buffer.

Carboxymethyl Cellulose Chromatography

The dialyzed DEAE pool is subjected to chromatography on a 3 × 10-cm column of carboxymethyl cellulose (Whatman CM-52) previously equilibrated in CM buffer containing no phosphatidylcholine. The DEAE pool is applied at a rate of 30 ml/hr then the column is washed with 100 ml CM buffer containing 0.5 mg/ml PC. Under these conditions, signal peptidase is retained by the column and the endogenous phospholipids and some protein contaminants flow through the column. After this step, PC is added to all chromatography buffers to stabilize enzyme activity. Signal peptidase is eluted from the column using a linear gradient of NaCl. The 200-ml gradient is formed using 100 ml CM buffer containing 0.5 mg/ml PC and 100 ml CM buffer containing 0.5 mg/ml PC and 0.4 *M* NaCl. Fractions containing 6 ml are collected during the gradient elution. Protein assays are performed using 20-μl aliquots of each fraction and signal peptidase assays are performed using 4-μl aliquots. Active fractions are pooled as indicated in the typical chromatogram shown in Fig. 3. The

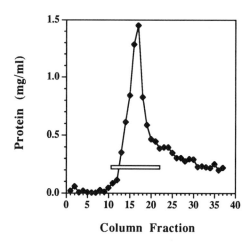

FIG. 3. Carboxymethyl cellulose chromatography of solubilized oviduct signal peptidase. Fractions (6 ml) containing signal peptidase were pooled, as indicated by the bar. Protein concentration was determined as described.[24]

CM pool is a very active signal peptidase preparation and can be used without further purification for many applications. The preparation is stable for months when stored at $-20°$.

Hydroxylapatite Chromatography

Chromatography on hydroxylapatite (HA) is used to change detergents from NP-40 to CHAPS prior to chromatography on concanavalin A–Sepharose because better results were obtained using CHAPS during the lectin affinity chromatography step. A 1.5 × 5-cm column is packed with 2.5 g hdyroxylapatite (Bio-Rad, Richmond, CA) previously equilibrated in HA buffer. The CM chromatography pool is dialyzed against HA buffer, applied to the column at a flow rate of approximately 25 ml/hr, before the column is washed with HA baseline buffer until the $A_{280\,nm}$ (due to the NP-40) reaches baseline. Signal peptidase is eluted using HA elution buffer. Protein eluting from the column is monitored using the UV absorbance monitor and the peak of $A_{280\,nm}$ is collected. The HA pool can be stored frozen at $-20°$.

Concanavalin A–Sepharose Chromatography

The signal peptidase complex contains a 23-kDa glycoprotein that has a single Asn-linked side chain containing α-D-mannopyranosyl residues,[4,5] so concanavalin A (Con A) affinity chromatography is used as a final purification step for signal peptidase. The HA pool is applied at a flow rate of 10 ml/hr to a column (1 × 17 cm) of Con A–Sepharose (Pharmacia, Piscataway, NJ) equilibrated in Con A buffer. The column is washed with Con A buffer then bound glycoproteins are released by elution with 0.25 M methyl α-D-mannopyranoside in Con A buffer. The eluted peak fractions, monitored by a UV absorbance monitor at 280 nm, are pooled and frozen at $-20°$. Figure 4 shows an SDS–PAGE gel of oviduct signal peptidase stained with Coomassie blue.

Physical Characterization

Microsomal signal peptidase was first purified from canine pancreatic ER microsomes as a complex of five proteins with masses of 25, 22/23 (a glycoprotein), 21, 18, and 12 kDa.[4] As described here, purification of signal peptidase from ER microsomes obtained from chicken oviduct tubular gland cells (the cells that synthesize egg white proteins) shows that a complex of only two proteins (Fig. 4), with molecular masses of 22/23 and 19 kDa, is sufficient for removal of signal peptides *in vitro*.[5] The 22/23-kDa bands contain a single glycoprotein that migrates as a doublet due

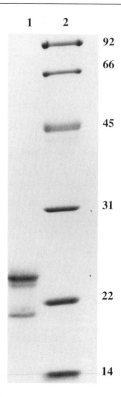

Fig. 4. Purified oviduct signal peptidase separated by SDS–PAGE and stained with Coomassie blue (lane 1). The masses of the standard proteins (lane 2) are indicated in kilodaltons.

to apparent differences in the carbohydrate side chains.[5] The proteins associated with purified microsomal signal peptidases are shown schematically in Fig. 5. cDNAs encoding the 22/23-kDa glycoprotein from dogs[25] and chickens[26] have revealed that these two gene products are close homologs, based on their amino acid sequences being 90% identical. Furthermore, the 21-[27] and the 18-kDa canine subunits are homologous gene products found in the same species.[28] Tryptic peptide analysis of the purified chicken 19-kDa subunit[5] and a partial cDNA clone encoding it[29]

[25] G. S. Shelness, Y. S. Kanwar, and G. Blobel, *J. Biol. Chem.* **263,** 17063 (1988).
[26] A. L. Newsome, J. W. McLean, and M. O. Lively, *Biochem. J.* **282,** 447 (1992).
[27] G. Greenberg, G. S. Shelness, and G. Blobel, *J. Biol. Chem.* **264,** 15762 (1989).
[28] G. S. Shelness and G. Blobel, *J. Biol. Chem.* **265,** 9512 (1990).
[29] S. J. Walker and M. O. Lively, unpublished data (1993).

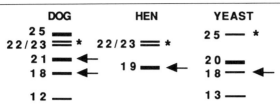

FIG. 5. Schematic representation of purified signal peptidase proteins from dog,[4] hen,[5] and yeast.[7] The glycoproteins are indicated by the asterisks and the SEC11 family of proteins is designated with arrows.

show that its amino acid sequence is closely related to the 21- and 18-kDa canine subunits. Although there is currently no evidence from tryptic peptide analysis for homologous 19-kDa proteins in purified chicken signal peptidase, DNA sequence analysis of amplified chicken cDNA suggests that a homologous mRNA is expressed.[29]

The 18- and 21-kDa canine and the 19-kDa chicken signal peptidase proteins are all homologs of the yeast SEC11 gene product, identified genetically as an essential gene required for processing precursor proteins in yeast.[30] Although the global alignments of the amino acid sequences of this SEC11 family of signal peptidase proteins with the bacterial leader peptidase sequences[10] do not provide convincing evidence that these two families of proteins are evolutionarily related, alignments of limited regions within these sequences suggest similarities in three regions.[31] One of these regions includes a Ser residue required for enzymatic activity of E. coli leader peptidase,[32] suggesting that the SEC11 homologs contain an active site that may be similar to bacterial signal peptidase I.

The mechanistic class of microsomal signal peptidase is unknown. There are no clearly defined irreversible, active site-directed inhibitors of the enzyme and the known amino acid sequences do not align with any of the established families of proteolytic enzymes. The individual functional roles of the signal peptidase subunits are unknown. Purified hen oviduct signal peptidase is proteolytically active as a complex that contains at least one 22/23-kDa glycoprotein subunit and one of the SEC11-like proteins, although the stoichiometry of the active complex is not known. The other proteins isolated as part of canine and yeast signal peptidases do not

[30] P. C. Böhni, R. J. Deshaies, and R. W. Schekman, J. Cell Biol. 106, 1035 (1988).
[31] R. E. Dalbey and G. von Heijne, Trends Biochem. Sci. 17, 474 (1992).
[32] W. R. Tschantz, M. Sung, V. M. Delgado-Partin, and R. E. Dalbey, J. Biol. Chem. 268, 27349 (1993).

appear to be required for proteolysis *in vitro*, but may exist is a complex *in vivo* that plays a role in protein translocation as well as proteolysis.

Acknowledgment

This work was supported by Grant GM32861 from the National Institutes of Health.

[23] Endopeptidase Clp: ATP-Dependent Clp Protease from *Escherichia coli*

By Michael R. Maurizi, Mark W. Thompson,
Satyendra K. Singh, and Seung-Ho Kim

Introduction

Escherichia coli Clp protease is a multicomponent protease that has an ATP-activated proteolytic activity and an ATPase activity that is activated by proteins and peptides.[1–6] Clp protease has also been called protease Ti.[2,4] Initially, two components of Clp protease were purified; the proteolytic activity and the ATPase activity reside on the separate subunits, ClpP and ClpA, which interact to form the active protease.[3,4,6] *In vitro*, the complex of ClpA and ClpP can degrade a number of proteins, such as α-casein, into small peptides and can hydrolyze ATP. Another component of Clp protease, called ClpX, has been genetically identified[7] and purified.[8] ClpX participates with ClpP in the rapid and specific degradation of the λ O protein, but the combination of ClpX and ClpP has very little activity against α-casein.[8] Sequence comparisons indicate that ClpA and ClpX are part of a family of ATPases that includes two other *E. coli*

[1] Y. Katayama-Fujimura, S. Gottesman, and M. R. Maurizi, *J. Biol. Chem.* **262**, 4477 (1987).

[2] B. J. Hwang, W. J. Park, C. H. Chung, and A. L. Goldberg, *Proc. Natl. Acad. Sci. U.S.A.* **84**, 5550 (1987).

[3] Y. Katayama, S. Gottesman, J. Pumphrey, S. Rudikoff, W. P. Clark, and M. R. Maurizi, *J. Biol. Chem.* **263**, 15226 (1988).

[4] B. J. Hwang, K. M. Woo, A. L. Goldberg, and C. H. Chung, *J. Biol. Chem.* **263**, 8727 (1988).

[5] S. Gottesman, W. P. Clark, and M. R. Maurizi, *J. Biol. Chem.* **265**, 7886 (1990).

[6] M. R. Maurizi, W. P. Clark, Y. Katayama, S. Rudikoff, J. Pumphrey, B. Bowers, and S. Gottesman, *J. Biol. Chem.* **265**, 12536 (1990).

[7] S. Gottesman, W. P. Clark, V. de Crecy-Lagard, and M. R. Maurizi, *J. Biol. Chem.* **268**, 22618 (1993).

[8] D. Wojtkowiak, C. Georgopoulos, and M. Zvlicz, *J. Biol. Chem.* **268**, 22627 (1993).

proteins, ClpB[9,10] and ClpY,[7] the proteolytic activities of which are unknown. ClpA acts as a specific ATP-dependent chaperone in vitro,[11] and ClpA-related ATPases may all be components of chaperone-linked proteases.

This chapter describes the purification and properties of ClpA and ClpP. The procedures presented here differ from those described previously.[3,4,6] *In vivo* and *in vitro* data indicate that these two components by themselves form an active complex, referred to as ClpAP protease, responsible for degradation of specific classes of proteins. The two proteins are purified separately and possess limited intrinsic enzymatic activities that have shed light on the mechanistic basis of specificity and the ATP dependence of the holoenzyme, ClpAP protease.

Methods

Purification of ClpA

The regulatory subunit of Clp protease, ClpA, can be overexpressed in mostly soluble form in *E. coli* cells, both under its own promoter and under very strong promoters such as p_L and p_{tac} on multicopy plasmids. When expressed under the wild-type *clpA* promoter on a multicopy plasmid, about 30–35 mg pure ClpA protein can be routinely obtained from 30 g of cells; however, the yield of purified ClpA can be increased severalfold using only half the amount of cells when ClpA is expressed from a p_L or p_{tac} promoter. The procedure routinely used in our laboratory[12] to purify ClpA is described below and summarized in Table I.

E. coli CSH100 cells transformed with pSK-39 (with *clpA* under the p_{tac} promoter),[13] are grown to an A_{600} of 5–6 at 37° in a 10-liter fermentor in superbroth medium (per liter: 12 g tryptone, 24 g yeast extract, 6.3 g glycerol, 12.5 g dipotassium phosphate, 3.8 g monopotassium phosphate, adjusted to pH 7.3) with 25 μg/ml chloramphenicol. Isopropyl β-D-thiogalactoside (IPTG) (1 mM) is added and growth is continued for 2 hr before harvesting the cells by centrifugation at 4°. Alternatively, *E. coli* MZ1 *clpP::CM* cells[14] transformed with pSK24 (a derivative of pRE1,[14] with

[9] C. Squires and C. L. Squires, *J. Bacteriol.* **174,** 1081 (1992).

[10] S. Gottesman, C. Squires, E. Pichersky, M. Carrington, M. Hobbs, J. S. Mattick, B. Dalrymple, H. Kuramitsu, T. Shiroza, T. Foster, W. P. Clark, B. Ross, C. L. Squires, and M. R. Maurizi, *Proc. Natl. Acad. Sci. U.S.A.* **87,** 3513 (1990).

[11] S. Wickner, S. Gottesman, D. Skowyra, J. Hoskins, K. McKenney, and M. R. Maurizi, submitted (1994).

[12] M. W. Thompson and M. R. Maurizi, *J. Biol. Chem.* **269,** 18201 (1994).

[13] S. K. Singh and M. R. Maurizi, submitted (1994).

[14] P. Reddy, A. Peterkofsky, and K. McKenney, *Nucleic Acids Res.* **17,** 10473 (1989).

TABLE I
PURIFICATION OF ClpA FROM CELLS CARRYING MULTICOPY *clpA*

Fraction[a]	Total protein (mg)	Total activity (units)[b]	Specific activity (units/mg)[b]	Purification (-fold)	Yield (%)
Crude extract	2000	4400	2.2	1	100
PEI/KCl extract	800	2600	3.3	1.5	60
40% (NH₄)SO₄	200	1850	9.3	4.2	42
Low-salt precipitate[c]	44	1030	24	11	—
Mono S	45	990	22	10	—
Mono Q	43	950	22	10	—
Low-salt precipitate[c] and Mono Q combined	87	2080	24	11	47

[a] See text for definitions of fractions.
[b] A unit of activity is defined as 1 mg α-casein degraded per hour.
[c] ClpA from 40% ammonium sulfate precipitation that did not go into solution in the low-salt phosphate buffer used for the Mono S column was solubilized in buffer B with 0.3 M KCl.

clpA under the p_L promoter)[13] are grown to an A_{600} of 2.0 at 32° in a 10-liter fermentor in superbroth medium with 100 μg/ml ampicillin. The temperature is increased to 42° for 2 hr and the cells are cooled and harvested by centrifugation at 4°. Cell pellets are frozen immediately in dry ice and stored at −70°.

In a typical purification procedure, 15–16 g of frozen cells are suspended in 60 ml of buffer B [50 mM Tris-HCl, pH 7.5, 2 mM EDTA, 2 mM DTT, and 10% (v/v) glycerol] and are broken by a single pass through an ice-chilled French pressure cell at 20,000 psi. The crude suspension is diluted to 100 ml with buffer B and spun at 30,000 g for 60 min at 4°. All purification steps are carried out at or below 5°. To the crude supernatant, 0.5 ml of a 10% (w/v) polyethyleneimine (PEI) solution (titrated to pH 8.0 with HCl) is added with thorough mixing to give a final concentration of 0.05% PEI, and the turbid suspension is centrifuged at 30,000 g for 30 min. PEI is then added to the supernatant solution to give a final concentration of 0.3% PEI, which precipitates most of the soluble ClpA. The 0.3% PEI pellet is collected by centrifugation at 30,000 g for 30 min and extracted with 400 ml of buffer B containing 0.4 M KCl. The KCl extract is centrifuged at 30,000 g for 30 min, and solid ammonium sulfate is added to the supernatant solution to a final concentration of 40% saturation. The precipitated protein is collected by centrifugation for 20 min at 20,000 g.

The ammonium sulfate pellets are dissolved in a volume (65–100 ml) of buffer P [20 mM potassium phosphate buffer, pH 7.5, 2 mM EDTA,

2 mM DTT, and 10% (v/v) glycerol] to obtain a solution with the equivalent conductivity of 0.1 M KCl. The solution is clarified by centrifugation at 20,000 g for 15 min. A portion of the ClpA is insoluble in low salt in buffer P but can be dissolved in buffer P or buffer B with high salt (0.3–0.4 M KCl). ClpA recovered by dissolving in high salt is more than 90% pure, fully active, and is used without further purification. The ClpA solubilized in low salt is purified further by chromatography on a Mono S HR (16/10) Sepharose column (LKB-Pharmacia, Piscataway, NJ). ClpA is loaded onto the Mono S column previously equilibrated with buffer P containing 0.1 M KCl, and ClpA is eluted at 0.3 M KCl within a 70-ml linear gradient from 0.1 to 0.4 M KCl. At this stage of purification ClpA is usually more than 90% pure as judged by SDS–PAGE, but trace protein and other contaminants can be removed and the buffer exchanged by an additional chromatographic step. The ClpA-containing fractions are pooled, precipitated with 40% saturated ammonium sulfate, and dissolved in buffer B to give a solution with the equivalent conductivity of 0.1 M KCl. The ClpA is loaded onto a Mono Q HR (10/10) column equilibrated with buffer B containing 0.1 M KCl, and ClpA is eluted at 0.28 M KCl in a 60-ml linear gradient from 0.2 to 0.4 M KCl. Using this procedure, the yield of purified ClpA typically ranges from 90 to 150 mg.

Solubility and Stability of ClpA

Purified ClpA is stable for extended periods when stored in buffer B with 10% glycerol at −70°. Repeated freezing and thawing and relatively short exposures to temperatures above 10° lead to losses of activity. ATP and nonhydrolyzable analogs of ATP stabilize ClpA.[15] ClpA has a tendency to lose activity when diluted, particularly in plastic tubes, but can be stabilized by including 0.05% (v/v) Triton X-100 and 0.1 M KCl in the buffer. In buffers containing Triton X-100, ClpA is stable for several days at 4° but is sensitive to freezing. ClpA is lost when exposed to various gel filtration media (silica-based TSK columns, polyacrylamide), but recoveries are improved in the presence of nucleotide or when Triton X-100 is included in the buffer.

ClpA has limited solubility in low-ionic-strength buffers but is quite soluble (≤15 mg/ml) in buffers with 0.3 M KCl. The protein displays hysteretic solubility properties in response to changes in temperature and ionic strength. Sudden decreases in ionic strength cause ClpA to precipitate, and the precipitated protein slowly redissolves as it becomes equilibrated to the new solution.

[15] M. R. Maurizi, M. W. Thompson, and S. K. Singh, unpublished (1994).

General Properties of ClpA

ClpA is a protein of 758 amino acids with a molecular mass of 83 kDa.[5] The pI calculated from the amino acid composition is 6.3. The absorption coefficient of ClpA at 279 nm is 0.4 (mg/ml)$^{-1}$. ClpA has an apparent native molecular mass of 120–140 kDa by gel filtration[4,16] and an $s_{20,w}$ of 9.0 by sedimentation velocity ultracentrifugation.[17] Both methods suggest that ClpA is purified as a mixture of monomeric and dimeric forms. Addition of ATP or a nonhydrolyzable analog of ATP (ATPγS or AMPPNP) induces self-assocation of ClpA to produce a monodisperse species with an apparent molecular mass and sedimentation coefficient of a hexamer. In the presence of nucleotide, the association constant between subunits in the hexamer is $>10^8 M^{-1}$.[17]

The translated amino acid sequence of ClpA reveals two ATP-binding consensus motifs.[5] These ATP-binding motifs occur in two separate regions of the protein sequence that are otherwise not homologous to each other.[5,10] Mutational analysis suggests that both ATP-binding sites of ClpA have ATPase activity,[13] and kinetic analysis of proteolytic and ATPase activities gave Hill coefficients between 1.3 and 1.6 for ATP, suggesting that more than one ATP site is involved in these activities (see below).[17] Two of the three cysteines in ClpA, which all reside in domain I, are accessible to sulfhydryl reagents such as the organomercurial, neohydrin, and some alkylating agents (N-methylmaleimide but not iodoacetamide).[16]

Purification of ClpP

ClpP has routinely been purified from wild-type E. coli cells transformed with the multicopy plasmid, pWPC9,[6] which carries ClpP under its natural promoter. In cells grown at 37° in 10 liters of superbroth medium containing 100 μg/ml ampicillin to an $A_{600} > 10$, ClpP constitutes about 2% of the cellular protein. ClpP can also be overproduced in soluble form in cells carrying plasmid pSK20[7,15] with clpP under the p$_L$ promoter, as described above for ClpA. clpP has been cloned under control of the T7 promoter in a pET3a plasmid[18] and expressed in E. coli strain SG1147 clpA::kan, which carries a prophage encoding T7 RNA polymerase. Cells grown overnight in 1% tryptone/0.5% NaCl, without induction, produce about 10–20% of their cellular protein as soluble ClpP. Cell pellets are stored at −70° until used.

[16] M. R. Maurizi, Biochem. Soc. Trans. 19, 719 (1991).
[17] M. R. Maurizi, M. W. Thompson, S. K. Singh, and A. Ginsburg, in preparation (1994).
[18] J. Shanklin, personal communication (1994).

The following purification procedure, summarized in Table II, has been used to prepare ClpP overproduced from the T7 expression system but is used essentially without change for smaller amounts of ClpP as well. All purification steps are carried out at <5°. About 16 g of frozen cells are suspended in 64 ml of buffer B (see above) and are broken by a single pass through an ice-chilled French pressure cell at 20,000 psi. The cell extract is centrifuged at 30,000 g for 50 min. A 10% (w/v) PEI solution is added to the supernatant solution to give a final concentration of 0.1% PEI, and the mixture is centrifuged at 30,000 g for 50 min. Solid KCl is added to the polyethyleneimine supernatant to give a final concentration of 0.1 M KCl. ClpP that precipitates with the 0.1% PEI can be extracted into 50 ml of buffer B containing 0.1 M KCl, centrifuged as described above, and combined with the PEI supernatant.

The combined PEI supernatants are passed over a Q Sepharose column (2.6 × 14 cm) previously equilibrated with buffer B containing 0.1 M KCl. Unbound protein is washed from the column with 150 ml of buffer B containing 0.1 M KCl, and ClpP is eluted with a 375-ml linear gradient of 0.1 to 0.4 M KCl in buffer B. ClpP activity is observed in fractions eluting from 0.32 to 0.37 M KCl. ClpP is precipitated from the Q Sepharose pool by adding two volumes of ice-cold acetone followed by centrifugation at 12,000 g for 10 min. The protein pellet is dissolved in 30 ml of buffer B (10.7 mg/ml). About 100 mg of protein is passed over a Mono Q HR 16/10 column previously equilibrated with buffer B containing 0.05 M KCl. The column is washed with 40 ml of buffer B containing 0.05 M KCl, followed by a 100-ml linear gradient from 0.05 to 0.4 M KCl. ClpP is eluted at 0.3 M KCl. ClpP pools from three separate columns are combined and the enzyme is precipitated using ice-cold acetone as described above. The

TABLE II
PURIFICATION OF ClpP FROM CELLS CARRYING MULTICOPY *clpP*

Fraction[a]	Total protein (mg)	Total activity (units)[b]	Specific activity (units/mg)	Purification (-fold)	Yield (%)
Crude extract	2820	11,100	3.9	1	100
0.1% PEI super/KCl extract of PEI pellet	1550	16,400	10.6	2.7	(148)
Q Sepharose	321	8920	27.8	7.1	80
Mono Q	189	5740	30.4	7.8	52
TSK 250	119	4910	41.3	10.6	44

[a] See text for identification of fractions.
[b] A unit of activity is 1 mg α-casein degraded per hour.

protein is dissolved in 4 ml of buffer B containing 0.2 M KCl (47.3 mg/ml protein), and 60-mg aliquots of the enzyme mixture are run over a Bio-Sil (Bio-Rad, Richmond, CA) TSK-250 gel filtration column (2.1 × 60 cm) in buffer B containing 0.2 M KCl. ClpP emerges in a single peak of essentially homogeneous protein.

General Properties and Stability of ClpP

ClpP is composed of 193 amino acids and has a molecular weight of 21,000.[6] ClpP is synthesized *in vivo* as a precursor with 207 amino acids and is cleaved in an autocatalytic (ClpP-dependent) process between Met-14 and Ala-15.[6] Purified ClpP has a native molecular weight of 230,000 in the presence of >0.1 M KCl and associates to a form with a molecular weight of 450,000 in low-salt buffer. Electron micrographs show that ClpP in high salt is composed of two superimposed rings of six subunits each. Particles that appeared to be two dodecamers stacked on each other, which presumably represent the low-salt form of ClpP, were also seen.[6] ClpP is stable when stored in buffer with 10% glycerol at −70°. The protein is also relatively stable at elevated temperatures and can be heated at 59° for 10 min with little loss of activity. ClpP has two cysteines that are not reactive with sulfhydryl reagents in the native protein but that are reactive when the protein is denatured. ClpP is a serine protease and has a serine residue (Ser-111) that is highly reactive with diisopropyl fluorophosphate.[2,6,19]

Association of ClpA and ClpP

Clp protease activity can be readily detected in extracts of *E. coli* cells by measuring the ATP-dependent degradation of α-casein.[1] Fractionation of extracts leads to loss of proteolytic activity against α-casein, because the components, ClpA and ClpP, have very little affinity for each other and are easily separated under purification conditions. Proteolytic activity of Clp is restored by addition of the two components in the presence of ATP. ClpA and ClpP form a tight complex in the presence of $MgCl_2$ and ATP or the nonhydrolyzable analog, ATPγS.[16] The ClpAP complex is composed of a dodecamer of ClpP and a hexamer of ClpA. The complex can be isolated by gel filtration or ultracentrifugation in the presence of nucleotide, but dissociates when Mg^{2+} or nucleotide is removed. Activity titrations indicate that the K_d for the complex is 5–8 nM for a complex of the dodecamer and hexamer.[17]

[19] M. R. Maurizi, W. P. Clark, S.-H. Kim, and S. Gottesman, *J. Biol. Chem.* **265**, 12546 (1990).

Assays of Enzymatic Activities of ClpA and ClpP

ClpP has peptidase activity against very short peptides (fewer than five amino acid residues) in the absence of ClpA and nucleotide.[4,12,20] When activated by ClpA, ClpP can degrade longer polypeptides and proteins. Activation of polypeptide degradation requires the presence of ATP or a nonhydrolyzable analog of ATP, whereas activation of protein degradation requires ATP hydrolysis and does not occur with nonhydrolyzable analogs. The exact length or composition of a polypeptide that imposes the requirement for ATP hydrolysis is not yet known. ClpA has a basal ATPase activity that is activated in the presence of proteins and peptides.[1,4,12] These activities of ClpA and ClpP are described in more detail below.

Cleavage of Fluorogenic Peptides by ClpP

ClpP alone hydrolyzes the short fluorogenic peptide, Succ-Leu-Tyr-aminomethylcoumarin (AMC).[20] Peptidase activity is usually determined in 50-μl solutions containing 50 mM Tris-HCl, pH 8.0, 0.1 M KCl, 1 mM dithiothreitol (DTT), 1 mM Succ-Leu-Tyr-AMC, and 0.5 μg ClpP. The fluorogenic peptide is added from a stock solution of 100 mM in dimethyl sulfoxide (DMSO); up to 5% dimethyl sulfoxide in the assay has no effect on enzyme activity. Reactions are run for 15–60 min at 37° in dim light and quenched with 2 ml of 0.2 M sodium borate, pH 9.0, containing 0.1% SDS. Fluorescence is measured in a Gilson Spectra/glo fluorometer (Middleton, WI) with 7–60× excitation and 3–73 emission filters or in a spectrofluorometer with excitation at 360 nm and emission at 455 nm.

Cautionary Note on Use of Fluorogenic Peptide Substrates

Multiple sites of cleavage in some simple peptides we examined (see below) prompted us to reexamine the cleavage of the fluorogenic peptides by separating the products on a C$_{18}$ column. ClpP cut the Leu-Tyr bond of Succ-Leu-Tyr-AMC at a significant rate, which was about 25% of the rate of cleavage of the amidomethylcoumarin bond. When other fluorogenic peptides that appeared to be poor substrates for ClpP, because little or no fluorescent product was obtained, were assayed by reversed-phase chromatography, several were found to be rapidly cleaved at other peptide bonds within the fluorogenic peptides (Fig. 1). Hydrolysis at the alternate sites was 10–100 times faster than cleavage of the amidomethylcoumarin. The use of fluorogenic peptides can result in a considerable underestimation of the enzymatic activity of the peptidase, because the shorter fluoro-

[20] K. M. Woo, W. J. Chung, D. B. Ha, A. L. Goldberg, and C. H. Chung, *J. Biol. Chem.* **264**, 2088 (1989).

	Site of cleavage	
	---	---
	Alternate peptide bond	Amidomethyl coumarin
Peptide	(relative rate)	
Succ-Leu+Tyr-AMC	0.2	0.8
Succ-Ala-Ala+Phe-AMC	0.9	0.1
Succ-Leu+Leu+Val-Tyr-AMC	0.99	0.01

FIG. 1. Cleavage of fluorogenic peptide substrates at peptide bonds that do not release the fluorophore. Fluorogenic peptides were incubated with ClpP under standard conditions for 1 hr at 37° and the degradation products were separated by C_{18} reversed-phase chromatography and identified by amino acid analysis and spectral properties. The relative yields of alternate cleavage products compared to the yield of aminomethylcoumarin are given. All three peptides were degraded at comparable rates. Sites other than the amidomethylcoumarin bond at which peptides were cleaved are shown by the crossbars.

genic product of the more rapid alternate cleavage may no longer be a substrate for the peptidase. We have also observed that rat liver proteasome[21,22] cleaves one of the commonly used substrates, Succ-Leu-Leu-Val-Tyr-AMC, at alternate sites.

Separation and Quantitation of Peptide Products by Reversed-Phase Chromatography

Synthetic peptides[12] can be degraded either by ClpP or by ClpP and ClpA in the presence of nucleotide. For short peptides (<6 amino acids), the assay buffer is the same as the one used for fluorogenic peptide cleavage. With longer peptides (6–30 amino acids), the following additions are required: 0.02% Triton X-100, 25 mM MgCl$_2$, 4 mM ATP, or 1 mM ATPγS, and 4 μg of ClpA. Reactions are quenched by addition of an equal volume of 7.4 M guanidine hydrochloride and injected onto a C_{18} reversed-phase column in a system equipped with a detector and integrator. Peptide products as well as the remaining substrate are detected by absorbance at 210 and 280 nm and relative peak areas are used for quantitation.

Degradation of α-Casein and Other Proteins

Protein degradation by Clp protease is usually measured by acid solubilization of radioactively labeled α-casein. The assay solution contains, in

[21] M. Orlowski, *Biochemistry* **29**, 10289 (1990).
[22] A. J. Rivett, *Arch. Biochem. Biophys.* **268**, 1 (1989).

250 μl, 50 mM Tris-HCl, pH 8.0, 0.1 M KCl, 1 mM dithiothreitol, 0.02% Triton X-100, 25 mM MgCl$_2$, 4 mM ATP, 8 μg of [^{3}H]α-casein (7000–17,000 cpm/μg), and a combination of ClpA and ClpP. When ClpA is limiting, assays contain 0.2 μg ClpA and 2–4 μg of ClpP; when ClpP is limiting, assays contain 0.2 μg ClpP and 4–8 μg ClpA. Assays are initiated by adding ClpP followed after 15 sec by ClpA. Reactions are run at 37° for 10–15 min and quenched with 0.31 ml of 9.6% (w/v) trichloroacetic acid and 40 μl of 10 mg/ml bovine serum albumin (BSA). The precipitated protein is removed by centrifugation in an Eppendorf microcentrifuge at 13,000 g, and 0.5 ml of the supernatant solution is counted in 7 ml of Aquasol. A unit of activity is defined as the degradation of 1 mg α-casein per hour. Specific activities, measured by keeping one component limiting and using a saturating amount of the other component, were 24 ± 4 U/mg for ClpA and 50 ± 5 U/mg for ClpP.

ATPase Activity of ClpA

Purified ClpA hydrolyzes ATP to yield ADP and phosphate. ATPase activity is measured in 50-μl solutions containing 50 mM Tris-HCl, pH 8.0, 0.1 M KCl, 1 mM dithiothreitol, 0.02% Triton X-100, 25 mM MgCl$_2$, 1–2 mM [γ-^{32}P]ATP [4400 disintegrations per minute (dpm)/nmol], and 0.2–0.5 μg ClpA. Solutions are incubated at 37° in a 1.5-ml microcentrifuge tube for 10–15 min and the reactions are stopped with 200 μl of an ice-cold quenching solution. The quenching solution is prepared fresh daily by combining equal volumes of 5% ammonium molybdate in 1 M H$_2$SO$_4$ and 5 mM silicotungstic acid in 1 mM H$_2$SO$_4$. The phosphomolybdate complex is extracted into 0.5 ml of an ice-cold 1 : 1 mixture of toluene and isobutanol by vigorous vortexing for 15–30 sec and the layers are separated by centrifuging the solution for 2 min in a microfuge at 12,000 g; 250 μl of the solvent (upper layer) is counted in 7 ml of Aquasol (DuPont/ NEN, Boston, MA). ATPase activity of ClpA is activated or inhibited by different peptides and proteins (see below).

Catalytic Properties of ClpP and ClpA

Specificity

ClpP alone can cleave short peptides such as the fluorogenic peptide, Succ-Leu-Tyr-AMC.[20] Synthetic pentapeptides, Xaa-Tyr-Leu-Tyr-Trp, are competitive inhibitors of Succ-Leu-Tyr-AMC degradation and are cleaved by ClpP.[12] Analysis of the degradation products by C$_{18}$ reversed-phase chromatography indicated that cleavage of Leu-Tyr-Leu-Tyr-Trp

TABLE III
KINETIC CONSTANTS FOR PROTEIN AND PEPTIDE SUBSTRATES

Substrate	K_m or $S_{0.5}$ (M)	Peptide turnover[a] (min⁻¹)
Succ-Leu-Tyr-AMC	$1.0 \pm 0.5 \times 10^{-3}$	2.5 ± 0.5[b]
Phe-Ala-Pro-His-Met-Ala-Leu-Val-Pro-Val	$4.6 \pm 0.5 \times 10^{-3}$	800[c]
Insulin B chain	$5.0 \pm 1 \times 10^{-5}$	10 ± 3[d]
α-Casein	$4.0 \pm 0.4 \times 10^{-7}$	15 ± 5[e]

[a] Moles of peptide bonds cleaved per mole of ClpP subunit.
[b] Cleavage of the amidomethylcoumarin bond only.
[c] Propeptide is cleaved between Met and Ala only.
[d] Calculated from an average of three bonds cleaved per molecule of oxidized insulin B chain.
[e] Calculated based on an average of 15 peptide bonds cleaved per molecule.

occurred primarily at the Leu³-Tyr⁴ bond but that significant cleavage also occurred at Tyr²-Leu³ and Leu⁴-Trp⁵. Moreover, the preferred site of cleavage was affected by succinylation of the amino terminus or by making the peptide amide.[12] Replacing Leu-1 with Phe, Thr, Gly, Lys, or Glu did not change the specificity of cleavage by ClpP but did alter the binding affinity.[12] Peptides with hydrophobic amino termini were better (5- to 10-fold) competitive inhibitors than those with amino-terminal charged amino acids.

Cleavage of peptides of more than five amino acid residues by ClpP requires activation by ClpA and either ATP or a nonhydrolyzable analog, such as ATPγS or AMPPNP.[12,23] ClpP cleaves polypeptides such as oxidized insulin B chain and glucagon at multiple sites. Cleavage occurs at a limited number of discrete sites, but the basis for selectivity of those sites is not obvious, because cuts are made between both hydrophobic and hydrophilic amino acid residues. The most specific and also the most rapidly degraded substrate found to date is a synthetic peptide that corresponds to the 10 amino acids surrounding the *in vivo* processing site in ClpP. This peptide, Phe-Ala-Pro-His-Met-Ala-Leu-Val-Pro-Val, and several related peptides, were degraded with a k_{cat} of about 800 min⁻¹ at 37° (Table III).[12] In contrast, degradation of other polypeptides, such as oxidized insulin B chain and glucagon, is much slower (<30 min⁻¹).[23] The propeptide derivatives are cut exclusively at the Met-Ala bond. Most substitutions for the Met block peptide cleavage, and tolerated substitu-

[23] M. W. Thompson, S. K. Singh, and M. R. Maurizi, *J. Biol. Chem.* **268,** 18209 (1994).

tions, such as Leu or Trp, reduce the rate of cleavage more than 80% without changing the site of cleavage.[12]

α-Casein and other proteins and polypeptides are degraded into small peptides by ClpAP protease. SDS gels of reaction mixtures during α-casein degradation reveal no degradation intermediates or products >8000 M_r, indicating that proteins are cleaved multiple times, possibly in all available sites, before being released from the enzyme. Between 20 and 30 peptide products from α-casein degradation were separated by reversed-phase chromatography.[23] Because there is no indication that degradation intermediates accumulate, these data suggest that α-casein may be degraded at 20–30 different peptide bonds.

Many protein and peptide substrates of Clp protease, which are cleaved in many sites, display normal Michaelis–Menten kinetics, probably because the processive mechanism ensures that complete degradation occurs each time a substrate protein encounters the protease. Thus proteins are turned over at a rate that is a partial sum of rates of individual cleavage reactions during a processive cycle. Table III shows the substrate K_m or $S_{0.5}$ values and the approximate k_{cat} values for ClpP and ClpAP.[12,23] The K_m values vary over a 100-fold range. α-Casein has a low K_m, perhaps indicating the multiple sites on α-casein that can interact with Clp protease, and the most rapidly degraded substrate, FAPHMALVPV, has the highest K_m.[12] It appears that the k_{cat} makes a significant contribution to the K_m and the protein and peptide substrates may not exchange rapidly in binding to the protease.

Protein and Peptide Interactions

Random peptides were screened for the ability to compete with Succ-Leu-Tyr-AMC in the oligopeptidase assay and with α-casein in the proteinase assay. A series of dipeptides, Ala-Xaa and Xaa-Ala, revealed little effect (<10% inhibition at 2 mM) on either activity. Of 120 longer peptides (10–50 amino acids), only a few showed any affinity for Clp protease. Table IV shows several of the peptides that inhibited either the oligopeptidase activity or the proteinase activity. As expected, peptides that inhibited oligopeptidase activity, presumably by interaction at the active site of ClpP, also inhibited proteinase activity. Interestingly, several peptides inhibited proteinase activity of ClpAP but had less effect on the oligopeptidase activity of ClpP. These peptides appear to interact primarily at a site that is present on the holoenzyme but interact less well at the active site on ClpP. These results point to the presence of an allosteric site on Clp protease that is involved in degradation of proteins. Of various proteins

TABLE IV
INHIBITION OF OLIGOPEPTIDASE AND PROTEINASE ACTIVITY OF ClpP AND
ClpAP BY VARIOUS PEPTIDES

	Inhibition of activity (%)	
Competitor peptide[a]	Oligopeptidase[b]	Proteinase[b]
QNGGVCVSYKYFSSIRRCSAPKKF	0	76
NGGTAVSNKYFSNIHWCN	16	40
YGIFQINSR	78	30
SFYILAHTEFTPTETD	5	50

[a] Peptide sequences are given in single-letter code. Peptide competitors were present at 100 μg/ml.

[b] Standard assays were run with Succ-Leu-Tyr-AMC at ~50% saturation in the peptidase assay and α-casein at ~80% saturation in the protease assay.

that were screened, κ-casein was a strong, competitive inhibitor of α-casein degradation (Fig. 2), but is not a substrate for Clp protease.

Nucleotide Requirement

Only ATP and dATP supported α-casein degradation; no other nucleotides or nucleotide analogs were effective.[1,2] The K_m values for ATP in

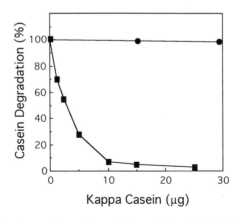

FIG. 2. Effects of κ-casein on oligopeptidase activity of ClpP (●) and proteinase activity of ClpAP (■). α-Casein degradation was measured by acid-solubilization of ^{3}H-labeled α-casein in the presence of limiting ClpP, saturating ClpA and MgATP, and increasing amounts of κ-casein. Separate analyses using SDS gel electrophoresis and reversed-phase chromatography showed no evidence for degradation of κ-casein by ClpAP under these or other conditions (data not shown).

the ATPase and proteolysis assays are identical (Table V), suggesting that the same sites for ATP binding are used for both activities.[17,23] Nonhydrolyzable ATP analogs inhibit ATPase activity and ATP-dependent α-casein degradation (Table V), but promote assembly of the ClpAP complex and activate cleavage of some polypeptides, such as insulin B chain and propeptide, by ClpP. Because binding of ATP (or a nonhydrolyzable analog) is sufficient to activate ClpP for degradation of polypeptides but not α-casein, there are at least two steps involved in ATP-dependent degradation. One step is an allosteric effect on ClpP produced by the interaction with ClpA. Substrates that can enter the active site of activated ClpP are cleaved. The second step, requiring ATP hydrolysis, is not completely understood at this time, but may involve a ClpA-dependent change in the larger protein substrates.

Stoichiometry of ATP Hydrolysis and Peptide Bond Cleavage

Basal ATPase activity of ClpA is 0.7–0.9 μmol/min/mg ClpA (turnover number of 70 $\pm$ 10 min^{-1}). Several substrates such as α-casein, β-casein, bovine serum albumin, and the propeptide Phe-Ala-Pro-His-Met-Ala-Leu-Val-Pro-Val activate the ATPase activity between 40 and 60%, whereas some proteins and peptides inhibit the ATPase activity. κ-Casein, which is a competitive inhibitor of α-casein degradation, has little effect on ClpA ATPase activity. The effects of peptides and proteins on ClpA are seen in the absence of ClpP and thus reflect direct interactions with binding

TABLE V
KINETIC CONSTANTS FOR NUCLEOTIDES

Substrate	$S_{0.5}$ (μM)	$I_{0.5}$ (μM)	Turnover of ATP^a (min^{-1})
ATP (proteolysis)^b	180 $\pm$ 50		130 $\pm$ 20
ATP (ATPase)^c	150 $\pm$ 20		70 $\pm$ 10
ATPγS^{d,e}		50 $\pm$ 10	
AMPPNP^{d,f}		1000 $\pm$ 200	
ADP^d		200 $\pm$ 50	

^{a} Turnover number was calculated from the V_{max} and is expressed as the moles of ATP hydrolyzed per mole of ClpA subunit. The individual rate constants at the two ATPase sites are not known.
^{b} The ATP dependence of activation of α-casein degradation.
^{c} The ATP dependence of hydrolysis of ATP by ClpA alone.
^{d} Inhibition of ATP-dependent α-casein degradation.
^{e} ATPγS is adenosine 5'-thiotriphosphate.
^{f} AMPPNP is 5'-adenylyl imidodiphosphate.

TABLE VI
EFFECTS OF PEPTIDES AND PROTEINS ON ATPase ACTIVITY
OF ClpA AND ClpB

Protein/peptide[b]	ATP hydrolysis (%)[a]	
	ClpA	ClpB
None	100	100
α-Casein	145	69
β-Casein	185	153
κ-Casein	107	285
CPSESLLERITRKLRDGWKRLIDIL[c]	124	198
DRGDQSTDTGIFQINSR[c]	91	47
GCQNGGVAVSYKYFSRIRRCS[c]	46	43
NRPILSLNRKPKS[c]	84	139

[a] ATPase activity assayed as described in this article.
[b] Peptides were present at 200 μg/ml.
[c] Peptide sequences are expressed in the single-letter amino acid code.

sites on ClpA. The diversity of the effects suggest that there are several peptide- and protein-binding sites on ClpA and that occupancy of those binding sites induces different conformational changes in ClpA. ClpB, a closely related protein from *E. coli,* also has ATPase activity that is affected by proteins and peptides.[9,10,24] However, the activities of ClpA and ClpB are affected to different extents and in some cases in opposite directions by different proteins and peptides (Table VI).

Addition of α-casein to the ClpAP complex produces a somewhat greater increase in the ATPase activity of ClpA than seen when α-casein is added to ClpA alone. ATPase activity under conditions of α-casein degradation is increased to 1.3–1.7 μmol/min/mg (turnover number of 130 ± 20 min⁻¹). The ratio of ATP hydrolysis to peptide bond cleavage when both reactions are operating at maximal velocity is about 6–8 molecules of ATP per peptide bond.

Metal Ion and Salt Effects and Other Properties

Proteolytic activity of ClpAP is dependent on MgCl$_2$, but concentrations higher than about 30 mM are inhibitory (Fig. 3). This effect of high Mg^{2+} may reflect the requirement for binding free nucleotide to one of

[24] K. M. Woo, K. I. Kim, A. L. Goldberg, D. B. Hai, and C. H. Chung, *J. Biol. Chem.* **267,** 20429 (1992).

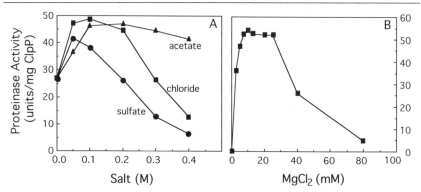

FIG. 3. Effects of Mg^{2+} and different anions on ClpAP proteinase activity. α-Casein degradation was measured by acid solubilization of ^3H-labeled α-casein as described above with ClpP limiting and 4 mM ATP in all assays. (A) The potassium salts were used in the assays shown; however, LiCl, NaCl, and KCl produced identical effects (data not shown). (B) Mg^{2+} is required for ClpAP protease but is inhibitory at high concentrations.

the sites on ClpA. Ca^{2+} and Zn^{2+} are poor substitutes for Mg^{2+} and, at a concentration of 5 mM, do not antagonize the effect of Mg^{2+}.[4,15] Clp protease activity is strongly inhibited by high salt concentrations (Fig. 3). The effect appears to be specific for the anion, because changing the cation had little effect on the inhibition. Chloride is much more inhibitory than acetate, and divalent anions are also very inhibitory. Because ClpA and ClpB can associate in high salt, it is likely that the inhibitory effects of high ionic strength influence another step such as substrate binding or catalysis at either the ATPase or the proteolytic active site. High salt and excess Mg^{2+} also inhibit the basal ATPase activity of ClpA.

The degradation of α-casein by ClpAP has a broad pH optimum between pH 7.5 and 9.5, and the pH optimum for the ATPase activity of ClpA is essentially the same.[4,15] Peptidase activity of ClpP against Succ-Leu-Tyr-AMC has an optimum at pH 7.0, although significant activity is observed up to pH 9.0.[15] ClpAP is active in degrading α-casein up to 42°, but activity begins to decline rapidly above that temperature and no proteinase activity is detectable at 55°.

Inhibitors

ClpP is a serine peptidase and is inhibited by diisopropyl fluorophosphate (DFP).[4,19] DFP inhibits both oligopeptidase activity of ClpP and proteinase activity of ClpAP, indicating that a single active site on ClpP is involved in both activities.[4,19,23] ClpP reacts slowly with DFP, and incubations of 1–3 hr with 2–5 mM DFP are required to obtain complete inhibition. The low reactivity of the active site serine with DFP may reflect a higher pK_a for the activated serine in ClpP than for the active site residue in classical serine proteases such as trypsin and subtilisin, and may be a characteristic of ATP-dependent proteases. The ATPase activity of ClpA and the ability to activate ClpP are inhibited by sulfhydryl reagents,[1,2,16] but the proteolytic activity is more sensitive than the ATPase activity.[16] Because all three sulfhydryl groups are in domain I of ClpA and may be located near the ATPase active site,[5] nucleotide binding or hydrolysis at that site is critical for activation of ClpP. Mutation studies have led to a similar conclusion.[13] ClpA is inhibited by the affinity reagent, fluorosulfonylbenzoyladenosine.[15]

Significance of Clp Protease and Other ATP-Dependent Endopeptidases

ATP-dependent proteases of *E. coli* fall into at least two families: Lon[25,26] and Clp[26,27] To date, only a single Lon protease has been identified in *E. coli,* although *Myxococcus* has two highly similar Lon proteases that function during vegatative growth or during development.[28–30] In *E. coli,* Lon targets a number of specific proteins and plays a role in regulatory pathways involved in capsular polysaccharide production and cell division.[27,31,32] Lon also helps degrade abnormal proteins *in vivo.*[33]

The Clp family was initially identified biochemically by *in vitro* proteolytic assays. Clp protease appears to have a modular design consisting of a proteolytic component, ClpP, that has little or no activity by itself, and at least two different activators, ClpA and ClpX, that direct its activity to different targets *in vivo*. It is possible that additional components of the Clp modular protease exist, because *E. coli* has close homologs of both ClpA and ClpX. ClpB is closely related to ClpA (60% identical/85% similar amino acid sequences),[10] although the purified protein does not interact with ClpP under the conditions tested. A gene coding for a protein, which we call ClpY,[7] has been identified in *E. coli.* It seems probable that ClpB and ClpX activate some ClpP-related proteases (if not ClpP) in energy-dependent degradation *in vivo*.

[25] A. L. Goldberg, *Eur. J. Biochem.* **203,** 9 (1992).
[26] M. R. Maurizi, *Experientia* **48,** 178 (1992).
[27] S. Gottesman and M. R. Maurizi, *Microbiol. Rev.* **56,** 592 (1992).
[28] R. E. Gill, M. Karlok, and D. Benton, *J. Bacteriol.* **175,** 4538 (1993).
[29] N. Tojo, S. Inouye, and T. Komano, *J. Bacteriol.* **175,** 4545 (1993).
[30] N. Tojo, S. Inouye, and T. Komano, *J. Bacteriol.* **175,** 2271 (1993).
[31] S. Mizusawa and S. Gottesman, *Proc. Natl. Acad. Sci. U.S.A.* **80,** 358 (1983).
[32] A. S. Torres-Cabassa and S. Gottesman, *J. Bacteriol.* **169,** 981 (1987).
[33] S. Gottesman, *Annu. Rev. Genet.* **23,** 163 (1989).

After the genes for the different Clp components were cloned and mutations introduced into the chromosomal genes, it was shown that the Clp proteases had only a small role in abnormal protein degradation.[3,7] However, a variety of specific protein targets for Clp proteases have been identified. ClpAP degrades several β-galactosidase fusion proteins[5] and may show specificity for the amino-terminal amino acid in at least one class of such substrates.[34] ClpXP degrades the highly unstable O protein of λ phage *in vivo* ($t_{1/2} \leq 2$ min)[7] and *in vitro*.[8] ClpXP appears to be involved in plasmid maintenance[35] and in phage mu virulence.[36] Data indicate that ClpP also contributes to degradation of proteins synthesized during carbon starvation,[37] and it is possible that ClpP and other Clp family members may have broader roles in both specific and nonspecific protein degradation in *E. coli* than is currently appreciated.

[34] J. W. Tobias, T. E. Shrader, G. Rocap, and A. Varshavsky, *Science* **254**, 1374 (1991).
[35] M. Yarmolinsky (personal communication).
[36] V. Geuskens, A. Mhammedi-Alaoui, L. Desmet, and A. Toussaint, *EMBO J.* **11**, 5121 (1992).
[37] K. Damerau and A. C. St. John, *J. Bacteriol.* **175**, 53 (1993).

[24] Multicatalytic Endopeptidase Complex: Proteasome

By A. Jennifer Rivett, Peter J. Savory, and Hakim Djaballah

Introduction

The multicatalytic endopeptidase complex (EC 3.4.99.46) is a 700-kDa multisubunit enzyme complex that is widely distributed in eukaryotic cells. It has been described as a high molecular mass protease under many different names[1] and is apparently identical to a variety of cylindrical particles of unknown function,[2] as well as to the prosome, a particle that was believed to play a role in the control of translation.[3] The complex is now commonly referred to as either the multicatalytic proteinase complex[1,4]

[1] A. J. Rivett, *Arch. Biochem. Biophys.* **268**, 1 (1989).
[2] P. E. Falkenberg, P. C. Haass, P. M. Kloetzel, B. Niedel, F. Kopp, L. Kuehn, and B. Dahlmann, *Nature (London)* **331**, 190 (1988).
[3] H. P. Schmid, O. Akhayat, C. Martins de Sa, F. Puvion, K. Koehler, and K. Scherrer, *EMBO J.* **3**, 29 (1984).
[4] M. Orlowski, *Biochemistry* **29**, 10289 (1990).

or the proteasome.[5–7] The multicatalytic endopeptidase complex, either by itself or as the catalytic core of the 26S proteinase complex,[8,9] is believed to play an important role in ubiquitin-dependent as well as ubiquitin-independent nonlysosomal pathways of protein turnover, including the degradation of regulatory proteins and the processing of antigens for presentation by the major histocompatibility complex (MHC) class I pathway.[6,7]

Assays of Multicatalytic Endopeptidase Activity

The term "multicatalytic" has been applied to the complex because it was realized from early work[10] that the broad specificity of the mammalian enzyme could be attributed to distinct types of catalytic sites, the activities of which can be distinguished using a variety of protease inhibitors and other effectors (see below). Activities responsible for cleavage on the carboxyl side of basic (usually Arg), hydrophobic (Leu, Tyr, Phe), and acidic (Glu) residues have been referred to as trypsin-like, chymotrypsin-like, and peptidylglutamyl-peptide bond hydrolase activities, respectively,[10] although these terms are not very accurate in describing the specificities. The enzyme can also catalyze cleavage after other residues, including Gln, Thr, and Ala.[10–12] Different substrates have been used to assay the endopeptidase activities of the complex purified from many different sources. Its substrates include proteins, peptides, and synthetic peptides (Table I) and the enzyme is active over a range of neutral to weakly alkaline pH values in a variety of different buffers.[11,13–15] However, there can be significant variations in activity in different buffers, with Tris and both Na^+ and K^+ ions having been found to inhibit some activities.[10]

Assaying multicatalytic endopeptidase activity is complicated by the fact that there are multiple distinct catalytic sites and interactions between them. The complex has some unusual kinetic properties (Table II) and there may be differences depending on the source of the enzyme and the

[5] K. Tanaka, T. Tamura, T. Yoshimura, and A. Ichihara, *New Biol.* **4**, 173 (1992).

[6] A. L. Goldberg and K. L. Rock, *Nature (London)* **357**, 375 (1992).

[7] A. J. Rivett, *Biochem. J.* **291**, 1 (1993).

[8] A. Hershko and A. Ciechanover, *Annu. Rev. Biochem.* **61**, 761 (1992).

[9] M. Rechsteiner, L. Hoffman, and W. Dubiel, *J. Biol. Chem.* **268**, 6065 (1993).

[10] S. Wilk and M. Orlowski, *J. Neurochem.* **35**, 1172 (1980).

[11] A. J. Rivett, *J. Biol. Chem.* **260**, 12600 (1985).

[12] H. Djaballah and A. J. Rivett, unpublished observations, (1992).

[13] K. Tanaka, T. Yoshimura, A. Kumatori, A. Ichihara, A. Ikai, M. Nishigai, K. Kameyama, and T. Takagi, *J. Biol. Chem.* **263**, 16209 (1988).

[14] R. W. Mason, *Biochem. J.* **265**, 479 (1990).

[15] H. Djaballah and A. J. Rivett, *Biochemistry* **31**, 4133 (1992).

TABLE I
SUBSTRATES OF MULTICATALYTIC ENDOPEPTIDASE COMPLEX

Type of substrate	Examples	Refs.
Proteins	Casein, oxidized enzymes, α-crystallin, lysozyme, serum albumin, hemoglobin	a–d
Peptides	Insulin B chain, substance P, glucagon, angiotensin I, and many others	a, e–g
Synthetic peptides— 7-amido-4-methyl- coumarin (AMC) derivatives	Suc-Leu-Leu-Val-Tyr-AMC (Suc)-Ala-Ala-Phe-AMC Boc-Leu-Ser-Thr-Arg-AMC Boc-Phe-Ser-Arg-AMC Boc-Leu-Arg-Arg-AMC Z-Gly-Gly-Arg-AMC	e, f, h, i
Synthetic peptides—2- naphthylamides	Z-Leu-Leu-Glu-2-naphthylamide Z-DAla-Leu-Arg-2-naphthylamide	e
Other synthetic peptides		
p-Nitroanilide	Z-Gly-Gly-Leu-p-nitroanilide	e, j
Methoxynaphthylamide	Z-Ala-Ala-Arg-methoxynaphthylamide	k
Thiobenzyl	Boc-Ala-Ala-Asp-S-benzyl	l
p-Aminobenzoate (pAB)	Z-Gly-Pro-Ala-Ala-Gly-pAB	m

[a] A. J. Rivett, *J. Biol. Chem.* **260,** 12600 (1985).
[b] A. J. Rivett, *Arch. Biochem. Biophys.* **243,** 624 (1985).
[c] K. Ray and H. Harris, *Proc. Natl. Acad. Sci. U.S.A.* **82,** 7545 (1985).
[d] M. Orlowski and C. Michaud, *Biochemistry* **28,** 9270 (1989).
[e] S. Wilk and M. Orlowski, *J. Neurochem.* **35,** 1172 (1980).
[f] A. J. Rivett, *J. Biol. Chem.* **264,** 12215 (1989).
[g] J. R. McDermott, A. M. Gibson, A. E. Oakley, and J. A. Biggins, *J. Neurochem.* **56,** 1509 (1991).
[h] K. Tanaka, K. Ii., A. Ichihara, L. Waxman, and A. L. Goldberg, *J. Biol. Chem.* **261,** 15197 (1986).
[i] S. Ishiura, T. Tsukahara, T. Tabira, and H. Sugita, *FEBS Lett.* **257,** 388 (1989).
[j] T. Achstetter, C. Ehmann, A. Osaki, and D. H. Wolf, *J. Biol. Chem.* **259,** 13344 (1984).
[k] J. Arribas and J. G. Castaño, *J. Biol. Chem.* **265,** 13969 (1990).
[l] H. Djaballah and A. J. Rivett, *Biochemistry* **31,** 4133 (1992).
[m] M. Orlowski, C. Cardozo, and C. Michaud, *Biochemistry* **32,** 1563 (1993).

purification procedure. There are latent and active forms of the complex (for references, see Table II) as well as the possibility of different subpopulations.[16,17]

A number of synthetic peptides have proved useful for assaying individual activities of the multicatalytic endopeptidase complex (examples in Table I). The 7-amino-4-methylcoumarin leaving group provides the most sensitive assay and these synthetic peptides are often used at concentra-

---

[16] P. E. Falkenburg and P. M. Kloetzel, *J. Biol. Chem.* **264,** 6660 (1989).
[17] M. G. Brown, J. Driscoll, and J. J. Monaco, *J. Immunol.* **115,** 1193 (1993).

TABLE II
KINETIC PROPERTIES OF PURIFIED EUKARYOTIC MULTICATALYTIC ENDOPEPTIDASE COMPLEXES

Kinetic property	Refs.
Broad specificity of bond cleavage due to multiple distinct catalytic components, which can be assayed using appropriate synthetic substrates and distinguished using inhibitors and other effectors (see Tables V–VII)	a–d
Can be isolated in "latent" form; activation by removal of glycerol, by dialysis, by heat treatment, or by addition of polylysine	e–h
Optimum pH range, pH 7–9, depending on substrate	f, i
No protease inhibitor yet found to block all peptidase activities; different catalytic components have very different reactivity with some protease inhibitors	c, d; see also Table VII
Addition of inhibitors of one activity can affect kinetic parameters at other catalytic sites	j–l
Activation (see Table VI for effectors) involves conformational changes, allosteric effects	m–o
Some peptidase activities show positive cooperativity, possible substrate channeling	i, o, p
Mechanism not established but believed to be an unusual type of serine peptidase	q, r
Some differences in kinetic characteristics of enzyme isolated from different sources	c, d, l

[a] S. Wilk and M. Orlowski, *J. Neurochem.* **35,** 1172 (1980).

[b] B. Dahlmann, L. Kuehn, M. Rutschmann, and H. Reinauer, *Biochem. J.* **228,** 161 (1985).

[c] H. Djaballah, J. A. Harness, P. J. Savory, and A. J. Rivett, *Eur. J. Biochem.* **209,** 629 (1992).

[d] M. Orlowski, C. Cardozo, and C. Michaud, *Biochemistry* **32,** 1563 (1993).

[e] K. Tanaka, K. Ii., A. Ichihara, L. Waxman, and A. L. Goldberg,, *J. Biol. Chem.* **261,** 15197 (1986).

[f] K. Tanaka, T. Yoshimura, A. Kumatori, A. Ichihara, A. Ikai, M. Nishigai, K. Kameyama, and T. Takagi, *J. Biol. Chem.* **263,** 16209 (1988).

[g] D. L. Mykles, *Arch. Biochem. Biophys.* **274,** 216 (1989).

[h] D. Weitman and J. D. Etlinger, *J. Biol. Chem.* **267,** 6977 (1992).

[i] H. Djaballah and A. J. Rivett, *Biochemistry* **31,** 4133 (1992).

[j] S. Wilk and M. Orlowski, *J. Neurochem.* **40,** 842 (1983).

[k] C. Cardozo, A. Vinitsky, M. C. Hidalgo, C. Michaud, and M. Orlowski, *Biochemistry* **31,** 7373 (1992).

[l] H. Djaballah, P. J. Savory, and A. J. Rivett, unpublished observations.

[m] H. Djaballah, A. J. Rowe, S. E. Harding, and A. J. Rivett, *Biochem. J.* **292,** 857 (1993).

[n] J. Arribas and J. G. Castaño, *J. Biol. Chem.* **265,** 13969 (1990).

[o] M. Orlowski, C. Cardozo, M. C. Hidalgo, and C. Michaud, *Biochemistry* **30,** 5999 (1991).

[p] L. R. Dick, C. R. Moomaw, G. N. DeMartino, and C. A. Slaughter, *Biochemistry* **30,** 2725 (1991).

[q] M. Orlowski, *Biochemistry* **29,** 10289 (1990).

[r] A. J. Rivett, *Biochem. J.* **291,** 1 (1993).

Phe-Val-Asn-Gln- His-Leu- Cya-Gly-Ser-His-Leu- Val-Glu- Ala-Leu- Tyr-Leu- Val-Cya- Gly-Glu-Arg-Gly-Phe-Phe-Tyr-Thr-Pro-Lys-Ala

(a) Rat liver

(b) Human erythrocyte

(c) *Xenopus laevis* oocytes

FIG. 1. Multicatalytic endopeptidase complex cleavage sites in oxidized insulin B chain. (*a*) A. J. Rivett, *J. Biol. Chem.* **260**, 12600 (1985); (*b*) L. R. Dick, C. R. Moomaw, G. N. DeMartino, and C. A. Slaughter, *Biochemistry* **30**, 2725 (1991); and (*c*) T. Takahashi, T. Tokumoto, K. Ishikawa, and K. Takahashi, *J. Biochem.* (*Tokyo*) **113**, 225 (1993).

tions below their K_m values, which are in the range of 0.1–1 mM.[18] Other fluorogenic or chromogenic leaving groups are also suitable (Table I). In particular, Z-Leu-Leu-Glu-2-naphthylamide is commonly used to assay peptidylglutamyl-peptide hydrolase activity. For some activities of the bovine pituitary enzyme a coupled assay with an aminopeptidase has been used.[19] Although we describe below stopped-fluorimetric assay procedures, it is necessary to establish that the production of product is linear with both time and enzyme concentration, because this is not always the case. Fluorimetric assays can of course be carried out continuously using a water-jacketed cell holder to maintain a constant temperature.

The degradation of peptide substrates such as oxidized insulin B chain (Table I) can be monitored by C_{18} reversed-phase high-performance liquid chromatography (HPLC). Some differences have been observed in the cleavage pattern obtained with multicatalytic endopeptidases from different sources (Fig. 1).[12] The degradation of protein substrates, of which casein has been the most widely used, is easily assayed by measuring acid-soluble counts released from radiolabeled protein (see below). Alternative methods for investigating protein degradation include sodium dodecyl sulfate (SDS)–polyacrylamide gel electrophoresis and the use of fluorescamine to determine the level of acid-soluble products from unlabeled proteins.[20] Not all proteins are substrates for the complex but it is not yet entirely clear what structural features of proteins determine proteolytic susceptibility.[21,22] Those proteins which are substrates are usually degraded to peptides.[11]

[18] A. J. Rivett, *J. Biol. Chem.* **264**, 12215 (1989).
[19] M. Orlowski, C. Cardozo, and C. Michaud, *Biochemistry* **32**, 1563 (1993).
[20] A. J. Rivett, *Arch. Biochem. Biophys.* **243**, 624 (1985).
[21] A. J. Rivett and R. L. Levine, *Arch. Biochem. Biophys.* **278**, 26 (1990).
[22] R. E. Pacifici, Y. Kono, and K. J. A. Davies, *J. Biol. Chem.* **268**, 15405 (1993).

Assays with 7-Amido-4-methylcoumarin Substrates

A variety of different commercially available synthetic peptides may be used to assay peptidase activities of the complex and the assay procedure is the same for all of them. Although the pH optimum varies for different substrates, we routinely use buffer of pH 7.5. Commonly used peptidyl-7-amino-4-methylcoumarin substrates are listed in Table I. We routinely use Ala-Ala-Phe-7-amido-4-methylcoumarin, Suc-Leu-Leu-Val-Tyr-7-amido-4-methylcoumarin, and Boc-Leu-Ser-Thr-Arg-7-amido-4-methylcoumarin.

Reagents

Buffer: 100 mM HEPES/KOH, pH 7.5, diluted to 50 mM in the assay.
Substrate stock solutions: These can be made at 2 or 10 mM in water, 50 mM acetic acid, or dimethyl sulfoxide (depending on the solubility of the substrate) and stored in aliquots at $-20°$. Dimethyl sulfoxide concentrations in the assays should not normally exceed 5%.
Stop mix: Sodium acetate trihydrate (0.25 g) is added to 4.375 ml 1 M acetic acid and made up to a volume of 25 ml with water.
Procedure. Assay mixtures containing approximately 1–2 μg enzyme, substrate (at a fixed concentration, e.g., 20 or 50 μM), and 50 mM HEPES/KOH, pH 7.5, are made up in a total volume of 200 μl in 4-ml disposable test tubes and then incubated at 37° for 15 or 30 min. Reactions are started by addition of either substrate or enzyme and are stopped by addition of 0.1 ml stop mix. H$_2$O (2 ml) is added to each tube prior to measuring fluorescence (excitation 370 nm, emission 430 nm). Blanks are prepared without the addition of enzyme and a standard curve is prepared with 7-amino-4-methylcoumarin.

Assay with Z-Leu-Leu-Glu-2-naphthylamide

The substrate is hydrolyzed to give 2-naphthylamine, which can be either assayed colorimetrically by coupling with a diazonium salt or measured directly in a fluorimeter. This product is probably carcinogenic and cannot easily be purchased, so we make just enough by an enzymatic procedure to produce a standard curve.

Reagents

Buffer: 100 mM HEPES/KOH, pH 7.5, diluted to 50 mM in the assay.
Substrate stock solution: 10 mM Z-Leu-Leu-Glu-2-naphthylamide in dimethyl sulfoxide (DMSO), stored in aliquots at $-20°$.

Procedure. Assay mixes are made to a volume of 200 μl in 4-ml disposable test tubes. Usually 1–2 μg of enzyme is used with a final buffer concentration of 50 mM. The amounts of substrate and water are varied. The amount of substrate is limited to a maximum of 0.6 mM by its solubility in aqueous solutions.[15] There are two distinct types of catalytic centers (LLE1 and LLE2), which possess peptidylglutamyl-peptide hydrolase activity.[15] We usually assay with substrate at concentrations of 0.1 mM (LLE1 activity) and 0.4 mM (LLE1 + LLE2 activity). Because the kinetic characteristics of the peptidylglutamyl-peptide hydrolase activities (see later) of the rat liver and bovine pituitary enzymes are not identical,[15,23] they should be tested in other systems.

Assay mixtures are incubated at 37° usually for 30 min and are then stopped by addition of 0.3 ml ethanol. Water (2 ml) is added prior to reading samples in a fluorimeter (excitation 333 nm, emission 450 nm). Blank samples are prepared at the appropriate substrate concentrations with no enzyme. 2-Naphthylamine is produced by complete digestion of a known amount of Z-Leu-Leu-Glu-2-naphthylamide (e.g., 0.6 mM) by *Staphylococcus aureus* V8 proteinase in 50 mM ammonium bicarbonate buffer, pH 7.8. A standard curve can then be prepared using the 50 mM HEPES/KOH, pH 7.5, assay buffer, and diluting samples with ethanol and water as described above.

Assay with Casein as Substrate

Protein substrates are usually degraded to low molecular mass trichloroacetic acid-soluble products. Therefore, assays with radiolabeled substrate are most convenient.

Reagents

Buffer: 100 mM HEPES/KOH, pH 8.0, diluted to 50 mM in the assay.
Substrate stock solution: Casein (^{14}C, ^{3}H, or ^{125}I labeled), stored in aliquots at −20°.
Others: 10% (w/v) trichloroacetic acid (TCA); 5% (w/v) bovine serum albumin (BSA) in 0.1 N HCl.

Procedure. Assays are carried out in a 100-μl volume using 1–2 μg enzyme with 10 μg casein in 50 mM HEPES/KOH buffer, pH 8.0, in 1.5-ml Eppendorf tubes. Incubations are at 37° for 1 hr. Assays are stopped by the addition of 0.5 ml of the 10% trichloroacetic acid followed by addition of 0.1 ml of the bovine serum albumin solution. Samples are left on ice for 10 min and then centrifuged in a microfuge for 5 min. Scintillant

[23] M. Orlowski, C. Cardozo, M. C. Hidalgo, and C. Michaud, *Biochemistry* **30**, 5999 (1991).

TABLE III
SOURCES USED FOR PURIFICATION OF MULTICATALYTIC
ENDOPEPTIDASE COMPLEX[a]

Source	Ref.
Animal tissues and cells	
Liver (e.g., mouse, rat, human)	b–e
Muscle (rat)	f
Lung (human)	g
Brain (human, bovine)	h, i
Pituitary (bovine)	j, k
Lens (bovine)	l
Placenta (human)	m
Erythrocytes (human)	n
Plants (tobacco, potato, mung bean, pea seed)	o, p
Frog (*Xenopus laevis* or *Rana pipiens*): oocytes	q, r
Lobster: muscle	s
Fish (carp, white croaker): muscle	t, u
Sea urchin: eggs, sperm	v, w
Yeast (*Saccharomyces cerevisiae*)	x
Archaebacteria (*Thermoplasma acidophilium*)	y

[a] The enzyme has been described under many different names [reviewed in A. J. Rivett, *Arch. Biochem. Biophys.* **268**, 1 (1989)].

[b] A. J. Rivett, *J. Biol. Chem.* **260**, 12600 (1985).

[c] K. Tanaka, K. Ii., A. Ichihara, L. Waxman, and A. L. Goldberg, *J. Biol. Chem.* **261**, 15197 (1986).

[d] R. W. Mason, *Biochem. J.* **265**, 479 (1990).

[e] H. Djaballah, and A. J. Rivett, *Biochemistry* **31**, 4133 (1992).

[f] B. Dahlmann, L. Kuehn, M. Rutschmann, and H. Reinauer, *Biochem. J.* **228**, 161 (1985).

[g] R. Zolfaghari, C. R. F. Baker, P. C. Canizaro, A. Amirgholami, and F. J. Behal, *Biochem. J.* **241**, 129 (1987).

[h] A. Azaryan, M. Banay-Schwartz, and A. Lajtha, *Neurochem. Res.* **14**, 995 (1989).

[i] J. R. McDermott, A. M. Gibson, A. E. Oakley, and J. A. Biggins, *J. Neurochem.* **56**, 1509 (1991).

[j] S. Wilk and M. Orlowski, *J. Neurochem.* **35**, 1172 (1980).

[k] M. Orlowski and C. Michaud, *Biochemistry* **28**, 9270 (1989).

[l] K. Ray and H. Harris, *Proc. Natl. Acad. Sci. U.S.A.* **82**, 7545 (1985).

[m] K. B. Hendil and W. Uerkvitz, *J. Biochem. Biophys. Methods* **22**, 159 (1991).

[n] M. J. McGuire and G. N. DeMartino, *Biochim. Biophys. Acta.* **873**, 279 (1986).

[o] M. Schliephacke, A. Kremp, H. P. Schmid, K. Kohler, and U. Kull, *Eur. J. Cell Biol.* **55**, 114 (1991).

TABLE III (*continued*)

p B. Skoda and L. Malek, *Plant Physiol.* **99**, 1515 (1992).

q K. Tanaka, T. Yoshimura, A. Kumatori, A. Ichihara, A. Ikai, M. Nishigai, K. Kameyama, and T. Takagi, *J. Biol. Chem.* **263**, 16209 (1988).

r Y. Azuma, T. Tokumoto, and K. Ishikawa, *Mol. Cell. Biochem.* **100**, 177 (1991).

s D. L. Mykles, *Arch. Biochem. Biophys.* **274**, 216 (1989).

t M. Kinoshita, H. Toyohara, and Y. Shimizu, *Comp. Biochem. Physiol.* **96B**, 565 (1990).

u E. J. Folco, L. Busconi, C. B. Martone, and J. J. Sanchez, *Arch. Biochem. Biophys.* **267**, 599 (1988).

v J. L. Grainger and M. M. Winkler, *J. Cell Biol.* **109**, 675 (1989).

w K. Matsumura and K. Aketa, *Mol. Reprod. Dev.* **29**, 189 (1991).

x T. Achstetter, C. Ehmann, A. Osaki, and D. H. Wolf, *J. Biol. Chem.* **259**, 13344 (1984).

y B. Dahlmann, F. Kopp, L. Kuehn, B. Niedel, G. Pfeifer, R. Hegerl, and W. Baumeister, *FEBS Lett.* **251**, 125 (1989).

(10 ml) is added to an aliquot of the supernatant (0.63 ml) and samples are then counted in a liquid scintillation counter.

Purification of the Multicatalytic Endopeptidase Complex

The multicatalytic endopeptidase complex is an abundant protein and can constitute up to 1% of the soluble cellular protein, with the highest levels in animal cells and tissues being found in the liver.[24,25] The enzyme has been purified from a wide variety of sources (Table III). Purification procedures often involve a combination of anion-exchange chromatography and gel filtration, often preceded by an ammonium sulfate fractionation, and usually involve additional chromatographic steps such as hydroxylapatite, hydrophobic, heparin–Sepharose or Affi-Gel blue chromatography or a second ion-exchange or gel-filtration step. Purification by immunoaffinity chromatography has also been described.[26] The yield of multicatalytic endopeptidase complex that can be expected depends on the source but is usually in the range of 1–10 mg/100 g tissue.[13,15] A low amount obtained from erythrocytes[27] reflects the relatively low level

[24] K. Tanaka, K. Ii., A. Ichihara, L. Waxman, and A. L. Goldberg, *J. Biol. Chem.* **261**, 15197 (1986).

[25] A. J. Rivett and S. T. Sweeney, *Biochem. J.* **278**, 171 (1991).

[26] K. B. Hendil and W. Uerkvitz, *J. Biochem. Biophys. Methods* **22**, 159 (1991).

[27] M. J. McGuire and G. N. DeMartino, *Biochim. Biophys. Acta* **873**, 279 (1986).

of the complex in these cells.[25] The enzyme is usually prepared from soluble extracts, but purification from erythrocyte membranes has also been reported.[28]

The enzyme can be purified in a latent form (references in Table II), the difference in procedure usually involving the inclusion of 20% glycerol in all the buffers for the latent enzyme. There are only minor differences between latent and active forms in the molecular masses and pI values of the associated polypeptides.[29,30] In the majority of cases (Table I) the enzyme has been purified in the active state. It is, however, quite difficult to compare the specific activities of the final preparations obtained using different procedures, for seveal reasons. First, the purification procedure may have an effect on activity; second, the activity can be markedly influenced by the composition of the buffer (see below); and third, the activities have often not been assayed using the same substrates under identical conditions.

The purification-fold can also be misleading. For example, it is not straightforward to measure activity in crude extracts of animal cells using protein substrates,[11,27] possibly due to the presence of endogenous inhibitor proteins[31,32] or competing protein substrates. With synthetic peptide substrates, the ratio of activity with different substrates varies at different stages of the purification[33] and some, if not all, of the usual substrates are also hydrolyzed by other peptidases. With apparently homogeneous enzyme preparations from different sources it is also difficult to make comparisons because of differences in assay conditions and possibly also some kinetic differences, the basis for which we do not yet understand. The kinetic characteristics of the purified multicatalytic endopeptidase complex (Table I) will be discussed in more detail below.

Purification of the Multicatalytic Endopeptidase from Rat Liver

The enyzme can be purified from fresh or frozen rat liver using the following procedure. All steps are carried out at 4° except for those involv-

[28] M. Kinoshita, T. Hamakubo, I. Fukui, T. Murachi, and H. Toyohara, *J. Biochem.* (*Tokyo*) **107**, 440 (1990).

[29] L. W. Lee, C. R. Moomaw, K. Orth, M. J. McGuire, G. N. DeMartino, and C. A. Slaughter, *Biochim. Biophys. Acta* **1037**, 178 (1990).

[30] M. E. Pereira, T. Nguyen, B. J. Wagner, J. W. Margolis, B. Yu, and S. Wilk, *J. Biol. Chem.* **267**, 7949 (1992).

[31] X. Li, M. Gu, and J. D. Etlinger, *Biochemistry* **30**, 9709 (1991).

[32] M. Chu-Ping, C. A. Slaughter, and G. N. DeMartino, *Biochim. Biophys. Acta* **1119**, 303 (1992).

[33] S. Wilk and M. Orlowski, *J. Neurochem.* **40**, 842 (1983).

ing fast protein liquid chromatography (FPLC), which are performed at room temperature. Column and buffer volumes are given for a preparation from 200 g rat liver. We recommend Ala-Ala-Phe-7-amino-4-methylcoumarin as the substrate for assays during the purification.

Step 1. Homogenization and Centrifugation. Rat livers are washed in homogenization buffer (20 mM HEPES, pH 8.0, containing 1 mM 2-mercaptoethanol and 1 mM EDTA) and then homogenized (3 × 30 sec in a Waring blendor) in 5 volumes of buffer. The homogenate is centrifuged at 27,000 g for 2.5 hr.

Step 2. Ammonium Sulfate Fractionation. Solid ammonium sulfate to give 35% saturation is added to the supernatant from step 1. After stirring for 30 min on ice, the precipitate is removed by centrifugation and ammonium sulfate is added to the supernatant to give 60% saturation. Proteins precipitated during the second ammonium sulfate precipitation (left stirring for 45 min after all the ammonium sulfate has dissolved) are collected by centrifugation and redissolved in approximately 200 ml of 10 mM Tris-HCl buffer, pH 7.2. The preparation is dialyzed against 2 × 5 liters 10 mM Tris-HCl buffer, pH 7.2, containing 50 mM KCl, 0.1 mM EDTA, 1 mM 2-mercaptoethanol.

Step 3. DEAE–Cellulose Chromatography. The dialyzed ammonium sulfate fraction is centrifuged at 45,000 g for 20 min and then loaded on to a DEAE–cellulose column (2.6 × 30 cm) equilibrated in 10 mM Tris-HCl, pH 7.2, containing 50 mM KCl at a flow rate of approximately 2 ml/min. After washing the column with equilibration buffer, a gradient of 50–350 mM KCl (2 × 1.5 liters) is applied. The enzyme elutes at approximately 150 mM KCl. Active fractions are pooled (approximately 400 ml) and concentrated to a volume of about 100 ml using an Amicon (Danvers, MA) ultrafiltration cell with an XM50 membrane. EDTA (0.1 mM) and 2-mercaptoethanol (1 mM) are added to the preparation after pooling fractions from this and all the subsequent chromatography steps.

Step 4. Mono Q Anion-Exchange Chromatography. The DEAE-cellulose pool is diluted with one-half volume of 20 mM Tris-HCl, pH 7.2, and then divided into at least two portions. The portions are loaded separately on to a Mono Q 10/10 FPLC column (Pharmacia, Piscataway, NJ) equilibrated in 20 mM Tris-HCl, pH 7.2, containing 100 mM KCl at a flow rate of 4 ml/min. The enzyme is eluted on a gradient (200 ml) of 100–500 mM KCl at a KCl concentration of about 400 mM (usually peak fraction 19 out of 25 × 8 ml fractions). Active fractions are pooled and concentrated to less than 10 ml using an Amicon ultrafiltration cell (XM50 membrane).

Step 5. Superose 6 Gel Filtration. The sample is filtered and loaded onto a preparative-size Superose 6 FPLC column (Pharmacia), equilibrated in 50 mM potassium phosphate buffer, pH 7.0, containing 0.1 M

KCl. Although the sample could be concentrated more and loaded for one column run, we find that the resolution and recovery are better if the sample is divided into two or three portions, loading 2 or 3 ml each time. The enzyme elutes at approximately the same volume as thyroglobulin.

Step 6. Mono Q Ion-Exchange Chromatography. We usually find it helpful to carry out a final ion-exchange (Mono Q 5/5) chromatography step. The pool from the gel-filtration step is therefore dialyzed against 20 m*M* Tris-HCl buffer, pH 7.2, and then run (<5 mg/run) on the Mono Q 5/5 column equilibrated in the same buffer and eluted at approximately 0.35 − 0.4 *M* KCl by applying a linear gradient to 500 m*M* KCl. The active fractions from this column are pooled and dialyzed against 50 m*M* potassium phosphate, pH 7.0, containing 1 m*M* dithiothreitol, 0.1 m*M* EDTA, and 10% glycerol.

Protein Determination. The amount of enzyme can be quantitated using the Bradford method[34] for protein determination with bovine serum albumin as standard, because this gives a very good estimate of actual protein concentration determined by amino acid analysis.[35]

Purity. The purity of the preparation is best assessed by PAGE carried out under nondenaturing conditions,[11] and the enzyme can be visualized under UV light in nondenaturing gels following incubation with fluorogenic peptide substrates (for example, at a concentration of 50–100 μ*M* in 50 m*M* HEPES buffer, pH 7.5, incubated at 37° for 1 hr). There has been a report suggesting multiple electrophoretic forms of the enzyme,[36] but these are not always observed.[11] SDS–PAGE gels can also be used to determine whether preparations contain any contaminating proteins in addition to the characteristic ladder of multicatalytic endopeptidase complex bands of M_r of 20–34 kDa.[11,15]

Specific Activity. The specific activities of the rat liver complex assayed with 50 μ*M* Ala-Ala-Phe-7-amido-4-methylcoumarin and 0.4 m*M* Z-Leu-Leu-Glu-2-naphthylamide are usually in the ranges of 12–25 nmol/min/mg protein and 100–150 nmol/min/mg, respectively.

Storage. The purified enzyme is stable for several months when stored frozen at −20° in 50 m*M* potassium phosphate buffer, pH 7.0, containing 10% glycerol, 1 m*M* DTT, 0.1 m*M* EDTA.

Purification of 26S Proteinase

The multicatalytic endopeptidase complex forms part of the 26S proteinase (also called ubiquitin-conjugate degrading proteinase), and several

[34] M. M. Bradford, *Anal. Biochem.* **72**, 248 (1976).
[35] P. J. Savory and A. J. Rivett, *Biochem. J.* **289**, 45 (1993).
[36] L. Hoffman, G. Pratt, and M. Rechsteiner, *J. Biol. Chem.* **267**, 22362 (1992).

purification protocols have been reported for this larger proteinase complex.[9,37–40] The 26S proteinase, which is precipitated at lower ammonium sulfate concentrations than the multicatalytic endopeptidase complex alone, appears to require ATP for its stability. Also, unlike purified multicatalytic endopeptidase complex preparations,[1] it catalyzes ATP-dependent protein degradation. 26S proteinase activity, as well as multicatalytic endopeptidase activity, can conveniently be assayed with Suc-Leu-Leu-Val-Tyr-7-amido-4-methylcoumarin or with casein as a substrate.

Properties of Purified Multicatalytic Endopeptidase Complex

The size, shape, and subunit structure of multicatalytic endopeptidase complexes from eukaryotic cells are broadly similar irrespective of the source of the enzyme, and are summarized in Table IV. The number of different types of polypeptide associated with the complex does seem to vary with the source and it is not yet clear whether these differences can be explained solely in terms of post-translational modification, such as phosphorylation, glycosylation, and proteolysis. cDNAs have been cloned for many yeast, rat, and human multicatalytic endopeptidase subunits.[5,7] All subunits of the complex are encoded by members of the same gene family, with different subunits from the same source having 18–40% identity. The individual subunit sequences are highly conserved between different mammalian species (rat/human 95–98% identity at the amino acid level) and are 40–70% identitical with the most closely related subunits in yeast.

The multicatalytic endopeptidase complex isolated from the archaebacterium *Thermoplasma acidophilum* is an interesting and useful one because it is a much simpler molecule than the eukaryotic complex. The overall structure is similar,[41] but it is composed of only two different types of subunit, α and β ($\alpha_{14}\beta_{14}$ stoichiometry). The subunit sequences are related to those of all the eukaryotic multicatalytic endopeptidase complex subunits.[42] The α subunits are located on the outer rings of the cylinder

---

[37] E. Eytan, D. Ganoth, T. Armon, and A. Hershko, *Proc. Natl. Acad. Sci. U.S.A.* **86**, 7751 (1989).

[38] J. Driscoll and A. L. Goldberg, *J. Biol. Chem.* **265**, 4789 (1990).

[39] H. Kanayama, T. Tamura, S. Ugai, S. Kagawa, N. Tanahashi, T. Yoshimura, K. Tanaka, and A. Ichihara, *Eur. J. Biochem.* **206**, 567 (1992).

[40] A. Azaryan, M. Banay-Schwartz, and A. Lajtha, *Neurochem. Res.* **14**, 995 (1989).

[41] G. Pühler, S. Weinkauf, L. Bachmann, S. Müller, A. Engel, R. Hegerl, and W. Baumeister, *EMBO J.* **11**, 1607 (1992).

[42] P. Zwickl, A. Grziwa, G. Pühler, B. Dahlmann, F. Lottspeich, and W. Baumeister, *Biochemistry* **31**, 964 (1992).

TABLE IV
PROPERTIES OF EUKARYOTIC MULTICATALYTIC ENDOPEPTIDASE COMPLEX

Property	Comments	Refs.
Occurrence	Found in all types of cells (see also Table III); usually in nucleus as well as in cytoplasm	a–c
Level	Can constitute up to 1% of cellular protein	d, e
Size	Has a molecular mass of around 700,000 Da; also forms part of 26S proteinase	f, g
Subunits	Many (14–20) different polypeptides M_r of 20,000–34,000, pI 4–9; some may be related by post-translational modification	h, i
RNA	Has very low amounts of small species (approximately 80 nucleotides) of RNA associated	j, k
Shape	Has a cylindrical structure (approximately 11 × 17 nm)	l
	Pseudo-helical arrangement of subunits	m
	Possibly a complex dimer	n
Genes	Subunits are encoded by members of same gene family, but can be divided into two groups, A (α) and B (β), related to archaebacterial α and β subunits	o, p
	Genes are located on different chromosomes; some are essential for cell proliferation	i, q, r
Activity	Mammalian complex contains at least five distinct peptidase sites (for substrates, see Table I; other kinetic properties, see Table II)	
Function	Ubiquitin-dependent and ubiquitin-independent nonlysosomal pathways of protein breakdown, including the degradation of regulatory proteins, and probably antigen processing for presentation by MHC class I pathway	i, o, p, s, t

[a] K. Tanaka, A. Kumatori, K. Ii, and A. Ichihara, *J. Cell. Physiol.* **139,** 34 (1989).
[b] A. J. Rivett, A. Palmer, and E. Knecht, *J. Histochem. Cytochem.* **40,** 1165 (1992).
[c] A. J. Rivett and E. Knecht, *Curr. Biol.* **3,** 127 (1993).
[d] L. Tanaka, K. Ii., A. Ichihara, L. Waxman, and A. L. Goldberg, *J. Biol. Chem.* **261,** 15197 (1986).
[e] K. B. Hendil, *Biochem. Int.* **17,** 471 (1988).
[f] M. Rechsteiner, L. Hoffman, and W. Dubiel, *J. Biol. Chem.* **268,** 6065 (1993).
[g] A. Hershko and A. Ciechanover, *Annu. Rev. Biochem.* **61,** 761 (1992).
[h] A. J. Rivett and S. T. Sweeney, *Biochem. J.* **278,** 171 (1991).
[i] W. Heinemeyer, J. A. Kleinschmidt, J. Saidowsky, C. Escher, and D. H. Wolf, *EMBO J.* **10,** 555 (1991).
[j] H. S. Skilton, I. C. Eperon, and A. J. Rivett, *FEBS Lett.* **279,** 351 (1991).
[k] H. G. Nothwang, O. Coux, G. Keith, I. Silva-Pereira, and K. Scherrer, *Nucleic Acids Res.* **20,** 1959 (1992).
[l] W. Baumeister, B. Dahlmann, R. Hegerl, F. Kopp, L. Kuehn, and G. Pfeifer, *FEBS Lett.* **241,** 239 (1988).
[m] H. Djaballah, A. J. Rowe, S. E. Harding, and A. J. Rivett, *Biochem. J.* **292,** 857 (1993).
[n] F. Kopp, B. Dahlmann and K. B. Hendil, *J. Mol. Biol.* **229,** 14 (1993).
[o] K. Tanaka, T. Tamura, T. Yoshimura, and A. Ichihara, *New Biol.* **4,** 173 (1992).

TABLE IV (continued)

p A. J. Rivett, *Biochem. J.* **291**, 1 (1993).
q T. Fujiwara, K. Tanaka, E. Orino, T. Yoshimura, A. Kumatori, T. Tamura, C. H. Chung, T. Nakai, K. Yamaguchi, S. Shin, A. Kakizuka, S. Nakanishi, and A. Ichihara, *J. Biol. Chem.* **265**, 16604 (1990).
r Y. Emori, T. Tsukahara, H. Kawasaki, S. Ishiura, H. Sugita, and K. Suzuki, *Mol. Cell. Biol.* **11**, 344 (1991).
s M. Orlowski, *Biochemistry* **29**, 10289 (1990).
t A. L. Goldberg and K. L. Rock, *Nature (London)* **357**, 375 (1992).

and the β subunits are in the inner rings.[43] The subunits of the eukaryotic multicatalytic endopeptidase complex can be divided into two groups, A (α) and B (β),[6] depending on whether they resemble the archaebacterial α or β subunit. The A group members are quite similar, with a very highly conserved region close to the N terminus, which is blocked. The B group members, on the other hand, are not so closely related to each other and have a variable region at the N terminus, which is often unblocked.[44]

Catalytic Mechanism and Identification of Catalytic Components

The enzyme is now widely believed to be an unusual type of serine peptidase based on its inhibitor specificity (Table V). However, a definite assignment must await the identification of catalytic residues. Although it seems likely that the different catalytic centers of the multicatalytic endopeptidase complex are mechanistically related, there are remarkable differences in reactivity of the different peptidase sites with some serine peptidase inhibitors.[45] For example, a given inhibitor can rapidly inactivate one activity while having no effect on another. Such observations can explain why in some of the earliest studies the enzyme was not thought to be a serine protease. The inhibition observed with thiol-reactive reagents is probably nonspecific, because activity of the *Thermoplasma* enzyme is unaffected by even the mercury-containing thiol reagents.[46] Because the primary structures of the multicatalytic endopeptidase complex subunits bear no resemblence to any other known peptidases, the catalytic components cannot be identified from sequence information. Also, dissociated

---

[43] A. Grziwa, W. Baumeister, B. Dahlmann, and F. Kopp, *FEBS Lett.* **290**, 186 (1991).
[44] K. S. Lilley, M. D. Davison, and A. J. Rivett, *FEBS Lett.* **262**, 327 (1990).
[45] H. Djaballah, J. A. Harness, P. J. Savory, and A. J. Rivett, *Eur. J. Biochem.* **209**, 629 (1992).
[46] B. Dahlmann, L. Kuehn, A. Grziwa, P. Zwickl, and W. Baumeister, *Eur. J. Biochem.* **208**, 789 (1992).

TABLE V
PROTEASE INHIBITORS EFFECTIVE ON MAMMALIAN MULTICATALYTIC
ENDOPEPTIDASE ACTIVITIES[a]

Type of inhibitor	Examples	Refs.
Organophosphorus compound	Diisopropyl fluorophosphate	b–d
Sulfonyl fluoride	4-(2-Aminoethyl)benzene sulfonyl fluoride	c
Isocoumarins	3,4-Dichloroisocoumarin; others	c, e, f
Peptide aldehydes	Leupeptin, antipain	g, h
	Chymostatin and analogs	c
	Z-Leu-Leu-PheH	i
Peptidyl chloromethanes	Tyr-Gly-Arg-CH$_2$Cl (some have no effect)	j, k
Peptidyl diazomethanes	Z-Leu-Leu-Tyr-CHN$_2$, Z-Phe-Gly-Tyr-CHN$_2$ (some have no effect)	j
Thiol-reactive reagents	N-Ethylmaleimide, p-hydroxymercuribenzoate, others	g, l

[a] There are marked differences in reactivity of different peptidase sites. None of the inhibitors here inhibit all activities of the mammalian multicatalytic endopeptidase complex (see also Table VII).
[b] K. Tanaka, K. Ii., A. Ichihara, L. Waxman, and A. L. Goldberg, J. Biol. Chem. 261, 15197 (1986).
[c] H. Djaballah, J. A. Harness, P. J. Savory, and A. J. Rivett, Eur. J. Biochem. 209, 629 (1992).
[d] M. Orlowski, C. Cardozo, and C. Michaud, Biochemistry 32, 1563 (1993).
[e] M. Orlowski and C. Michaud, Biochemistry 28, 9270 (1989).
[f] R. W. Mason, Biochem. J. 265, 479 (1990).
[g] S. Wilk and M. Orlowski, J. Neurochem. 40, 842 (1983).
[h] P. J. Savory and A. J. Rivett, Biochem. J. 289, 45 (1993).
[i] A. Vinitsky, C. Michaud, J. C. Powers, and M. Orlowski, Biochemistry 31, 9421 (1992).
[j] P. J. Savory, H. Djaballah, H. Angliker, E. Shaw, and A. J. Rivett, Biochem. J. 296, 601 (1993).
[k] C. Cardozo, A. Vinitsky, M. C. Hidalgo, C. Michaud, and M. Orlowski, Biochemistry 31, 7373 (1992).
[l] A. J. Rivett, J. Biol. Chem. 260, 12600 (1985).

subunits are inactive. It is possible that the enzyme has a novel type of catalytic mechanism.

The distinction between different catalytic activities of the eukaryotic complex is based on the use of different synthetic peptide substrates with a variety of protease inhibitors and other effectors (Table VI). The effects of activators such as low concentrations of SDS can be quite variable depending on the source of the enzyme, and, at least in some cases,

TABLE VI
OTHER EFFECTORS OF MULTICATALYTIC ENDOPEPTIDASE ACTIVITIES

Tested effector	Concentration	Effect on activity[a,b]
EDTA	1 mM	Little or no effect
$MnCl_2$, $MgCl_2$, $CaCl_2$	1 mM	Stimulation or inhibition
$ZnCl_2$	0.1 mM	Inhibition
KCl	25 mM	Inhibition or no effect
ATP	1–5 mM	No effect or slight inhibition
Polylysine	0.1 mg/ml	Stimulation or no effect
Linoleic acid, oleic acid	0.2–5 mM	Stimulation
Sodium dodecyl sulfate	0.01%	Stimulation, inhibition, or no effect
Nonidet P-40	0.5%	Stimulation, inhibition, or no effect
Guanidine hydrochloride	20 mM	Stimulation or inhibition

[a] Depends on source of multicatalytic endopeptidase complex and substrate.
[b] Many references, including the following sources: S. Wilk and M. Orlowski, *J. Neurochem.* **35**, 1172 (1980); B. Dahlmann, M. Rutschmann, L. Kuehn, and H. Reinauer, *Biochem. J.* **228**, 171 (1985); K. Tanaka, K. Ii., A. Ichihara, L. Waxman, and A. L. Goldberg, *J. Biol. Chem.* **261**, 15197 (1986); Y. Saitoh, H. Yokosawa, and S.-I. Ishii, *Biochem. Biophys. Res. Commun.* **162**, 334 (1989); J. Arribas and J. G. Castaño, *J. Biol. Chem.* **265**, 13969 (1990); R. L. Mellgren, *Biochim. Biophys. Acta* **1040**, 28 (1990); D. L. Mykles and M. F. Haire, *Arch. Biochem. Biophys.* **288**, 543 (1991); H. Djaballah, J. A. Harness, P. J. Savory, and A. J. Rivett, *Eur. J. Biochem.* **209**, 629 (1992); H. Djaballah, A. J. Rowe, S. E. Harding, and A. J. Rivett, *Biochem. J.* **292**, 857 (1993).

involve conformational changes within the complex.[47,48] It is difficult to define precisely the specificity of the individual peptidase activities of the complex because of the possibility of overlapping specificity and lack of suitable substrates/inhibitors. The number of distinct peptidase sites of the mammalian multicatalytic endopeptidase complex is probably greater than the five shown in Table VII. The original nomenclature for the different activities is clearly inadequate because there are at least two distinct chymotrypsin-like activities and two peptidylglutamyl-peptide hydrolase activities. Additional names such as "branched-chain amino acid preferring" and "small neutral amino acid preferring"[19] are cumbersome. Moreover, all of these names imply something about the specificity of the sites, of which we still know rather little. We therefore suggest a simple and short alternative nomenclature for the different sites (MEC 1, MEC 2, MEC 3, etc., as defined in Table VII), which can be added to as necessary.

[47] H. Djaballah, A. J. Rowe, S. E. Harding, and A. J. Rivett, *Biochem. J.* **292**, 857 (1993).
[48] Y. Saitoh, H. Yokosawa, and S.-I. Ishii, *Biochem. Biophys. Res. Commun.* **162**, 334 (1989).

TABLE VII
Catalytic Activities of Rat Liver Multicatalytic Endopeptidase Complex[a]

Activity	MEC1	MEC2	MEC3	MEC4	MEC5
Previous name:	Trypsinlike	Chymotrypsin-like		Peptidylglutamyl-peptide hydrolase	
Substrate:[b]	LSTR-AMC	AAF-AMC	LLVY-AMC	LLE1 LLE-NA	LLE2 LLE-NA
Protease inhibitors					
3,4-Dichloroisocoumarin (20 μM)	Stimulation	Inhibition	Inhibition	—	Inhibition
Diisopropyl fluorophosphate (5 mM)	Inhibition	—	Inhibition	Inhibition	—
4-(2-Aminoethyl)benzene sulfonyl fluoride	Inhibition	—	—	—	—
Leupeptin (20 μM)	—	—	—	—	—
Ala-Ala-Phe-CH$_2$Cl (10 μM)	—	Inhibition	Inhibition	—	—
Z-Phe-Gly-Tyr-CHN$_2$ (0.1 mM)	—	—	Inhibition	—	—
Other effectors					
KCl (25 mM)	—	—	—	—	—
Casein (0.1 mg/ml)	—	—	Inhibition	Inhibition	Inhibition
MnCl$_2$ (1 mM)	Inhibition	—	—	Stimulation	Stimulation

[a] From H. Djaballah, J. A. Harness, P. J. Savory, and A. J. Rivett, *Eur. J. Biochem.* **209**, 629 (1992); P. J. Savory, H. Djaballah, H. Angliker, E. Shaw, and A. J. Rivett, *Biochem. J.* **296**, 601 (1993).
[b] LSTR-AMC, Boc-Leu-Ser-Thr-Arg-7-amido-4-methylcoumarin; AAF-AMC, Ala-Ala-Phe-7-amido-4-methylcoumarin; LLVY-AMC, Leu-Leu-Val-Tyr-7-amido-4-methylcoumarin; LLE-NA, Z-Leu-Leu-Glu-2-naphthylamide.

Another difficulty in trying to define peptidase activities is that there are some substantial differences between the kinetic properties of the best studied, rat liver and bovine pituitary multicatalytic endopeptidase complexes.[19,45] For example, the trypsinlike activity, which in many ways is the easiest to distinguish, is inhibited by 3,4-dichloroisocoumarin in the case of the bovine pituitary enzyme, but not in the case of the rat liver enzyme. Therefore, the precise definition of the activities of the mammalian multicatalytic endopeptidase complex may ultimately depend on the identification of the catalytic subunits and residues. Some useful reagents have been identified,[45,49] but work with peptidyl chloromethane and peptidyl diazomethane inhibitors has provided some surprising results.[49] For example, Ala-Ala-Phe-chloromethane inhibits MEC3 but not MEC2, for which the substrate is Ala-Ala-Phe-7-amido-4-methylcoumarin. The catalytic subunits of the yeast multicatalytic endopeptidase complex have been investigated by the production of mutants defective in chymotrypsin-like or peptidylglutamyl-peptide hydrolase activity[50-52] and the data suggest that two subunits may be required to form a single catalytic site.

Functions of Multicatalytic Endopeptidase Complex and 26S Proteinase

It appears that the multicatalytic endopeptidase complex and the 26S proteinase constitute a major nonlysosomal proteolytic system. The 26S proteinase is believed to play a role in the ubiquitin system of protein degradation.[9] Nonlysosomal pathways, which can be either ubiquitin-dependent[8,53] or ubiquitin-independent, seem to account for the degradation of short-lived regulatory proteins and for the breakdown of abnormal proteins and damaged proteins produced under stress conditions.[8,54,55] The yeast mutants that are defective in multicatalytic endopeptidase complex activity have proved useful for establishing the role of the complex in ubiquitin-dependent proteolysis *in vivo*,[50] because these mutants have been found to be defective in the degradation of known substrates of the

[49] P. J. Savory, H. Djaballah, H. Angliker, E. Shaw, and A. J. Rivett, *Biochem. J.* **296**, 601 (1993).

[50] W. Heinemeyer, J. A. Kleinschmidt, J. Saidowsky, C. Escher, and D. H. Wolf, *EMBO J.* **10**, 555 (1991).

[51] W. Hilt, C. Enenkel, A. Gruhler, T. Singer, and D. H. Wolf, *J. Biol. Chem.* **268**, 3479 (1993).

[52] W. Heinemeyer, A. Gruhler, V. Möhrle, Y. Mahé and D. H. Wolf, *J. Biol. Chem.* **268**, 5115 (1993).

[53] S. Jentsch, *Trends Cell Biol.* **2**, 98 (1992).

[54] A. J. Rivett and E. Knecht, *Curr. Biol.* **3**, 127 (1993).

[55] M. Rechsteiner, *Cell* (*Cambridge, Mass.*) **66**, 615 (1991).

ubiquitin system.[56,57] Also, the discovery of two γ-interferon-inducible multicatalytic endopeptidase complex genes in the MHC class II region[58,59] has prompted the suggestion that the enzyme is involved in antigen processing for presentation by the MHC class I pathway.

[56] B. Richter-Ruoff, W. Heinemeyer, and D. H. Wolf, *FEBS Lett.* **302,** 192 (1992).
[57] S. Seufert and S. Jentsch, *EMBO J.* **11,** 3077 (1992).
[58] R. Glynne, S. H. Powis, S. Beck, A. Kelly, L. A. Kerr, and J. Trowsdale, *Nature (London)* **353,** 357 (1991).
[59] C. K. Martinez and J. J. Monaco, *Nature (London)* **353,** 664 (1991).

[25] ATP-Dependent Protease La (Lon) from *Escherichia coli*

By Alfred L. Goldberg, Richard P. Moerschell, Chin Ha Chung, and Michael R. Maurizi

Introduction

Protease La (endopeptidase La, EC 3.4.21.53), the product of the *lon* gene in *Escherichia coli,* is an ATP-dependent cytosolic protease that plays an important role in intracellular protein degradation.[1-4] In bacteria, as in eukaryotic cells, protein degradation is an energy-requiring process.[5] Studies of the biochemical basis for this energy requirement led to the discovery of protease La, a new type of proteolytic enzyme, whose activity is coupled to ATP hydrolysis.[6-9] This enzyme catalyzes the rate-limiting steps in the degradation of highly abnormal proteins in *E. coli*[7] and certain short-lived regulatory proteins.[10] It functions independently of ubiquitin; in fact, no factor similar to ubiquitin has been found in *E. coli* or other bacteria. Unlike proteases described previously, this enzyme has ATPase

[1] A. L. Goldberg, *Eur. J. Biochem.* **203,** 9 (1992).
[2] A. L. Goldberg, K. H. S. Swamy, C. H. Chung, and F. S. Larimore, this series, Vol. 80, p. 680.
[3] S. Gottesman, *in* "*Escherichia Coli* and *Salmonella Typhimurium*" (F. C. Neidhardt, J. L. Ingraham, K. B. Low, B. Magasanik, M. Schaechter, and H. E. Umbarger, eds.), p. 1308. American Society for Microbiology, Washinghton, D.C., 1987.
[4] S. Gottesman and M. R. Maurizi, *Microbiol. Rev.* **56,** 592 (1992).
[5] K. Olden and A. L. Goldberg, *Biochim. Biophys. Acta* **542,** 385 (1978).
[6] K. S. Swamy and A. L. Goldberg, *Nature (London)* **292,** 652 (1981).
[7] C. H. Chung and A. L. Goldberg, *Proc. Natl. Acad. Sci. U.S.A.* **78,** 4931 (1981).
[8] M. Charette, G. W. Henderson, and A. Markovitz, *Proc. Natl. Acad. Sci. U.S.A.* **78,** 4728 (1981).
[9] F. S. Larimore, L. Waxman, and A. L. Goldberg, *J. Biol. Chem.* **257,** 4187 (1982).
[10] S. Gottesman, *Annu. Rev. Genet.* **23,** 163 (1989).

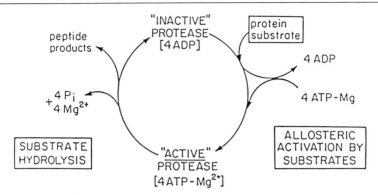

Fig. 1. Mechanism of ATP-dependent protein breakdown by protease La. This multistep model is based on published data and some unpublished findings. It is proposed that the interaction of the substrate with the allosteric site leads to the release of ADP and the partial activation of the protease (as monitored by its ability to hydrolyze oligopeptides). The binding of ATP–Mg to the enzyme leads to further activation. If the allosteric site is occupied by the protein substrate, ATP–ADP exchange and activation will occur. Hydrolysis of ATP to ADP somehow enables the enzyme to degrade proteins. Formation of the "active" protease allows maximal rates of breakdown of small peptides, but repeated rounds of this cycle would be required to degrade proteins completely to acid-soluble fragments. Adapted from Goldberg.[1]

as well as proteolytic activity[11] and is a multimeric structure of high molecular mass. Moreover, rapid degradation of protein substrates is dependent on concurrent hydrolysis of ATP or other nucleoside triphosphates. Protease La also has multiple modes of interaction with proteins such that the proteolytic active site remains in an inactive state until an appropriate substrate binds to an allosteric site on the enzyme. This binding step temporarily activates the enzyme and leads to rapid degradation of the bound protein. A model has been proposed to account for this behavior (Fig. 1; see below).

Major advances have been made in our understanding of protease La since the first discussion of this enzyme appeared in this series.[2] Several articles have appeared that review this novel enzyme, its mechanism, physiological roles, and intracellular regulation.[1,4,12] This chapter describes the isolation and assay of protease La from *E. coli* and summarizes some of its enzymatic properties. Protease La serves as a paradigm for other ATP-dependent proteases from prokaryotic and eukaryotic cells, all of which use the energy of ATP hydrolysis to accelerate the rate-limiting steps in the kinetically driven degradation of proteins.

[11] L. Waxman and A. L. Goldberg, *Proc. Natl. Acad. Sci. U.S.A.* **79**, 4883 (1982).
[12] M. R. Maurizi, *Experientia* **48**, 178 (1992).

Enzymes closely homologous to protease La (Lon protease) appear to be widespread in nature, although the enzymatic properties of these other ATP-dependent proteases have not been studied in depth. Proteases homologous to Lon are present in gram-negative[13] and gram-positive bacteria,[14-16] and, in bacilli and mycobacteria, these enzymes are necessary for cell differentiation.[14,15] Genes encoding the mitochondrial Lon protease homologs have been cloned from human cells[17] and yeast.[18-20] Figure 2 shows a comparison of the amino acid sequences of Lon proteases from gram-negative (*E. coli*)[21,22] and gram-positive (*Bacillus brevis*)[15] bacteria, yeast (*Saccharomyces cerevisiae*),[20] and humans.[17] Using the *E. coli* sequence as a reference, there is conservation of identical amino acids in 25% and retention of identical or highly similar amino acids in 53% of the positions in all four species. Conservation is highest in two regions, one containing a well-defined ATP-binding consensus motif and the other near the carboxy terminus containing the putative active site serine residue.

A protease resembling La in physical and enzymatic properties is found in the mitochondrial matrix in rats,[23-25] and this protease cross-reacts with antibodies against *E. coli* protease La (S. Kuzela, M. Desautels, and A. L. Goldberg, unpublished observations, 1992). The sequences of the eukaryotic proteins and the *E. coli* La protease have >25% identical and >50% identical or similar amino acid residues. The human gene expressed in *E. coli* produced an ATP-dependent protease with enzymatic properties consistent with those reported for the mitochondrial protease.[17] Mitochondrial La is essential for growth of yeast on nonfermentable carbon sources[19,20] and appears to catalyze the rapid degradation of abnormal matrix proteins in rat liver[26] and yeast[19] mitochondria.

[13] D. Downs, L. Waxman, A. L. Goldberg, and J. Roth, *J. Bacteriol.* **165**, 193 (1986).
[14] R. E. Gill, M. Korluk, and D. Benton, *J. Bacteriol.* **175**, 4538 (1993).
[15] K. Ito, S. Udaka, and H. Yamagato, *J. Bacteriol.* **174**, 2281 (1992).
[16] N. Tojo, S. Inouye, and T. Komano, *J. Bacteriol.* **175**, 4545 (1993).
[17] N. Wang, S. Gottesman, M. C. Willingham, M. M. Gottesman, and M. R. Maurizi, *Proc. Natl. Acad. Sci. U.S.A.* **90**, 11247 (1993).
[18] E. Kutejova, G. Durcova, E. Surovkova, and S. Kuzela, *FEBS Lett.* **329**, 47 (1993).
[19] L. Van Dyck, D. A. Pearce, and F. Sherman, *J. Biol. Chem.* **269**, 238 (1994).
[20] C. Suzuki, K. Suda, N. Wang, and G. Schatz, *Science* **264**, 273 (1994).
[21] D. T. Chin, S. A. Goff, T. Webster, T. Smith, and A. L. Goldberg, *J. Biol. Chem.* **263**, 11718 (1988).
[22] A. Y. Amerik, L. G. Christyakowa, N. I. Ostroumova, A. I. Gurevich, and V. K. Antonov, *Bioorg. Khim.* **14**, 408 (1988).
[23] M. Desautels and A. L. Goldberg, *J. Biol. Chem.* **257**, 11673 (1982).
[24] S. Watabe and T. Kimura, *J. Biol. Chem.* **260**, 5511 (1985).
[25] S. Watabe and T. Kimura, *J. Biol. Chem.* **260**, 14498 (1985).
[26] M. Desautels and A. L. Goldberg, *Proc. Natl. Acad. Sci. U.S.A.* **79**, 1869 (1982).

Protease La catalyzes the initial ATP-requiring steps in the degradation of polypeptides with highly abnormal conformations, as may result from nonsense or missense mutations, biosynthetic errors, or intracellular denaturation,[7] and also of several short-lived normal proteins whose instability is important for their regulatory function.[4,27,28] Mutants with a reduced capacity for degrading abnormal proteins (initially called *deg*) were first isolated by Bukhari and Zipser[29] in 1973, who subsequently found them to map in the *lon* locus[30] (*deg, lon,* and *capR* all refer to the gene for Lon protease; see Ref. 10 for a review). Such mutants have a variety of unusual phenotypic features, including mucoidy and increased sensitivity to DNA-damaging agents.[3,10] These properties result from the inability of the *lon* mutants to degrade rapidly critical regulatory proteins, such as SulA and RcsA.[10] In wild-type cells, these proteins have half-lives of less than 2 min, but are stabilized 10- to 20-fold in *lon* mutants.[4,10] In *lon* mutants, the degradation of abnormal polypeptides containing canavanine or puromycin occurs at 30–50% the rate in wild-type cells,[31] and the residual proteolysis seems to involve other ATP-dependent proteases (e.g., Clp protease; see [23] in this volume).[32] Thus, *lon* is not essential for normal growth,[33] but is necessary for viability in certain conditions, e.g., for recovery from DNA damage.

On the other hand, cells that carry multiple copies of the *lon* gene and have increased levels of this protease hydrolyze both abnormal and certain normal proteins at increased rates.[34] Protease La is one of the cellular heat-shock proteins expressed under the control of the *rpoH* operon (*htpR*).[35-37] Consequently, the transcription and content of protease La and the overall proteolytic capacity of the cell increase in wild-type cells during the heat-shock response, when cells accumulate large amounts of abnormal polypeptides.[35] Studies in which the cloned *lon* gene was put under the control of an inducible *lac* or *tac* promoter indicated that increasing the cellular content of protease La only 3- to 5-fold blocks growth within 20 min.[34]

[27] S. Misuzawa and S. Gotesman, *Proc. Natl. Acad. Sci. U.S.A.* **80,** 358 (1983).
[28] S. Gottesman, M. Gottesman, J. E. Shaws, and M. L. Pearson, *Cell (Cambridge, Mass.)* **26,** 223 (1981).
[29] A. I. Bukhari and D. Zipser, *Nature (London)* **243,** 238 (1973).
[30] S. Gottesman and D. Zipser, *J. Bacteriol.* **133,** 844 (1978).
[31] J. D. Kowit and A. L. Goldberg, *J. Biol. Chem.* **252,** 8350 (1977).
[32] M. R. Maurizi, M. W. Thompson, S. K. Singh, and S. H. Kim, this volume [24].
[33] M. R. Maurizi, P. Trisler, and S. Gottesman, *J. Bacteriol.* **164,** 1124 (1985).
[34] S. A. Goff and A. L. Goldberg, *J. Biol. Chem.* **262,** 4508 (1987).
[35] S. A. Goff and A. L. Goldberg, *Cell (Cambridge, Mass.)* **41,** 587 (1985).
[36] S. A. Goff, L. P. Casson, and A. L. Goldberg, *Proc. Natl. Acad. Sci. U.S.A.* **81,** 6647 (1984).
[37] A. D. Grossman, J. W. Erickson, and C. A. Gross, *Cell (Cambridge, Mass.)* **38,** 383 (1984).

```
         1
E.coli    .....MNPER SERIE...... ..IPVLPLRD VVVPHM...  ..........  VIPLFVGREK SIRCLEAAM. ....DHDKKI MLVAQKEAST
B.brevis  .....MGER  SGKRE.....  ..LPLPLRG  LLVPTM...  ..........  VLHLDVGREK SIRALEQAMV ....DDNKI  LLATQEEVHI
S.cerevi  RSSASGGQS  SSSRSDSGDG SSKQKPKDV  PEVPQMLAL  PIARRLFPG   FYKAVVISDE RVMKAIKEML DRQPYIGAF  MLKNSEDTD
H.sapien  GGAEEGAGGA GGSAGAGEGP VITALTPMTI PDVFPHLPLI AITRNPVFPR  FIKIIEVKNK KLVELLRRKV RLAQPYVGVF LKRDDSNESD

         91                                                                                               180
E.coli    DEPGVNDLFT VGTVASILQM LKLPD...... GTVKVLVEGL QRARIS....  ..........  ..........  ..........  ..........
B.brevis  EEPDAEQIYS IGTVARVKQM LKLPN...... GTIRVLVEGL QRAKIE....  ..........  ..........  ..........  ..........
S.cerevi  VITDKNDVYD VGVLAQITSA FPSKDEKTGT ETMTALLYPH RRIKIDELFP  PNEEKEKSKE QAKDTDTETT VVEDANNPED QESTSPATPK
H.sapien  VVESLDEIYH TGTFAQIHEM QDLGD..... .KLRMIVMGH RRVHISRQLE  VE........  ..........  ......PEE  PEAENKHKPR

         181                                                                                              270
E.coli    .......... .......... .......... .......... ..ALSDNGEH  FSAKAEYLES PTIDEREQE. .VLVRTAISQ FEGYIKLNKK
B.brevis  .......... .......... .......... .......... ..BYLQKEDY  FVVSITYLKE EKAEENEVE. .ALMRSLLTH FEQYIKLSKK
S.cerevi  LEDIVVERIP DSELQHHKRV EATEEESEEL DVAMEPTPEL DDIQEGEDIN  PTEFLKNYNV SLVNVLNLED EPFDRKSPVI NALTSEILKV FKEISQLNTM
H.sapien  RKSKRGKKEA EDELSARHPA DVAMEPTPEL PAEVL..... .MVEVENVVH  EDFQVTEEVK .ALTAEIVKT IRDIIALNPL

         271                                                                                              360
E.coli    IPPEVLTSLN SI........ DDPARLADTI AAHMPLKLAD KQSVLEMSDV  NERLEYLMAM MESEIDLLQV EKRIRNRVKK QMEKSQREYY
B.brevis  VSPETLTSVQ DI........ EEPGRLADVI ASHLPLKMKD KQEILETVNI  QERLEILLTI LNNEREVLEL ERKIGNRVKK QMERTQKEYY
S.cerevi  FREQIETFSA SIQSATTNIF EEPARLADFA AAVSAGEEDE LQDILSSLNI  EHRLEKSLLV LKKELMNAEL QNKISKDVET KIQKRQREYY
H.sapien  YRESVLQ... .MQQAGQRVV DNPIYILSDMG AALTGAESHE LQDVLEETNI  PKRLYKALSL LKKFEELSKL QQRLGREVEE KIKQTHRKYL

         361                                                                                              450
E.coli    LNEQMKAIQK ELG.EMDDAP DENEALKRKI DAAKMPKEAK EKAEAELQKL  KMMSPMSAEA TVVRGYIDWM VQVPWNARSK VKKDLRQAQE
B.brevis  LREQMKAIQK ELG.DKDGRQ GEVDELRAQL EKSDAPERIK AKIEKELERL  EKMPSTSAEG SVIRTYIDTL FALPWTKTTE DNLDIKHAEE
S.cerevi  LMERLKGIKR ELGID.DGRD KLIDTYKERI KSLKLPDSVQ KIFDDEITKL  STLETSMSEF GVIRNLDWL TSIPWGKHSK EQYSIPRAKK
H.sapien  LQEQLKIIKK ELGLEKDDKD AIEEKFRERL KELVVPKHVM DVVDEELSKL  GLLDNHSSEF DVTRNYLDWL TSIPWGKYSN ENLDLARAQA
                                                        ***  *****     ******

         451                                                                                              540
E.coli    ILDTDHYGLE RVKDRILEYL AVQSRVNKIK GPILCLVGPP GVGKTSLGQS  IAKATGRKYV RMALGGVRDE AEIRGHRRTY IGSMPGKLIQ
B.brevis  VLDEDHYGLE KPKERVLEYL AVQKLVNSMR GPILCLVGPP GVGKTSLARS  VARALGREFV RISLGGVRDE AEIRGHRRTY VGALPGRIIQ
S.cerevi  ILDEDHYGMV DVKDRILEFI AVGKLLGKVD GKIICFVGPP GVGKTSIGKS  IARALNRKFF RKSVGGMTDV AEIKGHRRTY IGALPGRVVQ
H.sapien  VLEEDHYGME DVKKRILEFI AVSQLRGSTQ GKILCFYGPP GVGKTSIARS  IARALNREYF RFSVGGMTDV TEIKGHRRTY VGAMPGKIIQ
```

```
                                       ******
         541                                                                              630
E.coli    KMAKVGVKNP LFLLDEIDKM SS.DMRGDPA SALLEVLDPE QNVAFSDHYL EVDYDLSDVM FVATSNSMN. IPAPLLDRME VIRLSGYTED
B.brevis  GMKQAGTINP VFLLDEIDKL AS.DFRGDPA SALLEVLDPN QNDKFSDHYI EETYDLSDVM FITTANSLDT IPRPLLDRME VISISGYTEL
S.cerevi  ALKKCQTQNP LILIDEIDKI GHGIHGDPS  AALLEVLDPE QNNSFLDNYL DIPIDLSKVL FVCTANSLET IPRPLLDRME VELTGYVAE
H.sapien  CLKKTKTENP LILIDEVDKI GR.GYQGDPS SALLELLDPE QNANFLDHYL DVPVDLSKVL FICTANVTDI IPEPLRDRME MINVSGYVAQ

         631                                                                              720
E.coli    EKLNIAKRHL LPKQIERNAL KKGELTVDDS AIIGIIRYYT REAGVRGLER EISKLCRKAV KQLLLDKSLK
B.brevis  EKLNILRGYL LPKQMEDHGL GKDKLQMNED AMLKLVRLYT REAGVRNLNR EAANVCRKAA KIIVGGEKKR
S.cerevi  DKVKIAEQYL VPSAKKSAGL ENSHVDMTED AITALMKYYC RESGVRNLKK HIEKIYRKAA LQVVKKLSIE DSPTSSADSK PKESVSSEEK
H.sapien  EKLAIAERYL VPQARALCGL DESKAKLSSD VLTLLIKQYC RESGVRNLQK QVEKVLRKSA YKIVSGEAES

         721                                                                              810
E.coli    .......... .......... .......... ..HIEINGD NLHDYLGVQR FDYGRADNEN RVGQVTGLAW TEVGGDLLTI ETACV......
B.brevis  .......... .......... .......... ...VVTAK  TLEALLGKPR YRYGLAEKKD QVGSVTGLAW TQAGGDTLNV EVSIL......
                                                      ●
S.cerevi  AENNEKSSSE KTKDNNSEKT SDDIEALKTS EKINVSIQOK NLQDFVGKPV YTTDRLYETT PPGVVMGLAW TNMGGCSLYV ESVLEQPLH.
H.sapien  .......... .......... .......... ...VEVTPE NLQDFVGKPV FTVERMYDVT PPGVVMGLAW TAMGGSTLFV ETSLRRPQDK

         811                                                                              900
E.coli    ...PGKGKL TYTGSLGEVM QESIQAALTV VRARAEKLGI NPDFYEKRDI HVHVPEGATP KDGPSAGIAM CTALVSCLTG NPVRADVAMT
B.brevis  ..AGKGKL  TLTGQLGDVM KESAQAAFSY IRSRASEWGI DPEFHEKNDI HIHVPEGAIP KDGPSAGITM ATALVSALTG IPVKKEVGMT
S.cerevi  ..NCKHPTF ERTGQLGDVM KESSRLAYSF AKMYLAQKFP ENRFEKASI  HLHCPEGATP KDGPSAGVTM ATSFLSLALN KSIDPTVAMT
H.sapien  DAKGDKDGSL EVTGQLGEVM KESARIAYTF ARAFLMQHAP ANDYLVTSHI HPHVPEGATP KDGPSAGCAI VTALLSLAMG RPVRQNLAMT

         901                                                                              990
E.coli    GEITLRGQVL PIGGLKEKLL AAHRGGIKTV LIPFENKRDL EEIPDNVIAD LDIHPVKRIE EVLTLALQNE PSGMQVVTAK*
B.brevis  GEITLRGRVL PIGGLKEKCM SAHRAGLTTI ILPKDNEKDI EDIPESVREA LTFYVEHLD  EVLRHALTKQ PVGDKK*
S.cerevi  GELTLTGKVL RIGGLRERAV AAKRSGAKTI IFPKDNLNDW EELPDNVKEG LEPLAADWYN DIFQKLFKDV NTKEGNSVWK AEFEILDAKK
H.sapien  GEVSLTGKIL PVGGIKEKTI AAKRAGVTCI VLPAENKKDF YDLAAFITEG LEVHFVEHYR EIFDIAFPDE QAEALAVER*
```

FIG. 2. Comparison of amino acid sequences of prokaryotic and eukaryotic protease La. The amino acid sequences were obtained by translation of the DNA sequences from the gram-negative bacterium E. coli,[21,22] the gram-positive bacterium Bacillus brevis,[15] the yeast Saccharomyces cerevisiae,[20] and human hippocampus.[17] The alignment was made using the program Pileup of GCG [J. Devereux, P. Haeberli, and O. Smithies, Nucleic Acids Res. 12, 387 (1984)], and adjustments were made by inspection. Homologies are emphasized by boldface letters for amino acids whenever at least three of the four amino acids are identical or when two identical amino acids occur in a position with four highly similar amino acids.

Such enhanced rates of proteolysis must be highly deleterious, because cells carrying additional copies of *lon* rapidly lose viability or acquire insertion elements on the plasmid that inactivate the cloned gene, especially at 37°. In fact, this property was long a barrier to cloning and production of large amounts of this protease until tightly regulated expression plasmids were used. Because both increasing and decreasing the level of protease La in cells affect proteolytic rates, this enzyme is rate-limiting for degradation of some classes of many unstable proteins *in vivo*.

Enzymatic Assays

Protein Degradation

Protease La is routinely assayed by measuring the hydrolysis of a radiolabeled protein, most frequently [³H]methylcasein, to acid-soluble radioactive peptides in the presence of ATP. This proteolytic activity requires ATP hydrolysis and, consequently, the enzyme is identified during purification as a peak of proteolytic activities seen with ATP present and not in its absence (Fig. 3). The typical reaction mixture (0.1 ml) contains 1–2 μg of the purified enzyme (or fractions generated during enzyme purification), 10 μg of [³H]methylcasein [5000–10,000 counts per minute (cpm)/μg], 50 mM Tris-HCl (pH 8), ATP (1 mM), and 5 mM MgCl$_2$. The ³H-labeled α- or β-casein is prepared by reductive methylation

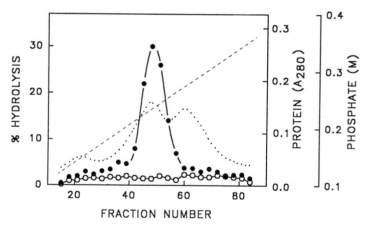

FIG. 3. Phosphocellulose chromatography. Fractions (5 ml) were collected at a flow rate of 50 ml/hr. Casein hydrolysis was assayed by incubating 10 μl of the fractions at 37° for 30 min in the presence (●) and absence (○) of 1 mM ATP. The dotted line indicates the protein profile and the dashed line shows the phosphate gradient.

using the method of Rice and Means.[38] Incubations are performed for 30 or 60 min at 37°, and the reactions are terminated by the addition of trichloroacetic acid (final concentration of 5%) plus 100 μg bovine serum albumin as a carrier. After cooling on ice for at least 1 hr, the samples are centrifuged to remove particulate material, and the radioactivity released into the acid-soluble fraction is determined by scintillation counting. The pure enzyme is routinely stored in 20–30% glycerol; however, the final concentration of glycerol in the assay should not exceed 5%, because high levels of glycerol can inhibit protein degradation.

Cleavage of Synthetic Fluorogenic Peptides

Another convenient assay method for this protease involves measuring the hydrolysis of the defined peptide substrates, glutaryl-Ala-Ala-Phe-methoxynaphthylamide (MNA), succinyl-Ala-Ala-Phe-MNA, or succinyl-Phe-Leu-Phe-MNA, which releases the highly fluorescent moiety, methoxynaphthylamine. The assay mixture (0.2 ml) typically contains 1–2 μg of protease La, 50 mM Tris-HCl (pH 7.9), 5 mM MgCl, 0.1–0.3 mM of the fluorogenic peptide dissolved in dimethyl sulfoxide (DMSO; final concentration of <5%), and 1 mM of ATP or AMPPNP (5-adenylylimidodiphosphate). This nonhydrolyzed ATP analog supports hydrolysis of small peptides (but not proteins) as well as or even more effectively than ATP, apparently because such analogs do not support self-hydrolysis of protease La. The reactions are carried out at 37° for 30–60 min and are terminated by the addition of 100 μl of 1% sodium dodecyl sulfate (SDS) and 1.2 ml of 0.1 M sodium borate (pH 9.1). Fluorescence is measured in a spectrofluorometer with an excitation wavelength of 335 nm and emission wavelength of 410 nm, and a standard curve is constructed with increasing amounts of methoxynaphthylamine hydrochloride.

Measurement of Generation of New Amino Groups

For studies of enzyme mechanism or specificity, it has also been useful to measure the hydrolysis of oligopeptides or proteins by quantitating the appearance of new amino groups with fluorescamine.[39] This approach allows quantitation of the exact number of peptide bonds cleaved, but necessitates the use of buffers lacking amino groups (50 mM HEPES, pH 7.9) in place of Tris-HCl. Otherwise, the reactions are run under conditions similar to those described above. The enzyme reaction is stopped by immersing tubes in ice and then adding 1.2 ml of borate buffer (pH 9.2)

---

[38] R. H. Rice and G. E. Means, *J. Biol. Chem.* **246,** 831 (1971).
[39] A. S. Menon, L. Waxman, and A. L. Goldberg, *J. Biol. Chem.* **262,** 722 (1987).

and 200 μl of fluorescamine dissolved in acetone (0.3 mg/ml). Fluorescence is measured with an excitation wavelength of 395 nm and an emission wavelength of 475 nm, and increasing concentrations of leucine are used to establish a standard curve. This assay is linear for at least 60 min with either large proteins (e.g., denatured bovine serum albumin, casein) or peptides (e.g., parathyroid hormone or glucagon) as substrate, and rates of cleavage are significantly faster with the peptide hormones. Similar data can be obtained using trinitrobenzene sulfonate in place of fluorescamine to monitor the appearance of free amino groups, but this approach is less sensitive and requires additional steps to block the preexisting lysine amino groups on protein substrates.[39] Use of ninhydrin or o-phthalaldehyde was found to be less reliable than fluorescamine due to its variable degree of reaction with diverse peptides.

Isolation of Degradation Products by Reversed-Phase Chromatography

Protease La degrades proteins and polypeptides to yield discrete peptide products about 5–15 residues long that are generated at a constant and reproducible rate during the initial time of incubation. Reaction mixtures (described above) can be quenched with an equal volume of 7.4 M guanidine hydrochloride and the peptide products separated by chromatography on a C_{18} reversed-phase column (see [23] in this volume for details).[32] Quantitation of the peptide products by integration of the absorbance signal allows precise measurement of the rate of peptide bond cleavage. Peptide products from substrates whose sequences are known can be isolated and identified by amino acid analysis, which allows positive identification of the sites of cleavage in protein and peptide substrates.

Measurement of ATPase Activity

Protease La also has an inherent ATPase activity, which releases orthophosphate from ATP and (less rapidly) from other nucleoside triphosphates. The enzyme shows no hydrolytic activity against ADP or AMP. The ATPase activity requires Mg^{2+} and is activated 2- to 5-fold by protein substrates [e.g., casein or reduced and alkylated bovine serum albumin (BSA)]. No such stimulatory effect is seen with tetrapeptide substrates, whose degradation also does not require ATP hydrolysis. ATPase activity can readily be measured in reaction mixtures under conditions identical to those used for protein or peptide hydrolysis. A variety of methods have been used to detect the products of hydrolysis. Measurement of hydrolysis of radiolabeled ATP allows for the greatest sensitivity, and the ATPase activity has been monitored by following the release of [^{32}P]phosphate from $[\gamma\text{-}^{32}P]ATP^{32}$ or by the conversion of $[^3H]ATP$ to $[^3H]ADP$, which can be resolved by thin-layer chromatography.[11]

A sensitive, nonradioactive assay that offers many advantages for routine use involves measurement of released inorganic phosphate by the method of Lanzetta et al.[40] This approach necessitates the use of buffers lacking phosphate. A malachite–molybdate solution is made fresh by mixing 600 μl of a 0.045% malachite green-HCl stock solution and 200 μl of a 4.2% (w/v) ammonium molybdate, 4 N HCl stock solution. This malachite–molybdate solution is allowed to stand at room temperature for 30 min, then filtered with Whatman (Clifton, NJ) paper prior to use. A 50-μl reaction mixture containing 50 mM Tris-HCl, pH 7.8, 5 mM MgCl$_2$, 1 mM ATP, 0.5 μg protease La is incubated at 37° for 20 min, then 800 μl of the malachite–molybdate solution is added, and the development of color is terminated by addition of 100 μl of 34% sodium citrate. The optical density at 660 nm is measured, and the amount of phosphate released is determined using a standard curve made with sodium phosphate. The degree of protein stimulation of the ATPase activity can easily be assayed by performing the ATPase reaction in the presence of 0.1 mg/ml β-casein.

Purification

Preparation of Cell Extracts

Initial methods for purification of this enzyme have been developed with wild-type E. coli, in which the protease is generally less than 1% of the cell protein.[41] However, enzyme isolation is most effective in cells carrying the lon gene on a multicopy plasmid. Because expression of high concentrations of the enzyme can reduce or prevent growth, derivatives of a plasmid have been made in which lon is under the control of a gratuitous inducer such as isopropylthiogalactoside (IPTG). The E. coli strain SG840 [F⁻ Δ(lac-pro)XIII recA56 argEam thi nal^r pSG10 (promoter^{tac}-lon, amp) pSG11 (lacI^q, tet)][34] is grown at 37° in Luria broth containing 50 μg/ml each of tetracycline and ampicillin until the culture attains an OD$_{600}$ of 1.5. IPTG is added to 1 mM and the culture is incubated for an additional 2 hr. The cells (10 g) are harvested, washed, and suspended in 30 ml of 0.1 M KH$_2$PO$_4$/K$_2$HPO$_4$ buffer (pH 6.5) containing 10% (v/v) glycerol, 1 mM dithiothreitol (DTT), and 1 mM EDTA. The

[40] P. A. Lanzetta, L. J. Alvarez, P. S. Reinach, and O. A. Candia, Anal. Biochem. **100**, 95 (1979).

[41] B. A. Zehnbauer, E. C. Foley, G. W. Henderson, and A. Markovitz, Proc. Natl. Acad. Sci. U.S.A. **78**, 2043 (1981).

cells are then frozen and stored at −70° until purification. The cells are then disrupted with a French press at 14,000 psi and centrifuged at 120,000 g for 4 hr. The resulting supernatant is dialyzed overnight at 4° against the same buffer.

Enzyme Purification

The crude cell extract (1.1 g) is loaded on a phosphocellulose column (5 × 15 cm) equilibrated with the phosphate buffer. After washing the column extensively with the buffer, proteins are eluted with a linear gradient of 0.1–0.4 M phosphate. The peak of ATP-dependent, casein-hydrolyzing activity is eluted at about 0.25 M phosphate (Fig. 3). The fractions with the highest activity are pooled, titrated to pH 7.5 with Tris–base, and dialyzed for 6 hr against 25 mM Tris-HCl buffer (pH 7.5) containing 20% (v/v) glycerol, 1 mM DTT, and 1 mM EDTA. The dialyzed proteins (9.7 mg) are loaded on a DEAE–Sepharose column (1 × 5 cm) equilibrated with the same buffer. After washing the column, proteins are eluted with a linear gradient of 0–0.3 M NaCl. The peak of activity is eluted at about 0.15 M NaCl (Fig. 4). The fractions with the greatest activity are pooled and concentrated by ultrafiltration on a YM30 membrane and then with a Centricon apparatus (Amicon, Danvers, MA).

The concentrated sample (5.6 mg) is then loaded on a Superose 6 column (1 × 30 cm) equilibrated with the Tris buffer. Fractions of 0.5 ml

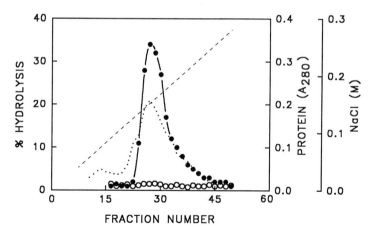

FIG. 4. DEAE–Sepharose chromatography. Fractions (1 ml) were collected at a flow rate of 20 ml/hr. Casein hydrolysis was assayed by incubating 5 μl of the fractions at 37° for 30 min in the presence (●) and absence (○) of 1 mM ATP. The dotted line indicates the protein profile; the dashed line shows the NaCl gradient.

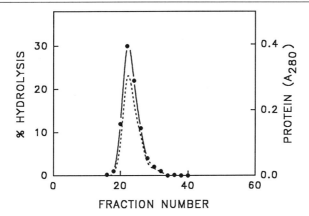

FIG. 5. Superose 6 chromatography. Fractions (0.5 ml) were collected at a flow rate of 20 ml/hr. Casein hydrolysis was assayed by incubating 3 μl of the fractions at 37° for 30 min in the presence of (●) and absence (○) of 1 mM ATP. The dotted line indicates the protein profile.

are collected at a flow rate of 20 ml/hr. A sharp, symmetric peak of activity is eluted in fractions at an apparent molecular weight of about 800,000 (Fig. 5). This peak of activity coincides exactly with the protein peak, as expected for a homogeneous preparation. The enzyme obtained appears as a single band of 87 kDa on SDS–polyacrylamide gels. Table I shows the summary of the purification protocol.

Enzyme Purification: Alternate Procedure

In *E. coli* bearing additional copies of *lon* on plasmids derived from pBR322 under the control of the *lon* promoter, this protease constitutes about 0.1% of the cell protein. Such cells do not grow well at 37° or above,

TABLE I
PURIFICATION PROTOCOL FOR PROTEASE La[a]

Purification step	Total protein (mg)	Total activity (units)	Specific activity (units/mg)	Recovery (%)
Crude extract	1100	—	—	—
Phosphocellulose	9.7	20.4	2.1	100
DEAE–Sepharose	5.6	15.1	2.7	74
Superose 6	4.3	13.3	3.1	65

[a] One unit is defined as 1 mg of casein hydrolyzed per hour.

but can be grown at 32° to an optical density of 1.0 at 600 nm. Protease La can be purified from frozen cells maintained at −70° by the following procedure. The frozen cells are thawed in 100 ml of cold 20 mM potassium phosphate, pH 7.2, 2 mM EDTA, 2 mM DTT, and 10% (v/v) glycerol. The cells are disrupted by a single pass through a French pressure cell at 20,000 psi and centrifuged at 30,000 g for 1 hr at 4°. The supernatant extract is treated with 0.16% polyethyleneimine with stirring and centrifuged at 20,000 g for 30 min to remove debris and nucleic acids. The supernatant solution is passed through a S-Sepharose column (5 × 15 cm). The ATP-dependent protease is eluted at about 0.25 M KCl in a 1500-ml linear gradient from 0 to 0.4 M KCl. The active fractions are pooled and the protease precipitated with 65% saturated ammonium sulfate. After dissolving the enzyme in 20 ml of Tris-HCl, pH 7.5, with EDTA, DTT, and glycerol as above, it is subjected to gel filtration on a Bio-Sil (Bio-Rad, Richmond, CA) TSK250 column (2.1 × 60 cm) in aliquots of 3 ml (about 50 mg total protein per run). The fractions containing maximal ATP-dependent protease activity are pooled and run over a Mono Q (Pharmacia, Piscataway, NJ) 10 HR anion-exchange column, and eluted with a linear gradient at about 0.3 M KCl. The most active fractions from the Mono Q column typically contain about 95% pure protease La.

Stability

The purified enzyme is stored at −70° in 20–30% glycerol, which is essential to prevent loss of activity. It may be stored in this fashion for many months without significant loss of activity, although with time it begins to exhibit ATP-independent activity, especially against small peptide substrates. For example, although ATP stimulates Glt-Ala-Ala-Phe-MNA hydrolysis 10- to 30-fold by the freshly purified enzyme, after storage for 6 months, only a 3- to 4-fold stimulation was found.[42] In the absence of glycerol, the pure enzyme loses activity rapidly at room temperature and especially rapidly at 37° due to denaturation (rather than autolysis). The presence of 1 mM ATP, AMPPNP, or ADP helps maintain the enzyme in a functional state for hours.

General Properties

Sequence of lon Gene

The amino acid sequence of protease La derived from the sequence of the *lon* gene[21,22,43] predicts a protein of 784–787 amino acids with a

[42] L. Waxman and A. L. Goldberg, *J. Biol. Chem.* **260,** 12022 (1985).
[43] H. Fischer and R. Glockshuber, *J. Biol. Chem.* **268,** 22502 (1993).

molecular mass of 88,000 Da (Fig. 2). The sequence exhibits no homologies to previously described families of proteases.[21] Because the enzyme appears to have serine at the active site,[44] this enzyme represents a novel family of serine proteases. The ATP-binding consensus sequence in protease La is similar to that found in domain I of the ATP-dependent Clp protease,[45] but these proteases are otherwise quite different in sequence. The sequence published by Chin et al.[21] appears to have contained errors leading to two frame shifts between residues 264–317 and 539–563,[22,43] and the revised sequences within the frame-shifted region are homologous to other bacterial ATP-dependent proteases. Other minor discrepancies in protease La sequences obtained by different investigators have not been resolved.

Protease Size and Structure

Protease La, with a predicted subunit size of 88,000 Da, migrates on SDS–polyacrylamide gels with an apparent molecular weight of 94,000. The native protein has displayed somewhat heterogeneous behavior on gel-filtration columns and behaves either as an apparent tetramer or as an octamer. The active form of the protease appears to be a multimer composed of at least four subunits, and many observations suggest that the subunits function cooperatively. For example, the inactive mutant protein, CapR9, can form a mixed multimer with the wild-type enzyme, and acts as a dominant inhibitor of both proteolytic and ATPase activities.[46] Early gel-filtration and sucrose gradient studies indicated that protease La from wild-type *E. coli* or cells carrying plasmid pJMC40 had a molecular mass of approximately 400–450 kDa, which suggested that La is a tetramer. However, more recent studies with protease La encoded on plasmid pSG11 indicate a much higher molecular weight of approximately 840 kDa, more consistent with an octameric structure (Fig. 6). The sedimentation coefficient of protease La determined by velocity sedimentation with exponential sucrose gradients (15–30%, w/v) was 24S. The Stokes radius, R_s, was determined from the elution pattern on Superose 6 by the method of Laurent and Killander[47] to be 8.5 to 9 nm. From the sedimentation coefficient and Stokes radius, and assuming a partial spe-

[44] A. Y. Amerik, V. K. Antonov, A. E. Gorbalenya, S. A. Kotova, T. V. Rotanova, and E. V. Shimbarevich, *FEBS Lett.* **287**, 211 (1991).

[45] S. Gottesman, C. Squires, E. Pickersky, M. Carrington, M. Hobbs, J. S. Mattick, B. Dalrymple, H. Kuramitsu, T. Shiroza, T. Foster, W. P. Clark, B. Ross, C. Squires, and M. R. Maurizi, *Proc. Natl. Acad. Sci. U.S.A.* **87**, 3513 (1990).

[46] C. H. Chung, L. Waxman, and A. L. Goldberg, *J. Biol. Chem.* **258**, 215 (1983).

[47] T. C. Laurent and J. Killander, *J. Chromatogr.* **14**, 317 (1964).

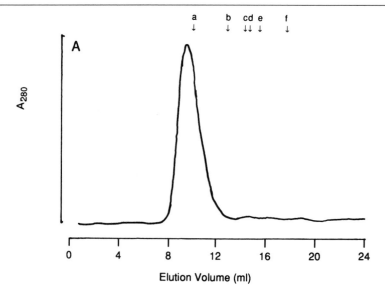

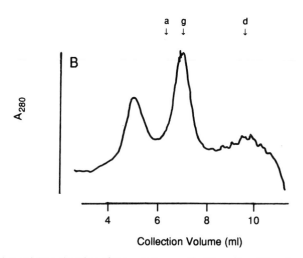

FIG. 6. Molecular weight estimation of the protease purified from *E. coli* bearing pSG11. The molecular weight of protease La was calculated from the observed Stokes radius and sedimentation coefficient. The standards used for calibration are (a) thyroglobulin; (b) apoferritin; (c) β-amylase; (d) alcohol dehydrogenase; (e) BSA; (f) carbonate dehydratase; (g) β-galactosidase. (A) Protease La, 4-mg aliquot, was eluted on a Superose 6 gel filtration column (1 × 30 cm) at 4° with 25 m*M* Tris-HCl, pH 7.5, 1 m*M* EDTA, 1 m*M* DTT, 20% glycerol. (B) A 15–30% (w/v) exponential sucrose gradient was prepared with the same buffer in a 12-ml centrifuge tube using a gradient maker with a 12-ml mixing chamber. A 100-μl sample

TABLE II

EFFECT OF NUCLEOTIDES ON HYDROLYSIS OF ^{125}I-LABELED BSA, [³H]CASEIN,
AND Glt-Ala-Ala-Phe-MNA BY PROTEASE La

Addition (0.5 mM)	Relative hydrolysis (%)		
	^{125}I-Labeled BSA	[³H]Casein	Glt-Ala-Ala-Phe-MNA
ATP	100	100	100
dATP	79	113	125
ATPγS	3	18	96
AMP-PCP	0	13	52
AMP-PNP	0	20	81
Diol ATP	—	10	88
CTP	23	32	68
UTP	16	8	56
GTP	2	12	52
None	0	5	12
ADP	0.3	—	5

cific volume of 0.73 cm³/ml, a molecular mass of 840 to 900 kDa was calculated by the method of Siegel and Monty,[48] which would suggest an octamer (or possibly a larger multimer). Possibly the discrepancy in the apparent size of the enzyme prepared from different strains may indicate a tendency of this enzyme to dimerize under certain conditions.

Enzymatic Activities

In the presence of ATP or another nucleotide, La hydrolyzes peptide bonds in a variety of proteins (especially unfolded polypeptides) as well as certain small peptides. In the absence of nucleotide, the enzyme has limited activity on very short peptides, but rapid degradation of both peptides and proteins requires ATP.[42,49] Nonhydrolyzable analogs of ATP, especially AMPPNP and to a lesser extent ATPγS and AMPPCP, also can support peptide degradation (Table II), presumably because they lock the enzyme into a fully active conformation, which is only temporary in the

[48] L. M. Siegel and K. J. Monty, *Biochim. Biophys. Acta* **112**, 346 (1966).
[49] A. L. Goldberg and L. Waxman, *J. Biol. Chem.* **260**, 12029 (1985).

---

containing 50 μg protease La and 50 μg β-galactosidase was layered on top of the gradient, and the tube centrifuged at 176,000 g for 16 hr at 4°. The sucrose gradient was then drawn from the bottom of the tube and 0.4-ml fractions were collected. The peak of peptidase activity corresponds to protease La, the first peak. The second peak is β-galactosidase.

ATP-stimulated reaction cycle (see below). That these nonhydrolyzable analogs are much less effective than ATP in supporting degradation of large proteins (Table II) is a function of several variables, not all of which are completely understood. For example, breakdown of denatured BSA appears to be completely dependent on ATP hydrolysis, whereas degradation of λ N protein[50] or partial degradation of β- or α-casein[51] occurs in the presence of AMPPNP at 20–40% of the rate observed with ATP. However, this dependence of proteolysis on ATP hydrolysis varies with the enzyme purification method, the length of storage of the pure enzyme, the temperature, and the ionic conditions.[42] For example, during the course of purification, ATP-independent activity increases. The addition of physiological concentrations of ADP (<100 mM) blocks this ATP-independent activity (which presumably represents spontaneous transition of the enzyme into its activated form. Therefore it seems likely that, in vivo, ATP hydrolysis is tightly linked to protein degradation by protease La; however, the peptide bond and ATP hydrolysis can be uncoupled in vitro (see below).

Protease La has an intrinsic ATPase activity and binds ATP and ADP very tightly.[52] Protein substrates (e.g., casein, globin, or denatured albumin) allosterically stimulate ATP hydrolysis 2- to 4-fold, whereas proteins that are poor substrates (e.g., native albumin or hemoglobin) have no such effect.[42] Inactivation of the proteolytic site by mutagenesis of the active site serine residue[43,44] or by reaction with a peptidyl chloromethane (L. Engler and A. L. Goldberg, unpublished observations, 1991) blocks proteolysis, but does not affect the ATPase activity or its stimulation by proteins. Thus, a protein substrate does not have to be degraded to stimulate ATP hydrolysis by the protease. The K_m for ATP for both ATP hydrolysis and protein breakdown is less than 50 μM,[42,52] which is far below intracellular levels of ATP (approximately 3 mM) under normal growth conditions or in carbon starvation (>2 mM).[53] Thus, changes in the cellular level of ATP seen in different physiological conditions should never limit protein breakdown in vivo, and ATP must be serving as a cofactor rather than a regulator of proteolysis.

Metal Ion Requirements

Protease La requires Mg^{2+} for breakdown of both proteins and ATP.[11,52] Maximal binding of ATP (one per subunit) occurs at approxi-

[50] M. R. Maurizi, J. Biol. Chem. **262**, 2696 (1987).
[51] T. Edmunds and A. L. Goldberg, J. Cell. Biochem. **32**, 187 (1986).
[52] A. S. Menon and A. L. Goldberg, J. Biol. Chem. **262**, 14921 (1987).
[53] A. C. St. John and A. L. Goldberg, J. Biol. Chem. **253**, 2705 (1978).

mately 100 μM Mg^{2+}, but much higher Mg^{2+} concentrations (1–2 mM) are necessary for maximal protein and peptide hydrolysis. Thus, Mg^{2+} must serve two roles: it serves as a cofactor for ATP binding and, at a low-affinity site, it allows proteolysis. Both Ca^{2+} and Mn^{2+} can replace Mg^{2+} to activate cleavage of fluorogenic peptides. This process is stimulated to the same extent by the same concentrations of all three metal ions.[42] However, Mn^{2+} or Ca^{2+} is less effective than Mg^{2+} in activating casein degradation. Ca^{2+} at higher concentrations (i.e., ≥ 2 mM) activates about 70% less effectively than Mg^{2+}, so that at 15 mM Ca^{2+} casein degradation is about 25% as fast as with Mg^{2+}. The failure of Mn^{2+} or Ca^{2+} to support rapid protein degradation is apparently because neither ion activates ATPase as well as Mg^{2+} does.

Substrate Specificity

In the presence of ATP and Mg^{2+} protease La displays a broad specificity in degrading denatured proteins and polypeptide substrates. Protease La hydrolyzes certain hydrophobic peptide substrates, such as Glutaryl-Ala-Ala-Phe-methoxynaphthylamide and succinyl-Phe-Leu-Phe-aminomethylcoumarin (MCA), which are also substrates for chymotrypsin, but it does not digest the characteristic peptide substrates of the trypsinlike or elastase-like enzymes (Table III).[42] La has a lower turnover number in degrading these fluorogenic substrates than does chymotrypsin. In addition, changing amino acids in the P2, P3, and P4 positions or blocking the amino terminus of fluorogenic peptides alters the rate of cleavage. Thus, binding of substrates at the active site may involve multiple interactions for both protein and peptide substrates.

TABLE III
HYDROLYSIS OF FLUOROGENIC SUBSTRATES BY PROTEASE La AND CHYMOTRYPSIN[a]

Substrate	Concentration (μM)	Relative rates of hydrolysis by	
		Protease La	Chymotrypsin
Glt-Ala-Ala-Phe-MNA	300	100	100
Suc-Ala-Ala-Phe-MNA	300	75	90
Glt-Gly-Gly-Phe-MNA	300	6	4
Glt-Ala-Ala-Ala-Mna	300	3	1
Ala-Ala-Phe-MNA	300	<1	2
Suc-Phe-Leu-Phe-MNA	100	137	22
MeO-Glt-Ala-Ala-Phe-MNA	100	<1	69

[a] Data from Waxman and Goldberg.[42]

Protease La degrades proteins and longer polypeptides into short peptides of 5–15 amino acids. The enzyme does not show amino- or carboxypeptidase activity. Analysis of cleavage sites in insulin and glucagon as well as in the physiological substrate, λ N protein, also indicates that La prefers to cleave after hydrophobic and nonpolar residues, such as leucine and alanine.[50] Degradation of proteins occurs in a highly reproducible pattern, indicating that protein substrates bind and are positioned for cleavage by specific interactions between the substrate and the enzyme. Because there is no clear consensus in the amino acid residues surrounding the cleavage sites in protein substrates, and cleavage does not occur after all available hydrophobic residues in protein and polypeptide substrates, the specificity is not determined strictly by the amino acids at the site of cleavage. As discussed below, there is evidence that proteins interact at allosteric sites in addition to the proteolytic active site, and analysis of the specificity of binding at these sites is likely to shed more light on the selectivity of proteolysis by La.

Covalent Inhibitors

Protease La behaves as a novel type of serine protease, sensitive to diisopropyl fluorophosphate (DFP) and to 3,4-dichloroisocoumarin,[42] which covalently inactivate the enzyme (Table IV). Treatment with DFP

TABLE IV

EFFECTS OF PROTEASE INHIBITORS ON HYDROLYSIS OF [³H]METHYL-CASEIN AND Glt-Ala-Ala-Phe-MNA BY PROTEASE La AND OF CASEIN BY CHYMOTRYPSIN[a]

		Activity (% control)		
		Protease La		Chymotrypsin
Inhibitor	Concentration	Glt-Ala-Ala-Phe-MNA	Casein	Casein
Z-Gly-Leu-Phe-CH₂Cl	50 μM	5	11	2
Z-Gly-Gly-Phe-CH₂Cl	50 μM	60	57	2
Z-Phe-CH₂Cl	50 μM	73	90	7
Tos-Lys-CH₂Cl	50 μM	107	84	91
Chymostatin	50 μM	96	63	0
Tosyl-Phe-CH₂Cl	250 μM	63	61	3
Phenylmethysulfonyl fluoride	5 mM	38	48	0
Diisopropyl fluorophosphate	10 mM	31	32	0

[a] Data from Waxman and Goldberg.[42]

inactivates hydrolysis of both peptides and proteins. The slow rate of inactivation by DFP is consistent with the rather low turnover number for peptide substrates and suggests that the active site serine is not activated to the same extent as in classical serine endopeptidases. Mutant enzymes in which Ser-679 was replaced by Ala lacked all peptidase activity but retained ATPase activity.[22,43]

Certain peptidylchloromethanes containing hydrophobic peptides, which are substrate analogs (e.g., Z-Gly-Leu-Phe-CH$_2$Cl), can rapidly inactivate protease La[42]; basic peptidylchloromethanes that inhibit trypsinlike enzymes have no such effect (Table IV). These observations are consistent with the preference of the enzyme for hydrophobic peptides as substrates and the pattern of bonds cleaved in λ N protein.[50] These agents rapidly inactivate protease La with the covalent incorporation of Z-Gly-Leu-Phe-CH$_2$Cl into each subunit (A. L. Goldberg and L. Engler, unpublished results, 1991). However, these peptidylchloromethanes can inhibit only in the presence of ATP or a nonhydrolyzed ATP analog. By contrast, ADP completely prevents this inactivation reaction, just as it inhibits peptide hydrolysis.[42] Thus ADP seems to prevent peptides from binding to the active site (see below), while ATP binding causes the formation (or exposure) of the active sites on the protease.

Both the proteolytic and ATPase activities of protease La are sensitive to high concentrations (5 mM) of sulfhydryl blocking agents, iodoacetamide, N-ethylmaleimide, and dansyl fluoride.[7,11] However, the sensitivity of protease La to such agents is much less than that of typical cysteine proteases; so, presumably, the important cysteinyl residues are not found in the active sites.

DNA Binding

An intriguing feature of protease La is that it has a high affinity ($K_a = 10^{-9} M$) for DNA. In fact, Zehnbauer *et al.*[41] originally isolated this protein as a DNA-binding protein, assuming it to be a regulator of gene expression. DNA, especially single-stranded DNA, activates the proteolytic activity and the protein-activated ATPase of protease La.[54] mRNA and tRNA have no effect. No sequence specificity for activation by DNA has been found, although poly(dT), poly(rC), and poly(rU) are effective, but poly(rA) is not. Polynucleotide length is a factor because poly(dT), but not (dT)$_{10}$, stimulates casein hydrolysis.[54] A stoichiometric binding of La to plasmid DNA has been demonstrated and has been shown to increase the V_{max} of the protease against casein. Moreover, these protein

---

[54] C. H. Chung and A. L. Goldberg, *Proc. Natl. Acad. Sci. U.S.A.* **79**, 795 (1982).

substrates promote the dissociation of the enzyme from DNA. The physiological significance of the effect of DNA remains uncertain; possibly, some of the protease *in vivo* is associated with DNA in the bacterial chromosome, where rapid degradation of damaged polypeptides or normal regulatory proteins may be particularly important for cell homeostasis.

Effects of Nucleotide Binding on Protease La

Studies of ATP binding by Menon and Goldberg indicate that each subunit binds an ATP molecule.[52,55] Even though there is only one type of subunit, four binding affinities were observed, and each ATP bound seems to reduce the affinity of the enzyme for the next nucleotide. In the presence of Mg^{2+} (100 μM) the protease exhibited a set of high-affinity binding sites for ATP ($K_a < 10^{-6}$ M) and a set of lower affinity sites ($K_a < 50$ μM), possibly suggesting a flip-flop mechanism in proteolysis. These different sites also show distinct specificities for a variety of ATP analogs and different requirements for Mg^{2+}. Although the binding of an ATP–Mg complex or nonhydrolyzable analog to the high-affinity site is sufficient to allow maximal rates of peptide hydrolysis, ATP occupancy of all four sites is necessary for maximal rates of protein degradation.[52] Once bound, the ATP molecules are hydrolyzed rapidly to ADP, which remains tightly associated with the enzyme, while the inorganic phosphates are released. ATP hydrolysis probably also causes Mg^{2+} release, because the ion has little affinity for ADP. ADP actually binds with higher affinity than does ATP, and because ADP does not support proteolysis, its release from the enzyme must be a rate-limiting step in proteolysis.[52,55]

Protein degradation by protease La is accompanied by rapid hydrolysis of ATP. It also hydrolyzes CTP, GTP, UTP, and dATP (at 0.5 mM concentration) at rates comparable to those seen with ATP, although only ATP and dATP support protein degradation efficiently (Table V).[11] No evidence for protein kinase activity or phosphoprotein intermediates has been found and the enzyme does not appear to be phosphorylated or adenylylated.[11] The ATPase activity is sensitive to vanadate but the critical inhibitory form is decavanadate and not orthovanadate, which is ineffective when free of decavanadate contamination. Decavanadate at 15 μM almost completely blocks the binding of ATP or ADP to the protease, and thus it prevents proteolysis. These results are consistent with the lack of phosphoprotein intermediates during the function of this ATPase, in contrast to many enzymes sensitive to orthovanadate.

[55] A. S. Menon and A. L. Goldberg, *J. Biol. Chem.* **262,** 14929 (1987).

TABLE V
RELATIVE RATES OF HYDROLYSIS OF DIFFERENT
NUCLEOTIDES AND EFFECTS ON CASEIN DEGRADATION BY
PROTEASE La[a]

Nucleotide	Nucleotide hydrolysis (relative rate)
ATP	100
ATP in the absence of Mg^{2+}	0
ADP	0
dATP	100
p[CH₂]ppA	0
p[NH]ppA	0
CTP	82
GTP	113
UTP	77

[a] Data from Waxman and Goldberg.[11]

The amino acid sequences of the peptide substrates hydrolyzed by protease La rule out certain models that had been proposed to explain the involvement of ATP in proteolysis. For example, peptide substrates lack amino acids that can be phosphorylated or adenylylated; they also lack free amino groups that might be covalently modified in a reaction analogous to ubiquitin conjugation. Because these peptides lack secondary structure, it is unlikely that protease La utilizes ATP to unfold these substrates. Therefore, such mechanisms cannot entirely account for the stimulation of peptide hydrolysis by ATP.

ADP cannot support proteolysis and is a potent inhibitor of the peptidase activity of the enzyme.[55] Moreover, ADP binds to the enzyme with even higher affinity than ATP ($K_m = 10 \ \mu M$). The rate-limiting step in proteolysis appears to be ADP release (see below). Menon and Goldberg[55] showed that the binding of a protein substrate promotes the release of these ADP molecules from the protease and thereby allows binding of new ATP moieties. When protease La degrades casein[51] or λ N protein,[50] the substrate is totally converted to small products (9–20 amino acids), and no large polypeptide fragments accumulate. With nonhydrolyzable analogs (e.g., AMPPCP) cleavage of the substrate is slower and polypeptide intermediates may accumulate in significant amounts. Thus, La appears to function processively (A. S. Menon and A. L. Goldberg, unpublished, 1990),[50,51] and the energy from ATP hydrolysis may serve to increase the rate at which substrates are repositioned on the enzyme to move appropriate cleavage sites into the proteolytic active site.

Nonhydrolyzable analogs of ATP and even inorganic pyrophosphate or triphosphate strongly stimulate the breakdown of fluorogenic peptides and other oligopeptides, in some cases even better than ATP. ATP binding and hydrolysis must, therefore, serve distinct functions in the proteolytic mechanism. Apparently, binding of the nonhydrolyzable ATP analog induces an active conformation of the enzyme in which the active site is accessible to oligopeptides and to some regions of protein substrates. Unlike protein substrates, peptides do not stimulate ATP hydrolysis, but actually inhibit this process.[42] Thus, the protein-stimulated ATP hydrolysis seems to be required for the multiple peptide cleavage steps necessary to generate small peptides. These findings with peptide substrates also indicate that ATP hydrolysis and peptide cleavage do not occur simultaneously in a concerted reaction. Instead, these processes probably occur sequentially as part of an ordered reaction cycle in which the ATP-consuming step follows the peptide bond hydrolysis.[49]

Coupling of ATP Hydrolysis to Proteolysis

The use of peptide substrates and ATP analogs has allowed the dissection of partial reactions of the degradation cycle, suggesting the following multistep process (Fig. 1).[1,42,55] (1) ATP binding occurs and leads to the formation or exposure of the proteolytic site. (2) Peptide bond cleavage occurs. (3) ATP hydrolysis is triggered by the presence of a protein substrate on an allosteric regulatory site (see below). (4) Because ADP does not support proteolysis, ATP hydrolysis leads to a temporary inactivation of the protease until the ADP molecule is replaced by another ATP molecule. In the degradation of large proteins, this activation–inactivation cycle must occur repetitively and rapidly until the protein is converted to small peptides.

When the enzyme works optimally, about two ATP molecules are consumed for each peptide bond cleaved in proteins.[39] This apparent stoichiometry is similar with several different polypeptide substrates (Fig. 7), but the coupling between these reactions is not tight. With either proteins or peptides as substrates, the peptide bond cleavage activity is maximal at a pH of 9.6. Activity falls off sharply above pH 10 and more gradually below pH 8.6. However, ATPase activity is optimal at pH 7.5,[42] such that two to three times more ATP is consumed in cleavage of peptide bonds at pH 7.5 than at pH 7.8 (the usual pH for assay). Under nonoptimal conditions (e.g., decreased pH or magnesium concentrations above 10 mM), more molecules of ATP (up to 15) are consumed for each peptide bond cleaved. Even under optimal conditions, protease La consumes almost as many ATP molecules in cleaving peptide bonds in proteins as

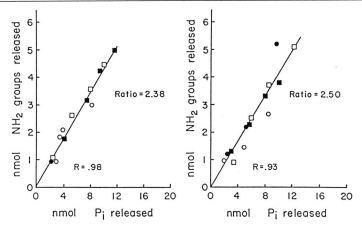

FIG. 7. Relationship between ATP and peptide bond hydrolysis of proteins by protease La. The stoichiometric ratio of ATP consumed per peptide bond cleaved was approximately 2 with different proteins. These data are with 10 mM Mg^{2+}. With 1 mM Mg^{2+} present, a ratio of 2.1 was found. Data from Menon *et al.*[39]

were utilized in the synthesis of the peptide bonds.[39] However, protease La generates oligopeptides generally greater than 1500 Da, and most of the subsequent degradative steps involved in the rapid breakdown of these oligopeptides to amino acids probably do not require ATP *in vivo*. Presumably, this large investment of ATP is advantageous in that it makes possible the rapid elimination of abnormal, potentially damaging polypeptides, and also helps prevent excessive degradation of desired cell proteins.[1]

Allosteric Activation by Protein Substrates

Probably, the most important regulatory feature of protease La is that it is activated on interaction with a protein substrate.[56] For example, protein substrates enhance the capacity of the enzyme to degrade small peptide substrates and to stimulate ATP hydrolysis, but nonsubstrate proteins (e.g., native BSA) do not. Because peptides and proteins are hydrolyzed at the same active site, the ability of proteins to increase activity against small peptide substrates suggests that the proteins interact allosterically with a regulatory region on the enzyme to cause activation of the peptidase site. Small peptide substrates that bind only to the active site have no such allosteric effect. A variety of kinetic studies indicates

[56] L. Waxman and A. L. Goldberg, *Science* **232**, 500 (1986).

that binding of a protein substrate leads to formation or exposure of active sites and is the critical step in triggering enzymatic attack on the bound proteins. This activation by potential substrates probably helps ensure that the protease remains in an inactive form *in vivo* until the enzyme binds an appropriate substrate.[56]

Protein substrates also activate protease La by another allosteric mechanism. As noted above, ADP has a higher affinity for La than does ATP[52] and, *in vivo,* an ADP molecule is probably bound to each subunit and thus inhibits proteolysis. Protein substrates, such as denatured albumin, induce the rapid release of bound ADP from the enzyme, whereas nondegraded polypeptides (e.g., native BSA) have no such activating effect. Thus, protein substrates not only block the inhibition of peptidase activity by ADP,[55] but also stimulate the binding of nucleotides to the protease. For example, in the absence of a protein, AMPPNP is associated with less than half of the binding sites, but in the presence of a protein substrate, this analog is bound to each subunit. This substrate-induced release of ADP and the stimulation of ATP binding constitute a protein-activated ATP–ADP exchange mechanism analogous to the GTP–GDP exchange mechanism of G proteins, although this regulatory ATP–ADP exchange is an inherent part of protease La and not a separate component. These enzymological properties seem to have evolved to increase the selectivity of intracellular proteolysis and to prevent inappropriate degradation of cell constituents.

Other Regulators of Protease Function

Although purified protease La catalyzes the degradation of model protein substrates to small peptides, proteolytic function *in vivo* against certain substrates requires the presence of certain heat-shock proteins that act as molecular chaperones. Degradation of abnormal puromycyl peptides and proteins containing canavanine *in vivo* is dependent on a functional *lon* gene and on expression of functional heat-shock genes, such as *dnaJ* and *dnaK*.[35,57,58] Specific abnormal proteins, e.g., a mutant form of alkaline phosphatase, PhoA61, is degraded rapidly *in vivo* by a process requiring *lon, dnaJ, dnaK,* and *grpE*.[59] Because purified La cannot digest this polypeptide, these chaperones may function as recognition factors for the protease or they may modify substrate conformation or aid the protease attack. Unfortunately, the way these chaperones and the protease

[57] D. B. Straus, W. A. Walter, and C. A. Gross, *Genes Dev.* **2,** 1851 (1988).
[58] T. A. Phillips, R. A. Van Bogelen, and F. C. Neidhardt, *J. Bacteriol.* **159,** 283 (1984).
[59] M. Y. Sherman and A. L. Goldberg, *EMBO J.* **11,** 71 (1992).

interact functionally has not been elucidated. Specific, highly unstable *in vivo* targets of protease La, such as SulA and λ N protein, do not require heat-shock proteins for rapid degradation.[60] Purified protease La can also recognize and completely degrade λ N protein *in vitro*.[50] The ability of the protease to interact with potential substrates must depend on the structure or flexibility of the target protein as well as the conformation of the protease. Auxillary proteins such as the heat-shock chaperones may be needed to promote a protease-susceptible conformation on certain proteins that are not in a form that can be directly recognized.

During the infection of *E. coli* with T4 phage, a specific inhibitor of protease La is produced, forming an inactive complex with the enzyme.[61-64] This protease inhibitor, PinA, is a polypeptide encoded on an early phage gene whose expression seems to account at least partially for reduction in the ability of the bacteria to degrade abnormal proteins after phage infection.[61] Purified PinA binds to protease La and inhibits basal ATPase as well as substrate-stimulated ATPase activity. The PinA–La complex is inactive against high molecular weight proteins but can degrade fluorogenic peptides or small proteins.[64] This failure to degrade large proteins may be due to the inhibition of the ATPase activity of the enzyme. Thus PinA defines another binding site on protease La distinct from the active site or the allosteric site at which protein substrates bind. No similar endogenous inhibitors of protease La have yet been found in exponentially growing *E. coli* cells.[65]

Acknowledgment

The authors are grateful to their many colleagues who have collaborated in studies of these enzymes, and to Mrs. Aurora P. Scott for assistance in the preparation of this manuscript. These studies were supported in part by grants from the National Institutes of Health (RO1 GM46147) and AMGEN to Alfred L. Goldberg, the Korea Science and Engineering Foundation (93-1-3) to Chin Ha Chung, and a fellowship from the National Institutes of Health (F32 GM16326) to Richard P. Moerschell.

[60] Y. Jubete, M. R. Maurizi, and S. Gottesman, submitted (1994).

[61] L. D. Simon, Tomezak, and A. L. St. John, *Nature (London)* **275**, 424 (1978).

[62] K. Skorupski, J. Tommaschewski, W. Ruger, and L. D. Simon, *J. Bacteriol.* **170**, 3016 (1988).

[63] J. Hilliard, M. R. Maurizi, and L. Simon, *J. Biol. Chem.* submitted (1994).

[64] J. Hilliard, M. R. Maurizi, and L. Simon, *J. Biol. Chem.* submitted (1994).

[65] C. H. Chung, H. E. Ives, S. Almeda, and A. L. Goldberg, *J. Biol. Chem.* **258**, 11032 (1983).

[26] Mitochondrial ATP-Dependent Protease from Rat Liver and Yeast

By STEFAN KUZELA and ALFRED L. GOLDBERG

Introduction

Proteins within mitochondria, like those in the cytosol, are subject to continual turnover, and this process appears important for both the formation and maintenance of these organelles.[1-4] Isolated mitochondria can degrade completely proteins found within this organelle by an ATP-dependent process.[1-4] Polypeptides with abnormal conformations and free subunits of multimeric enzymes are hydrolyzed particularly rapidly,[2] and the resulting amino acids exchange quickly with cytosolic amino acid pools.[2] A crucial enzyme in this process is an ATP-dependent, vanadate-sensitive endoprotease located in the mitochondrial matrix of mammals[5-7] and yeast.[8] In primary sequence[9,10] and catalytic mechanism,[12] this enzyme resembles closely the ATP-hydrolyzing protease La, the product of *lon* gene in *Escherichia coli*[13,14] and in other bacteria (for a review, see [25] in this volume).[14]

The mitochondrial protease is encoded on a nuclear gene. It is synthesized in the cytosol and taken up and processed to its mature form in the mitochondria.[11] The yeast[4,10] and human genes[11] have been sequenced. This enzyme appears necessary not only for rapid degradation of many organellar proteins,[10] but also for cell growth on fermentable substrates and for maintenance of mitochondrial DNA.[4,10] Like the *E. coli* prote-

---

[1] V. Luzikov, *Cell. Biol. Rev.* **25**, 249 (1991).
[2] M. Desautels and A. L. Goldberg, *Proc. Natl. Acad. Sci. U.S.A.* **79**, 1869 (1982).
[3] M. Desautels and A. L. Goldberg, *Proc. Biochem. Soc. Trans.* **13**, 290 (1985).
[4] C. Suzuki, K. Suda, N. Wang, and G. Schatz, *Science* **264**, 273 (1994).
[5] M. Desautels and A. L. Goldberg, *J. Biol. Chem.* **257**, 11673 (1982).
[6] S. Watabe and T. Kimura, *J. Biol. Chem.* **260**, 5511 (1985).
[7] S. Watabe and T. Kimura, *J. Biol. Chem.* **260**, 14498 (1985).
[8] E. Kutejova, G. Durcova, E. Surovkova, and S. Kuzela, *FEBS Lett.* **329**, 47 (1993).
[9] S. Kuzela and A. L. Goldberg, submitted for publication (1993).
[10] L. Van Dyck, D. A. Pearce, and F. Sherman, *J. Biol. Chem.* in press (1994).
[11] N. Wang, S. Gottesman, M. C. Willingham, M. M. Gottesman, and M. R. Maurizi, *Proc. Natl. Acad. Sci. U.S.A.* **90**, 11247 (1993).
[12] M. Desautels and A. L. Goldberg, submitted (1994).
[13] A. L. Goldberg, *Eur. J. Biochem.* **203**, 9 (1993).
[14] A. L. Goldberg, R. Moerschell, C. H. Chung, and M. Maurizi, this volume [25].

TABLE I
EFFECT OF INHIBITORS AND ADENINE NUCLEOTIDES ON CASEIN-DEGRADING ACTIVITY
OF YEAST AND RAT MITOCHONDRIAL ATP-DEPENDENT PROTEASES, AND
Escherichia coli PROTEASE La

Addition[a]	Concentration (mM)	2 mM ATP	Relative activity (%)		
			Yeast	Rat liver	*E. coli*
None		+	100	100	100
PMSF	5	+	48	51	55
NEM	1	+	<1.0	22.0	53
EDTA	25	+	<1.0	<0.1	<0.1
Orthovanadate	5	+	40	39	63
None		−	<1.0	<0.1	<0.1
ADP	2	−	<1.0	<0.1	<0.1
AMPCPP	2	−	<1.0	<0.1	10.0
AMPPCP	2	−	<1.0	<0.1	<0.1

[a] PMSF, Phenylmethylsulfonyl fluoride; NEM, *N*-ethylmaleimide; EDTA, ethylenediaminetetraacetic acid; AMPCPP, α,β-methylene adenosine 5'-triphosphate; AMPPCP, β,γ-methylene adenosine 5'-triphosphate.

ase,[15,16] the matrix enzyme is induced during heat shock[10] when cells and organelles accumulate damaged proteins. A number of early articles had reported other mitochondrial-associated proteases, but most of these activities were due to contaminating proteases that originate in lysosomes or mast cell granules.[2] Nevertheless, mitochondria do contain additional proteolytic activities[9] that are independent of ATP and that hydrolyze proteins, signal peptides, and peptides generated by the ATP-dependent protease. However, only the ATP-dependent protease has been characterized in depth and clearly shown to play a role in turnover of mitochondrial proteins *in vivo*.

The mitochondrial protease can be assayed by methods similar to those for the *E. coli* protease La, with which it shares many properties (Tables I and II). This enzyme was isolated first from rat liver[5] and then purified completely from bovine adrenal cortex.[7,8] However, this purification procedure is not applicable to the enzyme from mitochondria of yeast or rat tissues, which are the cells used most often in studies of mitochondrial physiology and biogenesis. Therefore, methods for purification to homogeneity of the ATP-dependent protease from *Saccharomyces cerevisiae*[9] and rat liver mitochondria[9] have recently been developed and are described

[15] S. A. Goff, L. D. Casson, and A. L. Goldberg, *Proc. Natl. Acad. Sci. U.S.A.* **81**, 6647 (1985).

[16] A. D. Grossman, J. W. Erickson, and C. A. Gross, *Cell (Cambridge, Mass.)* **38**, 383 (1984).

TABLE II
PROPERTIES OF ATP-DEPENDENT PROTEASE FROM LIVER MITOCHONDRIA AND PROTEASE
La FROM *Escherichia coli*[a]

Property	Liver mitochondria	*E. coli*
Action	Endoprotease	Endoprotease
pH optimum	7.8	7.8
ATP–Mg^{2+}	Essential	Essential
Inhibitors	DFP, PMSF, vanadate, SH blockers	DFP, PMSF, vanadate, SH blockers
ATPase activity	Essential, protein activated, vanadate sensitive	Essential, protein activated, vanadate sensitive
Glu-Ala-Ala-Phe-MNA	Hydrolyzed, protein activated	Hydrolyzed, protein activated
Ubiquitin	No effect	No effect
DNA activated	No	Yes
Multimeric mass	600 kDa	400 or 800 kDa
Subunit mass	106 kDa	87 kDa

[a] These findings on the mitochondrial enzyme are based on work of Desautels and Goldberg,[5] Watabe and Kimura,[6,7] and Kuzela and Goldberg.[9] For data on protease La, see Goldberg *et al.*[14] MNA, Methoxynaphthylamide; DFP, diisopropyl fluorophosphate; PMSF, phenylmethylsulfonyl fluoride.

below. The cloned human brain enzyme has been expressed in *E. coli* and was shown to function similarly as the rat and mitochondrial adrenal proteases.[11]

Isolation of Protease from Rat Liver Mitochondria

The following protocol was originally developed for large-scale isolation of the enzyme from mitoplasts. It can also be employed without modification if whole mitochondria devoid of most of the contaminating microsomes[17] are used as the starting material. If smaller amounts of the mitochondrial matrix (below 0.2 g) are processed, the Q Sepharose Fast Flow chromatography step can be omitted.

Enzyme Assays

Proteolytic activity is assayed by the conversion of ^{14}C-methylated casein to peptides soluble in trichloroacetic acid (see [25][14]). Enzyme preparations are incubated at 37° in a 0.15-ml mixture containing 50 mM Tris-HCl (pH 7.9), 10 mM MgSO$_4$, and 5 μg [^{14}C]methylcasein [15,000

[17] S. Kuzela, A. Mutvei, and B. D. Nelson, *Biochim. Biophys. Acta* **936**, 372 (1988).

counts per minute $(cpm)/\mu g]^{18}$ for 20–60 min, during which the reaction proceeds at a linear rate. The ATP-dependent proteolytic activity is determined by addition of 0.5 or 1.0 mM ATP. For enzyme preparations containing ATP, the ATP-independent activity is conveniently assayed by introducing 7 mU apyrase (Sigma, St. Louis, MO) into the reaction mixture to deplete it of ATP. Reactions are terminated by the addition of trichloroacetic acid and a carrier protein.[14] Alternatively, this protease can be assayed by using the fluorogenic peptide, glutaryl-Ala-Ala-Phe-methoxynaphthylamine[14] in the presence of ATP or β,γ-methylene adenosine 5′-triphosphate (AMPPCP). Its ATP-hydrolyzing activity can be assayed in a fashion similar to that for *E. coli* protease La.[14]

Preparation of Mitochondrial Matrix Fraction

Isolated rat liver mitochondria[17] or mitoplasts[19] are suspended in 220 mM mannitol, 1 mM EDTA, 20 mM Tris-HCl (pH 7.4) to give 80–100 mg protein/ml, and, after addition of glycerol to 20% (v/v), are stored at −70° (for up to several months). All of the subsequent operations are carried out at 0–4°. After thawing, the mitoplasts (6.5 g protein) are diluted to 50 ml with 20 mM Tris-HCl (pH 7.9), 20% glycerol, 1 mM dithiothreitol, 0.1 mM EDTA, and are disrupted by sonication for 5 min using a Branson Sonifier (Branson Ultrasonic Corp., Danbury, CT) at the maximum output. To reduce the ATP content in the resulting suspension of mitochondria, 2 mM MgSO$_4$ is added to the preparation to allow ATP hydrolysis. After 15 min at 4°, the Mg^{2+} is chelated by adding 5 mM EDTA, and the suspension centrifuged for 1 hr at 120,000 g. The supernatant (about 3.5 g protein) represents the mitochondrial matrix fraction.

Anion-Exchange Chromatography

The matrix is applied to a 50-ml Q Sepharose Fast Flow (Pharmacia LKB, Piscataway, NJ) column equilibrated with buffer A, which contains 10 mM Tris-HCl (pH 7.9), 20% (v/v) glycerol, 1 mM dithiothreitol, and 0.5 mM EDTA. The column is washed with 50 ml of this buffer and then eluted with a 350-ml linear NaCl gradient (0 to 0.5 M) in buffer A. The ATP-stimulated proteolytic activity is recovered between 230 and 300 mM NaCl. The active fractions are pooled and dialyzed overnight against buffer A. (This preparation can be stored for several days at −70° without appreciable loss of activity.) The pooled material (about 90 mg protein)

[18] N. Jentof and D. G. Dearborn, *J. Cell Biol.* **254**, 4359 (1979).
[19] C. A. Schnaitman and J. W. Greenawalt, *J. Cell Biol.* **38**, 158 (1969).

is then applied to a FPLC (fast protein liquid chromatography) Mono Q HR 10/10 column (Pharmacia LKB) equilibrated with buffer A. The bound material is eluted with a 20-ml linear NaCl gradient (0 to 0.25 M) followed by a 30-ml isocratic wash with 0.25 M NaCl and finally with a 40-ml linear NaCl gradient (0.25 to 0.5 M) in buffer A. The peak of the ATP-dependent protease activity is eluted at 375 mM NaCl. The active fractions (containing about 3 mg protein) are then pooled, dialyzed overnight against buffer A supplemented with 0.5 mM ATP, and, if necessary, stored at −70°. After this purification step, the casein-degrading activity is almost completely ATP dependent.

Phosphocellulose Chromatography

Potassium phosphate is added to the pooled fractions to give a final concentration of 10 mM (pH 7.9), and the sample is then applied to a 2-ml phosphocellulose column (PC 11, Whatman, Clifton, NJ) equilibrated with buffer A containing 0.5 mM ATP and 10 mM potassium phosphate (pH 7.9). The presence of ATP in these solutions is crucial, because the simplicity of this solution procedure is based on changes induced by this nucleotide in the behavior of the protease on phosphocellulose chromatography. The column is washed with 10 ml of the binding buffer, then by 7 ml of the binding buffer containing 0.1 M potassium phosphate, and finally eluted with a linear K_2HPO_4 gradient (0.1 to 0.4 M) in buffer A containing 0.5 mM ATP. The peak of the ATP-dependent protease activity is eluted at about 250 mM potassium phosphate. The most active fractions contain the pure enzyme, as shown by sodium dodecyl sulfate polyacrylamide gel electrophoresis (SDS–PAGE).

Activity and Composition of Rat Liver Protease

A quantity of 1 μg of pure protease hydrolyzes 30 nmol of ^{14}C-methylated casein[18] to trichloroacetic acid-soluble fragments in 1 hr, assayed as described previously.[2,5] The total yield of enzyme from 6.5 g of mitoplasts is about 0.2 mg. The molecular mass of the isolated enzyme estimated by gel filtration is about 600 kDa. In SDS–PAGE, the purified enzyme gives two protein bands corresponding to peptides of 100 and 80 kDa (Fig. 1, lane 3). The relative amounts of proteins in the two bands vary in different enzyme preparations. The smaller of the two polypeptides is absent in intact mitochondria, and it appears to be generated during the isolation of the enzyme by cleavage of the authentic 100-kDa subunit near its N terminus.[9] Thus, like the bovine adrenal protease,[6,7] rat liver mitochondrial enzyme is a hexamer consisting of identical 100-kDa subunits.

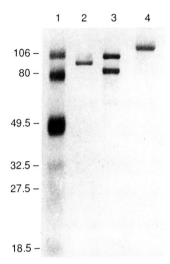

FIG. 1. Comparison of bacterial protease La (lane 2) with rat liver (lane 3) and yeast (lane 4) mitochondrial ATP-dependent proteases by SDS–polyacrylamide gel electrophoresis. Lane 1 is prestained molecular weight standards (Bio-Rad). Proteins were stained with silver.

Isolation of Protease from Yeast Mitochondria

Because *S. cerevisiae* cells contain potent vacuolar proteases that often contaminate and damage the preparations of isolated mitochondria, the ATP-dependent protease is best isolated from the mutant strain *pep949*, which is deficient in the major vacuolar proteases.[20] Also, the ATP-dependent activity of the matrix from poorly coupled mitochondria is very low; therefore, only organelle preparations exhibiting good respiratory control should be used for enzyme purification.

Preparation of Mitochondria and Matrix Fraction

Mitochondria are isolated[21] from spheroplasts[22] of yeast grown to early stationary phase in yeast extract peptone (YEP) medium[23] with 2% glucose. The mitochondria are suspended to about 40 mg protein per ml of the preparation medium supplemented with 20% (v/v) glycerol, frozen in liquid nitrogen, and stored at −70° for up to several months. All of the subsequent steps are carried out at 0–4°. The mitochondria (about 2 g

---

[20] E. W. Jones, this series, Vol. 194, p. 428.
[21] T. Yasuhara and A. Ohashi, *Biochem. Biophys. Res Commun.* **144,** 277 (1987).
[22] A. Ohashi and G. Schatz, *J. Biol. Chem.* **255,** 7740 (1980).
[23] B. Feinberg and C. S. McLaughlin "Yeast, a practical approach" (I. Campbell and J. H. Duffus, eds.), p. 147. IRL Press, Oxford, 1988.

protein) are thawed and diluted to about 20 mg protein/ml by buffer B [20 mM Tris-HCl, 0.1 mM EDTA, 1 mM dithiothreitol, 10% (v/v) glycerol (pH 7.9)]. After adding N-tosyl-1-phenylalanine chloromethyl ketone (TPCK) and N^α-p-tosyl-1-lysine chloromethyl ketone (TLCK) to 0.1 mM and aprotinin to 0.1 mg/ml, the matrix fraction is released by a 15-min treatment with 0.16 mg Lubrol WX per 1 mg protein. The suspension is centrifuged for 30 min at 150,000 g, and the pelleted membranes are twice washed with 50 ml of buffer B. The combined supernatants from the three centrifugations are referred to as the matrix fraction.

Q Sepharose Fast Flow and Hydroxylapatite Chromatography

The matrix fraction (about 0.7 g protein) is passed through a 50-ml column of Q Sepharose Fast Flow (Pharmacia LKB) equilibrated with buffer B. The flow-through fraction (about 150 mg protein), which contains the ATP-stimulated protease activity, is brought to 10 mM by addition of potassium phosphate. It is applied to a 20-ml column of hydroxylapatite (Bio-Rad, Richmond, CA) equilibrated with buffer B containing 10 mM potassium phosphate. The bound proteins are eluted from the column with a linear K_2PO_4 gradient (10 to 500 mM) in buffer. The ATP-stimulated protease peak is eluted between 0.2 and 0.3 M K_2HPO_4.

Chromatography on the FPLC Mono Q and Superose 6 Columns

The active fractions (about 40 mg protein) are dialyzed against buffer B and applied to the FPLC Mono Q HR10/10 column (Pharmacia LKB) equilibrated with the same buffer. The column is developed with 20 ml of a linear NaCl gradient (0 to 0.25 M) followed by a 30-ml isocratic elution with 0.25 M NaCl and then with a 30-ml linear NaCl gradient (0.25 to 0.5 M) in buffer B. The protease activity, which is completely ATP dependent at this stage, is eluted by about 0.4 M NaCl. The active fractions (0.5 to 0.6 mg protein) are concentrated by Centricon (Amicon, Danvers, MA) and applied to the FPLC Superose 6 column in the same buffer. The active fractions in the gel filtration step contain the pure protease, as confirmed by PAGE. Its elution volume corresponds to a globular protein with a molecular mass of 700 kDa.

Activities and Composition of Yeast Protease

Of the resulting protease, 1 μg under standard conditions hydrolyzes about 60 nmol of ^{14}C-methylated casein[18] to trichloroacetic acid-soluble fragments in 1 hr.[2] On SDS–PAGE, the purified enzyme contains a major protein band corresponding to a molecular mass of 120 kDa (Fig. 1, lane 4), plus a cross-reacting 100-kDa band, which appears to be derived from the larger band by proteolysis during the course of purification. This result, along with its behavior on gel filtration, indicates that the yeast mitochondrial ATP-dependent protease is a hexamer consisting of 120-kDa subunits.

Properties of Mitochondrial ATP-Dependent Proteases

The enzymatic properties of the bovine, rat, and yeast mitochondrial ATP-dependent proteases closely resemble each other and those of the homologous bacterial protease La (*lon*) (Tables I and II). These enzymes all show similar sensitivities to covalent inhibitors and a similar dependence on ATP hydrolysis for degradation of large polypeptides. All these proteases have an inherent ATPase activity that is activated allosterically by protein substrates[9] and is linked to proteolysis.[13,14] Casein hydrolysis by the mitochondrial proteases is not supported by ADP or by nonhydrolyzable ATP analogs[5,6] and requires Mg^{2+}.[5,6] In fact, protein degradation by the mitochondrial enzymes appears to be even more tightly linked to ATP hydrolysis than the bacterial enzyme. In addition, the rat liver and yeast enzymes rapidly degrade certain small hydrophobic substrates,[9] the same ones degraded by the E. coli protease[14] [e.g., glutaryl-Ala-Ala-Phe-methoxynaphthylamide (MNA) and succinyl-Phe-Leu-Phe-MNA], and the cleavage of these small peptides requires ATP binding, but not its hydrolysis. By contrast, the mitochondrial enzyme does not degrade typical peptide substrates of trypsin or elastase-like proteases. These fluorogenic peptides, which are commercially available, are very convenient for enzyme isolation or study (see Goldberg et al.[14]). The capacity of the enzyme to hydrolyze these hydrophobic peptides is stimulated severalfold on binding of a protein substrate to an allosteric regulatory site.[9,13]

One apparent difference between the mitochondrial enzymes and the E. coli protease is in their subunit composition. Unlike protease La, which contains four, or perhaps eight,[14] 87-kDa subunits, the mitochondrial enzymes act like hexamers composed of 100- or 120-kDa subunits. (Electrophoretic migration rates of the two mitochondrial ATP-dependent proteases and the bacterial protease in SDS–polyacrylamide gels are compared in Fig. 1.) The significance of this difference in subunit organization is uncertain, because the mitochondrial and E. coli enzymes are homologous, cross-react immunologically,[9] function through a similar reaction cycle, and play analogous roles *in vivo*.

Acknowledgments

This work was supported by grants from the National Institutes of Health (RO1 GM46147), Muscular Dystrophy Association, and National Aeronautics and Science Administration to ALG, and part of the work of SK was supported by the National Institutes of Health grant (RO3 TW169) to Dr. Fred Sherman.

[27] Omptin: An *Escherichia coli* Outer Membrane Proteinase That Activates Plasminogen

By WALTER F. MANGEL, DIANA L. TOLEDO, MARK T. BROWN, KIMBERLY WORZALLA, MIJIN LEE, and JOHN J. DUNN

Introduction

An *Escherichia coli* outer membrane proteinase that can activate human plasminogen to plasmin was discovered and partially characterized in 1980.[1] In 1988 the gene for an *E. coli* outer membrane proteinase that cleaves recombinant T7 RNA polymerase at dibasic residues was identified as *ompT*.[2] Previous studies had shown that OmpT, formerly called protein a[3] and protease VII,[4–6] is a trypsinlike protease with a narrow specificity. Interestingly, only OmpT$^+$ strains could activate plasminogen.[2] The nucleotide sequence of *ompT* and the N-terminal sequence of the mature OmpT protein indicated that OmpT is synthesized as a preprotein of 317 amino acids and during export to the outer membrane a signal sequence 20 amino acids long is removed.[7] Comparison of the nucleotide sequence of *ompT* to consensus sequences in various classes of proteases indicated there is no homology, and thus omptin must belong to a new class of proteinase. Inactivation of the *ompT* gene has no discernible effect on *E. coli*,[2,8] so the function of this gene is unknown. The *ompT* gene product has been shown to cleave several recombinant proteins expressed in *E. coli*, and thus OmpT$^-$ strains have been constructed that have proved very useful for the expression of recombinant proteins.[8,9]

There are other genes homologous to the *ompT* gene. Of the 278 amino acid residues encoded by the *Salmonella typhimurium* E protein gene,

[1] S. P. Leytus, L. K. Bowles, J. Konisky, and W. F. Mangel, *Proc. Natl. Acad. Sci. U.S.A.* **78,** 1485 (1981).
[2] J. Grodberg and J. J. Dunn, *J. Bacteriol.* **170,** 1245 (1988).
[3] W. C. Hollifield, Jr., E. H. Fiss, and J. B. Nielands, *Biochem. Biophys. Res. Commun.* **83,** 739 (1981).
[4] K. Sugimura and T. Nishihara, *J. Bacteriol.* **170,** 5625 (1988).
[5] K. Sugimura and N. Higashi, *J. Bacteriol.* **170,** 3650 (1988).
[6] K. Sugimura, *Biochem. Biophys. Res. Commun.* **153,** 753 (1988).
[7] J. Grodberg, M. D. Lundrigan, D. L. Toledo, W. F. Mangel, and J. J. Dunn, *Nucleic Acids Res.* **16,** 1209 (1988).
[8] F. Baneyx and G. Georgiou, *J. Bacteriol.* **172,** 491 (1990).
[9] F. W. Studier, A. H. Rosenberg, J. J. Dunn, and J. W. Dubendorff, this series, Vol. 185, p. 60.

135 (48.2%) are identical in OmpT.[10] Furthermore, the *Salmonella typhimurium* E protein can localize in the outer membrane of *E. coli* and can cleave the T7 RNA polymerase. The organism that causes plague, *Yersinia pestis*, had been shown to contain a plasmid whose expression correlates with virulence.[11,12] The plasmid contains a gene, *pla*, for a plasminogen activator activity.[13] The nucleotide sequence of the *pla* gene was determined and when translated into amino acids indicated 47.5% homology with OmpT and 71% homology with protein E from *S. typhimurium*.[14] Inactivation of the plasmid gene *pla* increased the median lethal dose of the bacteria for mice by 10^6-fold.[15] Studies are being done to determine whether the expression of omptin correlates with the pathogenic potential of strains of *E. coli* that infect humans.[16] Both human plasminogen[17,18] and plasmin[19,20] can bind specifically to the surface of bacteria, so an entire fibrinolytic enzyme system can assemble there, perhaps in ways similar to the assembly of plasminogen activation systems on the surface of eukaryotic cells and structures.[21,22]

The Nomenclature Committee of the IUBMB has recommended the name omptin (EC 3.4.21.87) be used in place of OmpT, protein a, protease VII, and Pla.

Materials and Methods

Human Plasminogen. Human plasminogen has been purified by the method of Deutsch and Mertz[23] with slight modification.[24]

[10] J. Grodberg and J. J. Dunn, *J. Bacteriol.* **171,** 2903 (1989).
[11] R. Ben-Gurion and A. Shafferman, *Plasmid* **5,** 183 (1981).
[12] D. M. Ferber and R. R. Brubaker, *Infect. Immun.* **31,** 839 (1981).
[13] E. D. Beesley, R. R. Brubaker, W. A. Jansen, and M. J. Surgalla, *J. Bacteriol.* **165,** 19 (1967).
[14] O. A. Sodeinde and J. D. Goguen, *Infect. Immun.* **57,** 1517 (1989).
[15] O. A. Sodeinde, Y. V. B. K. Subrahmanyam, K. Stark, T. Quan, Y. Bao, and J. D. Goguen, *Science* **258,** 1004 (1992).
[16] M. D. Lundrigan and R. M. Webb, *FEMS Microbiol. Lett.* **76,** 51 (1992).
[17] J. Parkkinen and T. K. Korhonen, *FEBS Lett.* **250,** 437 (1989).
[18] M. Ullberg, G. Kronvall, I. Karlsson, and B. Wiman, *Infect. Immun.* **58,** 21 (1990).
[19] T. A. Broeseker, M. D. P. Boyle, and R. Lottenberg, *Microb. Pathog.* **5,** 19 (1988).
[20] R. Lottenberg, C. C. Broder, and M. D. P. Boyle, *Infect. Immun.* **55,** 1914 (1987).
[21] M. P. Stoppelli, C. Tacchetti, M. V. Cubellis, A. Corti, V. J. Hearing, G. Cassani, E. Appella, and F. Blasi, *Cell (Cambridge, Mass.)* **45,** 675 (1986).
[22] W. F. Mangel, *Nature (London)* **344,** 488 (1990).
[23] D. G. Deutsch and E. T. Mertz, *Science* **170,** 1095 (1970).
[24] S. P. Leytus, G. P. Peltz, H.-Y. Liu, J. F. Cannon, S. W. Peltz, D. C. Livingston, J. R. Brocklehurst, and W. F. Mangel, *Biochemistry* **20,** 4307 (1981).

T7 RNA Polymerase. The T7 RNA polymerase, the T7 gene 1 protein, has been purified as described in Grodberg and Dunn.[2]

Protein Concentration. The concentration of protein has been determined from the absorbance at 260 and 280 nm using a nomograph.[25]

Molecular Mass of Omptin. The calculated molecular mass of pro-omptin is 35,562 Da. The calculated molecular mass of omptin, pro-omptin minus the first 20 NH_2-terminal amino acids, is 33,477 Da.

Omptin Molar Extinction Coefficient. A molar extinction coefficient at 280 nm of 74,960 has been calculated using the method of Gill and von Hippel.[26]

Units of Enzyme Activity. One unit of enzyme activity is the amount of omptin required to activate 1 μmol of plasminogen to plasmin per minute. Specific activity is units/milligram of protein.

Isoelectric Point. The pI has been calculated to be 5.62.

Substrate for Plasmin. (Ile-Pro-Arg-NH_2)$_2$-rhodamine has been synthesized as described in Leytus *et al.*[27] A stock solution 10 mM in dimethyl sulfoxide (DMSO) is stored at $-20°$.

Growth Medium for E. coli. The medium in which *E. coli* is grown is M9-TB medium containing 50 μg/ml kanamycin.

SDS–PAGE. Samples are diluted into $4\times$ SDS–PAGE sample buffer such that the final concentrations are 0.01 M Tris-HCl (pH 8.0), 0.001 M EDTA, 2.5% (w/v) SDS, 0.01% (w/v) bromphenol blue, and 5% (v/v) mercaptoethanol. Unless stated otherwise, the solutions are placed for 90 sec in a boiling water bath prior to electrophoresis on a Pharmacia (Piscataway, NJ) PhastSystem.

Assay Procedures

Assay for Plasminogen Activator Activity

The plasminogen activator activity in an extract of omptin is determined by incubating 5 μl of extract with 95 μl 0.01 M HEPES (pH 7.4) containing 2.7 μM plasminogen. After 5 min at 40°, 85 μl is removed and added to 915 μl of 25 μM (Ile-Pro-Arg-NH_2)$_2$-rhodamine in 0.01 M HEPES (pH 7.4). The increase in fluorescence with time is followed with an SLM 500C (SLM Instruments, Urbana, IL) spectrofluorometer. The excitation and emission wavelengths are 492 and 523 nm, respectively, both set with a bandwidth of 5 nm. The change in fluorescence (ΔF) is equal to the

[25] O. Warburg and A. Christian, *Biochem. Z.* **310,** 384 (1942).
[26] S. G. Gill and P. H. von Hippel, *Anal. Biochem.* **182,** 319 (1989).
[27] S. P. Leytus, D. L. Toledo, and W. F. Mangel, *Biochim. Biophys. Acta* **788,** 74 (1984).

magnitude of the fluorescence in the sample cuvette minus the magnitude of the fluorescence in a reference cuvette containing the same concentrations of (Ile-Pro-Arg-NH)$_2$-rhodamine and plasminogen, but in which the plasminogen has been incubated for 5 min without omptin.

Assay for T7 RNA Polymerase Cleavage

Assays are performed in 60 μl of reaction buffer consisting of 10 mM Tris-HCl (pH 8.0), 20 mM NH$_4$Cl, 10 mM MgCl$_2$, 5 μg of T7 RNA polymerase, and either 0.25 A_{600} units of live cells or aliquots of a protein fraction. After 1 hr at 37°, the reaction is chilled on ice and then centrifuged at 8000 g in an Eppendorf centrifuge for 2 min. A portion of the supernatant is mixed with SDS–PAGE sample buffer and the reaction mixture is fractioned by SDS–PAGE. The initial cleavage of T7 RNA polymerase by omptin occurs between Lys-172 and Arg-173 to yield fragments of 20 and 80 kDa.[2]

Assay for Inhibitors of Omptin

To 50-μl solutions of 0.05 M HEPES (pH 6.5) containing different concentrations of inhibitor are added omptin and either 0.05, 0.75, or 0.133 mg/ml human plasminogen. After 5 min at 40°, 35-μl aliquots are removed and added to 965-μl aliquots containing 0.05 M HEPES (pH 7.4) and 100 μM (Ile-Pro-Arg-NH)$_2$-rhodamine. The increase in fluorescence is then measured. Control experiments are performed to determine whether the putative inhibitor inhibits plasmin under the conditions of the assay. One-ml solutions of 0.05 M HEPES (pH 7.4), 1 mg/ml bovine serum albumin (BSA), 100 μM (Ile-Pro-Arg-NH)$_2$-rhodamine, plasmin, and various concentrations of inhibitor are incubated at room temperature while the increase in fluorescence is measured.

Assays for Cleavage of Synthetic, Fluorogenic Substrates

These new assays are based on the observations that omptin cleaves between basic amino acids and that aminopeptidases will not cleave after a blocked amino acid. Thus the compounds benzyloxycarbonylarginylarginine 4-methyl-7-coumarylamide or Cbz-Arg-Arg-NHMec and Cbz-Ala-Lys-Arg-NHMec will not be cleaved by aminopeptidase M. In the presence of omptin, either substrate is cleaved to Arg-NHMec. When Arg-NHMec is incubated with aminopeptidase M, the highly fluorescent cleavage product aminomethylcoumarin is formed. This two-step procedure to measure the activity of omptin can be combined into a coupled assay or separated in an uncoupled assay.

In the coupled assay, 5 μl of omptin is added to 995 μl of 0.05 M HEPES (pH 7.4) containing 0.02 U/ml aminopeptidase M from hog kidney and either 0.02 mM Cbz-Arg-Arg-NHMec or Cbz-Ala-Lys-Arg-NHMec. The increase in fluorescence with time is followed at room temperature with an SLM 500C spectrofluorometer. The excitation and emission wavelengths are 380 and 440 nm, respectively, both set with a bandwidth of 5 nm. The change in fluorescence (ΔF) is equal to the magnitude of the fluorescence in the sample cuvette minus the magnitude of the fluorescence in a reference cuvette treated similarly but without omptin. Under conditions in which the concentration of Arg-NHMec is much less than its K_m for aminopeptidase M, its rate of hydrolysis is directly proportional to its concentration, which is directly proportional to the concentration of omptin.

An example of the data from a coupled assay and the analysis is shown in Fig. 1. The primary data, ΔF versus t, yield a parabolic curve (Fig. 1A). The rates of formation of Arg-NHMec are obtained from the slopes of plots of $\Delta F'/\Delta t$ versus $\bar{t}$ (Fig. 1B), where $\Delta F'/\Delta t$ represents an increase in F ($\Delta F' = \Delta F_2 - \Delta F_1$) over a given fixed time interval ($\Delta t = t_2 - t_1$) and where $\bar{t}$ is the mean of the time interval $(\bar{t}) = (t_1 + t_2)/2$.[28] When the slopes of these straight lines are plotted versus the concentration of omptin, a straight line passing through points 0, 0 is obtained (Fig. 1C). An alternative to the plot in Fig. 1B is a plot of ΔF versus t^2.[29]

In the uncoupled assay, 50 μl 0.05 M 2-(N-Morpholino) ethanesulfonic acid (MES) (pH 6.0) is incubated with omptin and 0.4 mM Cbz-Arg-Arg-NHMec or Cbz-Ala-Lys-Arg-NHMec. After 10 min at 37°, 40 μl is removed and added to 960 μl of 0.05 M HEPES (pH 7.4) containing 0.02 U/ml aminopeptidase M and 50 mM octylglucoside. The increase in fluorescence with time is followed at room temperature. The octylglucoside is present in the second part of the assay to inhibit omptin so that in this part of the assay no additional Arg-NHMec is being created.

Comments on Assays for Omptin

The simplest assay for omptin is cleavage of T7 RNA polymerase. However, this is not a quantitative assay. There are many sites on T7 RNA polymerase that are cleaved by omptin; some of them are cleaved early in an incubation, others late. The assay for the plasminogen activator activity of omptin is highly quantitative and sensitive. However, this assay is dependent on the homogeneity of a plasminogen preparation. There are several different forms of plasminogen and each may interact with omptin

[28] L. C. Petersen, J. Brender, and E. Suenson, *Biochem. J.* **225**, 149 (1985).
[29] U. Christensen and S. Mullertz, *Biochim. Biophys. Acta* **480**, 275 (1977).

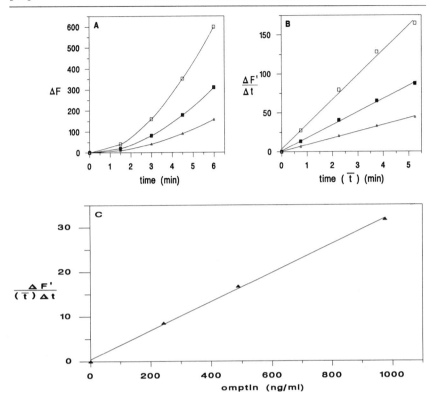

FIG. 1. Coupled assay for omptin using a synthetic, fluorogenic substrate. One-ml solutions of 0.05 M HEPES (pH 7.4) with 0.02 mM Cbz-Ala-Lys-Arg-NHMec and 0.02 U/ml aminopeptidase M were incubated with either 244 (△), 488 (■), or 976 (□) ng/ml omptin and the increase in fluorescence with time was determined using an excitation wavelength of 380 nm and an emission wavelength of 440 nm. (A) The increase in fluorescence, ΔF, is plotted versus time. (B) $\Delta F'/\Delta t$ is plotted versus $\bar{t}$, where $\Delta F'/\Delta t$ represents an increase in F ($\Delta F' = \Delta F_2 - \Delta F_1$) over a given fixed time interval ($\Delta t = t_2 - t_1$) and where $\bar{t}$ is the mean of the time interval $(\bar{t}) = (t_1 + t_2)/2$. (C) The slopes of the lines in (B), $\Delta F/(\bar{t})\Delta t$, are plotted versus the concentration of omptin.

with different kinetic parameters, as they do with human urokinase.[30] When the plasminogen activation assay is used to screen for inhibitors of omptin, interaction of the inhibitors with plasmin must also be assessed. The assay with synthetic, fluorogenic substrates is very sensitive. Tenfold less omptin is required in this assay to obtain a quantitative signal of enzyme concentration compared to that required in a plasminogen activation assay.

[30] S. W. Peltz, T. A. Hardt, and W. F. Mangel, *Biochemistry* **21**, 2798 (1982).

Cloning of Omptin

In order to obtain large quantities of omptin, we decided to clone the entire *E. coli* omptin gene in a pET plasmid vector, thereby placing the gene under control of a T7 promoter and efficient translation signals from phage T7 (Fig. 2).[9] However, because the omptin coding sequence[7] does not begin with a restriction site that could be used to fuse it directly to the ATG initiation start codon in the vector, we used the polymerase chain reaction (PCR) to introduce convenient restriction enzyme sites at both the 5' and 3' ends of the gene. This approach would also provide a construct that retains the intrinsic amino-terminal signal sequence of omptin, a signal that is necessary for transport of the proprotein to the outer membrane of the cell and proteolytic processing by signal peptidase I. Most DNA manipulations are performed as described in either Maniatis

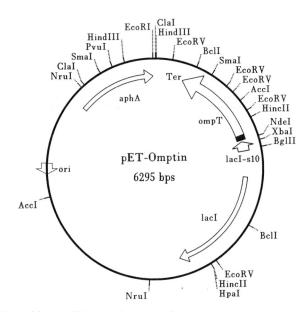

FIG. 2. Recombinant pET translation vector for expression of omptin. The details of the construction are described in the text. Open arrows depict genes and direction of their expression: the lac repressor (*lacI*), the Tn5-derived gene conferring kanamycin resistance (*aphA*), and the gene for pro-omptin (*ompT*). The origin of replication and diagnostic restriction enzyme sites are indicated. T7*lac*-s10 and Ter are, respectively, the T7*lac* promoter–translational start signal directing omptin expression and the T7 transcriptional terminator. The amino-terminal signal peptidase I signal sequence of pro-omptin is indicated by the solid box.

et al.[31] or Ausubel *et al.*[32] DNA modification enzymes are obtained from several sources and are used according to the instructions of the manufacturers. Oligonucleotide primers complementary to noncoding (N-terminal) and coding (C-terminal) strands of omptin are synthesized in a Milligen (Millipore, Bedford, MA) DNA synthesizer and purified using Poly-Pak purification cartridges (Glen Research Corp., Herndon, VA) according to the manufacturers' specifications. Primer 1 → 13 (5'-CCGGGATCCAT ATGCGGGCGAAAC-3') is used to redesign the 5' end of the gene and introduce a unique *Nde*I restriction site (CATATG), whereas primer 942 ← 954 (5'-GATATCTAGATCTTAAAATGTGTAC-3') is used to redesign the 3' end and introduce a unique recognition site for *Bgl*II (the oligonucleotide numbers and the regions underlined indicate the specific nucleotide positions of the omptin coding sequence to which they are complementary).[7]

Polymerase chain reaction amplifications are carried out in 50-μl reaction volumes containing 1 U AmpliTaq DNA polymerase (Perkin-Elmer Cetus, Norwalk, CT), each primer at 1 μM and ≈0.1 μg of purified pML19 plasmid DNA as template. The reaction mix also contains 10 mM Tris-HCl (pH 8.0), 50 mM KCl, 1.5 mM MgCl$_2$, 0.05% (v/v) Tween 20, 0.05% (v/v) Nonidet P-40 (NP-40), and 0.25 mM each dNTP. Samples are overlaid with mineral oil and amplification is carried out for 25 cycles in a DNA thermal cycler (Perkin-Elmer Cetus), with each cycle consisting of 1 min at 94°, 1 min at 47°, and 3 min at 72°. Amplification is completed by a final incubation at 72° for 10 min. The amplified products are extracted with phenol, ethanol precipitated, and then purified by electrophoresis on a 1% low-melting-point agarose gel (Bethesda Research Laboratories, Gaithersburg, MD). Prior to electrophoresis, the amplified DNA is incubated with *Nde*I and *Bgl*II restriction enzymes to generate ends suitable for cloning into the T7 base pET expression vector,[9] which has been digested with *Nde*I and *Bam*HI. We use pET13A as the vector.[33] This vector has an *aphA* gene conferring resistance to kanamycin as its selective marker and it also carries a *lac*-repressible T7 promoter (T7*lac*) upstream of the strong translational start from the T7 gene 10 capsid protein (s10). In addition, pET13A has its own copy of the *lac* repressor gene, *lacI*, to prevent the *lac* operator on the plasmid from titrating out all the *lac* repressor. In this configuration, the *lac* repressor blocks transcription of the omptin target sequence by any T7 RNA polymerase in the expression

[31] T. Maniatis, E. F. Fritsch, and J. Sambrook, "Molecular Cloning: A Laboratory Manual." Cold Spring Harbor Laboratory, Cold Spring Harbor, New York, 1982.

[32] F. M. Ausubel, R. Brent, R. E. Kingston, D. D. Moore, J. G. Seidman, J. A. Smith, and K. Struhl, "Current Protocols in Molecular Biology." Wiley, New York, 1989.

[33] J. J. Dunn, unpublished data (1993).

host, BL21(DE3), before induction with isopropylthiogalactoside (IPTG). Another expression host is BL26(DE3), which is a LacY⁻ deletion derivative of BL21(DE3), which is an omptin null strain. Because it lacks an active lac permease, the extent of derepression of the lacUV5 promoter used to direct expression of T7 RNA polymerase from the chromosome, as well as the rate of expression of target genes under control of the T7lac hybrid promoter, can be regulated by varying the concentration of IPTG in the medium. Synthesis of omptin in *E. coli* cells that were transformed with pro-omptin was detected by SDS–PAGE following induction of cultures with IPTG. The induced protein comigrated with natural omptin synthesized from pML19.

Cell Growth and Induction

BL26(DE3) 13A KanR pro-omptin cells are stored in a frozen glycerol stab. An overnight culture is initiated by inoculating 2 ml of growth medium and allowing it to stand at 37°. The next day 25 μl of the culture is diluted 1 : 10^3, 1 : 10^6, 1 : 10^9, and 1 : 10^{12} in growth medium and the cultures are grown without shaking overnight at 37°. The next day, 5 ml of the lowest dilution still containing growing cells is added to 1 liter of the growth medium. The flask is shaken vigorously at 37°. When the A_{600} reaches 0.978, IPTG is added to 0.4 mM. When the A_{600} reaches 1.4, the cells are harvested by centrifugation at 13,180 g for 10 min. The pellets are washed once with phosphate-buffered saline (PBS; pH 7.4) and then frozen. The induction of omptin in equal numbers of BL26 DE3 13A KanR pro-omptin cells as a function of time after the addition of IPTG is shown in the gel in Fig. 3.

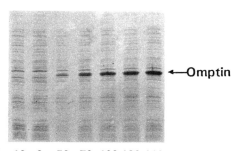

Time (min) -10 0 50 70 100 120 140

Fig. 3. Induction of omptin in BL26 DE3 13A KanR preomptin cells. The growth of BL26 cells was measured by the absorbance at 600 nm. Induction by IPTG was at time 0. At the indicated times before and after induction, the optical densities of aliquots from the culture were determined. Equal numbers of cells were then pelleted by centrifugation and their proteins were fractionated by SDS–PAGE.

Purification Procedures

Extraction of Omptin from Membrane

The pellet from 4 liters of cells is suspended in 200 ml 0.03 M Tris-HCl (pH 8.0) with 20% (w/v) sucrose and stirred at 4° for 30 min. EDTA to 0.75 mM and lysozyme to 0.015 mg/ml are then added and stirred for another 30 min. The suspension is then shell frozen in a dry ice–ethanol bath and thawed three times. After adding MgCl$_2$ to 0.01 M, bovine pancreatic DNase I (1680 Kunitz U/mg solid) is added to 0.038 mg/ml. The suspension is stirred for 30 min at 4°, centrifuged at 27,000 g for 30 min, and the pellet suspended in 200 ml 0.01 M HEPES (pH 7.4), 0.1 M NaCl, 5 mM MgSO$_4$, and 0.03 M octylglucoside. After stirring for 30 min at 4°, the suspension is centrifuged at 27,000 g for 30 min. The Mg supernatant is immediately frozen and the pellet is suspended in 200 ml 0.01 M HEPES (pH 7.4), 0.1 M NaCl, 0.01 M EDTA, and 0.03 M octylglucoside. After stirring for 30 min at 4°, the suspension is centrifuged at 27,000 g for 30 min. The EDTA supernatant is immediately frozen and the pellet is suspended in 200 ml 0.01 M HEPES (pH 7.4), 0.1 M NaCl, and 0.04 M octylglucoside. After stirring for 30 min at 4°, the suspension (remainder) is centrifuged at 27,000 g for 30 min.

Comments on Extraction from Membrane. A quantitative summary of the purification procedure is shown in Table I and a gel of extracts at various steps in the purification is shown in Fig. 4A. There must be an inhibitor of either plasminogen activation or of plasmin in the initial DNase extract because the total plasminogen activator activity measured is less than that found in subsequent steps. The pellet obtained after centrifuging

TABLE I
PURIFICATION OF OMPTIN

Purification step	Volume (ml)	Total protein (mg)	Total activity (units[a])	Specific activity (units/mg)	Yield (%)	Purification (-fold)
DNase fraction	200	10,200	4.37×10^{-2}	4.28×10^{-6}	100	1
DNase supernatant	200	480	8.54×10^{-3}	1.78×10^{-5}	19.5	4.16
Mg supernatant	200	480	8.21×10^{-2}	1.71×10^{-4}	188	40
EDTA supernatant	200	80	6.50×10^{-2}	8.12×10^{-4}	149	190
Remainder	200	3000	5.55×10^{-3}	1.85×10^{-6}	12.7	0.432
Remainder supernatant	200	64	1.22×10^{-2}	1.91×10^{-4}	27.9	44.6
Mg supernatant, boiled supernatant	200	80	5.57×10^{-2}	6.96×10^{-4}	127	163

[a] One unit corresponds to 1 μmol plasminogen converted to plasmin per minute.

A

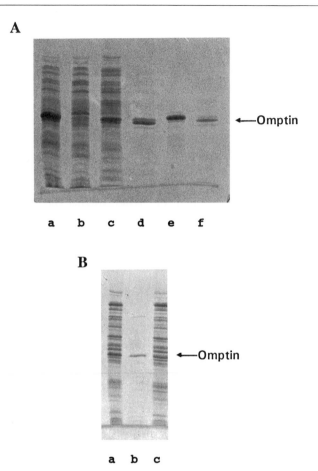

←—Omptin

a   b   c   d   e   f

B

←—Omptin

a   b   c

FIG. 4. SDS–PAGE of fractions during the purification of omptin. (A) Lane a, DNase fraction; lane b, DNase supernatant; lane c, Mg supernatant; lane d, EDTA supernatant; lane e, remainder; lane f, remainder supernatant. (B) Lane a, Mg supernatant; lane b, Mg boiled supernatant; lane c, Mg boiled pellet.

the suspension treated with DNase was sequentially extracted with solutions containing Mg, EDTA, and 40 mM octylglucoside. The distribution of plasminogen activator activity among those three fractions varied. Whereas in this enzyme preparation, 50% of the total activity in those three fractions was in the Mg fraction and 39% in the EDTA fraction, in two other enzyme preparations the Mg fractions contained 76 and 31% of the total activity compared to the EDTA fractions, which contained 21 and 65%, respectively. The gel in Fig. 4A indicates that most of the

omptin remained in the DNase pellet. The major band in the Mg supernatant contained a doublet, the predominant upper band being omptin. The
EDTA supernatant also contained the doublet, with the lower band being
predominant. There was a small amount of omptin in the remainder supernatant and very little in the remainder pellet.

Boiling of Omptin

Glass test tubes containing aliquots of the Mg supernatant are placed
in a boiling water bath for 45 sec and then quickly chilled to 4°. The
resultant cloudy solutions are centrifuged at 27,000 g for 20 min. The
supernatants contain almost pure opmtin (Fig. 4B).

Comments on Boiling Step. The boiling step resulted in the precipitation of most of the contaminating proteins with only a 32% loss in total
activity (Table I). The EDTA supernatant is much more pure in omptin
compared to the Mg supernatant. However, if the EDTA supernatant is
treated similarly to the Mg supernatant, the prominent lower band of the
doublet does not precipitate. Indeed, that band has proved extremely
difficult to remove from an omptin preparation. It also cofractionates with
omptin on DEAE-cellulose, S-Sepharose, benzamidine-Sepharose, and
HW75 (hydrophobic) chromatography.[34]

Anomalous Behavior of Omptin in SDS–PAGE

Omptin, when heated in a boiling water bath for 1.5 min in SDS–PAGE
sample buffer containing 5% mercaptoethanol and then fractionated by
SDS–PAGE, runs as a 42-kDa protein. However, if it is heated to less
than 42° for 5 min and then fractionated, it runs as a 28-kDa protein. When
omptin in SDS–PAGE sample buffer, 5% mercaptoethanol, and either
41.25% ammonium sulfate or 375 mM arginine is heated in a boiling water
bath for 1.5 min and is then fractionated by PAGE, it runs not as a
42-kDa protein but as a 28-kDa protein. Removal of the ammonium sulfate
or arginine by dialysis reverses the apparent decrease in molecular weight
in that it runs as a 42-kDa protein after heating in a boiling water bath for
1.5 min.

Self-Cleavage of Omptin

If omptin is placed in SDS–PAGE sample buffer with 5% mercaptoethanol, heated in a boiling water bath for 90 sec, and allowed to sit at room

[34] D. L. Toledo, M. Lee, M. T. Brown, and W. F. Mangel, unpublished observations (1993).

temperature overnight before SDS–PAGE, omptin migrates as a 31-kDa protein. This seems to be an irreversible, self-degradation product. A similar band is irreversibly formed if omptin is allowed to sit at room temperature, at pH 5, in the presence of a detergent above its critical micelle concentration.

Optimization of Assay Conditions for Plasminogen Activation

Assay

The plasminogen activator activity of omptin is optimized with a two-step assay. In the first step, omptin is incubated with plasminogen in a solution containing the variable for the experiment. After 5 min, the resultant plasmin is assayed by diluting the reaction mixture at least 10-fold in a solution containing the plasmin substrate (Ile-Pro-Arg-NH)$_2$-rhodamine.[27] The rate of hydrolysis of the plasmin substrate is then monitored under optimal conditions. Similar plasminogen activator assays are also performed with human urokinase.

pH, Temperature, and Ionic Strength

The pH optimum for omptin activating plasminogen is 5. At pH 4 there is less than 95% of the optimal activity and at pH 6, about 25% of the optimal activity. Human urokinase has less than 5% of its optimal activity at pH 5. Activity increases almost linearly from pH 5 to 9. At pH 7, it has about 50% of its optimal activity. The activity of both omptin and urokinase increases as the temperature is raised from 20° to 45°. The activity of omptin at 37° is 40% of its activity at 45°; for urokinase, 60% of its activity at 45°. The activation of plasminogen by urokinase is relatively insensitive to ionic strength at NaCl concentrations from 0 to 0.18 M; however, omptin plasminogen activator activity is quite sensitive. Optimal activity is in the absence of NaCl. Activity decreases to 40% at 0.04 M NaCl and to 20% at 0.15 M NaCl.

Michaelis–Menten Kinetics

Omptin exhibits Michaelis–Menten kinetics in activating human plasminogen (Table II). At pH 7.4 and 37°, the K_m is 3.6 μM. The K_m decreases to 58 nM at pH 5.0. The V_{max} is 600-fold greater at pH 5 than it is at pH 7.4. Human urokinase has a K_m for activating human Glu-plasminogen of 200 μM at pH 7.4 and room temperature.[30] The K_m for activation of human Glu-plasminogen by the omptin in outer membranes of *Yersinia pestis* is also quite low, 122 nM.[14]

TABLE II
KINETIC PARAMETERS OF OMPTIN

Plasminogen activator activity	K_m (nM)	V_{max} (relative)	K_i
pH dependence			
pH 7.4	3600	1	
pH 5.0	58	600	
Inhibitors			
Arginine-HCl	(competitive)		1.7 mM
Zinc Chloride	(uncompetitive)		9.9 μM

Substrate Specificity

Omptin cleaves between basic amino acids. Cleavage of the T7 RNA polymerase occurs at Lys-Arg[173],[35] Lys-Lys[180],[2] and Arg-Lys[392].[36] Sugimura and Nishihara[4] have shown that omptin cleaves between basic amino acids in recombinant human γ-interferon and in several biologically active peptides. In activating plasminogen, omptin produces a heavy and light chain of plasmin indistinguishable by SDS–PAGE from that produced by human urokinase. Furthermore, the NH$_2$ terminus of the light chain of human plasmin produced by the omptin of *Y. pestis* was shown to be identical to the NH$_2$ terminus of the light chain of human plasmin produced by human urokinase.[15] Thus cleavage of human plasminogen occurred at an Arg-Val bond.[37]

Inhibitors

Leytus *et al.*[1] reported that an *E. coli* outer membrane plasminogen activator activity is inhibited by diisopropyl fluorophosphate. Sugimura and Nishihara[4] reported that *E. coli* omptin is inhibited by diisopropyl fluorophosphate. They also reported that omptin is significantly inhibited by benzamidine and by the bivalent cations ZnCl$_2$, CuCl$_2$, and FeSO$_4$. We quantitatively characterized the inhibition by zinc and arginine of plasminogen activation by omptin. The type of inhibition with ZnCl$_2$ was uncompetitive (Fig. 5B) with a K_i of 9.9 μM (Table II). The second part of the assay, where the amount of plasmin was measured, contained 10 mM EDTA because ZnCl$_2$ is a competitive inhibitor of plasmin (Fig. 5A) with a K_i of 24 μM. Inhibition by ZnCl$_2$ is pH dependent. Although the enzyme could be significantly inhibited at pH 6.5, at pH 7.4 the same

[35] S. Tabor, and C. C. Richardson, *Proc. Natl. Acad. Sci. U.S.A.* **82,** 1074 (1985).
[36] D. L. Toledo, J. J. Dunn, and W. F. Mangel, unpublished data (1993).
[37] K. C. Robbins, L. Summaria, B. Hsieh, and R. Shah, *J. Biol. Chem.* **242,** 2333 (1967).

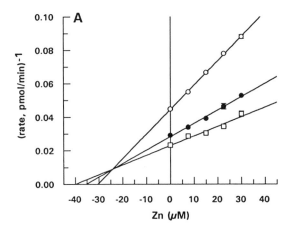

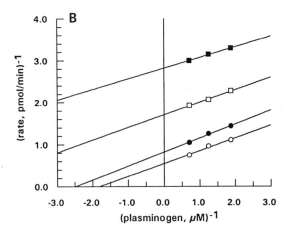

FIG. 5. Inhibition of plasmin (A) and plasminogen activation by omptin (B) by $ZnCl_2$. (A) One-ml solutions of 0.05 M HEPES (pH 7.4) with 1 mg/ml BSA and containing either 0, 7.5, 15, 22.5, or 30 μM $ZnCl_2$ and either 25 (○), 50 (●), or 75 (□) μM (Ile-Pro-Arg-NH)$_2$-rhodamine were incubated at 40° and the increase in fluorescence was measured. (B) To 50-μl solutions of 0.05 M HEPES (pH 6.5) containing either 0 (○), 25 (●), 50 (□), or 75 (■) μM $ZnCl_2$ were added either 0.05, 0.75, or 0.133 mg/ml human plasminogen, and omptin. After 5 min at 40°, 40-μl aliquots were removed and added to 960-μl aliquots containing 0.05 M HEPES (pH 7.4), 0.01 M EDTA, and 100 μM (Ile-Pro-Arg-NH)$_2$-rhodamine. The increase in fluorescence, ΔF, was then measured.

concentrations of $ZnCl_2$ had little or no effect. Plasminogen activation by omptin is also inhibited by arginine. The mode of inhibition is competitive, and the K_i is 1.7 mM.[38]

General Comments on Omptin

Omptin is a unique protease in that its sequence is not related to conserved sequences in other protease families. Thus, knowledge of the active site nucleophile and of the structure of the active site is eagerly awaited. The purification procedure presented here is simple; it utilizes three short centrifugations and no column chromatography. Hence large amounts of pure enzyme can readily be obtained. Because the gene for omptin can be deleted in laboratory strains of *E. coli* without any obvious consequences, OmpT⁻ strains of *E. coli* can be used for the expression and purification of recombinant proteins. That omptin can activate plasminogen to plasmin and that it is the virulence factor in the organism that causes plaque suggest that omptin may be utilized by other pathogenic organisms.

Acknowledgments

We thank William J. McGrath for valuable discussions. We also thank J. Coughlin and J. Buffett. This work was supported by the Office of Health and Environmental Research of the U.S. Department of Energy. Mijin Lee and Kimberly Worzalla were supported by the Department of Energy's Division of University and Industry Programs, Office of Energy Research, as Lab Co-op Program participants.

[38] M. T. Brown, D. L. Toledo, and W. F. Mangel, unpublished observations (1993).

[28] Transient Transfection Assay of the Herpesvirus Maturational Proteinase, Assemblin

By WADE GIBSON, ANTHONY R. WELCH, and JENNIFER M. LUDFORD

Introduction

Herpes group viruses are widespread in nature and cause serious infections in man, ranging in severity from bothersome and often painful cold sores caused by herpes simplex virus (HSV), to birth defects, blinding retinitis, and life-threatening pneumonitis caused by cytomegalovirus

(CMV).[1] Herpesviruses have recently been shown to encode a proteinase called assemblin that cleaves its own precursor and that of an abundant capsid phosphoprotein during assembly.[2,3,4] The capsid protein substrate is called the assembly protein precursor (pAP) and its carboxyl end is removed as a consequence of the cleavage event.[5] This cleavage is required for the production of infectious virus,[6] thereby making the viral proteinase a potential target for new antiviral drugs.

The molecular and biochemical characteristics of the simian cytomegalovirus (SCMV, strain Colburn) proteinase, illustrated in Figs. 1 and 2, can be generalized to the other herpesvirus proteinase homologs[7-13] and summarized as follows. The gene that encodes the herpesvirus proteinase (e.g., *APNG1* in SCMV) is the longest open reading frame in a family of in-frame, overlapping, 3'-coterminal genes. The gene that encodes its substrate, pAP, (e.g., *APNG.5* in SCMV) is a member of this family and constitutes the 3' half of the proteinase gene. All herpesviruses characterized encode a homologous proteinase having a similar nested genetic arrangement with its substrate, but the number of potential genes included in each nested family varies from two (e.g., HSV-1, EBV) to eight (e.g., VZV).[14,15]

The proteinase gene is transcribed into an ≈2-kb mRNA that is translated into a precursor form of the proteinase (e.g., pNP1 in SCMV). The

[1] B. Roizman, in "Fields Virology" (B. Fields and D. Knipe, eds.), p. 1787. Raven, New York, 1990.

[2] A. R. Welch, A. S. Woods, L. M. McNally, R. J. Cotter, and W. Gibson, *Proc. Natl. Acad. Sci. U.S.A.*, **88**, 10792 (1991).

[3] F. Liu and B. Roizman, *J. Virol.* **65**, 5149 (1991).

[4] V. G. Preston, F. J. Rixon, I. M. McDougall, M. McGregor, and M. F. Al Kobaisi, *Virology* **186**, 87 (1992).

[5] W. A. Gibson, I. Marcy, J. C. Comolli, and J. Lee, *J. Virol.* **64**, 1241 (1990).

[6] V. G. Preston, J. A. Coates, and F. J. Rixon, *J. Virol.* **45**, 1056 (1983).

[7] A. R. Welch, L. M. McNally, M. R. T. Hall, and W. Gibson, *J. Virol.* **67**, 7360 (1993).

[8] F. Liu and B. Roizman, *J. Virol.* **67**, 1300 (1993).

[9] M. C. Smith, J. Giordano, J. A. Cook, M. Wakulchick, E. C. Villarreal, G. W. Becker, K. Bemis, J. Labus, and J. S. Manetta, this volume [29].

[10] P. J. Burck, D. H. Bergf, T. P. Luk, L. M. Sassmannshausen, M. Wakulchik, G. W. Becker, D. P. Smith, H. M. Hsiung, W. Gibson, and E. C. Villarreal, *J. Virol.* **68**, 2937 (1994).

[11] B. Holwerda, A. Wittwer, L. Carr, R. Wiegand, M. Toth, C. Smith, K. Duffin, and M. Bryant, *J. Cell. Biochem.* (Suppl. 18D), 168 (Abstr. S414) (1994).

[12] C. L. DiIanni, J. T. Stevens, M. Bolgar, D. R. O'Boyle, S. P. Weinheimer, and R. J. Colonno, *J. B. C.*, **269**, 12672 (1994).

[13] R. R. Jones, L. Sun, G. A. Bebernitz, V. P. Muzithras, H.-J. Kim, S. H. Johnston, and E. Z. Baum, *J. Virol.* **68**, 3742 (1994).

[14] A. R. Welch, L. M. McNally, and W. Gibson, *J. Virol.* **65**, 4091 (1991).

[15] F. Liu and B. Roizman, *J. Virol.* **65**, 206 (1990).

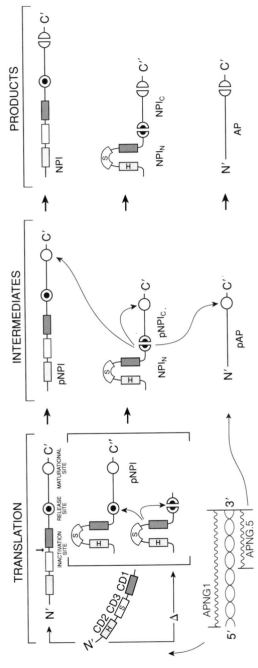

FIG. 1. A model of the synthesis and function of the herpesvirus proteinase, assemblin. Salient features of this model are described in the text and elsewhere.[2,7] The model is based on properties of these genes and proteins in transfection assays; additional factors may modulate their activity in the context of an infected cell. Rectangles indicate highly conserved domains CD1, CD2, and CD3 in the proteolytic portion of the molecule; H in CD2 and S in CD3 indicate the absolutely conserved and essential histidine and serine residues noted in the text. The maturational (○), release (◉), and "inactivation" (↓) sites are indicated in the "translation" panel. The main proteolytic cleavage products of the proteinase precursor pNP1 and the assembly protein precursor pAP are assemblin (NP1$_n$), the nonproteolytic portion of the proteinase precursor (NP1$_c$), the mature assembly protein (AP), and the tail fragment produced by cleavage of pNP1 and pAP at the maturation site (C'). The predicted sizes and other landmarks of these and additional proteins and fragments are shown in Fig. 2B.

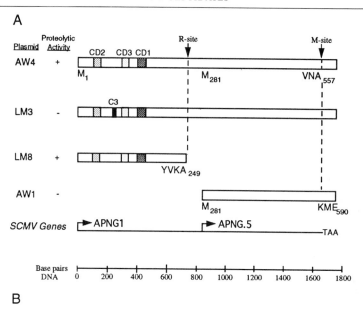

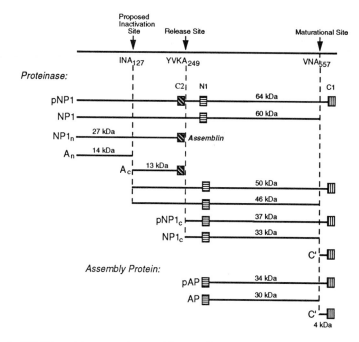

FIG. 2. SCMV proteinases and assembly protein precursor genes, and their protein products. (A) Four genes used to study the SCMV proteinase, their plasmid designations, and their proteolytic activity are depicted. The overlapping relationship of the genes encoding

precursor undergoes two principal cleavages: one at the maturational (M) site (Fig. 1, empty circles) that removes its carboxyl end (e.g., C'), and a second at the release (R) site (Fig. 1, circles containing dot) that separates the nonproteolytic domain (e.g., $NP1_c$ in SCMV) from the proteolytic portion of the molecule (e.g., $NP1_n$). Neither of these cleavages is absolutely essential for proteolytic activity per se.[7,8,13] The M and R sites have similar sequences that are highly conserved among the different herpes group viruses: V/L-X-A$\downarrow$S and Y-V/L-K/Q-A$\downarrow$S, respectively.[7] The proteinases of SCMV and human CMV (HCMV) have a third major cleavage site located near the midpoint of the proteolytic domain. Although this site has been called the inactivation (I) site, recent evidence indicates that cleavage at that position may not inactivate the proteinase.[11] The I site is not well conserved between SCMV and HCMV and may not have a counterpart among the assemblin homologs of other herpesviruses.

The proteolytic portion of the molecule was named assemblin after its substrate, the assembly protein precursor, and contains three highly conserved domains, CD1, CD2, and CD3 (Figs. 1 and 2A). The only absolutely conserved serine in assemblin resides in CD3,[7] and both site-directed mutagenesis[7] and affinity labeling[12] have identified it as the active site nucleophile. An absolutely conserved histidine in CD2 is also essential for proteolytic activity[7,8] and may also be a member of the active site triad. Because herpesvirus assemblins do not contain the characteristic sequence motifs found near the active site serines of the chymotrypsin-like or subtilisin-like proteinases (i.e., G-X-S/C-G-G or G-T-S-M/A), they would appear to represent a new subclass of serine proteinases.

the proteinase (i.e., *APNG1*) and the assembly protein precursor (i.e., *APNG.5*) is indicated at the bottom of the panel. The release (R) and maturational (M) cleavage sites are indicated by arrows; three of the five conserved domains (CD1–CD3) are indicated by shaded rectangles; a 14-amino-acid epitope (C3) inserted into the amino half of LM3[2] is indicated by the black rectangle; the translational start methionines of the proteinase (M_1) and the assembly protein precursor (M_{281}), and the carboxyl-terminal residues of assemblin (YVKA) and the mature assembly protein (VNA) and its precursor (KME), are also indicated. (B) The full-length proteinase precursor, pNP1, its three principal cleavage sites (i.e., inactivation, release, and maturational), and the protein products expected from cleavages at these sites are depicted. Also shown are the assembly protein precursor, pAP, and the two products of its cleavage at the maturational site. The designation of each protein is shown to the left of the line representing it, and the computer-predicted size of each is indicated. The carboxyl-terminal residues of the mature assembly protein (VNA), assemblin (YVKA), and the amino half of I-site-cleaved assemblin (INA) are indicated at the top. Antisera were made to synthetic peptides representing the amino (N1) and carboxyl (C1) ends of the assembly protein precursor (i.e., anti-N1 and anti-C1, respectively), and to the carboxyl end of assemblin ($NP1_n$) (C2) (i.e., anti-C2).

This paper presents a more detailed description of the transient trans-fection assay procedure that we have used to study the herpesvirus protein-ase. The following sections describe the methods that we have used to (1) prepare plasmids, (2) carry out transfections, (3) separate proteins by SDS–PAGE, and (4) identify proteinase and substrate proteins by Western and immunoprecipitation immunoassays. A short discussion section at the end considers some of the advantages and limitations of these techniques.

Genetic constructs, proteinase and substrate cleavage products, com-puter-predicted sizes, and other landmarks for the SCMV proteins dis-cussed in this paper are indicated in Fig. 2. We have found it convenient to refer to the proteins, peptides, domains, and cleavage sites of the proteinase and assembly protein by names that are applicable to all herpes-viruses, rather than by sizes or amino acid numbers, etc., which may differ for each virus.[2,7] A standardized nomenclature for the herpesvirus proteinases would be useful and aid in discussion of this new member of the serine proteinase super family.

Methods

General Procedure

The procedure that we have used to study the activity of the herpesvi-rus proteinase is based on transient transfection and, in outline, is done as follows. Human embryonal kidney (HEK) cells are transfected with a plasmid encoding the proteinase, alone or together with a plasmid encoding a substrate (Fig. 2A; AW4 and AW1, respectively); the transfected cells are harvested 2 or 3 days after adding the DNAs; the transfected-cell proteins are resolved by SDS–PAGE, and the proteins of interest are visualized by Western immunoassays. The procedures used are standard published methods and are described more fully below.

Proteolytic cleavages are monitored by the disappearance of the pre-cursor forms of the proteins and appearance of the appropriate product forms. These proteins are identifiable by their migration following SDS–PAGE and by their reactivity with specific antisera (e.g., produced against the N1, C1, or C2 peptides indicated in Fig. 2B.[7] Representative data from such an assay are shown in Fig. 3. Plasmid AW4 contains the full-length proteinase gene (*APNG1*, from SCMV) and gives rise to the full-length proteinase precursor pNP1 and its autoproteolytic products, including NP1 and $NP1_c$ seen in Fig. 3, lane 4. Plasmid AW1 contains the gene for pAP (*APNG.5* from SCMV) and gives rise to a predominant protein band corresponding to pAP (Fig. 3, lane 2). A comparatively weaker band just above pAP appears to be a posttranslationally modified

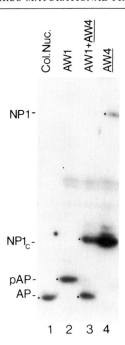

FIG. 3. Western immunoassay of HEK cells transfected with the gene for the SCMV proteinase precursor, alone or together with the gene for the SCMV assembly protein precursor. Shown here is a fluorogram of a Western immunoblot visualizing proteins reactive with anti-N1 (see Fig. 2B) in cells transfected with plasmids containing the gene for the assembly protein precursor pAP (i.e., AW1, lane 2), the proteinase precursor pNP1 (i.e., AW4, lane 4), or both (AW1 + AW4, lane 3). The nuclear fraction of SCMV-infected cells (Col. Nuc.) was used for comparison. Protein bands are designated as indicated in Fig. 2B.

form of pAP (e.g., Fig. 4, lane 13, asterisk) that accumulates with time (Fig. 4). When AW1 is cotransfected with AW4, pAP is cleaved to AP. The proportion of pAP converted to AP is influenced by the relative amount of each protein and can vary from essentially quantitative conversion at low ratios of pAP to proteinase (e.g., Fig. 3, lane 3), to 50% or less conversion at high ratios of pAP to proteinase (e.g., Fig. 4, lanes 11, 15, and 19).

Plasmid Construction

We have used the eukaryotic expression vector RSV.5(neo)[16] to express wild-type and mutant forms of cytomegalovirus assemblin, its pre-

[16] E. O. Long, S. Rosen-Bronson, D. R. Karp, R. Malnati, R. P. Sekaly, and D. Jaraquemada, *Hum. Immunol.* **31**, (1991).

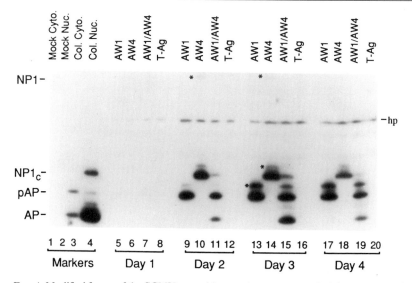

FIG. 4. Modified forms of the SCMV assembly protein precursor and of the corresponding nonproteolytic portion of proteinase precursor, showing increase in relative abundance with time following transfection. Shown here is a fluorogram of a Western immunoblot visualizing proteins reactive with anti-N1 (see Fig. 2B) in cells transfected for 1, 2, 3, or 4 days with plasmids containing the gene for the assembly protein precursor pAP (i.e., AW1, lanes 5, 9, 13, and 17), the proteinase precursor pNP1 (i.e., AW4, lanes 6, 10, 14, and 18), or both (AW1 + AW4, lanes 7, 11, 15, and 19), or with the SV40 T-antigen gene alone (T-Ag, lanes 8, 12, 16, and 20). Markers (lanes 1–4) were known viral proteins in the cytoplasmic (Cyto.) and nuclear (Nuc.) fractions of noninfected (Mock) and SCMV-infected (Col.) human foreskin fibroblasts; their designations and predicted sizes are as indicated in Fig. 2B. Asterisks near top of the day 2 and day 3 panels indicate the position of weakly visible NP1 band; asterisks above the pAP and NP1$_c$ bands in the day 3 panel indicate positions of modified proteins mentioned in text. An antigenically cross-reactive host protein (hp) is seen in the day 2, 3, and 4 panels.

cursor, pNP1, and its substrate, pAP.[7,14] The viral genes have been inserted at the SalI, BamHI, or XbaI cloning sites of the vector by standard techniques.[17] Mutations have been introduced by primer-directed mutagenesis[18] or by replacing small portions of the wild-type genes with mutation-containing oligomers. Mutations are verified by dideoxynucleotide sequence analysis.[19] Transfection plasmids were initially prepared by iso-

[17] J. Sambrook, E. F. Fritsch, and T. Maniatis, "Molecular Cloning: A Laboratory Manual." 2nd Ed. Cold Spring Harbor Laboratory, Cold Spring Harbor, New York, 1989.

[18] J. Taylor, J. Ott, and F. Eckstein, Nucleic Acids Res. 13, 8765 (1985).

[19] F. Sanger, S. Nicklen, and A. R. Coulson, Proc. Natl. Acad. Sci. U.S.A. 74, 5463 (1977).

pycnic banding in CsCl, but are now routinely prepared by using Qiagen columns (Cat No. 12162, Qiagen, Inc., Chatsworth, CA), as per instructions of the manufacturer.

Transfection. Plasmid DNA is transfected into HEK cells (line 293, ATCC, Rockville, MD) by a modified calcium phosphate precipitation technique.[20] Cells are propagated in 6- or 10-cm plastic petri dishes (e.g., Corning No. 25010, Corning Glass Works, Corning, NY) containing 5 or 10 ml, respectively, of Dulbecco's modified Eagle's medium (DMEM) (GIBCO, No. 430-2100EF, Grand Island, NY) with 10% fetal calf serum (HyClone Laboratories, Inc., Logan, UT) and the antibiotics penicillin (100 U/ml), streptomycin (100 μg/ml), and Nystatin (10 U/ml). When confluent, the loosely adherent cell layer is dispersed with trypsin (0.1%; GIBCO, No. 610-5090AG, in calcium- and magnesium-free PBS, CMF/ PBS) and reseeded at a 1 : 20 dilution into new petri dishes. Trypsin treatment reduces the tendency of the HEK cells to clump and results in more uniform monolayers.

Transfections typically are done in 24-well plates (e.g., Falcon No. 3047 multiwell plate, Becton Dickinson Labware, Oxnard, CA). Cells from one confluent 6-cm petri dish are dispersed with trypsin (0.1% in CMF/PBS) and suspended in 10 ml of medium; 2 ml of the suspension is combined with 22 ml of medium, and each 1.8 cm^2 recipient well receives 1 ml of cell suspension ($\approx 10^5$ cells/well). Cells are allowed to adhere to the plate and grow for about 24 hr at 37° in 5% CO_2, and are used when still sparse (i.e., $\approx$30% confluence).

Transfecting DNAs in 50 μl of water are combined sequentially with 5 μl of 2.5 M $CaCl_2$ and 50 μl of 2× BBS [55 mM N,N-bis-(2-hydroxyethyl)-2-aminoethanesulfonic acid, 280 mM NaCl, 1.5 mM Na_2HPO_4, pH 6.95], and are allowed to stand 15 min at room temperature ($\approx$25°). A 5-μl volume of each mixture is removed for analysis by agarose gel electrophoresis to verify the amount and integrity of the plasmid DNA, and the remaining 100-μl volume is slowly discharged into the culture medium of the cells being transfected. The total amount of DNA added per well is generally 2 μg. The molar ratio of proteinase plasmid to substrate plasmid is generally 1 : 4 in cotransfections. Varying this ratio from 1 : 1 to 1 : 16 gives similar results, in terms of the relative amount of assembly protein precursor cleaved; further dilution of the proteinase gene results in diminished cleavage of pAP. A plasmid encoding SV40 T-antigen is included (one-tenth molar amount of other plasmids, i.e., 0.2 μg) in most transfections to increase the copy number of the viral constructs by stimulating replica-

[20] C. Chen and H. Okayama, *Mol. Cell. Biol.* **7**, 2745 (1987).

tion of the RSV.5(neo) vector,[21] which contains an SV40 origin of DNA replication.

Transfected cells are incubated at 35° for ≈18 hr in 3% CO_2[20]; then the medium is removed, the cell layer is rinsed gently (cells detach easily) with PBS, fresh medium is added, and the cells are returned to incubation at 5% CO_2, 37°. Comparable levels of expression and cleavage are observed when the entire procedure is done at 33°, 37°, or 39°; however, longer incubations (i.e., 3 to 4 days) are done at the lower temperature and shorter incubations (i.e., 2 days) are done at the higher temperature to compensate for the differential rates of cell growth at the different temperatures. Transfections done at 37° are typically terminated 2 or 3 days following addition of DNA.

This procedure was also used to transfect a T-antigen gene-transformed African green monkey kidney cell line (COS), but the efficiency of transfection, as judged by Western immunoassay (see below), was ≥10-fold lower than with HEK cells. We were unsuccessful in several attempts to demonstrate transfection of human foreskin fibroblasts (5th to 15th passage from primary culture) by the same procedure.

To date there has been good agreement between results obtained from transfection experiments and those obtained from *in vitro* assays done using bacterially expressed proteinase. Two recent examples of this are (1) agreement between biochemical experiments done with bacterially synthesized HSV assemblin,[12] and site-directed mutagenesis/transfection experiments done with SCMV assemblin,[7] that CD3 serine is the active site nucleophile of the enzyme, and (2) agreement between *in vitro* proteinase assays and transfection experiments that a chelating peptide handle, added for purification purposes to the amino terminus of assemblin, is tolerated, whereas the same handle added to the carboxy terminus abolishes activity.[9]

Polyacrylamide Gel Electrophoresis

Analysis of the proteinase precursor, its substrate (the assembly protein precursor), and their multiple proteolytic products is complicated by the broad size range of the proteins (i.e., see Fig. 2B). Twelve percent acrylamide gels cross-linked with N,N'-methylenebisacrylamide (bis; 0.735 : 28 bis : acrylamide) provide good general-purpose separation and enabled all four of the principal proteolytic reactions to be monitored: (1) M-site cleavage products of the proteinase precursor (e.g., resolution of pNP1 from NP1, and $pNP1_c$ from $NP1_c$); (2) M-site cleavage of the sub-

---

[21] D. C. Rio, S. G. Clark, and R. Tijan, *Science* **227**, 23 (1985).

strate assembly protein precursor (e.g., resolution of pAP from AP); (3) R-site cleavage products of the proteinase precursor (e.g., identification of $NP1_n$, resolution of $pNP1_c$ from $NP1_c$); and (4) I-site cleavage of assemblin (e.g., resolution of 13-kDa A_c from 14-kDa A_n). Ten percent bis-linked gels give better resolution of the higher molecular weight forms of the proteinase and of pAP from AP, and also permit a more uniform electrotransfer of proteins from the gel in Western immunoassays (see below), but have the limitation of not retaining the 13- and 14-kDa I-site cleavage products (A_c and A_n, respectively) of CMV assemblin. Even better overall resolution and size range are provided by 18% acrylamide gels crosslinked with diallyltartardiamide (DATD, 1.09 : 28 DATD : acrylamide), and these are generally used to analyze immunoprecipitated proteins.[7] The principal limitation of DATD-linked, 18% gels for general purpose is that they expand significantly during the electrotransfer step of Western immunoassays and, because of the tighter acrylamide matrix, the higher molecular weight forms of the proteinase (e.g., pNP1 and NP1) are not efficiently electrotransferred from the gel.

These gel conditions are used routinely to analyze the proteinase precursors and substrates, and their respective cleavage products, from SCMV, HCMV, and HSV.

Western Immunoassays

In preparation for Western immunoassays, the medium is aspirated from the transfected cell cultures and the cell layer is rinsed with PBS (optional; may dislodge some cells); 75 μl (i.e., a volume similar to that of the cell pellet) of 2× SDS–PAGE sample buffer (4% SDS, 10% 2-mercaptoethanol, 20% glycerol, 50 mM Tris, pH 7, and 0.02% bromophenol blue) is added to the cell layer, and the viscous lysate is collected by "mopping" it around the bottom of the well with an Eppendorf tip and pipetting it out. The lysate is vortexed vigorously to help shear the DNA and disrupt the viscous aggregate, heated in a boiling water bath for 3 min, vortexed again, and frozen at −80° until analyzed.

Western immunoassays are done essentially as described by Towbin et al.[22] Electrotransfer of the proteins to Immobilon P (Millipore, Bedford, MA) is done using a semidry unit, an electrode buffer containing 50 mM Tris and 20% methanol, and a time of transfer calculated by the formula (gel width × height × 2.5 = mA per 30 min). The resulting membrane is blocked by gentle rocking in 5% BSA/TN (0.9% NaCl, 10 mM Tris, pH 7.4) for 60 to 90 min at room temperature, or 12 to 18 hr at 4°, incubated

[22] H. Towbin, T. Staehelin, and J. Gordon, *Proc. Natl. Acad. Sci. U.S.A.* **76**, 4350 (1979).

for 90 min at room temperature with appropriate dilutions of antisera in 5% BSA/TN, and rinsed thoroughly. It is then incubated with ^{125}I-labeled protein A (Cat. no. IM.144, New England Nuclear, Boston, MA) diluted 1 : 400 in 5% BSA/TN for 60 min at room temperature, rinsed thoroughly, dried, and exposed to X-ray film, usually with a calcium tungstate intensifying screen. When needed, direct autoradiographic exposure affords better resolution of the protein bands and more accurate quantification than is possible with fluorographic exposure (e.g., ^{125}I-labeled protein A with intensifying screen, or detection by chemiluminescence).

Immunoprecipitations

In preparation for immunoprecipitation, the medium is aspirated from the well and 150 μl of lysis buffer (1% NP-40, 0.5% deoxycholate, 0.5 M KCl, in ice-cold CMF/PBS) is added per 1.5-cm well. The resulting viscous cell lysate is collected by "mopping" it around the bottom of the well with an Eppendorf tip and pipetting it out. The lysate is vortexed vigorously, subjected to centrifugation ($\approx$12,000 g, 10 min at 4°) to remove large particulate material and much of the viscous DNA, and then frozen at $-80°$ until needed. Freezing prior to clarification makes it more difficult to free the lysate of viscous DNA.

Immunoprecipitations are done by combining 50 μl of clarified lysate (see above) with 20 μl of antiserum and incubating the mixture with gentle rocking for 90 min at room temperature. Fifty μl of protein A beads (Sigma No. P3391, St. Louis, MO; 100 μg/ml of CMF/PBS) is added to each tube and incubation with gentle rocking is continued for 60 min at room temperature. The reacted beads are collected by centrifugation (12,000 g, 30 sec at 4°), washed 4 times in immunoprecipitation buffer (lysis buffer with no KCl), transferred to a new tube during a fifth wash, and combined with an equal volume ($\approx$50 μl) of 2× SDS–PAGE sample buffer. This is heated in a boiling water bath for 3 min and stored at $-80°$ until analyzed by SDS–PAGE. Following SDS–PAGE, gels are stained in Coomassie brilliant blue, destained, and radiolabeled proteins are detected by fluorography following sodium salicylate enhancement.[23]

Discussion

The transfection procedure outlined above has been used to study the synthesis, processing, and structure of the herpesvirus proteinase. The technique has proved convenient, reproducible, and versatile. Because the

[23] J. P. Chamberlain, *Analyt. Biochem.* **98**, 132 (1979).

transfection procedure enables the proteins to be expressed in mammalian cells, there is less concern about unknown, but possibly important, host modifications, cofactors, etc., that might be lacking in prokaryotic or nonmammalian cells. For example, both the pAP substrate and the nonproteolytic domain (e.g., $NP1_c$) of the proteinase precursor are phosphorylated, and the influence of this modification of the proteolytic cleavage reactions, if any, is unknown. Coupled with the Western immunoassay, the procedure also provides a sensitive means of simultaneously analyzing multiple events (e.g., cleavage at the M-, R-, and I-sites) in a crude unfractionated lysate. Additionally, the procedure lends itself to analysis of nonenzymatic aspects of the proteinase, such as intracellular localization, that cannot be studied in bacteria.

The main limitation of the transfection assay is that of quantification. Although it has been possible to use the transfection procedure to investigate the substrate specificity of the enzyme and to identify specific amino acids and domains that are essential for proteolytic activity, it has not been possible with this approach to obtain the kinetic data needed to determine the substrate binding and catalysis rates of the proteinase. The transfection assay is also slower than experiments done *in vitro* with purified constituents, is by nature not a defined reaction system, and does not afford the most versatile screen for inhibitors because many such compounds or their solvents may be cytotoxic.

Although the different assay systems being used to study the herpesvirus proteinase (e.g., *in vitro* transcription and translation, transient transfection, bacterial transformation, baculovirus infection, *in vitro* reactions using purified proteinase and synthetic peptide substrates, and recombinant herpesviruses carrying mutated proteinase or substrate genes) have their own advantages and limitations, generally consistent and mutually complimentary results have been obtained for each. As biological, biochemical, and pharmacological studies of the herpesvirus proteinase progress, it will be important to exploit the advantages of each of these systems.

Acknowledgments

We thank Lisa McNally for superb technical assistance, and Shaneth Merbs and Jeremy Nathans for generously sharing transfection reagents and protocols. This work was aided by Public Health Service research grants AI13718, AI22711, and AI32957.

[29] Purification and Kinetic Characterization of Human Cytomegalovirus Assemblin

By Michele C. Smith, Joanna Giordano, James A. Cook,
Mark Wakulchik, Elcira C. Villarreal, Gerald W. Becker,
Kerry Bemis, Jean Labus, and Joseph S. Manetta

Introduction

Human cytomegalovirus (HCMV) proteinase, or assemblin (S21 in the serine assemblin family), is an attractive target for antiviral drug development. Assemblin cleaves the assembly protein precursor (pAP), a step required for encapsidation of progeny DNA.[1] Thus, inhibition of the proteinase offers a means of preventing viral replication and could lead to the potential treatment of CMV infection.[2] Simian CMV assemblin has been expressed transiently in mammalian cells,[3] and HCMV assemblin has been expressed in *Escherichia coli*.[4]

We report here on the expression and facile preparation of pure HCMV assemblin with the aid of a chelating peptide purification handle. Chelating peptide-immobilized metal ion affinity chromatography (CP-IMAC)[5,6] uses an engineered metal-binding site, or chelating peptide (CP) at either the N terminus or C terminus of a recombinant protein for a one-step affinity purification using IMAC.[7] These CP sequences are easily incorporated into the protein with recombinant DNA cloning techniques. The kinetic characterization of CP-assemblin was carried out using a high-performance liquid chromatography (HPLC) assay to measure the hydrolysis of a peptide that mimics the maturational cleavage site of pAP. A nonlinear regression analysis was used to determine the K_m and V_{max}. These data

---

[1] A. R. Welch, A. S. Woods, L. M. McNally, R. J. Cotter, and W. Gibson, *Proc. Natl. Acad. Sci. U.S.A.* **88**, 10792 (1991).

[2] D. W. Gibson and A. R. Welch, Intl. Pat. WO93-01291 (1992).

[3] W. Gibson, A. R. Welch, and J. Ludford, this volume [28].

[4] E. Z. Baum, G. A. Bebernitz, J. D. Hulmes, V. P. Muzithras, T. R. Jones, and Y. Gluzman, *J. Virol.* **67**, 497 (1993).

[5] M. C. Smith, T. C. Furman, T. D. Ingolia, and C. Pidgeon, *J. Biol. Chem.* **263**, 7211 (1988).

[6] M. C. Smith, J. A. Cook, T. C. Furman, P. D. Gesellchen, D. P. Smith, and H. Hsiung, *in* "Protein Purification" (M. R. Ladish, R. C. Willson, C. C. Painton, and S. E. Builder, eds.), p. 168. American Chemical Society, Washington, D.C., 1990.

[7] F. H. Arnold (ed.), *Methods (San Diego)* **4**, 1 (1992).

METHODS IN ENZYMOLOGY, VOL. 244

and the purified enzyme were used to develop a high-throughput assay to screen for inhibitors.

Plasmid Construction and Expression in *Escherichia coli*

The HCMV assemblin gene from the LM12 plasmid is amplified by polymerase chain reaction (PCR) to create a new plasmid, pCZR332.28K, for expression in *E. coli*.[8] This plasmid is further modified to include codons for a chelating peptide sequence at either the N terminus or C terminus of the enzyme. The oligonucleotides used to attach the codons for Met-His-Trp-His-Trp-His, to the 5′ end of the HCMV assemblin gene to express CP-assemblin with the CP at the N terminus of enzyme are as follows:

5′-TATGCATTGGCACTGGCA-3′
3′-ACGTAACCGTGACCGTAT-5′

The sequence includes *Nde*I restriction sites for cloning into the *Nde*I site of pCZR332.28K. The addition of the sequence encoding the CP results in the creation of an additional *Nsi*I site within the new plasmid pCZR332.28K/CN. The *Nsi*I site is used to identify clones that contained the CP. Sequencing of *Nsi*I-positive clones is carried out to identify those with only one copy of the CP in the correct orientation.

A 63-oligonucleotide sequence (shown below) is used to introduce the CP sequence into the 3′ end of the assemblin gene to express assemblin-CP with the CP at the C terminus of the protein. The oligonucleotide contains the 3′ end of the assemblin gene from the *Spe*I site, followed by the CP codon sequence and a *Bam*HI restriction site.

5′-CTAGTCGGCGTGACGGAGCGCGAGTCATACGTAAAGGCG
3′-AGCCGCACTGCCTCGCGCTCAGTATGCATTTCCGC
ATGCACTGGCACTGGCACTAATAG-3′
TACGTGACCGTGACCGTGATTATCCTAG-5′

The addition of the 63-nucleotide oligomer results in the creation of a unique *Sna*BI site within the plasmid pCZR332.28K/CC. The *Sna*BI site is used to identify positive clones that are then sequenced to select those with only one copy of the CP handle.

The new plasmids pCZR332.28K/CN, with the CP at the 5′ end, and pCZR332.28K/CC, with the CP at the 3′ end, are used to transform *E.*

[8] P. J. Burck, D. H. Berg, T. P. Luk, L. M. Sassmannshausen, M. Wakulchik, D. P. Smith, H. M. Hsiung, G. W. Becker, W. Gibson, and E. C. Villarreal, *J. Virol.* **68,** 2937–2946 (1994).

coli K12 RV308, and the transformants are selected by their ability to attain log growth in the presence of 10 μg/ml tetracycline. *Escherichia coli* cells containing plasmid pCZR332.28K/CN or pCZR332.28K/CC are grown to stationary phase in TY medium containing 10 μg/ml tetracycline. The cells are transferred to fresh medium and grown at 32° until they reach a culture density of 0.3–0.4 A_{550}. Production of the CP containing assemblin is induced by shifting the temperature to 42° and allowing the cells to grow for 3 hr in an air shaker incubator or in a BioFlo-III fermenter (New Brunswick Scientific Co., Edison, N.J.). The production of assemblin is monitored by 12% SDS–PAGE of cells lysed with sample buffer.

Purification of Assemblin

Cells are lysed with lysozyme and sonicated as described.[9] Inclusion bodies are collected by centrifugation, lyophilized, and 10 mg/ml of solids dissolved in 0.5 M Tris, 7 M urea, pH 8.2. The cysteine residues are converted to S-sulfonate groups by including 100 mM Na$_2$SO$_3$ and 10 mM Na$_2$S$_4$O$_6$ in the buffer.[9] Blocking the cysteine residues as S-sulfonates facilitates the purification by solubilizing more protein from the inclusion bodies and preventing the formation of intermolecular disulfide bonds and aggregates. We also expect the S-sulfonate form of the enzyme to be inactive, thus eliminating autocatalysis during purification. The S-sulfonate groups are removed after purification by reduction with dithiothreitol (DTT) during the activation procedure.

The crude mixture of sulfitolyzed proteins (Fig. 1A, lane 1) is filtered through a 0.45-μm filter and applied to a Poros MC/P column on a fast protein liquid chromatography system (FPLC, Pharmacia, Piscataway, NJ). Poros MC/P, from PerSeptive Biosystems (Cambridge, MA), is a polymeric packing designed for IMAC in the perfusion chromatography mode, wherein high linear flow rates between 200 and 5000 cm/hr are used.[10] This resin is used because perfusion chromatography shortens the time required to chromatograph a protein, but other commercially available resins, such as chelating Sepharose Fast Flow (Pharmacia) or Toyopearl AF-chelate-650 M (TosoHaas) (Philadelphia, PA), can also be used at slower flow rates. The breakthrough of HCMV CP-assemblin in the flowthrough peak is significant with the Poros column, but this material can be recycled by passing the flowthrough pool through the column again. Because the times required in the perfusion mode are so short (minutes),

[9] M. S. Kasher, M. Wakulchik, J. A. Cook, and M. C. Smith, *BioTechniques* **14**, 630 (1993).
[10] N. B. Afeyan, S. P. Fulton, N. F. Gordon, I. Mazsaroff, L. Várady, and F. E. Regnier, *Bio/Technology* **8**, 203 (1989).

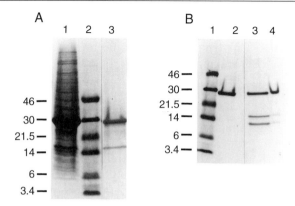

FIG. 1. Purification of assemblin. A. Lane 1, crude sulfitolyzed granules; lane 2, molecular weight markers; lane 3, IMAC pool of CP-assemblin S-sulfonate and its N-terminal degradation products. B. Lane 1, molecular weight markers; lane 2, MonoQ pool of CP-assemblin S-sulfonate in 7M urea; lane 3, MonoQ pool of CP-assemblin S-sulfonate after removal of urea by dialysis; lane 4, activated CP-assemblin in 50% glycerol containing buffer.

the total time for processing the crude mixture is considerably less than with the more conventional resins (hours).

A 1 × 10-cm HR column (Pharmacia) containing 8 ml Poros MC/P resin is equilibrated in Milli Q water, charged with 6 ml of 50 mM NiCl$_2$, and washed with water. The column is equilibrated with 10 mM imidazole in 50 mM NaH$_2$PO$_4$, 0.5 M NaCl, 7 M urea, pH 7.5, for 8 min at 3 ml/min (3 bed volumes). A 5-ml sample of the sulfitolyzed proteins (10 mg solids/ml) is applied to the column with the same buffer, washed for 1 min, followed by an 80-min (10 bed volumes) imidazole gradient from 10 to 250 mM imidazole in 50 mM NaH$_2$PO$_4$, 0.5 M NaCl, 7 M urea, pH 7.5. Fractions are collected and analyzed by SDS–PAGE with prepoured 10–20% gels in a Tricine buffer system (Novex, San Diego, CA). Pools of the eluted bound protein fractions containing the S-sulfonate form of HCMV CP-assemblin are dialyzed against 50 mM Tris, 7 M urea, pH 7.6.

SDS–PAGE (Fig. 1A, lane 1) and Western blots (data not shown) of the isolated granules reveal the presence of a 15-kDa species that cross-reacts with antibodies to HCMV assemblin. Degradation of the full-length 28-kDa form to a smaller fragment with the same N-terminal sequence as the 28-kDa form has been observed by others and proposed as a mechanism to inactivate the enzyme during the viral life cycle.[4,8] CP-IMAC[5,6] is independent of the protein and derives its specificity from the high-affinity metal-binding site, or CP, engineered into the recombinant protein. Therefore, CP-IMAC of CP-assemblin results in the copurification of the 15-kDa species, which also contains the CP (Fig. 1A, lane 3). An additional chromatography step is necessary to separate the full-length CP-assemblin from the 15-kDa species.

Anion-exchange chromatography is able to distinguish between the desired 28-kDa form and the N-terminal 15-kDa fragment. A 10/16 Mono Q column is equilibrated in 50 mM Tris, 7 M urea, pH 7.6 (buffer A), and 50 ml of the IMAC CP-assemblin pool is applied to the column at a 5-ml/min flow rate. After the sample is loaded, a gradient from 0 to 100 mM NaCl over 8 bed volumes (160 ml) is generated with buffer B (50 mM Tris, 1 M NaCl, 7 M urea, pH 7.6). The column is washed with half a bed volume of 1 M NaCl before equilibrating in buffer A for the next injection. Fractions are collected and analyzed by SDS–PAGE with pre-poured Novex 10–20% gels in a Tricine buffer system. Pools of fractions containing the 28-kDa S-sulfonate form of HCMV CP-assemblin are dialyzed against 20 mM NH$_4$HCO$_3$ for 3 days. This material is lyophilized and stored in the freezer at $-20°$.

SDS–PAGE of the Mono Q pool before and after dialysis (Fig. 1B) shows that in the absence of 7 M urea the S-sulfonate form of HCMV CP-assemblin is able to cleave itself. Two new forms of the enzyme (15 and 18 kDa) are generated (Fig. 1B, lane 3) from pure full-length HCMV CP-assemblin (Fig. 1B, lane 2). Because the cysteine groups are blocked but the protein has proteolytic activity, a catalytic role for cysteine, as postulated by others,[4] seems highly unlikely. The 28-, 18-, and 15-kDa forms of the enzyme have the same N-terminal sequence.[11] The absence of the corresponding C-terminal fragments suggests that they are digested into peptides too small to be observed by SDS–PAGE or are removed during dialysis. The appearance of the 18-kDa form also suggests that an additional internal cleavage site exists within the enzyme to inactivate the enzyme, because the 15-kDa form is inactive.[4,8] Further investigation of these cleavage sites is underway.

Dialysis of the Mono Q pool against a low-pH buffer (10 mM ammonium acetate, pH 4.3) prevents the autocatalytic internal cleavage (data not shown), because this pH is outside the pH optimum for the enzyme.[8] The protein also forms a precipitate slowly at this pH, which may protect it from autodigestion. The UV spectrum of this protein in 0.01 N HCl has a λ_{max} at 277 nm with an extinction coefficient of 1.14 ml mg^{-1} cm^{-1} based on the amino acid analysis of the same solution.

In an attempt to avoid the copurification of the full-length assemblin and its cleavage products on the IMAC column, we add the CP to the C terminus of the enzyme. The 15-kDa fragment would not bind because it

[11] Proteins separated by SDS–PAGE were transferred to ProBlot or Immobilon and N-terminal sequence analysis was performed by automated Edman degradation using either an Applied Biosystems (Foster City, CA) Model 477A or Model 470A sequencer.

would not contain a CP, and if the C-terminal fragments contain an intact CP, they would again be small enough to separate by dialysis. SDS–PAGE of the granules reveals the presence of the 28-kDa assemblin-CP and a smaller 15-kDa form (data not shown). The 28-kDa assemblin-CP is indeed purified in a single IMAC step as expected, because fragments lacking the CP do not bind the column whereas full-length assemblin-CP binds tightly.

Activation of HCMV CP-Assemblin S-Sulfonate

A 1-mg/ml solution of lyophilized CP-assemblin S-sulfonate in 100 mM HEPES, 1 M urea, 0.2 mM EDTA, 100 mM DTT, 200 mM NaCl, pH 7.5, is allowed to stir for 2 hr at room temperature under nitrogen. Protein dialyzed at low pH is first dissolved in 7.5 M urea and then diluted with a concentrated version of the buffer to achieve the final concentrations given above. The DTT concentration is lowered by desalting the reduced HCMV CP-assemblin (1 ml) on a 1 × 9.5-cm Sephadex G-25SF column equilibrated in 100 mM HEPES, 0.2 mM EDTA, 2 mM DTT, 200 mM NaCl, pH 7.5. The absorbance of the protein pool is measured at 282 nm, the new λ_{max} for the reduced protein in this buffer, and the extinction coefficient of 1.13 ml cm^{-1} mg^{-1} for the S-sulfonate form is used to determine the protein concentration. This method for determining protein concentration agrees with the results from the Coomassie Plus Protein Assay (Pierce, Rockford, IL). The pool is diluted 1 : 1 with glycerol and allowed to sit overnight at 4° under a blanket of nitrogen at a final concentration of 120 μg/ml in folding buffer (50 mM HEPES, 0.1 mM EDTA, 1 mM DTT, 100 mM NaCl, 50% (v/v) glycerol, pH 7.5). The activated HCMV CP-assemblin is stored in folding buffer at −20° and is stable for 8 months or more.

When this activation procedure is used with assemblin-CP (CP at the C terminus) it yields inactive assemblin that is unable to cleave the labeled peptide substrate. The inherent inactivity of this C-terminal modified form of the enzyme is confirmed in the transfection assay for assemblin activity.[3,12] These results agree with previous observations that deletions from or additions to the C terminus lead to reduced production of cleaved assemblin fragments in *E. coli*.[4]

Preparation of Substrate

The activity of HCMV CP-assemblin is measured by following the hydrolysis of biotinyl-R-G-V-V-N-A-S-S-R-L-A-K-FITC (FITC, fluorescein isothiocyanate), a labeled peptide mimic of the maturational site

---

[12] E. C. Villarreal and W. Gibson, unpublished data (1993).

of the assembly protein precursor. The pAP sequence mimicked by the substrate peptide is G-V-V-N-A-S-C-R-L-A.[1] The cysteine residue following the A-S cleavage site is substituted with serine, and arginine and lysine residues are added to the N and C termini of the peptide, respectively, to facilitate the addition of biotin and FITC to the peptide.

Biotinyl-R-G-V-V-N-A-S-S-R-L-A-K (95% pure) is purchased from American Peptide Co. (Sunnyvale, CA) and derivatized with fluorescein isothiocyanate using the following procedure. Peptide (800 mg, 0.54 mmol) is dissolved in 150 ml of 0.1 M sodium borate (pH 9.5) buffer with 15 ml of methanol. The solution is slightly turbid because the peptide has limited solubility at 5 mg/ml. FITC [1.5 g/7.5 ml dimethyl sulfoxide (DMSO), 3.85 mmol] is added to the peptide solution in 10 equal parts over 2 hr and the reaction is allowed to stir for an additional hour at room temperature. A precipitate forms during the course of the reaction and again after the pH is lowered to 7.5 with 5 N HCl at the end of the reaction. The suspension is centrifuged at 3000 g for 30 min at 22° and the supernatant and precipitate are frozen at $-20°$.

Purification of Substrate

HPLC analysis of the pellet and supernatant from the reaction shows that most of the product is in the precipitate. The precipitate is dissolved in two volumes of 7.5 M urea for every volume of solid, stirred for 15 min, and centrifuged at 2000 g for 15 min at 22°, which leaves behind a small, dark orange pellet. Preparative reversed-phase HPLC is used to purify biotinyl-R-G-V-V-N-A-S-S-R-L-A-K-(FITC), with a 2.2 × 30-cm Zorbax C_{18} column attached to the FPLC run at 4 ml/min. The column is equilibrated in 10% buffer B (buffer A is 0.1 M ammonium acetate, pH 7.5, and buffer B is 0.1 M ammonium acetate, pH 7.5, 50% acetonitrile) for one bed volume (113 ml); sample is applied (15- to 55-ml loads) and washed for 30 min with 10% buffer B. A gradient between 10 and 50% buffer B over 300 min separates the product from impurities. The column is recycled by washing with a steep gradient (50–100% buffer B over 80 min), washing at 100% buffer B for 15 min and then lowering the acetonitrile concentration back to 10% buffer B over 75 min. The product pool is lyophilized and analyzed by electrospray mass spectroscopy and amino acid analysis, giving the correct mass and composition.

An analytical HPLC system using the same buffer system with a Vydac C_{18} column (0.46 × 10 cm) is used to analyze individual fractions from the preparative HPLC runs and pool fractions. In this case the column is equilibrated with 0.1 M ammonium acetate, pH 7.5, 15% (v/v) CH_3CN. The sample is applied at 0.5 ml/min and washed with the same buffer for

5 min. A gradient from 15 to 25% CH_3CN in 0.1 M ammonium acetate, pH 7.5, over 60 min is used to separate the product from the contaminants. The column is then recycled by washing with 50% CH_3CN in 0.1 M ammonium acetate, pH 7.5, for 5 min and then reequilibrating in 15% CH_3CN in 0.1 M ammonium acetate, pH 7.5.

The high-pH buffers used to purify the labeled substrate shorten column lifetimes. Reduced performance is observed on the analytical column after about 200 injections and a new column has to be substituted.

HPLC Assay for Proteolytic Activity

An isocratic HPLC assay is used to measure substrate hydrolysis as a function of time and substrate concentration in order to determine the K_m and V_{max} for HCMV CP-assemblin with the labeled substrate. An enzyme concentration of 0.13 μM and nine different substrate concentrations between 30 and 374 μM in 50 mM HEPES, 0.1 mM EDTA, 1.0 mM DTT, 100 mM NaCl, 15% glycerol, pH 7.5, are used to follow the kinetics at room temperature. Triplicate samples (30 μl) are taken at each time point, quenched with 570 μl of 0.1% (v/v) trifluoroacetic acid (TFA), and then frozen in a dry ice/acetone bath. Freezing the samples is important to stop the reaction completely because samples left ar room temperature for a few days show additional hydrolysis of substrate, and it takes 5 to 10 days to analyze all the samples. Samples for an individual substrate concentration are thawed as a group, loaded into a Bio-Rad (Richmond, CA) autoinjector, and 500 μl of each is injected onto the HPLC column.

An analytical Brownlee (Applied Biosystems Inc., Foster City, CA) Spheripore C_{18} column (0.46 × 10 cm) in 0.1% TFA and 25% CH_3CN is able to separate the product S-S-R-L-A-K-(FITC) from the substrate with elution times of 8.4 and 14.1 min, respectively. Data are collected on a PE Nelson TurboChrom 3 chromatography system (Perkin-Elmer Corp. Cupertino, CA), which digitizes the analog output from the HPLC detector and records the retention time and peak height and integrates each peak area. Velocity is calculated as the amount of product formed as measured by peak height. This is found to give the most precise correlation with concentration of the product peptide.

Statistical Analysis of Kinetic Data

Data files containing the product peak height at different time points for the nine different substrate concentrations are transferred from the TurboChrom 3 system to a JMP template file on an Apple Macintosh computer. JMP (SAS Institute Inc., Cary, NC) is a statistical software package that allows for nonlinear regression analysis of data. Two different

TABLE I
KINETIC PARAMETERS FOR CP-ASSEMBLIN AND
BIOTINYL-R-G-V-V-N-A-S-S-R-L-A-K-(FITC)[a]

	Lineweaver–Burk		Nonlinear regression[b]		
Experiment	K_m (μM)	V_{max} $(\mu M/min)$	$K_m \pm SE$ (μM)	Confidence limits	V_{max} $(\mu M/min)$
1	194	nd[c]	127 ± 35	62–270	nd[c]
2	166	1.0	91 ± 32	41–206	1.2

[a] The buffer used was 50 mM HEPES, 0.1 mM EDTA, 1.0 mM DTT, 100 mM NaCl, 15% glycerol, pH 7.5, with 0.13 μM enzyme and between 30 and 374 μM substrate. The k_{cat} is the V_{max}/E_0, where E_0 is the initial enzyme concentration, and is 7.7 min^{-1} and 9.2 min^{-1} for the Lineweaver–Burk and nonlinear regression analysis, respectively.
[b] JMP was also used to calculate 95% confidence limits on the parameters fit (K_m) using nonlinear regression.
[c] Not determined.

methods are used to calculate a K_m and V_{max} value from the data. The first method measures the initial velocities using linear regression to obtain the slope of the linear portion of the time course. These velocities are then used in a Lineweaver–Burk plot to obtain the K_m and V_{max}. In the second method each time course for a given substrate concentration is fitted to Eq. (1),

$$P = P_\infty(1 - e^{-kt}) + P_{back} \qquad (1)$$

where P is the amount of product, P_∞ is the amount of product as time approaches infinity, k is the rate constant, t is time, and P_{back} is the amount of contaminating product or nonspecific background signal.[13] The initial velocities are obtained from Eq. (1) using Eq. (2). Equation (2) is derived

$$v = P_\infty k \qquad (2)$$

from Eq. (1) by taking the derivative at $t = 0$. The initial velocities obtained are then used to solve the Michaelis–Menten equation [Eq. (3)], using nonlinear regression analysis. Table I shows the K_m and V_{max} values obtained with these two methods.

$$v = V_{max}[S]/([S] + K_m) \qquad (3)$$

High-Throughput Assay for Inhibitors

The labeled substrate is used to develop a fluorescence-based high-throughput assay for identifying inhibitors of assemblin as potential

[13] R. J. Leatherbarrow, *Anal. Biochem.* **184,** 274 (1990).

HCMV antivirals. The high-throughput assay uses avidin-derivatized polystyrene beads to capture biotinylated peptides. The beads with bound biotinyl-substrate and biotinyl-product peptides are washed in special 96-well plates (IDEXX Fluoricon assay plate, IDEXX Co., 100 Four St., Portland, ME 04101) to remove the C-terminal cleavage product, S-S-R-L-A-K-(FITC). The beads are filtered, and the remaining fluorescence is read directly in an IDEXX Screen Machine. A decrease in fluorescence indicates that the substrate is cleaved and loses the FITC label when the beads are washed. Retention of the fluorescent signal in the presence of a test compound would be expected for a potential inhibitor of the enzyme.

The reaction is run in a Costar #3794 polypropylene microtiter plate (Costar Corp., Cambridge, MA) by combining 20 μl of 20 μg/ml of HCMV CP-assemblin in diluting buffer [50 mM HEPES, 0.1 mM EDTA, 1 mM DTT, 100 mM NaCl, 25% (w/v) glycerol, pH 7.5] with 10 μl of inhibitor in 2% DMSO. The plate is covered and the enzyme and inhibitor are incubated for 1 hr at 22°. At the end of 1 hr, 20 μl of a 40 μg/ml of the labeled peptide substrate in diluting buffer is filtered through a 0.22-μm filter and added to the reaction. These conditions give a final concentration of 0.28 μM enzyme and 8.5 μM substrate. The plate is covered and incubated overnight at 22°. The enzyme cleaves about 80% of the substrate under these conditions, which enhances the sensitivity of the assay for detecting inhibitors. The reaction is quenched by diluting with 200 μl of TBSA [0.02 M Tris, 0.15 M NaCl, 1 mg/ml bovine serum albumin (BSA), pH 7.5]. An aliquot (20 μl) of the diluted reaction is transferred to an IDEXX Fluoricon assay plate (IDEXX Co., Portland, ME) containing 25 μl of a 0.1% solution of avidin–polystyrene Fluoricon assay particles. The beads and diluted reaction aliquot are incubated at 22° for 10 min and then washed twice with TBSA in the IDEXX Screen Machine. After the addition of each new reagent, a pipette is used to mix the reaction thoroughly. The fluorescence is measured with excitation at 485 nm and emission at 535 nm and a gain setting of 1×. The coefficient of variation of the reaction of the CP-assemblin and labeled peptide substrate in the conditions stated above is under 16%. The coefficient of variation of the substrate filtered through a 0.22-μm filter and plated alone is 3–5%.

Discussion

The purification of HCMV assemblin was achieved by expressing the enzyme in *E. coli* as a fusion protein with a CP. This purification is the first example of CP-IMAC on a proteinase, where the enzyme was modified with a high-affinity metal-binding site to enhance its purification using IMAC. Our experience also highlighted some of the limitations inherent in modifying proteins to facilitate their purification. The internal cleavage sites in assemblin generated assemblin fragments during expression in *E. coli,* some of which also contained the N-terminal CP. These contaminants retained the high-affinity metal-binding site and were therefore co-purified with the full-length CP-assemblin. This limitation has not been observed before because other proteins that have been expressed with CP purification handles either are not enzymes or are not cleaved by host proteinases to generate such fragments.[14] A second anion-exchange step was sufficient to purify the material to a single band on SDS–PAGE. On the other hand, when the CP was attached to the C terminus of a proteinase, we were able to purify the protein in a single chromatographic step. Unfortunately, assemblin activity seems to be sensitive to changes at the C terminus, as this form of the enzyme was inactive. Other proteinases may not have internal cleavage sites or be as sensitive to the addition of five or six extra amino acids to either end of the protein, so the limitations we encountered may be unique to assemblin.

The complete lack of proteolytic activity of HCMV assemblin with a C-terminal chelating peptide agrees with previously published observations about assemblin.[4] Deletion of three or eight amino acids from the C terminus leads to a profound decrease in the amount of enzyme cleaved at the internal site during expression in *E. coli.*[4] Twelve additional C-terminal amino acids had the same effect. Our results extend these observations and showed that six additional amino acids prevented the enzyme from cleaving a maturational release site peptide substrate. Interestingly, we did observe the 15-kDa species in isolated granules (data not shown), which is in contrast to that observed by Baum *et al.,*[4] and suggests that an intact C terminus is critical for cleaving pAP but does not prevent the enzyme from hydrolyzing itself.

An unexpected finding during the purification was the apparent activity of the S-sulfonate form of CP-assemblin. All purification steps were carried out in buffers containing 7 M urea, which maintained the enzyme in an unfolded inactive conformation. We expected the blocked cysteine residues to prevent the enzyme from refolding when the urea was removed by dialysis at pH 8, but instead smaller molecular weight fragments of CP-assemblin were generated. These data provided additional evidence that assemblin is not a cysteine proteinase because the S-sulfonate form of cysteine is oxidized and cannot act as a nucleophile. The molecular weight of at least one of these fragments is larger than those reported,[4,8] suggesting that assemblin may have multiple internal cleavage sites.

The purified enzyme was used to measure a K_m and V_{max} for CP-assemblin with biotinyl-R-G-V-V-N-A-S-S-R-L-A-K-(FITC), a labeled

[14] M. C. Smith, *Ann. N.Y. Acad. Sci.* **646**, 315 (1991).

peptide mimic of the maturational release site in pAP. The K_m values obtained in the 100 to 200 μM range and V_{max} and k_{cat} values are reasonable for proteolytic enzymes and gave a specific activity of 0.32 U/mg.[15] More importantly, they allowed a high-throughput assay to be developed for identifying inhibitors of HCMV assemblin. Such compounds can then be used to determine if inhibiting this critical step in the viral life cycle is sufficient to inhibit viral replication.

Acknowledgments

The authors would like to thank Mel Johnson, Bob Ellis, John Richardson, Theresa Gygi, and Fred Chadwell for technical contributions and Dr. Laura Mendelsohn for helpful discussions.

[15] T. E. Barman, "Enzyme Handbook," Vol. 2, Springer-Verlag, New York, 1969.

[30] Amino Acid and Peptide Phosphonate Derivatives as Specific Inhibitors of Serine Peptidases

By Jozef Oleksyszyn and James C. Powers

Introduction

Serine peptidases are excellent targets for drug design because they are essential in many important physiological processes, including coagulation, fibrinolysis, and complement activation. In addition, serine peptidases are thought to be involved in disease states such as pulmonary emphysema, arthritis, inflammation, and tumor metastasis. Many reversible and irreversible inhibitors of serine peptidases have been developed.[1] These include affinity labels, transition-state inhibitors, and mechanism-based inhibitors, among others. Each class of compound has different degrees of specificity, stability, and utility for use in biological experiments. Many inhibitors, especially peptide-based inhibitors, are not easy to synthesize, often requiring multistep procedures. Amino acid and peptide organophosphate derivatives such as α-aminoalkyl phosphonate diphenyl esters (Fig. 1) comprise one class of inhibitor that is often highly specific for the target serine peptidase and is quite stable in physiological media. Because peptide phosphonates can often be synthesized without

[1] J. C. Powers and J. W. Harper, in "Proteinase Inhibitors" (A. J. Barrett and G. S. Salvesen, eds.), p. 56. Elsevier, Amsterdam and New York, 1986.

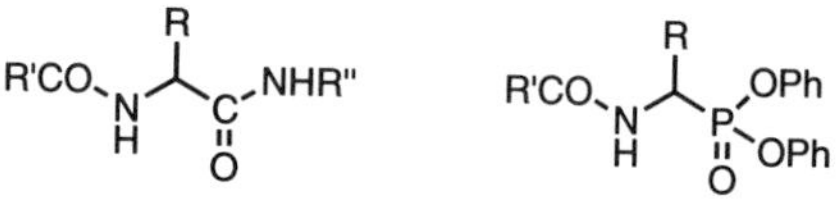

R'CO—N(H)—CR(—NHR")—C(=O)   R'CO—N(H)—CR(—P(=O)(OPh)(OPh))

FIG. 1. Relationship between an amino acid residue and the corresponding α-aminoalkyl phosphonate diphenyl ester derivative.

major problems, they should find wide applicability in biological experiments.

Synthetic Methods

General Considerations

The main advantage of the α-aminoalkyl phosphonate diphenyl ester derivatives over other types of serine peptidase inactivators is the relative ease of synthesis and use of these compounds. The parent diphenyl α-N-benzyloxycarbonylaminoalkyl phosphonates are synthesized in 35–70% yields by α-amidoalkylation of triphenyl phosphite with benzyl carbamate and the appropriate aldehyde (Fig. 2).[2] Isolation of the crystalline material is usually readily accomplished and additional recrystallization provides analytically pure compounds that can be kept in the lab for a few months without special precautions. The α-amidoalkylation reaction to produce the phosphonate inhibitors has broad scope, and many different variations in reactant structure are acceptable.

Carbamate

Simple alkyl carbamates undergo the α-amidoalkylation reaction; however, isolation of the product is difficult. Benzyl carbamate has been widely used and provides crystalline products in which the amino group of the product is blocked by a N-benzyloxycarbonyl (Cbz) protecting group.[3] This can easily be removed using standard methods used in peptide chemistry.

[2] J. Oleksyszyn, L. Subotkowska, and P. Mastalerz, Synthesis, 985 (1979).
[3] Abbreviations: Ac, acetyl; Boc, tert-butyloxycarbonyl; Cat G, cathepsin G; Cbz, benzyloxycarbonyl; CDI, carbonyldiimidazole; DCC, dicyclohexylcarbodiimide; DCU, dicyclohexylurea; DFP, diisopropyl fluorophosphate; Dpa, β,β-diphenylphenylalanine; DPP-IV, dipeptidyl-peptidase IV; Et, ethyl; HEPES, 4-(2-hydroxyethyl)1-piperazineethanesulfonic acid; HLE, human leukocyte elastase; Me, methyl; pNA, p-nitroanilide; Nva, norvaline; Ph, phenyl; PhGly, phenylglycine; RMCP II, rat mast cell protease II; Suc, succinyl; TLC, thin-layer chromatography.

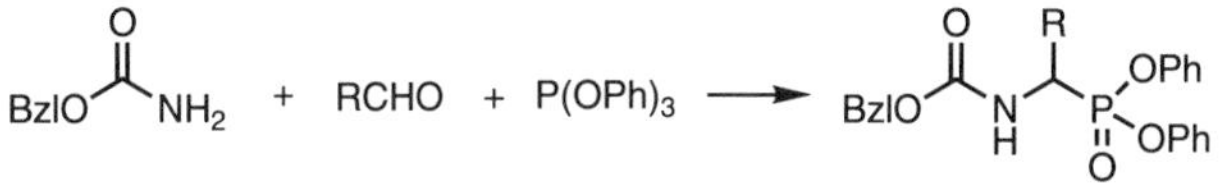

FIG. 2. Synthesis of α-aminoalkyl phosphonate diphenyl ester derivatives by reaction of Cbz-NH$_2$, an aldehyde, and triphenyl phosphite.

Organophosphorus Reagent

In general, any triester of phosphorous acid can be used if the ester groups are electron withdrawing. Trialkyl phosphites do not undergo the α-amidoalkylation reaction. In addition to triphenyl phosphite, we have used tri(4-chlorophenyl), tri(3-chlorophenyl), and tri(4-fluorophenyl) phosphites. The yields of crystalline products are comparable to those obtained with triphenyl phosphite.

Carbonyl Reagent

The α-amidoalkylation reaction works quite well with simple alkyl and aryl aldehydes. In addition, a number of other functional groups are tolerated in the carbonyl reagent. These include aryl nitriles (Ar-CN), aryl and alkyl halides (with the exception of 2-halogenated acetaldehydes), -O-CH$_2$Ph, -SCH$_3$, phthalyl-blocked amino groups [e.g., phthalyl-N(CH$_2$)$_3$CHO], Ar-NO$_2$, and -COOCH$_2$CH$_3$.

Peptide Synthesis

Most of the chemical procedures used in peptide synthesis can also be applied to the synthesis of peptides with a C-terminal diphenyl α-aminoalkyl phosphonate moiety.[4,5] The dicyclohexylcarbodiimide (DCC) method (in the presence of hydroxybenzotriazole hydrate), the mixed-anhydride method, and the carbonyldiimidazole (CDI) method are all effective methods for the synthesis of these peptide inhibitors. The yields are in the same range as for normal peptide syntheses. The peptide phosphonates survive classical peptide synthesis work-up conditions (sodium bicarbonate–citric acid washing) without any problems, but the phosphonate diphenyl esters are susceptible to hydrolysis in strong aqueous acidic or basic conditions. This sensitivity is similar to that of other amino acid ester derivatives used in peptide synthesis.

[4] J. Oleksyszyn and J. C. Powers, *Biochemistry* **30**, 485 (1991).
[5] C-L. J. Wang, T. L. Taylor, A. J. Mical, S. Spitz, and T. M. Reilly, *Tetrahedron Lett.* **33**, 7667 (1993).

The phosphonate diphenyl ester moiety survives a variety of other reaction conditions used during peptide synthesis. These include (a) catalytic hydrogenolysis in presence of a Pd/C catalyst, which is used to remove the benzyloxycarbonyl or benzyl ester groups (however, in the case of chlorophenyl esters, some loss of the chlorine atoms is observed); (b) anhydrous acidic conditions used to remove Boc groups, tert-butyl esters (CF_3COOH), or Cbz groups (HBr/acetic acid solution); (c) hydrazine, which is used to remove phthalyl groups; (d) ammonia used during the synthesis of amidino groups from iminoesters; (e) guanidinylation reaction conditions used for the synthesis of an arginine-related phosphonate diphenyl ester, and (f) oxidation (H_2O_2/AcOH) used in the synthesis of Met(O)^P from Met^P.[6]

We recommend that longer peptide phosphonates be synthesized stepwise by extension in the N-terminal direction, starting with diphenyl α-aminoalkyl phosphonates. The product peptide is then usually obtained as an equivalent mixture of diastereomers.[4] Only one diastereomer reacts with the enzyme and this is usually the diastereomer in which the ^{31}P NMR signal is shifted downfield. An alternative synthetic route involves coupling of diphenyl α-aminoalkyl phosphonates directly to longer peptides. In that case, the reaction is under kinetic control and the product is unlikely to be an equivalent mixture of diastereomers. For example, the coupling of MeO-Suc-Ala-Ala-Pro-OH with Phe^P(OPh)$_2$ gives MeO-Suc-Ala-Ala-Pro-Phe^P(OPh)$_2$ in a diasteomeric ration of 1:2.5. The less abundant isomer is the one that reacts with chymotrypsin. The correct pure diasteromer of a phosphonate inhibitor should be a much more effective inactivator than the diastereomeric mixture and it would be advantageous in certain cases to separate the mixture of stereoisomers. However, at present no general methods are available for either the synthesis of pure diastereomers or their separation from mixtures.

Synthetic Examples

Cbz-Val^P(OPh)$_2$

A mixture of benzyl carbamate (15.1 g, 0.1 mol), triphenyl phosphite (31 g, 0.1 mol), and isobutyraldehyde (10.8 g, 0.15 mol; 0.1 mol is sufficient

---

[6] Esters of α-aminoalkylphosphonic acids are analogs of natural α-amino acids and are designated by the generally accepted three-letter abbreviations for the amino acid residue followed by a superscript P. For example, diphenyl [α-(N-benzyloxycarbonylamino)ethyl] phosphonate, which is related in structure to alanine, is abbreviated as Cbz-Ala^P(OPh)$_2$. Other phosphonate abbreviations are (4-AmPhGly)^P(OPh)$_2$, diphenyl 1-amino(4-amidinophenyl)methane phosphonate, or diphenyl 4-amidinophenylglycyl phosphonate; (4-AmPhe)^P(OPh)$_2$, diphenyl 1-amino-2-(4-amidinophenyl)ethane phosphonate or diphenyl 4-amidinophenylalanyl phosphonate.

to obtain a satisfactory yield) is heated in 20 ml of glacial acetic acid at 80–85° for 1 hr.[2] The volatile products are removed on a rotary evaporator, and after cooling, 250 ml of methanol is added and the solution is allowed to stand at −20° for 2–12 hr. A white solid forms and is collected by filtration, washed with 20 ml of cold methanol, and air dried. For recrystallization, the solid is dissolved in a minimal amount of hot chloroform (about 10 g solid per 20 ml) and excess methanol (200 ml) is added. The solution is allowed to stand for 4 hr at −20°, and the product is collected by filtration and dried to give 22.8 g (52%), mp 104–105°. The product is analytically pure and is stable at room temperature in a dry place for several months.

Other amino acid derivatives such as Cbz-Ala^P(OPh)$_2$, Cbz-Leu^P(OPh)$_2$, Cbz-Nva^P(OPh)$_2$, Cbz-Met^P(OPh)$_2$, and Cbz-Phe^P(OPh)$_2$ have been synthesized using analogous reactions.[2,4] This α-amidoalkylation reaction has also been used for the synthesis of phosphonate diphenyl esters containing phthalimido-protected geminal amino groups. The phthalyl group could be removed by hydrazinolysis to produce derivatives related to ornithine, lysine, and homolysine.[7] Guanidinylation of the ornithine analog provided the arginine-related phosphonate diphenyl ester.[5]

Removal of Cbz Group Using Hydrobromic Acid in Acetic Acid Solution [HBr·Val^P(OPh)$_2$]

The diphenyl α-N-benzyloxycarbonylaminoalkyl phosphonate Z-Val^P(OPh)$_2$ is dissolved in a 45% solution of hydrogen bromide in acetic acid (about 15 ml per 0.1 mol; a 35% solution is equally efficient) and after 1 hr at room temperature, the solvent is removed. The oily residue is dissolved in a minimal amount of methanol and excess anhydrous diethyl ether is added. The solution is allowed to crystalize at −10° for a few hours. The crystalline hydrobromide is collected by filtration and recrystallized from methanol–anhydrous ethyl ether; the yield is >85%, mp 174–176°. It can be used directly in peptide synthesis reactions. Using these conditions the Cbz group can also be removed from other α-N-benzyloxycarbonylaminoalkyl phosphonates as well as longer peptide phosphonates.

Removal of Cbz Group by Catalytic Hydrogenolysis

The Cbz derivative of a diphenyl α-aminoalkyl phosphonate or a longer peptidyl phosphonate is dissolved in anhydrous ethanol and a few milliliters of a 4 N solution of HCl in dioxane (HBr/AcOH can also be used) and a 5% (w/w) Pd/C catalyst are added. The solution is stirred under a

---

[7] R. Hamilton, B. J. Walker, and B. Walker, *Tetrahedron Lett.* **34**, 2847 (1993).

hydrogen atmosphere until one equivalent of hydrogen is consumed. The catalyst is removed by filtration and the solvent is evaporated to dryness. The crude residue can be used directly in peptide synthesis reactions or it can be recrystallized from ethanol/ethyl ether to obtain analytically pure material.

Peptide Synthesis [Cbz-Pro-Val^P(OPh)$_2$]

Add 0.01 mol of DCC to a solution of 0.01 mol of Cbz-Pro-OH and 0.01 mol of 1-hydroxybenzotriazole hydrate in 20 ml of tetrahydrofuran. The solution is allowed to cool in a refrigerator for about 40 min before 0.01 mol of the hydrobromide of a diphenyl α_1-aminoalkyl phosphonate and 0.01 mol of triethylamine are added. The reaction mixture is then stirred overnight at room temperature. After the dicyclohexylurea (DCU) is removed by filtration, 100 ml of ethyl acetate is added and the solution is washed several times with 5% (w/v) sodium bicarbonate and 10% (w/v) citric acid until thin-layer chromatography (TLC) analysis shows one spot. The solution is dried over anhydrous maganesium sulfate and evaporated to dryness. The oily residue is dissolved in a minimal amount of methylene chloride, and the small amount of DCU, which usually precipitates, is removed by filtration. The solvents are removed and the oily residue is dried on a vacuum pump for several hours (yield 50–85%). In some cases, the oil solidifies and can be recrystallized from dichloromethane–pentane. The product is analytically pure and can be used in the next step without further purification.

Boc-Val-Pro-Val^P(OPh)$_2$

The Cbz group is removed from 0.005 mol of Cbz-Pro-Val^P(OPh)$_2$ dissolved in anhydrous ethanol containing a few milliliters of a 4 N solution of HCl in dioxane. The oily product can be used directly. A solution of 0.005 mol of Boc-Val, 0.005 mol of hydrobenzotriazole hydrate, and 0.005 mol of DCC in tetrahydrofuran is kept in a refrigerator for 40 min and added to the oily hydrogenolysis product. After addition of 0.006 mol of triethylamine, the mixture is stirred overnight at room temperature. After a standard peptide work-up, the product is dried on a vacuum line for several hours to obtain the solid analytically pure product (yield 72%; mp 62–66°).[4]

Cbz-(4-AmPhGly)^P(OPh)$_2$

Diphenyl N-benzyloxycarbonylamino(4-cyanophenyl)methane phosphonate [Cbz-(4-CN-PhGly)^P(OPh)$_2$] can be synthesized by reaction of

9.75 g of 4-cyanobenzaldehyde, 7.65 g of benzyl carbamate, and 13.5 ml of triphenyl phosphite in 20 ml of glacial acetic acid according to the procedure described for Cbz-Val^P(OPh)$_2$ (yield 70%, mp 135–138°).[8]

A solution of 7 g of Cbz-(4-CN-PhGly)^P(OPh)$_2$ in 150 ml of dry chloroform and 15 ml of absolute ethanol is saturated with dry HCl at 0° and kept in the refrigerator until the starting material is no longer detected by TLC analysis (~24 hr). Excess pentane is added and the precipitate is collected by filtration and dried under vacuum. The solid iminoester is then dissolved in 200 ml of dry methanol and dry ammonia gas is bubbled through the solution for ~20 min (better yields are obtained by the addition of 1 equivalent of ammonia using a standardized solution of ammonia). The methanol and excess ammonia are removed as fast as possible. Fresh methanol (100 ml) is added and the solution is heated at 50° for 8 hr until TLC analysis shows the absence of the imino ether. The solvent is evaporated and the resulting oil is dissolved in chloroform. Addition of ether causes the oil to solidify. The solid is again dissolved in chloroform and precipitated with ether to give the product (yield 70–80%; mp 154–158°). The product, diphenyl N-benzyloxycarbonylamino(4-amidinophenyl)methane phosphonate hydrochloride [Cbz-(4-Am-PhGly)^P(OPh)$_2$], contains a trace of ammonium chloride, but can be used directly in peptide synthesis or for enzyme inhibition experiments.

(4-AmPhGly)^P(OPh)$_2$ · 2HCl

A 1.8-g sample of Cbz-(4-AmPhGly)^P(OPh)$_2$ is hydrogenated in 150 ml of 2 N HCl–methanol using a 5% Pd/C catalyst to give on work-up a solid that is crystallized from ethanol–ethyl to give analytically pure product (yields 60–80%; mp 213–215°).

Boc-D-Phe-Pro-(4-AmPhGly)^P(OPh)$_2$

A solution of 1.0 mmol of Boc-D-Phe-Pro-OH and 1.05 mol of CDI in 2 ml dry dimethylformamide (DMF) is stirred at 0° for 1 hr before 1 mmol (4-Am-PhGly)^P(OPh)$_2$·2HCl is added and the solution is stirred for an additional 48 hr at 0°. Water (10 ml) is added and the oil is decanted and washed with distilled water. The oil residue is dissolved in chloroform and the solution is washed with 4% NaHCO$_3$, water, and 0.05 N HCl. After drying over MgSO$_4$ and removing the solvent, the resulting oil is dried under a vacuum for a few hours to give analytically pure Boc-D-Phe-Pro-(4-AmPhGly)^P(OPh)$_2$·HCl (yield 29%; mp 185–190°).[8]

---

[8] J. Oleksyszyn, B. Boduszek, C.-M. Kam, and J. C. Powers, *J. Med. Chem.* **337,** 226 (1994).

Inhibition Kinetics and Methods

General Methods

Stock solutions of the inhibitors (0.5–5 mM) are prepared in DMSO, although other solvents such as ethanol and acetonitrile can also be used. If the stock solution is too concentrated, precipitation can occur when the inhibitor solution is added to the reaction media. On the other hand, a stock solution that is too dilute may result in an increased concentration of organic solvent in the reaction medium when high inhibitor concentrations are required. In water or buffer solutions, peptidyl derivatives of diphenyl α-aminoalkyl phosphonates have limited but usually sufficient solubility, and we did not observe any solubility problems with most inhibitors at concentrations below 10 μM when the inhibition solution contained 2–5% organic solvent. In most cases, there is no need to use concentrations higher than 10 μM in the final reaction medium. A molar inhibitor : enzyme ratio of 5 is sufficient to obtain complete inhibition of enzyme activity after 10 min if $k_{obs}/[I] > 500 \ M^{-1} \ \sec^{-1}$, but investigators should account for the presence of the nonreacting stereoisomer. With less reactive inhibitors ($k_{obs}/[I] < 500 \ M^{-1} \ \sec^{-1}$) or in the presence of the subtrate, a higher inhibitor concentration and a longer reaction time may be required. If the enzyme concentration is approximately 0.1 μM, a final concentration of inhibitor of approximately 5 μM would be a reasonable starting point. For example, if we have 1 ml of a 0.1 μM solution of the enzyme and only 1% dimethyl sulfoxide (DMSO) is acceptable in assay, then 10 μl (1% by volume for 1 ml) of the inhibitor stock solution in DMSO could be added. This would require a stock solution of 0.5 mM to give a 5 μM final concentration in enzyme assay solution after the 100-fold dilution.

Inhibition Kinetics

Inhibition rate constants can be measured using the incubation method under pseudo-first-order reaction conditions (inhibitor concentration $\geq$ 10× enzyme concentration). An aliquot of inhibitor (25 or 50 μl) in DMSO is added to 0.3–0.6 ml of buffered enzyme solution (enzyme concentration 0.03–2.5 μM) to initiate the inactivation reaction. Aliquots (25–150 μl) are removed at various time intervals and residual enzyme activity is measured using specific chromogenic or fluorogenic substrates for the enzyme. The $t_{1/2}$ for the inhibition reaction is obtained from plots of ln v_t/v_0 vs. time. For first-order reactions, the pseudo-first-order inibition rate constant k_{obs} (in $\sec^{-1}$) can be calculated from the equation $t_{1/2} = 0.693/k_{obs}$. The apparent second-order inhibition rate constants $k_{obs}/[I]$ (in $M^{-1} \ \sec^{-1}$) are reported in various tables.

The rates of inhibition of chymotrypsin, PPE, human leukocyte elastase (HLE), cathepsin G (Cat G), rat mast cell peptidase (RMCP) II, and human thrombin are measured in a 0.1 M HEPES, 0.5 M NaCl (pH 7.5) buffer containing 9% Me$_2$SO at 25°. Inhibition of bovine trypsin and human plasma kallikrein are measured in a 0.1 M HEPES, 0.01 M CaCl$_2$ (pH 7.5) buffer. Chymotrypsin and Cat G are assayed with Suc-Val-Pro-Phe-pNA (0.476 mM),[9] HLE with MeO-Suc-Ala-Ala-Pro-Val-pNA (0.482 mM),[10] PPE with Suc-Ala-Ala-Ala-pNA (0.714 mM),[11] RMCP II with Suc-Ala-Ala-Pro-Phe-SBzl (88 μM),[12] trypsin with Z-Phe-Gly-Arg-pNA (100–120 μM),[13] and thrombin and kallikrein with Z-Arg-SBzl (80–90 μM).[14] Enzymatic hydrolysis rates of peptide p-nitroanilides are measured at 410 nm (ε_{410} = 8800 M^{-1} cm^{-1})[15] and hydrolysis of thioesters is measured at 324 nm in the presence of 4,4′-dithiodipyridine (ε_{324} = 19,800 M^{-1} cm^{-1}).[16]

Stability of Peptide Phosphonates

Peptidyl derivatives of diphenyl α-aminoalkyl phosphonates show remarkable chemical stability in buffer and in human plasma. For example, the UV spectrum of MeO-Suc-Ala-Ala-Pro-Leu^P(OPh)$_2$ (0.143 mM) in 0.1 M HEPES and 0.5 M NaCl, pH 7.5, at 25° shows broad absorbance at 250–280 nm, with maxima at 260 nm (ε = 1260 M^{-1} cm^{-1}) and 265 nm (ε = 1370 M^{-1} cm^{-1}). There is no change in the spectrum after incubation for 7 days. Addition of 0.395 mM glutathione, a compound that rapidly inactivates peptide chloromethyl ketone inhibitors, has no effect on the spectrum after incubation for at least 4 days. Similarly, the UV spectrum of Cbz-(4-AmPhGly)^P(OPh)$_2$ hydrochloride shows maximum absorbance at 246 nm at neutral pH in a 0.1 M HEPES, 0.5 M NaCl (pH 7.5) buffer, and no change was observed after incubation for 3 days at room temperature. The stability of MeO-Suc-Ala-Ala-Ala-Phe^P(OPh)$_2$ in a 0.5 M NaCl, 0.1 M HEPES (pH 7.5) buffer containing 20% DMSO-d_6 was studied by ^{31}P NMR (18.62 and 18.58 ppm) and no change was observed in the spectrum after incubation for 4 days.

---

[9] T. Tanaka, Y. Minematsu, C. F. Reilly, J. Travis, and J. C. Powers, *Biochemistry* **24**, 2040 (1985).

[10] K. Nakajima, J. C. Powers, B. Ashe, and M. Zimmerman, *J. Biol. Chem.* **254**, 4027 (1979).

[11] J. Bieth, B. Spiess, and C. G. Wermuth, *Biochem. Med.* **11**, 350 (1974).

[12] J. W. Harper, G. Ramirez, and J. C. Powers, *Anal. Biochem.* **118**, 382 (1981).

[13] T. Tanaka, B. J. McRae, K. Cho, R. Cook, J. E. Frake, D. A. Johnson, and J. C. Powers, *J. Biol. Chem.* **258**, 13552 (1983).

[14] R. R. Cook, B. J. McRae, and J. C. Powers, *Arch. Biochem. Biophys.* **234**, 82 (1984).

[15] B. F. Erlanger, N. Kokowsky, and W. Cohen, *Arch. Biochem. Biophys.* **95**, 271 (1961).

[16] D. R. Grassetti and J. F. Murray, Jr., *Arch. Biochem. Biophys.* **119**, 41 (1967).

The stability of Suc-Val-Pro-Phe^P(OPh)$_2$ in human plasma has been studied by measuring the ability of a plasma solution of the inhibitor to inactivate chymotrypsin. No change in the inhibition half-life was observed during the first 8 hr of incubation ($t_{1/2}$ values for inactivation after 0, 1, 2, 3, 4, and 8 hr were 0.16, 0.15, 0.12, 0.14, 0.17, and 0.19 min, respectively). However after a 24-hr incubation time, the $t_{1/2}$ had increased to 0.4 min, indicating partial destruction of the inhibitor. The estimated half-life for hydrolysis of inhibitor in human plasma is about 20 hr.[4]

Stability of Enzyme–Inhibitor Complex

Enzyme inhibitor complexes formed by reaction of α-aminoalkyl phosphonates with serine peptidases also exhibit remarkable stability, although only a limited number of studies have been reported in the literature. Trypsin and elastases inhibited by their specific inhibitors do not regain any activity on incubation for 48 hr at pH 7.5, after removal of excess inhibitor. Chymotrypsin inactivated by Suc-Val-Pro-Phe^P(OPh)$_2$ regained 50% activity after incubation for 7.5 hr at pH 7.5. Other chymotrypsin-specific inhibitors form somewhat more stable complexes and the $t_{1/2}$ for dephosphonylation of chymotrypsin inhibited by Cbz-Phe-Pro-Phe^P(OPh)$_2$ is 26 hr.[4] Inhibited chymotrypsin complexes are much more stable at pH < 4, suggesting that the active site histidine is involved in the dephosphonylation reaction. If the long-term stability of the enzyme–inhibitor complex is important, as in crystallographic studies, we suggest that the stability be measured. However, for most investigations, the enzyme–inhibitor complexes are completely stable for all practical purposes.

Discussion

Simple Organophosphorus Inhibitors

One of the most widely used organophosphorus serine peptidase inhibitors is diisopropyl fluorophosphate (DFP), which was discovered in 1949 to inactivate chymotrypsin and trypsin stoichiometrically.[17] DFP is frequently used diagnostically in the identication and classification of proteolytic enzymes. However, DFP and related organophosphorus derivatives are extremely toxic and have been used as chemical warfare agents.[18] These derivatives are also extremely sensitive to hydrolysis due to the highly electrophilic character of the phosphorus atom. The mechanism of inactivation of serine peptidases by DFP involves phosphorylation of the

[17] E. J. Jansen, M. D. F. Nutting, R. Jang, and A. K. Ball, *J. Biol. Chem.* **179,** 189 (1949).
[18] M. F. Sartori, *Chem. Rev.* **48,** 225 (1951).

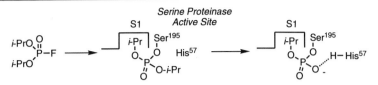

FIG. 3. Inhibition of a serine peptidase by diisopropyl fluorophosphate (DFP) to give a diisopropyl ester, which ages to a monoisopropyl ester.

active site serine (Fig. 3).[19] The catalytic apparatus of the enzyme is involved in the inhibition reaction due to the enhanced reactivity of DFP toward the active site serine compared to chemical hydrolysis or reaction with other serine residues. The phosphorylserine derivative formed on reaction with the serine oxygen is susceptible to an "aging" reaction that involves hydrolysis of one of the phosphate ester groups, as shown in Fig. 3. Bender and Wedler[20] have investigated the mechanism of "aging" of chymotrypsin inactivated by tris(4-nitrophenyl) phosphate. The initial inactivation process results in the rapid release of 1 equivalent of 4-nitrophenol; a second equivalent of nitrophenol is released more slowly in a first-order process to give the "aged" enzyme inhibitor complex. The imidazole of the active site histidine residue is probably assisting in the hydrolysis of the second nitrophenyl phosphate ester to give the aged product. The "aging" reaction also occurred during an X-ray crystallographic investigation of the diisopropyl–trypsin complex, wherein only the monoisopropyl derivative was observed.[19]

Many attempts have been made to improve the utility of simple organophosphorus inactivators by seeking a compromise that maintains their high reactivity while improving their specificity and reducing their instability and toxicity. To increase the specific interaction with the enzyme, Becker[21] synthesized a series of phosphonates with aromatic residues that were able to interact with the S_1 pocket[22] of chymotrypsin.[21] The compounds, which included structures such as 4-nitrophenyl ethyl 2-naphthylmethylene phosphonate, are more stable than phosphoryl fluorides and show significant specificity and reactivity toward chymotrypsin ($k_{inact} = 7600 \ M^{-1} \ sec^{-1}$). In a similar approach, Boter and Ooms[23] reported

[19] R. M. Stroud, L. M. Kay, and R. E. Dickerson, *J. Mol. Biol.* **83**, 185 (1974).

[20] M. L. Bender and F. C. Wedler, *J. Am. Chem. Soc.* **94**, 2101 (1972).

[21] E. L. Becker, *Biochim. Biophys. Acta* **147**, 289 (1967).

[22] The nomenclature of Schechter and Berger is used to designate the individual amino acid residues (P2, P1, P1′, P2′, etc.) of a peptide substrate and the corresponding subsites (S2, S1, S1′, S2′, etc.) of the enzyme. The scissile bond is the P1–P1′ peptide bond. I. Schechter and A. Berger, *Biochem. Biophys. Res. Commun.* **27**, 157 (1967).

[23] H. L. Boter and A. J. J. Ooms, *Biochem. Pharmacol.* **16**, 1563 (1967).

that 4-nitrophenyl ethyl 5-aminopentyl phosphonate is almost four orders of magnitude more reactive with trypsin ($5500\ M^{-1}\ \text{sec}^{-1}$) than with chymotrypsin ($1.5\ M^{-1}\ \text{sec}^{-1}$). Nayak and Bender[24] have synthesized a series of 4-nitrophenyl esters of alkyl phosphontaes as inhibitors of porcine pancreatic elastase (PPE). Due to specific interaction with the S_1 pocket of the enzyme (PPE prefers short aliphatic chain), these compounds are quite specific and inhibition rates with chymotrypsin and trypsin are three to four orders of magnitude slower. One of the best PPE inhibitors is ethyl 4-nitrophenyl pentyl phosphonate ($90\ M^{-1}\ \text{sec}^{-1}$).

Organophosphorus Analogs of Amino Acids

A variety of organophosphorus derivatives related in structure to amino acids and peptides have been synthesized as serine peptidase inhibitors. The structural relationship between an amino acid residue and the corresponding phosphonic analog is shown in Fig. 1. Lamden and Bartlett[25] reported organophosphorus compounds in which the phosphofluoridate moiety is incorporated into a blocked amino acid structure (Table I). The phenylalanine-related structure Cbz-Phe^P(O-iPr)F and the alanine-related structure Cbz-Ala^P(O-iPr)F were prepared as inhibitors of chymotrypsin and PPE.[25] The Phe derivative Cbz-Phe^P(O-iPr)F was remarkably reactive and specific. It inhibited chymotrypsin with a second-order rate constant $180,000\ M^{-1}\ \text{sec}^{-1}$, but is three orders of magnitude less reactive with PPE ($160\ M^{-1}\ \text{sec}^{-1}$). However, these compounds are extremely sensitive to hydrolysis.

The reactivity of various esters of simple N-blocked α-aminoalkylphosphonic acids are strongly dependent on the character of the ester group (Table I). Dimethyl esters such as Cbz-Phe^P(OMe)$_2$ do not react with chymotrypsin even after incubation for 22 hr at 0.5 mM.[26] The methoxy group is a very poor leaving group and replacement of only one methoxy group with a better leaving group, such as the thiomethoxy in Cbz-Phe^P(OMe)(SMe), gives an inhibitor with an inactivation rate constant of $120\ M^{-1}\ \text{sec}^{-1}$.[26] Replacing the methoxy group with the more reactive 4-nitrophenoxy leaving group yielded highly reactive chymotrypsin inhibitors such as Ac-Phe^P(OMe)(4-nitrophenoxy) ($k_{\text{inact}} = 50,000\ M^{-1}\ \text{sec}^{-1}$). However, these derivatives are very unstable ($k_{\text{hyd}} = 4.5 \times 10^{-2}\ \text{sec}^{-1}$).[21] The 3-chlorophenol esters are more stable [Ac-Phe^P(OEt) (3-chlorophenoxy), $k_{\text{hyd}} = 2.1 \times 10^{-5}\ \text{sec}^{-1}$], but they are also less reactive toward chymotrypsin ($k_{\text{inact}} = 1560\ M^{-1}\ \text{sec}^{-1}$).[21] The lysine derivative

[24] P. L. Nayak and M. L. Bender, *Biochem. Biophys. Res. Commun.* **83,** 1178 (1978).
[25] L. A. Lamden and P. A. Bartlett, *Biochem. Biophys. Res. Commun.* **112,** 1085 (1983).
[26] P. A. Bartlett and L. A. Lamden, *Bioorg. Chem.* **14,** 356 (1986).

TABLE I

IRREVERSIBLE INACTIVATION RATE CONSTANTS OF REPRESENTATIVE SERINE
PEPTIDASES BY SIMPLE ORGANOPHOSPHORUS INHIBITORS

	Rate constants ($M^{-1}\ \sec^{-1}$)		
Inhibitor	Chymotrypsin	PPE	Trypsin
Diisopropyl fluorophosphate[a]	250	72[b]	41
Dipropyl fluorophosphate[a]	33,300	—	300
Cbz-Ala^P(O-iPr)F[c]	8800	1300	—
Cbz-Phe^P(O-iPr)F[c]	180,000	160	—
Cbz-Phe^P(OMe)$_2$[d]	NI[e]	—	—
Cbz-Phe^P(OMe) (SMe)[d]	120	—	—
Cbz-Phe^P(OMe) (SPh)[d]	220	—	—
Ac-Phe^P(OMe)(p-nitrophenoxy)[f]	50,000	—	—
Ac-Phe^P(OEt)(m-chlorophenoxy)[f]	1560	—	—
Ac-Lys^P(OEt)(m-chlorophenoxy)[f]	—	—	2600
Cbz-Val^P(OPh)$_2$[g]	NI	9[h]	—
Cbz-Nva^P(OPh)$_2$[g]	NI	9.7[h]	—
Cbz-Met^P(OPh)$_2$[g]	30[h]	NI	—
Cbz-Leu^P(OPh)$_2$[g]	81[h]	NI	—
Cbz-Phe^P(OPh)$_2$[g]	1200[h]	NI	—
Cbz-(4-AmPhGly)^P(OPh)$_2$[i]	—	—	2000[h]

[a] J. A. Cohen, R. A. Oosterbaan, and F. Berends, this series, Vol. 11, p. 686.
[b] Estimated from M. A. Naughton and F. Sanger, *Biochem. J.* **78**, 156 (1961).
[c] L. A. Lamden and P. A. Bartlett, *Biochem. Biophys. Res. Commun.* **112**, 1085 (1983).
[d] P. A. Bartlett and L. A. Lamden, *Bioorg. Chem.* **14**, 356 (1986).
[e] NI, No inhibition observed after 22 hr of incubation of the inhibitor (0.5 mM) with the enzyme.
[f] J. Fastrez, L. Jespers, D. Lison, M. Renard, and E. Sonveaux, *Tetrahedron Lett.* **30**, 6861 (1989).
[g] J. Oleksyszyn, and J. C. Powers, *Biochemistry* **30**, 485 (1991).
[h] $k_{obs}/[I]$, [$M^{-1}\ \sec^{-1}$].
[i] J. Oleksyszyn, B. Boduszek, C.-M. Kam, and J. C. Powers, *J. Med. Chem.*, **37**, 226 (1994).

Ac-Lys^P(OEt) (3-chlorophenoxy) is a good inhibitor of trypsin and urokinase (Table I),[27] but the acyl moiety used as a nitrogen blocking group
does not allow these structures to be incorporated into longer peptide
sequences. More efficient synthetic routes to phosphonates related to
arginine, lysine, and ornithine and their peptide derivatives have been published.[5,7]

[27] J. Fastrez, L. Jespers, D. Lison, M. Rendard, and E. Sonveaux, *Tetrahedron Lett.* **30**, 6861 (1989).

Peptide Phosphonate Derivatives

A variety of peptidyl derivatives of α-aminoalkyl phosphonates have now been reported as inhibitors of serine peptidases. The majority of the derivatives are α-aminoalkyl phosphonate diphenyl esters, and derivatives that have Ala^P, Val^P, Nva^P, Leu^P, Met^P, $Met(O)^P$, Pro^P, Phe^P, Lys^P, Orn^P, Arg^P, 4-AmPhGly^P, and 4-AmPhe^P residues at the P_1 position in the inhibitor structure have been described. These compounds are selective inhibitors for porcine pancreatic elastase, human leukocyte elastase, human cathepsin G, bovine chymotrypsin, rat mast cell peptidase II, bovine trypsin, bovine and human thrombin, factor XII, plasmin, urokinase-type plasminogen activator, human plasma kallikrein, and dipeptidyl-peptidase IV. Representative inhibitor structures and second-order inactivation rate constants for elastases and chymotrypsin-like enzymes are shown in Table II, and those for trypsinlike series peptidases are shown in Table

TABLE II

INHIBITION OF ELASTASE AND CHYMASES BY PEPTIDYL DERIVATIVES OF DIPHENYL α-AMINOALKYL PHOSPHONATES[a]

Inhibitor	Inhibition rate constant $k_{obs}/[I]$ (M^- sec^{-1})				
	PPE[b]	HLE[c]	Chymotrypsin[d]	Cat G[e]	RMCP II
Cbz-Val^P(OPh)$_2$	9	280	NI[g]	NI	NI
Cbz-Ala-Ala-Val^P(OPh)$_2$	340	1300	NI	—	—
MeO-Suc-Ala-Ala-Nva^P(OPh)$_2$	4200	380	50	—	—
MeO-Suc-Ala-Ala-Ala-Phe^P(OPh)$_2$	NI	NI	12,000	370	6.8
MeO-Suc-Ala-Ala-Pro-Val^P(OPh)$_2$	7100	7100	21	—	—
Boc-Ala-Ala-Pro-Val^P(OMe)(SPh)	—	266[h]	—	—	—
MeO-Suc-Ala-Ala-Pro-Leu^P(OPh)$_2$	740	140	1500	—	—
MeO-Suc-Ala-Ala-Pro-Phe^P(OPh)$_2$	NI	NI	11,000	440	39
MeO-Suc-Ala-Ala-Pro-Met^P(OPh)$_2$	44	53	570	—	—
Boc-Val-Pro-Val^P(OPh)$_2$	11,000	27,000	NI	—	—
Cbz-Phe-Pro-Phe^P(OPh)$_2$	NI	NI	17,000	5100	32
Suc-Val-Pro-Phe^P(OPh)$_2$	NI	NI	44,000	36,000	15,000

[a] Data from J. Oleksyszyn and J. C. Powers, *Biochemistry* **30**, 485 (1991).
[b] Porcine pancreatic elastase.
[c] Human leukocyte elastase.
[d] Bovine chymotrypsin.
[e] Human cathepsin G.
[f] Rat mast cell protease II.
[g] NI, Less than 5% inhibition after a 30-min incubation with the inhibitor.
[h] Data recalculated (16,000 M^{-1} min^{-1}) from N. S. Sampson and P. A. Bartlett, *Biochemistry* **30**, 2255 (1991).

III. In general, tri- and tetrapeptide phosphonate derivatives are much more potent inhibitors compared to simple amino acid derivatives. The activity of a diaryl α-aminoalkyl phosphonate derivative toward a specific serine peptidase correlates with the specificity of that enzyme toward peptide substrates. For example, elastases are inhibited by phosphonates with C-terminal $Val^P(OPh)_2$ residues and not by those with $Phe^P(OPh)_2$ residues. In contrast, chymotrypsin and chymases are inhibited by phosphonates with C-terminal $Phe^P(OPh)_2$ residues and not by the Val phosphonates. Trypsinlike enzymes are inhibited by phosphonates containing $Arg^P(OPh)_2$ or the arginine analogs $(4-AmPhe)^P(OPh)_2$ and $(4-AmPhGly)^P(OPh)_2$. Interestingly, thrombin is inhibited not only by arginine-related phosphonates, but also by phosphonate with a 3-methoxypro-

TABLE III

INHIBITION OF TRYPSIN-LIKE ENZYMES BY PEPTIDYL α-AMINOALKYLPHOSPHONATE DIPHENYL ESTER INHIBITORS

Inhibitor	Inhibition rate constant $k_{obs}/[I]$ (M^{-1} sec^{-1})		
	Human thrombin	Human plasma kallikrein	Bovine trypsin
Cbz-(4-AmPhGly)^P(OPh)$_2$[a]	80	18,000	2000
Cbz-Pro-(4-AmPhGly)^P(OPh)$_2$[a]	20	60	100
Boc-D-Phe-Pro-(4-AmPhGly)^P(OPh)$_2$[a]	11,000	160	2200
D-Phe-Pro-(4-AmPhGly)^P(OPh)$_2$[a]	700	250	110
Cbz-(4-AmPhe)^P(OPh)$_2$[a]	0.2	—	24
Boc-D-Phe-Pro-(4-AmPhe)^P(OPh)$_2$[a]	0.7	—	130
D-Phe-Pro-(4-AmPhe)^P(OPh)$_2$[a]	24	—	50
Ac-D-Phe-Pro-Arg^P(OMe)$_2$[b]	NI[c]	—	—
Ac-D-Phe-Pro-Arg^P(OPh)$_2$[b]	1150[d]	—	—
Cbz-D-Phe-Pro-HNCH(n-pentyl)P(O)(OPh)$_2$[e]	19 nM	—	—
Dpa-Pro-HNCH(n-pentyl)P(O)(OPh)$_2$[e]	0.94 nM	—	—
D-Phe-Pro-HNCH(3-methoxypropyl)P(O)(OPh)$_2$[e]	12 nM	—	—
Dpa-Pro-HNCH(3-methoxypropyl)P(O)(OPh)2[e]	2.4 nM	—	—

[a] J. Oleksyszyn, B. Boduszek, C.-M. Kam, and J. C. Powers, *J. Med. Chem.* **37**, 226 (1994).

[b] C.-L. J. Wang, T. L. Taylor, A. J. Mical, S. Spitz, and T. M. Reilly, *Tetrahedron Lett.* **33**, 7667 (1993).

[c] NI, No inhibition after a 10-min incubation at an unknown inhibitor concentration.

[d] Recalculated from C.-L. J. Wang, T. L. Taylor, A. J. Mical, S. Spitz, and T. M. Reilly, *Tetrahedron Lett.* **33**, 7667 (1993).

[e] Data from L. Cheng, C. A. Goodwin, M. F. Scully, V. V. Kakkar, and G. Claeson, *Tetrahedron Lett.* **32**, 7333 (1991). The inhibitory potency is reported as IC_{50}, the inhibitor concentration required to inhibit 50% of the enzyme activity after preincubation of the enzyme and inhibitor at 37° for 1 hr. Other conditions were not specified.

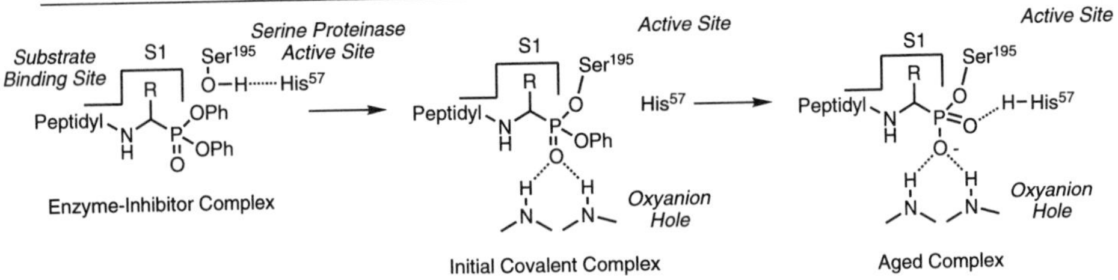

Serine Proteinase

Substrate   S1   Ser^{195} *Active Site*

Binding Site

Peptidyl — N — P(OPh)(OPh), O—H······His^{57}

Enzyme–Inhibitor Complex

Active Site

S1   Ser^{195}

Peptidyl — N — P, His^{57}, OPh, Oxyanion Hole

Initial Covalent Complex

Active Site

S1   Ser^{195}

Peptidyl — N — P=O, H—His^{57}, Oxyanion Hole

Aged Complex

FIG. 4. Mechanism of inhibition of a serine peptidase by an α-aminoalkyl phosphonate diphenyl ester derivative. The inhibitor first forms a reversible enzyme–inhibitor complex. Phosphonylation of the active site serine residue yields a covalent phosphonate diester, which ages to a monoester.

pyl substituent, such as D-Phe-Pro-HNCH(CH$_2$CH$_2$CH$_2$OCH$_3$)P(O)-(OPh)$_2$.[28]

Mechanism of Inhibition

The mechanism of inhibition of serine peptidases by peptidyl α-aminoalkyl phosphonate diphenyl ester derivatives is shown in Fig. 4. After initial formation of an E·I complex, the active site serine attacks the phosphorous atom via a pentacoordinate intermediate to form a tetrahedral inhibition product with the loss of a single phenoxy group. Subsequently, the second phenoxy group is lost in an aging process similar to that observed with DFP. Crystal structures of HLE, trypsin, and thrombin inhibited by three different phosphonate diphenyl ester derivatives have been completed, but not yet published. In each case, both phenoxy groups of the inhibitor have been lost, although it is not yet clear whether this is a rapid process or takes places slowly during the crystallization experiments.

The starting diphenyl α-aminoalkyl phosphonates used in peptide synthesis are racemates and thus the peptide derivatives derived from them are mixtures of diastereomers. It appears that only one diastereomer inhibits the enzyme, most likely the diastereomer with the configuration at the α-carbon identical to an L-amino acid, although this has not yet been demonstrated. For example, when chymotrypsin (1.3 μM) is allowed to react with Suc-Val-Pro-Phe^P(OPh)$_2$ (1.3 μM, diastereomeric ratio 1 : 1), only one diastereomer reacts. After 1.9 min, 50% inhibition is observed and no further inhibition is observed after incubation for an additional 60 min. When a higher concentration of this same inhibitor (2.6 μM) is

[28] L. Cheng, C. A. Goodwin, M. F. Scully, V. V. Kakklar, and G. Claeson, *Tetrahedron Lett.* **32**, 7333 (1991).

allowed to react with chymotrypsin (1.3 μM), complete inhibition is observed. The second-order inhibition rate is 146,000 M^{-1} sec^{-1}.[4]

In the design of new inhibitors, it is possible to imagine replacing the two phenoxy groups in the α-aminoalkyl phosphonate diphenyl ester moiety by two different ester groups. However, this would introduce chirality at the phosphorus atom and generate a mixture of diastereomers. This would complicate the inhibition reaction kinetics because each diastereomer would interact differently with enzyme.[20] It also appears that replacing the phenoxy groups with structures that can interact with the S_1'–S_2' subsites of the enzyme is not an advantageous change with phosphorus derivatives and results in decreased inactivation rate constants. For example, the extended peptide structure Boc-Ala-Ala-Pro-Val^P (SPh)[O-CH(CH$_3$)-CO-Ala-OMe][29] inhibits human leukocyte elastase with a rate constant of 9 M^{-1} sec^{-1}. In contrast, the diphenyl ester derivative MeO-Suc-Ala-Ala-Pro-Val^P(OPh)$_2$,[30] which has minimal interaction with the S_1' subsite of the enzyme, inhibits human leukocyte elastase with a rate constant of 7100 M^{-1} sec^{-1}. We believe the difference in reactivity of the two derivatives can be explained by differing stereoelectronic requirements for the transition state of nucleophilic substitution on the phosphorus atom compared to a carbonyl carbon atom (tetrahedron versus trigonal bipyramid). Strong interactions between the extended structure in the ideal phosphorus peptidomimetic with the extended substrate binding region in the active site of enzyme make nucleophilic substitution on the phosphorus atom very difficult, because the phosphorus atom already has a tetrahedral geometry resembling the transition state for peptide hydrolysis.

Inhibitor Design

Peptide derivatives of α-aminoalkyl phosphonate diphenyl esters appear to be an optimum compromise between chemical stability and inhibitory potency. They show remarkable chemical stability compared to other phosphorus analogs, such as phosphonate thiophenyl esters or phosphoryl fluorides. They bind to serine peptidases in a "substratelike" manner because inhibitors with a sequence identical to the sequence of the best substrate are usually the best inactivators for a particular enzyme. This indicates that the catalytic apparatus of the enzyme is needed for nucleophilic substitution on the phosphorus atom and phosphonylation of the active site serine residue. The inhibition product has an ideal tetrahedral

[29] N. S. Sampson and P. A. Bartlett, *Biochemistry* **30**, 2255 (1991).
[30] J. Oleksyszyn and J. C. Powers, *Biochem. Biophys. Res. Commun.* **161**, 143 (1989).

geometry around the phosphorus atom, as shown by a ^{31}P NMR study, and this is similar to the tetrahedral intermediate involved in peptide bound hydrolysis. As a result, it is often not necessary to increase the chemical reactivity or electrophilicity of the phosphorous atom in the inhibitor structures by the addition of better leaving groups. The inhibitor structures are chemically stable and very resistant to hydrolysis at neutral pH, properties that are lost if the phenoxy groups are replaced by better leaving groups.

Inhibitor specificity toward different serine peptidases can be controlled by changing the amino acid sequence of the inactivator. The interaction of the inhibitor with the S_1 pocket of the target enzyme is the predominant factor affecting potency and specificity with α-aminoalkyl phosphonate diphenyl ester derivatives. Thus, it is easy to obtain an inhibitor specific for elastase, chymotrypsin, or trypsin by altering the P_1 residue (Table II). Moreover, it is even possible to distinguish two closely related chymotrypsin-like enzymes, both with a preference for Phe as the P_1 residue, by changing the amino acid sequence of the inhibitor molecule. The optimal sequence of the inhibitor should be based first on an optimum substrate sequence and/or the sequence of other peptidyl inhibitors. This can be further improved by structure–activity studies. The P_5–P_4 positions in peptidyl inhibitors are often not critical for good interactions with enzymes and can tolerate a variety of functional groups, and thus these positions can be used to introduce other physiochemical properties (e.g., solubility) into the inhibitor structure.

Advantages and Summary

Peptidyl derivatives of α-aminoalkyl phosphonate diphenyl esters have a number of advantages for *in vitro* and *in vivo* experiments compared to other commonly used peptide serine peptidase inhibitors. They are easily synthesized, are chemically very stable, and are not alkylating agents such as the commonly used peptide chloromethyl ketone serine peptidase inhibitors. They are more stable than most other organophosphorus inhibitors, including peptidyl derivatives of the α-aminoalkyl phosphonates, where the phosphonate moiety is chemically activated by the presence of better leaving groups. The α-aminoalkyl phosphonate diphenyl esters have outstanding stability ($t_{1/2}$ usually greater than 4 days at pH 7.5; >24 hr in plasma). Thus, low inhibitor concentrations can effectively control unwanted serine peptidase activity with low inhibitor concentrations over long time periods, which makes them perfect tools for experiments involving cells. Because α-aminoalkyl phosphonate diphenyl esters are irreversible inhibitors, they offer real advantages in many experimental situations over reversible inhibitors in cases in which it may be necessary to maintain high concentrations of the reversible inhibitor for long time periods.

The second-order inhibition rate constants for phosphonate inhibitors are usually not as high as those observed with other types of peptidyl serine peptidase inhibitors. This is compensated for by their high stability and specificity. The irreversible character of the inhibition reaction allows effective inhibition even if the inactivation rate constant is not large. For example, Cbz-Val^P(OPh)$_2$ inhibits HLE with a rate constant of 260 M^{-1} sec^{-1}. Thus at an effective concentration of 10 μM, 50% of the enzyme is inactivated after 4.5 min, and almost no activity is detected after an 11-min incubation time.

Frequently there is a need to specifically inhibit serine peptidases *in vitro* during protein purification procedures or in biological experiments involving cells or tissue culture. Typically, peptide chloromethyl ketone derivatives are used. However, these inactivators are quite nonspecific alkylating agents and experimental results can be misleading. For example, the presence of a chymotrypsin-like enzyme activity on the neutrophil membrane was assumed when inhibition with Tos-Phe-CH$_2$Cl resulted in inhibition of the so-called oxidative burst of these cells.[31] However, it has been shown that the targeted protein is not a serine peptidase, and inhibition results from a nonspecific alkylation reaction.[32] As another example of the utility of phosphonates, dipeptide derivatives of α-aminoalkyl phosphonate diphenyl ester derivatives with a P$_1$ proline residue are effective inhibitors for dipeptidyl-peptidase IV.[33] The corresponding dipeptide boronic acid and chloromethyl ketone derivatives are unstable.

In summary, peptidyl derivatives of α-aminoalkyl phosphonate diphenyl esters are highly specific irreversible inhibitors of serine peptidases and are chemically stable and stable in plasma. They offer a number of advantages over other types of inhibitors currently in use in biological experiments. After reaction with the enzyme, they form very stable enzyme–inhibitor complexes, making them interesting tools for X-ray studies on the active site structure of new serine peptidases.

Acknowledgment

This work has been supported by Grants HL29307, HL34035, and GM42212 from the National Institutes of Health.

[31] S. Kitagawa, F. Takaku, and S. J. Sakamoto, *J. Clin. Invest.* **65,** 74 (1980).
[32] E. C. Conseiller, D. Schott, and F. Lederer, *Eur. J. Biochem.* **193,** 345 (1990).
[33] B. Boduszek, J. Oleksyszyn, C.-M. Kam, J. Selzler, R. E. Smith, and J. C. Powers, *J. Med. Chem.* unpublished results (1995).

[31] Isocoumarin Inhibitors of Serine Peptidases

By JAMES C. POWERS and CHIH-MIN KAM

Introduction

Isocoumarins are potent irreversible heterocyclic inhibitors of serine peptidases.[1] 3,4-Dichloroiscoumarin (DCI)[2] (Fig. 1) is a general serine peptidase inhibitor that inhibits most serine peptidases and many esterases. It is most effective against serine peptidases that prefer noncharged amino acid residues, such as human leukocyte elastase (HLE),[3] but it has inhibited all serine peptidases examined thus far with the exception of two complement proteins, C2a and Bb, and subtilisin, which turns over DCI. DCI is quite specific and will not inhibit metalloexopeptidases such as leucine aminopeptidase or β-lactamase. It does not appear to inhibit thiol peptidases belonging to the papain family, possibly due to rapid turnover of the acyl-enzyme derivatives, but does inhibit calpain.

Isocoumarins such as 3-alkoxy-4-chloro-7-substituted derivatives (Fig. 1) are more specific inhibitors of serine peptidases, including HLE, PPE, proteinase 3, chymotrypsin, cathepsin G, RMCP I, RMCP II, lymphocyte granzymes, and the trypsinlike enzymes that initiate blood coagulation and the complement cascade.[4-11] The 3-alkoxy group confers selectivity

[1] J. C. Powers and J. W. Harper, in "Proteinase Inhibitors" (A. J. Barrett and G. S. Salvensen, eds.), p. 55. Elsevier, Amsterdam and New York, 1986.

[2] Abbreviations: ADMP, 3,5-dimethylpyrazole-1-carboxamidine nitrate; Cat G, cathepsin G; ChyT, chymotrypsin; CIP, capillary injury-related proteases; CiTPrOIC, 4-chloro-3-(3-isothioureidopropoxy)isocoumarin; DCI, 3,4-dichloroisocoumarin; DFP, diisopropyl fluorophosphate; DMF, dimethylformamide; Gua, guanidino; HEPES, N-2-hydroxyethyl-piperazine-N'-2-ethanesulfonic acid; IC, isocoumarin; MS (FAB), fast atom bombardment mass spectroscopy; NA, 4-nitroanilide; PMSF, phenylmethylsulfonyl fluoride; PPE, porcine pancreatic elastase; RMCP I, rat mast cell peptidase I; RMCP II, rat mast cell peptidase II; Tris, tris(hydroxymethyl)aminomethane; Tos, toluenesulfonyl; SBzl, thiobenzyl ester; Z, benzyloxycarbonyl.

[3] J. W. Harper, K. Hemmi, and J. C. Powers, Biochemistry 24, 1831 (1985).

[4] J. W. Harper and J. C. Powers, Biochemistry 24, 7200 (1985).

[5] C.-M. Kam, K. Fujikawa, and J. C. Powers, Biochemistry 27, 2547 (1988).

[6] J. C. Powers, J. Oleksyszyn, S. L. Narasimhan, C.-M. Kam, R. Radhakrishnan, and E. F. Meyer, Jr., Biochemistry 29, 3108 (1990).

[7] S. Odake, C.-M. Kam, L. Narasimhan, M. Poe, J. T. Blake, O. Krahenbuhl, J. Tschopp, and J. C. Powers, Biochemistry 30, 2217 (1991).

[8] C.-M. Kam, T. J. Oglesby, M. K. Pangburn, J. E. Volanakis, and J. C. Powers, J. Immunol. 149, 163 (1992).

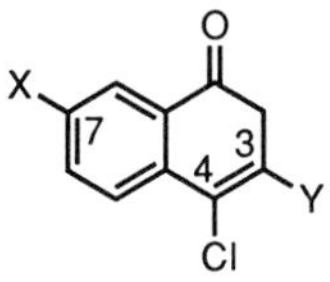

DCI	X = H	Y = Cl
3-Alkoxy-7-amino-4-chloroisocoumarin	X = NH$_2$	Y = OR
3-Alkoxy-4-chloro-7-guanidinoisocoumarin	X = NHC(=NH$_2^+$)NH$_2$	Y = OR
NH$_2$-CiTPrOIC	X = NH$_2$	Y = O(CH$_2$)$_3$-SC(=NH$_2^+$)NH$_2$

FIG. 1. Structures of DCI and substituted isocoumarins.

to the isocoumarins for various serine peptidases. For example, isocoumarins with 3-benzyloxy or 3-phenylethoxy substituents are potent inhibitors of chymotrypsin-like enzymes, and compounds with 3-methoxy or 3-ethoxy groups are effective inhibitors of elastases.[4] Isocoumarins containing basic substituents such as guanidino or isothioureidoalkoxy groups are potent inhibitors of trypsinlike enzymes.[5,7–9,11]

Isocoumarins are mechanism-based or "suicide" inhibitors wherein the isocoumarin ring is opened by the active site serine residue of a serine peptidase to form an acyl-enzyme derivative. Simultaneously, a new reactive structure is unmasked, which can react further with an active site nucleophile such as histidine-57 to form an alkylated acyl-enzyme derivative, a doubly covalent enzyme–inhibitor complex. This mechanism has been confirmed by several X-ray structures of complexes of serine peptidases with a variety of isocoumarins. Both acyl-enzyme derivatives and alkylated acyl-enzyme derivatives have been observed.

Synthesis

Materials

Homophthalic acid, PCl$_5$, POCl$_3$, ADMP, and 4,4'-dithiodipyridine (Aldrithiol-4) are obtained from Aldrich Chemical Co. (Milwaukee, WI). Thiourea is provided by Fisher Scientific (Norcross, GA), chlorine gas is

[9] M. Orlowski, M. Lesser, J. Ayala, A. Lasdun, C.-M. Kam, and J. C. Powers, *Arch. Biochem. Biophys.* **269**, 125 (1989).

[10] C.-M. Kam, J. E. Kerrigan, K. M. Dolman, R. Goldschmeding, A. Von dem Borne, and J. C. Powers, *FEBS Lett.* **297**, 119 (1992).

[11] C.-M. Kam, G. P. Vlasuk, D. E. Smith, K. E. Arcuri, and J. C. Powers, *Thromb. Haemostasis* **64**, 133 (1990).

obtained from Matheson Gas Products (Morrow, GA), and N-blocked amino acids are obtained from Bachem Bioscience Inc. (Philadelphia, PA) or other sources. HEPES is purchased from Research Organics Inc. (Cleveland, OH). Many of these materials are also available from other sources.

General Synthetic Procedures

Substituted isocoumarins are prepared by different synthetic methods, depending on the exact substitutents, and some detailed procedures are described below. DCI has been prepared by cyclization of homophthalic acid with PCl_5 in $POCl_3$ to give 3-chloroisocoumarin,[12] which is then converted to 3,3,4-trichloro-3,4-dihydroisocoumarin by reaction with Cl_2. Elimination of HCl from the trichloro derivative gives 3,4-dichloroiso-coumarin.[13] 3-Alkoxy-7-amino-4-chloroisocoumarins are prepared by cy-clization of the appropriate alkyl 2-carboxy-4-nitrophenylacetate deriva-tives with PCl_5 to give a 3-alkoxy-4-chloro-7-nitroisocoumarin followed by catalytic reduction of the nitro group.[4] 3-Alkoxy-4-chloro-7-guanidinoi-socoumarins are synthesized by hydrogenolysis of alkyl 2-carboxy-4-nitro-phenylacetates to give the corresponding amino compounds followed by guanidination with ADMP, cyclization with PCl_5, and chlorination with N-chlorosuccinimide.[5] 4-Chloro-3-isothioureidoalkoxyisocoumarins are prepared by the reaction of the corresponding 3-bromoalkoxyisocoumar-ins with thiourea.[5] Derivatives of 3-alkoxy-7-amino-4-chloroisocoumarin substituted at the 7-position with N-substituted ureas, thioureas, acyl groups, or carbonates are prepared by reaction of the isocoumarins with the appropriate isocyanate, isothiocyanate, N-blocked amino acid, or alkyl chloroformate.[6,14] The detailed syntheses of a few representative isocoum-arins are given below.

3,4-Dichloroisocoumarin

Homophthalic acid (20 g, 0.11 mol) is heated with PCl_5 (50 g, 0.24 mol) in $POCl_3$ (50 g) at 140–150° for 3 hr, and then $POCl_3$ is removed in vacuo.[12] Water is added to the residue and 3-chloroisocoumarin is steam distilled and further purified by silica gel column chromatography twice. The first column is eluted with CH_2Cl_2, and the second one is eluted with benzene

---

[12] W. Davies and H. G. Poole, J. Chem. Soc., 1616 (1928).
[13] V. B. Milevskaya, R. V. Belinskaya, and L. M. Yagupol'skii, J. Org. Chem. USSR (Engl. Transl.) 9, 2160 (1973).
[14] M. A. Hernandez, J. C. Powers, J. Glinski, J. Oleksyszyn, J. Vijayakshmi, and E. F. Meyer, Jr., J. Med. Chem. 35, 1121 (1992).

to give 3-chloroisocoumarin as a white solid (yield 2 g, 10%; mp 95–96°). Analysis (calculated) for $C_9H_5O_2Cl$ gives C (59.85), H (2.77), Cl (19.65); found: C (59.74), H (2.83), Cl (19.71).

Chlorine gas is bubbled through a solution of 3-chloroisocoumarin (0.45 g, 2.5 mmol) in CCl_4 at 20° for 50 min.[13] The solvent and Cl_2 are then removed using a water aspirator without heating to give 3,3,4-trichloro-3,4-dihydroisocoumarin [yield 0.58 g, 90%; TLC (benzene) R_f 0.73; ^1H NMR δ (CDCl$_3$) 8.1 and 7.6 (m, 4H), 5.5 (s, 1H)]. 3,3,4-Trichloro-3,4-dihydroisocoumarin (0.58 g, 2.3 mmol) is dissolved in a few milliliters of anhydrous ether, then Et$_3$N (0.23 g, 2.3 mmol) in 2 ml of ether is added and the precipitate Et$_3$N · HCl is removed by filtration. The solvent is removed and the residue is purified by silica gel chromatography using benzene as an eluant to give DCI as a white crystal solid [yield 0.25 g, 50%; mp 97–98°; ^1H NMR δ (CDCl$_3$) 8.3 and 7.8 (m, 4H)]. Analysis (calculated) for $C_9H_4O_2Cl_2$ give C (50.26), H (1.86), Cl (32.99); found: C (50.17), H (1.92), Cl (32.91).

7-Amino-4-chloro-3-methoxyisocoumarin

Nitrohomophthalic acid (18 g, 0.08 mol) is dissolved in 160 ml of methanol and 1 ml of conc. H_2SO_4 is added.[4] The mixture is stirred at room temperature for 48 hr. The solvent is removed *in vacuo*, 600 ml of 5% NaHCO$_3$ (w/v) is added to the residue, and the aqueous solution is extracted with 300 ml of benzene. The aqueous layer is cooled and acidified with 6 N HCl to pH 2–3; a white solid is formed, filtered, and dried to give 17.5 g (90%) of methyl nitrohomophthalate [mp 167–168°; TLC (CHCl$_3$: methanol, 5 : 1) R_f 0.25; ^1H NMR δ (acetone-d_6) 8.84 (d, 1H), 8.39 (dd, 1H), 7.70 (d, 1H), 4.23 (s, 2H), 3.64 (s, 3H)]. Methyl nitrohomophthalate (5 g, 0.02 mol) is mixed with 350 ml of benzene, PCl$_5$ (9.2 g, 0.04 mol) is added, and the mixture is refluxed for 3 hr. The solvent is removed *in vacuo* and the yellow residue is purified by silica gel chromatography using benzene as an eluant to give 1.8 g (34%) of 7-nitro-4-chloro-3-methoxyisocoumarin [mp 128–130°; TLC (benzene : hexane, 8 : 2) R_f 0.31; ^1H NMR δ (CDCl$_3$) 9.05 (d, 1H), 8.54 (dd, 1H), 7.83 (d, 1H), 4.19 (s, 3H)]. The nitro compound (1.1 g, 4.3 mmol) is dissolved in 50 ml of ethyl acetate : methanol (4 : 1), 5% of Pd/C (0.1 g) is added, and the mixture is hydrogenated. After hydrogenolysis, the catalyst is removed by filtration, the solvent is removed by evaporation, and the residue is purified by silica gel chromatography using CH$_2$Cl$_2$ as an eluant to give 0.34 g (35%) of the final product [mp 163–165°; TLC (CH$_2$Cl$_2$) R_f 0.19; ^1H NMR δ (CDCl$_3$) 7.56 (d, 1H), 7.46 (d, 1H), 7.13 (dd, 1H), 4.05 (s, 3H)].

7-(N-Tosylphenylalanylamino)-4-chloro-3-methoxyisocoumarin

The acid chloride of N-tosylphenylalanine (77 mg, 0.23 mmol) and 7-amino-4-chloro-3-methoxyisocoumarin (50 mg, 0.22 mmol) are mixed in CH_2Cl_2/tetrahydrofuran (THF) (1 : 1) with stirring and Et_3N (0.037 ml) in 2 ml CH_2Cl_2 is added dropwise.[4] The mixture is stirred at room temperature for 2 hr, the solvent is removed by evaporation, and the residue is dissolved in ethyl acetate, which is washed with 10% citric acid (3 × 30 ml) and 4% $NaHCO_3$ (2 × 30 ml), dried over $MgSO_4$, and concentrated in vacuo. The crude product is purified by silica gel chromatography using 1% methanol in CH_2Cl_2 as an eluant. The product (22 mg) is crystallized from methanol–isopropyl ether to give a pale yellow solid [mp 222–224°; TLC (CH_2Cl_2) R_f 0.3; MS m/e 527 (M^+ + 1)]. Analysis (calculated) for $C_{26}H_{23}Cl$-N_2O_6S · $0.5H_2O$ gives C (58.26), H (4.53); found: C (58.28), H (4.50).

7-Guanidino-4-chloro-3-methoxyisocoumarin

Hydrogenolysis of methyl 2-carboxy-4-nitrophenylacetate (described above) gives methyl 4-amino-2-carboxyphenylacetate (90%).[5] Methyl 4-amino-2-carboxyphenylacetate (2.2 g, 10 mmol), Et_3N (1.9 g, 19 mmol), and ADMP (3.0 g, 15 mmol) are heated in 50 ml of THF at reflux for 18 hr. The white precipitate is filtered and washed with cold methanol to give methyl 2-carboxy-4-guanidinophenylacetate nitrate [1.5 g, 46%; TLC R_f 0.6 (butanol : acetic acid : water : pyridine, 4 : 1 : 1 : 2); ^{1}H NMR δ (CH_3 COOH) 8.4, 7.7 (b, 4H), 6.6 (b, 4H), 4.4 (s, 2H), 4.1 (s, 3H)]. Analysis (calculated) for $C_{11}H_{13}N_3O_4$ · $0.5H_2O$ gives (50.77), H (5.42), N (16.15); found: C (51.03), H (5.38), N (16.19).

Methyl 2-carboxy-4-guanidinophenylacetate (0.9 g, 3 mmol) is heated with PCl_5 (1.5 g, 7.2 mmol) in a few milliliters of THF at 70–80° for 2 hr. A white precipitate is formed, filtered, and purified by silica gel chromatography using CH_2Cl_2 : methanol (5 : 1) as an eluant to give 7-guanidino-3-methoxyisocoumarin [0.5 g, 59%; mp 185–186°; TLC, R_f 0.7 (butanol : a-cetic acid : water : pyridine, 4 : 1 : 1 : 2); ^{1}H NMR δ (DMSO-d_6) 7.9, 7.6 (b, 3H), 7.7 (b, 4H), 6.1 (s, 1H), 3.9 (s, 3H); MS (FAB) m/e 234 (M^+ − Cl)]. Analysis (calculated) for $C_{11}H_{12}N_3O_3Cl$ · $0.5H_2O$ gives C (47.40), H (4.67), N (15.08), Cl (12.75); found: C (47.42), H (4.74), N (15.05), Cl (12.68).

7-Guanidino-3-methoxyisocoumarin (0.27 g, 1 mmol) is chlorinated with N-chlorosuccinimide (0.15 g, 1.1 mmol) in 5 ml of DMF at room temperature overnight. The solvent is evaporated and the residue is purified by silica gel chromatography using CH_2Cl_2 : MeOH (5 : 1) as an eluant to give 4-chloro-7-guanidino-3-methoxyisocoumarin [0.1 g, 34%; TLC R_f 0.75 (butanol : acetic acid : water : pyridine, 4 : 1 : 1 : 2); the NMR spectrum is similar to the precursor except for the absence of a peak at 6.1 ppm;

MS (FAB) m/e 268 (M^+ − Cl)]. Analysis (calculated) for $C_{11}H_{11}N_3O_3$ Cl_{12} · 0.5H_2O gives C (42.17), H (3.83), N (13.41), Cl (22.68); found: C (42.65), H (3.72), N (13.28), Cl (22.32).

Inhibition and Assay Kinetic Methods

Substituted isocoumarins inhibit serine peptidases irreversibly and the inhibition rates can be measured by the incubation method under pseudo-first-order conditions or by the progress curve method in the presence of the substrate.

Enzymes

Chymotrypsin, trypsin, and PPE are obtained from Sigma Chemical Co. (St. Louis, MO). HLE and human cathepsin G are obtained from Dr. James Travis of the University of Georgia or Athens Research and Technology, Inc. (Athens, GA). Bovine thrombin and factor Xa are provided by Drs. Kotoku Kurachi, Kazuo Fujikawa, and Earl Davie of the University of Washington (Seattle, WA). Human factor D is provided by Dr. John Volanakis of the University of Alabama (Birmingham, AL). Human C$\overline{1}$s and C$\overline{1}$r are provided by Dr. Michael Pangburn of the University of Texas (Tyler, TX). Mouse granzyme A is provided by Dr. Jürg Tschopp of the University of Lausanne (Epalinges, Switzerland).

Isocoumarin Solutions and Stability

Stock solutions of isocoumarins (5 mM) are prepared in DMSO and stored at −20°. Most isocoumarin stock solutions are stable for several weeks. The concentration of isocoumarins can be determined by the absorbance at λ_{max} 325–380 nm (isocoumarin chromophore) using the known extinction coefficients for the wavelength. The extinction coefficient of DCI at 325 nm is 3330 M^{-1} cm^{-1} in 0.1 M HEPES, 0.5 M NaCl, and 10% DMSO (pH 7.5) buffer.[3] The hydrolysis of DCI is measured spectrometrically by following the decrease in absorbance at 325 nm. DCI has half-lives ($t_{1/2}$) of 18 and 48 min for its hydrolysis in pH 7.5 HEPES and phosphate buffer, respectively. In the presence of 0.2 mM glutathione in HEPES buffer, the half-life of DCI becomes 1 min. Generally, 3-alkoxy-7-amino-4-chloroisocoumarins are more stable than DCI. For example, 7-amino-4-chloro-3-methoxyisocoumarin has $t_{1/2}$ values of 200 and 815 min in HEPES and phosphate (pH 7.5) buffer, respectively.[4] NH$_2$-CiT-PrOIC has a $t_{1/2}$ of 90 min in a pH 7.5 HEPES buffer.[5]

Incubation Method

An aliquot (15–100 μl) of inhibitor (≤5 mM) in DMSO is added to 0.25–1.0 ml of a buffered enzyme solution (0.06–2.3 μM) to initiate the inactivation reaction. Aliquots (25–100 μl) are withdrawn at various time intervals and the residual enzymatic activity is measured. The buffers used are 0.1 M HEPES, 0.01 M CaCl$_2$ (pH 7.5) for bovine trypsin and coagulation enzymes; and 0.1 M HEPES, 0.5–0.6 M NaCl (pH 7.5–7.6) for chymotrypsin, cathepsin G, HLE, PPE, factor D, C$\overline{1}$s, and granzyme A. The DMSO concentration in the final inhibition mixtures is 8–12% (v/v).

Chymotrypsin and cathepsin G are assayed with Suc-Val-Pro-Phe-NA (0.075–0.125 mM).[15] HLE and PPE are assayed with MeO-Suc-Ala-Ala-Pro-Val-NA (0.1–0.125 mM)[16] and Suc-Ala-Ala-Ala-NA (0.6–1.2 mM),[17] respectively. Trypsin is assayed with Z-Phe-Gly-Arg-NA · HCl (0.07 mM)[18] or Z-Arg-SBzl · HCl (0.07 mM),[19] and other trypsinlike enzymes are assayed with Z-Arg-SBzl · HCl (0.07 mM).[20] Peptide 4-nitroanilide hydrolysis is measured at 410 nm (ε_{410} = 8800 M^{-1} cm^{-1})[21] and peptide thioester hydrolysis rates are measured with assay mixtures containing the thiol reagent 4,4′-dithiodipyridine (ε_{324} = 19,800 M^{-1} cm^{-1} for 4-thiopyridone).[22,23] Pseudo-first-order inactivation rate constants (k_{obs}) are obtained from plots of ln v_t/v_0 vs. time, and the correlation coefficients are usually greater than 0.98.

Progress Curve Method

The inhibition rate constant (k_{obs}/[I]) can be measured in the presence of substrate by the method of Tian and Tsou.[24] This method is especially useful for fast inhibitors. For example, inhibition of HLE by 4-chloro-3-ethoxyisocoumarin is monitored by adding 25 μl of enzyme (final concentration 8 nM) to a buffered solution of MeO-Suc-Ala-Ala-Pro-Val-NA (0.171 mM) containing 0.2–0.6 μM inhibitor and 10% DMSO. The increase in absorbance at 410 nm is followed with time until no further increase is observed. The k_{obs}/[I] values are calculated from plots of log([P]$_\infty$ − [P]$_t$)

[15] T. Tanaka, Y. Minematsu, C. F. Reilly, J. Travis, and J. C. Powers, *Biochemistry* **24**, 2040 (1985).

[16] K. Nakajima, J. C. Powers, B. M. Ashe, and M. Zimmerman, *J. Biol. Chem.* **254**, 4027 (1979).

[17] J. Bieth, B. Spiess, and C. G. Wermuth, *Biochem. Med.* **11**, 350 (1974).

[18] K. Cho, T. Tanaka, R. R. Cook, W. Kiesiel, K. Fujikawa, K. Kurachi, and J. C. Powers, *Biochemistry* **23**, 644 (1984).

[19] C.-M. Kam, B. J. McRae, J. W. Harper, M. M. Niemann, J. E. Volanakis, and J. C. Powers, *J. Biol. Chem.* **262**, 3444 (1987).

[20] B. J. McRae, K. Kurachi, R. L. Heimark, K. Fujikawa, E. W. Davie, and J. C. Powers, *Biochemistry* **20**, 7196 (1981).

[21] B. F. Erlanger, N. Kokowsky, and W. Cohen, *Arch. Biochem. Biophys.* **95**, 271 (1961).

[22] D. R. Grassetti and J. F. Murray, Jr., *Arch. Biochem. Biophys.* **119**, 41 (1967).

[23] J. C. Powers and C.-M. Kam, this series, Vol. 245.

[24] W. Tian and C. Tsou, *Biochemistry* **21**, 1028 (1982).

vs. time, where $[P]_\infty$ and $[P]_t$ are the concentrations of 4-nitroaniline at the time approaching infinity and at time t, respectively.

Inhibition Kinetics

Irreversible inhibitors usually inactivate serine peptidases by first forming a reversible enzyme–inhibitor complex (E · I) followed by covalent bond formation. (Scheme I).

$$E + I \underset{}{\overset{K_I}{\rightleftharpoons}} E \cdot I \underset{}{\overset{k_2}{\rightleftharpoons}} E\text{–}I$$

$$k_{obs} = k_2[I]/(K_I + [I])$$

SCHEME I.

K_I is the dissociation constant for the E · I complex and k_2 is the rate of the inactivation step. The k_{obs} is the apparent pseudo-first-order rate constant for the inactivation reaction and can be obtained by measuring the residual enzyme activity at different time intervals of a reaction mixture of enzyme containing a large excess of inhibitor. Both k_2 amd K_I can be obtained by measuring k_{obs} values at various inhibitor concentrations and performing a double-reciprocal plot ($1/k_{obs}$ vs. $1/[I]$).[25] Alternately, if $K_I > [I]$, $k_{obs}/[I]$ will be equal to k_2/K_I. In most cases $k_{obs}/[I]$ values have been determined for the inhibition of serine peptidases by isocoumarins.

In some cases, the serine peptidases by isocoumarins are too fast to be measured using psuedo-first-order rate conditions ($[I] > [E]$) and conventional kinetic methods. Then the second-order inhibition reaction rate constant (k_{2nd}) can be obtained using equal concentrations of enzyme and inhibitor and the equation $1/[E] = 1/[E]_0 + k_{2nd}t$, where $[E]_0$ and $[E]$ are initial enzyme concentration and free enzyme concentration at time t, respectively. The half-life ($t_{1/2}$) can be measured and k_{2nd} is calculated from the equation $k_{2nd} = 1/(t_{1/2}[E]_0)$.

When using $k_{obs}/[I]$ values from the literature, the half-life of a reaction at any particular inhibitor concentration can be calculated from the equation $t_{1/2} = 0.693/[(\text{literature } k_{obs}/[I] \text{ value}) \times (\text{inhibitor concentration used})]$. In general, 10 half-lives should be sufficient for complete inhibition. However, many isocoumarins have limited stability in aqueous solution and it may be necessary to replenish the reaction mixture with fresh inhibitor solution to obtain complete inhibition, especially with the slower inhibitors.

[25] R. Kitz and I. B. Wilson, *J. Biol. Chem.* **237**, 3245 (1962).

TABLE I
Inhibition of Serine Peptidases, Esterases, and Other Enzymes by 3,4-Dichloroisocoumarin[a]

Enzymes	[I] (μM)	$k_{obs}/[I]$ (M^{-1} sec^{-1})
Serine peptidases		
Human leukocyte elastase	1.1	8920
Porcine pancreatic elastase	8.1	2500
Human proteinase 3[b]	3.6	2600
Bovine chymotrypsin A$_\alpha$	13	570
Human leukocyte cathepsin G	49	28
Rat mast cell protease I	38	260
Rat mast cell protease II	11	580
Human skin chymase	92	27
Streptomyces griseus protease A	136	310
Subtilisin	—	Substrate
Bovine trypsin	127	198
Human thrombin[c]	340	10
Bovine thrombin	127	25
Bovine factor Xa[c]	422	0.2
Bovine factor XIa[c]	239	27
Human factor VIIa[d]	44	31
Human factor XIIa[c]	135	64
Porcine pancreatic kallikrein[c]	127	27
Human factor D	109	192
Human C2a[e]	330	NI
Human Bb[e]	330	NI
Human C$\overline{1s}$[e]	44	170
Human C$\overline{1r}$[e]	470	42
Murine granzyme A[f]	45	50
Murine granzyme B[f]	4.2	4200
Staphylococcus aureus protease V-8	18	2770
Sheep lymph capillary CIP[g]	460	39
Protease La[h]	82	30
Dipeptidyl-peptidase IV[i]	50	18%
Esterases and other enzymes		
Bovine multicatalytic proteinase complex[j]		
Chymotrypsin-like activity	4	147
Glutamyl-hydrolyzing activity	12	32.9
Trypsinlike activity	40	11.6
Acetylcholinesterase[k]	157	<0.6
Influenza C virus esterase[l]	6.3	410
Glycogen phosphorylase b[m]	100	3.4
Papain[n]	422	NI
β-Lactamase	385	NI
Leucine aminopeptidase[k]	422	NI

[a] Inhibition rates were measured in 0.1 M HEPES, 0.5 M NaCl, 8–10% DMSO, pH 7.5, and at 25° unless otherwise noted. NI, No inhibition. Data obtained from J. W. Harper, K. Hemmi, and J. C. Powers, *Biochemistry* **24**, 1831 (1985), unless otherwise indicated.

Inhibitory Potency and Specificity

Inhibition rate constants of serine peptidases by DCI and various 7-substituted 3-alkoxy-4-chloroisocoumarins are shown in Tables I–III. DCI is a general serine peptidase inhibitor that inhibits all the serine peptidases that have been tested thus far, with the exception of C2a, Bb, and subtilisin, which turns over DCI. DCI inhibits elastases, chymotrypsin-like enzymes, and trypsinlike enzymes with a wide range of $k_{obs}/[I]$ values (0.2–8920 M^{-1} sec^{-1}).[3] DCI is most effective against elastases and murine granzyme B, but it also inhibits trypsinlike enzymes such as factor Xa and thrombin, although more slowly. In many cases where comparisons can be made, DCI reacts faster with a particular serine peptidase than other general inhibitors such as DFP and PMSF.

DCI is fairly specific for serine peptidases and will not inhibit leucine aminopeptidase (a metalloexopeptidase) and β-lactamase at 390–420 μM

[b] Data obtained from C.-M. Kam, J. E. Kerrigan, K. M. Dolman, R. Goldschmeding, A. E. G. Kr. Von dem Borne, and J. C. Powers, *FEBS Lett.* **297**, 119 (1992).

[c] Inhibition measured in 0.1 M HEPES, 5 mM CaCl$_2$, 8–10% DMSO, pH 7.5, at 25°.

[d] Inhibition measured in 0.05 M HEPES, 0.15 M NaCl, 0.05 M CaCl$_2$, 5% DMSO, pH 7.5, at 25°. Data obtained from C.-M. Kam, G. P. Vlasuk, D. E. Smith, K. E. Acuri, and J. C. Powers, *Thromb. Haemostasis,* **64**, 133 (1990).

[e] Data obtained from C.-M. Kam, T. J. Oglesby, M. K. Pangburn, J. E. Volanakis, and J. C. Powers, *J. Immunol.* **149**, 163 (1992).

[f] Inhibition measured in 0.1 M HEPES, 0.01 M CaCl$_2$, 8% DMSO, pH 7.5, at 25°. Data obtained from S. Odake, C.-M. Kam, L. Narashimhan, M. Poe, J. T. Blake, O. Krahenbuhl, J. Tschopp, and J. C. Powers, *Biochemistry* **30**, 2217 (1991).

[g] Data obtained from M. Orlowski, M. Lesser, J. Ayala, A. Lasdun, C.-M. Kam, and J. C. Powers, *Arch. Biochem. Biophys.* **269**, 125 (1989).

[h] Inhibition measured in 65 mM Tris, 6.5 mM MgCl$_2$, 0.36 mM ATP, 5.2% DMSO (pH 7.9) buffer at 25°.

[i] Percentage of inhibition was measured after incubation of enzyme and inhibitor in 50 mM Tris, 4% DMF (pH 7.5) buffer for 30 min at 25°.

[j] Inhibition measured in 0.01 M Tris-HCl, pH 8.0, at 25°. Data obtained from M. Orlowski and C. Michaud, *Biochemistry* **28**, 9270 (1989).

[k] Inhibition measured in 0.1 M phosphate, 10% DMSO, pH 7.0, at 25°.

[l] Inhibition measured in 10 mM phosphate, 1% DMSO, 1% CH$_3$CN, pH 7.1, at 25°. Data obtained from R. Vlasak, T. Muster, A. M. Laura, J. C. Powers, and P. Palese, *J. Virol.* **63**, 2056 (1989).

[m] Inhibition measured in 0.1 M maleate, pH 6.8, at 25°. Data obtained from N. M. Rusbridge and R. J. Beynon, *FEBS Lett.* **268**, 133 (1990).

[n] Recent results in the authors' laboratory indicate that papain may be turning over DCI.

TABLE II
INHIBITION RATES OF ELASTASES AND CHYMOTRYPSIN-LIKE ENZYMES
BY SUBSTITUTED ISOCOUMARINS[a]

Compounds	$k_{obs}/[I]$ (M^{-1} sec^{-1})			
	HLE	PPE	Cat G	ChyT
7-NH$_2$-4-Cl-3-OMeIC[b]	10,000	1040	17	110
7-NH$_2$-4-Cl-3-(OCH$_2$CH$_2$Br)IC[c]	200,000	1000	410	1200
7-NH$_2$-4-Cl-3-(OCH$_2$CH$_2$Ph)IC[b]	40	NI	70	1170
4-Cl-3-(OCH$_2$Ph)IC[b]	1500	6	1140	16,000
7-CH$_3$CONH-4-Cl-3-OMeIC[d]	30,000	2300	—	—
7-Ph$_2$CHCONH-4-Cl-3-OMeIC[d]	52,000	20	—	—
7-(N-Tos-Phe-NH)-4-Cl-3-OMeIC[b]	190,000	6500	NI	150

[a] Inhibition rates were measured in 0.1 M HEPES, 0.5–0.6 M NaCl, 8–10% DMSO, pH 7.5–7.6, at 25°. NI, No inhibition.
[b] Data obtained from J. W. Harper and J. C. Powers, *Biochemistry* **24**, 7200 (1985).
[c] Data obtained from J. Vijayakshmi, E. F. Meyer, Jr., C.-M. Kam, and J. C. Powers, *Biochemistry* **30**, 2176 (1991), and also from J. E. Kerrigan, J. Oleksyszyn, C.-M. Kam, J. Selzler, and J. C. Powers, unpublished results (1994).
[d] Data obtained from M. A. Hernandez, J. C. Powers, J. Glinski, J. Oleksyszyn, J. Vijayakshmi, and E. F. Meyer, Jr., *J. Med. Chem.* **35**, 1121 (1992).

concentrations. DCI will inhibit several esterases (Table I) as expected, because these enzymes are mechanistically quite similar to serine peptidases. It also inhibits three of the enzymatic activities of the multicatalytic proteinase complex. The situation with cysteine peptidases is not clear at present. Papain is not inhibited by DCI at 422 μM, but results in the authors' laboratory indicate that papain may be turning over DCI. DCI will inhibit calpain[26] and there is an unpublished report that interleukin 1β-converting enzyme (ICE) is inhibited by DCI. In the case of calpain, an unstable acyl-enzyme is formed and, in the presence of thiols, DCI is probably an effective inhibitor only as long as excess DCI is present.

Isocoumarins with a small alkoxy group such as methoxy or ethoxy at the 3-position are potent inhibitors of HLE and moderate inhibitors of PPE (Table II). Both elastases prefer peptide substrates with small aliphatic side chains at the P$_1$ site.[16,27] Like DCI, substituted isocoumarins are most effective against HLE when compared to PPE. Isocoumarins

[26] J. E. Foreman, N. T. Luu, J. C. Powers, and D. D. Eveleth, *FASEB J.* **7**, 763 (1993).
[27] The nomenclature for the individual amino acid residues (P$_2$, P$_1$, P$_1'$, etc.) of a substrate and for the subsites (S$_2$, S$_1$, S$_1'$, etc.) of an enzyme is that of I. Schechter and A. Berger, *Biochem. Biophys. Res. Commun.* **27**, 157 (1967).

with aromatic 3-substituents such as benzyloxy or phenylethoxy groups are good inhibitors of chymotrypsin and moderate inhibitors of cathepsin G (Table II). These two enzymes prefer peptide substrates with large hydrophobic aromatic side chains at the P_1 site.[16] These inhibitors show some specificity. For example, 7-amino-4-chloro-3-methoxyiscoumarin inhibits HLE more potently than chymotrypsin by 90-fold and 3-benzyloxy-4-chloroisocoumarin inhibits chymotrypsin more potently than HLE by 10-fold.

Isocoumarins with a basic substituent such as a 7-guanidino group or an isothioureidoalkoxy at the 3-position inhibit trypsin, thrombin, factor Xa, and granzyme A more effectively than DCI (Tables I and III). These isocoumarins also show some specificity toward various trypsinlike enzymes. For example, the 7-guanidinoisocoumarins inhibit thrombin better than $C\overline{1}s$ by 80- to 440-fold and NH_2-CiTPrOIC inhibits $C\overline{1}s$ more effectively than thrombin by 40-fold. The inhibition of other enzymes such as proteinase 3,[10] RMCP I,[4] RMCP II,[4] human skin chymase,[4] lymphocyte granzymes,[7] sheep lymph CIP,[9] complement proteins,[8] and coagulation serine peptidases[5,11] by 7-substituted-4-chloro-3-alkoxyisocoumarins is not dealt with here.

Inhibition Mechanism and Structures of Enzyme–Inhibitor Complexes

The mechanism of inhibition of serine peptidases by DCI involves the acylation of the active site serine by DCI, with the formation of an acid

TABLE III
INHIBITION RATES OF TRYPSINLIKE ENZYMES BY SUBSTITUTED ISOCOUMARINS

$k_{obs}/[I]$ (M^{-1} sec^{-1})

Compounds	Bovine trypsin[a]	Bovine thrombin[a]	Bovine factor Xa[a]	Human factor D[b]	Human $C\overline{1}s$[b]	Murine granzyme A[c]
CiTPrOIC	46,000	1400	220	150	130,000	18,000
NH₂-CiTPrOIC	410,000	630	1600	55	23,000	3000
7-Gua-4-Cl-3-OMeIC	310,000	290,000	3100	250	660	15,000
7-Gua-4-Cl-3-OEtIC	>110,000	55,000	27,000	190	690	26,000
7-Gua-4-Cl-3-(OCH₂CH₂Ph)IC	>110,000	30,000	96,000	90	90	6400

[a] Inhibition rates were measured in 0.1 M HEPES, 0.01 M CaCl₂, 8–12% DMSO, pH 7.5, at 25°. Data were obtained from C.-M. Kam, K. Fujikawa, and J. C. Powers, *Biochemistry* **27**, 2547 (1988).
[b] Inhibition measured in 0.1 M HEPES, 0.5 M NaCl, 8–10% DMSO, pH 7.5, at 25°. Data were obtained from C.-M. Kam, T. J. Oglesby, M. K. Pangburn, J. E. Volanakis, and J. C. Powers, *J. Immunol.* **149**, 163 (1992).
[c] Inhibition measured in 0.1 M HEPES, 0.01 M CaCl₂, 8% DMSO, pH 7.5, at 25°. Data obtained from S. Odake, C.-M. Kam, L. Narasimhan, M. Poe, J. T. Blake, O. Krahenbuhl, J. Tschopp, and J. C. Powers, *Biochemistry* **30**, 2217 (1991).

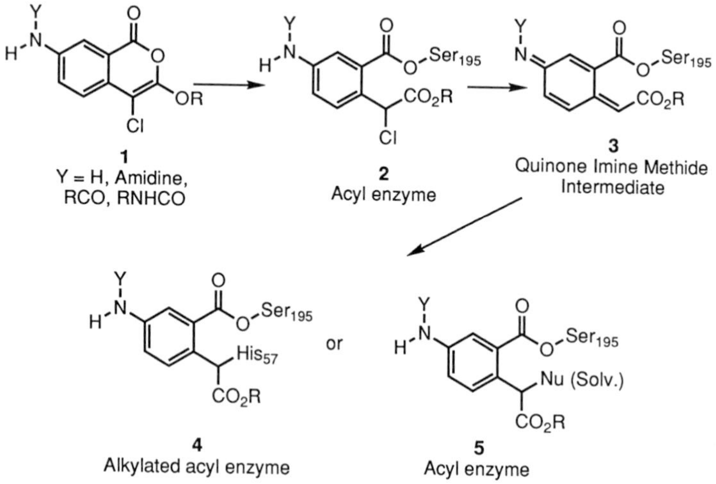

Fig. 2. Inhibition mechanism of serine peptidases by 7-substituted 3-alkoxy-4-chloroisocoumarins.

chloride (or ketene) functional group in the active site of the enzyme.[3] Isocoumarin ring opening catalyzed by the enzyme occurs concurrently with the enzyme inactivation reaction. Either a diacylated or monoacylated enzyme structure is consistent with the available evidence for the initial enzyme–inhibitor complex. An acyl-enzyme that is stabilized by a salt link between the carboxylate of the inhibitor and protonated His-57 has been detected by electrospray mass spectrometry.[28] The initial acyl derivative is reactivatable by NH_2OH treatment. In some cases, an aging process appears to take place and the enzyme is no longer completely reactivatable.

The inhibition mechanism of serine peptidases by substituted isocoumarins is shown in Fig. 2.[4,5] Isocoumarins (1) react with the active site Ser-195 initially to form an acyl-enzyme derivative (2), which can deacylate to regenerate the active enzyme. Alternately, the acyl-enzyme can eliminate the chlorine to form a quinone-imine methide intermediate (3), which then reacts with a nearby enzyme nucleophile such as His-57 to give an alkylated acyl-enzyme derivative (4) or with a solvent molecule to give an acyl-enzyme derivative with solvent replacing the chlorine atom (5). The alkylated acyl-enzyme 4 is stable and cannot be reactivated by NH_2OH. The acyl-enzymes 2 and 5 can be reactivated by NH_2OH. PPE, HLE, and chymotrypsin inhibited by 3-alkoxy-7-amino-4-chloro-

[28] R. T. Aplin, C. V. Robinson, C. J. Schofield, and N. J. Westwood, J. Chem. Soc., Chem. Commun., 1650 (1992).

isocoumarin can be reactivated with NH_2OH and only 15–43% of the enzyme activities are regained.[4] Thus, the majority of the inhibited enzyme is in the form of a nonreactivatable alkylated acyl-enzyme derivative **4**.

The X-ray structures of PPE inhibited with 7-amino-4-chloro-3-methoxyisocoumarin,[29]   4-chloro-3-ethoxy-7-guanidinoisocoumarin,[6] 7-amino-3-(2-bromoethoxy)-4-chloroisocoumarin,[30] and 7-[(N-tosylphenylalanyl)amino]-4-chloro-3-methoxyisocoumarin,[14] and one complex of trypsin inhibited with 4-chloro-3-ethoxy-7-guanidinoisocoumarin,[31] have been solved to atomic resolution. In the complex of PPE with 7-amino-4-chloro-3-methoxyisocoumarin, an acyl-enzyme at Ser-195 is formed; an acetate from solvent has displaced the chlorine and occupies the S_1 subsite. Hydrogen bonds are formed between the imidazole ring of His-57 and the carbomethoxy group. In the complex of PPE with the 7-guanidinoisocoumarin, the chlorine is still present in the acyl-enzyme and H bonds are formed between the guanidino group and Thr-41 at the S_2' subsite. The acyl carbonyl group is twisted out of the oxyanion hole and the ethoxy group occupies the S_1 pocket. In the complex of PPE with the 3-bromoethoxyisocoumarin, the inhibitor forms two covalent bonds with the enzyme at Ser-195 and His-57; H bonds are formed between Gln-192 of the enzyme and the bromine atom of the inhibitor, and also between Thr-41 and the 7-amino group. In the complex of PPE with the Tos-Phe derivative, the chlorine atom is still present in the acyl-enzyme and occupies the S_1 pocket. His-57 is not covalently linked to the inhibitor. Hydrogen bonds are formed between the 7-amino group and Thr-41. Tos-Phe also makes a few hydrophobic contacts with the S_1' subsites. In the complex of trypsin with the guanidino compound, both the chloroacyl-enzyme derivative at Ser-195 and the double covalent adduct with His-57 exist in the crystal, and the guanidino group interacts with Asp-189 in the S_1 pocket. The crystal structures are consistent with the proposed inhibition mechanism based on observations of the acyl-enzyme **2**, the alkylated acyl-enzyme **4**, and the solvated derivative **5**.

Advantages and Limitations

Substituted isocoumarins are potent mechanism-based inhibitors of serine peptidases and show different specificities toward individual en-

[29] E. F. Meyer, Jr., L. G. Presta, and R. Radhakrishnan, *J. Am. Chem. Soc.* **107**, 4091 (1985).
[30] J. Vijayakshmi, E. F. Meyer, Jr., C.-M. Kam, and J. C. Powers, *Biochemistry* **30**, 2176 (1991).
[31] M. M. Chow, E. F. Meyer, Jr., W. Bode, C.-M. Kam, R. Radhakrishnan, J. Vijayashmi, and J. C. Powers, *J. Am. Chem. Soc.* **112**, 7783 (1990).

zymes. Isocoumarins containing basic substituents are potent inhibitors of trypsinlike enzymes and those with small 3-alkoxy groups inhibit elastases potently. In contrast, DCI is a general serine peptidase inhibitor and inhibits all serine peptidases tested except for C2a, Bb, and subtilisin, which is turned over. DCI has several advantages as a general serine peptidase inhibitor. DCI is more stable than DFP and PMSF, which are often used to identify serine peptidases. DCI is also less toxic and much easier to handle than DFP. DCI is fairly specific toward serine peptidases, although it will inhibit esterases and may inhibit some cysteine peptidases.

Substituted isocoumarins have not been widely studied, although they have the potential to be developed as antiinflammatory and antithrombotic agents. For example, NH_2-CiTPrOIC inhibits several coagulation enzymes and is an effective anticoagulant both *in vitro* and *in vivo*.[11,32] Isocoumarins have a limited stability and lifetime in aqueous buffers and plasma, and they are hydrolytically destroyed by physiological nucleophiles such as glutathione. Thus their applicability is limited to situations in which the inhibitor is effective for only a short, defined time period. One such example would involve the use of isocoumarins as anticoagulants in extracorporeal circuits. It would be effective only when the isocoumarin was given by infusion and normal coagulation would be reestablished once the infusion was terminated.

Isocoumarins have other biological applications. For example, DCI and several substituted isocoumarins have been used to characterize the biological role of lymphocyte granule proteases. DCI inhibits all five granule protease activities and inactivates cytolysis.[33,34] When the inhibited proteases and reactivated by NH_2OH treatment, the lysis is also restored, indicating that proteases are essential for cytolysis. DCI and NH_2-CitPrOIC have also been used to identify an additional DCI-resistant component of the multicatalytic proteinase complex, which also exhibits three proteolytic activities inhibited by DCI.[35] Biotinylated isocoumarin derivatives have been synthesized. These derivatives should be useful for the

---

[32] S. W. Oweida, D. N. Ku, A. B. Lumsden, C.-M. Kam, and J. C. Powers, *Thromb. Res.* **58,** 191 (1990).
[33] D. Hudig, N. J. Allison, C.-M. Kam, and J. C. Powers, *Mol. Immunol.* **26,** 793 (1989).
[34] D. Hudig, N. J. Allison, T. M. Pickett, U. Winkler, C.-M. Kam, and J. C. Powers, *J. Immunol.* **147,** 1360 (1991).
[35] C. Cardozo, A. Vinitsky, M. C. Hidalgo, C. Michaud, and M. Orlowski, *Biochemistry* **31,** 7373 (1992).

isolation of new serine peptidases and for the histochemical localization of serine peptidases in biological systems.[36]

Acknowledgment

This work has been supported by Grants HL29307, HL34035, and GM42212 from the National Institutes of Health.

[36] C. M. Kam, A. S. Abuelyaman, Z. Li, D. Hudig, and J. C. Powers, *Bioconjugate Chem.* **4,** 560 (1993).

Section II

Cysteine Peptidases

[32] Families of Cysteine Peptidases

By NEIL D. RAWLINGS and ALAN J. BARRETT

Introduction

About 20 families of peptidases dependent on a cysteine residue at the active center can be recognized (Table I). As far as is known, the activity of all cysteine peptidases depends on a catalytic dyad of cysteine and histidine. The order of the cysteine and histidine residues (Cys/His or His/Cys) in the linear sequence differs between families (Table I), and this is among the lines of evidence suggesting that cysteine peptidases have had many separate evolutionary origins. As yet, tertiary structures are available for members of only two families of cysteine peptidases, so relationships between the families are not clear.

The order in which the families of cysteine peptidases are described in the present chapter is shown in Table I. The families C1, C2, and C10 can be loosely described as "papainlike," and form clan CA. Nearly half of the known families of cysteine peptidases (C3, C4, C18, C5, C6, C7, C8, C9, C16, and C21) are represented only in viruses, and the description of these forms the middle section of this chapter. The viral cysteine peptidases include families with both His/Cys and Cys/His orders of catalytic residues, and the distribution of the families among types of viruses is given in [2] in this volume. The final section of the present chapter deals with an additional eight families, three of which (C11, C15, and C20) are from bacteria, whereas the others are from eukaryotic organisms.

Papain Family (C1)

The best known family of cysteine peptidases is that of papain, often seen to be so typical of cysteine endopeptidases that a newly discovered enzyme of this type is automatically described as "papainlike" whether or not it is a homolog! The papain family contains peptidases with a wide variety of activities, including endopeptidases with broad specificity (such as papain), endopeptidases with very narrow specificity (such as glycyl endopeptidase), aminopeptidases, a dipeptidyl-peptidase, and peptidases with both endopeptidase and exopeptidase activities (such as cathepsins B and H) (Table II). There are also family members that show no catalytic activity.

Enzymes of the papain family are found in a wide variety of life forms:

TABLE I
CLANS AND FAMILIES OF CYSTEINE PEPTIDASES[a]

Clan	Family	Representative enzyme	Identified catalytic residues
CA	C1	Papain	Cys/His
CA	C2	Calpain	Cys
CA	C10	Streptopain	Cys/His
CB	C3	Polio virus picornain 3C	His/Cys
CB	C4	Tobacco etch virus NIa endopeptidase	His/Cys
—	C18	Hepatitis C virus endopeptidase 2	His/Cys
—	C5	Adenovirus endopeptidase	His/Cys
CC	C6	Tobacco etch virus HC-proteinase	Cys/His
CC	C7	Chestnut blight virus p29 endopeptidase	Cys/His
—	C8	Chestnut blight virus p48 endopeptidase	Cys/His
—	C9	Sindbis virus nsP2 endopeptidase	Cys/His
—	C16	Mouse hepatitis virus endopeptidase	Cys/His
—	C21	Turnip yellow mosaic virus endopeptidase	Cys/His
—	C11	Clostripain	Cys
—	C12	Deubiquitinating peptidase Yuh1	Cys/His
—	C19	Deubiquitinating peptidase Ubp1	Cys/His
—	C13	Hemoglobinase	—
—	C14	Interleukin 1β converting enzyme	Cys
—	C15	Pyroglutamyl-peptidase I	Cys
—	C17	Microsomal ER60 endopeptidase	—
—	C20	Type IV prepilin leader peptidase	—

[a] The order in which the families are listed here is that in which they are to be found in the text.

baculovirus,[1] eubacteria (*Porphyromonas* and *Lactococcus*), yeast,[2] and probably all protozoa, plants, and animals. In the present volume, articles [33]–[38] deal with enzymes of the papain family.

Papain homologs are generally either lysosomal (vacuolar) or secreted proteins. In plants, they are found in the vacuoles, but are also extracellular in the latex of papaya or fig, and in arthropods such as lobsters and mites they are among the digestive enzymes.[3,4] Exceptionally, bleomycin hydrolase is a cytosolic enzyme in fungi and mammals.[5]

[1] N. D. Rawlings, L. H. Pearl, and D. J. Buttle, *Biol. Chem. Hoppe-Seyler* **373**, 1211 (1992).
[2] C. Enenkel and D. H. Wolf, *J. Biol. Chem.* **268**, 7036 (1993).
[3] M. V. Laycock, R. M. MacKay, M. Di Fruscio, and J. W. Gallant, *FEBS Lett.* **292**, 115 (1991).
[4] K. Y. Chua, G. A. Stewart, W. R. Thomas, R. J. Simpson, R. J. Dilworth, T. M. Plozza, and K. J. Turner, *J. Exp. Med.* **167**, 175 (1988).
[5] S. M. Sebti, J. C. Deleon, and J. S. Lazo, *Biochemistry* **26**, 4213 (1987).

The catalytic residues of papain are Cys-25 and His-159, and these are conserved in all members of the family that are peptidases. Other residues important for catalysis include Gln-19, which helps form the "oxanion hole," and Asn-175, which orientates the imidazolium ring of His-159 (see [33]). Bromelain is reported to lack Asn-175.[6] There is strong conservation of sequence in the vicinity of the essential Cys and His residues and Asn-175 (Fig. 1).

Members of the papain family in which Cys-25 has been replaced include the soya bean oil-body-associated protein (P34_SOYBN) (Cys25→ Gly). In the *Plasmodium* surface protective protein (SERA_PLAFG) and a protein from *Schistosoma japonicum* (EMBL: X70969), Cys-25 has been converted to Ser.

Papain, like most other members of the family, shows a preference for a bulky hydrophobic residue occupying the S2 subsite, whereas cathepsin B prefers arginine (at least in small substrates). This can be explained by different amino acid side chains in the S2 binding pocket. In papain, Ser-205 lies at the bottom of the pocket, and this is replaced by Glu in cathepsin B, which is well suited to make a salt bridge with an arginine side chain.[7,8] Other homologs that contain Glu in this position include enzymes from *Brassica* (CYS4_BRANA), tomato (CYSL_LYCES), barley (CYS1_HORVU and CYS2_HORVU), *Plasmodium* (CYSP_PLAFA), the baculovirus from *Autographa* (CATV_NPVAC), lobster (CYS3_HOMAM), and rice (ORYA_ORYSA). Cysteine endopeptidases from *Entamoeba* that also prefer arginine in P2 have Ser-205 replaced by Asp (see [35]), as do stem bromelain and lobster digestive proteinase 1 (CYS1_HOMAM).

Being secreted or lysosomal enzymes, peptidases of the papain family are synthesized with signal peptides, and there are also propeptides at the N terminus. Proteolytic cleavage of the propeptides is necessary for activation of the proenzymes. The majority of the propeptides are homologous to that of papain. Although this group of propeptides does not contain any residues that are completely conserved, Glu-64, Arg-68, Phe-72, Asn-75, Asn-83, Phe-96, Asp-98, and Glu-103 (numbering according to prepropapain) are present in all but a few sequences. The first five of these amino acids are part of the "ERFNIN motif" that has been used to identify propeptides related to papain, and may have some structural significance.[9]

[6] A. Ritonja, A. D. Rowan, D. J. Buttle, N. D. Rawlings, V. Turk, and A. J. Barrett, *FEBS Lett.* **247,** 419 (1989).

[7] A. J. Barrett, M. J. H. Nicklin, and N. D. Rawlings, *Symp. Biol. Hung.* **25,** 203 (1984).

[8] D. Musil, D. Zucic, D. Turk, R. A. Engh, I. Mayr, R. Huber, T. Popovic, V. Turk, T. Towatari, N. Katunuma, and W. Bode, *EMBO J.* **10,** 2321 (1991).

[9] K. M. Karrer, S. L. Peiffer, and M. E. DiTomas, *Proc. Natl. Acad. Sci. U.S.A.* **90,** 3063 (1993).

TABLE II
Peptidases of Papain, Calpain, and Streptopain Families (Clan CA)[a]

Peptidase	EC	Database code
Family C1: Papain		
Actinidain	3.4.22.14	ACTN_ACTCH
Aleurain (barley)	-	ALEU_HORVU
Allergen (*Dermatophagoides*)	-	MMAL_DERPT
Allergen (*Euroglyphus*)	-	EUM1_EURMA
Bleomycin hydrolase	-	BLMH_RABIT, BLH1_YEAST
Calotropin (*Calotropis*)	-	CAL1_CALGI
Caricain	3.4.22.30	PAP3_CARPA
Cathepsin B	3.4.22.1	CATB_*, CYSP_SCHMA, (M75822), (X73074), (X70968),[b]
Cathepsin H	3.4.22.16	CATH_*
Cathepsin L	3.4.22.15	CATL_*
Cathepsin S	3.4.22.27	CATS_*
Chymopapain	3.4.22.6	PAP2_CARPA
Cysteine aminopeptidase (*Lactococcus*)	-	(M86245)
Cysteine endopeptidases 2 and 3 (barley)	-	CYS1_HORVU, CYS2_HORVU
Cysteine endopeptidase (*Brassica napus*)	-	CYS4_BRANA
Cysteine endopeptidase (*Caenorhabditis*)	-	CYS1_CAEEL
Cysteine endopeptidases 1 and 2 (*Dictyostelium*)	-	CYS1_DICDI, CYS2_DICDI
Cysteine endopeptidase (*Entamoeba*)	-	(M27307), (M64712), (M64721), (M94163)
Cysteine endopeptidases 1 and 2 (*Haemonchus*)	-	CYS1_HAECO, CYS2_HAECO, (M80385), (M80386)
Cysteine endopeptidase (*Hemerocallis*)	-	(X74406)
Cysteine endopeptidases 1, 2 and 3 (*Homarus*)	-	CYS1_HOMAM, CYS2_HOMAM, CYS3_HOMAM
Cysteine endopeptidase (*Leishmania*)	-	LCPA_LEIME, (M97695), (Z14061)
Cysteine endopeptidase (mung bean)	-	CYSP_VIGMU
Cysteine endopeptidase (*Ostertagia*)	-	(M88505)
Cysteine endopeptidase (pea)	-	CYSP_PEA, (X66061)
Cysteine endopeptidase (*Plasmodium*)	-	CSP_PLACM, CYSP_PLAFA, (L08500), (L26362)
Cysteine protease tpr (*Porphyromonas*)	-	TPR_PORGI
Cysteine endopeptidase (*Tetrahymena*)	-	(L03212)
Cysteine endopeptidase (*Theileria*)	-	CYSP_THEPA, CYSP_THEAN
Cysteine endopeptidase (tobacco)	-	(Z13959), (Z13964)
Cysteine endopeptidase (tomato)	-	CYSL_LYCES, (Z14028)
Cysteine endopeptidase (*Trypanosoma*)	-	CYSP_TRYBR, (L25130), (M90067)
Dipeptidyl peptidase I	3.4.14.1	CATC_RAT
Endopeptidase (baculovirus of *Autographa*)	-	CATV_NPVAC
Endopeptidase EP-C1 (*Phaseolus vulgaris*)	-	CYSP_PHAVU
Glycyl endopeptidase	3.4.22.25	PAP4_CARPA
Oryzain (includes forms α, β and γ) (rice)	-	ORYA_ORYSA, ORYB_ORYSA, ORYC_ORYSA
Papain	3.4.22.2	PAPA_CARPA
Stem bromelain	3.4.22.32	BROM_ANACO
Thaumatopain (*Thaumatococcus*)	-	THPA_THADA

TABLE II (*continued*)

Peptidase	EC	Database code
Family C2: Calpain		
Calpain (*Schistosoma*)	3.4.22.17	(M67499)
Calpain I	3.4.22.17	CAP1_CHICK, CAP1_HUMAN, CAP1_RABIT
Calpain II	3.4.22.17	CAP2_HUMAN, CAP2_RABIT
Calpain P94	3.4.22.17	CAP3_HUMAN, CAP3_RAT
Calcium-binding protein PMP41	3.4.22.17	CAP4_MOUSE
Sol gene product (*Drosophila*)	-	(M64084)
Family C10: Streptopain		
Cysteine endopeptidase (*Porphyromonas*)	-	(M83096)
Streptopain	3.4.22.10	STCP_STRPY

[a] EC is the enzyme nomenclature number (Nomenclature Committee of the International Union of Biochemistry and Molecular Biology, "Enzyme Nomenclature 1992." Academic Press, Orlando, 1992, and Supplement); where there is no EC number, none has been assigned. Literature references to the individual proteins are generally to be found in the database entries for which the codes are given; these codes are from the Swiss-Prot database, except those in parentheses, which are from the EMBL database.
[b] F. J. Cejudo, G. Murphy, C. Chinoy, and D. C. Baulcombe, *Plant J.* **2**, 937 (1992).

The papain propeptide is also homologous to mouse proteins (CT2A_MOUSE, CT2B_MOUSE) of unknown function from activated T cells.[10] These proteins are more closely related to cathepsin L and cathepsin S propeptides than to those from plants or protozoa, and thus seem to be derived from an ancestral proenzyme.

Not all of the propeptides in family C1 are related to that of papain. Among the exceptions are the propeptides of cathepsin B, dipeptidylpeptidase I, and bleomycin hydrolase, which are also unrelated to each other. The propeptides of peptidases from the nematodes *Haemonchus* and *Ostertagia* are homologous, but unrelated to other members of family C1, and the nonenzymatic surface antigen from *Plasmodium* (SERA_PLAFA) has an N-terminal extension unrelated to any cysteine peptidase propeptide, but similar to other *Plasmodium* proteins. The enzymes with the propapain-like propeptides are seen to be more closely related to each other in overall structure than to other members of the family, and thus form a natural subfamily.

In the chymotrypsin family of serine peptidases (see [2]), propeptides show even greater variation than in the papain family, and the mechanism

[10] F. Denizot, J.-F. Brunet, P. Roustan, K. Harper, M. Suzan, M.-F. Luciani, M.-G. Mattei, and P. Golstein, *Eur. J. Immunol.* **19**, 631 (1989).

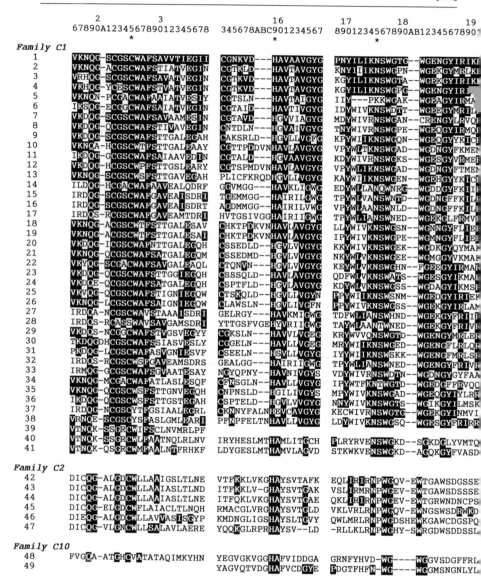

FIG. 1. Conservation of sequence around Cys-25, His-159, and Asn-175 in the papain family, and comparisons for the families of calpain and streptopain. Residues are numbered according to papain, and residues identical to those in papain are shown in white on black. Asterisks mark the catalytic residues. Key to sequences: 1, papain; 2, chymopapain; 3, caricain; 4, glycyl endopeptidase; 5, stem bromelain; 6, actinidain; 7, tomato cysteine endopeptidase; 8, mung bean cysteine endopeptidase; 9, pea cysteine endopeptidase; 10, aleurain; 11, oryzain α; 12, oryzain γ; 13, tobacco cysteine endopeptidase; 14, wheat cathepsin B

of exon shuffling has been invoked to explain this.[11] In the papain family, however, there is considerably less conservation of exon/intron junction positions or phasing,[12,13] so that exon shuffling appears to be an unlikely mechanism. Analysis of the GC content of the cathepsin H gene led Ishidoh *et al.*[12] to propose that the modern gene arose by fusion of several genes, of which that for the propeptide was one.

It has been shown by Fox *et al.*[14] that the propeptide of cathepsin B is a very potent inhibitor of the enzyme, and this presumably is part of the mechanism for the catalytic inactivity of the proenzyme. Comparable inhibitory activity of propeptides has been seen previously in a number of peptidases of other catalytic types, including members of the families of α-lytic endopeptidase (S2), subtilisin (S8), carboxypeptidase C (S10), pepsin (A1), and carboxypeptidase A (M14).

Many members of family C1 contain a proline residue at position 2 in the mature enzyme, which may serve to prevent unwanted N-terminal proteolysis. In addition to the N-terminal peptides, peptidases from tomato, pea, *Leishmania,* and *Trypanosoma* also have long C-terminal extensions.

---

[11] L. Patthy, *Semin. Thromb. Hemostasis* **16,** 245 (1990).
[12] K. Ishidoh, E. Kominami, N. Katunuma, and K. Suzuki, *FEBS Lett.* **253,** 103 (1989).
[13] M. Ferrara, F. Wojcik, H. Rhaissi, S. Mordier, M.-P. Roux, and D. Béchet, *FEBS Lett.* **273,** 195 (1990).
[14] T. Fox, E. De Miguel, J. S. Mort, and A. C. Storer, *Biochemistry* **31,** 12571 (1992).

(F. J. Cejudo, G. Murphy, C. Chinoy, and D. C. Baulcombe, *Plant J.* **2,** 937, (1992)]; 15, human cathepsin B; 16, mouse cathepsin B; 17, *Schistosoma japonicum* cathepsin B; 18, rat cathepsin H; 19, human cathepsin H; 20, chicken cathepsin L; 21, human cathepsin L; 22, human cathepsin S; 23, American lobster digestive cysteine endopeptidase 1; 24, American lobster digestive cysteine endopeptidase 3; 25, *Trypanosoma brucei* cysteine endopeptidase; 26, *Leishmania mexicana* cysteine endopeptidase; 27, *Haemonchus contortus* cysteine endopeptidase 1; 28, *Schistosoma japonicum* "serine endopeptidase"; 29, *Theileria parva* cysteine endopeptidase; 30, *Theileria annulata* cysteine endopeptidase; 31, *Plasmodium falciparum* cysteine endopeptidase; 32, *Schistosoma mansoni* cathepsin B; 33, house dust mite digestive cysteine endopeptidase; 34, baculovirus of alfalfa looper moth cysteine endopeptidase; 35, *Dictyostelium* cysteine endopeptidase 1; 36, *Dictyostelium* cysteine endopeptidase 2; 37, *Entamoeba histolytica* cysteine endopeptidase; 38, human dipeptidyl-peptidase I; 39, rabbit bleomycin hydrolase; 40, yeast bleomycin hydrolase; 41, *Lactococcus lactis* aminopeptidase; 42, chicken calpain; 43, human calpain I; 44, human calpain II; 45, *Schistosoma mansoni* calpain; 46, human calpain P94; 47, *sol* gene product, *Drosophila melanogaster;* 48, streptopain; 49, *Porphyromonas gingivalis* putative cysteine endopeptidase. Database codes for the above sequences may be found in Table II.

Three-dimensional structures have been eludicated for papain,[15] actinidain,[16] calotropin,[17] and cathepsin B,[8] and show bilobed molecules in which the catalytic site is located in a cleft between the lobes.

In an attempt to discover something of the evolution of the papain family, a dendrogram has been constructed using the KITSCH program of the PHYLIP suite.[18] This was calibrated by taking the divergence between cathepsin H and aleurain (the biochemically similar barley vacuolar enzyme) as 1000 million years ago. On this basis, the divergence between the *Lactococcus* aminopeptidase and the yeast bleomycin hydrolase homolog is found to be 2500 million years ago, which predates the origin of mitochondria, suggesting that papainlike cysteine peptidases were present in the organisms ancestral to eubacteria and eukaryotes, and that evolution of the papain family has not involved a horizontal transfer of genes.

The synthetic inhibitors of cysteine peptidases of the papain family have been reviewed by Rich,[19] Shaw,[20] and in articles [46]–[48] of the present volume. The most potent naturally occurring inhibitors are those of the cystatin family (see [49]).

Calpain Family (C2)

The calpain family includes the calcium-dependent cytosolic endopeptidase calpain, which is known from birds and mammals, and the product of the *sol* gene in *Drosophila*.[21] Calpain is a complex of two peptide chains. There are at least two variants of the enzyme in mammals, differing in calcium requirement (millimolar or micromolar concentrations) and with different heavy chains, but the same light chain. The heavy chain is a mosaic of four domains, in which domain 2 contains the catalytic cysteine and domain 4 binds calcium and regulates activity. The sequence of domain 2 shows some similarities to that of papain (Fig. 1), but the relationship is not statistically provable. The putative catalytic histidine residue in calpain has not been identified biochemically. Despite these uncertainties,

[15] I. G. Kamphuis, K. H. Kalk, M. B. A. Swarte, and J. Drenth, *J. Mol. Biol.* **179,** 233 (1984).

[16] E. N. Baker, *J. Mol. Biol.* **141,** 441 (1980).

[17] U. Heinemann, G. P. Pal, R. Hilgenfeld, and W. Saenger, *J. Mol. Biol.* **161,** 591 (1982).

[18] J. Felsenstein, *Evolution* **39,** 783 (1985).

[19] D. H. Rich, *in* "Proteinase Inhibitors" (A. J. Barrett and G. Salvesen, eds.), p. 153. Elsevier, Amsterdam, 1986.

[20] E. Shaw, *Adv. Enzymol.* **63,** 271 (1990).

[21] S. J. Delaney, D. C. Hayward, F. Barleben, K. F. Fischbach, and G. L. G. Miklos, *Proc. Natl. Acad. Sci. U.S.A.* **88,** 7214 (1991).

it seems very likely that the calpain and papain families are related, and belong in a single clan (CA). Domain 4 contains four calcium-binding EF-hand structures, and is homologous to sorcin, another EF-hand structure protein, although not to calmodulin. Domain 4 also is homologous to the calcium-binding domain of the calpain light chain.

The product of the *Drosophila sol* gene is also chimeric, but with a 1000-residue N-terminal domain containing six possible zinc fingers, and a C-terminal 300-residue domain of unknown function. Only the putative peptidase domain of the *sol*-encoded protein shows evidence of relationship to calpain.

Streptopain Family (C10)

Streptopain is the cysteine proteinase from a group A strain of *Streptococcus*,[22] probably *Streptococcus pyogenes*. Although there is no statistically significant relationship between the sequences of streptopain and papain, there are some similarities in structure and in properties.[23] In the primary structure, the catalytic residues of streptopain (Cys-46, the only cysteine in the protein, and His-194) occur in the same order as those in papain, and with some identical residues nearby. Asn-175 appears to be replaced by Asp (Fig. 1). The specificity of streptopain is also similar to that of papain, with preference for a hydrophobic residue at P2.[24] Streptopain is inactivated by E64 much more slowly than is papain.[25]

Streptopain is synthesized as a precursor with an 85 amino acid propeptide. The activation cleavage is at a Lys+Gln bond.[26]

The *prtT* gene of *Porphyromonas gingivalis* is reported to encode an endopeptidase of novel sequence.[27] However, we have noticed that the strand complementary to this gene contains a sequence that is clearly homologous to that of streptopain. If some sequencing errors are assumed (as is indicated by results of analysis of the data by the NIP program of the Staden package[28]), a sequence of 69 amino acids over 50% identical to the C-terminal part of streptopain is revealed. This seems more than a coincidence, and merits further investigation.

[22] S. D. Elliott and T.-Y. Liu, this series, Vol. 19, p. 252.
[23] J. Y. Tai, A. A. Kortt, T.-Y. Liu, and S. D. Elliott, *J. Biol. Chem.* **251,** 1955 (1976).
[24] A. A. Kortt and T.-Y. Liu, *Biochemistry* **12,** 328 (1973).
[25] A. J. Barrett, A. A. Kembhavi, M. A. Brown, H. Kirschke, C. G. Knight, M. Tamai, and K. Hanada, *Biochem. J.* **201,** 189 (1982).
[26] K. Yonaha, S. D. Elliott, and T.-Y. Liu, *J. Protein Chem.* **1,** 317 (1982).
[27] J. I. Otogoto and H. K. Kuramitsu, *Infect. Immun.* **61,** 117 (1993).
[28] R. Staden, this series, Vol. 183, p. 163.

TABLE III
VIRAL CYSTEINE PEPTIDASES OF PICORNAIN 3C, TOBACCO ETCH VIRUS NIa
ENDOPEPTIDASE, ADENOVIRUS ENDOPEPTIDASE, AND HEPATITIS C VIRUS
ENDOPEPTIDASE 2 FAMILIES[a]

Peptidase	EC	Database code
Family C3: Picornain		
Picornain 2A	3.4.22.29	POLG_POL1M, POLG_COXA2,
		POLG_SVDVH, POLG_BOVEV,
		POLG_HRV14
Picornain 3C	3.4.22.28	POLH_POL1M, POLG_COXA2,
		POLG_SVDVH, POLG_BOVEV,
		POLG_HPAV1, POLG_HRV14,
		POLG_ECHO9, POLG_TMEVD
Aphthovirus endopeptidase	-	POLG_FMDVD
Cardiovirus endopeptidase	-	POLG_EMCV
Comovirus endopeptidase	-	VGNB_CPMV, (D00657)
Nepovirus endopeptidase	-	POL1_GCMV, POL1_TBRVS,
		(D00915)
Family C4: Tobacco etch virus NIa		
endopeptidase		
NIa endopeptidase	-	POLG_PPVD, POLG_PPVYN,
		POLG_TEV, POLG_TVMV,
		POLG_WMV2, POLG_OMV,
		(D11118), (X68509)
Family C18: Hepatitis C virus		
endopeptidase 2		
Endopeptidase 2 (hepatitis C virus)	-	POLG_HCV1
Family C5: Adenovirus endopeptidase		
Adenovirus endopeptidase	-	VPRT_ADEB3, VPRT_ADEB7,
		VPRT_ADE02, VPRT_ADE03,
		VPRT_ADE04, VPRT_ADE05,
		VPRT_ADE12, VPRT_ADE40,
		VPRT_ADE41, VPRT_ADEM1,
		(L13161), (M72715)

[a] See Table II for general explanations.

Polio Virus Picornain 3C Family (C3)

Picornains are a family of polyprotein-processing endopeptidases from single-stranded RNA viruses. Examples are known from picornaviruses such as polio virus and coxsackie virus, as well as from plant viruses such as cowpea mosaic virus (a comovirus) and grapevine fanleaf virus (a nepovirus) (Table III). Each picornavirus has two picornains (known as 2A and 3C), whereas only one is known for the plant viruses. Cardioviruses and aphthoviruses have only one picornain, the second proteinase being unrelated.[29] Picornains have been most thoroughly characterized from polio virus (see [40]).

An early cleavage in the processing of the polio virus polyprotein is that of a Tyr+Gly bond by which picornain 2A releases the capsid protein precursor. With 161 amino acid residues, picornain 2A is one of the smallest cysteine peptidases. Picornain 3C performs the other nine cleavages of the polyprotein, mostly at Gln+Gly bonds.[30] The Gln+Gly specificity is also seen in picornain 3C from encephalomyocarditis virus and Mengo virus. The specificities of picornain 3C from other picornaviruses are less strict, but the P1 residue is generally glutamine.

The three dimensional structure of a mutant form of picornain 3C from a hepatitis A virus has recently been reported.[31] In the numbering of Fig. 2, His-40 and Cys-147 (replaced by Ala in the mutant) are located appropriately to form a catalytic dyad, confirming mutational studies.[32,33] However, a catalytic triad including Glu-71 (or Asp-85 in the coxsackie virus peptidase[33]) that had been proposed earlier was not seen. The tertiary fold of picornain is similar to those of chymotrypsin and α-lytic endopeptidase, and clearly unrelated to that of papain, confirming molecular modelling studies.[34] His-40 and Cys-147 in picornain 3C are functionally equivalent to His-57 and Ser-195 in chymotrypsin, and mutants of picornain 3C with the active site Cys replaced by Ser have been shown to retain some activity.[32,33,35] This is the only instance so far discovered among the many families of peptidases in which an evolutionary relationship crosses the boundary of catalytic type, and is presumed to have arisen from a single base change that converted a catalytic Ser to a Cys residue. In the Sindbis virus, the core protein is an endopeptidase also structurally related to chymotrypsin, but with the catalytic Ser retained (Chapter [2], this volume).

Tobacco Etch Virus NIa Endopeptidase Family (C4)

Tobacco etch virus is one of the potyviruses. These are plant viruses in which the single-stranded RNA encodes a large polyprotein that is

[29] A. C. Palmenberg, G. D. Parks, D. J. Hall, R. H. Ingraham, T. W. Seng, and P. V. Pallai, *Virology* **190,** 754 (1992).

[30] K. M. Kean, N. Teterina, and M. Girard, *J. Gen. Virol.* **71,** 2553 (1990).

[31] M. Allaire, M. M. Chernaia, B. A. Malcolm, and M. N. G. James, *Nature* **369,** 72 (1994).

[32] K. Miyashita, M. Kusumi, R. Utsumi, S. Katayama, M. Noda, T. Komano, and N. Satoh, *Protein Eng.* **6,** 189 (1993).

[33] J. T. Dessens and G. P. Lomonossoff, *Virology* **184,** 738 (1991).

[34] J. F. Bazan and R. J. Fletterick, *Proc. Natl. Acad. Sci. U.S.A.* **85,** 7872 (1988).

[35] M. A. Lawson and B. L. Semler, *Proc. Natl. Acad. Sci. U.S.A.* **88,** 9919 (1991).

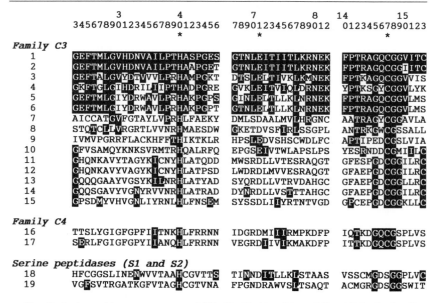

```
         3         4            7        8     14        15
3456789012345678901234 56    789012345678 9012   01234567890123
         *                       *                  *
Family C3
  1   GEFTMLGVHDNVAILPTHASPGES   GTNLEITIITLKRNEK   FPTRAGQCGGVITC
  2   GEFTMLGVHDNVAILPTHAAPGET   GTNLEITIITLKRNEK   FPTRAGQCGGIITC
  3   GEFTALGVYDTVVVLPRHAMPGKT   DTSLELTIVKLKMNEK   FPTKAGQCGGVVIS
  4   GKFTGLGIHDRILIIPTHADPGRE   GVKLEITVIQLDRNEK   YPTKSGYCGGVLYK
  5   GEFTMLGIYDRWAVLPRHAKPGPS   GINLEITLLKLNRNEK   FPTRAGQCGGVLMS
  6   GEFTMLGIYDRWAVLPRHAKPGPT   GTNLELTLLKLNRNEK   FPTRAGQCGGVLMS
  7   AICCATGVFGTAYLVPRHLFAEKY   DMLSDAALMVLHRGNC   AATRAGYCGGAVLA
  8   STQTCILVRGRTLVVNRHMAESDW   CKETDVSFIRLSSGPL   ANTRKGWCGSSALL
  9   IVMVPGRRFLACKHFFTHIKTKLR   HPSLEDVSHSCWDLFC   APTIPEDCGSLVIA
 10   GFVSAMQYKNKSVRMTRHQALRFQ   EPGSEIVTWLAPSLPS   YESRNDDCGMILLC
 11   GHQNKAVYTAGYKICNYHLATQDD   MWSRDLLVTESRAQGT   GFESPGDCGGILRC
 12   GHQNKAVYTAGYKICNYHLATPSD   LWDRDLMVVESRAQGT   GFAEPGDCGGILRC
 13   CQQQGAAYVGSYKIINRHLATYAD   SYQRDLLVTRVDAHGC   GFAEPGDCGGLLRC
 14   CQQSGAVYVGNYRVVNRHLATRAD   DYNRDLLVSTTTAHGC   GFAEPGDCGGILRC
 15   GPSDMYVHVGNLIYRNLHLFNSEM   SYSSDLIIYRTNTVGD   GPCEPGDCGGKLLC

Family C4
 16   TTSLYGIGFGPFIITNKHLFRRNN   IDGRDMIIIRMPKDFP   IQTKDGQCGSPLVS
 17   SERLFGIGFGPYIIANQHLFRRNN   VEGRDIIVIKMAKDFP   ITTKDGQCGSPLVS

Serine peptidases (S1 and S2)
 18   HFCGGSLINENWVVTAAHCGVTTS   TINNDITLLKLSTAAS   VSSCMGDSGGPLVC
 19   VGFSVTRGATKGFVTAGHCGTVNA   FPGNDRAWVSLTSAQT   ACMGRCDSGGSWIT
```

FIG. 2. Amino acid sequences around His-40, Glu/Asp-71, and Cys-147 in the families C3 and C4 of viral cysteine endopeptidases (clan CB), with sequences from the chymotrypsin clan (SA) for comparison. Residues are numbered according to polio virus (Mahoney strain) picornain 3C, and residues identical to those in polio virus picornain 3C are shown in white on black. Key to sequences: 1, polio virus picornain 3C; 2, coxsackie virus picornain 3C; 3, cattle enterovirus picornain 3C; 4, human rhinovirus picornain 3C; 5, echo 9 virus picornain 3C; 6, pig vesicular virus picornain 3C; 7, foot-and-mouth disease virus picornain 3C; 8, encephalomyocarditis virus picornain 3C; 9, cowpea mosaic virus picornain 3C; 10, tomato black ring virus picornain 3C; 11, polio virus picornain 2A; 12, coxsackie virus picornain 2A; 13, cattle enterovirus picornain 2A; 14, pig vesicular virus picornain 2A; 15, human rhinovirus picornain 2A; 16, tobacco etch virus NIa endopeptidase; 17, tobacco vein mottling virus NIa endopeptidase; 18, bovine chymotrypsin; 19, *Lysobacter* α-lytic endopeptidase.

processed by at least three peptidases, all of which are virally encoded. Two of these enzymes are cysteine endopeptidases—NIa endopeptidase (48 kDa) and HC-proteinase—whereas the third is a serine endopeptidase (family S30, see [2]). The HC-proteinase falls into family C6 (see below).

The catalytic cysteine and histidine residues of the NIa endopeptidase have been identified by site-directed mutagenesis, and are found to occur in the order His/Cys. A catalytic triad formed by His-234, Asp-269, and Cys-339 has been proposed.[36] Slight similarities in the sequences around the catalytic residues (Fig. 2), together with a similar specificity for

[36] W. G. Dougherty, T. D. Parks, S. M. Cary, J. F. Bazan, and R. J. Fletterick, *Virology* **172,** 302 (1989).

FIG. 3. Conservation of sequences around the catalytic residues in the families of adenovirus endopeptidase (C5) and hepatitis C virus endopeptidase 2 (C18). Residues are numbered according to human adenovirus type 40 endopeptidase. Residues identical to human adenovirus type 40 endopeptidase are shown in white on black. Key to sequences: 1, human adenovirus type 40 endopeptidase; 2, human adenovirus type 4 endopeptidase; 3, cattle adenovirus type 7 endopeptidase; 4, mastadenovirus mus1 endopeptidase; 5, hepatitis C virus endopeptidase 2.

Gln+Gly cleavages, suggest that families C3 and C4 are members of a clan, CB.

The NIa endopeptidase releases itself from the viral polyprotein by cleavages at Gln+Gly bonds at its N and C termini.[37] It then processes the viral polyprotein by cleavage at five similar sites in the C-terminal half of the polyprotein.[38]

The NIa endopeptidase is a bifunctional molecule; the C-terminal domain is the proteinase whereas the N-terminal domain functions as the VPg protein, which is attached covalently to the viral mRNA.[39]

Hepatitis C Virus Endopeptidase 2 Family (C18)

Hepatitis C virus is a flavivirus encoding a single polyprotein. As in other flaviviruses, the NS3 protein is a serine peptidase that is probably structurally related to chymotrypsin ([2], family S29). A second proteinase from this virus has been described, cleaving the Leu+Ala bond between the NS2 and NS3 proteins.[40] The limits of this second peptidase within the polyprotein have not been completely established, but deletion studies have shown that portions of the NS2 and NS3 proteins are essential for activity. Site-directed mutagenesis has identified His-952 and Cys-993 as essential for catalytic activity, and both of these reside in the NS2 protein (Fig. 3).

[37] K. Rorrer, T. D. Parks, B. Scheffler, M. Bevan, and W. G. Dougherty, *J. Gen. Virol.* **73,** 775 (1992).
[38] C.-S. Oh and J. C. Carrington, *Virology* **173,** 692 (1989).
[39] J. F. Murphy, R. E. Rhoads, A. G. Hunt, and J. G. Shaw, *Virology* **178,** 285 (1990).
[40] A. Grakoui, D. W. McCourt, C. Wychowski, S. M. Feinstone, and C. M. Rice, *Proc. Natl. Acad. Sci. U.S.A.* **90,** 10583 (1993).

Adenovirus Endopeptidase Family (C5)

Adenoviruses are double-stranded DNA, nonenveloped viruses that cause tumors in mammals. Unlike potyviruses, the adenoviruses do not encode a polyprotein, but have different genes for different proteins, as would be expected in a DNA virus. However, several adenovirus proteins are synthesized as precursors, which have to be processed before the virion is assembled.[41] Temperature-sensitive mutants fail to process the proteins, implying that the virus encodes its own proteinase (see [41]).

Although the endopeptidase has now been sequenced, there has been confusion about its catalytic type. The current thinking is that this is a cysteine peptidase, but for many years the enzyme was considered to be a serine peptidase, on the basis of reported inhibition by standard inhibitors of such enzymes.[42] Inhibition by cysteine proteinase inhibitors has been described, and the catalytic residues are thought to be His-54 and Cys-104 (Fig. 3) (see [41]).

The endopeptidase is synthesized as an active enzyme, not needing proteolytic processing.[43] This is unusual for an endopeptidase, but is perhaps explained by the very strict specificity of the enzyme.

Tobacco Etch Virus HC-Proteinase Family (C6)

A second potyvirus proteinase, helper component proteinase (HC-Pro), is known, and is responsible for only one cleavage, that of a Gly↓Gly bond that releases the precursor of HC-Pro. Further processing of the precursor is mediated by the third potyvirus proteinase (see [2]; family S30).[44] HC-Pro is probably a two-domain protein, the C-terminal domain carrying the endopeptidase activity and the N-terminal part being the helper component required for virus transmission from plant to plant by aphids.[38]

Site-directed mutagenesis of HC-Pro from tobacco etch virus has indicated that Cys-303 and His-376 are the essential catalytic residues, and these are completely conserved in all members of the family (Fig. 4; Table IV).[38]

Chestnut Blight Virus p29 and p48 Endopeptidase Families (C7 and C8)

The chestnut blight fungus *Cryphonectria parasitica,* which causes canker on chestnut trees, exhibits reduced pathogenicity if a viruslike,

[41] H.-G. Kräusslich and E. Wimmer, *Annu. Rev. Biochem.* **57,** 701 (1988).
[42] A. R. Bhatti and J. M. Weber, *Virology* **96,** 478 (1979).
[43] A. Webster and G. Kemp, *J. Gen. Virol.* **74,** 1415 (1993).
[44] Y. Stram, A. Chetsrony, H. Karchi, M. Karchi, O. Edelbaum, E. Vardi, O. Livneh, and I. Sela, *Virus Genes* **7,** 151 (1993).

FIG. 4. Conservation of sequences around the catalytic residues in clan CC. Residues are numbered according to tobacco etch virus HC-proteinase. Residues identical to those in tobacco etch virus HC-proteinase are shown in white on black, and asterisks indicate the catalytic residues identified in both families. Key to sequences: 1, tobacco etch virus HC-proteinase; 2, tobacco vein mottling virus HC-proteinase; 3, plum pox virus HC-proteinase; 4, potato virus Y HC-proteinase; 5, chestnut blight virus p29 endopeptidase.

double-stranded RNA element is present in the fungal hyphae. During anastomosis (joining) of hyphae, this hypovirulence particle is transmissible to other fungal strains lacking the particle. The hypovirulence particle encodes two polyproteins, both of which are proteolytically processed. The processing of the smaller polyprotein yields two proteins of 29 and 48 kDa, known as p29 and p48, respectively. The p29 component has been shown to be a proteinase, performing the single cleavage in the polyprotein at a Gly+Gly bond.[45] Site-directed mutagenesis has identified the possible catalytic residues in the p29 protein as Cys-162 and His-215. The postulated active site residues in the endopeptidases of family C6 (above) occur within similar motifs (Fig. 4), and a common ancestor has been postulated.[46] This suggests that families C6 and C7 comprise a clan (CC).

The second, larger polyprotein is also proteolytically processed, at a Gly+Ala bond,[47] and the proteinase responsible has been identified as the p48 protein. The catalytic residues have been identified by site-directed mutagenesis as Cys-341 and His-388 (Fig. 5). The p29 and p48 proteins show no sequence similarity, and are considered here to be representatives of separate families.

Sindbis Virus nsP2 Endopeptidase Family (C9)

Togaviruses such as the Sindbis virus produce two polyproteins, and these are processed by a combination of virally encoded peptidases and cellular peptidases. The p130 polyprotein is processed by a serine endo-

[45] G. H. Choi, R. Shapira, and D. L. Nuss, *Proc. Natl. Acad. Sci. U.S.A.* **88,** 1167 (1991).
[46] G. H. Choi, D. M. Pawlyk, and D. L. Nuss, *Virology* **183,** 747 (1991).
[47] R. Shapira and D. L. Nuss, *J. Biol. Chem.* **266,** 19419 (1991).

TABLE IV
OTHER CYSTEINE PEPTIDASES[a]

Peptidase	EC	Database code
Family C6: Tobacco etch virus HC-proteinase		
HC-Proteinase	-	POLG_PPVD, POLG_PVYN, POLG_TEV, POLG_TVMV
Family C7: Chestnut blight virus p29 endopeptidase		
p29 Endopeptidase	-	(M57938)
Family C8: Chestnut blight virus p48 endopeptidase		
p48 Endopeptidase	-	(M57938)
Family C9: Sindbis virus nsP2 endopeptidase		
Togavirus cysteine endopeptidase	-	POLN_SINDV, POLN_RRVN, POLN_SFV,POLN_ONNVG, V180_CGMVS, (J02246), (L01443),(X63135)
Family C16: Mouse hepatitis virus endopeptidase		
Mouse hepatitis virus endopeptidase	-	RRPA_CVMJH
Avian infectious bronchitis virus endopeptidase	-	VGF1_IBVB
Family C21: Turnip yellow mosaic virus endopeptidase		
Turnip yellow mosaic virus endopeptidase	-	POLR_TYMV
Family C11: Clostripain		
α-Clostripain	3.4.22.8	CLOS_CLOHI
Family C12: De-ubiquitinating peptidase Yuh1		
Ubiquitin carboxyl-terminal hydrolase	-	UBL1_*, UBL3_HUMAN, {8676}
Family C19: De-ubiquinating peptidase Ubp1		
Deubiquinating enzyme (DOA4 protein)	-	SSV7_YEAST, (U02518)
Ubiquitin-specific processing peptidase 1	-	UBP1_YEAST
Ubiquitin-specific processing peptidase 2	-	UBP2_YEAST
Ubiquitin-specific processing peptidase 3	-	UBP3_YEAST
tre oncogene protein (human)	-	(X63547)
unp protein (mouse)	-	(L00681)
Family C13: Hemoglobinase		
Hemoglobinase (*Schistosoma*)	-	HGLB_SCHMA, (X70967)
Legumain (jack bean)	-	[b]
Family C14: Interleukin-1β converting enzyme		
Interleukin-1β converting enzyme	-	I1BC_*
Family C15: Pyroglutamyl-peptidase I		
Pyroglutamyl-peptidase I	3.4.19.3	PCP_*, (X75919)

TABLE IV (*continued*)

Peptidase	EC	Database code
Family C17: Microsomal ER60 endopeptidase Microsomal ER60 protein	-	ER60_*
Family C20: Type IV-prepilin leader peptidase Prepilin leader peptidase	-	TCPJ_VIBCH, COMC_BACSU, XCPA_PSEAE, PULO_KLEPN, HOPO_ECOLI, (L11715)

[a] See Table II for general explanations.
[b] Y. Abe, K. Shirane, H. Yokosawa, H. Matsushita, M. Mitta, I. Kato, and S. Ishii, *J. Biol. Chem.* **268**, 3525 (1993).

peptidase (see [2], family S3). The p270 polyprotein contains nonstructural proteins, among which nsP2 is the cysteine endopeptidase that processes the polyprotein. The active site residues have been identified by site-directed mutagenesis as Cys-481 and His-558 in the Sindbis virus (Fig. 5).[48]

Protein nsP2 is a bifunctional, mosaic molecule, with the cysteine endopeptidase restricted to the C-terminal domain. The N-terminal domain is probably involved in RNA-binding during replication and is homologous to some plant virus proteins, such as the 180 kDa protein of cucumber green mottle mosaic virus, which is an unrelated tobamovirus. There is no sequence relationship to the endopeptidase domain.

Mouse Hepatitis Virus Endopeptidase Family (C16)

Mouse hepatitis is caused by one of the coronaviruses, which are single-stranded RNA viruses that encode several polyproteins. The polyprotein encoded by gene A of the virus is known to be autolytically processed to release the 28-kDa N-terminal p28 protein.[49] Site-directed mutagenesis has identified Cys-1137 and His-1288 as the catalytic dyad in this peptidase (Fig. 5).[50]

[48] E. G. Strauss, R. J. De Groot, R. Levinson, and J. H. Strauss, *Virology* **191**, 932 (1992).
[49] S. C. Baker, C. K. Shieh, L. H. Soe, M. F. Chang, D. M. Vannier, and M. M. C. Lai, *J. Virol.* **63**, 3693 (1989).
[50] S. C. Baker, K. Yokomori, S. Dong, R. Carlisle, A. E. Gorbalenya, E. V. Koonin, and M. M. C. Lai, *J. Virol.* **67**, 6056 (1993).

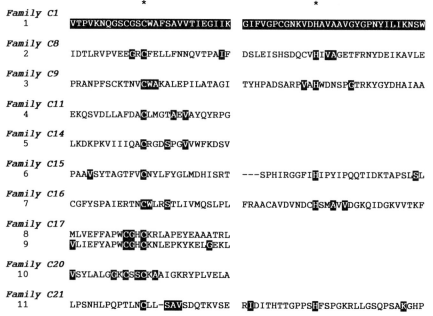

FIG. 5. Comparison of sequences in the vicinity of the catalytic residues of cysteine peptidases from families C8, C9, C11, C14, C15, C16, C17, C20, and C21 with those in papain. Residues identical to those in papain are shown in white on black. Key to sequences: 1, papain; 2, chestnut blight virus p48 endopeptidase; 3, Sindbis virus nsP2 endopeptidase; 4, clostripain; 5, human interleukin 1β converting enzyme; 6, *Bacillus subtilis* pyroglutamyl-peptidase I; 7, mouse hepatitis virus endopeptidase; 8, rat ER60 endopeptidase repeat 1; 9, rat ER60 endopeptidase repeat 2; 10, *Pseudomonas aeruginosa* type IV prepilin leader peptidase; 11, turnip yellow mosaic virus endopeptidase. Asterisks indicate identified catalytic residues (predictions only for ER60 endopeptidase, and His in pyroglutamyl-peptidase I).

Turnip Yellow Mosaic Virus Endopeptidase Family (C21)

Turnip yellow mosaic virus is a single-stranded RNA virus in which two polyproteins are encoded. The larger polyprotein (206 kDa) includes an endopeptidase that cleaves the 206-kDa polyprotein at a single bond to release an N-terminal 150-kDa protein, which contains a helicase, and a C-terminal 70-kDa protein, which includes a polymerase. The endopeptidase has been delimited to residues 731–885 (within the 150-kDa protein), and site-specific mutagenesis has identified Cys-783 and His-869 as the catalytic dyad (Fig. 5).[51] Homologous sequences occur in the ononis yellow

[51] K. L. Bransom and T. W. Dreher, *Virology* **198,** 148 (1994).

mosaic virus (POLR_OYMV), the eggplant mosaic virus (POLR_EPMV), the kennedya yellow mosaic virus (EMBL: D00637), and the erysimum latent virus.[52]

Clostripain Family (C11)

Clostripain, from the anaerobic bacterium *Clostridium histolyticum,* is a cysteine endopeptidase with strict specificity for the cleavage of arginyl bonds, only rarely cleaving lysyl bonds.[53] The two-chain enzyme is synthesized as a precursor protein that is processed to yield (from the N terminus) a 50-residue propeptide, the light chain of the mature enzyme, a nonapeptide, and the heavy chain. The C-terminal residues of both the light chain and nonapeptide linker are arginine, suggesting that their processing is autolytic, but the activation cleavage at the N terminus is of a Lys+Asn bond, and is probably mediated by another enzyme.[54]

Clostripain differs from the enzymes of the papain family both in its calcium dependence and its inhibition characteristics. The enzyme is more rapidly inactivated by iodoacetamide than by iodoacetate, and E64 gives only reversible inhibition.[55] The active site cysteine has been identified as Cys-181,[56] and occurs in a sequence unrelated to those in family C1 (Fig. 5). The dependence of catalytic activity on a histidine residue has been demonstrated by use of diethyl pyrocarbonate,[55] but the histidine remains unidentified.

Deubiquitinating peptidase Yuh1 Family (C12)

Ubiquitin is a protein of 76 amino acids that exists in eukaryotic cells and commonly occurs conjugated to ubiquitin or other proteins through the carboxyl group of its C-terminal glycine residue. The link may be to the N terminus of another polypeptide or to the ε-amino group of a lysine, in which case an isopeptide bond is formed.[57] Ubiquitination can act as a signal to mark proteins for rapid degradation, or may have a chaperone function in the assembly of oligomeric proteins and ribosomes. Whatever the function of the attachment of ubiquitin, the ubiquitin molecule is

[52] P. Srifah, P. Keese, G. Weller, and A. Gibbs, *J. Gen. Virol.* **73,** 1437 (1992).

[53] B. Keil, "Specificity of Proteolysis." Springer-Verlag, Berlin, 1992.

[54] H. Dargatz, T. Diefenthal, V. Witte, G. Reipen, and D. Von Wettstein, *Mol. Gen. Genet.* **240,** 140 (1993).

[55] A. A. Kembhavi, D. J. Buttle, P. Rauber, and A. J. Barrett, *FEBS Lett.* **283,** 277 (1991).

[56] A.-M. Gilles, A. De Wolf, and B. Keil, *Eur. J. Biochem.* **130,** 473 (1983).

[57] K. D. Wilkinson, K. Lee, S. Deshpande, P. Duerksen-Hughes, J. M. Boss, and J. Pohl, *Science* **246,** 670 (1989).

eventually released by the action of a peptidase hydrolyzing the glycyl bond at the C terminus, and is recycled.

There are a number of distinct deubiquitinating peptidases,[58] and they belong to at least two families. Nevertheless, they have important properties in common. They are activated by thiol compounds[59] and are inhibited by thiol-blocking reagents, but not by inhibitors of other classes of peptidases. They are also inhibited by ubiquitin aldehyde.[60] In these respects, the deubiquitinating enzymes have properties expected of cysteine peptidases. Also, they are all specific for the C terminus of ubiquitin, but seem to show almost no selectivity for residues on the prime side of the scissile bond (only proline being unacceptable).[61] A generic assay for the deubiquitinating enzymes is made with ubiquitin ethyl ester as substrate.[62]

The deubiquitinating peptidases generally fall into one of two molecular mass ranges, 20–30 or 100–200 kDa.[58] The small enzymes are those of the Yuh1 family (C12), and the large ones belong to the Ubp1 family (C19).

Family C12 comprises the product of the yeast *YUH1* gene and its mammalian counterparts. The Yuh1 protein is known to cleave only short ubiquitin conjugates, being inactive against ubiquitinated β-galactosidase.[63] One mammalian homolog, known as ubiquitin conjugate hydrolase (Uch) isozyme L1 or PGP 9.5, is among the most abundant proteins of the brain (1–5% of total soluble protein), and has been described as "neurone-specific", although it does occur in other tissues at lower concentrations. The other, Uch-L3, is the predominant form of ubiquitin C-terminal hydrolase in bovine thymus.[57]

The alignment of sequences[57] shows only one cysteine residue conserved among the members of this family, Cys-100 (Fig. 6). There are two conserved histidine residues, His-107 and His-181, the second of which has the spacing from the Cys that is more consistent with participation in a catalytic dyad.

Deubiquitinating Peptidase Ubp1 Family (C19)

Work on the high molecular mass deubiquitinating peptidases of yeast led to the sequencing of Ubp1,[64] Ubp2, and Ubp3.[61] The Ubp2 protein can

[58] A. N. Mayer and K. D. Wilkinson, *Biochemistry* **28,** 166 (1989).
[59] C. M. Pickart and I. A. Rose, *J. Biol. Chem.* **261,** 10210 (1986).
[60] A. Hershko and I. A. Rose, *Proc. Natl. Acad. Sci. U.S.A.* **84,** 1829 (1987).
[61] R. T. Baker, J. W. Tobias, and A. Varshavsky, *J. Biol. Chem.* **267,** 23364 (1992).
[62] K. D. Wilkinson, M. J. Cox, A. N. Mayer, and T. Frey, *Biochemistry* **25,** 6644 (1986).
[63] H. I. Miller, W. J. Henzel, J. B. Ridgway, W. J. Kuang, V. Chisholm, and C. C. Liu, *Bio/Technology* **7,** 698 (1989).
[64] J. W. Tobias and A. Varshavsky, *J. Biol. Chem.* **266,** 12021 (1991).

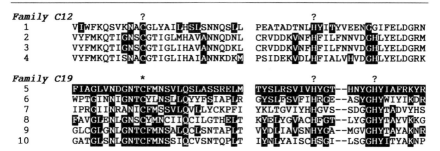

FIG. 6. Conservation of sequences around the potential catalytic cysteine and histidine residues of deubiquitinating peptidases. Residues identical to yeast deubiquitinating peptidase Ubp1 are shown in white on black, and the asterisk marks the cysteine residue shown to be involved in the activity of Doa4. Key to sequences: 1, yeast deubiquitinating peptidase Uch; 2, rat deubiquitinating peptidase Uch-L1; 3, human deubiquitinating peptidase Uch-L1; 4, human deubiquitinating peptidase Uch-L3; 5, yeast deubiquitinating peptidase Ubp1; 6, yeast deubiquitinating peptidase Ubp2; 7, yeast deubiquitinating peptidase Ubp3; 8, yeast deubiquitinating enzyme Doa4; 9, mouse protooncoprotein Unp; 10, human *tre* oncogene protein.

deubiquitinate any size of ubiquitinated protein, including polyubiquitin, whereas Ubp1 and Ubp3 do not act on polyubiquitin.[61] Subsequently, a fourth homolog (Doa4) was discovered in yeast. Doa4 appears to be involved in the later stages of ubiquitin recycling, perhaps working in conjunction with the 26S ubiquitin–conjugate-degrading enzyme.[65]

Site-directed mutagenesis has identified Cys-571 as catalytically important in Doa4,[65] and the closest similarities between the sequences of these proteins occur in the vicinity of this residue and a more C-terminal region containing two histidines, either of which may play a role in activity (see Fig. 6).

Mammalian proteins in the family include the products of the human *tre-2* oncogene, which is also a deubiquitinating enzyme,[65] and the mouse *unp* gene (Table IV). Also related is dog mucin (EMBL: L03387), which overlaps only the C-terminal half of the Doa4 protein and does not contain the active site Cys.

Hemoglobinase Family (C13)

Schistosoma mansoni, a blood fluke, is a human parasite causing schistosomiasis. The adult worms burrow through the skin and take up residence in the bloodstream, where they feed on hemoglobin. There are two cysteine endopeptidases in the parasite digestive tract, one a cathepsin

[65] F. R. Papa and M. Hochstrasser, *Nature (London)* **366,** 313 (1993).

B-like enzyme of family C1, and a second, termed hemoglobinase,[66] which shows no homology to papain. Hemoglobinase has proved difficult to isolate free from the cathepsin B, and consequently rather little is known about its catalytic activity, although it is trapped by α_2-macroglobulin and is inactivated by Z-Tyr-Ala-CHN$_2$ (C. L. Chappell, personal communication, 1991). An attempt at expressing hemoglobinase in *Escherichia coli* did not produce a product with detectable peptidase activity,[67] and it may be that the enzyme acts synergistically with the cathepsin B. A second clone has recently been sequenced from *Schistosoma japonicum* (Table IV).

It has been discovered (see [42]) that legumain, an atypical cysteine endopeptidase from legume seeds, is a homolog of hemoglobinase. Legumain is a very strict asparaginyl endopeptidase that acts on both small substrates and proteins. This enzyme has been assigned two separate functions in the legume seed: first, posttranslational splicing of seed proteins, including concanavalin A during maturation of the seed, and second, a role in the degradation of seed proteins during germination (see Ref. 68 and [42]). It was first suggested by Csoma and Polgár[69] that the bean asparaginyl endopeptidase might be a member of a family of cysteine peptidases different from that of papain, because it has the unusual characteristic of reacting more rapidly with iodoacetamide than with iodoacetate. No catalytic residues have been identified in the hemoglobinase family. A sequence tag from *Arabidopsis thaliana* is homologous to the N terminus of hemoglobinase (EMBL: Z17798).

Interleukin 1β Converting Enzyme Family (C14)

The precursor of the cytokine, interleukin 1β, is a 31- to 33-kDa protein synthesized by monocytes, and the active 17.5-kDa molecule is released by cleavage of an Asp+Ala bond. The interleukin 1α precursor is not processed by this enzyme.[70]

The interleukin 1β converting enzyme (ICE) has been purified, cloned, and sequenced (see [43]). The enzyme shows strict specificity for the cleavage of aspartyl bonds.

[66] A. H. Davis, J. Nanduri, and D. C. Watson, *J. Biol. Chem.* **262**, 12851 (1987).

[67] B. Götz and M.-O. Klinkert, *Biochem. J.* **290**, 801 (1993).

[68] A. A. Kembhavi, D. J. Buttle, C. G. Knight, and A. J. Barrett, *Arch. Biochem. Biophys.* **303**, 208 (1993).

[69] C. Csoma and L. Polgár, *Biochem. J.* **222**, 769 (1984).

[70] N. A. Thornberry, H. G. Bull, J. R. Calaycay, K. T. Chapman, A. D. Howard, M. J. Kostura, D. K. Miller, S. M. Molineaux, J. R. Weidner, J. Aunins, K. O. Elliston, J. M. Ayala, F. J. Casano, J. Chin, G.J.-F. Ding, L. A. Egger, E. P. Gaffney, G. Limjuco, O. C. Palyha, S. M. Raju, A. M. Rolando, J. P. Salley, T.-T. Yamin, and M. J. Tocci, *Nature (London)* **356**, 768 (1992).

ICE is synthesized as a 45-kDa precursor. Four peptide bond cleavages occur in the formation of the active enzyme, which is a heterodimer of a heavy chain (22 kDa) and a light chain (10 kDa). All of the cleavages in the maturation of ICE are of aspartyl bonds, and it is very probable that they are mediated by preexisting molecules of the active enzyme.

Although ICE has little reactivity with most inhibitors that are effective against enzymes of the papain family, it is inhibited by aldehyde, diazomethane, and (acyloxy)methane compounds of appropriate structure (see [43]). ICE is also inhibited by a protein inhibitor of the serpin family that is encoded by the cowpox virus.[71]

A broader physiological role for ICE has been suggested by the discovery that the product of the *ced-3* gene involved in programmed cell death in *Caenorhabditis elegans* is a homolog of ICE,[72] and that either this protein or ICE can cause programmed cell death in transfected cells.[73]

The catalytic residue of ICE has been identified as Cys-285 (Fig. 5), and the only His residues conserved in the family are N-terminal to this.

Pyroglutamyl-peptidase I Family (C15)

Pyroglutamyl-peptidase I removes an N-terminal pyroglutamyl residue from a polypeptide. Sequences are known from eubacteria only, although a mammalian enzyme with similar activity exists.[74] The enzyme from *Bacillus amyloliquefaciens* (23-kDa monomer mass) is inhibited by diazomethane and chloromethane inhibitors,[75] and site-directed mutagenesis has identified Cys-144 as the catalytic residue.[76] There are two conserved His residues, His-168 and His-215, the former the more likely to be part of a catalytic dyad because the latter forms the C terminus (Fig. 5).

Microsomal ER60 Endopeptidase Family (C17)

ER60 endopeptidase is a protein from the mammalian rough endoplasmic reticulum that is believed to degrade other endoplasmic proteins,

[71] C. A. Ray, R. A. Black, S. R. Kronheim, T. A. Greenstreet, P. R. Sleath, G. S. Salvesen, and D. J. Pickup, *Cell* (*Cambridge, Mass.*) **69**, 597 (1992).

[72] J. Yuan, S. Shaham, S. Ledoux, H. M. Ellis, and H. R. Horvitz, *Cell* (*Cambridge, Mass.*) **75**, 641 (1993).

[73] M. Miura, H. Zhu, R. Rotello, E. A. Hartwieg and J. Yuan, *Cell* (*Cambridge, Mass.*) **75**, 653 (1993).

[74] J. K. McDonald and A. J. Barrett, "Mammalian Proteases: A Glossary and Bibliography. Volume 2: Exopeptidases." Academic Press, London, 1986.

[75] K. Fujiwara, E. Matsumoto, T. Kitagawa, and D. Tsuru, *Biochim. Biophys. Acta* **702**, 149 (1980).

[76] T. Yoshimoto, T. Shimoda, A. Kitazono, T. Kabashima, K. Ito, and D. Tsuru, *J. Biochem.* (*Tokyo*) **113**, 67 (1993).

such as protein disulfide-isomerase and calreticulin.[77] Degradation of these proteins is inhibited by leupeptin and E64. These compounds also inhibit calpain, as well as the lysosomal cysteine endopeptidases that pass through the endoplasmic reticulum in the course of their biosynthesis. However, the ER60 activity has a neutral pH optimum, and is unaffected by EGTA (which inhibits calpain) and by other inhibitors that might affect the cathepsins.

The sequence reported for ER60 endopeptidase is that of a member of a large family of proteins related to thioredoxin. Thioredoxin acts as a protein disulfide oxidoreductase, and catalyzes dithiol–disulfide exchanges. The three-dimensional structure of *E. coli* thioredoxin has been resolved,[78] and a catalytic mechanism has been proposed in which the proximity of two Cys residues in a sequence -Cys-Xaa-Xaa-Cys- permits the reversible formation of a disulfide bond. The ER60 endopeptidase contains two thioredoxin-like sequences separated by an unrelated segment of 240 residues. The Cys residues involved in the catalytic activity of thioredoxin are conserved in the ER60 endopeptidase. It is notable that a -Cys-Xaa-Xaa-Cys- sequence contains the catalytic Cys in papain, and is suggested to contain that of the type IV prepilin leader peptidase also (Fig. 5).

Type IV Prepilin Leader Peptidase Family (C20)

Some of the biosynthetic precursors of proteins secreted by bacteria have a special leader peptide that is removed by type IV prepilin leader peptidase. Pili are hairlike structures that occur on the surface of bacteria, and each is assembled from one or more protein subunits known as pilins. Type IV pili are found in the gram-negative pathogens, including *Pseudomonas aeruginosa* and *Neisseria gonorrhoeae,* and are thought to be responsible for attaching the organism to the surface of host epithelial cells. The prepilin subunits are synthesized with six- to eight-residue leader peptides that contain charged amino acids, unlike the leader peptides removed by leader peptidase 1. All mature type IV pilins have a methylated N-terminal Phe residue. N-Methylation is unusual in bacteria, but is a prerequisite for assembly of pilin subunits.[79]

The type IV prepilin leader peptidase is found in the inner membrane, and cleavage and methylation of the pilin precursors appear to occur on

---

[77] R. Urade, M. Nasu, T. Moriyama, K. Wada, and M. Kito, *J. Biol. Chem.* **267,** 15152 (1992).

[78] H. J. Dyson, G. P. Gippert, D. A. Case, A. Holmgren, and P. E. Wright, *Biochemistry* **29,** 4129 (1990).

[79] M. S. Strom and S. Lory, *J. Biol. Chem.* **266,** 1656 (1991).

the cytoplasmic face of the membrane.[80] In strains of *Pseudomonas* in which activity is deficient, there is accumulation not only of prepilins, but also of proteins that are normally secreted, such as exotoxin A, phospholipase C, alkaline phosphatase, and pseudolysin.[81] The accumulation of these proteins occurs because type IV prepilin leader peptidase is required to process proteins that mediate secretion of the affected proteins.[82]

The prepilin leader peptidase cleaves Gly+Phe bonds, and specificity is for residues in the prime sites. Thus, the peptidase recognizes a -Gly+Phe-Thr-Leu- (or -Ile-) -Glu consensus in which the Gly in P1 is obligatory.[79] The peptidase is bifunctional, however, for not only is the leader peptide removed, but the newly exposed N terminus is methylated.[83]

Type IV prepilin leader peptidase is a product of the *pilD* gene in *Pseudomonas,* and homologs are known from other bacteria (Table IV). The enzyme is sensitive to thiol-blocking reagents such as *N*-ethylmaleimide, iodoacetamide, and *p*-mercuribenzoate, and the inhibition by *p*-mercuribenzoate is reversible by dithiothreitol. Site-directed mutagenesis has implicated four Cys residues in both peptidase and methylase activities to differing extents,[80] although one of these (Cys-97) is naturally replaced by Ser in the *Klebsiella* homolog.[84] However, there is no conserved His residue in an alignment of the sequences of members of the family.[80]

Conclusions

In the past, the cysteine proteinases of the papain family have been given so much more attention than the others that it has been easy to assume that all such enzymes were essentially "papainlike", but it is now abundantly clear that that is not the case. In the near future, this point seems likely to be brought into sharp focus by the elucidation of three-dimensional structures for cysteine peptidases of other families. In reviewing the structures and mechanisms of the catalytic sites of peptidases generally, James[85] has observed that for serine peptidases (e.g., chymotrypsin and subtilisin), aspartic endopeptidases (e.g., pepsin), and metallopeptidases (e.g., carboxypeptidase A and thermolysin), the attack on the

[80] M. S. Strom, P. Bergman, and S. Lory, *J. Biol. Chem.* **268,** 15788 (1993).
[81] D. N. Nunn and S. Lory, *Proc. Natl. Acad. Sci. U.S.A.* **88,** 3281 (1991).
[82] M. S. Strom and S. Lory, *J. Bacteriol.* **174,** 7345 (1992).
[83] M. S. Strom, D. N. Nunn, and S. Lory, *Proc. Natl. Acad. Sci. U.S.A.* **90,** 2404 (1993).
[84] M. R. Kaufman, J. M. Seyer, and R. K. Taylor, *Genes Dev.* **5,** 1834 (1991).
[85] M. N. G. James, *in* "Proteolysis and Protein Turnover" (J. S. Bond and A. J. Barrett, eds.), p. 1. Portland Press, London, 1993.

peptide bond that leads to hydrolysis of the substrate is from the *re* face of the bond. In contrast, the active site of papain attacks its substrate from the *si* face. As James points out, it will be of the greatest interest to learn, as more structures become available, whether other families of cysteine peptidases share this unusual characteristic of papain, or resemble the majority of peptidases of other catalytic types. It may well be that some of the other families will have geometries more like that of chymotrypsin. As we have seen, the picornains have been suggested to be distantly related to the chymotrypsin family. Also, Salvesen[86] has pointed out that the reactivity of interleukin 1β converting enzyme with a serpin suggests that it too may have geometry more like that of chymotrypsin than that of papain.

[86] G. Salvesen, *in* "Proteolysis and Protein Turnover" (J. S. Bond and A. J. Barrett, eds.), p. 57. Portland Press, London, 1993.

[33] Catalytic Mechanism in Papain Family of Cysteine Peptidases

By ANDREW C. STORER and ROBERT MÉNARD

Introduction

Cysteine peptidases are a class of enzymes that have been widely studied over the years. The overall principles of substrate recognition, catalysis, and inhibition are now reasonably well documented. However, the molecular basis of these properties is still not clearly established. For example, although it has formed the subject of numerous reviews (see Refs. 1–5), the mechanism by which cysteine peptidases hydrolyze their substrates is still poorly defined at the atomic level. By far the bulk of the literature reports dealing with enzymes in this class describe results

[1] L. Polgar and P. Halasz, *Biochem. J.* **207**, 1 (1982).

[2] E. N. Baker and J. Drenth, *in* "Biological Macromolecules and Assemblies, Volume 3—Active Sites of Enzymes" (F. A. Jurnak and A. McPherson, eds.), p. 314. Wiley, New York, 1987.

[3] K. Brocklehurst, *in* "Enzyme Mechanisms" (M. I. Page and A. Williams, eds.), p. 140. Royal Society of Chemistry, London, 1987.

[4] K. Brocklehurst, F. Willenbrock, and E. Salih, *in* "Hydrolytic Enzymes" (A. Neuberger and K. Brocklehurst, eds.), p. 39. Elsevier, Amsterdam, 1987.

[5] L. Polgar, *in* "Mechanisms of Protease Action," p. 123. CRC Press, Boca Raton, Florida, 1990.

obtained with the plant peptidase, papain. Consequently, this enzyme is considered to be the archetype of cysteine peptidases, and as such to constitute a good model for this family of enzymes (the papain sequence numbering will be used throughout this review). In this paper the main features of the catalytic mechanism for cysteine peptidases of the papain family that have been previously discussed will only be summarized. However, the most recent advances, which contribute significantly to increasing our understanding of catalysis by cysteine peptidases, will be covered in detail.

General Catalytic Mechanism

Cysteine peptidases of the papain family catalyze the hydrolysis of peptide, amide, ester, thiol ester, and thiono ester bonds.[4] The basic features of the mechanism include the formation of a covalent intermediate, the acyl-enzyme, resulting from nucleophilic attack of the active site thiol group on the carbonyl carbon of the scissile amide or ester bond of the bound substrate. The overall mechanism of hydrolysis involves a number of steps that are represented schematically in Fig. 1 for the hydro-

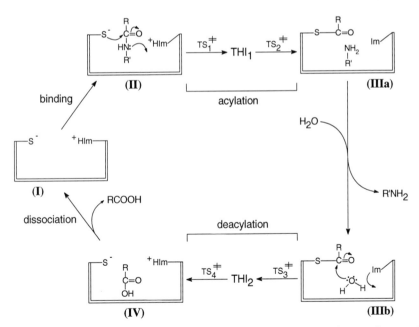

FIG. 1. Schematic representation of the various steps and putative intermediates and transition states involved in the reaction pathway for hydrolysis of an amide substrate by a cysteine peptidase. Details are given in the text.

lysis of an amide substrate. The first step in the reaction pathway corresponds to the association (or noncovalent binding) of the free enzyme **(I)** and substrate to form the Michaelis complex **(II)**. This step is followed by acylation of the enzyme **(IIIa)**, with formation and release of a first product from the enzyme, the amine $R'NH_2$. In the following step, the acyl-enzyme **(IIIb)** reacts with a water molecule to form the second product (deacylation step). Release of this product results in the regeneration of the free enzyme. Many intermediates and/or transition states are believed to exist along this pathway, i.e., based on limited experimental data and by analogy with the serine peptidases it has been proposed that the acylation and deacylation steps involve the formation of transient tetrahedral intermediates (THI_1 and THI_2). In addition, four transitions states ($TS_1\ddagger$, $TS_2\ddagger$, $TS_3\ddagger$, and $TS_4\ddagger$) separate the Michaelis complex, acyl-enzyme, and enzyme–product complex from the two tetrahedral intermediates.

Structural Features of Cysteine Peptidases

By definition, cysteine peptidases require the thiol group of a cysteine residue for their activity. Before the advent of structural data, it had been suggested that a histidine located in the vicinity of the active site cysteine residue was probably also involved in the catalytic process.[6,7] The Cys^{25}-His^{159} pairing in papain became evident when the crystal structure of the enzyme was solved.[8-10] The enzyme constitutes a single protein chain folded to form two domains delimiting a cleft into which the substrate can bind. The active site residues Cys-25 and His-159 are located at the interface of this cleft on opposite domains of the enzyme. Other residues near the active site were identified from the structural data and have been suggested to play various roles in the catalytic mechanism of the enzyme (see below). The structures of several other cysteine peptidases are now available [actinidain,[11] calotropin DI,[12] cathepsin B,[13] and caricain (papaya

---

[6] G. Lowe and A. Williams, *Biochem. J.* **96**, 194 (1965).
[7] S. S. Husain and G. Lowe, *Biochem. J.* **108**, 855 (1968).
[8] J. Drenth, J. N. Jansonius, R. KoeKoek, H. M. Swen, and B. G. Wolthers, *Nature (London)* **218**, 929 (1968).
[9] J. Drenth, J. N. Jansonius, R. KoeKoek, and B. G. Wolthers, *Adv. Protein Chem.* **25**, 79 (1971).
[10] I. G. Kamphuis, K. H. Kalk, M. B. A. Swarte, and J. Drenth, *J. Mol. Biol.* **179**, 233 (1984).
[11] E. N. Baker, *J. Mol. Biol.* **141**, 441 (1980).
[12] U. Heinemann, G. P. Pal, R. Hilgenfeld, and W. Saenger, *J. Mol. Biol.* **161**, 591 (1982).
[13] D. Musil, D. Zucic, D. Turk, R. A. Engh, I. Mayr, R. Huber, T. Popovic, V. Turk, T. Towatari, N. Katunuma, and W. Bode, *EMBO J.* **10**, 2321 (1991).

proteinase Ω)[14]] and their active site residues can be superimposed onto that of papain with an accuracy close to that of the atomic coordinates (see Baker and Drenth[2] for a review of the structural aspects of cysteine peptidases).

Thiolate–Imidazolium Ion Pair

Probably the aspect of cysteine peptidases of primary importance for their catalytic activity is the high nucleophilicity of the active site thiol group. It is now generally accepted that the active form of papain and of cysteine peptidases in general consists of a thiolate–imidazolium ion pair.[15–18] The existence of two ionizable groups as essential catalytic residues in papain is consistent with the bell-shape form for the pH dependency of activity with this enzyme. The acid limb, with a pK_a of ~4, is usually attributed to the ionization of the active site Cys-25, whereas the basic limb, with a pK_a of ~8.5, is considered to reflect ionization of His-159. The ionization pathways of the two active site residues are illustrated in Fig. 2.[19] In this model, which is a simplification of the actual situation, only the ion pair form of the enzyme is considered to be active and K_5 represents the equilibrium constant between this ion pair and the neutral form of the active site residues. Results from various experimental techniques have evaluated the percentage of papain present as its ion pair form at neutral pH to be in the range of at least 50% to almost 100% of the total enzyme,[1] and equilibrium constants (i.e., K_5) of 2^{20} and 12^{17} have been reported. The equilibrium therefore seems to lie in favor of the thiolate–imidazolium form. In reality, the experimentally observed pK_a values describing the pH–activity profile of papain are apparent pK_a values that depend on the position of the equilibrium between the neutral and ion pair form of the active site residues.[19,21] The acid limb apparent pK_a approximates the intrinsic pK_a of Cys-25 (pK_1^{int}) because $K_5 > 1$ for papain. If the stability of the ion pair form relative to that of the neutral form was lower ($K_5 \ll 1$), the apparent pK_a in the acid limb would approach the

[14] R. W. Pickersgill, P. Rizkallah, G. W. Harris, and P. W. Goodenough, *Acta Crystallogr. Sect. B: Struct. Sci.* **B47**, 766 (1991).

[15] L. Polgar, *FEBS Lett.* **47**, 15 (1974).

[16] L. A. Æ. Sluyterman and J. Wijdenes, *Eur. J. Biochem.* **71**, 383 (1976).

[17] S. D. Lewis, F. A. Johnson, and J. A. Shafer, *Biochemistry* **15**, 5009 (1976).

[18] S. D. Lewis, F. A. Johnson, and J. A. Shafer, *Biochemistry* **20**, 48 (1981).

[19] R. Ménard, H. E. Khouri, C. Plouffe, R. Dupras, D. Ripoll, T. Vernet, D. C. Tessier, F. Laliberté, D. Y. Thomas, and A. C. Storer, *Biochemistry* **29**, 6706 (1990).

[20] D. J. Creighton and D. J. Schamp, *FEBS Lett.* **110**, 313 (1980).

[21] R. Ménard, H. E. Khouri, C. Plouffe, P. Laflamme, R. Dupras, T. Vernet, D. C. Tessier, D. Y. Thomas, and A. C. Storer, *Biochemistry* **30**, 5531 (1991).

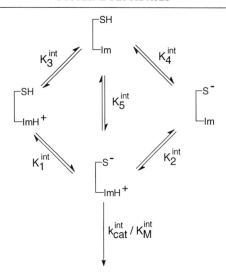

FIG. 2. The ionization pathways of the cysteine and histidine active site residues. Only the ion pair form of the enzyme is considered to be active and K_5 represents the equilibrium constant between this ion pair and the neutral form of the enzyme.

pK_3^{int} of the His-159 residue. The same comments apply to the apparent pK_a in the basic limb. An important consequence of the existence of two ionization pathways for the active site residues is that the width of a pH–activity profile reflects the stability of the ion pair form relative to that of the neutral form.[21] Ménard et al.[21] proposed that this could be used to divide the effects of mutations on measured activity, introduced in the enzyme via site-directed mutagenesis, between a direct influence on the intrinsic activity and an indirect influence through a change in the equilibrium between active (i.e., ion pair) and inactive forms of the enzyme. Using this approach, a value of K_5 = 4.4 was determined for papain,[22] which is within the range of K_5 values obtained using various other experimental techniques.[17,20]

The model illustrated in Fig. 2 reflects only the ionization of the active site residues Cys-25 and His-159, but other pH-dependent events are known to affect the activity of cysteine peptidases. For example, it is well known that the acid limb of the pH–activity profile of papain against substrates is modulated not by one but by two ionizable groups with pK_a

[22] R. Ménard, C. Plouffe, H. E. Khouri, R. Dupras, D. C. Tessier, T. Vernet, D. Y. Thomas, and A. C. Storer, *Protein Eng.* **4,** 307 (1991).

values of approximately 4,[23-25] and Brocklehurst and collaborators have shown that a number of ionizable groups can affect the activity of both papain and actinidain against pyridyl disulfides.[25-27] In addition to the two pK_a values defining the acid limb, it was shown using pyridyl disulfides that the basic limb of pH–activity profiles for papain also displays two pK_a values, indicating that a second ionizable group in addition to His-159 modulates the activity of the enzyme in this pH range.[28] This second ionization in the basic limb was also observed for substrate hydrolysis by papain under certain conditions (A. C. Storer and R. Ménard, unpublished work). In some cases, they also observed that ionization of a group with a pK_a of 5.0 can modulate activity.[27] The approach used by Brocklehurst and collaborators offers the advantage of combining results from a variety of substrates and pyridyl disulfides to test the validity of a "complete" model that would take into account all possible pH-dependent events that affect the activity of cysteine peptidases. However, one drawback of this approach lies in the complexity of the model fitting required. A complete understanding of the catalytic mechanism of cysteine peptidases will require knowledge of the ionizable groups that modulate enzymatic activity and the mechanisms by which such modulation occurs. At the moment, site-directed mutagenesis offers the best hopes for providing positive identification of these groups.

The enzyme features that stabilize the thiolate–imidazolium ion pair have been the subject of several theoretical studies over the years.[29-35] As evidenced from structural data, the catalytic thiol group of cysteine peptidases is located at the end of an α-helix and calculations indiate that

[23] M. R. Bendall and G. Lowe, *Eur. J. Biochem.* **65**, 481 (1976).

[24] S. D. Lewis, F. A. Johnson, A. K. Ohno, and J. A. Shafer, *J. Biol. Chem.* **253**, 5080 (1978).

[25] E. Salih, J. P. G. Malthouse, D. Kowlessur, M. Jarvis, M. O'Driscoll, and K. Brocklehurst, *Biochem. J.* **247**, 181 (1987).

[26] D. Kowlessur, C. M. Topham, E. W. Thomas, M. O'Driscoll, W. Templeton, and K. Brocklehurst, *Biochem. J.* **258**, 755 (1989).

[27] K. Brocklehurst, S. M. Brocklehurst, D. Kowlessur, M. O'Driscoll, G. Patel, E. Salih, W. Templeton, E. Thomas, C. M. Topham, and F. Willenbrock, *Biochem. J.* **256**, 543 (1988).

[28] G. W. Mellor, E. W. Thomas, C. M. Topham, and K. Brocklehurst, *Biochem. J.* **290**, 289 (1993).

[29] R. Lavery, A. Pullman, and Y. K. Wen, *Int. J. Quantum Chem.* **24**, 352 (1983).

[30] P. Th. van Duijnen, B. Th. Thole, R. Broer, and W. C. Nieuwpoort, *Int. J. Quantum Chem.* **17**, 651 (1980).

[31] W. G. J. Hol, P. Th. van Duijnen, and H. J. C. Berendsen, *Nature (London)* **273**, 443 (1978).

[32] P. Th. van Duijnen, B. Th. Thole, and W. G. J. Hol, *Biophys. Chem.* **9**, 273 (1979).

[33] J. P. Dijkman, R. Osman, and H. Weinstein, *Int. J. Quantum Chem.* **35**, 241 (1989).

[34] J. P. Dijkman and P. Th. van Duijnen, *Int. J. Quantum Chem., Quant. Biol. Symp.* **18**, 49 (1991).

[35] J. A. C. Rullman, M. N. Bellido, and P. Th. van Duijnen, *J. Mol. Biol.* **206**, 101 (1989).

the helix field could provide enough stabilization for the ion pair to exist over a wide pH range.[30,32] It has also been argued that the electrostatic interaction between the catalytic Cys and His residues could be enough to ensure that an ion pair form exists as a stable species at neutral pH and that this ion pair should be extremely sensitive to the geometry of the active site region.[35] Aside from the helix dipole and the geometry of the active site, specific neighboring residues have been considered to play a significant role in the stabilization of the thiolate–imidazolium ion pair. These residues and their possible roles will be discussed in the next section.

Residues in Proximity to Active Site Cysteine and Histidine Residues

Of possible importance in the catalytic mechanism of cysteine peptidases is the proximity to the catalytic histidine of a conserved asparagine residue. The amide oxygen $O_{\delta 1}$ of the Asn-175 side chain is hydrogen bonded to the $N_{\epsilon 2}$ atom of His-159, creating a Cys-His-Asn triad that is often considered analogous to the Ser-His-Asp arrangement of serine peptidases. The Asn[175]-His[159] hydrogen bond is approximately colinear with the His-159 C_{β}–C_{γ} bond, allowing the imidazole side chain to rotate about the C_{β}–C_{γ} bond without disruption of the Asn[175]-His[159] hydrogen bond. This, combined with the observation from crystallography studies that the orientation of the His-159 side chain can vary for different complexes of papain, has led to the proposal that the role of Asn-175 might be to direct the imidazole in positions optimum for the differing steps in the catalytic mechanism.[36] As suggested by theoretical studies,[35] Asn-175 might also assist in the stabilization of the thiolate–imidazolium ion pair by keeping the imidazole ring in a favorable orientation relative to the cysteine residue. The role of Asn-175 in catalysis by cysteine peptidases is presently under investigation by site-directed mutagenesis and initial results indicate that an Asn-to-Gln substitution can be tolerated in the active site of papain.[37]

The hydrogen bond between the side chains of residues His-159 and Asn-175 is buried in a hydrophobic region that, in the case of papain, is composed mainly of the side chains of residues Phe-141, Trp-177, and Trp-181. In particular, Trp-177 is in a position to interact positively with the protonated His-159 and to shield the His-Asn hydrogen bond from solvent. It has been shown by Loewenthal et al.[38] for the enzyme barnase,

---

[36] J. Drenth, H. M. Swen, W. Hoogenstraaten, and L. A. Æ. Sluyterman, *Proc. K. Ned. Akad. Wet., Ser. C* **78,** 104 (1975).

[37] T. Vernet, J. Chatellier, D. C. Tessier, and D. Y. Thomas, *Protein Eng.* **6,** 213 (1993).

[38] R. Loewenthal, J. Sancho, and A. R. Fersht, *J. Mol. Biol.* **224,** 759 (1992).

a ribonuclease, that the interaction of a protonated histidine side chain with a tryptophan residue results in an increase of the pK_a of the former. This suggests that the protonated form of the histidine can be preferentially stabilized by the adjacent aromatic group. Thus, this His-Trp pairing in cysteine peptidases is probably of importance for the ion pair stability and/or the catalytic activity of the enzyme through its combined effect of increasing the pK_a of the histidine and shielding the His[159]-Asn[175] hydrogen bond from external solvent.

Asp-158 is the only residue bearing a charged side chain located within 10 Å of the active site ion pair in papain. This residue has had several mechanistic roles suggested based on experimental data as well as theoretical studies, and it has been proposed to contribute significantly to the stabilization of the active site thiolate–imidazolium ion pair (see Ménard et al.[19] for a review). One of the two pK_a values defining the acid limb of pH–activity profiles with papain has often been attributed to the ionization of Asp-158, and this has been considered as experimental evidence that the side chain of Asp-158 has a direct influence on the catalytic activity of the enzyme.[23–25,39] The role of Asp-158 in papain was clarified when site-directed mutagenesis was used to replace the aspartic acid by a residue with a side chain that cannot ionize (Asn).[19] The mutant Asp158Asn displayed a k_{cat}/K_m value against the substrate CBZ-Phe-Arg-MCA only six-fold lower than that of wild-type papain. In addition, the pH–activity profile of the Asp158Asn mutant was shifted to lower pH values relative to that of papain by approximately 0.3 pH unit, but retained the two pK_a values in the acid limb. These results clearly demonstrated that Asp-158 is not a residue essential for catalytic activity in papain and that the negative charge of the side chain does not contribute significantly to the stability of the thiolate–imidazolium ion pair. Any effect that the negative charge could have on the ion pair stability is likely to be significantly screened by solvent molecules, as suggested by recent theoretical studies.[34] The shift in the pH–activity profile, however, is compatible with the assumption that electrostatic interactions exist between the negatively charged side chain of Asp-158 and the active site ion pair. Other mutations at that position indicate, however, that the network of hydrogen bonds involving the side chain of Asp-158 (Asp-158 to Ala-136, His-159, and a water molecule) might contribute to ion pair stability and catalytic activity.[21] The role of the Asp-158 hydrogen bonding network could be to maintain an optimum orientation of residues in the active site, especially considering that one of the H-bond partners of the side chain is the main chain amide of His-159. However, the fact that Asp-158 does not play an

[39] K. Brocklehurst, E. Salih, and T. S. Lodwig, *Biochem. J.* **220**, 609 (1984).

essential role in papain does not come as a surprise, because this residue is not conserved in all cysteine peptidases.[2]

From an observation of the crystal structures of cysteine peptidases and their complexes with acyl-chloromethane inhibitors,[36,40] other potentially important enzyme residues can be identified. For example, the side chain of Gln-19 and the amide group of Cys-25 have been proposed to be of importance to catalysis through their participation in an oxyanion hole similar to that found in serine peptidases. This enhancement of catalytic activity would originate from a stabilized transition state, and this aspect is discussed below. An extensive network of interactions at the interface of the two domains in papain and other cysteine peptidases is also evident from examination of the available structures and could play a role in catalysis.[11] The active site residues Cys-25 and His-159 are located on opposite walls of the active site cleft and have a low accessibility in papain.[10] Widening of the cleft through domain movement is suggested to be required for substrate binding and also to increase the nucleophilicity of the thiolate group of Cys-25 during acylation.[36,40] Interdomain interactions therefore might be important for the control of active site geometry and mobility during catalysis. However, mutation of residues involved in interdomain hydrophobic contacts (Val-32, Ala-162) and the hydrogen bond (Ser-176) failed to produce any significant effect on activity.[22,41] The narrowing of the pH–activity profile on mutation of Ser-176 to an alanine was interpreted to reflect a perturbation of the Cys^{25}-His^{159} ion pair stability.[22]

Substrate Binding and Acylation

Papain has a large binding site and there are a number of interactions that exist between the enzyme and the substrate over an extended region. Coupling of these substrate binding interactions to the hydrolytic process occurring at the active site is an important aspect of catalysis that has received relatively little attention so far. Brocklehurst and collaborators have made an effort to unravel proposed signaling mechanisms by which binding and catalysis are linked.[26,42–45] They have used substrate-derived

[40] J. Drenth, K. H. Kalk, and H. M. Swen, *Biochemistry* **15**, 3731 (1976).
[41] T. Vernet, D. C. Tessier, H. E. Khouri, and D. Altschuh, *J. Mol. Biol.* **224**, 501 (1992).
[42] K. Brocklehurst, D. Kowlessur, G. Patel, W. Templeton, K. Quigley, E. W. Thomas, C. W. Wharton, F. Willenbrock, and R. J. Szawelski, *Biochem. J.* **250**, 761 (1988).
[43] W. Templeton, D. Kowlessur, E. W. Thomas, C. M. Topham, and K. Brocklehurst, *Biochem. J.* **266**, 645 (1990).
[44] M. Patel, I. S. Kayani, G. W. Mellor, S. Sreedharan, W. Templeton, E. W. Thomas, M. Thomas, and K. Brocklehurst, *Biochem. J.* **281**, 553 (1992).
[45] M. Patel, I. S. Kayani, W. Templeton, G. W. Mellor, E. W. Thomas, and K. Brocklehurst, *Biochem. J.* **287**, 881 (1992).

pyridyl disulfides to study the relationship between substrate binding and catalytic site reactivity in a variety of cysteine peptidases. Results of these experiments have been interpreted as indicating that binding of a substrate in the S1–S2 subsite region causes a conformational change in the transition state, allowing the imidazolium group of His-159 to protonate the leaving group and thus facilitating the acylation reaction. Important interactions identified are between the P1–P2 amide bond and the main chain carbonyl of Asp-158, the P2 NH and P2 C=O, with Gly-66 and a hydrophobic side chain at P2 with the S2 subsite. It is suggested that these interactions function cooperatively to provide signaling to the catalytic site. Similar observations of cooperative enzyme–substrate interactions were made by Berti et al.[46] for the hydrolysis of ester substrates by papain. However, in this study the effect was suggested to be due to a decreased entropic penalty with the successive addition of favorable interactions, rather than a signaling mechanism per se.

Following the binding of the substrate to a cysteine peptidase, the carbonyl carbon of the scissile bond undergoes nucleophilic attack by the Cys-25 thiolate anion of the enzyme. This is facilitated by the carbonyl oxygen binding in the oxyanion hole that is defined by the Cys-25 NH and the side chain amide of Gln-19. As a result, the first anionic tetrahedral intermediate (THI_1, Fig. 1) is formed. It has been pointed out by Polgar and Asboth[47] that formation of this intermediate in cysteine peptidases requires only a migration of charge, whereas a full charge separation is required for serine peptidases and as a consequence catalysis by the former relies less heavily on the presence of an oxyanion hole. This suggestion has been supported by subsequent site-directed mutagenesis experiments,[48–50] i.e., the H bond formed with the side chain of Gln-19 contributes approximately 1 kcal less to the stabilization of the transition state in papain-catalyzed hydrolysis than does the equivalent H bond in subtilisin.[51] Following the formation of the charged THI_1 in cysteine peptidases, the protonated active site histidine, His-159, rotates in order to donate its proton (general acid catalysis) to the amide nitrogen, promoting the breakdown of the tetrahedral intermediate to give free amine product and an acyl-enzyme intermediate (Fig. 1). Experimental support for the rotation

[46] P. J. Berti, C. H. Faerman, and A. C. Storer, *Biochemistry* **30**, 1394 (1991).
[47] L. Polgar and B. Asboth, *J. Theor. Biol.* **121**, 323 (1986).
[48] R. Ménard, J. Carrière, P. Laflamme, C. Plouffe, H. E. Khouri, T. Vernet, D. C. Tessier, D. Y. Thomas, and A. C. Storer, *Biochemistry* **30**, 8924 (1991).
[49] P. Bryan, M. W. Pantoliano, S. G. Quill, H. Y. Hsiao, and T. Poulos, *Proc. Natl. Acad. Sci. U.S.A.* **83**, 3743 (1986).
[50] P. Carter and J. A. Wells, *Proteins: Struct. Funct. Genet.* **7**, 335 (1990).
[51] R. Ménard and A. C. Storer, *Biol. Chem. Hoppe-Seyler* **373**, 393 (1992).

of the histidine side chain can be found in the X-ray structures of the enzymes and their derivatives, in that, in oxidized forms of the enzyme and also acyl-chloromethane derivatives, the histidine ring is in an "upward" position, whereas in activatable papain it is rotated to a position, "downward," in which it is more capable of interacting with the Cys-25 thiolate anion.[2] Whether the histidine ring is free to move in the free and complexed forms of the enzyme or if the movement of the histidine results from a substrate signaling mechanism has yet to be determined. However, it should be noted that the X-ray structures of the acyl-chloromethane derivatives (acyl-enzyme analog) of papain[40] and of a leupeptin–papain complex[52] (tetrahedral intermediate analog) differ only slightly from that of the uncomplexed enzyme, indicating that any signaling mechanism, if present, must involve subtle changes in the structure of the enzyme.

An extensive spectroscopic study of cysteine peptidases has provided information on the conformation of the P2 and P1 residues of the acyl-enzyme (see Ref. 53 and references therein). Although the results have been obtained exclusively with thiono ester substrates, evidence suggests that observations can be transferred to the more natural oxygen ester substrates.[54] For thiono ester substrates with a glycine residue in P1, the acyl-enzyme formed assumes a conformation such that the glycine nitrogen atom is close to the thiol sulfur atom, with the distance between the atoms being less than the sum of their van der Waals radii. This interaction between the nitrogen and sulfur atoms involves a HOMO-to-LUMO donation of electrons from the nitrogen atom to the sulfur atom.[53] It has been suggested that this interaction could assist catalysis by stabilizing the acyl-enzyme and hence promote the partitioning of THI_1 toward the acyl-enzyme. It has also been proposed that the HOMO–LUMO interaction facilitates catalysis by providing a conduit for electron delocalization in the transition state, THI_2, of the deacylation step.[55] An observation that thiono ester substrates containing an amino acid other than glycine at P1 produce acyl-enzymes in which the nitrogen of the P1 residue is not in close contact with the sulfur of the active center thiol group and yet deacylates at a rate similar to that of the glycine-based substrates[53,56] is difficult to reconcile with the latter explanation.

[52] E. Schröder, C. Phillips, E. Garmen, K. Harlos, and C. Crawford, *FEBS Lett.* **315**, 38 (1993).

[53] P. J. Tonge, R. Ménard, A. C. Storer, B. P. Ruzsicska, and P. R. Carey, *J. Am. Chem. Soc.* **113**, 4297 (1991).

[54] A. C. Storer and P. R. Carey, *Biochemistry* **24**, 6808 (1985).

[55] G. D. Duncan, C. P. Huber, and W. J. Welsh, *J. Am. Chem. Soc.* **114**, 5784 (1992).

[56] A. C. Storer, R. H. Angus, and P. R. Carey, *Biochemistry* **27**, 264 (1988).

Concomitant with the formation of the acyl-enzyme on the cysteine peptidase reaction pathway, the pK_a of His-159 drops from its value of approximately 8.5 in the free enzyme to approximately 4.0 in the acyl-enzyme. This drop in pK_a enables the histidine to provide general base catalysis for the deacylation process, i.e., it has been postulated, based on kinetic isotope effects, that His-159 abstracts a proton from the attacking water molecule during tetrahedral intermediate (THI_2) formation (Ref. 57 and references therein). Breakdown would then involve a collapse of THI_2, resulting in the expulsion of the thiolate anion and, following the dissociation of the acid product, regeneration of the free enzyme.

Tetrahedral Intermediates and Transition States

As discussed above, by analogy with the serine peptidases, it is widely accepted that both the acylation and deacylation steps on the cysteine peptidase reaction pathway involve the formation of an anionic tetrahedral intermediate. It is interesting to note that, as determined by the geometries of their respective active sites, the intermediates formed in cysteine peptidases must be of the opposite hand as that in the serine peptidases,[58] which may reflect differences between the mechanisms of the two enzyme classes.

The experimental support for the formation of tetrahedral intermediates largely rests on the potent inhibition of these enzymes by peptide aldehydes, which form covalent tetrahedral adducts, hemithioacetals, with the active site cysteine residue. The potency of this inhibition is suggested to be due to the covalent intermediates mimicking the tetrahedral transition states that, in turn, are thought to closely resemble the tetrahedral intermediates on the enzyme reaction pathway.[59-61] Other evidence often quoted to be supportive of the formation of tetrahedral intermediates in the cysteine peptidase mechanism, such as Hammet ρ values obtained for the papain-catalyzed hydrolysis of phenyl esters[62] and anilides,[63] or nitrogen isotope effects on the papain-catalyzed hydrolysis of N-benzoyl-L-argininamide,[64]

---

[57] R. J. Szawelski and C. W. Wharton, *Biochem. J.* **199**, 681 (1981).

[58] R. M. Garavito, M. G. Rossmann, P. Argos, and W. Eventoff, *Biochemistry* **16**, 5065 (1977).

[59] M. R. Bendall, I. L. Cartwright, P. I. Clark, G. Lowe, and D. Nurse, *Eur. J. Biochem.* **79**, 201 (1977).

[60] C. A. Lewis and R. Wolfenden, *Biochemistry* **16**, 4890 (1977).

[61] A. Frankfater and T. Kuppy, *Biochemistry* **20**, 5517 (1981).

[62] G. Lowe and A. Williams, *Biochem. J.* **96**, 199 (1965).

[63] G. Lowe and Y. Yuthavong, *Biochem. J.* **124**, 117 (1971).

[64] M. H. O'Leary, M. Urberg, and A. P. Young, *Biochemistry* **13**, 2077 (1974).

is consistent with their formation but does not exclude the possibility of a concerted mechanism, i.e., a one-step acylation mechanism involving the direct displacement of the amine by the thiolate without the formation of an anionic tetrahedral intermediate.[65] Definitive proof of the existence of tetrahedral intermediates on the cysteine peptidase reaction pathway would be provided by the direct observation of such an intermediate. However, it has been estimated that the equilibrium constant for the formation of such an intermediate is approximately 10 orders of magnitude smaller than that for the corresponding hemithioacetal,[66] making the possibility of such an observation highly unlikely.

An increased stability of the enzyme-bound tetrahedral intermediates and hence of the structurally similar transition states relative to the catalytic ground state, i.e., the Michaelis complex, in part accounts for the catalytic enhancement of the enzyme. This increased stability is due to a greater degree of complementarity between the tetrahedral structures and the active site of the enzyme. This can be achieved by a better steric fit of the intermediates to the shape of the catalytic cleft and also by an increase in both the number and the strength of positive interactions between substrate and enzyme on moving along the reaction path from the Michaelis complex to the transition states and the tetrahedral intermediates. In addition to the covalent bond either partially or fully formed in the tetrahedral species between the thiol sulfur atom of the active site cysteine residue and the carbonyl carbon of the substrate scissile bond, other noncovalent interactions are formed or strengthened. It has been suggested, based on model building, that the optimal alignment of the tetrahedral intermediate THI_1 in the catalytic site requires a rotation of the substrate about its long axis.[44] On its formation, the geometry of the tetrahedral center is such that the oxyanion formed can interact more effectively with the H-bonding donors, the side chain of Gln-19 and the main chain NH of Cys-25, which comprise the oxyanion hole,[40] and in addition the NH of the scissile bond can now form a hydrogen bond with the main-chain carbonyl of Asp-158.[2]

The positive role of the oxyanion hole in stabilization of the tetrahedral species has been demonstrated using site-directed mutagenesis.[48] A mutant form of papain in which residue 19 has been converted to an alanine, thus removing one of the oxyanion hole proton donors, exhibits reduced catalytic activity. Additional evidence for the role of the oxyanion hole in the binding of tetrahedral species to the enzyme is found in the X-ray crystallographic structure of a papain–leupeptin complex.[52] Leupeptin is

[65] T. C. Curran, C. R. Farrar, O. Niazy, and A. Williams, *J. Am. Chem. Soc.* **102**, 6828 (1980).
[66] J. Fastrez, *Eur. J. Biochem.* **135**, 339 (1983).

a peptide aldehyde inhibitor of cysteine peptidases that forms a hemithioacetal with Cys-25. In the structure of this complex the hemithioacetal oxygen atom sits within the oxyanion hole of the enzyme. Although this observation is supportive of the formation and stabilization of tetrahedral species on the catalytic pathway of cysteine peptidases, it is inconsistent with the observation that the mutation of Gln-19 to Ala in papain results not in a decrease but a slight increase in affinity of the enzyme for aldehyde-based inhibitors.[48] Also, this latter result does not seem to support the suggestion that the cysteine peptidase mechanism operates via tetrahedral intermediates; however, additional experimentation is required before this issue can be clarified.

In the absence of direct experimental evidence for the existence of a stepwise process and tetrahedral intermediate formation, the acylation mechanism has been dissected in a theoretical study.[67] This study, which is further supported by calculations,[68] suggests that the transfer of the proton from the histidine to the amine leaving group is concerted with the attack of the cysteine nucleophile, i.e., it is suggested that an anionic tetrahedral intermediate is not formed. A similar conclusion was previously reached for the chymotrypsin-catalyzed hydrolysis of amide substrates.[69] The exclusively tetrahedral nature of the transition state(s) on the cysteine peptidase reaction pathway has been questioned by experimental evidence[70] that supports previous suggestions that peptide nitriles are also transition-state analog inhibitors of cysteine peptidases.[71–73] The evidence indicates that peptide nitriles, which are very potent inhibitors of this class of enzymes, form covalent thioimidate adducts to the active site thiol group and interact noncovalently with the enzyme in essentially the same way as do the hemithioacetals, i.e., a linear free energy correlation of 0.99 was found for the binding of 11 pairs of peptide aldehydes and nitriles. This similarity in binding exists in spite of the differing geometries of the inhibitor carbon atoms attached to the thiol sulfur, i.e., sp^2 for the thioimidate versus sp^3 for the hemithioacetal.

Unanswered Questions

Although the study of cysteine peptidases has been extensive, possibly second only in extent to that of the serine peptidases, several important

[67] D. Arad, R. Langridge, and P. A. Kollman, *J. Am. Chem. Soc.* **112,** 491 (1990).

[68] D. Arad, R. Kreisberg, and M. Shokhen, *J. Chem. Inf. Comput. Sci.* **33,** 345 (1993).

[69] M. Komiyama and M. L. Bender, *Proc. Natl. Acad. Sci. U.S.A.* **74,** 557 (1979).

[70] R. P. Hanzlik, S. P. Jacober, and J. Zygmunt, *Biochim. Biophys. Acta* **1073,** 33 (1991).

[71] J. R. Brisson, P. R. Carey, and A. C. Storer, *J. Biol. Chem.* **261,** 9087 (1986).

[72] T. C. Liang and R. Abeles, *Arch. Biochem. Biophys.* **252,** 626 (1987).

[73] R. P. Hanzlik, J. Zygmunt, and J. B. Moon, *Biochim. Biophys. Acta* **1035,** 62 (1990).

mechanistic issues have yet to be resolved. Is substrate binding and catalysis linked via a defined signaling mechanism? Are the acylation and deacylation processes stepwise, involving tetrahedral intermediates, or are they concerted processes? What are the origins of the pK_a values that influence the catalytic rate? Some progress toward answering these questions has been made, but much remains to be done. Developments in molecular biology offer hope that answers to these questions will be found soon.

Issued as NRCC publication number 36838.

[34] Cathepsin S and Related Lysosomal Endopeptidases

By HEIDRUN KIRSCHKE and BERND WIEDERANDERS

Introduction

Cathepsin S (EC 3.4.22.27) was the name given to a cysteine proteinase purified from bovine lymph nodes.[1] Although cathepsin S exhibited properties very similar to those of cathepsin L (EC 3.4.22.15), it was shown by biochemical and immunological methods and analysis of the amino acid sequences that cathepsin S and cathepsin L are distinct enzymes. The distribution in organs and cells of cathepsin S at the protein as well as at the mRNA level seems to be quite different from that of the other well-known lysosomal cysteine proteinases, such as cathepsins L, B (EC 3.4.22.1), and H (EC 3.4.22.16).

Assay Procedures

Assay of Cathepsin S

The close relationship, shown by amino acid sequence identities, between cathepsins L, S, H, and B is paralleled by their similar substrate specificities. Substrates containing the -Arg-Arg- or -Lys-Lys- sequences have been identified as very sensitive to cathepsin B.[2,3] These substrates are resistant to cathepsins S,[4,5] L,[4,6] and H[5-7] and as such they are the

[1] T. Turnšek, I. Kregar, and D. Lebez, Biochim. Biophys. Acta 403, 514 (1975).
[2] J. K. McDonald and S. Ellis, Life Sci. 17, 1269 (1975).
[3] A. J. Barrett and H. Kirschke, this series, Vol. 80, p. 535.
[4] H. Kirschke, P. Ločnikar, and V. Turk, FEBS Lett. 174, 123 (1984).
[5] X.-Q. Xin, B. Gunesekera, and R. W. Mason, Arch. Biochem. Biophys. 299, 334 (1992).
[6] H. Kirschke, J. Langner, S. Riemann, B. Wiederanders, S. Ansorge, and P. Bohley, in "Protein Degradation in Health and Disease" (D. Evered and J. Whelan, eds.), p. 15. Excerpta Medica, Amsterdam, Oxford, and New York, 1980.
[7] W. N. Schwartz and A. J. Barrett, Biochem. J. 191, 487 (1980).

TABLE I
KINETIC CONSTANTS FOR CATHEPSINS S, L, B, AND H

Substrate	pH	k_{cat} (sec^{-1})	K_m (μM)	k_{cat}/K_m (sec^{-1} mM^{-1})	Enzyme	Ref.
Z-F-R-NHMec[a]	6.5	4.7	14.7	320	cath.S$_{bov}$	b
	7.5	2.0	15.1	132	cath.S$_{bov}$	c
	6.5	1.9	22.4	85	rcath.S$_{hum}$	d
	5.5	17.0	2.4	7083	cath.L$_{hum}$	e
	5.5	17.6	2.8	6286	cath.L$_{rat}$	b
	6.0	364.0	223.0	1632	cath.B$_{rat}$	b
Z-V-V-R-NHMec	6.5	40.5	17.5	2314	cath.S$_{bov}$	b
	6.5	15.0	18.1	830	rcath.S$_{hum}$	d
	5.5	8.5	4.8	1771	cath.L$_{rat}$	b
	6.0	14.0	31.1	450	cath.B$_{rat}$	b
Bz-F-V-R-NHMec	6.5	13.0	8.1	1605	cath.S$_{bov}$	b
	7.5	1.6	4.6	348	cath.S$_{bov}$	c
	5.5	1.1	1.8	611	cath.L$_{rat}$	b
	6.0	17.5	29.0	603	cath.B$_{rat}$	b
	6.8	1.6	25.0	64	cath.H$_{bov}$	c
Z-F-V-R-NHMec	6.5	2.9	12.1	240	rcath.S$_{hum}$	d
Boc-F-L-R-NHMec	6.5	6.1	7.3	836	cath.S$_{bov}$	b
	5.5	7.9	1.5	5267	cath.L$_{rat}$	b
	6.0	46.2	107.0	432	cath.B$_{rat}$	b
Boc-F-F-R-NHMec	6.5	8.6	37.5	229	cath.S$_{bov}$	b
	6.5	3.3	48.0	69	rcath.S$_{hum}$	d
	5.5	7.5	1.4	5357	cath.L$_{rat}$	b
	6.0	22.7	114.0	199	cath.B$_{rat}$	b

[a] Standard abbreviations are used for common blocking groups and for the one-letter code for amino acid residues. NHMec, 7-(4-Methyl)coumarylamide; cath., cathepsin; rcath, recombinant cathepsin.
[b] D. Brömme, A. Steinert, S. Friebe, S. Fittkau, B. Wiederanders, and H. Kirschke, *Biochem. J.* **264**, 475 (1989).
[c] X.-Q. Xin, B. Gunesekera, and R. W. Mason, *Arch. Biochem. Biophys.* **299**, 334 (1992).
[d] D. Brömme, P. R. Bonneau, P. Lachance, B. Wiederanders, H. Kirschke, C. Peters, D. Y. Thomas, A. C. Storer, and T. Vernet, *J. Biol. Chem.* **268**, 4832 (1993).
[e] R. W. Mason, G. D. J. Green, and A. J. Barrett, *Biochem. J.* **226**, 233 (1985).

only specific substrates for lysosomal cysteine proteinases. Tests of a series of synthetic substrates revealed that all compounds were more or less sensitive to cathepsins S, L, and B (see Table I) and no peptide derivative has been described so far that is specific for cathepsin S at acidic pH values. However, changing the assay conditions to pH 7.5[8] and

[8] H. Kirschke, B. Wiederanders, D. Brömme, and A. Rinne, *Biochem. J.* **264,** 467 (1989).

using the most susceptible substrates,[9] such as Bz-Phe-Val-Arg-NHMec[10] or Z-Val-Val-Arg-NHMec,[10] allows the specific determination of cathepsin S. During a preincubation period up to 1 hr at pH 7.5 the activities of cathepsins L, B, and H are completely abolished, whereas cathepsin S retains 60–70% of its activity.[8] We have found that Z-Val-Val-Arg-NHMec is the best substrate to determine cathepsin S at pH 7.5 in the presence of cathepsins L, B, and H: cathepsin L exhibits a higher k_{cat}/K_m value with this substrate compared to cathepsin B,[9] no doubt, but on the other hand is more sensitive to inactivation at pH 7.5 than cathepsin B; cathepsin H hydrolyzes Z-Val-Val-Arg-NHMec fourfold less rapidly than Bz-Phe-Val-Arg-NHMec.[11]

Assay with Z-Val-Val-Arg-NHMec

Principle. The substrate Z-Val-Val-Arg-NHMec is hydrolyzed to liberate 7-amino-4-methylcoumarin, which is quantified continuously by its intense fluorescence, or in stopped assays after the termination of the enzymatic reaction by monochloroacetate (or simply by dilution with water—at least 1 : 15).[3,12]

Reagents

Buffer/activator: 0.1 *M* potassium phosphate buffer and 5 m*M* disodium EDTA, pH 7.5. The buffer is made 5 m*M* in dithiothreitol on the day it is to be used.

Substrate stock solution: A 10 m*M* solution of Z-Val-Val-Arg-NHMec in dimethyl sulfoxide, stored at −20°. When required, it is diluted to the working strength of 12.5 μ*M* with water (i.e., 10 μl to 8 ml).

Stopping reagent: 100 m*M* sodium monochloroacetate, 30 m*M* sodium acetate, 70 m*M* acetic acid, pH 4.3.

Diluent: Triton X-100 (0.01%) in 0.1 *M* potassium phosphate buffer containing 1 m*M* EDTA, pH 7.5.

Standard: 7-amino-4-methylcoumarin 10 m*M* in dimethyl sulfoxide is stored at −20°. Every 4 weeks a 10 μ*M* dilution in water is prepared and from this a 0.1 μ*M* solution in a 1 : 4 mixture of assay buffer and stopping reagent is prepared on the day it is to be used. The solutions are kept in brown bottles.

[9] D. Brömme, A. Steinert, S. Friebe, S. Fittkau, B. Wiederanders, and H. Kirschke, *Biochem. J.* **264,** 475 (1989).

[10] Available from Bachem Feinchemikalien AG, CH-4416 Bubendorf, Switzerland.

[11] H. Kirschke, unpublished results (1993).

[12] H. C. Blair, S. L. Teitelbaum, L. E. Grosso, D. L. Lacey, H.-L. Tan, D. W. McCourt, and J. J. Jeffrey, *Biochem. J.* **290,** 873 (1993).

Procedure. Assays are commonly done in 5-ml glass or polystyrene tubes. The enzyme sample is diluted to 100 μl with diluent, and 200 μl of buffer/activator is added. Preincubation for 60 min at 40° is necessary to destroy the activities of cathepsins L, B, and H. Thereafter, 200 μl of substrate solution is mixed in and after exactly 10 min, 2 ml of stopping reagent is added. The cathepsin S is used at a concentration of about 0.2 nM in the reaction mixture (500 μl).

The fluorescence of the free aminomethylcoumarin is determined by excitation at 360 nm and emission at 460 nm. If the instrument is zeroed against a 1 : 4 mixture of assay buffer and stopping reagent and set to read 1000 arbitrary fluorescence units with the 0.1 μM standard, an experimental reading (minus reaction blank) of 1000 fluorescence units corresponds to 25 μU of activity in the tube, for a 10-min incubation.

The above procedure is employed for the assay of cathepsin S in the presence of cathepsins L, B, and H. If the latter enzymes are absent, an assay buffer of pH 6.5 and a preincubation time of 5 min can be used with advantage. Control experiments confirming the activity of any cysteine proteinase being measured may be needed for both assay procedures, at pH 7.5 and pH 6.5.

Assay with Azocasein and Urea

The principle of the assay is the same as that published for the assay of cathepsin L.[3]

Here we describe a micromodification of the azocasein assay.

Reagents

Buffer/activator: 0.1 M acetate buffer containing 5 mM EDTA and 1 μg pepstatin/ml, pH 5.5. The buffer is made 5 mM in dithiothreitol on the day it is to be used.

Azocasein/urea solution: A solution of 1% (w/v) azocasein and 6 M urea is prepared in 0.1 M acetate buffer, pH 5.5.

Trichloroacetic acid: 10% (w/v) aqueous solution.

Procedure. A 50-μl enzyme sample in a 1.5-ml Eppendorf tube is mixed with 50 μl of buffer/activator and preincubated 5 min at 40°. A 100-μl volume of the azocasein/urea solution is added and after 15 min incubation at 40°, 200 μl of trichloroacetic acid is mixed in to stop the reaction. After centrifugation (e.g., 10 min at 10,000 g), the A_{366} of the supernatant is determined.

Cathepsin S is used at a concentration of about 30 nM in the reaction mixture (200 μl). The preparation of azocasein and standardization have been described in detail.[3] Cathepsin S is stable in the presence of 3 M urea,[4,8] like cathepsin L,[13] whereas cathepsin B is inactivated.[13] Changing the assay pH from 5.5 to 7.5 allows discrimination between cathepsin S and cathepsin L, because the latter is inactive at pH 7.5. It is necessary to confirm that the activity is due to a cysteine proteinase, for example, by showing inhibition by E-64 [*trans*-epoxysuccinyl-L-leucylamido(4-guanidino)butane].

Cathepsin S: Purification Procedures

We describe a purification method for cathepsin S from frozen or fresh cattle spleen.[5,8,14] This method is also applicable to the spleen of several other species (e.g., rats and humans). The whole procedure takes place at 4°, unless indicated otherwise.

Extraction, Autolysis, $(NH_4)_2SO_4$ Fractionation

After removing the capsule, about 300 g of spleen tissue is pushed through a sieve and homogenized in 2 volumes of 50 mM acetate buffer, 150 mM NaCl, 1 mM disodium EDTA, pH 4.2, using an Ultraturrax. The homogenate is centrifuged (20 min at 8000 g) and the supernatant is adjusted to pH 4.2, subjected to autolysis for 1 hr at 40°, and then to $(NH_4)_2SO_4$ fractionation. The 20–75% saturation fraction is dialyzed overnight against 20 mM sodium acetate buffer containing 1 mM disodium EDTA, pH 5.5.

CM-Sephadex C-50 Chromatography, Gel Filtration

The dialyzed fraction is centrifuged and then applied to a column (2.6 × 28 cm) of CM-Sephadex C-50 equilibrated in 20 mM sodium acetate buffer containing 1 mM disodium EDTA, pH 5.5. The column is washed at a flow rate of 11 cm/hr with 4 bed volumes of the buffer and then eluted with a linear gradient (0–0.3 M) of NaCl in the buffer (2 × 300 ml). The fractions containing cathepsins S and B elute before cathepsin H (bovine) or after cathepsin H (human and rat); these are pooled, concentrated by ultrafiltration on a YM10 membrane (Amicon, Danvers, MA), and applied to a column (2.5 × 90 cm) of Sephacryl S-200 equilibrated with 0.1 M sodium acetate buffer, containing 1 mM disodium EDTA, 0.5 M NaCl. The column is run at 4 cm/hr. Active fractions in the M_r range 15,000–30,000 are combined.

---

[13] B. Wiederanders, H. Kirschke, and S. Schaper, *Biomed. Biochim. Acta* **45**, 1477 (1986).
[14] H. Kirschke, I. Schmidt, and B. Wiederanders, *Biochem. J.* **240**, 455 (1986).

Chromatofocusing, Phenyl-Sepharose, CM-Sephadex Chromatography

The enzyme pool from the gel filtration step is concentrated and equilibrated with 25 mM Tris–acetate buffer, pH 6.0, and is then applied to a column (0.9 × 28 cm) of chromatofocusing gel PBE 94 or to a Mono P chromatofocusing column of the FPLC (fast protein liquid chromatography) system (Pharmacia, Piscataway, NJ) equilibrated with 25 mM Tris–acetate buffer, pH 8.3. The eluent is Polybuffer 96 (diluted 1 : 12 with water), pH 6.0.

Cathepsin S is eluted at pH 7.1–6.8, whereas the main part of cathepsin B can be obtained by subsequent elution with acetate buffer, pH 4.5. Polybuffer is removed from the pool of cathepsin S by hydrophobic chromatography on phenyl-Sepharose (0.9 × 4 cm), the sample being applied in 25% (w/v) $(NH_4)_2SO_4$ in 20 mM sodium malonate buffer containing 1 mM disodium EDTA, pH 5.0. Cathepsin S is eluted with 10% (v/v) ethylene glycol in the same buffer. For final purification the sample is run on a column (0.9 × 15 cm) of CM-Sephadex C-50 equilibrated with 20 mM sodium malonate buffer containing 1 mM disodium EDTA, pH 5.0. Cathepsin S is eluted at a flow rate of 11 cm/hr at about 0.18 M NaCl during a linear gradient (0–0.3 M NaCl) in the buffer (2 × 25 ml).

If the chromatofocusing is followed by purification on the Mono S column of the FPLC system, it may be observed that cathepsin S is eluted as early in the gradient as 50 mM[5] or 30 mM[15] NaCl instead of 150 mM NaCl.[5] This unusual behavior may be due to ampholines adsorbed to cathepsin S.[5]

The yield is about 3 mg cathepsin S/kg of bovine spleen. The enzyme is electrophoretically pure[14] and contains 30–40% active enzyme molecules as revealed by titration with E-64.[3] By inclusion of an affinity chromatography step in the procedure (e.g., activated thiol-Sepharose 4B[15,16] or activated thiopropyl-Sepharose 6B[17]), the inactive molecules are removed and the resulting cathepsin S preparation is 100% catalytically active.

Storage of Cathepsin S

Cathepsin S can be stored in 0.1 M sodium acetate buffer containing 1 mM disodium EDTA, pH 5.5, at −20°. Under these conditions the

[15] I. Dolenc, A. Ritonja, A. Čolić, M. Podobnik, T. Ogrinc, and V. Turk, *Biol. Chem. Hoppe-Seyler* **373,** 407 (1992).

[16] P. Ločnikar, T. Popović, T. Lah, I. Kregar, J. Babnik, M. Kopitar, and V. Turk, *in* "Proteinases and Their Inhibitors" (V. Turk and L. Vitale, eds.), p. 109. Mladinska Knjiga and Pergamon, Ljubljana and Oxford, 1981.

[17] D. Brömme, P. R. Bonneau, P. Lachance, B. Wiederanders, H. Kirschke, C. Peters, D. Y. Thomas, A. C. Storer, and T. Vernet, *J. Biol. Chem.* **268,** 4832 (1993).

enzyme activity remains stable for several months. Storage of cathepsin S in 0.1 M acetate buffer, pH 5.5, containing 0.5 mM HgCl$_2$ and 1 mM EDTA causes irreversible loss of activity, in contrast to cathepsins B, H, and L, which can successfully be stored as their mercury derivatives.[3]

Properties of the Enzymes

Well-characterized monomeric lysosomal cysteine proteases are the cathepsins B, L, H, and S, having M_r values of less than 30,000, whereas the related cathepsin C (dipeptidyl-peptidase I) forms oligomers of around 200,000.[18]

The enzymes are synthesized as precursor molecules. The primary translation products consist of NH$_2$-terminal signal peptides of 15 (cathepsin S) to 28 (cathepsin C) amino acid residues, which are cleaved off during the transport through the membrane of the endoplasmic reticulum (ER). The signal sequence is followed by a propeptide region of around 100 amino acid residues in cathepsins B, H, L, and S and a much longer propeptide of 201 residues in cathepsin C. Such precursor molecules of the cathepsins do not exhibit any measurable enzymatic activity toward common protease substrates at pH 6 and above, and they are stable at pH 7.[17,19,20] Like other lysosomal proteins, in the ER, they receive carbohydrate moieties directing them to the lysosomes via binding to the mannose 6-phosphate receptor. The proteolytic processing to active, mature enzymes has been described as cathepsin D dependent.[21] However, in vitro, autolytic activation below pH 5 of procathepsins L,[19] B,[20] and S[17] has also been observed.

The amino acid sequences of mature human cathepsins S[22] and L[23] share 57% identity, showing that the two cathepsins are more closely related than human cathepsins S[22] and H[24] (41% identical amino acid residues), rat cathepsins S[25] and C[18] (27%), and human cathepsins S[22] and B[26] (16%).

[18] K. Ishidoh, D. Muno, N. Sato, and E. Kominami, *J. Biol. Chem.* **266**, 16312 (1991).
[19] R. W. Mason, S. Gal, and M. M. Gottesman, *Biochem. J.* **248**, 449 (1987).
[20] A. D. Rowan, P. Mason, L. Mach, and J. S. Mort, *J. Biol. Chem.* **267**, 15993 (1992).
[21] Y. Nishimura, T. Kawarata, K. Furuno, and K. Kato, *Arch. Biochem. Biophys.* **271**, 400 (1989).
[22] B. Wiederanders, D. Brömme, H. Kirschke, K. v. Figura, B. Schmidt, and C. Peters, *J. Biol. Chem.* **267**, 13708 (1992).
[23] L. J. Joseph, L. C. Chang, D. Stamenkovich, and V. P. Sukhatme, *J. Clin. Invest.* **81**, 1621 (1988).
[24] R. Fuchs, W. Machleidt, and H.-G. Gassen, *Biol. Chem. Hoppe-Seyler* **369**, 469 (1988).
[25] S. Petanceska and L. Devi, *J. Biol. Chem.* **267**, 26038 (1992).
[26] A. Ritonja, T. Popović, V. Turk, K. Widenmann, and W. Machleidt, *FEBS Lett.* **181**, 169 (1985).

Karrer *et al.*[27] have proposed two distinct gene subfamilies within the family of cysteine protease genes based on a 20-residue-long sequence motif located in the propeptide region. Some plant, parasite, and animal cysteine proteases share this "E-X_3_R-X_3_F/W-X_2_N-X_3_I/V-X_3_N" motif, but in both cathepsins B and C, the motif is absent from the propeptide region, indicating that they do not belong to the "ERFNIN family".

Cathepsin S

Cathepsin S has been purified from bovine,[5,8,15,16] rabbit,[28] rat,[29] and human[29] tissues, and from *Saccharomyces cerevisiae* expressing the human enzyme.[17] The amino acid sequences have been reported from the bovine,[30,31] human,[22,32] and rat[25] enzyme (Fig. 1). cDNAs have been cloned from bovine,[30] human[22,32] (GenBank/EMBL: M90696), and rat cathepsin S[25] (GenBank/EMBL: L03201).

Mature active cathepsin S is a single-chain polypeptide comprising 217 amino acid residues. The NH_2-terminal prepro extension is 114 and 112 amino acids long in human[22,32] and rat[25] enzymes, respectively. The only potential glycosylation sites of cathepsin S are located at position N^{-11}-T^{-9} in the human enzyme and at position N^{-13}-S^{-11} in the rat enzyme. Thus, they are located in the propeptide region, which indicates that maturation of procathepsin S occurs after delivery to the final compartment.

Normally, *in vivo* maturation of procathepsin S leads to an enzyme with proline at position 2 from the NH_2 terminus, as is frequently found in lysosomal proteins. Expression of human preprocathepsin S cDNA in transfected BHK cells resulted in active enzyme with a M_r of 24,000 on SDS–PAGE, which is identical to that of purified cathepsin S,[22] whereas in COS cells, the recombinant enzyme was processed to M_r 28,000 active enzyme form.[32] Expressing human procathepsin S in *Saccharomyces cerevisiae*, Brömme *et al.*[17] described autoactivation at pH 4.5 forming active cathepsin S with six-amino-acid N-terminal extension. From these results

[27] K. M. Karrer, S. L. Pfeiffer, and M. E. DiTomas, *Proc. Natl. Acad. Sci. U.S.A.* **90**, 3063 (1993).

[28] R. A. Maciewicz and D. J. Etherington, *Biochem. J.* **256**, 433 (1988).

[29] H. Kirschke, unpublished results (1993).

[30] B. Wiederanders, D. Brömme, H. Kirschke, N. Kalkkinen, A. Rinne, T. Paquette, and P. Toothman, *FEBS Lett.* **286**, 189 (1991).

[31] A. Ritonja, A. Čolić, I. Dolenc, T. Ogrinc, M. Podobnik, and V. Turk, *FEBS Lett.* **283**, 329 (1991).

[32] G.-P. Shi, J. S. Munger, J. P. Meara, D. H. Rich, and H. A. Chapman, *J. Biol. Chem.* **267**, 7258 (1992).

```
                    ↘
rat MAVLGAPGVL CDNGHTAER    PTLDHHWDL WKKTRMRRNT DQNEEDVRRL IWEKNLKFIM
hum MKRLVCVLLV CSSAVAQLHK    DPTLDHHWHL WKKTYGKQYK EKNEEAVRRL IWEKNLKFVM
                    ↗

                                                            1
rat LHNLEHSMGM HSYSVGMNHM GDMTPEEVIG YMGSLRIPRP WNRSGTLKSS SNQTLPDSVD
hum LHNLEHSMGM HSYDLGMNHL GDMTSEEVMS LMSSLRVPSQ WQRNITYKSN PNRILPDSVD
bov                                                         LPDSMD


rat WREKGCVTNV KYQGSCGSCW AFSAEGALEG QLKLKTGKLV SLSAQNLVDC STEEKYGNKG
hum WREKGCVTEV KYQGSCGACW AFSAVGALEA QLKLKTGKLV SLSAQNLVDC ST EKYGNKG
bov WREKGCVTEV KYQGACGSCW AFSAVGALEA QVKLKTGKLV SLSAQNLVDC ST AKYGNKG

rat CGGGFMTEAF QYIIDTS ID SEASYPYKAM DEKCLYDPKN RAATCSRYIE LPFGDEEALK
hum CNGGFMTTAF QYIIDNKGID SDASYPYKAM DQKCQYDSKY RAATCSKYTE LPYGREDVLK
bov CNGGFMTEAF QYIIDNNGID SEASYPYKAM DGKCQYDVKN RAATCSRYIE LPFGSEEALK

rat EAVATKGPVS VGIDDASHSF FLYQSGVYDD PSCTENMNHG VLVVGYGTLD GKDYWLVKNS
hum EAVANKGPVS VGVDARHPSF FLYRSGVYYE PSCTQNVNHG VLVVGYGDLN GKEYWLVKNS
bov EAVANKGPVS VGIDASHSSF FLYKTGVYYD PSCTQNVNHG VLVVGYGNLD GKDYWLVKNS

rat WGLHFGDQGY IRMARNNKNH CGIASYCSYP EI
hum WGHNFGEEGY IRMARNKGNH CGIASFPSYP EI
bov WGLHFGDQGY IRMARNSGNH CGIANYPSYP EI
```

FIG. 1. Amino acid sequences of rat, human, and bovine cathepsin S. The ERFNIN sequence is printed in italics, the potential glycosylation sites are underlined, and the bold letters indicate identical residues in all three enzyme species. Arrows show the end of the signal peptide; the numeral 1 is above the NH_2-terminal leucine of the mature enzyme. Rat cathepsin S [S. Petance and L. Devi, *J. Biol. Chem.* **267**, 26043 (1992)]; human cathepsin S [B. Wiederanders, D. Brömme, H. Kirschke, K. von Figura, B. Schmidt, and C. Peters, *J. Biol. Chem.* **267**, 13708 (1992); G.-P. Shi, J. S. Munger, J. P. Meara, D. H. Rich, and H. A. Chapman, *J. Biol. Chem.* **267**, 7258 (1992)]; bovine cathepsin S [A. Ritonja, A. Čolić, I. Dolenc, T. Ogrinc, M. Podobnik, and V. Turk, *FEBS Lett.* **283**, 329 (1991); B. Wiederanders, D. Brömme, H. Kirschke, N. Kalkkinen, A. Rinne, T. Paquette, and P. Toothman, *FEBS Lett.* **286**, 189 (1991)].

it becomes evident that it is not necessary to remove the complete propeptide from the precursor molecule to produce active enzyme.

The enzyme is unevenly distributed between organs. On the protein basis we have detected high cathepsin S concentrations in spleen and lung[8,33]; similar results have been described by Qian et al.[34] on the mRNA basis. Additionally, Petanceska and Devi[25] presented results on the expression at high levels of cathepsin S mRNA in ileum, brain, thyroid, and ovary.

[33] H. Kirschke, D. Brömme, and B. Wiederanders, in "Proteolysis and Protein Turnover" (J. S. Bond and A. J. Barrett, eds.), p. 33. Portland Press, London and Chapel Hill, 1993.
[34] F. Qian, A. S. Bajkowski, D. F. Steiner, S. J. Chan, and A. Frankfater, *Cancer Res.* **49**, 4870 (1989).

TABLE II
SPECIFIC ACTIVITIES OF CATHEPSINS S AND L WITH PROTEIN SUBSTRATES

Substrate	pH	Cathepsin S (unit^a/μmol)	Cathepsin L (unit^a/μmol)	Ref.
Azocasein^b	6.0	382	294	f
	7.5	348	0	f
Azocasein–3 M urea^b	5.5	775	710	f
	7.5	234	0	f
[^{14}C]Hemoglobin^c	5.0	43	44	f
	7.5	42	0	f
[^{14}C]Albumin^c	5.0	34	25	f
	7.5	36	0	f
[^{3}H]Elastin (insoluble)^d	5.5	70	23	g
	7.0	39	0	g
Collagen (insoluoble)^e	3.5	201	569	f

^{a} One unit corresponds to 1 mg of substrate degraded per min.
^{b} Incubation, 15 min at 40°; enzyme concentration, 31 nM
^{c} Incubation, 30 min at 40°; enzyme concentration, 200 nM.
^{d} Incubation, 60 min at 37°; 3 mg elastin/ml; enzyme concentration, 10 nM.
^{e} Incubation, 30 min at 30°; enzyme concentration, 100 nM cathepsin S, 43 nM cathepsin L.
^{f} H. Kirschke, B. Wiederanders, D. Brömme, and A. Rinne, *Biochem. J.* **264**, 467 (1993).
^{g} H. Kirschke, unpublished results (1993).

Cathepsin S is active toward proteins and synthetic peptide substrates over the range of pH 5.0 to 8.0.[5,8,9,17] The stability of active enzyme above pH 7.0[17] is a remarkable property of cathepsin S. The isoelectric points for bovine cathepsin S have been measured in the range 6.3–7.0.[8,15,16]

The use of synthetic peptide substrates and low M_r inhibitors allows the characterization of substrate-binding sites. Cathepsin S does not prefer bulky hydrophobic residues in the S2 and S3 binding subsites as cathepsin L does (Table I). Instead, less bulky hydrophobic residues increase the specificity constant k_{cat}/K_m.

Bulky hydrophobic residues such as phenylalanine in the P2′ position of peptidyl acylhydroxamate-type irreversible inhibitors[35] reveal high k_2/K_i (8400 M^{-1} sec^{-1}) values in comparison to glycine, for example, in this position (780 M^{-1} sec^{-1}),[36] indicating strong hydrophobic interactions at the S2′ subsite of the cathepsin S active center.

Cathepsin S is capable of hydrolyzing protein substrates[8] as fast as cathepsin L (Table II). It shows collagenolytic[8] and elastinolytic[5,33] activi-

[35] D. Brömme and H.-U. Demuth, this volume [48].
[36] D. Brömme and H. Kirschke, *FEBS Lett.* **322**, 211 (1993).

TABLE III
KINETIC CONSTANTS FOR INHIBITION OF CATHEPSINS S AND L BY CYSTATINS

Enzyme	K_i (nM)			Ref.
	Cystatin A	Cystatin B	Cystatin C	
Cahepsin S	0.05	0.07	0.008	a
Cathepsin L	1.30	0.23	<0.005	b

[a] D. Brömme, R. Rinne, and H. Kirschke, *Biomed. Biochim. Acta* **50**, 631 (1991).
[b] A. J. Barrett, E. M. Davies, and A. Grubb, *Biochem. Biophys. Res. Commun.* **120**, 631 (1984).

ties. But cathepsin S has the unique property among the lysosomal cysteine proteinases of degrading soluble proteins as well as insoluble elastin at acid and neutral pH values. The action of cathepsin S on the oxidized B chain of insulin is very similar to that of cathepsin L with one exception: the Tyr^{26}-Thr^{27} bond is resistant to cathepsin S, even after 10 hr incubation.[9]

Cathepsin B

Cathepsin B is one of the well-characterized lysosomal cysteine proteinases and has been the subject of several reviews.[3,37] Here, only recent important results are mentioned, strengthening our knowledge of the biological significance of the enzyme.

Human cathepsin B was the first lysosomal cysteine proteinase whose crystal structure was elucidated at a 21.5-nm level by X-ray crystal structure analysis.[38] The enzyme shows a two-domain structure, the L domain containing residues 13–147 and 251–254 and the R domain containing residues 1–12 and 148–250. This explains the stability of cathepsin B after reductive cleavage of the disulfide bridge, which normally connects the N-terminal light chain [amino acids (AA) 1–47] and the C-terminal heavy chain (AA 50–254) of the enzyme. The V-shaped active site cleft between the two domains is partially occluded by a loop not present in other cysteine proteinases. On the basis of these data, the well-known peptidyl-dipeptidase activity of cathepsin B can be explained.

Cathepsin B can be selectively inhibited by a peptide derived from its own precursor peptide.[39] A synthetic peptide comprising residues -62 to

[37] H. Kirschke and A. J. Barrett, *in* "Lysosomes. Their Role in Protein Breakdown" (H. Glaumann and F. J. Ballard, eds.), p. 193. Academic Press, London 1987.
[38] D. Musil, D. Zucic, D. Turk, R. A. Engh, I. Mayr, R. Huber, T. Popović, V. Turk, T. Towatari, N. Katunuma, and W. Bode, *EMBO J.* **9**, 2321 (1991).
[39] T. Fox, E. de Miguel, J. S. Mort, and A. C. Storer, *Biochemistry* **31**, 12571 (1992).

−7 of procathepsin B showed a K_i value of 0.4 nM at pH 6.0 toward cathepsin B but of 5600 nM toward papain. The very specific interaction of the two molecules is pH dependent, because at pH 4.0, the K_i values are 64 and 2800 nM for cathepsin B and papain, respectively.

The gene of human cathepsin B has been assigned to chromosome 8p22-p23.1.[40] Gong *et al.*[41] reported on splice variants of cathepsin B premRNA leading to enzyme forms lacking the signal peptide and/or parts of the propeptide. *In vitro*, the authors found the splice variants translated at higher rates than normal messages. The variants occurred much more frequently in some human malignant tumors compared to normal tissues. The findings can partially explain cathepsin B forms in tumors showing aberrant localization within the cells or higher stability to mild alkaline treatment.[42]

Inhibitors

Several classes of irreversible inhibitors, such as *N*-peptidyldiazomethanes, chloromethanes, fluoromethanes, and *O*-acylhydroxylamines, inhibit the three lysosomal cysteine proteinases, cathepsins B, L, and S, more or less as expected from the substrate specificity.[35,43]

The cystatins, endogenous cysteine proteinase inhibitors,[44] react especially fast with cathepsin S.[45] The dissociation constant of cystatin A with cathepsin S[45] is one order of magnitude lower than that of cystatin A with cathepsin L.[46] These results imply that extralysosomal spaces have to be protected effectively against the destructive activities of cathepsin S, a lysosomal cysteine proteinase both catalytically active and stable at physiological pH values.

[40] D. Fong, M. M.-Y. Chan, W.-T. Hsieh, J. C. Menninger, and D. C. Ward, *Hum. Genet.* **89,** 10 (1992).

[41] Q. Gong, S. J. Chan, A. S. Bajkowski, D. F. Steiner, A. Frankfater, *DNA Cell Biol.* **12,** 299 (1993).

[42] B. F. Sloane, *Cancer Biol.* **1,** 137 (1990).

[43] E. N. Shaw, this volume [46].

[44] M. Abrahamson, this volume [49].

[45] D. Brömme, R. Rinne, and H. Kirschke, *Biomed. Biochim. Acta* **50,** 631 (1991).

[46] A. J. Barrett, E. M. Davies, and A. Grubb, *Biochem. Biophys. Res. Commun.* **120,** 631 (1984).

[35] Cysteine Endopeptidases of *Entamoeba histolytica*

By HENNING SCHOLZE and EGBERT TANNICH

Introduction

Entamoeba histolytica is a parasitic protozoon residing in the human gut. It is the etiological trigger of amebiasis, which is a widespread disease in tropical countries. Clinical symptoms include amebic dysentery and in some cases lethal organ abscesses, in particular in liver and spleen (for review, see Ref. 1). Penetration of intestinal epithelial tissue involves at least three steps: (1) recognition and adhesion to host cells and tissue, (2) destruction of cell membranes, and (3) digestion of the extracellular matrix. As to the latter process a series of proteases have been described to be responsible for the digestion of proteins of the extracellular matrix, some of which have been identified as cysteine proteinases (Table I): a 26- to 29-kDa cysteine proteinase named histolysain [EC 3.4.22.35],[2] a 22- to 27-kDa cysteine proteinase (amoebapain),[3] a cathepsin B-like 16 kDa protease (cytotoxic protease),[4] a 56 kDa cysteine proteinase,[5] and a collagenase.[6] Although these proteins differ somewhat in their molecular and enzymological properties (in particular, the 56-kDa protease is supposed to be a precursor form of a smaller intracellular enzyme), it seems probable that all of them are related and that there is a direct relationship between the *in vivo* activity of the cysteine proteases and the pathogenicity of the ameba isolate investigated.[7]

Cultivation of Pathogenic Trophozoites of *Entamoeba histolytica*

During its life cycle, *E. histolytica* is orally ingested in its quadrinucleated cyst form. In the large intestine each cyst produces eight mononucleate trophozoites by a series of nuclear divisions. So far, only the

---

[1] J. A. Walsh, *Rev. Infect. Dis.* **8,** 228 (1986).
[2] A. L. Luaces and A. J. Barrett, *Biochem. J.* **250,** 903 (1988).
[3] H. Scholze, in "Biochemical Protozoology" (G. Coombs and M. North, eds.), p. 251. Taylor & Francis, London and Washington, D.C., 1991.
[4] W. B. Lushbaugh, A. F. Hofbauer, and F. E. Pittman, *Exp. Parasitol.* **59,** 328 (1985).
[5] W. E. Keene, M. G. Petitt, S. Allen, and J. H. McKerrow, *J. Exp. Med.* **163,** 536 (1986).
[6] Ma. de L. Muñoz, J. Calderon, and M. Rojkind, *J. Exp. Med.* **155,** 42 (1982).
[7] H. Gadasi and D. Kobiler, *Exp. Parasitol.* **55,** 105 (1983).

TABLE I
MOLECULAR AND ENZYMOLOGICAL PROPERTIES OF PURIFIED CYSTEINE PROTEINASES OF
Entamoeba histolytica

Proteinase	Molecular weight ($\times 10^{-3}$)	Isolectric point	pH optimum	Favorite cleavage site	Ref.
Amoebapain	22–27	4.9	5–8.5	Arg-X-Y	a
Histolysain	26–29	n.d.[f]	5.5–9.5	Arg-X-Y	b
Cathepsin B	16	4–5	5–6	n.d.[f]	c
Neutral Cys-Proteinase	56	6.0	4.5–9.5	Arg-X-Y	d, e

[a] H. Scholze, in "Biochemical Protozoology" (G. Coombs and M. North, eds.), p. 251. Taylor & Francis, London and Washington, D.C., 1991.
[b] A. L. Luaces and A. J. Barrett, *Biochem. J.* **250**, 903 (1988).
[c] W. B. Lushbaugh, A. F. Hofbauer, and F. E. Pittman, *Exp. Parasitol.* **59**, 328 (1985).
[d] W. E. Keene, M. G. Petitt, S. Allen, and J. H. McKerrow, *J. Exp. Med.* **163**, 536 (1986).
[e] S. L. Reed, W. E. Keene, J. H. McKerrow, and I. Gigli, *J. Immunol.* **143**, 189 (1989).
[f] n.d., Not determined.

trophozoite form has been successfully cultured *in vitro* (Fig. 1A). About 25 years ago, Diamond developed a medium for axenic cultivation of *E. histolytica*.[8] This medium, called TYI-S-33, is rather complex, containing trypticase, yeast extract, a number of vitamins, and 10–15% (v/v) inactivated bovine serum [for detailed composition, see medium #1141 of ATCC catalog of protists (Rockville, MD), 17th edition, 1991]. Note that the brew TYI-S-33 (without serum and antibiotics) should be prepared and sterilized by filtration separately and stored at $-20°$. The other contents should be added under a lamina flow hood directly before inoculation of cells. The amebas can be cultured in plastic tissue culture flasks (50- or 250-ml content, several suppliers) or in 50-ml borosilicate glass flasks with screw caps. A prerequisite for proper growth is that the cells adhere to the vessel wall. Under these conditions the pathogenic form HM1 : IMSS divides every 6–9 hr, and an inoculate of 10^5 cells in 1000 ml of medium will have reached the end of the logarithmic growth phase after 3–4 days at 36.5°. The flasks are chilled on ice to release the cells from the walls, then the suspension is centrifuged at 700 g for 3 min at 4°. The cell pellet is washed 3 times with cold phosphate-buffered saline (PBS #8, 360 mOsm per kg[9]). The final yield is about 10^8 cells, corresponding to 2 ml packed cells or 200 mg total protein. For the characterization of intracellular protease activity, cells are disrupted by sonication on ice. For the isolation

[8] L. S. Diamond, *J. Parasitol.* **54**, 1047 (1968).
[9] L. S. Diamond, in "Amebiasis" (J. I. Ravdin, ed.), p. 38. Wiley, New York, 1988.

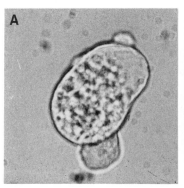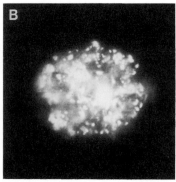

FIG. 1. Trophozoites of *Entamoeba histolytica*. (A) Light microscopy of strain HM1 : IMSS. (B) Fluorescence microscopy of the same isolate in the presence of protease substrate Arg-Arg-4-methoxy-2-naphthylamide.

of soluble proteases the homogenate is centrifuged at 50,000 g for 30 min at 4°.

Proteinase Assays

Azocasein Assay

For the routine determination of cysteine proteinase activity, azocasein is used as a substrate.[10,11] In a typical assay the following components are pipetted together: appropriate amounts of sample (x μl), 200 − x μl H$_2$O, and 100 μl 1 M phosphate buffer, pH 5.0, containing 2 mM dithiothreitol. To activate cysteine protease activity, the mixture is preincubated for 10 min at 37° in a water bath. Subsequently, 100 μl of a 2% (w/v) solution of azocasein is added and the mixture incubated at 37° for 30–60 min. The reaction is stopped by the addition of 1 ml 6% (w/v) trichloroacetic acid and undigested azocasein is removed by centrifugation at 5000 g. The absorption of the colored azopeptides in the supernatant fraction is measured photometrically at 366 nm. One enzyme unit is defined as the amount of proteinase releasing 1 μg soluble azopeptides per minute. The specific absorption, $A_{366}^{1\%}$, of the azocasein solution used may be calculated by measuring its absorption after total digestion and addition of the same amount 6% TCA as under test conditions.

[10] A. J. Barrett and H. Kirschke, this series Vol. 80, p. 535.
[11] P. M. Starkey, *in* "Proteinases in Mammalian Cells and Tissues" (A. J. Barrett, ed.), p. 57. North-Holland Publ., Amsterdam, 1977.

Assay with Small Chromogenic or Fluorogenic Substrates

For determination of kinetic parameters of the protease a well-defined substrate has to be employed. For this purpose small synthetic peptide analogs containing chromophores or fluorophores are available. These include blocked and unblocked synthetic peptide analogs such as Arg-Arg-4-methoxy-2-naphthylamide and Z-Arg-Arg-4-methoxy-2-naphthylamide. The release of naphthylamine can be measured photometrically after coupling with freshly diazotized *o*-aminoazotoluene (Fast Garnet GBC base) (Sigma).[10] To obtain a calibration curve, increasing amounts (10–50 nmol) of 4-methoxy-2-naphthylamine are coupled with Fast Garnet. The enzymatic reaction is allowed to proceed in a phosphate buffer at pH 5.0 and is stopped by the addition of mersalylic acid (final concentration, 5 mM) containing 2% (w/v) Brij 35 (final concentration). The presence of Brij 35 increases the solubility of the azo dye after coupling. For continuous measurements, the direct fluorometric assay with the commercially available, fluorogenic peptide analog Z-Arg-Arg-4-trifluoromethylcoumarinyl-7-amide is recommended. This substrate is used in a concentration of 5 nmol per ml and its hydrolysis followed in a spectrofluorometer set for excitation at 365 nm and emission at 495 nm. The buffer is the same as above without Brij 35. Substrate turnover is quantitated by measuring the fluorescence intensity of 0.5 nmol 4-trifluoromethyl-7-aminocoumarin per ml.

Gelatinase Assay in SDS Gel

The presence of cysteine proteinase activity can also be assessed by gelatin substrate sodium dodecyl sulfate (SDS) gel electrophoresis.[12] Samples of homogenates from different clinical amebic isolates or cell fractions are electrophoresed in 10–15% (w/v) SDS–polyacrylamide gels copolymerized with 0.1% (w/v) gelatin. After removal of SDS by shaking the gel in 0.1% (w/v) Triton X-100 and subsequent incubation of the gel in 0.1 M phosphate buffer, pH 6.0, containing 2 mM dithiothreitol and 5 mM EDTA overnight at 37°, cysteine proteinase activity is detected as a clear band after staining the gels with Coomassie blue.

Isolation of Amoebapain from Pathogenic *Entamoeba histolytica*

In general, the purification procedure for cysteine proteinases of *E. histolytica* follows the method of Lushbaugh *et al.*[4] For the example of

[12] S. L. Reed, W. E. Keene, and J. H. McKerrow, *J. Clin. Microbiol.* **27**, 2772 (1989).

TABLE II

PURIFICATION OF AMOEBAPAIN FROM PATHOGENIC *Entamoeba histolytica*[a]

Procedure	Total protein (mg)	Activity (U)	Specific activity (U/mg)	Purification (-fold)	Recovery (%)
1. Homogenate	130.0	8424	64.8	1.0	100
2. 60,000 g supernatant	62.4	6402	102.6	1.6	76.0
3. Sephadex G-100	12.3	4618	375.4	5.8	54.8
4. Organomercury–Sepharose 6B	0.3	2229	7430.0	114.7	26.5
5. DEAE-cellulose	0.09	841	9344.4	144.2	10.0
6. Sephadex G-75	0.03	457	13,848.5	213.7	5.4

[a] H. Egbringhoff, Diploma Thesis, University of Osnabrueck, Osnabrueck, Germany (1989).

amoebapain the purification procedure is summarized in Table II; protease activity is monitored with azocasein as substrate (see above) throughout the isolation procedure. The cell homogenate supernatant is first chromatographed on Sephadex G-100 (3 × 120 cm), equilibrated with 0.1 M phosphate buffer, pH 6.2, containing 0.1 M KCl and 1 mM mercury(II) chloride to prevent autoproteolytic digestion. Fractions with proteolytic activity (inhibition is reversed in the presence of 5 mM dithiothreitol during the test), which are eluted at about half the column volume, are pooled, concentrated by ultrafiltration in an Amicon (Danvers, MA) cell with a YM10 membrane, and further purified by covalent chromatography on an organomercury column after reduction of the active thiol group with dithiothreitol. Protease activity is eluted as a sharp peak with 50 mM sodium acetate, pH 5.5, containing 0.3 M KCl and 5 mM cysteine. The active fractions are equilibrated in 20 mM Tris-HCl, pH 7.8, by ultrafiltration and subjected to ion-exchange chromatography on DEAE-cellulose equilibrated with the same buffer. The enzyme is desorbed with a linear gradient of 20–200 mM sodium chloride in the same buffer. After concentration of active fractions and a final gel filtration on Sephadex G-75 in 0.1 M phosphate buffer, pH 6.2, containing 0.1 M KCl and 0.5 mM HgCl$_2$, the cysteine proteinase is eluted as a single protein peak. The protease solution is concentrated to 0.2 ml in an Amicon cell and stored in the presence of 0.5 mM HgCl$_2$. One liter of cell suspension containing about 10^8 cells yields about 20 μg of active amoebapain according to thiol titration with the specific inhibitor and sulfhydryl reagent E-64 [*trans*-epoxysuccinyl-L-leucylamido (4-guanidino)butane].[13]

[13] H. Scholze and W. Schulte, *Biomed. Biochim. Acta* **47**, 115 (1988).

Effectors of Amebic Cysteine Proteinase Activity

The effects of SH reagents on amebic protease activities are consistent with the properties of typical papainlike cysteine proteinases. Reducing agents with free thiol groups, such as cysteine, 2-mercaptoethanol, or dithiothreitol, and the chelator of divalent cations, EDTA, exhibit a stimulating effect on protease activity, whereas heavy metals with high affinity to sulfide ions, such as mercury or cadmium in millimolar concentrations, abolish activity completely. This inhibitory effect is reversed by excessive thiol groups. Also, active site inhibitors of cysteine proteinases, such as iodoacetamide, N-ethylmaleimide, and E-64, inhibit amoebapain activity, but because they bind covalently, their effect is only partially reversible by thiol-containing agents. E-64 and iodoacetic acid inactivate histolysain with second-order rate constants of 1100 and 180 M^{-1} sec^{-1}, respectively.[2] Cystatin C from chicken egg white, a specific protein inhibitor of cathepsins B and L, shows strong inhibitory effects on amoebapain; a concentration of 10 μM is sufficient to inhibit its activity completely.[13] The K_i value for the histolysain–cystatin interaction is approximately 10^{-11} M. The small natural microbial inhibitors antipain and leupeptin also inhibit histolysain activity, but are 50- to 100-fold less effective.[2] Specific reagents such as peptidyl diazomethanes and chloromethanes have been designed to identify and localize active forms of cysteine proteinases in cells and tissues.[14] Z-Leu-Met-CHN$_2$ is the most effective irreversible inhibitor for histolysain, followed by Z-Leu-Lys-CHN$_2$. The chloromethanes Pro-Phe-Arg-CH$_2$Cl and Z-Leu-Leu-Phe-CH$_2$Cl cause reversible inhibition of histolysain with K_i values of about 1.5 μM. Growth of *E. histolytica* trophozoites is inhibited by 50% at micromolar concentrations of benzyloxycarbonylleucyltyrosyldiazomethane (Z-Leu-Tyr-CHN$_2$).[15] Soluble fractions of trophozoites grown in the presence of sublethal concentrations of radiolabeled Z-Leu-[^{125}I]Tyr-CHN$_2$ exhibit two radioactive bands at 29 and 27 kDa after sodium dodecyl sulfate–polyacrylamide gel electrophoresis. In membrane fractions, additional bands are observed, among others a 56-kDa band (see Keene *et al.*[5]) and bands of 35 and 39 kDa, which may correspond to proforms of the mature proteases.

---

[14] R. W. Mason, D. Wilcox, P. Wikstrom, and E. N. Shaw, *Biochem. J.* **257**, 125 (1989).

[15] F. De Meester, E. Shaw, H. Scholze, T. Stolarsky, and D. Mirelman, *Infect. Immun.* **58**, 1396 (1990).

Substrate Specificity of Amebic Cysteine Endopeptidases

For *Entamoeba*, a correlation between virulence on the one hand and proteolytic and collagenolytic activities on the other hand was observed in the early 1980s.[7,16] These findings led to the hypothesis that cysteine proteinases contribute to pathogenicity by digesting proteins of the extracellular matrix during tissue penetration of the host. Within this framework, the activity of amebic lysates and purified proteases toward a series of proteins of the extracellular matrix has been investigated in several laboratories. Purified histolysain degrades basement membrane collagen and releases proteoglycan from bovine nasal cartilage.[2] The purified 56-kDa cysteine proteinase degrades purified laminin, fibronectin, and type I collagen, releasing a multiple pattern of protein fragments.[5] Amoebapain degrades collagen types I, IV, and V as well as fibronectin and laminin under *in vitro* conditions.[13,17] In all cases digestion does not proceed to completion and no discrete cleavage products are obtained. Amoebapain degrades human laminin and fibronectin much more rapidly than the collagens. Whereas the breakdown of laminin does not generate distinct fragments, the digestion of fibronectin yields a defined degradation pattern right from the beginning.[17] Another important question has been whether the parasite may possibly evade the immune system of the host by proteolytically interfering with the complement system. In support of this notion, the 56-kDa neutral cysteine proteinase has been shown to cleave the α chain of the native complement component C3 very specifically at a defined cleavage site one residue distal to the natural site for the C3 convertase.[18] Similar results have been obtained with amoebapain.[19] In both cases the β chain remains intact.

The substrate specificity of amoebapain against small synthetic peptides and peptide analogs has been studied in great detail. The kinetics of cleavage has been recorded by reversed-phase high-performance liquid chromatography (HPLC) employing C_{18} columns.[20–22] The protease does not cleave unblocked di- and tripeptides. Because the smallest cleavable peptide consists of four amino acid residues, the enzyme exhibits both dipeptidyl-peptidase and peptidyl-dipeptidase activity. In any case, an arginine in the P2 position is a prerequisite for proper turnover rates of

[16] H. Gadasi and E. Kessler, *Infect. Immun.* **39,** 528 (1983).
[17] W. Schulte and H. Scholze, *J. Protozool.* **36,** 538 (1989).
[18] S. L. Reed, W. E. Keene, J. H. McKerrow, and I. Gigli, *J. Immunol.* **143,** 189 (1989).
[19] H. Scholze, unpublished data (1991).
[20] H. Scholze, J. Otte, and E. Werries, *Mol. Biochem. Parasitol.* **18,** 113 (1986).
[21] W. Schulte, H. Scholze, and E. Werries, *Mol. Biochem. Parasitol.* **25,** 39 (1987).
[22] J. Otte and E. Werries, *Mol. Biochem. Parasitol.* **33,** 257 (1989).

a small synthetic peptide substrate, whereas the identity of the P1 position is rather marginal. Replacing arginine by lysine in the otherwise identical peptide sequence decreases its turnover rate to less than 20%.[22] Similar results have been obtained with fluorogenic substrates, where the blocking of the N-terminal amino group of arginine increases the susceptibility of the peptide about threefold, an additional arginine in P1 nearly twofold. The same specificity is exhibited toward natural peptides. Thus, amoebapain splits the α1-CB2 peptide of collagen type I exclusively at a unique Gly-Leu bond and the oxidized insulin B chain predominantly at a Gly-Phe bond, both of these bonds preceded by an arginine residue.[13] The same applies to the action of the 56-kDa protease on the complement component C3. Here, cleavage takes place at a Ser-Asn bond, one position behind the arginyl residue, which is the natural P1 site for the C3-convertase.[18] The preference for arginine in the P2 position seems to be a general property of the amebic cysteine proteinases, and the exact positioning of the guanidino group in the binding pocket may be of more importance than the basic character of this residue.

Cellular Localization of Cysteine Proteinase Activity

About 40% of the cell volume of *E. histolytica* is occupied by lysosome-like pinocytic vesicles, the contents of which are in continuous exchange with the extracellular space. In order to determine whether the cysteine proteinase activity is localized in these vesicles in analogy to those of higher eukaryotes, different histological methods can be employed. For immunohistochemistry, amebas are fixed for 20 min at 4° in PBS containing 1.5% (v/v) glutaraldehyde and embedded in Epon; 1-μm sections are prepared and etched for 30 min at 60° with sodium methoxylate (saturated solution in methanol). To block nonspecific antibody-binding sites, the sections are preincubated for 1 hr at room temperature with 1% (w/v) bovine serum albumin (BSA) in NET buffer [150 mM NaCl, 50 mM Tris, 5 mM EDTA, 0.05% (w/v) Nonidet P-40 (NP-40), 0.25% (w/v) gelatin, brought to pH 7.5 with HCl]. They are then incubated overnight in NET buffer containing antiserum against the cysteine proteinase in a final dilution of 1 : 100, washed, and incubated for 2 hr with fluorescein isothiocyanate (FITC)-conjugated goat antirabbit antiserum in a dilution of 1 : 2000. The sections are viewed under a fluorescence microscope at λ_{ex} 450–490 nm and $\lambda_{em} \geq$ 520 nm.

For activity staining of cysteine proteinases in living amebas, cells are incubated at room temperature in phosphate-buffered saline (PBS #8), pH 7.0, containing 5 mM of the substrate Arg-Arg-4-methoxy-2-naphthylamide and 2.5 mM 5-nitro-2-salicylaldehyde. The latter compound forms an insoluble fluorescent adduct with the proteolytically released naphthylamine derivative. After different intervals, the cells are spun down and washed four times in PBS #8 and viewed under a fluorescent microscope (λ_{ex} 360–430 nm, λ_{em} 550–600 nm). Both experiments give rise to similar results (Fig. 1B): fluorescence is clearly localized within subcellular vesicles, and, to a small extent, at the outer surface of the amebas.[23] These findings can be confirmed by electron microscopy of ultrathin sections after exposure to antiproteinase antiserum and subsequent immunogold staining.

Analyses of Cysteine Proteinase Genes of Entamoeba histolytica

Two different laboratories have reported the molecular cloning of cDNA or genomic sequences encoding *E. histolytica* cysteine proteinases. So far, three different cysteine proteinase genes have been identified in pathogenic *E. histolytica* (Eh-CPp1, Eh-CPp2, and Eh-CPp3, respectively) and one in nonpathogenic amebas (Eh-CPnp).[24–26] A cDNA clone coding for a cysteine proteinase (Eh-CPp1) has been identified by screening a λgt11 cDNA library derived from the pathogenic *E. histolytica* isolate HM1 : IMSS using an antiserum raised against amoebapain from the same isolate.[25] Using this clone an additional gene coding for a cysteine proteinase (Eh-CPp2) has been identified by screening a genomic library of pathogenic *E. histolytica* under low-stringency conditions. For Eh-CPp1 and Eh-CPp2 the complete coding regions have been analyzed.[26] They differ in 15% of the nucleotide sequences. As the two genes are derived from a cloned culture of the pathogenic isolate HM1 : IMSS, they must occur together within a single organism. The primary structure as deduced from the DNA sequences reveals that both enzymes are synthesized as prepro forms of 34.5 and 34.1 kDa, respectively; the mature proteins of 24.0 and 23.6 kDa differ by 13.5% in amino acid sequence. Alignments of the N-terminal amino acid residues determined by protein sequencing of purified *E. histolytica* cysteine proteinases suggest that Eh-CPp1 codes for amoebapain, whereas Eh-CPp2 encodes histolysain (Table III). A third cysteine proteinase gene identified in pathogenic *E. histolytica* (Eh-CPp3, ACPI according to Reed *et al.*[24]) has been obtained by gene amplification using

[23] H. Scholze, C. L. Baigent, G. Müller, and T. Bakker-Grunwald, *Arch. Med. Res.* **23**, 105 (1992).
[24] S. L. Reed, J. Bouvier, A. S. Pollack, J. C. Engel, M. Brown, K. Hirata, X. Que, A. E. Eakin, P. Hagblom, F. Gillin, and J. H. McKerrow, *J. Clin. Invest.* **91**, 1532 (1993).
[25] E. Tannich, H. Scholze, R. Nickel, and R. D. Horstmann, *J. Biol. Chem.* **266**, 4798 (1991).
[26] E. Tannich, R. Nickel, H. Buss, and R. D. Horstmann, *Mol. Biochem. Parasitol.* **54**, 109 (1992).

TABLE III

SEQUENCES OF HOMOLOGOUS CYSTEINE PROTEINASES FROM *Entamoeba histolytica*[a,b]

Sequence

Proteinase					*				40
		10		20		30			
Eh-CPp1	APKAV	DWRKK	GKVTP	IRDQG	NCGSC	YTFGS	IAALE	GRLLI	EKGGD
Eh-CPp2	APESV	DWRKE	GKVTP	IRDQA	QCGSC	YTFGS	LAALE	GRLLI	EKGGD
Eh-CPnp	APESV	DWRAQ	GKVTP	IRDQA	QCGSC	YTFGS	LAALE	GRLLI	EKGGN
Eh-CPp3[c]	APESV	DWRSI	MN--P	AKDQG	QCGSC	WTFCT	TAVLE	GRVNK	DLGKL

Proteinase	50				70		80		90
		60							
Eh-CPp1	SETLD	LSEEH	MVQCT	REDGN	NGCNG	GLGSN	VYNYI	MENGI	AKESD
Eh-CPp2	ANTLD	LSEEH	MVQCT	RDNGN	NGCNG	GLGSN	VYDYI	IEHGV	AKESD
Eh-CPnp	ANTLD	LSEEH	MVQCT	RDNGN	NGCNG	GLGSN	VYDYI	IQNGV	AKESD
Eh-CPp3	YS---	FSEQQ	LVDCD	ASD--	NGCER	GP-SN	SLKFI	QENNGL	GLESD

$S_1$140

Proteinase		100		110		120			130
Eh-CPp1	YPYTG	SDSTC	RSDVK	AFAKI	KSYNR	VARNN	EVELK	AAISQ	GLVDV
Eh-CPp2	YPYTG	SDSTC	KTNVK	SFAKI	TGYTK	VPRNN	EAELK	AALSQ	GLVDV
Eh-CPnp	TPYTG	TDSTC	KTNVK	AFAKI	TGYNK	VPRNN	EAELK	AALSQ	GLVDV
Eh-CPp3	YPYKA	VAGTC	KK-VK	NVATV	TGSRR	VTDGS	ETGLQ	TIIAENGPVAV	

Proteinase	150				160	*		170	180
Eh-CPp1	SIDAS	SVQFQ	LYKSG	AYTDT	QCKNN	YFALN	HEVCA	VGYGV	ADGKE
Eh-CPp2	SIDAS	SAKFQ	LYKSG	AYTDT	KCKNN	YFALN	HEVCA	VGYGV	ADGKE
Eh-CPnp	SIDAS	SAKFQ	LYKSG	AYSDT	KCKNN	FFALN	HEVCA	VGYGV	VDGKE
Eh-CPp3	GMDAS	RPSFQ	LYKKGTIYSDT		KCRSR	--MMN	HCVTA	VGYGSNSNGK-	

Proteinase	*		200			210		S_2	220
		190							
Eh-CPp1	CWIVR	NSWGT	GWGEK	GYINM	VIEGN	T---C	GVATD	PLYPT	GVEYL
Eh-CPp2	CWIVR	NSWGT	GWGDK	GYINM	VIEGN	T---C	GVATD	PLYPT	GVQYL
Eh-CPnp	CWIVR	NSWGT	GWGDK	GYINM	VIEGN	T---C	GVATD	PLYPT	GVQYL
Eh-CPp3	YWIIR	NSWGT	SWGDA	GYFL-	-LARD	SNNMC	GIGRD	SNYPT	GVKLI

[a] E. Tannich, H. Scholze, R. Nickel, and R. D. Horstmann, *J. Biol. Chem.* **266**, 4798 (1991).

[b] E. Tannich, R. Nickel, H. Buss, and R. D. Horstmann, *Mol. Biochem. Parasitol.* **54**, 109 (1992).

[c] S. L. Reed, J. Bouvier, A. S. Pollack, J. C. Engel, M. Brown, K. Hirata, X. Que, A. E. Eakin, P. Hagblom, F. Gillin, and J. H. McKerrow, *J. Clin. Invest.* **91**, 1532 (1993).

the polymerase chain reaction (PCR) and generic probes as primers.[27] Template DNA has been obtained from uncloned *E. histolytica* isolate HM1 : IMSS. Eh-CPp3 differs substantially from the other proteases; identity does not exceed 45% (Table III). Interestingly, it has been reported that Eh-CPp3 may be unique for pathogenic *E. histolytica*, because Southern blot analyses reveal that in contrast to Eh-CPp1 and Eh-CPp2, Eh-CPp3 genes hybridize with DNA of pathogenic amebas but not with DNA of nonpathogenic amebas.[24] The coding region of Eh-CPp1 has been used to identify a homologous sequence in a cDNA library from the nonpathogenic *Entamoeba* isolate SAW-1734.[25] DNA sequence analysis and comparison of the predicted amino acid sequences reveal a divergence of 15% from amoebapain, but only of 3.6% from histolysain.

The differences in primary structure of amebic cysteine endopeptidases are of interest with regard to possible functional differences among these enzymes and their role in the cytopathogenic process. The relationship between the structure and function of cysteine proteinases has been studied in great detail with the plant enzyme papain,[28] which corresponds to the ameba enzymes in 34% of the primary sequence. X-ray crystallography has revealed that papain consists of two domains separated by a deep cleft. A number of residues facing the cleft are essential for its proteolytic function. Alignments of the sequences of amebic cysteine proteinases with the three-dimensional structure of papain indicate that all amino acid residues forming and neighboring the putative catalytic site are conserved in the enzymes from both pathogenic and nonpathogenic amebas. Moreover, those residues facing the substrate-binding pocket, which are supposed to be responsible for proteolytic specificity (as deduced from known three-dimensional structures of several cathepsins[29]), are identical in all amebic cysteine proteinases analyzed so far; i.e., aspartate residues at the base of the S2 specificity pocket and at the position contacting the P1 residue, respectively. This suggests that no evidence is found for structural divergences likely to alter functional properties of the homologous enzymes from pathogenic and nonpathogenic *E. histolytica*. By contrast, there is a striking difference in the expression of cysteine proteinase genes between pathogenic and nonpathogenic isolates. Specifically, Northern blot analyses indicate that the amount of proteinase mRNA is consistently 10- to 100-fold higher in the former.[25] This finding is in good agreement with the observation that cysteine proteinase activity is much higher in

[27] A. E. Eakin, J. Bouvier, J. A. Sakanari, C. S. Craik, and J. H. McKerrow, *Mol. Biochem. Parasitol.* **39,** 1 (1990).
[28] G. Lowe, *Tetrahedron* **32,** 291 (1976).
[29] I. G. Kamphuis, J. Drenth, and E. N. Baker, *J. Mol. Biol.* **182,** 317 (1985).

pathogenic than in nonpathogenic *E. histolytica*.[12] At present, it is not known whether the higher proteolytic activity of pathogenic amebas is due to increased expression of one particular protease gene, or whether all genes are expressed at higher levels compared to the related homogeneous genes of nonpathogenic forms.

[36] Cysteine Endopeptidases of Parasitic Protozoa

By MICHAEL J. NORTH

Introduction

In recent years there has been an increasing interest in the role of proteolytic enzymes in parasites and especially in the possibility that some of the enzymes might prove to be appropriate targets for novel chemotherapeutic agents. Among parasitic protozoa, much of the focus has been on their cysteine endopeptidases.[1-3] These are often the most active proteolytic enzymes present, and most protozoa produce cysteine endopeptidases during at least one stage of their life cycle. Cysteine endopeptidase inhibitors have already been shown to be effective against a number of species of protozoa *in vitro*.[1,3]

A number of cDNAs and a few genomic DNA fragments encoding protozoan cysteine endopeptidases have now been cloned. In all cases the sequences of the predicted products confirm that the enzymes are members of the papain superfamily. Most of these show closest identity to mammalian cathepsin L, and only one cathepsin B-like sequence has so far been reported among protozoa.[4] In view of their similarity to cysteine endopeptidases of higher organisms, it is not surprising that the protozoan enzymes have been studied employing adaptations of techniques used initially with mammalian enzymes. Many of the protozoan enzymes do, however, share certain properties that distinguish them from most mammalian enzymes and have allowed the application of some methods less used in the analysis of the other cysteine endopeptidases. For example, the ability of many protozoan cysteine endopeptidases to retain activity after SDS–PAGE under partially denaturing conditions has led to the

[1] M. J. North, J. C. Mottram, and G. H. Coombs, *Parasitol. Today* **6**, 270 (1990).
[2] M. J. North, *Biol. Chem. Hoppe-Seyler* **373**, 401 (1992).
[3] J. H. McKerrow, E. Sun, P. J. Rosenthal, and J. Bouvier, *Annu. Rev. Microbiol.* **47**, 821 (1993).
[4] C. D. Robertson and G. H. Coombs, *Mol. Biochem. Parasitol.* **62**, 271 (1993).

widespread use of electrophoretic techniques as a basis for detection, identification, and partial characterization of these enzymes. The main aim of most studies has been to identify differences, often likely to be subtle, between the specificity of the parasite enzymes and that of their mammalian counterparts. Consequently, attention has often been focused on substrate preferences and inhibitor sensitivities. The significance of some of the more distinctive features of some of the protozoan cysteine endopeptidases, such as the C-terminal extension found in cruzipain, the cysteine endopeptidase of *Trypanosoma cruzi,*[5,6] and predicted to be present in the precursors of enzymes of other trypanosomatids,[7-9] has yet to be fully explored.

In this chapter I describe techniques that are likely to be generally applicable in studies of protozoan cysteine endopeptidases, but ones that differ from those generally used with the enzymes of higher organisms. The detailed methods and examples given are all drawn from work on the trichomonads *Tritrichomonas foetus*, a parasite of the genitourinary tract of cattle, and *Trichomonas vaginalis*, which infects the human genitourinary tract. They have, however, in most cases also been used for analysis of proteolytic enzymes of a number of other parasite species. For details of methods for *Entamoeba* cysteine endopeptidases the reader should consult [35] in this volume. Many of the methods described in other chapters on cysteine endopeptidases will also be applicable to protozoan enzymes.

Detection Methods

Because of the difficulties of culturing some parasite stages it has often been necessary to analyze cysteine endopeptidases using relatively small amounts of starting material. This has made the purification of individual endopeptidases problematic, and much of the initial characterization of cysteine endopeptidases has relied on the deployment of methods that allow the detection and analysis of enzymes without the need for purification. The presence in many species of multiple forms of endopeptidase also creates a need for methods that allow these to be identified before detailed analysis is undertaken. This has led to the widespread use of

[5] L. Åslund, J. Henriksson, O. Campetella, A. C. C. Frasch, U. Pettersson, and J. J. Cazzulo, *Mol. Biochem. Parasitol.* **45,** 345 (1991).
[6] U. Hellman, C. Wernstedt, and J. J. Cazzulo, *Mol. Biochem. Parasitol.* **44,** 15 (1991).
[7] J. C. Mottram, M. J. North, J. D. Barry, and G. H. Coombs, *FEBS Lett.* **258,** 211 (1989).
[8] A. E. Souza, S. Waugh, G. H. Coombs, and J. C. Mottram, *FEBS Lett.* **311,** 124 (1992).
[9] Y. M. Traub-Cseko, M. Duboise, L. K. Boukai, and D. McMahon-Pratt, *Mol. Biochem. Parasitol.* **57,** 101 (1993).

electrophoretic methods for simultaneous separation and detection. The methods can be used to look at the enzymes present in parasite lysates and growth medium and they have also been used to examine clinical material.[10] They have revealed differences between isolates and parasite stages and can be used to derive information on substrate specificity and inhibitor sensitivity.

Sample Preparation

Sample preparation will vary from parasite to parasite, but in general cells are lysed in detergent-containing solutions, although repeated freezing and thawing can also be used to release cysteine endopeptidases. Examples of detergents used are Nonidet P-40 (0.5–0.75%), Triton X-100 (0.1–0.25%), and Zwittergent 3–12 (0.05%). In general it is not necessary to add reducing agent to the lysing solution. For example, for trichomonads and *Leishmania mexicana* a solution of 0.25% Triton X-100 in 0.25 *M* sucrose is appropriate. These simple treatments usually release the cysteine endopeptidases, the majority of which are likely to be lysosomal. Cell debris and unbroken cells are removed by centrifugation. Samples of medium and other extracellular fluids can be used directly, although those containing high concentrations of serum may pose problems.

Protein–SDS–PAGE

Background. The method depends on the inclusion of a protein in the acrylamide gel. Following electrophoresis the protein can be hydrolyzed by endopeptidases and individual enzymes revealed as clear bands after staining the protein remaining. Initial analysis of protozoan endopeptidases involved the use of hemoglobin-containing gels and nondenaturing conditions.[11–14] It was these analyses that first revealed the complexity of the proteolytic system in many parasites and also showed that many of the enzymes involved were cysteine endopeptidases. Subsequently it was shown that samples could be analysed by SDS–PAGE, which in general gave sharper bands and more reproducible results. The method adopted by most workers has been adapted from that first described by Heussen and Dowdle.[15] It must be emphasized that the method is not specific for cysteine endopeptidases. In protozoa the majority of the enzymes detected

[10] J. F. Alderete, E. Newton, C. Dennis, and K. A. Neale, *Genitourin. Med.* **67**, 469 (1991).
[11] M. J. North and G. H. Coombs, *Mol. Biochem. Parasitol.* **3**, 293 (1981).
[12] G. H. Coombs and M. J. North, *Parasitology* **86**, 1 (1983).
[13] M. J. North, G. H. Coombs, and J. D. Barry, *Mol. Biochem. Parasitol.* **9**, 161 (1983).
[14] B. C. Lockwood, M. J. North, and G. H. Coombs, *Exp. Parasitol.* **58**, 245 (1984).
[15] C. Heussen and E. B. Dowdle, *Anal. Biochem.* **102**, 196 (1980).

are of this type, but the method has also been used to identify metalloendopeptidases and some putative serine endopeptidases: establishing the type of endopeptidase depends on the use of inhibitors (see below). Neither is its use confined to protozoa and it has been used to analyze the endopeptidases of a variety of organisms, including another major group of parasites, the helminths. Its success is dependent on the stability of the endopeptidases under the conditions used for electrophoresis (alkaline pH, presence of SDS, and usually a reducing agent), which must reflect a greater stability of these enzymes compared with their mammalian counterparts.

Gelatin (at 0.2%, w/v) is the most frequent choice of protein for inclusion in the gel, but bovine fibrinogen (at 330 μg/ml) has been used for the analysis of cysteine endopeptidases of African trypanosomes.[16] The method described below is as used for the analysis of trichomonad enzymes.[17]

Reagents

Acrylamide solution: 30% (w/v) acrylamide, 0.8% N,N'-methylenebisacrylamide.

Resolving gel buffer: 3 M Tris-HCl, pH 8.8.

Stacking gel buffer: 0.5 M Tris-HCl, pH 6.8.

10% (w/v) Sodium dodecyl sulfate (SDS).

1.5% (w/v) Ammonium persulfate (freshly prepared).

$N,N,N,'N'$-Tetramethylenediamine (TEMED).

Gelatin (Sigma, St. Louis, MO; porcine skin, 300 bloom): 2% (w/v) aqueous solution, stored at 4° and heated gently prior to use.

Reservoir buffer (10× final concentration): 0.25 M Tris, 1.92 M glycine, 1% (w/v) SDS.

Stain: 0.1% Coomassie Brilliant Blue R-250 in 40% (v/v) methanol/ 10% (v/v) acetic acid.

Destain: 10% (v/v) acetic acid.

Electrophoresis sample buffer: 1 ml stacking gel buffer, 0.8 ml glycerol, 1.6 ml 10% (v/v) SDS, 0.4 ml mercaptoethanol, 0.2 ml 0.05% bromphenol blue, 4 ml water.

2.5% (v/v) Triton X-100.

Gel incubation buffer: usually 0.1 M sodium acetate, pH 5.5, containing 1 mM dithiothreitol (DTT) (alternative buffers in the range from pH 3 to pH 8 have been used for other species).

---

[16] Z. R. Mbawa, I. D. Gumm, W. R. Fish, and J. D. Lonsdale-Eccles, *Eur. J. Biochem.* **195,** 183 (1991).

[17] M. J. North, C. D. Robertson, and G. H. Coombs, *Mol. Biochem. Parasitol.* **39,** 183 (1990).

Procedure. The method has been used with a variety of different electrophoresis systems but that described here is suitable for the preparation of two gels for the Bio-Rad (Richmond, CA) Mini-PROTEAN II. The acrylamide concentration can be varied, but for routine analyses either 7.5 or 10% (w/v) acrylamide gels are most appropriate. The following procedure describes the method for the preparation of two 10% acrylamide gels.

The separating gel is prepared by mixing 3.3 ml acrylamide solution, 1.25 ml resolving gel buffer, 0.1 ml SDS, 1.0 ml gelatin solution, 3.85 ml water, 0.5 ml ammonium persulfate, and finally 0.005 ml TEMED. Gels are 0.75 mm thick, poured to a standard height of 4 cm, and allowed to set under water-saturated 1-butanol. The stacking gel is prepared by mixing 0.625 ml acrylamide, 1.2 ml stacking gel buffer, 0.05 ml SDS, 2.625 ml water, 0.25 ml ammonium persulfate, 0.004 ml TEMED, and 0.2 ml 0.05% bromphenol blue. Usually a 10-well comb is used.

Parasite samples are mixed with an equal volume of electrophoresis buffer without heating. A maximum of 20 μg protein is loaded, and for the procedure given above this should be contained in a volume of less than 30 μl. When too much protein is loaded, endopeptidase bands are more difficult to detect because endogenous proteins interfere with the background staining. The gels are run at 16 mA per gel for approximately 40 min until the dye front has reached the bottom of the gel. They are then transferred to 10 ml 2.5% (v/v) Triton X-100 for 30 min with gentle shaking and incubated, with gentle shaking, in 10 ml gel incubation buffer at the required temperature (25° or 37°) for a period of a few hours (4 hr to overnight). The gels are stained for 30 min and destained to reveal bands of digested gelatin corresponding to individual endopeptidases. Molecular weight markers can be run and are visible through the background of stained gelatin.

Figure 1 provides an example of a gelatin gel in which samples of three different isolates of *T. foetus* have been analyzed. All of the major bands visible are cysteine endopeptidases [all are inhibited by *trans*-epoxysuccinyl-L-leucylamido(4-guanidino)butane (E-64)].[17] The analysis reveals some quantitative differences between the three isolates for some of the enzymes, notably in the fastest band (apparent M_r of 18,000), but the overall patterns are similar to one another.

SDS–PAGE with Fluorogenic Substrates

Background. The ability of the protozoan cysteine endopeptidases to hydrolyze peptidyl 7-amidomethylcoumarins has allowed the substrates to be used in combination with SDS–PAGE as an alternative to protein,

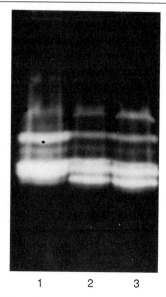

FIG. 1. Gelatin–SDS–PAGE analysis. The gel shows the endopeptidase patterns obtained at pH 5.5 for three isolates of *Tritrichomonas foetus*: (1) clone F2, (2) Manley strain, and (3) Belfast strain.

providing a sensitive method for detecting multiple forms of proteolytic enzymes in parasite samples and yielding information on the specificity of the individual enzymes without the need for purification. Initially the fluorogenic substrates were used with gelatin-containing gels. This allowed enzymes responsible for the bands on gelatin gels to be matched with those hydrolyzing the peptide derivatives. However, the presence of gelatin reduces the activity toward the fluorogenic substrates, and for greatest sensitivity it is preferable not to include protein in the electrophoresis gel. Proteolytic enzymes that fail to hydrolyze gelatin have also been revealed, in particular the *L. mexicana* type D enzymes, which have a narrow specificity, hydrolyzing Bz-Phe-Val-Arg-NHMec and Boc-Phe-Ser-Arg-NHMec.[18]

The method described below is that used for analyzing trichomonad cysteine endopeptidase specificity.[17]

Reagents. These are exactly as shown for gelatin–SDS–PAGE, with the addition of peptidyl amidomethylcoumarins (NHMecs). These are prepared as 5 mM stock solutions. The list includes those most useful for analysis of protozoan cysteine endopeptidases, but a much wider range

[18] J. C. Mottram, C. D. Robertson, G. H. Coombs, and J. D. Barry, *Mol. Microbiol.* **6**, 1925 (1992).

of substrates is available. They are available from a number of suppliers: all those listed can be obtained from Sigma and Bachem. The solvent for stock solutions is given in brackets.

Leu-Val-Tyr-NHMec [water].
Boc-Val-Leu-Lys-NHMec [water].
Bz-Phe-Val-Arg-NHMec [50% (v/v) acetonitrile].
Suc-Leu-Tyr-NHMec [20% (v/v) acetonitrile].
Z-Arg-Arg-NHMec [water].
Z-Phe-Arg-NHMec [20% (v/v) acetonitrile].
Gel incubation buffer: 0.1 M sodium acetate or 0.1 M sodium phosphate or 0.1 M Tris-HCl, all including 1 mM DTT.

Procedure. Sample and gel preparation are exactly as described above for gelatin–SDS–PAGE, except that gelatin may be omitted from the gels. If gelatin is included more sample protein may be required for detection of activity. When different substrates are tested on the same sample the stacking gel is prepared without wells and the sample is applied along the length of the gel: following electrophoresis the gel can be cut into strips for incubation with each substrate. After treatment with 2.5% (v/v) Triton X-100 the gels are incubated in prewarmed buffer for 5 min, although we have previously noted[19] that in the case of the trichomonad enzymes (and also cellular slime mold cysteine endopeptidases) the Triton X-100 treatment is not essential. The gels are then transferred to the same prewarmed buffer containing 0.05 mM substrate added from the stock solution. Gels are incubated without shaking at room temperature. High-activity bands appear within minutes and are detected by placing the gels on a Transilluminator, e.g., UVP (San Gabriel, CA) Chromato Vue TL33. The band pattern is recorded immediately using a Polaroid camera fitted with a Kodak (Rochester, NY) Wratten gelatin filter number 2E. With 665 (positive/negative) film an exposure of 10–20 sec at f11 is most appropriate. A shorter exposure (1 sec) is required when using 667 (positive only) film. Gels containing gelatin can be transferred to fresh buffer and incubated for longer periods so that bands of gelatin digestion can also be detected and subsequently matched with the fluorescent bands recorded photographically. It should be noted that the method is not suitable for enzymes with low activity because the fluorescent product (7-amino-4-methylcoumarin) diffuses too rapidly to maintain sharp bands during incubation periods longer than 1 hr.

Figure 2 provides an example involving four substrates and demonstrates the superiority of one of these substrates, Boc-Val-Leu-Lys-NH-Mec with all the *T. foetus* enzymes detectable. The bands seen in Fig. 2 correspond to the gelatin-digesting bands in Fig. 1.

[19] M. J. North and K. Nicol, unpublished (1993).

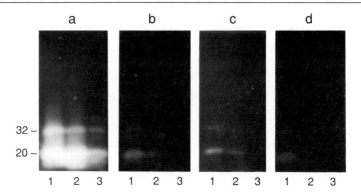

Fig. 2. SDS–PAGE analysis using fluorogenic substrates. The gels show the patterns of activity on four peptidyl amidomethylcoumarins at pH 6.0 of three isolates of *Tritrichomonas foetus*: (1) clone F2, (2) Manley strain, and (3) Belfast strain. (a) Boc-Val-Leu-Lys-NHMec; (b) Z-Arg-Arg-NHMec; (c) Suc-Leu-Tyr-NHMec; (d) Z-Phe-Arg-NHMec. The gels were incubated with the substrates for 20 min before being photographed. The numbers on the left indicate the apparent molecular weights ($\times 10^3$) of the major bands as judged by comparison with molecular weight markers.

Fluorogenic substrates have also been used with overlays in a more classical zymogram procedure, e.g., Z-Phe-Arg-7-amino-4-trifluoro-methylcoumarin used with methyl cellulose paper for the detection of an enzyme from *Trypanosoma brucei*.[20] The method described here is recommended, however, because it is simpler and allows the latter enzyme to be matched directly with a gelatin-hydrolyzing endopeptidase.[21]

Adaptations of SDS–PAGE

Two-Dimensional Gels. Detection of endopeptidases using SDS–PAGE can be combined with an isoelectric focusing step to increase the potential to detect multiple forms of endopeptidases and provide information on the isoelectric points of individual enzymes. Conventional two-dimensional electrophoresis (O'Farrell method[22]) has been adapted by the inclusion of gelatin or fibrinogen in the gel used for the electrophoresis

[20] E. G. Pamer, M. So, and C. E. Davis, *Mol. Biochem. Parasitol.* **33**, 27 (1989).
[21] C. D. Robertson, M. J. North, B. C. Lockwood, and G. H. Coombs, *J. Gen. Microbiol.* **136**, 921 (1990).
[22] P. H. O'Farrell, *J. Biol. Chem.* **250**, 4007 (1975).

step for analysis of enzymes from *T. vaginalis*,[23] *Trypanosoma congolense*,[24] and *T. cruzi*.[25]

Preparative isoelectric focusing can also be used in combination with SDS–PAGE, either with substrate-containing gels or fluorogenic substrates. Samples from the Bio-Rad Rotofor system, which generates 20 fractions, can be run directly on gels to generate the equivalent of a two-dimensional endopeptidase pattern.[26]

Acid Activation. A chance observation made during the analysis of cysteine endopeptidases of the cellular slime mold *Dictyostelium discoideum* revealed that additional enzymes can sometimes be detected by SDS–PAGE analysis if the gels are treated with acetic acid immediately after the electrophoretic separation.[27] The method simply involves the additional step of transferring a gel to 10% (v/v) acetic acid for 30 sec immediately after electrophoresis and before it is incubated in Triton X-100. This has allowed the detection of two additional enzymes in *L. mexicana*.[28] The activation of the enzymes results in a significant increase in electrophoretic mobility, indicating that they are likely to be inactive precursors and that activation might involve the removal of an N-terminal pro region from an inactive proenzyme. This mechanism is not universal, because a shift in electrophoretic mobility has not been noted with the *D. discoideum* enzymes.[27]

Immunoprecipitation. Some protozoan cysteine endopeptidases are sufficiently robust to be detectable using SDS–PAGE after immunoprecipitation. *Trichomonas vaginalis* endopeptidases have been immunoprecipitated using antibody raised against purified enzyme[29] or against parasite lysates[23] or antibody present in sera and vaginal washes of patients.[10,30,31] The method uses antibody bound to *Staphylococcus aureus* to precipitate endopeptidases from *T. vaginalis* lysates. Following precipitation the bacteria are suspended in electrophoresis sample buffer and the enzymes are released by heating at 37° for 3 min. The bacteria are pelleted and the supernatant is used for the standard gelatin–SDS–PAGE procedure.

[23] K. A. Neale and J. F. Alderete, *Infect. Immun.* **58**, 157 (1990).
[24] E. Authié, D. K. Muteti, Z. R. Mbawa, J. D. Lonsdale-Eccles, P. Webster, and C. W. Wells, *Mol. Biochem. Parasitol.* **56**, 103 (1992).
[25] M. C. Bonaldo, L. N. D'Escoffier, J. M. Saller, and S. Goldenberg, *Exp. Parasitol.* **73**, 44 (1991).
[26] M. J. North, M. M. Buchan, and K. Nicol, unpublished (1993).
[27] M. J. North, D. A. Cotter, and T. W. Sands, unpublished (1993).
[28] C. D. Robertson and G. H. Coombs, *FEMS Microbiol. Lett.* **94**, 127 (1992).
[29] M. J. North and C. Howe, unpublished (1993).
[30] J. F. Alderete, E. Newton, C. Dennis, and K. A. Neale, *Genitourin. Med.* **67**, 331 (1991).
[31] P. Bózner, A. Gomboscová, M. Valent, P. Demes, and J. F. Alderete, *Parasitology* **105**, 387 (1992).

General Comments on Electrophoretic Detection Methods

To detect endopeptidases by virtue of their activity, it is necessary to employ conditions that do not result in the complete denaturation of the protein. In most cases it is possible to use reducing conditions, although there are some enzymes for which the reducing agent must be omitted from the electrophoresis buffer. For example, inclusion of mercaptoethanol results in the autodegradation of a 33-kDa cysteine endopeptidase from *T. congolense*.[24] Two questions relating to the electrophoretic mobility arise. First, is the mobility affected by the presence of copolymerized protein in the gel? Comparisons of mobilities of enzymes detected with fluorogenic substrates on gels with and without gelatin suggest that the gelatin has no significant effect on the endopeptidase mobility. Second, can the electrophoretic mobility be related directly to the molecular mass of the enzyme? In most cases in which it has been possible to compare the mobility of cysteine endopeptidases run under totally denaturing conditions, i.e., after heating (purified protein detected by staining or protein in unfractionated samples detected after Western blotting), with that of the same enzyme run under the conditions used for activity gels, any differences have been small or not detectable.[18,24,32,33] Radiolabeling of cysteine endopeptidases of *Entamoeba histolytica*[34] and *Leishmania amazonensis*[35] with Z-[^{125}I]Tyr-Ala-CHN$_2$ has revealed bands (under denaturing conditions) with molecular masses similar to those expected from activity methods. Only in the case of cruzipain, the enzyme from *T. cruzi*, has anomalous electrophoretic behavior been reported, but this is so even with denaturing conditions.[36] In most cases it has been safe to conclude that an apparent molecular mass calculated from the electrophoretic mobility provides a reasonable estimate of the actual size.

In some species high apparent molecular mass (slow running) forms of cysteine endopeptidases have been detected. These have yet to be analyzed in detail and the possibility that they represent aggregates of enzymes or complexes of endopeptidases with other proteins, e.g., endogenous inhibitors[37] or serum proteins[38] cannot be eliminated. It is now

[32] J. W. Irvine, G. H. Coombs, and M. J. North, *FEMS Microbiol. Lett.* **110**, 113 (1993).
[33] Z. R. Mbawa, P. Webster, and J. D. Lonsdale-Eccles, *Eur. J. Cell Biol.* **56**, 243 (1992).
[34] F. De Meester, E. Shaw, H. Scholze, T. Stolarsky, and D. Mirelman, *Infect. Immun.* **58**, 1396 (1990).
[35] S. C. Alfieri, E. M. F. Pral, E. Shaw, C. Ramazeilles, and M. Rabinovitch, *Exp. Parasitol.* **73**, 424 (1991).
[36] J. Martinez and J. J. Cazzulo, *FEMS Microbiol. Lett.* **95**, 225 (1992).
[37] J. W. Irvine, G. H. Coombs, and M. J. North, *FEMS Microbiol. Lett.* **96**, 67 (1992).
[38] J. D. Lonsdale-Eccles, *in* "Biochemical Protozoology" (G. H. Coombs and M. J. North, eds.), p. 200. Taylor & Francis, London, 1991.

clear, however, that in many species in which multiple forms of cysteine endopeptidase are apparent, these can be accounted for, at least in part, by multiple genes and that multiplicity of electrophoretic forms is not simply an artifact.

Assays

The similarity of most of the well-characterized protozoan enzymes to mammalian cysteine endopeptidases has meant that the commonly used assay methods are very similar to those widely used for enzymes such as cathepsin L and cathepsin B. Proteins are rarely used as substrates in routine assays of protozoan enzymes. Continuous assays based on chromogenic and fluorogenic substrates are most appropriate. As with mammalian cathepsins, the most widely used substrates have an arginine residue at the P1 position, although substrates containing arginine alone are not the best substrates for cysteine endopeptidases and are as likely to be hydrolyzed by other parasite enzymes (probably serine peptidases). Many of the protozoan enzymes are able to hydrolyze substrates with a bulky residue at P2, and so Bz-Pro-Phe-Arg-*p*-nitroanilide (-Nan) and Z-Phe-Arg-NHMec are frequently used for chromogenic and fluorogenic assays, respectively. Some enzymes are also active on substrates with arginine at the P2 position, notably those from *E. histolytica* (see article [33]) and some of the trichomonad enzymes,[17,32] and Z-Arg-Arg-Nan and Z-Arg-Arg-NHMec are both used. A wide range of peptidyl nitroanilides and peptidyl amidomethylcoumarins is now available, although it is important to note that in unfractionated samples there may be many enzymes that contribute to the hydrolysis of any one of these substrates. Some indication of specificity might be gained from an electrophoretic analysis (see above).

The assay methods used are very similar to those used for mammalian cathepsins (see article [33]) and are not given in detail here. The following method is one we have used for assaying large numbers of samples, either to get an indication of specificity by using different substrates or for assaying activity during purification procedures.

Assay Using Peptidyl Nitroanilides

This assay is conveniently carried out in microtiter plates, but the volumes given below can be scaled up and the assay conducted in a cuvette and followed in spectrophotometer. The basis of the assay is the release from a peptidyl nitroanilide of nitroaniline, which can be followed at 405 nm.

Reagents

Nitroanilide stock solution: 1 mM. The most frequently used are Bz-Pro-Phe-Arg-Nan and Z-Arg-Arg-Nan, both of which can be prepared in water. Both are available from a number of suppliers, including Novabiochem (Nottingham, UK) and Bachem. Bz-Pro-Phe-Arg-Nan is also supplied by Sigma.

Buffer: 0.1 M sodium phosphate, pH 6.0.

Substrate/buffer solution: 4.0 ml buffer, 0.8 ml substrate, 3.2 ml water, and 0.08 ml 0.1 M DTT (prepared fresh). This is sufficient for assaying 30 samples.

Samples (15 μl) are placed in microtiter wells and the reaction is started by addition of 0.15 ml of the substrate/buffer solution. Final substrate concentration is 0.09 mM. The reaction is followed by regular readings in a microtiter plate reader, e.g., Bio-Rad Model 2550 EIA Reader fitted with a 405-nm filter. A change in optical density of 0.1 corresponds to 39.3 nmol nitroaniline and represents an activity of 2.62 μmol nitroaniline released per milliliter of sample.

Assays Using Peptidyl Amidomethylcoumarins

The use of fluorogenic substrates provides a more sensitive assay, with Z-Phe-Arg-NHMec frequently employed. The assay conditions reported for protozoan enzymes vary but are generally similar to those for the nitroanilide assay (0.05 mM substrate, pH 6.0, inclusion of DTT). The release of 7-amino-4-methylcoumarin is monitored in an appropriate spectrofluorometer with an excitation wavelength of 350 nm and an emission wavelength of 440 nm. An example used for African trypanosome cysteine endopeptidase activity is described by Mbawa *et al.*[33] Assays with amidomethylcoumarins are the most appropriate for determining the kinetic constants of purified enzymes.

Inhibitor Sensitivity

Determining the sensitivity of protozoan endopeptidases to various inhibitors usually has two aims. First, it allows the type of endopeptidase to be established, and second, it provides an indication of the features that might be appropriate for putative antiparasite agents targeted at proteolytic enzymes. Sensitivity to an inhibitor can be determined by the inclusion of the compound in an assay mix or by assaying samples preincubated with the compound. For endopeptidases detectable by SDS–PAGE methods the sensitivity of the various enzymes in unfractionated samples can be distinguished by treatment with test compounds at appropriate stages of the electrophoretic procedure. With irreversible inhibitors, samples should be preincubated for a short period (5 to 10 min) with the inhibitor before subjecting them to SDS–PAGE. The inhibitor can also be added to the gel incubation buffer but this may be prohibitively expensive in some cases. Inhibitors for which the binding to the enzyme is reversible have to be tested by inclusion in the gel incubation buffer.

The following three inhibitors are generally available and are particularly useful for studies on cysteine endopeptidases. The details in brackets indicate the concentration and solvent for the stock solution and the final concentration to be used. E-64 [2.80 mM (1 mg/ml) in water; 0.028–0.14 mM (10–50 μg/ml)] is a specific irreversible inhibitor of cysteine endopeptidases and is the best diagnostic compound for cysteine endopeptidases. Peptidyl diazomethanes [1 mM in acetonitrile; 10–100 μM] are irreversible inhibitors that are generally most effective toward cysteine endopeptidases. Z-Phe-Ala-CHN$_2$ and Z-Phe-Phe-CHN$_2$ are both commercially available but have varying effects on different protozoan enzymes and are not totally effective against all of them (e.g., see North et al).[17] A range of other peptidyl diazomethanes has been synthesized and an example of the use of some of these in combination with gelatin–SDS–PAGE is shown in Fig. 3. This shows differential effects dependent on the nature of the amino acids at the P1 and P2 positions.

Purification Methods

In spite of the interest in the protozoan enzymes, relatively few have been purified completely. Successful purification schemes have largely involved combinations of conventional techniques. For example, L. mexicana cysteine endopeptidases have been purified by a combination of affinity chromatography (concanavalin A–Sepharose), gel filtration (Superose 12), and anion-exchange chromatography (Mono Q, Pharmacia, Piscataway, NJ).[39] The T. congolense cysteine endopeptidase has been purified using affinity chromatography (thiopropyl Sepharose 6B or cystatin–Sepharose 4B) and gel filtration (Sephacryl S-200)[16,33] and the T. cruzi enzyme cruzipain by a combination of ion-exchange chromatography (DEAE-Sephacel and Mono Q) and gel filtration (Sephadex G-200 and Superose 6).[40] No one multistep scheme is likely to be generally applicable for the purification of a wide range of proteins, and we have sought to evaluate a number of affinity methods for one-step purification procedures

[39] C. D. Robertson and G. H. Coombs, Mol. Biochem. Parasitol. **42**, 269 (1990).
[40] J. J. Cazzulo, R. Couso, A. Raimondi, C. Wernstedt, and U. Hellman, Mol. Biochem. Parasitol. **33**, 33 (1989).

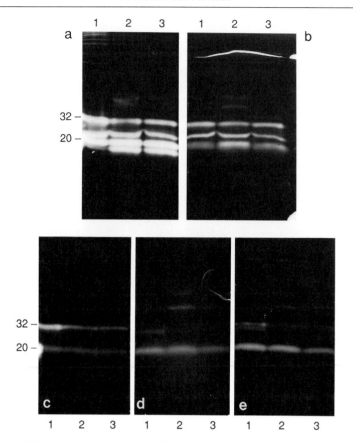

FIG. 3. Inhibition of cysteine endopeptidases of *Tritrichomonas foetus* by peptidyl diazomethanes. Samples of *T. foetus* (1) clone F2, (2) Manley strain, and (3) Belfast strain were
preincubated with 0.1 mM concentrations of peptidyl diazomethanes before analysis by
gelatin–SDS–PAGE. (a) Control; (b) Z-Phe-Ala-CHN$_2$; (c) Z-Leu-Ala-CHN$_2$; (d) Z-Leu-
Met-CHN$_2$; (e) Z-Leu-Nva-CHN$_2$. The latter three compounds were kindly provided by
Dr. E. Shaw (Basel). Nva, Norvaline. The numbers on the left indicate the apparent molecular
weights ($\times 10^3$) of the major bands as judged by comparison with molecular weight markers.

using trichomonads as the source of cysteine endopeptidases. The possibility of using single-step procedures was indicated by the very effective
purification of histolysain from *E. histolytica*, which uses immobilized
phenylalanyl(2-phenyl)aminoacetaldehyde   semicarbazone   (Phe-Phe-
Sc).[41] In summary, we have had limited success with thiopropyl, cystatin,
and Gly-Phe-Gly-Sc columns, all of which suffered from low binding ca-

[41] A. L. Luaces and A. J. Barrett, *Biochem. J.* **250,** 903 (1988).

pacities. Yields of cysteine endopeptidases using organomercurial columns (Affi-Gel 501) were variable. One method, bacitracin affinity chromatography, did prove successful and provided sufficient material from both *T. foetus* and *T. vaginalis* to obtain N-terminal sequences.[32]

Bacitracin Affinity Chromatography

Background. The use of the antibiotic cyclopeptide bacitracin as a ligand in affinity chromatography was first introduced by Stepanov and Rudenskaya[42] for the purification of endopeptidases from various sources. Bacitracin is a weak endopeptidase inhibitor of broad specificity and bacitracin–Sepharose chromatography can be applied to the purification of enzymes of different catalytic types, including plant cysteine endopeptidases.[42,43] In spite of its lack of specificity for cysteine endopeptidases, we have found this to be one of the more successful methods for purification of trichomonad cysteine endopeptidases.

The protocol given below is for a small-scale preparation suitable for N-terminal sequence determination.[32]

Reagents

Coupling buffer: 0.1 M NaHCO$_3$, 0.5 M NaCl, pH 8.3.
Bacitracin (Sigma).
Cyanogen bromide-activated Sepharose (Sigma) washed with 1 mM HCl.
20 mM Acetate buffer, pH 4.0 (buffer A).
0.1 M Tris-HCl, pH 7.0, 1.0 M NaCl, 25% (v/v) propanol (buffer B).

Procedure. Bacitracin–Sepharose is prepared as follows. Bacitracin (300 mg) in 25 ml coupling buffer is mixed with approximately 10 ml of activated Sepharose and shaken overnight at 4°. Further reaction is blocked by washing the Sepharose in 0.1 M Tris-HCl, pH 8.0, for 2 hr at room temperature. This is followed by three washing cycles with 0.1 M acetic acid/sodium acetate, pH 4.0, containing 0.5 M NaCl and 0.1 M Tris-HCl, pH 8.0, containing 0.5 M NaCl.

Parasites (2 × 10^9 trichomonad cells) are lysed in 2 ml buffer A by freeze-thawing (×4). The soluble fraction is obtained by centrifugation for 5 min at 11,600 g and applied to a 8 × 2-cm bacitracin–Sepharose column equilibrated with buffer A (flow rate 0.5 ml/min). The column is washed in buffer A until all unbound protein has been eluted (determined by A_{280} readings) and then a gradient of 0–100% buffer B is applied over 10 ml. Activity can be determined using Z-Arg-Arg-Nan as a substrate.

[42] V. M. Stepanov and G. N. Rudenskaya, *J. Appl. Biochem.* **5**, 420 (1983).
[43] J. M. van Noort, P. van den Berg, and I. E. Mattern, *Anal. Biochem.* **198**, 385 (1991).

The active fractions can be pooled and concentrated using Amicon (Danvers, MA) Centricon Microconcentrator (10-kDa cutoff).

In a typical purification from lysates of *T. foetus*, 1.35% of the total protein applied was recovered in the active fraction, which contained 157% of the activity initially present in the lysate (the increase in activity may be due to the removal of endogenous inhibitors[37]). This represents a purification of 126-fold. The material contained two major protein bands (by Coomassie blue staining) that appeared to correspond to two of four major protein-digesting bands. One of these had the following N-terminal sequence:

A D S L D W R E K G V V N S I K D Q A Q X G S

This sequence shows significant identity to the N terminus of most cysteine endopeptidases.

The method has also been used to purify a cysteine endopeptidase from *T. vaginalis* and enzymes from the medium of *T. foetus* cultures. Because it has also proved successful in the purification of cysteine endopeptidases from *D. discoideum*[44] it seems likely that this method may find general application.

Preparation of Recombinant Enzyme

The problems associated with the availability of limited amounts of starting material can be overcome by the use of recombinant enzymes. Cysteine endopeptidase genes have now been cloned from a number of species of parasitic protozoa[1,3] and expression of active recombinant enzymes in *Escherichia coli* have been achieved for enzymes from *T. brucei*[45] and *T. cruzi*.[46] Cruzain, one of the forms of the major cysteine endopeptidase of *T. cruzi*, has now been obtained in crystalline form.[47] Details of the molecular techniques used for the cloning and expression of the protozoan genes in *E. coli* and of the purification procedures for the recombinant enzyme are outside the scope of this chapter. However, it seems likely that this will prove to be the most appropriate approach for obtaining sufficient quantities of some of the protozoan enzymes for detailed structural and kinetic studies. This may be especially so for parasites that produce multiple forms of closely related enzymes, which have

[44] A. Champion, G. Harrison, V. Sham, A. A. Golley, M. Wilkins, M. J. North, and K. L. Williams, unpublished (1993).
[45] E. G. Pamer, C. E. Davis, and M. So, *Infect. Immun.* **59**, 1074 (1991).
[46] A. E. Eakin, A. A. Mills, G. Harth, J. H. McKerrow, and C. S. Craik, *J. Biol. Chem.* **267**, 7411 (1992).
[47] A. E. Eakin, M. E. McGrath, J. H. McKerrow, R. J. Fletterick, and C. S. Craik, *J. Biol. Chem.* **268**, 6115 (1993).

proved extremely difficult to obtain pure in large quantities from parasite sources.

Acknowledgments

This chapter is dedicated to the memory of Barbara Lockwood, who provided much of
the spark for our early studies on protozoan endopeptidases. Work in the author's laboratory
is supported by grants from the Wellcome Trust held jointly with G. H. Coombs (Glasgow)
and J. W. Irvine. I wish to thank Dr. J. W. Irvine and Dr. D. A. Scott for their help in the
preparation of this manuscript.

[37] Glycyl Endopeptidase

By David J. Buttle

The protein-digesting property of the juice of the unripe paw-paw, or
papaya fruit (*Carica papaya*), was first reported in 1750[1] (for an historical
account of research into the cysteine endopeptidases, see Brocklehurst
et al.[2]). Unlike some other fruits, such as the pineapple, most of the
proteolytic activity is found not in the pulp of the ripe fruit, but only in
the milky latex that is located in a network of laticifers just under the
epidermis of all parts of the plant except the ripe fruit and taproot.[3]

The preparation of dried papaya latex, mainly for export, is a large-
scale industry in some tropical countries. This is primarily because the
product provides a very rich and relatively cheap source of proteolytic
activity for use in the food, leather, and pharmaceutical industries. Good-
quality dried latex is more than 40% protein, and of this almost 70% is
made up of cysteine endopeptidases.[4]

The crude latex was known as "papain," and Balls *et al.*[5] gave the
name "crystalline papain" to the first single endopeptidase isolated from
the latex. The use of the name "papain" for both a purified product
(EC 3.4.22.2) and the crude starting material has led to some confusion,
however. Suspicions that crude papaya latex contained more than one

[1] G. Hughes, "The Natural History of Barbados," Book 7, p. 181. London, 1750.
[2] K. Brocklehurst, F. Willenbrock, and E. Salih, *in* "Hydrolytic Enzymes" (A. Neuberger
and K. Brocklehurst, eds.), p. 39. Elsevier, Amsterdam, 1987.
[3] M. L. Tainter and O. H. Buchanan, *Ann. N.Y. Acad. Sci.* **54,** 147 (1951).
[4] D. J. Buttle, P. M. Dando, P. F. Coe, S. L. Sharp, S. T. Shepherd, and A. J. Barrett,
Biol. Chem. Hoppe-Seyler **371,** 1083 (1990).
[5] A. K. Balls, H. Lineweaver, and R. R. Thompson, *Science* **86,** 379 (1937).

proteolytic enzyme were aroused in the early 1900s,[6,7] although this was not confirmed until the isolation of chymopapain (EC 3.4.22.6) in 1941.[8] A third endopeptidase was later isolated from crude papain, papaya peptidase A (now known as caricain; EC 3.4.22.30).[9] All three endopeptidases, although differing in primary structure, have very similar substrate specificities and are generally assayed with synthetic substrates with Arg in P1 (in the terminology of Ref. 10).[11,12] The similarities in enzymatic properties of these enzymes, together with their heterogeneity on cation-exchange columns, and perplexing results of studies of the reactivities of the enzymes with disulfides, led to continued confusion about the numbers and forms of cysteine endopeptidases in papaya latex.[11,13–17]

The first indications that papaya latex may contain an endopeptidase with a specificity markedly different from those of the three previously purified enzymes came with the use of blocked -Gly-nitrophenyl ester substrates. Lynn[18] and Polgár[19] detected a fraction that demonstrated activity against the glycine ester but not against Bz-Arg-NHPhNO$_2$ (NHPhNO$_2$, p-nitroanilide). This fraction was named papaya peptidase B, although its relationship to either chymopapain or the enzyme originally isolated by Schack (and called papaya peptidase A[9]) remained unclear. Confirmation of the existence of a fourth papaya endopeptidase awaited the serendipitous discovery that an endopeptidase with no activity on Bz-Arg-NHPhNO$_2$ could be separated with high yield from the other papaya endopeptidases by a simple purification procedure including affinity chromatography.[20] This enzyme, originally called papaya proteinase IV and now recommended to be called glycyl endopeptidase (EC 3.4.22.25), is

[6] S. H. Vines, *Ann. Bot.* **19**, 149 (1905).
[7] L. B. Mendel and A. F. Blood, *J. Biol. Chem.* **8**, 177 (1910).
[8] E. F. Jansen and A. K. Balls, *J. Biol. Chem.* **137**, 459 (1941).
[9] P. Schack, *C. R. Lab. Carlsberg* **36**, 67 (1967).
[10] Enzyme Nomenclature. "Recommendations of the Nomenclature Committee of the International Union of Biochemistry and Molecular Biology," p. 371. Academic Press, San Diego, 1992.
[11] B. S. Baines and K. Brocklehurst, *J. Protein Chem.* **1**, 119 (1982).
[12] S. Zucker, D. J. Buttle, M. J. H. Nicklin, and A. J. Barrett, *Biochim. Biophys. Acta* **828**, 196 (1985).
[13] D. J. Buttle and A. J. Barrett, *Biochem. J.* **223**, 81 (1984).
[14] A. J. Barrett and D. J. Buttle, *Biochem. J.* **228**, 527 (1985).
[15] K. Brocklehurst, E. Salih, R. McKee, and H. Smith, *Biochem. J.* **228**, 525 (1985).
[16] K. Brocklehurst, B. S. Baines, and C. Hatzoulis, *Biochem. J.* **221**, 553 (1984).
[17] L. Polgár, *Biochem. J.* **221**, 555 (1984).
[18] K. R. Lynn, *Biochim. Biophys. Acta* **569**, 193 (1979).
[19] L. Polgár, *Biochim. Biophys. Acta* **658**, 262 (1981).
[20] D. J. Buttle, A. A. Kembhavi, S. Sharp, R. E. Shute, D. H. Rich, and A. J. Barrett, *Biochem. J.* **261**, 469 (1989).

the subject of this chapter. The enzyme is commercially available as proteinase Gly-C from Calbiochem-Novabiochem (UK) Ltd. (Nottingham NG7 2QJ, UK).

Assay Procedures

Glycyl endopeptidase can be assayed with Z-Gly-OPhNO$_2$[19] (OPhNO$_2$, p-nitrophenyl ester) or Boc-Gly-OPhNO$_2$,[20] but the rapid spontaneous hydrolysis of these substrates makes quantification difficult and they are therefore not recommended.

Assay with Boc-Ala-Ala-Gly-NHPhNO$_2$

Principle. The substrate is hydrolyzed, releasing 4-nitroaniline, which is measured spectrophotometrically [$\varepsilon_{410\ nm}$ = 8800 M^{-1} cm^{-1} (Ref. 21)] after the enzyme is inactivated by alkylation of the active site thiol with chloroacetate.

Reagents

Buffer/activator: 200 mM sodium dihydrogen phosphate/200 mM disodium hydrogen phosphate/4 mM disodium EDTA, pH 6.8, made 8 mM with respect to cysteine by addition of the free base. The buffer can be made and stored for a few days but the cysteine must be added freshly.

Diluent: water.

Substrate stock solution: Boc-Ala-Ala-Gly-NHPhNO$_2$ is obtained from Bachem Feinchemikalien AG, Bubendorf, Switzerland. A stock solution (50 mM) in dimethyl sulfoxide (DMSO) can be stored at 4°.

Stopping reagent: 100 mM sodium chloroacetate/30 mM sodium acetate/70 mM acetic acid, pH 4.3.

Procedure. The assays can be made in disposable polystyrene tubes. The enzyme sample, containing between 3 and 60 pmol (70 ng–1.4 μg) of active glycyl endopeptidase, is placed in the tube together with 250 μl of the buffer/activator. The volume is made up to 975 μl with water and the mixture is incubated at 40° for 5 min to allow full enzyme activation. The reaction is then started by the addition of 25 μl of the substrate stock solution (to give a final concentration of 1.25 mM) and the reaction is allowed to proceed at 40°. After exactly 10 min, 1 ml of the stopping reagent is added and A_{410} is determined. Blanks and controls include

---

[21] B. F. Erlanger, N. Kokowsky, and W. Cohen, *Arch. Biochem. Biophys.* **95**, 271 (1961).

omitting the substrate from the reaction mixture, or adding the enzyme sample only after the addition of the stopping reagent, or making the assay in the presence of 10 μM E-64 (L-3-carboxy-2,3-*trans*-epoxypropionylleu-cylamido(4-guanidino)butane; see below).

If the spectrophotometer is equipped with a temperature-controlled cuvette holder and is linked to either a computer or a chart recorder, the reaction can be monitored and recorded continuously, without the addition of the stopping reagent.

The sensitivity of the assay can be improved by increasing the incubation time; the enzyme is stable under the conditions of the assay for at least 2 hr.

Assay with Boc-Ala-Ala-Gly-NHMec

Principle. A sensitive fluorimetric assay, detecting 20 fmol (500 pg) of active glycyl endopeptidase, involves the replacement of the spectrophotometric leaving group with NHMec [-7-(4-methyl)coumarylamide], so that hydrolysis results in release of the highly fluorescent 7-amino-4-methyl-coumarin. Boc-Ala-Ala-Gly-NHMec is not commercially available but can be synthesized quite easily as described below (C. Graham Knight, personal communication).

Synthesis. Dissolve H-Gly-NHMec·HBr (100 mg, 0.32 mmol), Boc-Ala-Ala-OH (92 mg, 0.35 mmol), and PyBOP (benzotriazol-1-yloxytripyr-rolidinophosphonium hexafluorophosphate) (156 mg, 0.35 mmol) in dry dimethylformamide (3 ml) with stirring at room temperature. Add diisopropylethylamine (0.12 ml, 0.7 mmol) and continue stirring overnight. Slowly add water (25 ml) to precipitate the product and triturate if necessary until fine white crystals are obtained. Stir the suspension for 1 hr and cool in ice before filtering. Wash the crystals with ice-cold water (100 ml) and dry *in vacuo* over P_2O_5. The yield of Boc-Ala-Ala-Gly-NHMec [homogeneous by high-performance liquid chromatography (HPLC)] is about 135 mg (90%).

Reagent Sources. H-Gly-NHMec (Sigma, St. Louis, MO), Boc-Ala-Ala-OH (Bachem Feinchemikalien), PyBOP (Calbiochem-Novabiochem).

Assay Reagents. The buffer/activator and stopping reagent are the same as those described above for assay with Boc-Ala-Ala-Gly-NHPhNO$_2$, except that the buffer/activator also contains 0.005% Brij 35.

Substrate stock solution: this can be made about 5 mM in DMSO. The exact concentration is best determined by total hydrolysis of a sample by papain or glycyl endopeptidase and quantitation of the amount of 7-amino-4-methylcoumarin released (see below), then adjustment of the stock solution to 2 mM by the addition of DMSO. The resulting stock solution can be stored at 4°.

Diluent: the nonionic detergent Brij 35 (0.005%) in water.

Aminomethylcoumarin standard: 7-amino-4-methylcoumarin (0.25 mM) in DMSO, stored in the dark at 4°.

Procedure. The enzyme sample, containing a maximum of 800 fmol (19 ng) of active glycyl endopeptidase, is placed in the tube along with 250 μl of the buffer/activator. The volume is made up to 975 μl with diluent and the mixture is incubated at 40° for 5 min to allow enzyme activation. The reaction is started by the addition of 25 μl of the substrate stock solution (to give a final concentration of 50 μM; the limit of solubility is about 80 μM) and the reaction is allowed to proceed at 40°. After 10 min, 2 ml of the stopping reagent is added.

The fluorescence of the liberated aminomethylcoumarin is determined in the fluorimeter by excitation at 360 nm and emission at 460 nm. The instrument is first standardized such that 10% of total hydrolysis corresponds to 1000 fluorescence units. This is done by zeroing the machine with a blank prepared by adding the enzyme sample after the stopping reagent. The aminomethylcoumarin standard (20 μl, giving a final concentration of aminomethylcoumarin of 1.67 μM) is added and mixed thoroughly by aspiration or stirring. The fluorimeter is then adjusted to read 1000 fluorescence units, equivalent to 500 pmol/min of substrate hydrolyzed.

If the fluorimeter is fitted with a temperature-controlled cuvette holder and linked to a chart recorder, the hydrolysis of the substrate by glycyl endopeptidase can be monitored and recorded continuously. A link with a computer allows kinetic analyses to be carried out conveniently by the use of software packages for recording and analyzing the data, such as FLUSYS[22] and ENZFITTER,[23] respectively.

Active Site Titration of Glycyl Endopeptidase

This method is a modification of the original description of E-64 titration, of cathepsins B and L.[24]

Principle. The amount of active glycyl endopeptidase is determined by stoichiometric inactivation by increasing amounts of E-64 in a series of tubes. The amount of activity remaining is then measured by cleavage of a substrate. The degree of substrate hydrolysis is plotted against E64

---

[22] N. D. Rawlings and A. J. Barrett, *Comput. Appl. Biosci.* **6,** 118 (1990).
[23] Biosoft, Cambridge CB2 1LR, UK.
[24] A. J. Barrett and H. Kirschke, this series, Vol. 80, p. 535.

concentration. Extrapolation of the straight line back to zero substrate hydrolysis gives the E-64 concentration at which all enzyme was inactivated. The stoichiometric nature of the inhibition means that this is also the concentration of active enzyme in the tubes (see elsewhere in this series).[24a]

Reagents, Buffer/Activator. This is the same as that described for the assay with Boc-Ala-Ala-Gly-NHPhNO$_2$.

Diluent: 1 mM EDTA.

Substrate stock solution: 50 mM Boc-Ala-Ala-Gly-NHPhNO$_2$ in DMSO.

E-64 stock solution: E-64 is obtained from Sigma Chemical Company Ltd. The dry powder is dissolved in methanol at 1 mM concentration. The solution is split into 20-μl aliquots and dried down in microcentrifuge tubes in a desiccator or by freeze-drying. The aliquots can then be stored dry at 4° until required. Before use, E-64 in one aliquot is dissolved in 10 ml of water, to give a concentration of 2 μM.

Enzyme stock solution: using the molar extinction coefficient (below), a 2 μM solution of glycyl endopeptidase in 1 mM EDTA is prepared.

Stopping reagent: as described for the assay with Boc-Ala-Ala-Gly-NHPhNO$_2$.

Procedure. Into each of 11 tubes, dispense 25 μl of the enzyme solution, followed by 0–25 μl of the E-64 stock solution, in 2.5-μl increments, and 75 μl of the buffer/activator. The volume is made up to 125 μl by the addition of diluent, where necessary. The tubes are incubated at 40° for 15 min, during which time the buffer/activator, diluent, and substrate stock solutions are warmed to 40°. Add 175 μl of the buffer/activator and 675 μl of diluent to each of the tubes, and the amidolytic reaction is started by the addition of 25 μl of the substrate stock solution. After 5 min the reaction is stopped by the addition of 1 ml of the stopping reagent. The degree of substrate hydrolysis is determined by measuring A_{410}. To find the operational molarity of active glycyl endopeptidase at the intersection of the slope with the abscissa, ΔA_{410} is plotted against E-64 concentration.

Assay with Polyclonal Antisera

Principle. Despite the structural similarities between the papaya endopeptidases (see below), they do not normally exhibit immunological cross-reactivity. Polyclonal antisera raised against one or the other of them can therefore be used to quantify the amount of a particular endopeptidase.

[24a] Knight, this series.

Such an approach is complementary to the use of enzyme–substrate assays and active site titration (above); these measure the amount of active enzyme whereas the antisera measure the amount of total enzyme.

Preparation of Antisera. Purified glycyl endopeptidase is used as the antigen. This can be dissolved at 1 mg/ml in phosphate-buffered saline (PBS) (see below), made 2 mM cysteine by addition of the free base. E-64 is added to 100 μM to inactivate glycyl endopeptidase. An equal volume of Freund's complete adjuvant is added and a stable emulsion prepared by aspiration through a syringe needle. Antisera can then be raised in rabbits[20] or sheep[4] by standard methods and immunoglobulins prepared by ammonium sulfate fractionation.[25] Cross-reacting antibodies to papain, chymopapain, and caricain can be removed by passing through columns of the immobilized enzymes.[4]

Single Radial Immunodiffusion. This procedure is carried out as described.[26] A convenient modification is to pour the antibody-containing agarose solution on to GelBond (FMC Corp., Rockland, ME).[20] This facilitates subsequent handling and storage of the gel. The limit of detection of the method is about 1 pmol (25 ng) of glycyl endopeptidase/well, and it takes about 2 days to complete. Enzyme samples should be inactivated, preferably by reaction with a molar excess of E-64 as described for preparation of antisera, before they are allowed to interact with antibody. This precludes digestion of the antibody by glycyl endopeptidase.

Dot–Blot Enzyme-Linked Immunosorbent Assay (ELISA). A rather more rapid method for the immunochemical detection of glycyl endopeptidase, with a limit of sensitivity of about 2 pmol (50 ng)/well, utilizes binding of the enzyme to nitrocellulose, amplification of the signal by use of a biotinylated second antibody, and detection via activity of an enzyme linked to avidin.[27]

Reagents

Microfiltration apparatus: this is the Bio-Dot apparatus, obtained from Bio-Rad (Richmond, CA).

Nitrocellulose sheets: the experiments described below were made using Trans-Blot no. 162-0114, from Bio-Rad, which is the correct size to fit in the microfiltration apparatus, but it is likely that nitrocellulose sheets from any source would be adequate.

PBS: 137 mM sodium chloride/8.1 mM disodium hydrogen phosphate/1.1 mM potassium dihydrogen phosphate, pH 7.4.

[25] K. Heide and H. G. Schwick, in *Handb. Exp. Immunol* **1**, 7.1 (1978).
[26] G. Mancini, A. O. Carbonara, and J. F. Heremans, *Immunochemistry* **2**, 235 (1965).
[27] P. M. Dando, S. L. Sharp, D. J. Buttle, and A. J. Barrett, unpublished.

Antibodies: Rabbit antibodies to glycyl endopeptidase (10 mg/ml in PBS) are stored at −20°.

Blocking buffer: PBS containing 10% (v/v) horse serum (Imperial Laboratories, Andover, Hampshire, UK).

Washing buffer: PBS containing 0.1% Tween 20 (Koch-Light Limited, Haverhill, Suffolk, UK).

Dilution buffer: PBS containing 0.1% Tween 20, 2% (v/v) horse serum, 10 mM EDTA, and 50 μg/ml heparin (Koch-Light). The inclusion of heparin inhibits nonspecific binding and adsorption of the basic papaya endopeptidases.[28]

Biotinylated antibodies to immunoglobulin G (IgG) and avidin–horseradish peroxidase: these are obtainable from many laboratories; we use the reagents from Vector Laboratories (Peterborough, Cambs, UK).

Substrate solution: PBS containing 60 μg/ml 4 chloro-1-naphthol (Aldrich Chemical Company, Milwaukee, WI) and 0.012% H_2O_2.

Procedure. All solutions are filtered (0.2 μm pore size) prior to use. A nitrocellulose sheet is placed in the microfiltration apparatus and washed with PBS. Serial twofold dilutions are made of the samples containing glycyl endopeptidase, and these are spotted (50 μl/well) onto the membrane and allowed to soak into it, before being drawn off using the vacuum manifold. Each of the following treatments is for 10 min at 20°, and the membrane is washed with washing buffer between each step. The remaining protein-binding sites on the membrane are blocked with blocking buffer. The antibody is diluted 1 : 500 in dilution buffer prior to use, and 200 μl/well is applied. The biotinylated antibody to rabbit IgG is diluted to 1 μg/ml in dilution buffer and applied (100 μl/well), followed by avidin–horseradish peroxidase (100 μl/well), at the concentration recommended by the suppliers. Then 600 μl of the substrate solution is applied, and color development is assessed by eye. The reaction is stopped by washing the membrane in distilled water and drying in air. The color intensity of stained dots is then quantified by reflectance densitometry, and compared with the values obtained from a standard curve generated by use of known quantities of glycyl endopeptidase.

Purification of Glycyl Endopeptidase

Two methods are described for the purification of glycyl endopeptidase, both of which rely heavily on affinity chromatography on Sepharose-Ahx-Gly-Phe-NHCH₂CN (Ahx-, 6-aminohexanoyl-). The ligand is not

---

[28] A. J. Pesce, R. Apple, N. Sawtell, and J. G. Michael, *J. Immunol. Methods* **87**, 21 (1986).

commercially available, but its synthesis has been described in detail.[20] It is a reversible inhibitor of papain and glycyl endopeptidase with only low affinity for chymopapain and caricain.[20] Method I incorporates cation-exchange chromatography to separate glycyl endopeptidase from papain. Affinity-purified papain is therefore obtained as a by-product if this method is used. The second method takes advantage of the very slow reactivity of glycyl endopeptidase with iodoacetate (see below).

Method I: Purification by Affinity Chromatography and
Cation-Exchange Chromatography

This is the original method described by Buttle *et al.*[20] A column (4-ml bed volume) of Sepharose-Ahx-Gly-Phe-NHCH$_2$CN is washed with 12 ml of 50 mM sodium citrate in water/ethanediol (2 : 1, v/v) (pH 4.5) (elution buffer), then with 12 ml of 50 mM sodium phosphate buffer containing 1 mM EDTA in water/ethanediol (2 : 1, v/v) (pH 6.8) (application buffer). Ethanediol is included in both buffers because it has been found to reduce the nonspecific binding of papaya latex proteins to the gel. Commercially available dried papaya latex is a convenient starting material. We have used the products of J. E. Siebel Sons Co. Inc. (Chicago, IL) and Powell and Scholefield Ltd. (Liverpool, UK), but it is likely that material from other suppliers, e.g., crude papain from Sigma Chemical Co., could be used. The latex (0.5 g) is dissolved in 10 ml of application buffer, clarified by filtration (0.2 μm pore size), and the approximate protein concentration is determined (an $A_{280,1\ cm}$ of 1 is roughly equivalent to 0.5 mg protein/ml). An 80-mg portion of the latex protein is taken and dithiothreitol added to 2 mM final concentration. The mixture is kept at 0° for 20 min, then applied to the column at 20°, at a flow rate of 35 ml/hr/cm². This is followed by a 8 ml of application buffer and then 8 ml of elution buffer (or more if the A_{280} has not returned to zero). One bed volume (4 ml) of elution buffer containing 50 mM hydroxyethyl disulfide is run on to the column, the flow is then stopped, and the column is left at 20° for about 18 hr. Glycyl endopeptidase and papain are then eluted with another 10 ml of elution buffer containing hydroxyethyl disulfide, after which the column can be reequilibrated in application buffer.

The mixture of glycyl endopeptidase and papain eluted in hydroxyethyl disulfide is applied to the Mono S HR 5/5 (cation-exchange) column of the fplc (fast protein liquid chromatography) apparatus (Pharmacia, Piscataway, NJ) preequilibrated in 50 mM sodium acetate/acetic acid containing 1 mM EDTA (pH 5.0). The column is washed with this buffer (1 ml/min) until A_{280} returns to zero. The bound proteins are then eluted with a gradient (21.5 mM Na$^+$/ml) to 1 M sodium acetate, and collected in 1 ml fractions. Papain (demonstrating activity against Bz-Arg-NHPhNO$_2$) is eluted at about 0.17 M Na$^+$, followed by glycyl endopeptidase (showing activity against Boc-Ala-Ala-Gly-NHPhNO$_2$ but not Bz-Arg-NHPhNO$_2$) at about 0.38 M Na$^+$. The yield is approximately 4 mg of glycyl endopeptidase and 2 mg of papain. Glycyl endopeptidase can be dialyzed into 1 mM EDTA, freeze-dried, and stored at $-20°$. Papain can be stored at 4° as a precipitate in 2 M (NH$_4$)$_2$SO$_4$, because the enzyme is not stable to freeze-drying.

Analysis of glycyl endopeptidase thus obtained, by single radial immunodiffusion using specific antisera to papain, chymopapain, and caricain,[4] has demonstrated levels of contamination of up to 3% by chymopapain, and 1% by each of papain and caricain. Both papain and glycyl endopeptidase are recovered with the active site thiol protected as the mercaptoethanol mixed disulfide, thus reducing autolysis and improving stability. They are very easily reactivated by exposure to millimolar concentrations of cysteine. Titration with E-64 has shown that both glycyl endopeptidase and papain obtained by this method are about 90% active.

Method II: Purification by Treatment with Iodoacetate Followed by Affinity Chromatography

In this method, papain, chymopapain, and caricain are selectively inactivated by reaction of the papaya latex with iodoacetate. Pure, active glycyl endopeptidase is then eluted from the affinity column as the mercury derivative in a one-step purification.

The column of Sepharose-Ahx-Gly-Phe-NHCH$_2$CN is pretreated exactly as described in Method I. About 200 mg of the latex protein is dissolved in 12 ml of application buffer, and clarified by centrifugation (8000 g, 10 min, 4°). The supernatant is made 4 mM with respect to cysteine by addition of the free base, and 500 μM iodoacetate, using the sodium salt. This leads to the loss of more than 99% of the Bz-Arg-NHPhNO$_2$-hydrolyzing activity, whereas the bulk of activity against Boc-Ala-Ala-Gly-NHPhNO$_2$ remains. After 30 min at 4° the latex solution is applied to the affinity column, followed by 12 ml of application buffer and 40 ml of elution buffer, by which time A_{280} should have returned to zero. Elution of purified glycyl endopeptidase is achieved in elution buffer containing 10 mM HgCl$_2$. An overnight incubation is not essential to dislodge the enzyme from the column, but is often convenient. The yield of glycyl endopeptidase is about 10 mg, with about 90% of the enzyme being active, as judged by active site titration. Levels of contamination by the other papaya endopeptidases are <0.01% for chymopapain and caricain, and 0.07% for papain, determined by single radial immunodiffusion. The mercurial form of glycyl endopeptidase is generally very stable and is rapidly activated by cysteine.

Properties of Glycyl Endopeptidase

Physicochemical Properties

The amino acid sequence of glycyl endopeptidase has been determined.[29] The enzyme exists as a single unglycosylated polypeptide of M_r 23,313. The $A_{280\ nm}^{1\%}$ of 16.5[20] therefore gives an extinction coefficient at 280 nm of $3.85 \times 10^4\ M^{-1}\ cm^{-1}$. Glycyl endopeptidase is clearly a member of the papain family of cysteine peptidases.[30] Alignment with the sequences of the other cysteine endopeptidases from papaya demonstrates that its closest homolog is caricain (81% identical residues), with chymopapain and papain also showing a large degree of identity (70 and 67%, respectively). The N-terminal 17 residues are identical to those published by Lynn and Yaguchi[31] for the enzyme they called papaya peptidase B. This information, together with the lack of cleavage of Bz-Arg-NHPhNO$_2$ and hydrolysis of a Gly-OPhNO$_2$ substrate,[18] leaves little doubt that papaya peptidase B and glycyl endopeptidase are the same enzyme. The C-terminal 62 residues are identical to those that are encoded by a partial cDNA sequence from a papaya leaf tissue library.[32]

Glycyl endopeptidase is an abundant protein in the latex of papaya. Quantification, based on single radial immunodiffusion[4] or dot–blot ELISA,[27] shows that the enzyme constitutes 23–28% of total protein.

Allergenic Properties

For some unknown reason enzymes of the papain family can be potent allergens. A major allergen of the house dust mite *Dermatophagoides pteronyssinus* is a papain homolog,[33] with about 30% sequence identity to glycyl endopeptidase. The first report of anaphylactic reactions to crude papain in animals appeared as early as 1912,[34] and a review of early reports

[29] A. Ritonja, D. J. Buttle, N. D. Rawlings, V. Turk, and A. J. Barrett, *FEBS Lett.* **258**, 109 (1989).
[30] N. D. Rawlings and A. J. Barrett, *Biochem. J.* **290**, 205 (1993).
[31] K. R. Lynn and M. Yaguchi, *Biochim. Biophys. Acta* **581**, 363 (1979).
[32] R. A. McKee, S. Adams, J. A. Matthews, C. J. Smith, and H. Smith, *Biochem. J.* **237**, 105 (1986).
[33] K. Y. Chua, G. A. Stewart, W. R. Thomas, R. J. Simpson, R. J. Dilworth, T. M. Plozza, and K. J. Turner, *J. Exp. Med.* **167**, 175 (1988).
[34] E. Seligmann, *Z. Immunitaetsforsch.* **14**, 419 (1912).

in humans was compiled by Osgood.[35] Glycyl endopeptidase[27] has been identified as perhaps the major allergen among the endopeptidases in papaya latex. Care should therefore be taken when dealing with crude papaya latex or preparations of glycyl endopeptidase. Perhaps the most likely route of sensitization is via the airways. Powders, in particular, should therefore be handled carefully, preferably in a fume cupboard.

Enzymatic Properties

Action on Small Substrates. Glycyl endopeptidase appears to hydrolyze only glycyl bonds efficiently. The kinetic constants for the cleavage of Boc-Ala-Ala-Gly-NHPhNO$_2$ and Boc-Ala-Ala-Gly-NHMec at pH 6.8 have been determined, and gave k_{cat}/K_m values of 4.23 and 31.2 sec^{-1} mM^{-1}, respectively. Both substrates are cleaved more efficiently by glycyl endopeptidase than by the other papaya cysteine endopeptidases, largely through differences in k_{cat}. The marked specificity for glycyl bonds was demonstrated by the substitution of Ala for Gly in the -NHMec substrate, the presence of the methyl side chain of Ala leading to a 60-fold reduction in k_{cat}.[36] Against Boc-Ala-Ala-Gly-NHPhNO$_2$ glycyl endopeptidase has a pH optimum of 7.0–7.5, but still retains about 75% of the optimal activity at pH 9.5, and 15% at pH 3.0 (Fig. 1).

Action on Protein Substrates. The marked preference for the hydrolysis of glycyl bonds also holds true with protein substrates. This makes the enzyme a useful tool in the analysis of protein primary structure. Table I is a compilation of known cleavage sites of proteins by glycyl endopeptidase. Of the 41 peptide bonds cleaved, 37 are glycyl bonds. Some specificity with regard to preferred amino acids in positions P3 and P2 is also apparent. In 24 cases (59%) an uncharged aliphatic residue is found in P2, most commonly Val or Ser. Uncharged aliphatic residues are also found 24 times in P3. Charged residues in the vicinity of the scissile bond do not appear to be favored, only negatively charged carboxymethyl-Cys and Glu being found, each on two instances, in P2, and positively charged Arg and Lys, both on three occasions, in P1'. Pro occurs only twice in each of P3 and P1', and once in P2. Carboxymethyl-Cys is not found in P1', and is found only twice in P2. Glycyl endopeptidase appears to cleave efficiently in the vicinity of disulfide bonds, cystine being accommodated well in P2. The enzyme is therefore a very useful reagent for the designation of disulfide bridges in proteins.[37] The four cleavages that did not take place at glycyl bonds occurred after prolonged

[35] H. Osgood, *J. Allergy* **16**, 245 (1945).

[36] D. J. Buttle, A. Ritonja, L. H. Pearl, V. Turk, and A. J. Barrett, *FEBS Lett.* **260**, 195 (1990).

[37] V. D. Bernard and R. J. Peanasky, *Arch. Biochem. Biophys.* **303**, 367 (1993).

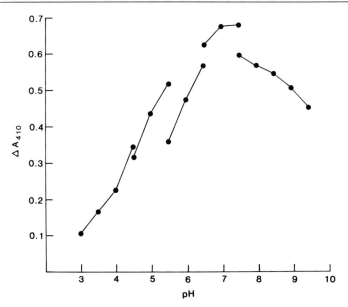

FIG. 1. The pH–activity profile of glycyl endopeptidase. The enzyme (100 nM final concentration) was allowed to activate in buffer (975 μl) containing 4 mM cysteine for 5 min at 40°. The reaction was started by the addition of 25 μl of 50 mM Boc-Ala-Ala-Gly-NHPhNO$_2$ and allowed to continue at 40° for 5 min. The reaction was then stopped by the addition of 1 ml of stopping reagent and A_{410} measured. The A_{410} of a control mixture without enzyme for each condition was subtracted. The values were not corrected for variation in A_{410} of 4-nitroaniline with pH, as this was found to vary by less than 7% over the range studied. Each point is the average of three determinations. Buffers were 50 mM sodium formate/formic acid (pH 3–4.5); 50 mM sodium acetate/acetic acid (pH 4.5–5.5); 50 mM Mes/NaOH (pH 5.5–6.5); 50 mM Na$_2$HPO$_4$/NaH$_2$PO$_4$ (pH 6.5–7.5); 50 mM Tris-HCl (pH 7.5–9.5).

reaction times with denatured substrate. They may demonstrate an ability of glycyl endopeptidase to cleave bonds other than glycyl with low efficiency, or they might have been made by the action of low levels of contaminating endopeptidases.

Inhibitors. Certain aspects of the interaction of glycyl endopeptidase with inhibitors have been studied and documented.[38] As a member of the papain family, glycyl endopeptidase would be expected to be inhibited by transition-state analogs such as peptidyl aldehydes, and by alkylating agents such as haloacetates, peptidyl diazomethanes, and epoxy-pep-

---

[38] D. J. Buttle, A. Ritonja, P. M. Dando, M. Abrahamson, E. N. Shaw, P. Wikstrom, V. Turk, and A. J. Barrett, *FEBS Lett.* **262**, 58 (1990).

TABLE I
KNOWN CLEAVAGE SITES IN PROTEINS FOR GLYCYL ENDOPEPTIDASE[a]

Cleavage number	P4	P3	P2	P1 P1'	P2'	P3'	P4'	Ref.[b]
1	A	S	A	G‡T	Q	C	L	1
2	F	C	A	G‡Y	L	E	G	1
3	E	A	A	G‡A	M	F	L	4
4		I	V	G‡G	Y	T	C	1
5	A	K	V	G‡V	V	I	Q	5
6	R	L	V	G‡G	P	M	D	3
7	S	A	V	G‡A	L	E	A	6
8	V	S	V	G‡I	D	A	S	6
9	G	A	L	G‡G	D	V	Y	5
10	R	L	L	G‡A	P	V	P	3
11	V	R	L	G‡Q	D	N	I	1
12	P	N	L	G‡G	F	S	D	7
13	I	S	I	G‡R	G	A	L	5
14	C*	S	P	G‡R	G			2
15	S	N	F	G‡Y	N	L	L	5
16	A	E	F	G‡K	Y	I	C	8
17	F	Q	F	G‡N	M	I	N	9
18	S	G	W	G‡N	T	Y	S	1
19	V	S	W	G‡S	G	C	A	1
20	C	G	W	G‡G	Q	G	K	9
21	F	C	G	G‡S	L	I	N	1
22	G	P	S	G‡T	P	V	R	5
23	G	D	S	G‡G	P	V	V	1
24	K	S	S	G‡T	S	Y	P	1
25	L	N	S	G‡Y	H	F	C	1
26	P	L	S	G‡G	I	F	E	5
27	V	C	S	G‡K	L	Q	G	1
28	E	C*	T	G‡C*	E			2
29	M	M	C	G‡G	T	S	A	10
30	I	V	C	G‡N	K	Y	G	9
31	E	S	C*	G‡P	N	E	V	2
32	K	P	C*	G‡K	N	E	V	2
33	L	K	C*	G‡Q	D	E	N	2
34		K	C*	G‡P	D	E	N	2
35	C	C	Y	G‡R	A	N	G	9
36	V	V	E	G‡N	Q	Q	F	1
37	Y	L	E	G‡G	K	D	S	1
38	P	I	L	S‡N	S	S	C	1
39	Y	S	S	S‡N	C	Q	E	9
40	P	V	V	C‡S	G	K	L	1
41	Q	D	N	I‡N	V	V	E	1

[a] The nomenclature relating to the amino acids around the scissile bond follows Ref. 10. The single-letter code for amino acids is used. The experimental conditions during hydrolysis varied. For instance, in some cases reduced and alkylated

tides.[39] Among the protein inhibitors, α_2-macroglobulin[40] and cystatins would be expected to inhibit.[41] On the whole, the spectrum of inhibition is consistent with glycyl endopeptidase as a papain homolog, with the important exception of lack of inhibition by cystatins.

The kinetic constants for the inhibition of glycyl endopeptidase by some small-molecule inhibitors are compared with those for the inhibition of papain in Table II. The low rates of inactivation by iodoacetate and iodoacetamide are an unusual, but not unique, property of glycyl endopeptidase. Other cysteine endopeptidases demonstrating low reactivity of the active site thiol include those from beans,[42,43] stem bromelain,[44] and clostripain.[45] The faster rate of inactivation with iodoacetate, as compared to iodoacetamide, appears to be a consistent feature of enzymes related to papain, however.

[39] D. H. Rich, in "Proteinase Inhibitors" (A. J. Barrett and G. Salvesen, eds.), p. 153. Elsevier, Amsterdam, 1986.

[40] A. J. Barrett, this series, Vol. 80, p. 737.

[41] A. J. Barrett, N. D. Rawlings, M. E. Davies, W. Machleidt, G. Salvesen, and V. Turk, in "Proteinase Inhibitors" (A. J. Barrett and G. Salvesen, eds.), p. 515. Elsevier, Amsterdam, 1986.

[42] C. Csoma and L. Polgár, Biochem. J. 222, 769 (1984).

[43] A. A. Kembhavi, D. J. Buttle, C. G. Knight, and A. J. Barrett, Arch. Biochem. Biophys. 303, 208 (1993).

[44] A. Ritonja, A. D. Rowan, D. J. Buttle, N. D. Rawlings, V. Turk, and A. J. Barrett, FEBS Lett. 247, 419 (1989).

[45] A. A. Kembhavi, D. J. Buttle, P. Rauber, and A. J. Barrett, FEBS Lett. 283, 277 (1991).

protein was used as substrate, whereas in others the protein was in its native conformation; C* denotes cystine.

[b] Key to references: (1) D. J. Buttle, A. Ritonja, L. H. Pearl, V. Turk, and A. J. Barrett, FEBS Lett. 260, 195 (1990); (2) V. D. Bernard and R. J. Peanasky, Arch. Biochem. Biophys. 303, 367 (1993); (3) D. J. Buttle, A. Ritonja, P. M. Dando, M. Abrahamson, E. N. Shaw, P. Wikstrom, V. Turk, and A. J. Barrett, FEBS Lett. 262, 58 (1990); (4) A. E. Mast, J. J. Enghild, and G. Salvesen, Biochemistry 31, 2720 (1992); (5) A. Ritonja, I. Krizaj, P. Mesko, M. Kopitar, P. Lucovnik, B. Strukelj, J. Pungercar, D. J. Buttle, A. J. Barrett, and V. Turk, FEBS Lett. 267, 13 (1990); (6) A. Ritonja, A. Colic, I. Dolenc, T. Ogrinc, M. Podobnik, and V. Turk, FEBS Lett. 283, 329 (1991); (7) B. Turk, I. Krizaj, B. Kralj, I. Dolenc, T. Popovic, J. G. Bieth, and V. Turk, J. Biol. Chem. 268, 7323 (1993); (8) B. Lenarcic, A. Ritonja, B. Turk, I. Dolenc, and V. Turk, Biol. Chem. Hoppe-Seyler 373, 459 (1992); (9) I. Krizaj, J. Siigur, M. Samel, V. Cotic, and F. Gubensek, Biochim. Biophys. Acta 1157, 81 (1993); (10) I. Krizaj, B. Turk, and V. Turk, FEBS Lett. 298, 237 (1992).

TABLE II
KINETIC CONSTANTS FOR INHIBITION OF
GLYCYL ENDOPEPTIDASE AND PAPAIN BY
SMALL-MOLECULE INHIBITORS[a]

	Kinetic constants	
Inhibitor	Glycyl endopeptidase	Papain
	K_i (nM)	
Leupeptin	20	1.6
Chymostatin	53	0.73
	k_2 (M^{-1} sec^{-1})	
Iodoacetate	2.2	1000
Iodoacetamide	0.5	46
E-64	58,000	1,160,000
Z-Phe-Phe-CHN$_2$	<10	5500
Z-Phe-Ala-CHN$_2$	60	43,000
Z-Gln-Gly-CHN$_2$	100	50
Boc-Val-Gly-CHN$_2$	5000	12,000
Z-Leu-Gly-CHN$_2$	5500	1600
Z-Met-Gly-CHN$_2$	5500	1700
Z-Phe-Gly-CHN$_2$	14,500	12,000
Z-Leu-Gly-Gly-CHN$_2$	<50	<50
Z-Leu-Phe-Gly-CHN$_2$	630,000	62,000
Z-Leu-Val-Gly-CHN$_2$	3,500,000	460,000

[a] All measurements were made at 30°, in 100 mM sodium phosphate buffer (pH 6.8) containing 1 mM EDTA, 0.005% Brij 35, and 2 or 4 mM cysteine. Continuous rate assays were made with Boc-Ala-Ala-Gly-NHMec (50 μM) as substrate for glycyl endopeptidase and Z-Phe-Arg-NHMec (10 μM) for papain. Reproduced from Ref. 38, with permission.

The rates of inactivation with a series of peptidyl diazomethanes are generally consistent with the substrate specificity of glycyl endopeptidase, the most rapid inactivation being achieved by a diazomethane with uncharged aliphatic residues in P3 and P2, as well as Gly in P1. Inactivation is very slow when either Phe or Ala are in P1.

Like most members of the papain family, glycyl endopeptidase is rapidly inactivated by the epoxy peptide E-64. This allows the concentration of active glycyl endopeptidase to be determined by stoichiometric titration (above).

Glycyl endopeptidase is bound by α_2-macroglobulin, although the reaction is much slower than the binding of papain. Binding to α_2-macroglobulin leads to steric inhibition against protein substrates, but not against small-molecule substrates.[40]

The lack of inhibition of glycyl endopeptidase by cystatins is of some interest. The enzyme cleaves chicken cystatin and human cystatin C (cleavages 6 and 10 in Table I), and in so doing inactivates them. It is unlikely that the lack of inhibition is due to cleavage, however. Glycyl endopeptidase does not cleave cystatin A, and is not inhibited by it.

Structure–Activity Studies. The three-dimensional structure of glycyl endopeptidase has been modeled on that of papain.[36] The model predicts that the S1 substrate-binding site of glycyl endopeptidase is altered by the substitution of Gly-23 and Gly-65 (papain numbering) with Glu and Arg, respectively. Their side chains and associated hydrogen bonds form a barrier to entry into the binding pocket, thus excluding substrate residues with long side chains. In addition to offering an explanation for the restricted substrate specificity of glycyl endopeptidase, the model also provides a reason for the lack of inhibition by cystatins. The X-ray crystal structure of the papain–cystatin B complex has been solved,[46] and shows that the inhibitor forms a wedge that fits into the active site cleft of papain, apparently without a great deal of distortion of either molecule. Any restriction of the active site cleft of the enzyme could therefore preclude binding to the inhibitor.

[46] M. T. Stubbs, B. Laber, W. Bode, R. Huber, R. Jerala, B. Lenarcic, and V. Turk, *EMBO J.* **9,** 1939 (1990).

[38] Pineapple Cysteine Endopeptidases

By Andrew D. Rowan and David J. Buttle

Nomenclature

The enzymes of the pineapple plant (*Ananas comosus*) have been known as the bromelains. The name was originally applied to "any protease from any plant member of the family Bromeliaceae."[1] The proteolytic fraction of the pineapple stem has been called stem bromelain, whereas that present in the fruit has been known as fruit bromelain (first described as "bromelin"[2]). At one time these two enzyme fractions were assigned separate systematic numbers, although more recently they were both

[1] R. M. Heinicke, *Science* **118,** 753 (1953).
[2] R. H. Chittenden, *J. Physiol.* (*London*) **15,** 249 (1894).

grouped together. It is evident that there has been considerable confusion as to whether these two enzymes are distinct proteins[3–5] or represent two forms of a single enzyme,[6,7] and the confusion has been exacerbated by the use of the same name, bromelain, for the commercially available dried powder prepared from waste pineapple stem material.[8] Many contradictory reports have described up to six different components in the stem,[4,5,9] and at least two components in the fruit,[3–5,10,11] possibly depending on the geographical location.[10,12,13] The bromelains have been reviewed previously in this series[14,15] and the confusion surrounding their identity discussed.

More recently the pineapple plant has been shown to contain at least four distinct cysteine endopeptidases.[16,17] The major endopeptidase present in extracts of plant stem is stem bromelain, and fruit bromelain is the major endopeptidase in the fruit. Two additional cysteine endopeptidases, ananain[16] and comosain,[17] have been detected only in the stem. The amino acid sequence of stem bromelain has been determined[18] and has shown this enzyme to be a member of the papain family.[19] The clarification of the identity of the pineapple enzymes has led to their reclassification.[20] Stem bromelain is now EC 3.4.22.32, fruit bromelain is EC 3.4.22.33 (also under this number is pinguinain, the major cysteine endopeptidase of the related bromeliad, *Bromelia pinguin*,[17,21–23] and ananain is EC 3.4.22.31.

[3] S. Ota, K. Horie, F. Hagino, C. Hashimoto, and H. Date, *J. Biochem.* (*Tokyo*) **71**, 817 (1972).
[4] S. Ota, E. Muta, Y. Katahira, and Y. Okamoto, *J. Biochem.* (*Tokyo*) **98**, 219 (1985).
[5] F. Yamada, N. Takahashi, and T. Murachi, *J. Biochem.* (*Tokyo*) **79**, 1223 (1976).
[6] S. Iida, M. Sasaki, and S. Ota, *J. Biochem.* (*Tokyo*) **73**, 377 (1973).
[7] M. Sasaki, T. Kato, and S. Iida, *J. Biochem.* (*Tokyo*) **74**, 635 (1973).
[8] J. C. Caygill, *Enzyme Microb. Technol.* **1**, 233 (1979).
[9] R. M. Heinicke and W. A. Gortner, *Econ. Bot.* **11**, 225 (1957).
[10] S. Ota, T. H. Fu, and R. Hirohata, *J. Biochem.* (*Tokyo*) **49**, 532 (1961).
[11] S. Ota, *J. Biochem.* (*Tokyo*) **59**, 463 (1966).
[12] S. Ota, K. Horie, and F. Hagino, *J. Biochem.* (*Tokyo*) **66**, 413 (1969).
[13] S. Ota, S. Moore, and W. H. Stein, *Biochemistry* **3**, 180 (1964).
[14] T. Murachi, this series, Vol. 19, p. 273.
[15] T. Murachi, this series, Vol. 45, p. 475.
[16] A. D. Rowan, D. J. Buttle, and A. J. Barrett, *Arch. Biochem. Biophys.* **267**, 262 (1988).
[17] A. D. Rowan, D. J. Buttle, and A. J. Barrett, *Biochem. J.* **266**, 869 (1990).
[18] A. Ritonja, A. D. Rowan, D. J. Buttle, N. D. Rawlings, V. Turk, and A. J. Barrett, *FEBS Lett.* **247**, 419 (1989).
[19] N. D. Rawlings and A. J. Barrett, *Biochem. J.* **290**, 205 (1993).
[20] Nomenclature Committee of the IUBMB, "Enzyme Nomenclature," p. 371. Academic Press, London, 1992.
[21] R. A. Messing, A. F. Santoro, and A. Bloch, *Enzymologia* **22**, 110 (1960).
[22] E. Toro-Goyco, A. Maretzki, and M. L. Matos, *Arch. Biochem. Biophys.* **126**, 91 (1968).
[23] E. Toro-Goyco and I. Rodriguez-Costas, *Arch. Biochem. Biophys.* **175**, 359 (1976).

Assay Procedures

Assay of Stem Bromelain

In the past, both stem and fruit bromelain have been assayed with protein substrates such as casein[24] and hemoglobin.[25] Typical synthetic substrates have included Bz-Arg-OEt[a] and Bz-Arg-NH$_2$,[26] although, due to the insensitivity of these methods, enzyme concentrations greater than 1 mg/ml were required.[14] Substrates containing longer peptide sequences and more sensitive leaving groups have been employed.[4,27] A sensitive, safe, and convenient leaving group for substrates of cysteine endopeptidases is 7-amino-4-methylcoumarin (-NHMec). We have found Z-Arg-Arg-NHMec to be an excellent and sensitive substrate for stem bromelain, the value of which is increased still further by its relative resistance to both ananain and fruit bromelain.[16,17] The sensitivity of the -NHMec substrates often necessitates considerable dilution of enzyme samples. It is therefore sometimes more convenient to use the less sensitive chromogenic nitroanilide (-NHPhNO$_2$) substrate. Procedures for both types of assay are described below. The -NHMec and -NHPhNO$_2$ substrates are available from Bachem Feinchemikalien AG (Bubendorf, Switzerland).

Assay with Z-Arg-Arg-NHMec

PRINCIPLE. The substrate, Z-Arg-Arg-NHMec, which is scarcely fluorescent, is hydrolyzed to liberate 7-amino-4-methylcoumarin, and this is quantified fluorimetrically once the enzyme has been inactivated with chloroacetate.

REAGENTS

Buffer/activator: 200 mM Na$_2$HPO$_4$/200 mM NaH$_2$PO$_4$/4 mM disodium EDTA, pH 6.8. Dithiothreitol (8 mM) or cysteine base (16 mM) is added on the day of use.

Substrate stock solution: a 2 mM solution of Z-Arg-Arg-NHMec·2HCl in DMSO (dimethyl sulfoxide) is made. The exact concentration is determined by total hydrolysis of a sample by crude pineapple stem extract and quantitation of the released 7-amino-4-methylcoumarin (see below), after which the substrate stock solution can be stored at 4° in the dark. A working-strength substrate solution (20 μM) is prepared, as required, by dilution of the stock with water.

[24] A. K. Balls, R. R. Thompson, and M. W. Kies, *Ind. Eng. Chem.* **33**, 950 (1941).
[25] T. Murachi and H. Neurath, *J. Biol. Chem.* **235**, 99 (1960).
[26] T. Inagami and T. Murachi, *Biochemistry* **2**, 1439 (1963).
[27] C. J. Gray, J. Boukouvalas, R. J. Szawelski, and C. W. Wharton, *Biochem. J.* **219**, 325 (1984).

Stopping reagent: 100 mM sodium monochloroacetate/30 mM sodium acetate/70 mM acetic acid, pH 4.3.

Diluent: Brij 35, 0.01% (w/v) in water.

Aminomethylcoumarin standard: 7-amino-4-methylcoumarin (1 mM) in DMSO, stored at 4° in the dark, and diluted to 0.5 μM in a 1 : 1 mixture of assay buffer and stopping reagent on the day of use.

PROCEDURE. The enzyme sample, containing 0.5–2.0 ng (20–80 fmol) of stem bromelain, is diluted to 500 μl with prewarmed diluent, and 250 μl of assay buffer is added. About 2 min is allowed for activation of the enzyme and temperature equilibration in a bath at 40°; prolonged activation times, particularly in the presence of dithiothreitol (DTT), can lead to loss of activity. The reaction is started by the addition of 250 μl of the 20 μM substrate solution. After exactly 10 min, 1 ml of stopping reagent is added.

The fluorescence of the free aminomethylcoumarin is determined by excitation at 360 nm and emission at 460 nm. The fluorimeter is zeroed against a reaction blank prepared by omitting enzyme from a reaction tube, and then set to read 1000 arbitrary fluorescence units with the 0.5 μM aminomethylcoumarin standard (equivalent to 10% of total substrate hydrolysis). When the fluorimeter is provided with a temperature-controlled cell and connected to a recorder, continuous rate assays can easily be made. The fluorescence of aminomethylcoumarin is scarcely affected by pH in the range 4–7, so pH-dependent measurements pose little problem.

Assay with Z-Arg-Arg-NHPhNO$_2$

PRINCIPLE. On substrate hydrolysis, 4-nitroaniline is liberated and measured colorimetrically ($\varepsilon_{410\ nm} = 8800\ M - 1\ cm^{-1}$)[28] following inactivation of the enzyme by alkylation with chloroacetate.

REAGENTS. The buffer/activator, diluent, and stopping reagent are the same as those described above for the assay with Z-Arg-Arg-NHMec.

Substrate stock solution: a 50 mM solution of Z-Arg-Arg-NHPhNO$_2$ in DMSO, which is stored at 4°.

PROCEDURE. The enzyme sample, containing between 0.2 and 2.0 μg (8 and 80 pmol) of stem bromelain is placed in a tube with 250 μl of buffer/activator and prewarmed diluent to a final volume of 975 μl. This mixture is incubated at 40° for 2 min to allow full enzyme activation, and the reaction is then started by the addition of 25 μl of the substrate stock solution (final concentration 1.25 mM). After exactly 10 min, 1 ml of stopping reagent is added and A_{410} determined. Blanks are prepared by

---

[28] B. F. Erlanger, N. Kokowsky, and W. Cohen, *Arch. Biochem. Biophys.* **95,** 271 (1961).

omission of the enzyme or adding the enzyme following addition of the stopping reagent.

Assay of Fruit Bromelain

Fruit bromelain has a markedly different substrate specificity compared to that of stem bromelain,[17] and is best assayed as described below.

Assay with Bz-Phe-Val-Arg-NHMec

PRINCIPLE. The principle of the assay is precisely analogous to that described for stem bromelain with Z-Arg-Arg-NHMec.

REAGENTS. The buffer/activator, diluent, stopping reagent, and standard are as described for the assay with Z-Arg-Arg-NHMec.

Substrate stock solution: A 1 mM solution of Bz-Phe-Val-Arg-NH-Mec · HCl in DMSO, which is diluted with water to a working strength of 20 μM when required.

PROCEDURE. This assay is performed as described for stem bromelain with Z-Arg-Arg-NHMec. Typically, 0.2–1.0 ng (8–40 fmol) of fruit bromelain is used in the assay.

Assay with Bz-Phe-Val-Arg-NHPhNO$_2$. The spectrophotometric assay can be very convenient, although a little less sensitive than the fluorimetric equivalent (above).

REAGENTS. The buffer/activator, diluent, and stopping reagent are as described for the assay of stem bromelain with Z-Arg-Arg-NHMec.

Substrate stock solution: a 10 mM solution in DMSO.

PROCEDURE. This is as described for stem bromelain with Z-Arg-Arg-NHPhNO$_2$.

Assay with Azocoll

PRINCIPLE. Crude denatured collagen (gelatin) in the form of hide powder is a good substrate for many endopeptidases. The products of hydrolysis are detected spectrophotometrically after separation from the substrate by centrifugation, by A_{520} of the azo dye coupled to the protein.

REAGENTS. The buffer/activator and diluent are as above.

Substrate stock solutions: a 3% (w/v) suspension of azocoll, obtainable from Sigma (St. Louis, MO), in 0.6 M sucrose, made fresh on the day of use. It has been recommended that azocoll be washed by preincubation in assay buffer for 90 min prior to use; this improves the linearity of the assay.[29]

Stopping reagent: a 20% (w/v) aqueous solution of trichloroacetic acid.

[29] R. Chavira, T. J. Burnett, and J. H. Hageman, *Anal. Biochem.* **136,** 446 (1984).

PROCEDURE. This assay is routinely performed in 1.5-ml minicentrifuge tubes. The enzyme sample containing 0.5–5.0 μg (20–200 pmol) of fruit bromelain is allowed to activate for 2 min in 125 μl of the activation buffer. Prewarmed diluent is then added to a final volume of 400 μl and the reaction started by the addition of 100 μl of the substrate suspension. After a 20-min incubation in a 40° shaking water bath or end-over-end mixer (agitation improves linearity of the assay), the reaction is stopped by the addition of 500 μl of the stopping solution. A precipitate is allowed to form for 5 min at 4°, and then removed by centrifugation (10,000 g, 2 min, 4°). The A_{520} of the supernatant is determined.

Assay of Ananain

Ananain is the second most abundant endopeptidase of pineapple stem extract. It was first detected by virtue of its activity against Z-Phe-Arg-NHMec, which is scarcely hydrolyzed by stem bromelain.[16]

Assay with Z-Phe-Arg-NHMec

PRINCIPLE. The principle of the assay is precisely analogous to that described for stem bromelain with Z-Arg-Arg-NHMec.

REAGENTS. The buffer/activator, diluent, stopping reagent, and coumarin standard are as described for the assay with Z-Arg-Arg-NHMec.

Substrate stock solution: a 1 mM solution of Z-Phe-Arg-NHMec · HCl in DMSO, which is diluted with water to a working strength of 20 μM when required.

PROCEDURE. This assay is performed exactly as described for stem bromelain with Z-Arg-Arg-NHMec. Typically, 1–5 ng (40–200 fmol) of ananain is used in the assay.

Assay with Z-Phe-Arg-NHPhNO$_2$. Again, the use of the colorimetric substrate can be very convenient.

REAGENTS. The buffer/activator, diluent, and stopping reagent are as described for the assay with Z-Arg-Arg-NHMec.

Substrate stock solution: a 10 mM solution in DMSO.

PROCEDURE. This is as described for stem bromelain with Z-Arg-Arg-NHPhNO$_2$.

Assay with Azocasein

PRINCIPLE. The His and Tyr side chains of casein are derivatized with diazotized sulfanilic acid in alkali. This azo coupling confers an intense yellow color on the protein ($A_{366\ nm}^{1\%}$ of about 40). Proteolytic degradation yields trichloroacetic acid-soluble peptides that are quantified by A_{366}. This method is very resistant to interference and is therefore particularly suitable for the detection of proteolytic activity in crude samples.

PREPARATION OF AZOCASEIN. The method we recommend for the preparation and standardization of azocasein[30] leads to a product with a higher degree of derivatization than was achieved with previously published methods. The use of sulfanilic acid rather than sulfanilamide[31,32] confers multiple negative charges on the substrate and thus increases its solubility in neutral and basic media.

REAGENTS. The buffer/activator, diluent, and stopping solution are as described for the assay with azocoll.

Substrate stock solution: a 3% (w/v) solution of azocasein as prepared above.

PROCEDURE. The enzyme sample containing 0.25–2.5 µg (10–100 pmol) of ananain is added to 125 µl of activation buffer and incubated at 40° for 2 min. Prewarmed diluent is added to 400 µl and the reaction is started by the addition of 100 µl of the substrate solution. The reaction is stopped after 20 min (though longer times may increase sensitivity) at 40°, with 500 µl of the stopping solution. After 5 min at 4° insoluble material is removed by centrifugation (10,000 g for 2 min) and A_{366} of the supernatant is measured.

Assay with Hide Powder Azure

PRINCIPLE. As with azocoll (above), this is a preparation of hide powder that has been derivatized with a dye. The susceptibility to hydrolysis by endopeptidases is therefore similar to that of azocoll, but the intensely blue Remazol Brilliant Blue R and insolubility of the substrate may add to its convenience.

REAGENTS. The buffer/activator, diluent, and stopping reagent are as described for the assay with Z-Arg-Arg-NHMec.

Substrate stock solution: Freshly prepared 3% (w/v) suspension of hide powder azure, obtainable from Sigma Chemical Co., in 0.6 M sucrose.

PROCEDURE. The enzyme sample containing between 0.5 and 5.0 µg (20 and 200 pmol) of ananain is added to 250 µl of activation buffer and left for 2 min at 40°. Prewarmed diluent is then added to 700 µl, and the reaction is started by the addition of 300 µl of the substrate suspension. Tubes are incubated with agitation of 40° for 20 min, although sensitivity might be increased by longer incubation times, after which 1 ml of stopping reagent is added. Following centrifugation (1200 g, 5 min), A_{595} of the supernatant is determined.

[30] A. J. Barrett and H. Kirschke, this series, Vol. 80, p. 535.
[31] J. Charney and R. M. Tomarelli, *J. Biol. Chem.* **171,** 501 (1947).
[32] J. Langner, A. Wakil, M. Zimmerman, S. Ansorge, P. Bohley, H. Kirschke, and B. Wiederanders, *Acta Biol. Med. Ger.* **31,** 1 (1973).

Purification Procedures

We have found the following purification procedures for the various pineapple endopeptidases to yield highly pure and active enzyme in sufficient quantities for most kinetic analyses.

Purification of Endopeptidases Present in Pineapple Stem Extract

We describe the purification of stem bromelain, ananain, and comosain (partial purification) from crude pineapple stem extract. All procedures, unless stated otherwise, are performed at 4°.

Starting Material. The protein components of pineapple stem juice, which is usually obtained with a press, can be concentrated by a variety of methods,[8,33] although acetone precipitation is the most commonly used method. The preparation is generally referred to as "bromelain" and can be obtained as a dried powder from several sources. We have had good results with the product containing 50% protein from Sigma Chemical Co. The dried powder is dissolved in sodium acetate/acetic acid (0.05 M, pH 5.0) containing 1 mM disodium EDTA, and insoluble material is removed by passing through a 0.22-μm filter or by centrifugation.

Cation-Exchange Chromatography at pH 5.0. Filtered pineapple stem extract (approximately 10 mg protein) is applied to a Mono S HR 5/5 cation-exchange column of the Pharmacia (Piscataway, NJ) FPLC (fast protein liquid chromatography) system. The column is first equilibrated with stock buffer (below) diluted to 0.05 M Na$^+$. Proteins are eluted with a linear gradient (0.0215 M Na$^+$/ml) from 0.05 to 1.00 M Na$^+$ using a 1.00 M sodium acetate/acetic acid, pH 5.0, stock buffer containing 1 mM disodium EDTA, and 0.01% NaN$_3$ (pump B), and 1 mM disodium EDTA, 0.01% NaN$_3$ (pump A), to create the gradient at a flow rate of 1 ml/min. Fractions are collected into tubes containing sufficient 100 mM hydroxyethyl disulfide to give a final concentration of 2 mM. The presence of disulfide ensures reversible inactivation of the enzyme, precluding autocatalytic degradation and irreversible oxidation. Stem bromelain elutes predominantly as two protein peaks containing immunologically identical enzyme, between 0.26 and 0.39 M Na$^+$.[16] The molecular basis for the charge heterogeneity of stem bromelain is not known. Storage of the early-eluting peak after dialysis into 1 mM disodium EDTA and freeze-drying, and subsequent rechromatography under the same conditions, results in a chromatographic shift to the same position as the second peak, suggesting either the loss of negative charges or an increase in positive charges with

[33] J. C. Caygill, D. J. Moore, and L. Kanagasabapathy, *Enzyme Microb. Technol.* **5**, 365 (1983).

time. Chromatographically purified, freeze-dried stem bromelain from Sigma Chemical Co. is also the late-eluting form.[34] Ananain and comosain elute essentially together between 0.6 and 0.7 M Na$^+$.

AFFINITY PURIFICATION OF STEM BROMELAIN. The fractions eluting between 0.30–0.32 M Na$^+$ and 0.35–0.39 M Na$^+$ containing activity against Z-Arg-Arg-NHMec are combined and dialyzed against 10 volumes of 50 mM sodium phosphate, 1 mM disodium EDTA in 33% (v/v) ethanediol, pH 6.8 (application buffer). Ethanediol has been found to be effective at preventing nonspecific interactions between plant cysteine endopeptidases and chromatographic media. An aliquot containing approximately 10 mg protein is taken and made 2 mM with respect to dithiothreitol and left at 4° for 10 min to allow enzyme activation. This sample is then applied at a flow rate of approximately 40 ml/hr/cm^2 to a column (5.0 × 0.5 cm) of Sepharose-Ahx-Gly-Phe-GlySc[35] (Ahx, 6-aminohexanoyl) (see [45] in this volume) preequilibrated in the application buffer at room temperature. The column is washed with 10 ml of application buffer followed by 1 ml of 50 mM sodium acetate/acetic acid in 33% (v/v) ethanediol, pH 4.5. Inactive enzyme passes unadsorbed through the column. Active enzyme is eluted as a sharp peak by application to the column of 10 mM HgCl$_2$ in the acetate buffer. This material can be further purified by reapplication to the Mono S column at pH 5.0. The major protein peak, active against Z-Arg-Arg-NHMec, is then dialyzed into 1 mM disodium EDTA and freeze-dried. Stem bromelain purified in this way is typically >80% active by titration with E-64[16] (L-3-carboxy-2,3-*trans*-epoxypropionylleucylamido(4-guanidino)butane), with a yield of about 2.5 mg/100 mg of the dried starting material.

PURIFICATION OF ANANAIN. The fractions eluting between 0.6 and 0.7 M Na$^+$ from the Mono S column (above), active against Z-Phe-Arg-NHMec, are combined, dialyzed, activated, and affinity-purified as described for stem bromelain. The affinity-purified material is diluted with 2 volumes of 25 mM sodium tetraborate buffer, pH 9.0, containing 1 mM disodium EDTA and applied to a Mono S column equilibrated with 25 mM sodium tetraborate buffer, 100 mM NaCl, pH 9.0, containing 1 mM disodium EDTA. Protein is eluted with a linear gradient (0.01 M Na$^+$/ml) to 0.3 M Na$^+$ in the same buffer, at a flow rate of 0.5 ml/min. Fractions (0.2 ml) containing the major peak of activity against Z-Phe-Arg-NHMec are combined, diluted with the 25 mM sodium tetraborate buffer (above) and reapplied to the Mono S column and eluted in the same manner. The fractions active against Z-Phe-Arg-NHMec are pooled, dialyzed, and

[34] A. D. Rowan, Ph.D. Thesis. The British Library, London (1989).
[35] D. H. Rich, M. A. Brown, and A. J. Barrett, *Biochem. J.* **235,** 731 (1986).

freeze-dried as described for stem bromelain. This purification procedure results in a 19-fold increase in specific activity with a 20% yield of enzyme >70% active by titration with E-64.[16] We usually obtain about 0.5 mg of ananain/100 mg of starting material.

PURIFICATION OF COMOSAIN. The fractions eluting between 0.6 and 0.7 M Na$^+$ from the Mono S column at pH 5.0 (above) are also the source of comosain, active against Z-Arg-Arg-NHMec.[17] This enzyme is purified essentially as described for ananain, except that the ligand used is -Phe-GlySc^{35} (GlySc, glycinaldehyde semicarbazone) instead of Gly-Phe-GlySc (see [45]). Enzyme activity is monitored with Z-Arg-Arg-NHMec exactly as described for stem bromelain. The affinity-purified material is subjected to three cycles of FPLC on the Mono S column at pH 9.0 as described for ananain, except that fractions containing activity against Z-Arg-Arg-NHMec (preceding the major peak of ananain) are combined, diluted, and rerun. Following the third cycle of FPLC, the fraction with the highest specific activity against Z-Arg-Arg-NHMec is dialyzed against 1 mM disodium EDTA containing 0.05% Brij 35 and kept at 4°. This procedure produces the purest enzyme, but trace contamination by ananain is still often detected.[17]

Purification of Fruit Bromelain

Fresh pineapple fruit contains fruit bromelain[3–5,10,11,13] and also stem bromelain in much smaller amounts.[17] Here we shall describe the purification of fruit bromelain from fresh pineapple fruit purchased from a local supermarket (variety unknown; country of origin, Costa Rica; varieties may differ in their enzyme content[10,12,13]).

Acetone Fractionation. The leaves and outer husk are removed from the fruit and the main fruit portion (850 g) is diced into small cubes and minced in a blender. This is then mixed with 500 ml of 1 mM disodium EDTA and filtered, under vacuum, through Whatman (Clifton, NJ) type 540 paper. The resulting extract (930 ml) is cooled to 0° by stirring in a beaker surrounded by an ice–NaCl freezing mixture. Acetone (1 volume) cooled to −5° in the same manner is added, with stirring over a 10-min period, and then left stirring for a further 10 min. The precipitate, which contains stem bromelain,[17] is removed by centrifugation (1500 g, 15 min, 0°) and discarded. A further 2 volumes of ice-cold acetone are then added as before. The second precipitate is collected by filtration through Whatman type 540 paper in the presence of a filter aid, Hyflo Super-Cell (Koch-Light) (1 g/100 ml of extract). The filtrate is discarded and the filter cake is extracted three times with 5 mM disodium EDTA (10 ml/100 ml of extract, each time); the extracts are combined and stored at 4°.

Cation-Exchange Chromatography. An aliquot of the extract of the second precipitate, containing approximately 10 mg of protein, is clarified and chromatographed on the Mono S column at pH 5.0 as described above for pineapple stem extract, except that the buffers contain 0.1% w/v Brij 35. Fruit bromelain elutes as a double peak between 0.12 and 0.18 M Na$^+$; active fractions in this region are combined.

Affinity Chromatography. Fruit bromelain is then affinity-purified on a column of Sepharose-Ahx-Phe-GlySc exactly as described for ananain. The enzyme is eluted with 10 mM mercuric chloride and then subjected to two cycles of FPLC on the Mono S column at pH 5.0 as described for stem bromelain. Fractions active against Bz-Phe-Val-Arg-NHMec (or Z-Phe-Arg-NHMec) are combined, dialyzed against 1 mM disodium EDTA, and freeze-dried. Fruit bromelain purified by this method is >85% active enzyme by titration with E-64.

Storage of Purified Enzymes

All enzymes purified by active-site-directed affinity chromatography will have been reversibly blocked by mercuric chloride or a disulfide. Because no further activation is performed and disulfide is routinely added to column fractions, the enzymes remain in the reversibly blocked state following freeze-drying. We have found such enzyme preparations to remain stable when stored dry at −20°.

Properties of Enzymes

Amino Acid Sequence

Only stem bromelain has been sequenced completely at the amino acid level[18] and shown to be a member of the papain family. A partial N-terminal sequence has been reported for fruit bromelain,[5] and recently the first 20 N-terminal residues of ananain and comosain have been determined.[36] The N-terminal sequences of the pineapple endopeptidases, along with that of papain,[37] are shown in Fig. 1. It seems clear that the pineapple endopeptidases are closely related in both an evolutionary and structural sense, no unequivocal sequence differences being found between those available for fruit bromelain and stem bromelain, and only 2 different residues out of 20 between ananain and comosain, and stem bromelain.

[36] A. D. Napper, S. P. Bennett, M. Borowski, M. B. Holdridge, M. J. C. Leonard, E. E. Rogers, Y. Duan, R. A. Laursen, B. Reinhold, and S. L. Shames, *Biochem. J.* **301,** 727 (1994).

[37] L. W. Cohen, V. M. Coghlan, and L. C. Dihel, *Gene* **48,** 219 (1986).

```
S.Brom   AVPQSIDWRDYGAVTSVKNQNPCGACW

F.Brom   AVPQSIDWRDYGA    NZNPCGAC

Anan     VPQSIDWRDSGAVTSVKNQG

Como     VPQSIDWRNYGAVTSVKNQG

Pap      IPEYVDWRQKGAVTPVKNQGSCGSCW
```

FIG. 1. N-Terminal amino acid sequences of the pineapple endopeptidases and papain. Residues identical to those of stem bromelain are shown as white-on-black. S. Brom, Stem bromelain[18]; F. Brom, fruit bromelain[5]; Anan, ananain[36]; Como, comosain[36]; Pap, papain.[37]

The available evidence therefore suggests that the pineapple endopeptidases are evolutionarily more closely related to each other than to other members of the papain family, suggesting relatively recent divergence.

Physicochemical Properties

These are summarized in Table I.[38–41]

Catalytic Behavior

All of the pineapple endopeptidases exhibit broad pH–activity profiles, optimal near neutral pH.[34] In terms of specificity for the hydrolysis of amide bonds, ananain and fruit bromelain are typical members of the papain family, preferring hydrophobic residues in P2 (in the recognized terminology[42]) (Table II). Stem bromelain is unusual, apparently requiring Arg in both P1 and P2 for the efficient cleavage of small substrates, even di-Lys-containing substrates being unaffected.[17] It is not known if this rigid and unexplained specificity applies also to protein substrates.

Inhibition Profiles

Both stem and fruit bromelain are unusual among cysteine endopeptidases of the papain family in their resistance to inhibition by chicken cystatin (Table II), glycyl endopeptidase being the only other known exam-

---

[38] Y. Minami, E. Doi, and T. Hata, *Agric. Biol. Chem.* **35**, 1419 (1971).
[39] T. Murachi and M. Yasui, *Biochemistry* **11**, 2275 (1965).
[40] H. Ishihara, N. Takahashi, S. Oguri, and S. Tejima, *J. Biol. Chem.* **254**, 10715 (1979).
[41] T. Murachi, M. Yasui, and Y. Yasuda, *Biochemistry* **3**, 48 (1964).
[42] Enzyme Nomenclature. "Recommendations of the Nomenclature Committee of the International Union of Biochemistry and Molecular Biology," p. 371. Academic Press, San Diego, 1992.

TABLE I
PHYSICOCHEMICAL PROPERTIES OF PINEAPPLE ENDOPEPTIDASES

Endopeptidase	M_r	pI	$A_{280\,nm}^{1\%}$	Glycoprotein
Stem bromelain	26,000[16]	9.45[38]	20.1[39]	Yes[18,40]
	23,800[18]	9.55[41]	—	—
	27,000[4]	—	—	—
Fruit bromelain	25,000[17]	4.6[5]	19.2[5]	Yes[13]
	27,000[4]	4.7[38]	—	No[5]
	31,000[5]	—	—	—
Ananain	25,000[16]	>10	16.5[16]	No[16]
	23,420–23,580[36]	—	—	No[36]
Comosain	24,509–24,568[36]	>10	?	Yes[36]

ple (see [37]). Stem bromelain is also unusual in its very slow rate of inactivation by E-64 (Table II), and in this respect it appears to be unique among the papain homologs.

Immunological Identity

Double-immunodiffusion studies have shown that stem and fruit bromelain and ananain are totally distinct with respect to precipitating reactions with polyclonal antisera.[16,17,34] Earlier reports of cross-reaction between stem and fruit bromelains[6,7] are probably explained by the presence,

TABLE II
ENZYMATIC PROPERTIES OF PINEAPPLE ENDOPEPTIDASES[a]

Endopeptidase	Preferred substrate [K_m (μM); k_{cat} (sec^{-1})]	E-64 k_2 (M^{-1} sec^{-1})	Chicken cystatin K_i (nM)
Stem bromelain	Z-Arg-Arg-NHMec [15.4; 27.3]	678	36,000
Fruit bromelain	Bz-Phe-Val-Arg-NHMec [4.0; 18.8]	3385	>1100
Ananain	Bz-Phe-Val-Arg-NHMec [13.1; 63.8]	302,297	1.1
Comosain	Z-Arg-Arg-NHMec [1.4; ?]	8900	?

[a] The data are taken from Ref. 17, except for the value of k_2 for E-64 and comosain.[36]

at the time not known, of stem bromelain in pineapple fruit and traces of fruit bromelain in the stem.

Acknowledgment

The authors would like to thank Dr. Andrew D. Napper and colleagues at Genzyme Corporation for allowing us to see data prior to publication.

[39] Cancer Procoagulant

By STUART G. GORDON

Introduction

Prior to 1970, because there was a well-documented association between hypercoagulation and cancer, several investigators evaluated procoagulant activity associated with malignant tissue. For the most part, they found the procoagulant to be tissue thromboplastin (tissue factor) or "thromboplastin-like" activity. The notable exception was a group headed by O'Meara.[1,2] This team found an activity that they believed was not tissue factor and it activated the coagulation cascade by an undefined mechanism. They concluded that the procoagulant was serum albumin associated with fatty acids[3]; they were unable to delineate further the properties of this substance. Although their conclusions were incorrect, much of their work provided an important basis for the discovery of cancer procoagulant (CP). Their observations included the extraction of procoagulant activity from tumor tissue in veronal buffer, several useful purification steps, stability data on the procoagulant, and a protein with a molecular weight of 68,000. As will be described below, cancer procoagulant has been purified to apparent homogeneity and many of its physical and chemical properties have been determined.

An important element of the discovery of CP was the timing and the development of coagulation studies by many other investigators from Europe and the United States. Until the early 1970s, purified coagulation factors were scarce and an understanding of how to characterize specific coagulation factors was limited—in particular, there was an inability to distinguish between tissue factor and other procoagulant activities (e.g.,

[1] R. A. Q. O'Meara and R. D. Thornes, *Ir. J. Med. Sci.* **423**, 106 (1961).
[2] R. A. Q. O'Meara, *Ir. J. Med. Sci.* **394**, 474 (1958).
[3] W. W. Fullerton, W. A. Boggust, and R. A. Q. O'Meara, *J. Clin. Pathol.* **20**, 624 (1967).

METHODS IN ENZYMOLOGY, VOL. 244

cancer procoagulant). One of the early efforts to show that CP was not tissue factor focused on the ability to inhibit CP with diisopropyl fluorophosphate (DFP),[4,5] a serine proteinase inhibitor. Tissue factor is insensitive to DFP. Then we showed that CP could activate coagulation in the absence of factor VII by using either genetically deficient citrated human plasma or DFP-treated bovine plasma (bovine factor VII/VIIa but not human factor VII/VIIa is inhibited by DFP). With these tools, the distinction between tissue factor or tissue factor/factor VII and CP was established. All of the early studies relied on the clotting time of recalcified, citrated plasma as a means for quantitating CP activity. Now more sensitive, specific, and quantitative assays have been developed (see below).

CP has been identified in a broad spectrum of malignant tissues but it has not been found in normally differentiated tissues.[6] In addition, following leads provided by O'Meara and co-workers,[7] CP has been extracted from human placental amnion–chorion tissue, suggesting that CP is an oncofetal protein. One of the unique features of this protein is that there is a high degree of enzymatic and structural similarity between CP isolated from rabbit V2 carcinoma, rat Walker 256 carcinosarcoma, mouse B16 melanoma, and human amnion–chorion tissue. CP from all these sources was immunopurified on an immunoaffinity chromatography column in which polyclonal anti-CP V2 carcinoma antibodies were coupled to resin; they all had the same properties, including molecular mass, isoelectric point, activation of human factor X, inhibition characteristics, and, of course, immunoreactivity to anti-CP antibodies. The same similarity in properties occur with CP from different human tumor tissues. These data suggest that CP is highly conserved across species lines and tissue (tumor) types, an uncommon characteristic for any protein and an interesting and provocative feature of CP. CP may have a primary function in the undifferentiated or dedifferentiated cell and its procoagulant activity may be coincidental and of secondary importance.

Measurement of Cancer Procoagulant Activity

The classic methods for measuring CP activity employed measurement of the recalcified clotting time of citrated plasma. Because CP directly activates factor X, several other more sensitive and reproducible chromo-

[4] S. G. Gordon, J. J. Franks, and B. Lewis, *Thromb. Res.* **6,** 127 (1975).
[5] S. G. Gordon and B. A. Cross, *J. Clin. Invest.* **67,** 1665 (1981).
[6] S. G. Gordon, *Semin. Thromb. Hemostasis* **18,** 424 (1992).
[7] W. A. Boggust, D. J. O'Brien, R. A. Q. O'Meara, and R. D. Thornes, *Ir. J. Med. Sci.* **477,** 131 (1963).

genic assays have been developed that measure the generation of factor Xa.

Recalcified Clotting Time Assay

Reagents. Rabbit brain thromboplastin, referred to as rabbit brain tissue factor (RBTF), Russell's viper venom (RVV), papain, citrated normal and factor VII-deficient bovine plasma, Bis–Tris propane and Tris-HCl buffer, sodium barbital, iodoacetamide, mercuric chloride, and *trans*-epoxysuccinyl-L-leucylamido(4-guanidino)butane (E-64) are purchased from Sigma Chemical Co. (St. Louis, MO). $CaCl_2$ and dimethyl sulfoxide (DMSO) can be purchased from Baker (Phillipsburg, NJ). If fresh plasma (human or bovine) is collected, 9 parts of blood are drawn into 1 part of 3.8% sodium citrate (Sigma) and the blood is centrifuged twice at 1600 g for 10 min. A Fibrometer (Bioquest) is used to time the coagulation reaction.

Assay. The recalcification clotting time of citrated normal or factor VII-deficient plasma, using any accurate clot timer, has been a standard assay of the coagulation laboratory. In this laboratory, 0.1 ml of citrated plasma, 0.1 ml of veronal buffer (pH 7.6) or procoagulant sample veronal buffer are prewarmed in the fibrometer reaction cup for 2 min at 37° and the reaction is initiated with 0.1 ml of 30 mM $CaCl_2$; the 37° reaction is timed until a clot (fibrin formation) is detected and the timer is stopped.[4,5] Other coagulation timers should work as well. For the analysis of factor VII-deficient plasma, RVV (from 10 to 1000 ng/ml 0.1 mM Tris buffer) is used as a positive control and RBTF (1 to 100 µg/ml Tris buffer) is used as a negative control.

Chromogenic Measurement of Factor X Activation

Reagents. Purified bovine factor X and bovine prothrombin can be purchased from Hematologic Technologies (Essex Junction, VT) or from Sigma. Chromogenic substrate for factor Xa, S-2222, Bz-Ile-Glu-Gly-Arg-pNA, and for thrombin, S-2238, Sar-Pro-Arg-pNA, can be purchased from Kabi Pharmacia, (Stockholm, Sweden) or Sigma. Other reagents and their sources are identified above.

Direct Assay for Factor Xa. This method was developed and championed by Colucci *et al.*[8] CP (in the form of cell suspensions) is incubated at 37° with 10 µg of purified factor X and an equal volume of 25 mM $CaCl_2$. After the predetermined time interval, the reaction is terminated by mixing a 300-µl sample of the reaction mixture with 400 µl of Tris-

---

[8] M. Colucci, L. Curatolo, M. B. Donati, and N. Semeraro, *Thromb. Res.* **18**, 589 (1980).

buffered saline (pH 8.4) containing 7.5 mM Na$_2$EDTA. The factor Xa generated by the first stage of the assay is measured by adding 100 μl of 4 mM S-2222; the reaction is incubated for 5 min at 37° followed by 200 μl of 40% trichloroacetic acid to stop the reaction. After centrifugation and addition of an equal volume of NaOH to return the pH to 7 ± 1, the absorbance is read at 405 nm on a spectrophotometer. Our studies find that this direct measurement of factor Xa is somewhat less sensitive than other assays (see "Active Site Configuration" below for possible explanation).

Indirect, Coupled Assay for Factor Xa: Method 1. To improve the sensitivity of the assay, a coupled assay was developed in which factor Xa generated by CP is used to activate prothrombin; the thrombin generated is measured by its activity on Sar-Pro-Arg-pNA.[9] This assay is very sensitive and specific for CP and has a high degree of precision. The assay is performed in a cuvette as follows: a sample containing CP (100 μl), or a veronal buffer blank, is mixed with 10 μl of bovine factor X (100 μg/ml) and 30 μl of 25 mM CaCl$_2$ in 50 mM Bis–Tris propane buffer (pH 6.7), such that the final pH is 7.2, and the first-phase reaction mixture is incubated at 25° for 30 min. The second phase of the reaction (amplification phase) contains 10 μl of bovine prothrombin (1 mg/ml) and 30 μl of RBTF/Ca^{2+} mixture [the RBTF mixture contains 1 part RBTF, prepared according to manufacturer's instructions, diluted 1 : 10 in water, 1 part of 50 mM CaCl$_2$ in water and 2 parts of 100 mM Bis–Tris propane buffer (pH 7.8)]; it is incubated for 30 min at 25°. In the third phase of the assay, 200 μl of 50 mM Tris buffer (pH 7.8) and 50 μl of 2 mM Sar-Pro-Arg-pNA substrate in 10% DMSO is added to the reaction and the increase in absorbance is monitored at 405 nm on a spectrophotometer. Generally, the absorbance is read at 5, 35, and 65 min and the change in absorbance with time is noted.

Indirect, Coupled Assay for Factor Xa: Method 2. Using Immunocaptured CP. An interesting and very specific variation on the solution-phase assay (Method 1) described above is an immunocapture enzyme (ICE) assay.[10] In this assay, CP is captured from a pure or crude solution of enzyme by a murine monoclonal anti-CP immunoglobulin M (IgM) antibody that is adsorbed onto the surface of microtiter plate wells. The IgM–CP complex formation does not interfere with the proteinase activity of CP. Following capture of the CP onto the surface of the well, other unbound proteinases, proteinase inhibitors, or other interfering substances in the solution can be removed by washing.

[9] W. P. Mielicki and S. G. Gordon, *Blood Coagulation Fibrinolysis* **4**, 441 (1993).
[10] W. P. Mielicki, M. Tagawa, and S. G. Gordon, *Thromb. Hem.* **71**, 1 (1994).

This assay is performed as follows: 50 μl of murine monoclonal anti-CP IgM (1 μg/ml) in 100 mM phosphate-buffered saline (PBS, pH 8.0) is added to the wells of a Costar 96-well $\frac{1}{2}$ area microtiter plates (Cambridge, MA) and incubated at 5° for 15 hr (overnight). The plates are washed with Tris-buffered saline (pH 7.4) containing 0.05% Tween 20 (TBS-T). The wells are blocked three times with 100 μl/well of Superblock (Pierce Chem Co., Rockford, IL) according the manufacturer's instruction. The plates are washed eight times with TBS-T, and 100 μl of the CP samples, blanks, or standards diluted in veronal-buffered saline (pH 7.6) (VBS) is put in the microtiter wells and incubated for 2 hr at 37°. The wells are washed three times with TBS-T to remove unbound sample and contaminants. Bovine factor X (50 μl; 7 μg/ml) in 10 mM Bis–Tris buffer (pH 7.2) containing 7 mM MnCl$_2$ is added to the wells and incubated for 1 hr at 37°. Bovine prothrombin (30 μl/well) at a concentration of 140 μg/ml in RBTF/Ca^{2+} mixture [RBTF stock solution diluted 1 : 100 with 10 mM CaCl$_2$ in 50 mM Bis–Tris buffer (pH 7.4)] is added to the wells and incubated for 30 min at 37°. Chromogenic substrate (Sar-Pro-Arg-pNA) (20 μl; 0.8 mg/ml in 100 mM Bis–Tris buffer, pH 7.9) is added to the wells and the absorbance at 405 nm is monitored at timed intervals for 30 min at room temperature on a UVmax Microtiter Plate Reader (Molecular Devices Corp., Menlo Park, CA).

General Comments. Figure 1 shows standard curves from the three different assays using the same CP sample for the purpose of comparing sensitivity and reproducibility of the assays. However, it is difficult to compare the sensitivities because of the different scales for the different assays. The clotting assay (open circles) shows greater variability and poorer linearity than the chromogenic assays. The three-stage solution phase assay (solid circles) is about 25% less sensitive than the ICE assay (open triangles) at the low concentrations of CP, but it has about five times greater linear range than the ICE assay; the ICE assay begins to saturate at about 3 μg/ml.

It has not been possible to analyze CP activity directly with a chromogenic substrate because the specificity of such an assay is uncertain. Many potential proteinases might hydrolyze a synthetic substrate and lead to erroneous conclusions. The use of the coupled assays has provided important information about cofactors for CP activity. The pH optimum for the activation of factor X by CP is 7.2.[9] Phospholipid, such as rabbit brain cephalin, is not necessary for factor X activation by CP.

Without calcium, CP has no activity; the optimum level of Ca^{2+} is 7 mM. Magnesium appears to substitute for Ca^{2+} and has an optimum concentration of 10 mM.[11] Manganese activates CP by binding at a different

---

[11] W. P. Mielicki, D. L. Kozwich, L. C. Kramer, and S. G. Gordon, *Arch. Biochem. Biophys,* in press (1994).

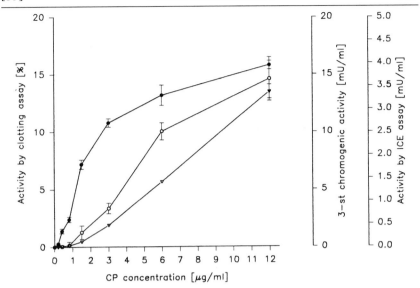

FIG. 1. Comparison of the sensitivity and reproducibility of three of the assays described in the text for analysis of a purified CP sample: (○) The recalcification time of citrated plasma [S. G. Gordon, J. J. Franks, and B. Lewis, *Thromb Res.* **6**, 127 (1975)]. (●) The three-stage, solution-phase chromogenic assay [W. P. Mielicki and S. G. Gordon, *Blood Coag. Fibrinol.* **4**, 441 (1993)]. (△) The ICE chromogenic assay [W. P. Mielicki, M. Tagawa, and S. G. Gordon, *Thromb. Hem.* **71**, 1 (1994)]. The data show the mean ± the standard error of the mean of three measurements. Note the difference in the scales for the vertical axis.

site than Ca^{2+} and Mg^{2+} and has maximum activation at 20 μM Mn^{2+}. When CP is bound to IgM in the ICE assay, neither calcium nor magnesium is needed as a cofactor. In the ICE assay, the optimum level of Mn^{2+} is 10 mM. Thus, binding CP to a surface changed the divalent ion requirements for the proteinase. In the solution phase assay, the optimum condition for the measurement of CP activity is 7 mM Ca^{2+}, 20 μM Mn^{2+} in 50 mM Bis–Tris buffer (pH 7.2).

The active site of CP appears to be easily oxidized. The use of a reduced environment (e.g., 1 mM 2-mercaptoethanol and 5 mM potassium cyanide[12] or 5 μM cysteine and 4 mM KCN[10]) has beeen shown to regenerate activity in partially purified samples or facilitate the inhibion of CP by some cysteine proteinase inhibitors. Frequently, buffers will be flushed with nitrogen or helium gas to reduce the amount of dissolved oxygen and, thus, decrease the apparent oxidative loss of activity.

Inhibitors. The appropriate use of inhibitors of CP and other procoagulants has provided a powerful tool for the characterization of these factors.

---

[12] A. Falanga and S. G. Gordon, *Biochemistry* **24**, 5558 (1985).

Original studies used DFP to distinguish CP from TF (see above) and, based on this observation, it appeared that CP was a serine proteinase. When it was discovered that CP was a cysteine proteinase, we were able to demonstrate that there was an impurity in the commercial DFP that was blocking the proteinase activity without forming an diisopropyl phosphoserine ester bond.[5] Subsequently, many cysteine proteinase inhibitors have been shown to be effective inhibitors of CP, including iodoacetamide, $HgCl_2$,[5,12] E-64,[13] and peptidyl diazomethanes (PDK).[14] The general procedure for studying CP inhibition is as follows: a sample of CP (purified or crude extract) is prepared in veronal (preferable) or Bis–Tris buffer (pH 7.2) containing 5 μM cysteine and 4 mM KCN. A 100-fold concentration of the inhibitor is prepared in water or other appropriate solvent (e.g., 20% (v/v) DMSO) and is added to the CP sample to achieve the desired final concentration. The sample is incubated at room temperature for 30 min and assayed for CP activity by one of the methods described above. Appropriate controls for these experiments include papain, as a factor X-activating cysteine proteinase to verify the effective inhibition of the inhibitor in the experimental system, and RVV, as a serine proteinase activator to verify the absence of an effect of the inhibitor, buffer, solvents, etc. on the experimental system. With crude extracts or partially purified CP preparations, nonspecific inhibitors such as Hg^{2+} or iodoacetamide may react with other sulfhydryls in the mixture and decrease the effective concentration of the inhibitor with respect to CP. The inhibitors of CP and their effective concentration are listed in Table I.

Immunoassay for Measurement of Cancer Procoagulant Antigen

A double-antibody enzyme-linked immunosorbent assay (ELISA) has been developed for CP.[15] Its primary focus has been for the analysis of CP in human serum samples, as a tumor marker assay. The presence of CP in the serum of an individual predicts, at the level of approximately 85%, whether or not the individual has cancer. CP appears to be particularly good for identifying individuals with stage I or II cancer.

Reagents. Polyclonal antibodies against CP are raised in a goat by standard methods. IgG is purified from the goat serum by $(NH_4)_2SO_4$ precipitation and DEAE ion-exchange chromatography. The IgG is further purified by affinity chromatography to remove antibodies that could cross-react with antigens present in human and rabbit serum. The affinity column

[13] W. R. Moore, *Biochem. Biophys. Res. Commun.* **184**, 819 (1992).
[14] A. Falanga, E. Shaw, M. B. Donati, R. Consonni, T. Barbui, and S. G. Gordon, *Thromb. Res.* **54**, 389 (1989).
[15] S. G. Gordon and B. A. Cross, *Cancer Res.* **50**, 6229 (1990).

TABLE I
INHIBITORS OF CANCER PROCOAGULANT[a]

Inhibitor	Concentration	Inhibition (%)	Effect of reducing agent	Ref.[b]
Iodoacetamide	1	100	n.d.	A
	1	82	+3.5×	B
	0.5	85	n.d.	C
Mercuric chloride	0.1	100	n.d.	A
	0.1	75	−0.15×	B
	0.5	95	n.d.	C
trans-epoxysuccinyl-L-leucylamido	0.1	86	+5.6×	B
(4-guanidino)butane (E-64)	0.1	80	n.d.	C
Leupeptin	0.1	71	+3.5×	B
	0.1	71	n.d.	C
Antipain	0.1	87	+5.4×	B
Cystatin	0.1 mg/ml	0	0	B
α_2-Macroglobulin	20 μg/ml	19	+1.2×	B
α_1-Antichymotrypsin	20 μg/ml	27	+1.3×	B
α_1-Antiprotease	20 μg/ml	0	0	B
Antithrombin III	5 μg/ml	0	0	B
Diisopropyl fluorophosphate (DFP)	5	100	n.d.	A, D
Phenylmethylsulfonyl fluoride (PMSF)	0.1	80	n.d.	A
Peptidyl diazomethanes	0.2	95–100	n.d.	E
	0.2	95–98	n.d.	C
Peptidyl chloromethanes	0.2	92–100	n.d.	C
Peptidyl dimethyl sulfonium salts	0.4	60–70	n.d.	E

[a] Final concentration of each inhibitor that has been used in studies of CP. (Concentration is millimolar unless otherwise specified.) As noted in the text, the inhibition by DFP was due to impurities in the commercial preparation. The inhibition by peptidyl diazomethanes, peptidyl chloromethanes, and peptidyl dimethyl sulfonium salts depends on the amino acid sequence of the peptide and is expressed as a range of inhibition. n.d., Not done.

[b] Key to references: (A) S. G. Gordon and B. A. Cross, *J. Clin. Invest.* **67**, 1665 (1981); (B) W. P. Mielicki, M. Tagawa, and S. G. Gordon, *Thromb. Hem.* **71**, 1 (1994); (C) W. R. Moore, *Biochem. Biophys. Res. Commun.* **184**, 819 (1992); (D) S. G. Gordon, J. J. Franks, and B. Lewis, *Thromb. Res.* **6**, 127 (1975); (E) A. Falanga, E. Shaw, M. B. Donati, R. Consonni, T. Barbui, and S. G. Gordon, *Thromb. Res.* **54**, 389 (1989).

is constructed by coupling human and rabbit serum proteins to CNBr-activated Sepharose. The immunoreactivity of the IgG against CP and normal human and rabbit serum is measured by ELISA using purified CP coated in the wells of a microtiter plate and antigoat IgG alkaline phosphatase conjugate as secondary antibody. The polyclonal anti-CP antibodies are conjugated to alkaline phosphatase by Lampire Biological

Laboratories Inc. (Pipersville, PA) by a single-step glutaraldehyde labeling method at a 4 : 1 (v/v) enzyme : protein ratio.

Hybridomas are generated by standard methodology. Hybrids are screened for production of anti-CP antibodies by ELISA utilizing purified CP coated microtiter wells and antimouse IgG or IgM labeled with alkaline phosphatase. Ascites fluid is produced in pristane-treated BALB/c mice. The hybridoma also synthesizes CP, necessitating a purification step to remove the contaminating CP from the antibody. Ascites are precipitated with 10% (v/v) polyethylene glycol 8000, centrifuged, and resuspended in 0.2 M phosphate-buffered saline containing 300 mM NaCl (pH 7.4), (2 × PBS), this procedure is repeated once more. The final preparation is diluted with an equal volume of ethylene glycol stored for at least 2 hr at 4° and applied to a 10 × 30-cm Superdex 2000 gel filtration column (Pharmacia LKB Biotechnology, Piscataway, NJ) run at 8 ml/min. Fractions are collected, concentrated, and diluted into 2 × PBS in a Millipore (Bedford, MA) Minitan Ultrafiltration System using 100,000 molecular weight retentate separators. Purified IgM is quantitated and tested for CP contamination, immunoreactivity, and nonspecific binding.

Assay. The ELISA is performed as follows[16]: murine monoclonal anti-CP IgM (100 μl of 1 μg/ml) in 0.1 M potassium phosphate buffer containing 0.15 M NaCl (pH 8.0) is adsorbed onto ELISA wells (Corning, 1 × 8 strips) for 15 hr at 4°, aspirated, and washed three times (Bio-Tek Instruments EL 403E Microplate AutoWasher) with 50 mM Tris buffer (pH. 7.2.) containing 150 mM NaCl and 0.05% Tween 20 (TBS-T). The wells are blocked with 150 μl/well of Superblock (Pierce Chem Co.) according to manufacturer's instructions. The plates are stable for months if stored desiccated at 4°. CP standards, control sera, and patient sera (50 μl/well) are added to monoclonal antibody (MAb)-coated wells in triplicate (Beckman Biomek 1000 Automated Laboratory Workstation, Fullerton, CA) and incubated at 37° for 2 hr. The serum is removed from the wells by aspiration and the wells are washed three times with TBS-T. Goat polyclonal anti-CP alkaline phosphatase conjugate is diluted (1 : 4000) in conjugate diluent buffer (50 mM Tris buffer, 150 mM NaCl, 1 mM MgCl$_2$, 1 mM ZnCl$_2$, 0.1% sucrose, 0.001% BSA, 0.001% NaN$_3$). The diluted conjugate (50 μl/well) is added and incubated for 1 hr at 37°. Conjugate is removed by aspiration and the wells are washed eight times with TBS-T. p-Nitrophenyl phosphate substrate (50 μl/well) is added to the wells and incubated at 22° for 20 min before adding 1 M NaOH (50 μl/well) stopping reagent. The absorbance is read on a microtiter plate spectrophotometer (Molecular Devices UVmax Microtiter Plate Reader) at a wavelength of 405 nm.

---

[16] D. Kozwich, L. Kramer, W. Mielicki, and S. Gordon, *Cancer*, in press (1993).

Purification of Cancer Procoagulant

Cancer procoagulant has been purified in several ways. The following procedures will highlight the various techniques used and comment on their application.

Reagents

Many of the reagents for purification studies are identified in earlier sections of this chapter. DE-53 and DEAE-cellulose were purchased from Whatman (Maidstone, UK) Other chromatography materials include CNBr-activated Sepharose 4B and phenyl-Sepharose (Pharmacia, Uppsala, Sweden), 1.5 M agarose, and Minitan II SDS–PAGE (Bio-Rad Laboratories, Richmond, CA) Amicon Ultrafiltration systems (Beverly, MA).

Procedures

Extraction. Sliced (to increase the surface area) human or animal tumor tissue[4] or amnion–chorion tissue from human placenta[17] is placed in cold (5°) 20 mM veronal (barbital) buffer (pH 7.8) (VB) containing 0.001% Na azide for 3 hr. The buffer is exchanged three times and the individual extracts are pooled. The pooled extract is concentrated about 10-fold on an Amicon PM10, PM30, or XM50 ultrafiltration membrane. The concentrated extract is centrifuged at 5000 rpm (about 2500 rcf) at 5° for 30 min to removed cellular debris.

Fractional Precipitation. Fractional precipitation has been used for partial purification and concentration of CP extracts.[4,5,9–12] Three precipitation procedures have been used with CP; the basic procedure is the same for each precipitant. (1) Crystalline ammonium sulfate is added to an extract or partially purified CP sample to 40% saturation. (2) An equal volume of 100% ethanol is added to an extract or partially purified CP sample. (3) Crystalline polyethylene glycol 8000 is added to an extract or partially purified CP sample to a concentration of 10%. Regardless of the precipitating agent, the precipitated sample is stirred at 5° for from 2 to 4 hr and the precipitate is removed by refrigerated centrifugation at 5000 rpm for 30 min. The precipitate is redissolved in from 10 to 20% of the initial volume of veronal buffer and undissolved material is removed by refrigerated centrifugation. Frequently, the precipitation procedure is repeated to improve the purity of the final sample.

Gel Filtration, Sizing Chromatography. A 1.5 × 100-cm column is packed with 1.5 m agarose resin that is equilibrated in 10 mM veronal

[17] S. G. Gordon, U. Hasiba, B. A. Cross, M. A. Poole, and A. Falanga, *Blood* **66**, 1261 (1985).

buffer.[5,12] The column is then equilibrated with 0.5 mg/ml of crude phospholipid in 10 mM veronal buffer and excess phospholipid is washed off the resin. No more than 15 ml (10% of resin volume) of a concentrated CP sample in veronal buffer is applied to the column and elutes in 10 mM veronal buffer at 1 ml/min. About 60–70% of the CP elutes in the high molecular weight-included peak, a small amount is in the void volume peak, and about 20% is in various lower molecular weight peaks. We speculate that in concentrated samples CP elutes as an aggregate, complexed with itself and other proteins. Moore[13] found that CP eluted from a Sephacryl 200 column in fractions comparable to a 10-kDa protein and suggests that CP may interact with the Sephacryl like an affinity resin.

Affinity Chromatography. Several affinity chromatography procedures have been used that exploit different characteristics of CP. These include benzamidine affinity, *p*-chloromecuribenzoate (PCMB) affinity, immunoaffinity, and phenyl-Sepharose hydrophobic interaction chromatography resins.

BENZAMIDINE AFFINITY CHROMATOGRAPHY. A benzamidine-Sepharose affinity resin is prepared with a aminohexanoic acid spacer arm coupling the CNBr-activated Sepharose and the *p*-aminobenzamidine.[12] The resin is packed in a 1 × 11-cm column and equilibrated in 10 mM veronal buffer (pH 7.8) containing 50 mM NaCl and 1 mM EDTA. A sample is applied to the column and unbound protein is washed off in the same 10 mM VB/50 mM NaCl–1 mM EDTA and then with 0.1% (v/v) Triton in 10 mM VB (pH 7.8). Proteinases are eluted from the resin with 0.05 and 0.5 M propionic acid. The acid is immediately neutralized and the sample is dialyzed against 20 mM VB and concentrated on an Amicon ultrafiltration membrane.

PCMB-AFFINITY CHROMATOGRAPHY. The PCMB resin is equilibrated in 20 mM Bis–Tris propane (BTP) buffer (pH 6.5) and packed in a 1 × 10-cm column.[5,12] After each run, the resin is regenerated with 100 ml of 10 mM HgCl$_2$ and 20 mM EDTA in 50 mM acetate buffer (pH 4.8) and 100 ml of 0.2 M NaCl and 1 mM EDTA in acetate buffer (pH 4.8), followed by reequilibration in 20 mM BTP buffer (pH 6.5). A partially purified sample is applied to the column and unbound protein is washed from the resin with 20 mM BTP buffer and then BTP buffer containing 1 M urea and 1% (v/v) Tween 20. Bound protein is eluted with 100 ml of 1 mM HgCl$_2$, 100 ml of 2 mM HgCl$_2$, and then 50 ml of 50 mM glutathione to strip the resin. The Hg^{2+}-eluted sample requires reactivation by making the samples 10 mM KCN, 2 mM DTT, and 2 mM EDTA. The reactivation mixture is incubated at 5° for 30 min and dialyzed overnight against 20 mM BTP buffer.

IMMUNOAFFINITY CHROMATOGRAPHY. Goat polyclonal anti-CP IgG antibodies are coupled to CNBr-Sepharose at a concentration of 5 mg/ml of resin; additional sites are blocked with 1 M ethanolamine and the resin is equilibrated in 20 mM veronal buffer (pH 7.8).[12] A CP sample is applied to the column in 20 mM VB and the column is washed with VB until the A_{280} is the same as the buffer. Then 100 ml of 1 M urea and 1% Tween 20 in 20 mM VB is put over the column to remove nonspecifically bound protein. CP is eluted with 100 ml of 3 M NaSCN in 40 mM VB; the eluate is immediately dialyzed against 20 mM BTP buffer (pH 6.5) and concentrated 10-fold on an Amicon PM10 ultrafiltration membrane. Other dissociating agents such as 5–8 M guanidine hydrochloride or 8 M urea may also be used to elute CP from the immunoaffinity column.

PHENYL-SEPHAROSE HYDROPHOBIC INTERACTION CHROMATOGRAPHY. Phenyl-Sepharose resin is packed in a 1 × 5-cm column and equilibrated in 10 mM VB (pH 7.8) containing 0.5 mg/ml of crude phospholipid; excess phospholipid is washed off with 10 mM VB and then with 10% (v/v) dimethyl sulfoxide (DMSO) in VB.[12] A CP sample is applied to the column in 20 mM VB, unbound protein is eluted with 10 mM VB (pH 7.8), and CP is eluted with 10% DMSO in 20 mM VB. The eluted sample is concentrated and dialyzed against 20 mM VB and the CP is stabilized by adding 0.1 volume of 2 mg/ml of crude phospholipid in 20 mM VB.

Ion-Exchange Chromatography. DE-53 resin is prepared according to the manufacturer's instructions, suspended in 20 mM VB (pH 7.8), and packed into a 2.5 × 10-cm column.[9] The CP sample is applied in 20 mM VB and unbound protein is washed from the column with 20 mM VB. A stepwise elution of CP is done using about 180 ml each of 0.15, 0.26, 0.5, and 1.0 M NaCl in VB. Each sample is immediately desalted and concentrated by Amicon ultrafiltration. CP is eluted primarily in the 0.5 M NaCl fraction.

Sequence of Purification Procedures

Crude extracts are effectively concentrated and partially purified by ammonium sulfate or ethanol precipitation. A large-capacity purification step, like DE-53 ion-exchange chromatography or phenyl-Sepharose chromatography, is a good second step with good yields. Affinity chromatography with benzamidine of PCMB resins provide selectivity and generally yields a pure or almost pure CP preparation.

Molecular Properties

Hydrophobic Properties

Cancer procoagulant is an unstable, hydrophobic cysteine proteinase. Due to its hydrophobic nature, CP associates with other proteins in tissue extracts and in serum. This has been the source of significant problems in the purification of this enzyme. In high concentrations, CP probably self-associates or associates with other proteins, so it appears to be a high molecular weight aggregate and elutes from a sizing column in the 10^7 molecular weight region. CP does not like high-salt conditions and generally, when such procedures are used [e.g. $(NH_4)_2SO_4$ precipitation], it is best to get the protein into low-salt conditions as soon as possible. It likes a lipid environment and early studies utilized crude phospholipid as a means of stabilizing CP activity; currently, it is not used as extensively. PEG precipitation of CP yields a protein that is very stable and maintains activity, apparently because PEG coats the hydrophobic regions of the peptide and diminishes the denaturation process without affecting the active site. Unfortunately, PEG coating of CP renders it less immunoreactive because, we think, the same hydrophobic regions that are covered by PEG are epitopes for anti-CP antibodies.

Active Site Oxidation

The active site of CP oxidizes easily, and frequently loss of activity during purification procedures can be recovered by using the reduction "reactivation" procedure described in the discussion of PCMB-affinity chromatography. In addition, there is substantial sensitivity of inhibitors to the reduced environment; iodoacetamide, antipain, E-64, and leupeptin have from three- to fivefold greater inhibition when CP is treated with 5 μM cysteine and 4 mM KCN. $HgCl_2$ is about 30% as effective in the reduced environment as in the nonreduced environment (buffer without the reducing agents), probably because Hg^{2+} binds nonspecifically to the excess -SH groups in the reduced conditions.[10] The activity of CP on factor X is not enhanced by the reduced conditions, a somewhat surprising finding.

Composition of CP

Although most of the early studies were performed on CP from rabbit V2 carcinomas, most of the more recent research has focused on CP purified from human amnion–chorion.

Amino Acids. The amino acid composition of CP purified from rabbit V2 carcinoma and human amnion–chorion tissue was determined by standard methodology.[12,18] The amino acid composition of CP from both tissues

---

[18] A. Falanga and S. G. Gordon, *Biochim. Biophys. Acta* **831**, 161 (1985).

TABLE II
AMINO ACID COMPOSITION OF CANCER PROCOAGULANT FROM RABBIT
AND HUMAN TISSUE[a]

| Component[b] | Number of residues/ mole of CP | | Difference (rabbit − human) |
	Rabbit V2	Human A/C	
Lysine	53	47	6
Histidine	26	21	5
Arginine	15	16	−1
Aspartic acid	46	49	−3
Threonine	33	35	−2
Serine	125	99	26
Glutamic acid	82	91	−9
Proline	24	19	5
Glycine	123	139	−16
Alanine	42	48	−6
Cystine	8	10	−2
Valine	20	28	−8
Methionine	11	9	2
Isoleucine	17	14	3
Leucine	22	27	−5
Tyrosine	13	13	0
Phenylalanine	14	14	0
	674	679	

[a] CP purified from rabbit V2 carcinoma and human amnion–chorion tissue and the difference in the number of residues (rabbit minus human). From A. Falanga and S. G. Gordon, *Biochim. Biophys. Acta* **831**, 161 (1985).
[b] Carbohydrate, <1 mol hexose and sialic acid/mole CP.

is shown in Table II. Note that, with the exception of serine and glycine, there is very little difference in the amino acid composition of CP from significantly different species. About 40% of the amino acids from both sources is hydrophobic and 33% of the amino acids in both human and rabbit CP consists of basic and acidic amino acids. The amino acid sequence of CP has not been completed.

For several years substantial data have been gathered regarding the existence of γ-carboxylation of glutamic acid to form the calcium-binding site in CP. Warfarin treatment of animals yields a protein that immunoreacts with anti-CP antibodies but has no procoagulant activity.[19] Furthermore, preliminary data show about five Gla residues/mole of CP. It is

[19] F. Delaini, M. Colucci, G. De Bellis Vitti, D. Locati, A. Poggi, N. Semeraro, and M. B. Donati, *Thromb. Res.* **24**, 263 (1981).

clearly established that CP requires Ca^{2+} for activity. In the absence of direct chemical evidence, it is probable that CP contains Gla in its calcium-binding sites.

Carbohydrate. Hexose and sialic acid contents were determined in purifed CP; neither of these sugars was detected.[12] Based on the sensitivity of the methods for analysis of these sugars in the controls, it seems likely the CP is not a glycoprotein.

Active Site Configuration of Cancer Procoagulant

The active site of CP was evaluated by examining the cleavage site of the heavy chain of factor Xa.[20] In this experiment factor X was activated by CP and the peptides of the activated factor X were separated by a 10% Tris–Tricine SDS–PAGE, electroeluted onto Immobilon, and stained with Coomassie blue; the bands containing the activation peptides were cut out and sequenced on a Applied Biosystems 477A (Foster City, CA) amino acid sequencer. There was one primary cleavage site and two secondary cleavage sites in the factor X heavy chain: the Tyr^{21}-Asp^{22} bond was the primary cleavage site, yielding an amino-terminal peptide Try-Pro-Lys-Trp $\cdots$; the secondary cleavage sites were the peptide bonds between Asp-14 and Ser-15 and Thr-18 and Glu-19, yielding amino-terminal peptides Asp-Pro-Ala-Glu $\cdots$ and Thr-Pro-Asp-Leu $\cdots$. In all three peptides, proline is the penultimate amino acid and appears to play a significant role in directing the cleavage of the substrate proteins by CP. The usual cleavage site of factor X by factors VIIa and IXa and Russell's viper venom is at Arg-52.

Because the cleavage site of factor X is different than the usual cleavage site, the configuration of the active site of the resulting factor Xa may be different also. If this is true, it might explain why tripeptide synthetic substrates for factor Xa (e.g., S-2222) may have a lower affinity/reactivity with CP-generated factor Xa than with naturally occurring substrates (e.g., prothrombin), which should have many more points of interaction with the factor Xa active site.

Summary and Conclusions

Cancer procoagulant is a unique cysteine proteinase. The enzyme has been purified by several procedures and many of its characteristics and enzymatic properties have been determined. Several sensitive and repro-ducible assays are now available. Many proteinase inhibitors have been evaluated for their effect on CP; most low molecular weight inhibitors

---

[20] S. G. Gordon and A. M. Mourad, *Blood Coagulation Fibrinolysis* **2,** 735 (1991).

work well in a reduced environment. In the foreseeable future, protein and gene sequence information, expression vectors, molecular probes, and highly specific antibodies and inhibitors should provide the research tools to delineate a functional understanding of CP at the molecular and cellular level.

[40] Picornains 2A and 3C

By TIM SKERN and HANS-DIETER LIEBIG

Nomenclature

Picornains 2A and 3C are synthesized in cells following infection by viruses of the family Picornaviridae. The genera of this family are the enteroviruses [polio viruses, coxsackie viruses (CVs), and echoviruses], human rhinoviruses (HRV; the major cause of the common cold), cardioviruses [encephalomyocarditis virus (EMCV), mengovirus, and Theiler's virus], aphthoviruses [foot-and-mouth disease virus (FMDV)], and hepatitis A virus. Viral proteins of this family are named according to their position on the viral polyprotein using the system of Rueckert and Wimmer[1] (see Fig. 1); the protein designated picornain 3C (EC 3.4.22.28; picornavirus endopeptidase 3C) contains about 180–220 amino acids and acts as a proteinase in all picornaviruses.[2] The viral 2A proteins of rhinoviruses and enteroviruses have also been shown to be proteinases and are designated picornain 2A (EC 3.4.22.29; picornavirus endopeptidase 2A). Furthermore, the 2A proteins of cardiovirus and aphthovirus are also now known to be proteinases[3,4]; however, at the amino acid level they show neither similarity to the 2A proteins of rhinoviruses and enteroviruses nor to any other known proteinases; these proteins are therefore referred to as cardiovirus and aphthovirus 2A proteinases in this article. Proteolytic activity has not yet been identified for the 2A protein of hepatitis A virus.

Primary Structure of Picornains 2A and 3C

The sequences of the picornains were deduced from genomic sequences long before their purification. Comparison of these sequences

[1] R. R. Rueckert and E. Wimmer, *J. Virol.* **50,** 957 (1984).
[2] C. U. T. Hellen, H.-G. Kraeusslich, and E. Wimmer, *Biochemistry* **28,** 9881 (1989).
[3] A. C. Palmenberg, G. D. Parks, D. J. Hall, R. H. Ingraham, T. W. Seng, and P. V. Pallai, *Virology* **190,** 754 (1992).
[4] M. D. Ryan, A. M. Q. King, and G. P. Thomas, *J. Gen Virol.* **72,** 2727 (1991).

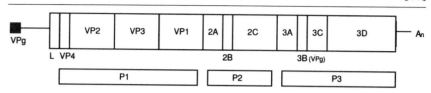

FIG. 1. Schematic representation of the picornavirus genome. The virally encoded protein VPg is covalently linked to the 5′ end of the RNA. The capsid proteins are VP1–VP4; the remainder are nonstructural. The L protein is found only in aphthoviruses and cardioviruses.

with protein data banks revealed similarity to sequences of trypsin and chymotrypsin-like serine proteinases.[5–7] However, the active site nucleophile of the picornains is cysteine, not serine, as borne out by the inhibitor profiles described below. Based on this, Gorbalenya et al.[6,8] and Brenner[9] have suggested that picornains may represent evolutionary intermediates between serine and cysteine proteinases.

This sequence similarity, together with the presence of conserved Asp and His residues in all picornains 2A and 3C, has led to attempts to predict the structure by modeling them onto known proteinase structures. The model of Bazan and Fletterick[10] proposes that the picornains 3C have a structure similar to that of the chymotrypsin-like serine proteinases, whereas the picornains 2A can be fitted to the structure of smaller bacterial serine proteinases such as α-lytic proteinase. The model of Gorbalenya et al.[6] is similar; however, they assume that the equivalent of chymotrypsin Asp-102 in the active site triad is Glu-71 and not Asp-85 in polio virus 3C. Site-directed mutagenesis experiments favor this hypothesis.[11] Nevertheless, the relevance of both models awaits the determination of the three-dimensional structures of the picornains 2A and 3C.

The 2A proteinases of cardioviruses and apthoviruses consist of 135–145 and 16 amino acids, respectively; at the carboxy terminus, the apthovirus 2A proteins show strong homology to the cardiovirus proteins.[3,4]

[5] P. Argos, G. Kamer, M. J. H. Nicklin, and E. Wimmer, Nucleic Acids Res. 12, 7251 (1984).
[6] A. E. Gorbalenya, A. P. Donchenko, V. M. Blinov, and E. V. Koonin, FEBS Lett. 243, 103 (1989).
[7] W. Sommergruber, M. Zorn, D. Blaas, F. Fessl, P. Volkmann, I. Mauer-Fogy, P. Pallai, V. Merluzzi, M. Matteo, T. Skern, and E. Kuechler, Virology 169, 68 (1989).
[8] A. E. Gorbalenya, V. M. Blinov, and A. P. Donchenko, FEBS Lett. 194, 253 (1986).
[9] S. Brenner, Nature (London) 334, 528 (1988).
[10] J. F. Bazan and R. J. Fletterick, Proc. Natl. Acad. Sci. U.S.A. 85, 7872 (1988).
[11] K. M. Kean, N. .L. Teterina, D. Marc, and M. Girard, Virology 181, 609 (1990).

Role of Picornains 2A and 3C in Viral Replication

The genetic information encoded in picornaviral genomes is expressed as a single polyprotein of about 2200 amino acids, which undergoes proteolytic processing to give the mature viral proteins. Under normal conditions, this polyprotein is not observed because processing begins before its synthesis is complete; in rhinoviruses and enteroviruses, the picornain 2A cleaves as it is synthesized.[12] Cleavage occurs between the C terminus of VP1 and the N terminus of picornain 2A in what is presumed to be an intramolecular reaction. This cleavage separates the capsid protein precursor from the noncapsid protein precursor. In some rhinoviruses and enteroviruses, the picornains 2A also carry out an intermolecular cleavage on the protein 3CD to yield 3C' 3D'; however, because this cleavage does not occur in all viruses and the cleavage site can be destroyed by site-directed mutagenesis without affecting viral replication, the relevance of this cleavage is not clear.[13]

The cleavage sites recognized by the rhinovirus and enterovirus picornains 2A are characterized by a P1' Gly and a preference for P2 Thr or Asn and P2' Pro or Phe. P1 is variable; Ala, Thr, Phe, Tyr, and Val have been observed.[14,15]

In cardioviruses and aphthoviruses, the primary cleavage is also carried out by the 2A proteinase, but between the C terminus of 2A and the N terminus of 2B. Cleavage is always that of the glycyl bond in the sequence -Asn-Pro-Gly↓Pro-.[3,4] All of the remaining cleavages except that between VP4 and VP2 (for which no proteinase has as yet been identified) are performed by picornain 3C or by precursors of the mature form of 3C such as 3CD or 3ABC. These forms of 3C can have different specificities; for instance, the cleavage site between VP2 and VP3 in polio virus can be processed by the 3CD form of picornain 3C but not by the 3C form.[16,17] It has been suggested that the picornains 3C and their precursors should be viewed as a family of proteinases with differential specificities that play a role in controlling processing rates during viral replication.[18]

Picornain 3C cleavage sites usually show Glu or Gln at the P1 site and

[12] H. Toyoda, M. J. H. Nicklin, M. G. Murray, C. W. Anderson, J. J. Dunn, F. W. Studier, and E. Wimmer, Cell (Cambridge, Mass.) **45**, 761 (1986).
[13] C.-K. Lee and E. Wimmer, Virology **166**, 405 (1988).
[14] T. Skern, W. Sommergruber, H. Auer, P. Volkmann, M. Zorn, H.-D. Liebig, F. Fessl, D. Blaas, and E. Kuechler, Virology **181**, 46 (1991).
[15] C. U. T. Hellen, C.-K. Lee, and E. Wimmer, J. Virol. **67**, 2110 (1992).
[16] M. J. H. Nicklin, K. S. Harris, P. V. Pallai, and E. Wimmer, J. Virol. **62**, 4586 (1988).
[17] M.-F. Ypma-Wong, P. G. Dewalt, V. H. Lamb, and B. L. Semler, Virology **166**, 265 (1988).
[18] A. C. Palmenberg, Annu. Rev. Microbiol. **44**, 603 (1990).

Gly, Ser, or Ala at P1'; however, the enzymes of the various genera differ in their cleavage specificities. For example, polio virus picornain 3C sites are exclusively Glu+Gly pairs whereas those of EMCV picornain 3C include Gln+Gly, Glu+Ser and Glu+Ala. Further characteristics are the tendency toward an aliphatic amino acid at P4 and Pro at P2' in rhinovirus and enterovirus sites and Pro at P2 or P2' in those of cardioviruses.[2,18] Not all such amino acid pairs with such a context on the polyprotein are, however, cleaved by picornains 3C. Although 13 Gln+Gly amino acid pairs are present on the polio virus polyprotein, only 9 are cleaved. The topology and accessibility of the potential cleavage sites are a further determinant; those pairs that lie between β barrels are cleaved efficiently whereas those inside a β barrel are not.[19] Primary, secondary, and tertiary structure constraints are most probably responsible for the difference in the rates at which the cleavage sites are processed *in vivo*; in polio virus, the picornain 3C sites between 2C and 3A and 2B and 2C occur early in replication whereas those between 3A and 3B and/or 3C and 3D take place much later.[20]

Picornaviral infection also leads to changes in cellular rates of replication, transcription, and translation. Specific proteolysis of cellular proteins by picornains 2A and 3C appears to be responsible. For instance, the expression of polio virus picornain 2A in COS-1 monkey kidney cells led to 4-fold reduction in DNA synthesis, a 22-fold reduction in transcription, and a 3-fold reduction in translation.[21] The reduction in translation appears to be due to the limited specific proteolysis of eukaryotic translation initiation factor (eIF)-4γ by picornains 2A.[22,23]

FMDV picornain 3C has been shown to cleave specifically histone protein H3 in baby hamster kidney cells,[24] whereas polio virus 3C cleaves the activated form of transcription factor TFIIIC in HeLa cells, thus inactivating it and leading to the shutoff of host–cell transcription.[25]

[19] M.-F. Ypma-Wong, D. J. Filman, J. M. Hogle, and B. L. Semler, *J. Biol. Chem.* **263**, 17846 (1988).

[20] P. V. Pallai, F. Burkhardt, M. Skoog, K. Schreiner, P. Bax, K. A. Cohen, G. Hansen, D. E. H. Palladino, K. S. Harris, M. J. H. Nicklin, and E. Wimmer, *J. Biol. Chem.* **264**, 9738 (1989).

[21] M. V. Davies, J. Pelletier, K. Meerovitch, N. Sonenberg, and R. J. Kaufman, *J. Biol. Chem.* **266**, 14714 (1991).

[22] E. E. Wyckoff, R. E. Lloyd, and E. Ehrenfeld, *J. Virol.* **66**, 2943 (1992).

[23] H.-D. Liebig, E. Ziegler, R. Yan, K. Hartmuth, H. Klump, H. Kowalski, D. Blaas, W. Sommergruber, L. Frasel, B. Lamphear, R. Rhoads, E. Kuechler, and T. Skern, *Biochemistry* **32**, 7581 (1993).

[24] M. Tesar and O. Marquardt, *Virology* **174**, 364 (1990).

[25] M. E. Clark, T. Haemmerle, E. Wimmer, and A. Dasgupta, *EMBO J.* **10**, 2941 (1991).

Assay Procedures

Polyprotein Substrates for Intermolecular Cleavage

Translation of cloned cDNA fragments of picornaviral genomes in rabbit reticulocyte lysates (RRLs) has allowed the production of radiolabeled proteins corresponding to various parts of the polyprotein. In order to produce substrates for intermolecular cleavage, regions containing the picornain genes are excluded or only partially present. Efficient translation of subgenomic RNAs is achieved by placing the cDNA fragment of interest downstream from the EMCV 5' untranslated region; the RNA is produced by *in vitro* transcription from a T7 or SP6 promoter at the 5' side of the EMCV sequence.[26] The translation extracts are then incubated with picornain preparations and the products of cleavage are separated by SDS–PAGE and detected by fluorography. Given the problem in ascertaining the effective concentration of protein produced, quantification of results obtained is difficult. Nevertheless, this method is effective in detecting picornain activity in extracts of infected cells and bacterial extracts.[15,27]

Reagents

5× Incubation buffer: 250 mM NaCl, 250 mM Tris-HCl, pH 8.0, 25 mM Dithiothreitol (DTT), 5 mM EDTA.
Substrate: ^{35}S-labeled protein prepared by translating *in vitro* transcribed RNA in RRLs.

One should verify beforehand that translation of the *in vitro* transcribed RNA generates sufficient labeled polyprotein for assay. After translation, RRLs may be stored at −20°, although precipitation of the labeled protein may reduce the quality of the substrate.

Procedure. Mix 1 μl RRL containing ^{35}S-labeled polyprotein, (7 − x) μl H$_2$O, 2 μl 5× assay buffer, and x μl picornain preparation.

Incubate at 30° for 1 hr. Add 10 μl 2× Laemmli sample buffer, heat at 95° for 2–3 min, and analyze by SDS–PAGE[28] and fluorography.[29]

Polyprotein Substrates for Picornain 2A Intramolecular Cleavage

The intramolecular cleavage by picornain 2A has been monitored by expression of cDNA encoding the region between VP3 or VP1 and 2A

[26] S. K. Jang, H.-G. Kraeusslich, M. J. H. Nicklin, G. M. Duke, A. C. Palmenberg, and E. Wimmer, *J. Virol.* **62**, 2636 (1988).
[27] M. J. H. Nicklin, H.-G. Kraeusslich, H. Toyoda, J. J. Dunn, and E. Wimmer, *Proc. Natl. Acad. Sci. U.S.A.* **84**, 4002 (1987).
[28] U. K. Laemmli, *Nature (London)* **291**, 547 (1970).
[29] J. P. Chamberlain, *Anal. Biochem.* **98**, 132 (1979).

in bacteria[14] or by *in vitro* translation of this region in RRLs[15,30,31] as described above.

Procedure. For assay *in vitro* in RRLs, an RNA bearing at least 100 amino acids of VP1 and all of 2A is translated using standard methods and analyzed as above. DTT is usually added to 5 mM to ensure complete reduction of all SH groups. The translation products are separated and analyzed as above; three products are normally visible, corresponding to the N-terminal cleavage product (containing VP1), the C-terminal product (2A), and uncleaved material (VP1/2A). For an RNA bearing a wild-type 2A, generally only about 5% uncleaved material is present; mutations in the 2A active site or at the VP1/2A boundary or the presence of specific inhibitors can increase the amount of the precursor protein to 100%. Exact values are determined by densitometric scanning of the autoradiograms.

Oligopeptide Substrates

All picornains 2A and 3C so far examined are capable of cleaving oligopeptides (synthesized and purified using standard techniques[32]) with sequences derived from viral cleavage sites. Peptides should contain at least 12 amino acids (6 at either side of the cleavage site) for efficient cleavage. The standard method of monitoring the reaction is by separation of the cleavage products from uncleaved material by reversed-phase high-performance liquid chromatography (HPLC).[16,20,33,34] The nature of the cleavage products can be determined by N-terminal sequencing of the products or the measurement of their mass or by comigration with reference peptides. The following protocol is for determination of HRV2 picornain 2A activity in bacterial extracts and during purification.[23]

Reagents

Peptide (TRPIITTA↓GPSDMYVH): 20 mg/ml in buffer A (50 mM NaCl, 50 mM Tris-HCl (pH 8.0 at 20°), 1 mM EDTA, 5 mM DTT, and 5% glycerol).
0.125 M Perchloric acid.

Procedure. The picornain 2A fraction (brought to 80 μl with buffer A and heated at 25°) is mixed with 20 μl of the peptide (also at 25°); 10-μl

[30] C. U. T. Hellen, M. Faecke, H.-G. Kraeusslich, C.-K. Lee, and E. Wimmer, *J. Virol.* **65**, 4226 (1991).
[31] S. F. Yu and R. R. Lloyd, *Virology* **182**, 615 (1991).
[32] R. B. Merrifield, *J. Am. Chem. Soc.* **85**, 2149 (1963).
[33] W. Sommergruber, H. Ahorn, A. Zoephel, I. Maurer-Fogy, F. Fessl, G. Schnorrenberg, H.-D. Liebig, D. Blaas, E. Kuechler, and T. Skern, *J. Biol. Chem.* **267**, 22639 (1992).
[34] M. G. Cordingley, P. L. Callahan, V. V. Sardana, V. M. Garsky, and R. J. Colonno, *J. Biol. Chem.* **63**, 5037 (1990).

aliquots are removed aftere 1, 2, 3, 4, 5, 7, 10, 15, and 30 min, the reaction is stopped by the addition of 80 μl perchloric acid and the samples are placed on ice. Following centrifugation through Millipore (Bedford, MA) ultra-free-MC NMWL filter units at 14,000 rpm in a Sigma (St. Louis, MO) microcentrifuge for 30 min at 4° to remove precipitated protein (W. Sommergruber, personal communication, 1993) the sample is applied to either a Merck supersphere (Gibbstown, N.J.) or a Bakerbond (J. T. Baker, Phillipsburg, NJ) WP C_{18} column; uncleaved substrate and products are eluted with a linear gradient of 10–40% (v/v) acetonitrile in 0.1% trifluoroacetic acid (TFA). Peptide elution is monitored by measuring the UV absorption at 210 and 280 nm; the peak areas can be used to determine substrate conversion and hence the initial rate of cleavage. One unit of HRV2 2A proteinase is the amount of enzyme that cleaves 1 μmol of the above peptide per minute at 25°.

   Alternative Method. Polio virus picornain 2A has been assayed[35] using its esterase activity on the peptide GlyLeuGlyGlnMet-OCH$_3$ to generate GlyLeuGlyGlnMet-COOH. Separation of processed from unprocessed material was achieved on high-performance thin-layer chromatography silica gel 60 F_{254} plates (Gibbstown, N.J.) followed by detection with ninhydrin.

Competition Assays

   Comparison of the ability of peptides to serve as substrates for picornains 2A and 3C has generally been performed by assaying the cleavage of the peptide under test in the presence of a reference peptide.[20,33,34] A protocol for HRV2 picornain 2A is described here.[33]

Reagents

   Picornain 2A: 0.5 μM in buffer A.
   Standard and reference peptides: 4 mg/ml in buffer A.
   0.5 M Perchloric acid
   Procedure. 500 μl of the picornain solution and 25 μl of each peptide are brought to 34° and then mixed. 100-μl aliquots are removed after 2, 4, 8, 16, and 32 min; the reaction is stopped by the addition of 200 μl perchloric acid and placed on ice. Preparation for and analysis by HPLC is as above.
   Calculation. Provided that all substrates and products can be separated from another, the extent of substrate conversion of each peptide can be determined from the analysis of peak areas. $(V_{max}/K_m)_{rel}$ values can be calculated from the formula $(V_{max}/K_m)_1/(V_{max}/K_m)_2 = \log(1 - F_1)/\log(1 - F_2)$, where F is the fraction of substrate converted.[20]

[35] H. Koenig and B. Rosenwirth, *J. Virol.* **62**, 1243 (1988).

Purification of Picornains 2A and 3C from Virally Infected Cells

Purification to homogeneity of polio virus picornain 2A from polio virus-infected HeLa cells was carried out by solubilization using different detergents followed by gel filtration on a Superose 12 column[35]; approximately 30 μg of pure polio virus picornain 2A was obtained from 2×10^9 HeLa cells. Picornain 3C activity of polio virus has been enriched from an S-300 membrane-free supernatant from infected HeLa cell homogenates by concentrating the proteins by ammonium sulfate precipitation and gel filtration on a Superose 12 column.[16] However, neither yield nor purity was determined.

Purification of Recombinant Picornains 3C from Bacteria

Expression

Two methods have been described. The first produced polio virus picornain 3C in a soluble form,[16] whereas the second resulted in HRV14 3C appearing in the membrane fraction.[36] Because the first method has also been applied to HRV14 3C,[37] it is described in detail. Polio virus picornain 3C is expressed in the *Escherichia coli* strain BL21 (DE3) from a plasmid containing a T7 RNA polymerase promoter followed by an initiation codon, the gene for the picornain 3C of the Sabin strain of polio virus type 2, and a termination codon. For expression, bacteria are grown (0.5-liter culture) in M9 medium[38] containing per liter, 2 g NH_4Cl, 2 g casamino acids, 10 g glucose, 20 μmol $FeCl_3$, and 0.1 g sodium ampillicin to an A_{600} of 0.5 and are injected into 9.5 liters of the same medium in a 14-liter fermentor (New Brunswick, Edison, NJ). The culture is stirred (900 rpm) and aerated (16 liters/min); at an A_{600} of 2.5, isopropyl-β-D-thiogalactoside (IPTG) is added to 0.4 mM and incubation is continued for 2.5 hr. Bacteria are harvested by centrifugation at 3500 g for 15 min, washed with buffer B (100 mM NaCl, 40 mM Tris-HCl, pH 7.9), and the pellet stored at $-80°$.

Purification

All steps should be performed at 4°. After thawing, 15 g of cell paste is resuspended in 40 ml buffer B and lysed in a French press by two

[36] R. T. Libby, D. Cosman, M. K. Cooney, J. E. Merriam, C. J. March, and T. P. Hopp, *Biochemistry* **27**, 6262 (1988).
[37] M. G. Cordingley, R. B. Register, P. L. Callahan, V. M. Garsky, and R. J. Colonno, *J. Virol.* **63**, 5037 (1989).
[38] J. Sambrook, E. Fritsch, and T. Maniatis, "Molecular Cloning: A Laboratory Manual," 2nd Ed., Cold Spring Harbor Laboratory, Cold Spring Harbor, New York, 1987.

passages at 70 MPa. 40 ml of buffer C (100 mM KCl, 20 mM Tris-HCl, 10 mM MgCl$_2$, 5 mM DTT, 1 mM EDTA, pH 7.9) is added and the diluted lysate in centrifuged at 360,000 g for 2 hr. The supernatant is diluted to 200 ml with water and the pH is adjusted to 8.3 by the addition of 1 M Tris base. In contrast to most $E.$ $coli$ proteins, polio virus picornain 3C fails to bind to DEAE-cellulose at this pH; the sample is therefore applied to a 180-ml Whatman (Clifton, NJ) DE-52 DEAE column preequilibrated with buffer D (40 mM Tris-HCl, 1 mM DTT, pH 7.9). Buffer D is also used for elution; the first 130 ml of flow-through are discarded, the next 320 ml being collected. Then 2.5 ml of 0.5 M EDTA (trisodium salt) and 5 ml of 0.5 M morpholinoethanesulfonic acid are added dropwise with stirring followed by the addition of solid ammonium sulfate to 0.55 g/ml over a 15-min period. After 12 hr, proteins are pelleted at 6000 g for 30 min and resuspended in less than 4 ml of buffer E (100 mM NaCl, 20 mM HEPES–NaOH, 1 mM EDTA, 1 mM DTT, pH 7.4). Following a clearing spin at 4000 g for 5 min, protein is precipitated by the addition of 2 g solid ammonium sulfate and centrifugation at 10,000 g for 30 min. The pellet is resuspended in 1 ml of buffer E and then chromatographed on a 55 × 1.6-cm column of superfine Sephadex G-75 (Pharmacia, Piscataway, NJ) in buffer E at 6 ml/hr. Peak fractions are identified by SDS–PAGE and the protein is concentrated as above with ammonium sulfate and redissolved in 1 ml of buffer E. This procedure yields about 5 mg of 98% pure picornain 3C.

Expression and Purification of Recombinant Picornain 3CD
  from Bacteria

Expression

Production of picornain 3CD is difficult because the molecule slowly undergoes autocatalytic processing to generate 3C and 3D. For polio virus picornain 3CD, this was overcome by using site-directed mutagenesis to substitute the P4Thr residue of the 3CD cleavage site with Lys; this mutation almost abolishes self-processing.[39] Expression of this picornain 3CD variant in bacteria was as above for polio virus picornain 3C[39]; however, the temperature of induction was reduced to 25°. Nevertheless, much picornain 3CD remained insoluble.

Purification

All operations except the gel filtration step are performed at 4°. Fifteen grams of induced cell paste is suspended in 40 ml buffer F [50 mM Tris-

---

[39] K. S. Harris, S. R. Reddigari, M. J. H. Nicklin, T. Haemmerle, and E. Wimmer, *J. Virol.* **66**, 7481 (1992).

HCl (pH 8.0), 100 mM NaCl, 10 mM dithiothreitol (DTT), 1 mM EDTA, and 5% glycerol] and the cells are lysed by three passages through a French pressure cell. 40 ml of buffer F is added and the lysate is centrifuged at 10,000 g for 20 min. The supernatant from this centrifugation step is spun at 100,000 g for 1 hr. EDTA is added to a final concentration of 5 mM. Subsequently, ammonium sulfate is slowly added with stirring to 35% saturation and the precipitated proteins are collected by centrifuging at 10,000 g for 15 min. The pellet is dissolved in 15 ml of buffer H [50 mM Tris-HCl (pH 8.0), 1 mM DTT, 1 mM EDTA, and 5% glycerol] containing 30 mM NaCl and dialyzed twice against 1 liter of the same buffer overnight at 4°.

To recover 3CD from the membrane fractions, the pellets from the 10,000 g and the 100,000 g centrifugations that followed lysis are separately suspended in 20 ml of buffer G [50 mM Tris-HCl (pH 8.0), 1 M NaCl, 10 mM. dithiothreitol, 1 mM EDTA, and 5% glycerol] and extracted for 30 min at 4°. The suspensions are then centrifuged at 100,000 g and the supernatants are dialyzed twice against 1 liter of buffer C containing 30 mM NaCl at 4°.

The next day, the dialyzed fractions are pooled and loaded onto a 70-ml DEAE-cellulose column equilibrated with buffer H containing 30 mM NaCl. The column is washed with 280 ml of this buffer and bound proteins are eluted with 140 ml of buffer H containing 300 mM NaCl. This fraction is then dialyzed twice against 2 liters of buffer H containing 50 mM NaCl at 4° for 2 hr and the retentate is then loaded onto a 15-ml phosphocellulose p11 column equilibrated with buffer H containing 50 mM NaCl. After a wash with four column volumes of this buffer, the bound proteins are eluted with 30 ml of buffer H containing 200 mM NaCl. The eluted protein is precipitated by the addition of ammonium sulfate to 50% saturation and is collected by centrifugation at 10,000 g for 15 min at 4°. The pellet is dissolved in 2 ml of buffer J (100 mM sodium phosphate, 100 mM sodium sulfate, 1 mM DTT, 1 mM Na$_3$EDTA, 100 μM GTP, and 5% (v/v) glycerol) and spun at 14,000 rpm in an Eppendorf microfuge for 5 min to remove any particulate matter. The supernatant is loaded onto a TosoHaas G3000SW (Mahwah, NJ) HPLC sizing column equilibrated with buffer J at room temperature. Peak fractions are pooled; the yield is 0.4 mg (at 86% homogeneity) from 15 g of cell paste.[39]

Expression and Purification of Recombinant Picornain 2A from Bacteria

Expression

Optimization of the expression vector and of the induction conditions has led to a protocol that allows the production of 5 mg of pure HRV2 picornain 2A or CV serotype B4 2A picornain from 5 g $E.$ $coli$ cell paste.[23]

The picornains 2A are expressed as fusion proteins containing 10 amino acids of the bacteriophage T7 gene 10 protein, about 30 amino acids of VP1, and the 2A protein followed by a termination codon. The 28 amino acids of VP1 are enough to allow picornains 2A to cleave themselves off the growing polypeptide chain, resulting in the accumulation of mature picornain 2A in the bacterial cell. For expression, a 400-ml stationary-phase culture of *E. coli* BL21 (DE3) pLysE bearing a 2A expression plasmid is diluted 1 : 12 in 5 liters of fresh medium. After a 3-hr incubation with shaking (160 rpm) at 34° (A_{590} 0.5), IPTG is added to 0.3 mM; the cultures are incubated for a further 7 hr at 34°. The bacteria are harvested by centrifugation at 6000 g for 10 min; after rinsing with 200 ml of washing buffer (150 mM NaCl, 50 mM Tris-HCl, pH 8.0, 1 mM EDTA), the cell pellets are stored at −80°.

Purification

Typically, the cell pellet from 5 liters of culture is resuspended in 30 ml of buffer A; cell lysis is completed by sonication using three to six bursts of 5 sec each using an MSE ultrasonic power unit. Insoluble material is removed by centrifugation (39,000 g, 30 min; all centrifugation steps are performed at 4°); 8 ml of a saturated ammonium sulfate solution (adjusted to pH 8.0 with Tris-HCl) is added with stirring to the 32 ml of supernatant (20% saturation). Precipitation is for 12 hr at 4°. Precipitated proteins are removed by centrifugation (39,000 g, 30 min), the supernatant (40 ml) is saved, and a further 13 ml of ammonium sulfate is added (40% saturation). After at least 4 hr at 4°, the precipitated proteins are collected by centrifugation at 15,000 g for 30 min and are dissolved in 31 ml buffer A. This solution is loaded onto a Pharmacia FPLC (fast protein liquid chromatography) HR 10/10 Mono Q column equilibrated with buffer A (all chromatography steps are performed at room temperature). After washing the column with four bed volumes of buffer A, the following gradient composed of buffer A and buffer K (as buffer A, except for 1 M NaCl) is applied: 20 ml of 0–30% K, 120 ml of 30–60% K, and 20 ml of 60–100% K. HRV2 picornain 2A elutes at around 35% K whereas picornain CVB4 2A elutes at 27% K. Peak fractions are identified by SDS–PAGE and cleavage activity (as detailed below) and are pooled and applied directly to a Pharmacia Superdex 75 Hiload 26/60 column equilibrated with buffer A. The gel filtration column is developed with buffer A. HRV2 2A proteinase elutes with an apparent M_r of 30,000 whereas the CVB4 2A elutes at M_r 17,000. Fractions are examined by SDS–PAGE and those containing 2A proteinase at a purity of more than 98% are pooled (total volume, 18 ml) and concentrated in an Amicon (Danvers, MA) Centriprep 10 cell.

Properties of Enzymes

Both picornains 2A and 3C have a pH optimum between pH 7 and pH 8.[33,37] HRV2 picornain 2A is active over a temperature range from 4° to 37° and is largely unaffected by ionic strength. Cleavage of capsid protein precursors by picornain 3CD is inhibited by the presence of 0.1% Triton X-100.[16,17]

Inhibitors

Picornains 2A and 3C are inhibited by typical cysteine proteinase inhibitors such as *N*-ethylmaleimide (NEM), iodoacetamide, and mercuric(II) ions, but are characteristic in not being inhibited by the papain inhibitor E-64.[33,35,40] Certain serine proteinase inhibitors (e.g., chymostatin, elastatinal) are also effective against the picornains. In general, the inhibitor profiles reflect the presence of cysteine as the active site nucleophile, acting in the environment of a serine proteinase.

Specificity and Kinetic Properties

The intramolecular reaction characterized by HRV2 picornain 2A has an absolute requirement for glycine at P1′ and a strong preference for Thr at P2; a variety of amino acids (Ala, Tyr, Phe, Leu) can be tolerated at P1. Similar results were obtained with polio virus picornain 2A, though mutations at P1′ did not abolish cleavage completely; furthermore, certain substitutions resulted in cleavage at other sites.[15]

HRV 2A picornain has a K_m of 0.5 mM on a 16-residue peptide derived from the sequence at the VP1/2A boundary (TRPIITTA⊣GPSDMVY). As in the case of intramolecular cleavage, substitutions at P1′ and P2 are detrimental to cleavage whereas most substitutions at P1 are well tolerated. Use of truncated peptides indicates an extended substrate-binding pocket; the smallest cleavable symmetrical peptide as a dodecapeptide (P6–P6′). The asymmetric peptide P8–P1′ was still cleaved, albeit 30-fold less efficiently than the P8–P8′ peptide; in contrast, the cleavage of P8–P2′ was almost as efficient as the P8–P8′ peptide.[33] Cleavage of radiolabeled polyprotein substrates by polio virus picornain 2A was also reduced by changes at P2 and P1′ but was tolerant to changes at P1.[15]

On peptides derived from their polyprotein cleavage sites, HRV14 picornain 3Cs have K_m values ranging from 0.25 to 2.6 mM and V_{max} values ranging from 1.0 to 7 μmol/min/μg.[40] Generally, the efficiency of cleavage of a peptide correlates well with the rate of cleavage at that

---

[40] D. C. Orr, A. C. Long, J. Kay, B. M. Dunn, and J. M. Cameron, *J. Gen. Virol.* **70,** 2931 (1989).

particular site *in vivo*. Thus, cleavage by polio virus and HRV14 picornains 3C is most efficient on peptides derived from the cleavage site 2C/3A and least efficient on that from 3C/3D.[20,34]

Cleavage of substituted peptides and mutational analysis of cleavage sites *in vivo* indicate that the occupation of positions P4, P1, P1′, and P2′ are critical for picornain 3C cleavage. Optimal occupancy for polio virus picornain 3C would be P4 Ala, P1 Gln, P1′ Gly and P2 Pro[20]; this pattern is that found for HRV14, except that Thr and Val are as well accepted as Ala at P4.[34] Nevertheless, mutant polio viruses containing Ala or Ser at the P1′ position of the 3C/3D junction were still viable.[41]

[41] K. M. Kean, N. Teterina, and M. Girard, *J. Gen. Virol.* **71**, 2553 (1990).

[41] Adenovirus Endopeptidases

By JOSEPH M. WEBER and KAROLY TIHANYI

Adenoviruses, like many complex viruses, encode an endopeptidase that is required for virion maturation and infectivity. This was initially demonstrated by the isolation of the *Ad2ts1* temperature-sensitive mutant, which is defective for proteolytic cleavages at the nonpermissive temperature, and its mapping to the L3 23-kDa protein.[1,2] Formal proof of the proteinase came with the demonstration of the *in vitro* cleavage of viral precursor proteins with *Escherichia coli*-expressed and purified enzyme.[3,4] The gene has been sequenced in 12 different adenovirus serotypes, and, although the amino acid sequence is highly conserved, it lacks typical proteinase motifs and shows no significant homology to known proteins.[5] The enzyme appears to be a monomer and it shows no changes at its N terminus or in its molecular weight, such as many other proteinases show during activation.[6]

[1] J. M. Weber, *J. Virol.* **17**, 462 (1976).
[2] L. Yeh-Kai, G. Akusjarvi, P. Alestrom, U. Pettersson, M. Tremblay, and J. M. Weber, *J. Mol. Biol.* **167**, 217 (1983).
[3] A. Houde and J. M. Weber, *Gene* **88**, 269 (1990).
[4] C. W. Anderson, *Virology* **177**, 259 (1990).
[5] F. Cai and J. M. Weber, *Virology* **196**, 358 (1993).
[6] K. Tihanyi, M. Bourbonniere, A. Houde, C. Rancourt, and J. M. Weber, *J. Biol. Chem.* **268**, 1780 (1993).

Expression of Recombinant Endopeptidases

The endopeptidase coding sequence was subcloned in pLAM, a modified pRIT2T expression vector under the control of the pL promoter.[6] Recombinant plasmids were transfected and selected on N99CI$^+$ cells, which express the wild-type λ CI repressor constitutively. The selected plasmids were then transferred into AR120 cells for expression. AR120 is derived from N99 and is CI$^+$.[7] Induction through the inactivation of CI$^+$ can be promoted with nalidixic acid (80 μg/ml) or via the SOS pathway when cells attain high density.[7] This second alternative was found equally efficient, if not more so, as nalidixic acid induction. Induction by means of inactivation of a temperature-sensitive repressor was found to promote the formation of inclusion bodies and was therefore abandoned.

The adenovirus endopeptidase has also been successfully expressed in another *E. coli* vector system[4] as well as in insect cells using a baculovirus expression system.[8]

Proteinase Assays

Protein Substrate Assay

Proteinase activity is measured by the cleavage of core protein pVII to VII as detected by SDS–PAGE by autoradiography.[6] The substrate consists of [^{35}S]methionine-labeled purified, disrupted Ad2 ts1-39C virions. Virions are disrupted in 2 *M* urea/50 m*M* Tris-HCl/1 m*M* EDTA/5 m*M* dithiothreitol (DTT), at 4° overnight. This preparation is dialyzed to remove the urea and centrifuged at 100,000 *g* for 1 hr. The supernatant contains the proteinase activity and most of the viral proteins. The reaction mixture (final pH 7.5) contains 20 μl of substrate, 20 μl of enzyme, and is incubated at 37° for 18 hr.

Peptide Substrate Assay

The reaction mixture contains 5–40 n*M* protease in 0.1 *M* phosphate buffer, pH 6.5. The protease is activated with 100-fold molar excess of peptide pVI-c (GVQSLKRRRCF) for 60 min prior to the addition of the 100 μ*M* octapeptide substrate [LAGG4-F(NO$_2$)RHR], which starts the cleavage reaction. The decrease in absorbance is monitored at 310 nm.[9]

[7] J. E. Mott, R. A. Grant, Y.-S. Ho, and T. Platt, *Proc. Natl. Acad. Sci. U.S.A.* **82,** 88 (1985).
[8] A. Webster, R. T. Hay, and G. Kemp, *Cell (Cambridge, Mass.)* **72,** 97 (1993).
[9] A. D. Richards, L. H. Phylip, W. G. Farmerie, P. E. Scarborough, A. Alvarez, B. Dunn, Ph.-H. Hirel, J. Konvalinka, P. Strop, L. Pavlickova, V. Kostka, and J. Kay, *J. Biol. Chem.* **265,** 7733 (1990).

Cleavage of the peptide substrate can also be monitored by HPLC.[8] Mangel *et al.*[10] have employed a rhodamine-tagged tetrapeptide as a fluorogenic substrate.

Purification of Endopeptidase

Bacteria (AR120) bearing the pLPV expression vector are harvested from 1 liter of medium and suspended in 3 volumes of 20 mM Tris-HCl buffer containing 10% glycerol, 10 mM 2-mercaptoethanol, and 0.5 mM EDTA, pH 8.5 (buffer A), and lysozyme added at a final concentration of 30 μg/ml. Inclusion bodies are sedimented by centrifugation at 10,000 g for 15 min and the supernatant is kept. The sedimented inclusion bodies are dissolved in buffer A (2 volumes) saturated with urea and then diluted slowly with the supernatant to promote refolding.

DEAE-Sephacel Batch Treatment. The above solution is batch treated with 10 ml of DEAE-Sephacel equilibrated with buffer A. The support is washed twice with 20 ml of buffer A and the wash is combined with the batch-treated solution.

Hydroxyapatite Batch Treatment. The DEAE-Sephacel-treated solution is then added to 15 ml hydroxyapatite equilibrated with 20 mM phosphate buffer containing 10% glycerol (v/v), 10 mM 2-mercaptoethanol, 0.5 mM EDTA, pH 6.5 (buffer B). After washing the hydroxyapatite three times with buffer B, the proteins are eluted by stepwise washing with increasing concentrations of phosphate in buffer B. The protease elutes in the 150–200 mM phosphate fraction.

Carboxymethyl-cellulose Batch Treatment. The endopeptidase-containing solution is added to 20 ml carboxymethyl-cellulose equilibrated with buffer B. After 1 hr cold shaking, the resin is washed three times with buffer B. The endopeptidase is eluted with 0.1 M Tris buffer, pH 8.5, containing 0.1 mM DTT and 0.5 mM EDTA.

After concentrating the eluted pure endopeptidase on Centricon 10 microconcentrator (Amicon, Danvers, MA), 10% glycerol is added to the endopeptidase solution. This procedure generally yields 4 mg of enzyme at better than 98% purity. The major contaminant is lysozyme. Extraction procedures that avoid lysozyme may be expected to result in better purity. Three cycles of freeze-thawing followed by sonication can be substituted for lysozyme treatment. Alternative purification procedures from bacteria and insect cells (baculovirus expression vector) have also been reported.[8,10]

[10] W. F. Mangel, W. J. McGrath, D. L. Toledo, and C. W. Anderson, *Nature (London)* **361**, 274 (1993).

Western Blotting

SDS–PAGE gels are electroblotted onto nitrocellulose C-extra membranes (Amersham) as described,[11] and antigen–antibody complexes are detected with protein A labeled with ^{125}I by the chloramine-T method. A polyclonal rabbit serum prepared against purified recombinant endopeptidase as well as a serum against the 15 N-terminal residues both give excellent results.

Affinity Labeling

2.5 μCi of 50–60 mCi/mmol iodo[2-^{14}C]acetate or 2.5 μCi of 3 Ci/mmol [1,3-^{3}H]diisopropyl fluorophosphate is added to various amounts of endopeptidase and incubated for 60 min at room temperature in the dark under nitrogen. The reaction is stopped by adding 100 mM 2-mercaptoethanol. The amount of incorporated radioactivity can be determined three ways: (1) The endopeptidase is precipitated with 10% final concentration of trichloroacetic acid (TCA), centrifuged, and washed with TCA three times. The washed proteins are dissolved in 0.1 M Tris buffer, pH 8.0, containing 1% sodium dodecyl sulfate (SDS) and 10 mM 2-mercaptoethanol and are mixed with scintillation cocktail. (2) 1 M NaOH (98:2) is added to the reaction mixture and centrifuged for 10 min at 4°. This procedure is repeated three times and the pellets are dissolved and added to a scintillation cocktail. (3) The reaction mixture is loaded on a cellulose thin-layer plate and developed in n-butanol:pyridine:distilled water (4:3:3) mixture. Following autoradiography the spots corresponding to the labeled protein are scraped off the plate and placed in a scintillation cocktail.

Endopeptidase Sequences

Figure 1 shows an alignment of the 12 known adenovirus endopeptidase sequences. All of the sequences are translations from DNA sequences. Direct verification by protein sequencing has been partially accomplished for Ad2 only.[4] The amino acid homology, as defined by the percentage similarity among the 12 sequences, varies from 93.6% for the closest pair (H40 and H41) to 54.5% for the most distant pair of enzymes (H4 and Bav7). Only 17.6% of the residues are identical across the 12 sequences in this particular alignment. Homology searches of data banks with the entire Ad2-EP sequence, or with three highly conserved segments individ-

[11] W. N. Burnette, *Anal. Biochem.* **112**, 195 (1981).

ually (residues 21–33, 41–62, and 68–97), using TFASTA and BLAST failed to identify any significant relationship to known sequences.

Properties

Activation

Studies with recombinant endopeptidase have shown that the protein requires a cofactor for enzymatic activity.[8,10] The cofactor is an 11-amino-acid cleavage fragment from viral protein pVI: GVQSLKRRRCF. It is not known how this peptide activates the enzyme, nor whether it is absolutely required for the cleavage of other adenoviral substrate proteins. Optimal activation requires oxidized peptide, suggesting that the disulfide-linked dimer of the 11-amino-acid peptide is the active form.[8] This region of the pVI protein has been sequenced in several adenovirus serotypes and appears to be highly conserved (Table I). Negatively charged polymers appear to enhance enzyme activity by about 10-fold.[10]

Active Site

The sequence alignment of the endopeptidase from 12 different adenoviruses show one conserved histidine, H54, and two conserved cysteines, C104 and C122[5] (Fig. 1). Mutations introduced in these residues (as well as others) combined with studies of the binding of E-64 [*trans*-epoxysuccinyl-L-leucylamido-(4-guanidino)butane] are consistent with H54 and C104 as two residues of the active site.[12] A candidate for a third residue has not been proposed.

Substrate Specificity

Using a variety of octapeptides, Webster *et al.* proposed a consensus cleavage site that can be updated to (M,I,L)XGG+X or (M,I,L)XGX+G.[13] Table II shows a compilation of adenovirus proteins for which the cleavage site has been inferred or confirmed. All cleaved sites conform to the proposed consensus. Adenovirus proteins contain a number of consensus sites that are known not to be cleaved, presumably because they are inaccessible to the enzyme.[14] Table III shows the distribution of amino acids in the P4 to P4' positions taken from data in Table II. The enzyme is also capable of cleaving nonviral proteins.[6,14]

[12] C. Rancourt, K. Tihanyi, M. Bourbonniere, and J. M. Weber, *Proc. Natl. Acad. Sci. USA.* **91**, 844 (1994).
[13] A. Webster, S. Russell, W. C. Russell and G. Kemp, *J. Gen. Virol.* **70**, 3225 (1989).
[14] J. M. Weber, *Semin. Virol.* **1**, 379 (1990).

```
                                          26
H40         ......MGSS EQELvAIVRE LGCGPYFLGT FDKRFPGFma PHKLACAIVN
H41         ......MGSS EQELvAIaRD LGCGsYFLGT FDKRFPGFma PnKLACAIVN
H12         ......MGSS EQELtAIVRD LGCGPYFLGT FDKRFPGFVS rdrLsCAIVN
H2          ......MGSS EQELkAIVkD LGCGPYFLGT YDKRFPGFVS PHKLACAIVN
H3          mtcgsgnGSS EQELkAIVRD LGCGPYFLGT FDKRFPGFma PdKLACAIVN
H4          ....maaGSg EQELRAIIRD LGCGPYFLGT FDKRFPGFma PHKvACAIVN
Bav3        ......MGSr EeELRfIlhD LGvGPYFLGT FDKhFPGFIS kdrMsCAIVN
Cav1        ...maegGSS EeELRAIVRD LavtPFFLGT FDKRFPGFIS sqritCAVVN
Mav1        ......MGSS EtqELRqlVaD LGiGs.FLGi FDKhFPGFIS vnKpACAIVN
Aav1        .....MsGtt EtqLRdllss MhlrhrFLGv FDKsFPGFld PHvpAsAIVN
Bav7        .....MsGlS EkEvflllss LqCthgFLGT FDcRFPGFIn kvKvqtAIIN
Consensus         MGSS EQELRAIVRD LGCGPYFLGT FDKRFPGFIS PHKLACAIVN

                 54              67
H40         TAGRETGGVH WLALAWNPkn rTCYLFDPFG FSDeRLKQIY QFEYEGLLkR
H41         TAGRETGGVH WLALAWNPkS hTCYLFDPFG FSDeRLKQIY QFEYEGLLkR
H12         TAGRETGGVH WLAFgWNPkS hTCYLFDPFG FSDQRLKQIY QFEYEsLLRR
H2          TAGRETGGVH WMAFAWNPrS kTCYLFEPFG FSDQRLKQVY QFEYEsLLRR
H3          TAGRETGGeH WLAFgWNPry nTCYLFDPFG FSDeRLKQIY QFEYEGLLRR
H4          TAGRETGGeH WLAFAWNPrS nTCYLFDPFG FSDQRLKQIY QFEYEGLLRR
Bav3        TAGRETGGVH WLAMAWhPaS qTfYMFDPFG FSDQkLKQIY nFEYqGLLkR
Cav1        TAGRETGGVH WLAMAWNPrS kTfYMFDPFG FSDskLKQVY sFEYEGLLRR
Mav1        TAsRETGGVH WLAMAWyPtS sTfYLFDPFG FSDrkLqQVY kFEYErLLkR
Aav1        TgsRasGGmH WigFAFdPaa grCYMFDPFG WSDQkLwelY rvkYnaFMRR
Bav7        TgpREqGGIH WiALAWdPkS yqmFiFDPLG WkndqLmkyY kFsYsnLikR
Consensus   TAGRETGGVH WLAFAWNP-S -TCYLFDPFG FSDQRLKQIY QFEYEGLLRR

                 104                  122              ↓ts1
H40         SALASTPDHC ITLIKSTQT. ....VQGPFS AACGLFCCMF LHAFVnWPts
H41         SALASTPDHC ITLVKSTQT. ....VQGPFS AACGLFCCMF LHAFIhWPsN
H12         SALAaTkDRC VTLeKSTQT. ....VQGPFS AACGLFCCMF LHAFthWPdh
H2          SAiASsPDRC ITLeKSTQs. ....VQGPnS AACGLFCCMF LHAFanWPqt
H3          SALA.TkDRC ITLeKSTQs. ....VQGPrS AACGLFCCMF LHAFVhWPdr
H4          SALA.TkDRC VTw.KShQTc rvrvgrcgFS AACstaCa.. ......WPt.
Bav3        SALtSTaDRC lTLIqSTQs. ....VQGPnS AACGLFCCMF LHAFVrWPlr
Cav1        SAiASTPDRC VTLaKSneT. ....IQGPnS AACGLFCCMF LHAFVnWPdN
Mav1        SAvsSssskC VTLVKShQT. ....VQGPhS AACGLFCvLF LaAFgkYPqN
Aav1        tgL.rqPDRC fTLVrSTea. ....VQcPcS AACGLFsaLF ivsFdrYrsk
Bav7        SAL.SsPDkC VkvIKnsQs. ....VQctca gsCGLFCvFF LycFykYksN
Consensus   SALASTPDRC VTL-KSTQT- ----VQGPFS AACGLFCCMF LHAFV-WP-N

                 160
H40         PMErNPTMDL lTGVPNSMLQ SPQVvPTLRh NQERLYRFLa qrSPYFqrHc
H41         PMEqNPTMDL lTGVPNSMLQ SPQVePTLRR NQERLYRFLt qHSPYFRrHR
H12         PMDkNPTMDL lTGVPNcMLQ SPQVvgTLqR NQneLYkFLN slSPYFRhnR
H2          PMDhNPTMnL iTGVPNSMLn SPQVQPTLRR NQEqLYsFLe rHSPYFRSHs
H3          PMDgNPTMkL vTGVsNSMLQ SPQVQPTLRR NQEvLYRFLN tHSsYFRSHR
H4          PMDkNPTMnL lTGVPNgMLQ SPQVePTLRR NQEaLYRFLN sHSaYFRSHR
Bav3        aMDNNPTMnL ihGVPNnMLe SPssQnvFlR NQqnLYRFLr rHSPhFvkHa
Cav1        PfnhNPTMgp lksVPNykLy dPtVQhvLWe NQEkLYkFLe KnSaYFRaHa
Mav1        PMnNNPiMgp ieGVPNdqMf nPcytkTLYR NQqWvYsYLN KnSlYFRlHv
Aav1        PMDgNPviDt vvGVkhenMn SPpyrdiLhR NQERtYwWWt KnSaYFRaHq
Bav7        afkNclfqsL ygsIPs...l tPpnptnLhk NQDFLYkFFk ekSlYFRqne
Consensus   PMDNNPTMDL -TGVPNSMLQ SPQVvPTLRR NQERLYRFLN KHSPYFRSHR

H40         ERIkKATAFD qMKNNM.... ......  205
H41         ERIEKATAFD qMKNaqVlfh nkify  214
H12         ERIEKATsFt KMqNgLk... ......  206
H2          aqIrsATsFc hLKNm..... ......  204
H3          aRIErATAFD rMdmq..... ......  209
H4          aRIEKATAFD rMnqdM.... ......  201
Bav3        aqIEadTAFD KMltN..... ......  204
Cav1        aaIktrTAFn KLKq...... ......  206
Mav1        ElIkKnTAFD KLlvrk.... ......  204
Aav1        EelrreTALn aLpeNhV... ......  206
Bav7        EyIvsnTkig liKshi.... ......  202
Consensus   ERIEKATAFD KMKNNMV--- -----  202
```

TABLE I
SEQUENCES OF PROTEINASE-ACTIVATING
PEPTIDES (PVI-c)

Virus	Peptide
H2	GVQSLKRRRCF
H5	GVQSLKRRRCF
H12	GVKSLKRRRCY
H41	GVKSLKRRRCY
Mav1	GLQPIKRRRCF[a]

[a] J. M. Weber, F. Cai, R. Murali, and R. M. Burnett, *J. Gen. Virol.* **75,** 141 (1994).

Physical–Chemical Properties

The specific activity of the endopeptidase was 12.9 nmol^{-1} min^{-1} nmol^{-1} enzyme. It is a monomer protein of 24,838 Da carrying no reported posttranslational modifications and no measurable disulfide bonds.[6] The p*I* is 10.59 and the pH optimum is between 7–8. The protein binds to DNA (unpublished observations) and appears to be found in association with the nuclear matrix in infected cells.[15]

Inhibitors

The enzyme is inhibited by *N*-ethylmaleimide, iodoacetate, dithiopyridine, *p*-chloromercuribenzoate, leupeptin, and E-64.[6,14,16] Addition of dithiothreitol abrogates the inhibition by phenylmethyl sulfonate. Earlier reports of inhibition by diisopropyl fluorophosphate were not reproduced once purified recombinant enzyme was used in the assay systems.

[15] G. Khittoo, L. Delorme, C. V. Dery, M. L. Tremblay, J. M. Weber, V. Bibor-Hardy, and R. Simard, *Virus Res.* **5,** 391 (1986).
[16] A. Webster, W. C. Russell, and G. D. Kemp, *J. Gen. Virol.* **70,** 3215 (1989).

Fig. 1. Multiple sequence alignment of the human (H), bovine (Bav), canine (Cav), murine (Mav), and avian (Aav) endopeptidases. The alignment was constructed by the PILEUP program and displayed with PRETTY (University of Wisconsin Genetics Computer Group package). Numbering is according to the Ad2 (H2) sequence. Because Ad5 differs from Ad2 only in having an H, instead of an R, at position 63, it is not included. Conserved residues are shown in capital letters; nonconserved, in lowercase letters. For easier reference, some of the residues are numbered. The two active site residues H54 and C104 are boxed in. The location of the Ad2 *ts1* mutation (P137L) is indicated.

TABLE II
ADENOVIRUS PROTEINS CLEAVED BY PROTEINASE

Virus	Protein[a]	Cleavage site		P1	Peptide cleaved[b]	
					In vitro	In vivo
H2[c]	pVII (197)	MFGG	AKKR	24	Yes	Yes
	pVIII (227)	LAGG	FRHR	111	Yes	—
	pVI (250)	MSGG	AFSW	33	Yes	Yes
	pVI (250)	IVGL	GVQS	239	—	Yes
	11K (79)	MAGH	GLTG	26	—	—
		LTGG	MRRA	31	—	Yes
		MRGG	ILPL	50	—	Yes
	pTP (668)	MRGF	GVTR	172	Yes	—
		MGGR	GRHL	180	Yes	—
		LGGG	VPTQ	314	No	—
		MTGG	VFQL	346	No	—
	pIIIa (585)	LGGS	GNPF	570	—	—
	L-1-52K (415)	LAGT	GSGD	351	—	—
H12	pVI (265)[d]	LNGG	AFNW	33	—	Yes
	pVI (265)[d]	IVGL	GVKS	254	—	Yes
	11K (72)[d]	LTGN	GRFR	27	—	Yes
	11K (72)[d]	MKGG	VLPF	42	—	Yes
	pVIII[d]	—	AKTR	—	—	Yes
	pIIIa[d]	—	GNPF	—	—	Yes
H41	pVIII (233)[e]	LAGG	SRHV	111	—	—
	pVI[f]	IVGL	GVKS	—	—	—
CAN1	pVII (132)[g]	LFGG	AKQK	23	—	—
MAV1	pVI[h]	IMGL	GLQP	—	—	—

[a] The number of amino acids in the precursor protein is given in parentheses.
[b] A dash indicates that the cleavage site has not been confirmed by direct means, such as sequencing or in vitro cleavage of the cognate peptide.
[c] The cleavage sites in H5 proteins pVI, pVII, pVIII, and L-1-52K are identical to those in H2 [J. Chroboczek, F. Bieber, and B. Jacrot, Virology 186, 280 (1992)].
[d] P. Freimuth and C. W. Anderson, Virology 193, 348 (1993).
[e] N. Pieniazek, J. Velarde, D. Pieniazek, and R. B. Luftig, Nucleic Acids Res. 17, 5398 (1989).
[f] C. I. A. Toogood, R. Murali, R. Burnett, and R. T. Hay, J. Gen. Virol. 70, 3203 (1989).
[g] F. Cai and J. M. Weber, Virology 193, 986 (1993).
[h] J. M. Weber, F. Cai, R. Murali, and R. M. Burnett, J. Gen. Virol. 75, 141 (1994).

Role in Virus Infection

The adenovirus proteinase is encapsidated during virus assembly and it is thought that the precursor proteins are cleaved during or subsequent to virus assembly.[14] Six of the currently known substrate proteins, i.e.,

TABLE III
AMINO ACIDS AT PROTEINASE CLEAVAGE SITE

N P4	P3	P2	P1	P1'	P2'	P3'	P4' C
A	G	G	A		S	A	Small, hydrophilic
G		S	G	S	T	S	
S		T	S	P	P	P	
T					G	G	
						A	
N	N		N		N	N	D	Acid, acid amide, hydrophilic
					Q	Q	
R	R		R		R	R	R	Basic
K	K		H		K	K	K	
						H	
M	V		L	M	V		L	Small, hydrophobic
I	M			I	L		V	
L				V				
F	F		F	F	F	F	F	Aromatic
							W	

IIIa, pTP, 11K (also known as mu, μ), pVI, pVII, and pVIII, are present in virions. The scaffolding protein L1-52K is present in assembly intermediate.[22] The requirement for the peptide cofactor may be a critical step in delaying enzyme activity until virus assembly. Although the precise function of these maturation cleavages remains unknown, studies with *ts1* have shown that cleavage is not required for virus assembly, but such virions are not infectious because they fail to uncoat.[1,17–19] Therefore, one possible function of maturation cleavages is the preparation of virions for uncoating in a subsequent infection. A second function might be to release completed virions from their anchorage on the nuclear matrix via the cleavage of the precursor to the terminal protein (pTP), which is covalently linked to the viral DNA destined to be packaged.[15,20,21] The proteinase

[17] M. A. A. Mirza and J. M. Weber, *Intervirology* **13**, 307 (1980).
[18] M. A. A. Mirza and J. Weber, *J. Virol.* **30**, 462 (1979).
[19] B. D. Miles, R. B. Luftig, J. A. Weatherbee, R. R. Weihing, and J. M. Weber, *Virology* **105**, 265 (1980).
[20] M. D. Challberg and T. J. Kelly, *J. Virol.* **38**, 272 (1981).
[21] J. N. Fredman and J. A. Engler, *J. Virol.* **67**, 3384 (1993).
[22] T. B. Hasson, D. A. Ornelles and T. Shenk, *J. Virol.* **66**, 6133 (1992).

also cleaves cytokeratin 18 and possibly cytokeratin 7, which may facilitate virus release and spread.[23]

Acknowledgment

This research was supported by Grant MT4164 from the Medical Research Council of Canada.

[23] P. H. Chen, D. A. Ornelles, and T. Shenk, *J. Virol.* **67**, 3507 (1993).

[42] Legumain: Asparaginyl Endopeptidase

By SHIN-ICHI ISHII

The presence of cysteine proteases in legume seeds has frequently been reported. Shutov and Vaintraub[1] suggested that the seeds contain two types of cysteine proteases, an enzyme of broad specificity, which they term proteinase A, and another that is selective for asparaginyl bonds, termed proteinase B. For example, SH-EP, which has been purified by Minamikawa and colleagues[2] from cotyledons of germinating *Vigna mungo* as an enzyme responsible for storage protein degradation, seems to correspond to proteinase A. They cloned the gene, as well as cDNA, for the precursor of this enzyme and showed a close relation[3] of the primary structure deduced for this precursor to those for precursors of papain-type proteases. On the other hand, proteases purified from cotyledons of germinating beans by Baumgartner and Chrispeels,[4] Shutov *et al.*,[5] and Csoma and Polgár[6] may be classified in proteinase B. These enzymes have also been proposed to play a role in degradation of seed storage proteins. Several papers, however, describe the isolation of proteinase B-like enzymes that seem to be responsible for limited proteolysis during maturation of storage protein in beans, including *Canavalia ensiformis*,[7,8] *Ricinus*

[1] A. D. Shutov and I. A. Vaintraub, *Phytochemistry* **26**, 1557 (1987).
[2] W. Mitsuhashi, T. Koshiba, and T. Minamikawa, *Plant Physiol.* **80**, 628 (1986).
[3] D. Yamauchi, H. Akafuku, and T. Minamikawa, *Plant Cell Physiol.* **33**, 789 (1992).
[4] B. Baumgartner and M. J. Chrispeels, *Eur. J. Biochem.* **77**, 223 (1977).
[5] A. D. Shutov, D. N. Lanh, and I. A. Vaintraub, *Biochemistry (Engl. Transl.)* **47**, 678 (1982).
[6] C. Csoma and L. Polgár, *Biochem. J.* **222**, 769 (1984).
[7] S. Ishii, Y. Abe, H. Matsushita, and I. Kato, *J. Protein Chem.* **9**, 294 (1990).
[8] Y. Abe, K. Shirane, H. Yokosawa, H. Matsushita, M. Mitta, I. Kato, and S. Ishii, *J. Biol. Chem.* **268**, 3525 (1993).

communis,[9] and *Glycine max*.[10] In this chapter, the recommended name legumain (EC 3.4.22.34) in the latest edition of Enzyme Nomenclature[11] will be adopted, instead of "proteinase B," as the general term for these asparaginyl bond-specific cysteine proteases found in legume seeds, regardless of their proposed physiological roles.

This chapter deals mainly with asparaginyl endopeptidase[8] from jack bean (*C. ensiformis*), because this is one of the best-characterized legumains. Carrington *et al.*[12] and Bowles *et al.*[13] have reported that concanavalin A, a lectin of jack bean, is produced in growing seeds by posttranslational proteolysis and transpeptidation (ligation) on the carboxyl side of Asn residues of its precursor. Based by these reports, Ishii and colleagues started their work and succeeded in purifying an enzyme that seems to be responsible for these posttranslational reactions. Strict hydrolytic specificity to asparaginyl bonds of this enzyme was confirmed by the use of various polypeptide substrates. Studies on the inhibitor susceptibility and the primary structure of this enzyme suggest that legumain may constitute a unique group of cysteine proteases distinct from the papain family.

Assay of Asparaginyl Endopeptidase Activity

During the characterization and purification of legumain from jack bean seeds, DNP-Pro-Glu-Ala-Asn-Val-Ile-Arg-NH$_2$ (DNP, dinitrophenyl) and DNP-Pro-Glu-Ala-Asn-NH$_2$, which contain a sequence similar to that around one of the processing sites in proconcanavalin A,[13] were used as substrates. These peptides were synthesized with an Applied Biosystems Model 430A Peptide Synthesizer and their NH$_2$ termini were blocked by manual treatment with 1-fluoro-2,4-dinitrobenzene. The presence of the DNP group with absorption near 370 nm allows easy monitoring of separation between an enzymatic reaction product, DNP-Pro-Glu-Ala-Asn, and either of the substrates by high-performance liquid chromatography (HPLC), even when the enzyme in a crude extract of seeds is assayed. Because the two substrates showed almost equal sensitivity in the assay, the shorter one was adopted for routine use in the purification study.

[9] I. Hara-Nishimura, K. Inoue, and M. Nishimura, *FEBS Lett.* **294**, 89 (1991).
[10] M. P. Scott, R. Jung, K. Muntz, and N. C. Nielsen, *Proc. Natl. Acad. Sci. U.S.A.* **89**, 658 (1992).
[11] Nomenclature Committee of the International Union of Biochemistry and Molecular Biology, *in* "Enzyme Nomenclature" Academic Press, San Diego, 1992.
[12] D. M. Carrington, A. Auffret, and D. E. Hanke, *Nature (London)* **313**, 64 (1985).
[13] D. J. Bowles, S. E. Marcus, D. J. C. Pappin, J. B. C. Findlay, E. Eliopoulus, P. R. Maycox, and J. Burgess, *J. Cell Biol.* **102**, 1284 (1986).

An enzyme sample is incubated with DNP-Pro-Glu-Ala-Asn-NH$_2$ (20 μM) in 20 mM sodium acetate, pH 5.0, containing 1 mM EDTA and 10 mM 2-mercaptoethanol (AEM buffer) for 10 min at 35°. The reaction mixture is then analyzed by HPLC on a reversed-phase column (0.6 × 10 cm) (ERC-ODS1161, Erma, Tokyo) in 45 mM ammonium acetate, pH 4.5, containing 22.5% (v/v) acetonitrile at a flow rate of 1 ml/min. One of the reaction products, DNP-Pro-Glu-Ala-Asn, is determined by monitoring the column effluent at 370 nm and using DNP-Ser as an internal standard. DNP-Pro-Glu-Ala-Asn-NH$_2$, DNP-Pro-Glu-Ala-Asn, and DNP-Ser are separately eluted within 8 min, as shown in Fig. 1. One unit is defined as the amount of the enzyme needed to yield 1 μmol of the product per minute. Specific activities of enzyme preparations are expressed in units per milligram of protein, where the protein concentration is estimated by absorption at 280 nm by using a value of 10 for the absorbance at 1%. The addition of 2-mercaptoethanol (or dithiothreitol) to the incubation mixture is essential to yield the full activity.

Purification

Most of the chromatographic media used for enzyme purification are obtained from Tosoh Co. (Tokyo). Organomercurial-Sepharose is prepared by immobilizing p-chloromercuribenzoic acid on an aminoethyl derivative of Sepharose 4B (Pharmacia LKB Biotechnology Inc., Piscataway, NJ), as described previously.[14]

Jack bean meal (300 g) purchased from Sigma (St. Louis, MO) is defatted by acetone and homogenized with AEM buffer containing 100 μM leupeptin (twice, 1 liter and then 800 ml) to extract the enzyme. Ammonium sulfate is added until the concentration reaches 1 M, and the supernatant is run on a column (3.1 × 28 cm) of TSKgel phenyl-Toyopearl 650M equilibrated with AEM buffer containing 10 μM leupeptin and 2 M ammonium sulfate. The enzyme adsorbed on the column is eluted with the same buffer containing 0.01% (w/v) Brij 35 but no ammonium sulfate. The enzyme is then adsorbed on a column (4.1 × 11 cm) of organomercurial-Sepharose previously equilibrated with 20 mM sodium acetate, pH 5.0, containing 1 mM EDTA, 10 μM leupeptin, and 0.01% Brij 35 (AELB buffer) and eluted with a linear gradient of 0–50 mM 2-mercaptoethanol in the same buffer (1.8 liters). TSKgel phenyl-Toyopearl 650S is the adsorbent used for the next purification step. The chromatographic conditions are the same as for Phenyl-Toyopearl 650M except that the column size is 1.9 × 9 cm, and the elution is carried out with a linear gradient of 2–0

[14] H. Yokosawa, S. Ojima, and S. Ishii, *J. Biochem. (Tokyo)* **82**, 869 (1977).

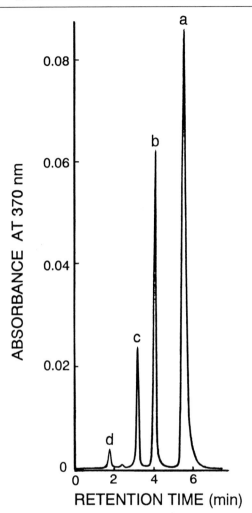

FIG. 1. Determination of asparaginyl endopeptidase activity with DNP-Pro-Glu-Ala-Asn-NH₂ as a substrate. The substrate is incubated with the enzyme, and the mixture is analyzed by reversed-phase HPLC. The substrate (a), DNP-Pro-Glu-Ala-Asn as a reaction product (b), and DNP-Ser as an internal standard (c) are separated by isocratic elution. Peak d is of 2-mercaptoethanol added to the incubation mixture. The text gives a full detail of experimental conditions.

TABLE I

PURIFICATION OF LEGUMAIN FROM JACK BEAN MEAL[a]

Step	Protein (mg)	Activity (mU)	Specific activity (U/mg)	Yield (%)
Extract	—	6800	—	100
Phenyl-Toyopearl 650M	6380	5680	0.00089	84
Organomercurial-Sepharose	138	2330	0.0169	34
Phenyl-Toyopearl 650S	84.2	1640	0.0195	24
SP-Toyopearl 650M	11.8	1900	0.161	28
TSKgel phenyl-5PW	3.68	1590	0.432	23
TSKgel G3000SW (1st)	0.105	443	4.22	7
TSKgel G3000SW (2nd)	0.072	352	4.90	5

[a] 300 g.

M ammonium sulfate (300 ml). The enzyme fraction in the effluent is dialyzed against AELB buffer containing 10 mM 2-mercaptoethanol and chromatographed on a column (1.9 × 7.4 cm) of TSKgel SP-Toyopearl 650M. A linear gradient of 0–0.2 M NaCl in the same buffer (200 ml) is applied to elute the enzyme. HPLC on a column (0.75 × 7.5 cm) of TSKgel phenyl-5PW is used next. After being loaded on this column in AELB buffer containing 10 mM 2-mercaptoethanol and 1 M ammonium sulfate, the enzyme is eluted with a linear gradient of 1–0 M ammonium sulfate in the same buffer (25 ml). The enzyme is further purified by gel-permeation HPLC on a column (0.75 × 60 cm) of TSKgel G3000SW in AEM buffer containing 0.01 M Brij 35 and 0.1 M NaCl. The HPLC is repeated under the same conditions. The addition of leupeptin to the media used for the extraction of meal and for the next several steps of purification is useful to protect the enzyme against the harmful action of contaminating protease(s) with papain-like activity. Table I summarizes the purification of the enzyme from an extract of 300 g of jack bean meal.

Kembhavi et al.,[15] who purified legumain from moth bean (Vigna aconitifolia), reported that the enzyme activity in a homogenate of cotyledons from the germinating seeds was increased by 65% after an initial acid treatment (at pH 3). This may be due to the removal of an inhibitor (or competing substrate), as they suggested. In the purification of jack bean legumain, too, 16% increase of the activity is observed after the step of SP-Toyopearl chromatography (see Table I), probably due to separation

[15] A. A. Kembhavi, D. J. Buttle, C. G. Knight, and A. J. Barrett, Arch. Biochem. Biophys. 303, 208 (1993).

from an inhibitor. The initial acid treatment of jack bean extract might improve the purification yield of legumain.

Properties

Purity and Fundamental Activity

The final preparation of jack bean legumain obtained after the second gel-permeation chromatography on G3000SW gives a single peak in HPLC on a reversed-phase column (TSKgel octadecyl-NPR) and a single protein band with mobility corresponding to a molecular mass of about 37,000 Da on SDS–polyacrylamide gel after electrophoresis under reducing conditions (see Ref. 8). Also, NH_2-terminal sequence analysis of the material in the HPLC peak fraction (with an Applied Biosystems Model 477A/ 120A protein sequencer/phenylthiohydantoin analyzer system) indicates the presence of a single polypeptide chain. This highly purified enzyme shows maximal activity toward DNP-Pro-Glu-Ala-Asn-NH_2 in the pH range 5.0–6.5, and gives Michaelis constants of 23 μM at pH 5.0 and 33 μM at pH 5.9 (at 35°). The addition of the nonionic detergent Brij 35 (0.005–0.05%), as well as that of 2-mercaptoethanol or dithiothreitol, is beneficial in stabilizing the enzyme. Even with these additives, the enzyme is labile above pH 7.5.

Substrate Specificity

Substrate specificity of the enzyme is examined with various polypeptide substrates containing Asn residue(s). These include pancreatic polypeptide (PP), parathyroid hormone [1–34] (PTH), neuropeptide Y (NPY), gastric inhibitory polypeptide (GIP), vasoactive intestinal peptide (VIP), tyrosine kinase substrate (TKS), gastrin-releasing peptide (GRP), [Asn[1],Val[5]]angiotensin I, neuromedin C, mastoparan, oxidized insulin B chain (insulin B), reduced S-pyridylethylated erabutoxin c (RPE-erabutoxin), reduced S-carboxymethylated ribonuclease A and B (RCM-RNase A and B), and reduced S-3-(trimethylamino)-propylated lysozyme (RTP-lysozyme).

Each substrate (10 nmol) is digested with the enzyme (0.2 mU for RTP-lysozyme and RCM-RNases and 0.1 mU for the other peptides) at 37° for 15 hr in 100 μl of 20 mM sodium acetate, pH 5.0, containing 1 mM EDTA and 10 mM dithiothreitol. After the reaction is terminated by the addition of 10 μl of formic acid, peptide fragments in the digests are separated by reversed-phase HPLC in 0.1% trifluoroacetic acid with an acetonitrile gradient on a column (0.39 × 15 cm) of μBondasphere C_4

TABLE II

SITES OF CLEAVAGE BY JACK BEAN LEGUMAIN IN VARIOUS PEPTIDE SUBSTRATES[a]

Site	Substrate	Site	Substrate
-G-N╀G-	Lysozyme[b] (102–104)	-V-N╀Q-	Insulin B (2–4)
-K-N╀G-	RNase A and B (66–68)	-C-N╀Q-	RNase A and B (26–28)
-M-N╀A-	Lysozyme (105–107)	-A-N╀K-	RNase A and B (102–104)
-D-N╀A-	PP (10–12)	-T-N╀R-	Lysozyme (43–45)
-K-N╀V-	RNase A and B (61–63)	-R-N╀R-	Lysozyme (112–114)
-H-N╀L-	PTH (9–11)	-V-N╀C-	Lysozyme (92–94)
-R-N╀L-	Lysozyme (73–75)	-T-N╀C-	RNase A and B (70–72)
-R-N╀L-	RNase A (33–35)	-P-N╀C-	RNase A and B (93–95)
-I-N╀L-	NPY (28–30)	-I-N╀M-	PP (28–30)
-I-N╀L-	Erabutoxin (50–52)	-H-N╀F	PTH (32–34)
-H-N╀I-	GIP (38–40)	-S-N╀F-	Lysozyme (36–38)
-L-N╀S-	VIP (23–25)	-D-N╀Y-	VIP (8–10)
-L-N╀S-	PTH (15–17)	-D-N╀Y-	Lysozyme (18–20)
-I-N╀S-	Lysozyme (58–60)	-S-N╀Y-	RNase A and B (23–25)
-F-N╀T-	Lysozyme (38–40)	-G-N╀W-	Lysozyme (26–28)
-R-N╀T-	Lysozyme (45–47)	-V-N╀W-	GIP (23–25)
-V-N╀T-	RNase A and B (43–45)	-G-N╀H-	GRP (18–20)
-C-N╀D-	Lysozyme (64–66)	-F-N╀H-	Erabutoxin (4–6)
-K-N╀D-	GIP (33–35)	-G-N╀P-	RNase A and B (112–114)
-D-N╀E-	TKS (6–8)	-L-N╀NH₂	VIP (27–29)
-C-N╀N	Erabutoxin (60–62)		

[a] Cleavage sites are marked "╀."

[b] Lysozyme, RNase, and erabutoxin denote RTP-lysozyme. RCM-ribonuclease(s), and RPE-erabutoxin c, respectively. Abbreviations for other peptides are defined in the text.

(Millipore). The peptides recovered from the effluent peaks are identified by analyzing their amino acid sequences (or amino acid compositions in cases of small peptides). The sites of enzymatic cleavage are assigned in this way, and the results are summarized in Table II.

As shown in Table II, the cleavable bonds are exclusively those possessing Asn as the P1 residue (according to the nomenclature in Ref. 16). By contrast, there are 20 varieties of the P1' residue, including S-alkylated cysteines. (The peptides containing intact cysteine or cystine residues have not been used as substrates.) In a few cases, the enzyme does not hydrolyze the peptide bonds with Asn in the P1 position. For example, Asn^7-Pro^8 in neuropeptide Y is not cleaved, whereas Asn^{113}-Pro^{114} in RCM-RNases is cleavable. Asn^{39}-Ile^{40} of gastric inhibitory polypeptide is split, but Asn^{77}-Ile^{78} of RTP-lysozyme is not. The reason for these findings is unclear. Properties of the residues surrounding P1 and P1' positions could

[16] I. Shechter and A. Berger, *Biochem. Biophys. Res. Commun.* **27,** 157 (1967).

modify the susceptibility of some peptide bonds to the enzyme. The enzyme seems to be excluded from the Asn residue that exists at the NH_2 terminus or at the second position from the NH_2 terminus of peptides, because Asn^1-Arg^2 in [Asn^1,Val^5]angiotensin I, Asn^2-His^3 in neuromedin C, and Asn^2-Leu^3 in mastoparan are not cleavable. On the other hand, the linkage with Asn at the penultimate position is split, as shown in the cases of Asn^{33}-Phe^{34} in parathyroid hormone [1–34], Asn^{61}-Asn^{62} in RPE-erabutoxin c, and Asn^{28}-NH_2^{29} in vasoactive intestinal peptide. Whereas Asn^{34}-Leu^{35} in RCM-RNase A is easily cleaved, the corresponding bond with N-glycosylated Asn in RCM-RNase B is not.

The strict specificity toward asparaginyl bonds thus confirmed promises the high utility of jack bean legumain for protein sequence analysis and for discrimination between glycosylated and nonglycosylated Asn residues in proteins. Examples of successful application have been reported.[17,18] A specimen of jack bean legumain, which is still heterogeneous as a protein, but contains no other protease activities, is now available from Takara Shuzo Co., Otsu.

Susceptibility to Inhibitors

The sensitivity of the purified enzyme to various peptidase inhibitors has been examined. The enzyme (0.14 mU) is incubated with inhibitors in 20 μl of 50 mM sodium phosphate, pH 6.0, containing 0.01% Brij 35 and 5 mM 2-mercaptoethanol at 35° for 30 min. Some experiments are carried out in the absence of 2-mercaptoethanol. In the case of pepstatin, the medium contains 5% (v/v) dimethyl sulfoxide. Then 10 μl of 240 μM DNP-Pro-Glu-Ala-Asn-NH_2 dissolved in the phosphate buffer is added to the enzyme–inhibitor mixture. After 10 min incubation at 35°, the amount of DNP-Pro-Glu-Ala-Asn produced is analyzed by HPLC as described above to determine the residual enzyme activity. The results are shown in Table III. The enzyme is inhibited strongly with p-chloromercuribenzenesulfonic acid (PCMBS) and N-ethylmaleimide and weakly with peptide aldehydes such as leupeptin. All the other small compounds listed in Table III show almost no effect. Among the protein protease inhibitors examined, substantial inhibition is observed only with cystatin EW (chicken egg white) and high molecular weight kininogen (human), both of which have been reported to be specific inhibitors of cysteine peptidases. SSI (*Streptomyces* subtilisin inhibitor) is known as a potent inhibitor

[17] T. Kumazaki, N. Urushibara, and S. Ishii, *J. Biochem.* (*Tokyo*) **112**, 11 (1992).
[18] Y. Abe, M. Tokuda, K. Azumi, and H. Yokosawa, *Seikagaku* **65**, 1021 (1993).

TABLE III
EFFECTS OF VARIOUS REAGENTS ON ACTIVITY OF JACK BEAN LEGUMAIN

Reagents	Concentration (μM)	Inhibition (%)
In absence of 2-mercaptoethanol		
PCMBS	100	100
N-Ethylmaleimide	100	99
E-64	500	7
Leupeptin	1000	62
Diisopropyl fluorophosphate	1000	4
Bestatin	1000	13
In presence of 5 mM 2-mercaptoethanol		
Leupeptin	1000	16
Chymostatin	500	4
Elastatinal	1000	36
Diisopropyl fluorophosphate	1000	0
Phenylmethylsulfonyl fluoride	5000	3
Pepstatin	500	0
Phosphoramidon	250	0
Bestatin	1000	2
Benzylsuccinic acid	1000	0
High molecular weight kininogen	5	32
Cystatin EW	5	26
Cystatin EW	10	47
Cystatin α	5	5
Cystatin A	40	2
Calpastatin	5	0
Aprotinin	5	0
Lima bean trypsin inhibitor	5	0
Bowman–Birk inhibitor	5	0
Chicken ovomucoid	5	0
Turkey ovomucoid	5	0
SSI (Met73Asn)	5	0

for a variety of serine proteases[19] and some metalloproteases.[20] Even recombinant SSI, whose reactive site Met-73 has been replaced with Asn,[21] does not inhibit jack bean legumain.

[19] U. Christensen, S. Ishida, S. Ishii, Y. Mitsui, Y. Iitaka, J. McClarin, and R. Langridge, J. Biochem. (Tokyo) 98, 1263 (1985).
[20] K. Kajiwara, A. Fujita, H. Tsuyuki, T. Kumazaki, and S. Ishii, J. Biochem. (Tokyo) 110, 350 (1991).
[21] S. Kojima, S. Obata, I. Kumagai, and K. Miura, Bio/Technology 8, 449 (1990).

Features of Legumain

The inhibition profile of jack bean legumain shown in Table III suggests that the enzyme is a cysteine-type endopeptidase. Some differences are observed, however, between the jack bean enzyme and papain-type proteases. For example, lower susceptibilities are seen in the former toward high molecular weight kininogen and cystatins. In particular, cystatin α and cystatin A show almost no inhibitory effect. E-64 [*trans*-epoxysuccinyl-L-leucylamido(4-guanidino)butane] behaves only as a poor inhibitor. Kembhavi *et al.*[15] have reported that legumain from *Vigna aconitifolia* has similarly low sensitivity to cystatin egg white and E-64, whereas vignain, another enzyme isolated from the same bean, resembles papain-type proteases in its higher sensitivity to these inhibitors. Vignain seems to correspond to Shutov's proteinase A. They also pointed out that legumain from *V. aconitifolia*, as well as that from *Phaseolus vulgaris*,[5] has the unusual characteristic of reacting more rapidly with iodoacetamide than with iodoacetate. This feature has never been seen in papain-type proteases, including vignain, but is seen with clostripain.

As mentioned at the beginning of this chapter, *Vigna mungo* SH-EP, an enzyme that may correspond to proteinase A, has the primary structure (deduced from the gene structure) homologous to those of papain-type proteases.[3] On the other hand, the NH$_2$-terminal sequence determined for jack bean legumain,[8] H-Glu-Val-Gly-Thr-Arg-Trp-Ala-Val-Leu-Val-Ala-Gly-Ser-Asn-Gly-Tyr-Gly-Asn-Tyr-Arg-His-Gln-Ala-Asp-Val-, has no counterpart in the primary structures of proteins that have been reported to be cysteine proteases. The only exception is one of the hemoglobinases of *Schistosoma mansoni*; the primary structure of its precursor has been deduced by the analysis of a cDNA.[22] Nineteen identical residues are detectable between the NH$_2$-terminal sequence of legumain and that of mature hemoglobinase. Takeda *et al.*[23] isolated a cDNA clone for jack bean legumain with a degenerate DNA probe representing all possible base sequences predicted from the NH$_2$-terminal amino acid alignment described above. The cDNA clone was that encoding a putative precursor of this enzyme, the deduced sequence for which consisted of 475 amino acid residues and contained the NH$_2$-terminal sequence of the enzyme in a range starting from residue 36. After a thorough survey of the NBRF database, prohemoglobinase was picked up again as the sole protein with a sequence homologous to prolegumain. The homology was marked (50%

---

[22] M. A. Meanawy, T. Aji, N. F. B. Phillips, R. E. Davis, R. A. Salata, I. Malhotra, D. McClain, M. Aikawa, and A. H. Davis, *Am. J. Trop. Med. Hyg.* **43**, 67 (1990).
[23] O. Takeda, Y. Miura, M. Mitta, H. Matsushita, I. Kato, Y. Abe, H. Yokosawa, and S. Ishii, *J. Biochem.* (Tokyo) **116**, in press (1994) [Database Accession No. D31787].

identical) in the range corresponding to the mature form (residues 32–291) of prohemoglobinase, as shown in the dot matrix (Fig. 2). Although detailed information about the substrate specificity concerning cleavable peptide bonds and the inhibitor susceptibility is not available yet for schistosome hemoglobinase, this protease and legumain seem to constitute a new family of cysteine peptidases (see [32] this volume).

Physiological Roles

Two roles have been suggested for legumain, as mentioned above. The researchers[4–6] who purified the enzyme from cotyledons of bean seedlings proposed the role of degradation of storage proteins. Actually, Kembhavi *et al.*[15] have observed marked increase in the activity of legumain, as well as of vignain, during germination of moth beans. On the other hand, some reports[9,10] attributed the catalysis of limited proteolysis essential for maturation of storage proteins to legumain from premature seeds. The ability of jack bean legumain to convert proconcanavalin A

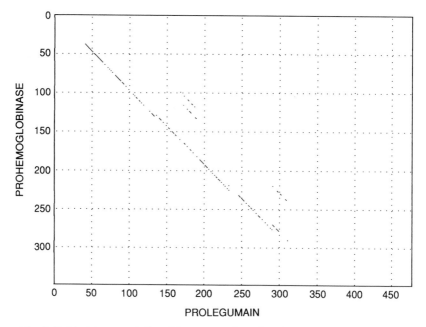

FIG. 2. Dot matrix presentation of the sequence homology between jack bean prolegumain and schistosome prohemoglobinase. In this plot, each dot represents 7 agreements out of 20 continuous amino acid residues. The part of the matrix corresponding to the C-terminal region of prohemoglobinase (residues 351–429) is omitted because no dot appears in this area.

to concanavalin A has not yet been verified. It has been demonstrated in an artificial system,[8] however, that the enzyme can also catalyze traspeptidation, which has been reported to be involved in the maturation process of concanavalin A.[13] Although jack bean legumain was purified from the mature dry seeds (meal) as described already, an asparaginyl endopeptidase activity that behaves similarly during purification has been found in the immature seed extract. Furthermore, the mRNA mixture employed to construct a cDNA library for the isolation of prolegumain cDNA clone was one prepared from immature jack bean seeds. Comparison of the primary structure of legumain from germinating seeds with that from immature seeds could offer a definite answer to the question whether the same molecular species of legumain plays the two roles. In either case, careful assay for legumain(s) in active and latent forms with seeds of various stages should resolve the uncertainties about the physiological roles.

Acknowledgments

I am grateful to Drs. Y. Abe, K. Shirane, H. Matsushita, M. Mitta, and I. Kato for their enthusiastic cooperation in the study of purification and characterization of jack bean legumain.

[43] Interleukin-1β Converting Enzyme

By Nancy A. Thornberry

Introduction

Interleukin-1 (IL-1) has been implicated in the pathogenesis of several acute and chronic inflammatory diseases, including rheumatoid arthritis, inflammatory bowel disease, and septic shock.[1] Evidence that IL-1 is an excellent target for therapeutic intervention has been provided by a naturally occurring receptor antagonist, IL-1ra, which is efficacious in several clinical models.[1,2] Two forms of IL-1 have been identified, IL-1α and IL-1β, the latter being the predominant form of this cytokine released by human monocytes in culture. IL-1β is synthesized as a 31 kDa inactive precursor (pro-IL-1β), which is activated to the 17.5 kDa active form by a highly selective proteinase termed IL-1β converting enzyme (EC

[1] C. A. Dinarello, *Blood* **77**, 1627 (1991).
[2] C. A. Dinarello and R. C. Thompson, *Immunol. Today* **12**, 404 (1991).

3.4.22.36),[3,4] Although the relative importance of IL-1α and IL-1β in inflammation remains to be established, two recent findings suggest a major role for IL-1β. First, neutralizing antibodies to IL-1β are effective in murine models of collagen induced arthritis.[5] Second, cowpox virus facilitates infection through inhibition of the host inflammatory response, at least in part by encoding a potent inhibitor of IL-1β converting enzyme, cytokine response modifier A (crmA).[6]

The purification, cloning, and characterization of this proteinase have recently been described.[7,8] The active enzyme is a heterodimer composed of two subunits of 20 and 10 kDa, both of which are required for catalytic activity. Both subunits are derived from a single 45 kDa proenzyme by a process that involves autoproteolysis. Regarding substrate specificity, the enzyme has a stringent requirement for Asp in the P1 position. This unusual specificity has enabled the design of potent, selective inhibitors which have been employed to prove that the enzyme is essential to the production of IL-1β in monocytes.

IL-1β converting enzyme appears to be the first member of a new cysteine proteinase family. Although it is not related to any known cysteine proteinase, it has recently been shown to share homology with a gene required for programmed cell death in C. elegans, CED-3, leading to the proposal that members of this family might function in programmed cell death in vertebrates.[9]

Distribution and Localization

Cellular Distribution

Human IL-1β converting enzyme activity has been detected in monocytes and the human monocytic cell line THP.1.[3,4] Although stimulation

[3] R. A. Black, S. R. Kronheim, and P. R. Sleath, *FEBS Lett.* **247**(2), 386 (1989).

[4] M. J. Kostura, M. J. Tocci, G. Limjuco, J. Chin, P. Cameron, A. G. Hillman, N. A. Chartrain, and J. A. Schmidt, *Proc. Natl. Acad. Sci. U.S.A.* **86**, 5227 (1989).

[5] T. Geiger, H. Towbin, A. Cosenti-Vargas, O. Zingel, J. Arnold, C. Rordorf, M. Glatt, and K. Vosbeck, *Clinical and Experimental Rheumatology*, **11**, 515–522 (1993).

[6] C. A. Ray, R. A. Black, S. R. Kronheim, T. A. Greenstreet, P. R. Sleath, G. S. Salvesen, and D. J. Pickup, *Cell (Cambridge, Mass.)* **69**, 597 (1992).

[7] N. A. Thornberry, H. G. Bull, J. R. Calaycay, K. T. Chapman, A. D. Howard, M. J. Kostura, D. K. Miller, S. M. Molineaux, J. R. Weidner, J. Aunins, K. O. Elliston, J. M. Ayala, F. J. Casano, J. Chin, G. J.-F. Ding, L. A. Egger, E. P. Gaffney, G. Limjuco, O. C. Palyha, S. M. Raju, A. M. Rolando, J. P. Salley, T. T. Yamin, T. D. Lee, J. E. Shively, M. MacCross, R. A. Mumford, J. A. Schmidt, and M. J. Tocci, *Nature (London)* **356**, 768 (1992).

[8] D. P. Cerretti, C. J. Kozlosky, B. Mosley, N. Nelson, K. Van Ness, T. A. Greenstreet, C. J. March, S. R. Kronheim, T. Druck, L. A. Cannizzaro, K. Huebner, and R. A. Black, *Science* **256**, 97 (1992).

[9] J. Yuan, S. Shaham, S. Ledoux, H. M. Ellis, and H. R. Horvitz, *Cell*, **75**, 641–652 (1993).

of cells by lipopolysaccharide or other stimuli is required for secretion of IL-1β, the enzyme activity has been identified in both stimulated and unstimulated cells. The mRNA for the enzyme has been detected in several nonmonocytic cell types, including T cells, B cells, and neutrophils.[7,8] Because some of these cell types do not contain detectable IL-1β mRNA, it has been suggested that the enzyme may have other substrates in addition to pro-IL-1β.[8]

The murine IL-1β converting enzyme also has broad tissue distribution. Activity has been detected in murine peritoneal exudate cells, and the murine macrophage cell lines IC21 and J774.[10] Convertase mRNA is constitutively expressed in IC21, J774, WeHI-3 (murine myelomonocyte), RAW8.1 (murine lymphoma), and RLM-11 (mature, murine T-cell) cells.[10,11] In addition, mRNA has been detected in murine spleen, heart, brain, adrenal glands, and liver tissue.[11]

Subcellular Localization

Enzyme activity was first localized to the cytosolic fraction of THP.1 cells.[4] However, this result must be reevaluated in light of the recent finding that the proenzyme is the major form of the enzyme in these cells, and is efficiently converted to the mature, active enzyme on cell lysis and incubation.[12] Thus, it is probable that the enzyme detected in, and purified from, cell cytosol was derived from the proenzyme following cell lysis. Consequently, the location of the active enzyme in intact cells remains unknown.

Substrate Specificity

The substrate specificity of the enzyme has been defined using both macromolecular and peptide substrates.[7,13,14] The enzyme catalyzes the cleavage of pro-IL-1β at Asp^{116}-Ala^{117} to generate the mature, biologically active product. (Pro-IL-1β is the only macromolecular substrate so far

[10] S. M. Molineaux, F. J. Casano, A. M. Rolando, E. P. Peterson, G. Limjuco, J. Chin, P. R. Griffin, J. R. Calaycay, G. J.-F. Ding, T.-T. Yamin, O. C. Palyha, S. Luell, D. Fletcher, D. K. Miller, A. D. Howard, N. A. Thornberry, and M. J. Kostura, *Proc. Natl. Acad. Sci. U.S.A.* **90**, 1809 (1993).

[11] M. A. Nett, D. P. Cerretti, D. R. Berson, J. Seavitt, D. J. Gilbert, N. A. Jenkins, N. G. Copeland, R. A. Black, and D. D. Chaplin, *J. Immunol.* **149**(10), 3254 (1992).

[12] J. M. Ayala, T.-T. Yamin, L. A. Egger, J. Chin, M. J. Kostura, and D. K. Miller, *J. Immunol.* in press, (1994).

[13] P. R. Sleath, R. C. Hendrickson, S. R. Kronheim, C. J. March, and R. A. Black, *J. Biol. Chem.* **265**(24), 14526 (1990).

[14] A. D. Howard, M. J. Kostura, N. Thornberry, G. J. F. Ding, G. Limjuco, J. Weidner, J. P. Salley, K. A. Hogquist, D. D. Chaplin, R. A. Mumford, J. A. Schmidt, and M. J. Tocci, *J. Immunol.* **147**(9), 2964 (1991).

identified, apart from the proenzyme of the converting enzyme; see below.) The natural substrate is also cleaved by the enzyme at Asp^{27}-Gly^{28}, although the physiological relevance of this cleavage is not known. Both cleavage sites are evolutionarily conserved in mammalian species. The most distinguishing feature of the enzyme is its stringent requirement for Asp in the P1 position; any substitution of this residue in pro-IL-1β and peptide substrates leads to >100-fold reduction in the rate of catalysis.

To be hydrolyzed efficiently, peptide substrates must contain at least four amino acids to the left of the cleavage site. Regarding substitutions in these positions, hydrophobic residues are favored in P4, Val is preferred in P3, and liberal substitutions are tolerated in P2. Not surprisingly, the optimal sequence to the left of the cleavage site, Ac-Tyr-Val-Ala-Asp, closely matches the corresponding sequence in the natural human substrate, Tyr^{113}-Val^{114}-Cys^{115}-Asp^{116}.

In contrast to the specificity requirements of the enzyme for the P1–P4 positions, residues beyond P1' are not required for catalytic recognition. Small, hydrophobic amino acids (Gly,Ala) are preferred in the P1' position of both pro-IL-1β and peptide substrates. Indeed, the best peptide substrate yet identified for the enzyme contains methylamine in this position. Incorporation of relative bulky photometric leaving groups [e.g., amino-4-methylcoumarin (AMC), p-nitroanilide (pNA)] in P1' results in substrates with low values for k_{cat} (Table I); however, this is compensated by low K_m values. Consequently, compounds with the general structure, Ac-Tyr-Val-Ala-Asp-X, where X is a photometric leaving group, are excellent substrates for the enzyme, as described in detail below.

Assay Methods

General Considerations

Three methods for assay of IL-1β converting enzyme, and their respective virtues, are described. The standard reaction conditions, defined as

TABLE I
KINETIC CONSTANTS FOR SUBSTRATES

Substrate	k_{cat} (sec^{-1})	K_m (μM)	k_{cat}/K_m (M^{-1} sec^{-1})
Ac-Tyr-Val-Ala-Asp-AMC	0.89	14	6.4×10^4
Ac-Tyr-Val-Ala-Asp-pNA	2.6	16	1.6×10^5
Pro-IL-1β	n.d.[a]	n.d.	1.5×10^5

[a] n.d., Not determined.

100 mM HEPES, pH 7.5, 10% (w/v) sucrose, 0.1% (w/v) 3-[(3-cholamido-propyl) dimethylammonio]-1-propane sulfonate (CHAPS), 10 mM dithio-threitol (DTT), at 25°, are selected for the following reasons. First, the enzyme has a broad pH optimum between 6.5 and 7.5. Second, the enzyme is stabilized by 10% sucrose and 0.1% CHAPS, for reasons described later. Finally, it is important to include DTT in all reaction mixtures because the catalytic cysteine is rather susceptible to oxidation.

Pro-IL-1β Cleavage Assay

Description. Assays using pro-IL-1β as a substrate are useful for assessing enzyme activity in crude cell lysates, where peptide-based substrates are cleaved by contaminating proteinases. The assay used in this laboratory, described below, employs radiolabeled pro-IL-1β, followed by SDS–PAGE and fluorography.[4] An alternative assay has been developed that involves incubation of enzyme samples with recombinant pro-IL-1β, followed by SDS–PAGE and detection with an IL-1β-specific antibody.[15]

Reagents. Substrate: [^{35}S]Methionine-labeled pro-IL-1β. Radiolabeled pro-IL-1β is synthesized by addition of 10 μg of 1 mg/ml pro-IL-1β RNA[4], 16 μl [^{35}S]methionine (Amersham; 15 μCi/μl), and 4 μl methionine-free amino acid mixture (Promega, Madison, WI) to 70 μl of rabbit reticulocyte lysate (Promega), followed by incubation for 90 min at 30°. The reticulocyte lysate is stable for 1–2 months at −70°. Buffer: 100 mM HEPES, 10% sucrose, 0.1% CHAPS, pH 7.5, 10 mM DTT.

Procedure. A typical assay contains 1 nM enzyme, buffer, and 1 μl substrate in a total volume of 20 μl. Under these conditions 50% of the substrate is converted to product in 77 min. Reactions are quenched by addition of 20 μl of 2× Laemmli sample buffer containing 10 mM DTT, and boiling for 3–5 min. The [^{35}S]Met-labeled substrate and product are then separated by SDS–PAGE (NOVEX 16% Tris–glycine; NOVEX, San Diego, CA). Gels are fixed in 40% (v/v) methanol, 10% (v/v) glacial acetic acid for 30 min at 25°, enhanced for fluorography by incubation in Amplify (Amersham) for 30 min at 25°, and vacuum dried prior to exposure to X-ray film. For quantitative analysis of reaction rates, the substrate and product bands are counted using a radioanalytical imaging system (e.g., Ambis Systems, San Diego, CA). It is necessary to multiply the counts in the 17.5-kDa band by 1.86 to correct for the loss of Met residues on conversion of pro-IL-1β to mature IL-1β.

---

[15] S. R. Kronheim, A. Mumma, T. Greenstreet, P. J. Glackin, K. Van Ness, C. J. March, and R. A. Black, *Arch. Biochem. Biophys.* **296,** 698 (1992).

Precautions. In crude cell lysates, pro-IL-1β is susceptible to cleavage by proteinases other than IL-1β converting enzyme.[16] Thus, it is advisable to include proteinase inhibitors [1 mM EDTA, 2 mM phenylmethylsulfonyl fluoride (PMSF), 100 μM pepstatin, 10 μg/ml leupeptin] in crude enzyme preparations. Also, cleavage of pro-IL-1β by the enzyme is inhibited by NaCl with an IC_{50} of approximately 20 mM. This inhibition is not observed with other substrates.

Fluorometric Assay

Description. A continuous fluorometric assay has been developed with the substrate Ac-Tyr-Val-Ala-Asp-AMC, which is cleaved by the enzyme to release the fluorescent leaving group. The kinetic parameters for this substrate are shown in Table I. This assay has been used in this laboratory routinely for studies of enzyme and inhibitor mechanism, screening, and analysis of purification fractions. A unit is defined as the amount of enzyme required to produce 1 pmol of AMC/min at 25° at saturating substrate concentrations.

Reagents. Substrate: Ac-Tyr-Val-Ala-Asp-AMC[17]. Stock solution is prepared at 0.14 M in dimethyl sulfoxide (DMSO) and is stable for at least 1 year at -20°. The substrate is soluble in assay buffer to at least 1.4 mM (with vigorous mixing). Buffer: 100 mM HEPES, 10% sucrose, 0.1% CHAPS, pH 7.5, 10 mM DTT. Fluorometric standard: amino-4-methylcoumarin. Stock solution is 0.1 M in DMSO and is stable for at least 1 year at -20°.

Procedure. Liberation of AMC from the substrate can be monitored continuously in a fluorometer using an excitation wavelength of 380 nm and an emission wavelength of 460 nm. Routine assays in our laboratory contain approximately 25 U enzyme (0.9 nM), 14 μM substrate (1 × K_m), and buffer in a total reaction volume of 500 μl. Reactions are placed in a Gilford Fluoro IV spectrofluorometer (Ciba-Corning, Oberlin, Ohio) calibrated such that 1 μM AMC produces 100% relative fluorescence. Under these conditions, 1 μM AMC is produced in 40 min at 25°. The assay has also been adapted to 96-well plate format using a Titertek Fluoroskan II detector (ICN, Costa Mesa, CA) and Micro FLUOR "W" Plates (Dynatech Laboratories, Chantilly, VA).

This assay can also be run discontinuously which facilitates processing of large numbers of samples and uses less enzyme. A typical assay contains 1 U enzyme (0.2 nM), 14 μM substrate, and buffer in a total volume of

---

[16] R. A. Black, S. R. Kronheim, M. Cantrell, M. C. Deeley, C. J. March, K. S. Prickett, J. Wignall, P. J. Conlon, D. Cosman, T. P. Hopp, and D. Y. Mochizuki, *J. Biol. Chem.* **263**(19), 9437 (1988).

[17] Available from Peptides International [Louisville, KY, (800)777-4779].

100 μl. Following incubation for 120 min at 25°, assays are quenched with 400 μl 0.2% trifluoroacetic acid (TFA). Under these conditions, 0.6 μM AMC is produced. The sample is applied to a Vydac C_{18} reversed-phase high-performance liquid chromatography (HPLC) column (300 Å pore size, 5-μm particle size, 4.6 mm × 25 cm) equilibrated with 25% CH_3CN/ 75% H_2O containing 0.1% TFA at a flow rate of 1 ml/min. Substrate and product, which are separated isocratically in less than 10 min, are detected using a fluorometer equipped with a flow cell.

Precautions. This substrate is not suitable for assay of crude cell lysates, due to hydrolysis by contaminating proteinases. However, it is possible to measure the enzyme activity in a crude sample if a selective IL-1β converting enzyme inhibitor is used to assess the contribution of the enzyme to the overall rate. Partial purification by anion-exchange chromatography (described below) removes the interfering proteinases.

Spectrophotometric Assay

Description. A spectrophotometric assay has been developed using the substrate Ac-Tyr-Val-Ala-Asp-pNA. The kinetic constants for this substrate, shown in Table I, are similar to those for the fluorogenic substrate described above. The extinction coefficient for the reaction is 9160 ± 226 cm^{-1} M^{-1} at 400 nm. This assay is less sensitive than the fluorometric assay described above, requiring approximately 10-fold more enzyme in a typical assay.

Reagents. Substrate: Ac-Tyr-Val-Ala-Asp-pNA. Stock solution is prepared at 10 mM in DMSO and is stable for at least 1 year at −20°. Buffer: 100 mM HEPES, 10% sucrose, 0.1% CHAPS, pH 7.5, 10 mM DTT

Procedure. Liberation of pNA from the substrate is monitored continuously in a spectrophotometer using a wavelength of 400 nm. A reaction mixture containing 7 nM enzyme, 16 μM substrate ($1 \times K_m$), and buffer in a total reaction volume of 500 μl will produce an optical density change of 0.1 in 20 min at 25°.

Precautions. This assay is subject to the same limitations as the fluorometric assay.

Concentration Dependence of Activity

Prior to any description of enzyme purification, it is important to note that catalytic activity of IL-1β converting enzyme is a function of enzyme concentration. Evidence for this is shown in Fig. 1, where dilution of enzyme (5 U/μl, 93 nM) 1000-fold resulted in complete loss of catalytic activity with a half-life of 2.7 ± 0.4 hr. Activity was fully restored by

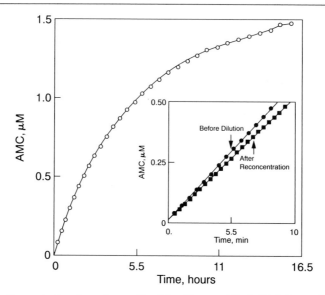

FIG. 1. Concentration dependence of activity. Enzyme (50 μl, 5 U/μl) was diluted 1000-fold into reaction mixtures containing 14 μM Ac-Tyr-Val-Ala-Asp-AMC under standard reaction conditions. The solid line is theoretical for first-order loss of activity with a rate constant of 0.004 min^{-1} ($t_{1/2}$ = 2.9 hr). Less than 2% of the total substrate was consumed over the reaction period. Complete recovery of activity is achieved on 1000-fold reconcentration of the enzyme by ultrafiltration. The inset compares the initial velocities of samples assayed before dilution (v_0 = 0.057 μM AMC/min) and after reconcentration (v_0 = 0.051 μM AMC/min). Taken from Thornberry et al.,[7] with permission.

reconcentration of the enzyme. This phenomenon is believed to result from dissociation of the enzyme to inactive subunits.[7] The enzyme is stabilized by saturating levels of substrate or competitive inhibitor. Including 10% sucrose and 0.1% CHAPS in the reaction or chromatography buffer also decreases the rate of dissociation by approximately 4-fold.

There are several practical implications of this property of the enzyme. First, the assay described in Fig. 1 results in incomplete conversion of substrate to product at low enzyme concentration and extended incubation time. Second, while loss of activity is observed at low substrate concentrations, saturating substrate drives the reassociation of inactive monomers, resulting in an increase in activity during the assay. Finally, because purification tends to generate concentrated enzyme preparations, this frequently results in an increase in total activity, as was observed in the purification described below.

TABLE II
PURIFICATION OF IL-1β CONVERTING ENZYME

Step	Volume (ml)	Protein[a] (mg)	Enzyme (units)[b]	Specific activity (U/mg)	Purification (-fold)	Recovery overall (%)
THP.1 cell supernatant	1000	5000	45,000	9	1	100
DEAE-5PW chromatography	9	63	100,000	1587	176	220
Affinity chromatography	2.7	0.051	60,750	1,200,000	130,000	135

[a] The protein concentration of affinity-purified enzyme was determined by densitometry of silver-stained SDS–PAGE gels using an IL-1β converting enzyme standard. Otherwise, protein concentration was determined from absorbance at 280 nm. The extinction coefficient of homogeneous enzyme at 280 nm, determined by amino acid analysis, is 1.3 $(mg/ml)^{-1} cm^{-1}$.
[b] Units are based on cleavage of Ac-Tyr-Val-Ala-Asp-AMC.

Purification

The purification of catalytically active IL-1β converting enzyme to homogeneity by conventional chromatography has not yet been achieved. This is probably due to the low abundance of the activated enzyme in monocytic cells, and the facile dissociation to catalytically inactivate subunits that occurs with dilution. Fortunately, a potent and highly selective tetrapeptide aldehyde inhibitor[18] is an exceptionally good affinity chromatography ligand. The affinity purification described below, and summarized in Table II, produces homogeneous, catalytically active enzyme from THP-1 cells. Although the affinity column can be used to purify the enzyme directly to homogeneity from a crude cell cytosol, for practical reasons it is preferable to include an anion-exchange chromatography step,[19] as described below.

Cells

Human THP.1 cells are obtained from the American Type Culture Collection (Rockville, MD) and are grown in batch fermenters in Iscove's modified Dulbecco's medium supplemented with 9% (v/v) horse serum

---

[18] K. T. Chapman, *Bioorg. Med. Chem. Lett.* **2**(6), 613 (1992).
[19] D. K. Miller, J. M. Ayala, L. A. Egger, S. M. Raju, T.-T. Yamin, G. J.-F. Ding, E. P. Gaffney, A. D. Howard, O. C. Palyha, A. M. Rolando, J. P. Salley, N. A. Thornberry, J. R. Weidner, J. H. Williams, K. T. Chapman, J. Jackson, M. J. Kostura, G. Limjuco, S. M. Molineaux, R. A. Mumfore, and J. R. Calaycay, *J. Biol. Chem.* **268**(24), 18062 (1993).

and 0.1–0.3% F68 pluronic (JRH Biosciences, Lenexa, KS). Stimulation of the cells is not required to produce active enzyme.

Preparation of Cell Extracts

Cells (1×10^{11}) are harvested by centrifugation (1500 g, 10 min) and are washed three times by suspension in phosphate-buffered saline (250 ml) and centrifugation for 10 min at 1500 g at 4°. The cells are suspended at 10^8 cells/ml in a hypotonic buffer containing 25 mM HEPES, pH 7.5, 5 mM MgCl$_2$, and 1 mM EDTA, and are placed on ice for 20 min. Following addition of proteinase inhibitors (1 mM PMSF, 10 μg/ml pepstatin, 10 μg/ml leupeptin), the cells are broken with 25 strokes in a tight-fitting Dounce homogenizer. Nuclei are removed by centrifugation at 2000 g for 10 min at 4°. Following addition of 6 mM EDTA, the supernatant is centrifuged at 20,000 g for 20 min (to remove organelles) and 200,000 g for 60 min at 4° (to sediment membranes). The supernatant is dialyzed overnight at 4° against 25 mM HEPES, pH 7.5, 10% sucrose, 0.1% CHAPS, and 2 mM DTT. As noted above, the majority of the enzyme protein in THP.1 cells is present as the inactive proenzyme. Thus, the yield of active enzyme/cell has been variable in our laboratory, depending on the extent of conversion of proenzyme to the mature, active form during the preparation of cell lysates. Dialysis has been found to facilitate the conversion of the proenzyme to the mature, active form and generally results in a large increase in activity.[12]

Anion-Exchange Chromatography

The cell supernatant (1000 ml, 5 g protein, 45,000 U enzyme) is clarified by passage through a 0.22-μm hollow fiber membrane and applied to a TosoHaas DEAE-5PW anion-exchange HPLC column (475 ml, 20 $\times$ 5.5 cm) equilibrated with 20 mM Tris, pH 7.8, 10% sucrose, 0.1% CHAPS, and 2 mM DTT, at a flow rate of 25 ml/min at 4°. After washing with 1 column volume of equilibration buffer, the enzyme is eluted with a linear gradient of 0–0.4 M NaCl and 20–240 mM Tris-HCl (pH 7.8) in 10% sucrose, 0.1% CHAPS, and 2 mM DTT. The majority of the activity chromatographs in a single peak eluting at approximately 30 mM NaCl. The active fractions (25 ml each) are pooled and concentrated approximately 16-fold to 9 ml with an Amicon (Beverly, MA) ultrafiltration apparatus using a YM3 membrane.

Affinity Chromatography

Preparation of Affinity Resin. Selection of the affinity chromatography ligand, Ac-Tyr-Val-Lys-Asp-CHO, is based on substrate specificity studies (described above) that indicate that liberal substitutions are tolerated in the P2 position. The affinity resin is prepared by combining 5 ml of a 4 mM methanol solution of the aldehyde protected as its aspartyl-*tert*-butyl ester and dimethyl acetal, 5 ml of a 400 mM solution of potassium carbonate adjusted to pH 11.00 with HCl, and 10 g of suction-dried epoxy-activated Sepharose CL-4B[20] (prepared from 1,4-butanediol diglycidyl ether as described[21]). This slurry was stirred by rotation for 3 days at 37°. The resulting affinity matrix is washed thoroughly with 1 M KCl and water, and stored as a slurry at 4°. This procedure gives the dimethylacetal of Acetyl-Tyr-Val-Lys-Asp-CHO coupled to the spacer arm, the *tert*-butyl protecting group on the aspartate residue having been lost during the coupling conditions. Activation of this matrix to the aldehyde is carried out in the affinity column prior to use, by equilibrating the matrix with 0.1 N HCl and letting it stand overnight at 25°. The capacity of the column is approximately 0.1 μmol enzyme bound/ml packed affinity matrix.

Chromatography. The activated affinity column (2 ml, 5.5 × 0.7 cm) and a guard column of native Sepharose CL-4B (10 ml, 6.5 × 1.5 cm) are equilibrated with 10 column volumes of the chromatography buffer (100 mM HEPES, 10% sucrose, and 0.1% CHAPS at pH 7.50) supplemented with 1 mM DTT. The enzyme solution (9 ml, 100,000 U enzyme, 63 mg protein) is applied through the guard column to the affinity column at a flow rate of 0.3 ml/min at 4°, and washed through with an additional 4 guard column volumes of chromatography buffer at the same flow rate. During loading >90% of the enzymatic activity is retained. The guard column is removed and the affinity column washed with 50 column volumes of buffer at a flow rate of 0.3 ml/min at 4°, followed by 50 column volumes at 25° at a faster flow rate of 1 ml/min. Prior to elution, the column is washed with 10 column volumes of chromatography buffer lacking DTT to facilitate removal of inhibitor (described below). To elute the enzyme, the column is flooded with 1 column volume of this buffer containing 100 μM acetyl-Tyr-Val-Ala-Asp-CHO, and left for 24 hr at room temperature to achieve dissociation of the matrix-bound enzyme. The free enzyme–inhibitor complex is then recovered from the affinity column by washing with approximately 1 column volume of buffer at a flow rate of 0.02 ml/min.

Removal of Bound Inhibitor. Removal of the inhibitor by standard dialysis is not feasible, because practical concentrations of enzyme are well above K_i, forcing the equilibrium to lie in favor of enzyme–inhibitor complex. Instead, the enzyme is reactivated using two synergistic chemi-

---

[20] L. Sundberg and J. Porath, *J. Chromatogr.* **90**, 87 (1974).
[21] H. G. Bull, N. A. Thornberry, and E. H. Cordes, *J. Biol. Chem.* **260**(5), 2963 (1985).

cal approaches: conversion of the inhibitor to its oxime with hydroxyl-amine, and oxidation of the active site thiol to its mixed disulfide with glutathione disulfide (GSSG) by thiol–disulfide interchange, as shown in Scheme I.

The rate of reactivation of enzyme is governed by the rate of dissociation of E–I complex to form free enzyme and inhibitor ($k_2 = 3 \times 10^{-4}$ sec^{-1}), the rate of oxime formation ($k_4 = 7.7 \pm 1.7 \times 10^{-2}$ M^{-1} sec^{-1}), and the rate of thiol–disulfide interchange between the free enzyme and glutathione disulfide ($k_3 = 3$ M^{-1} sec^{-1}). Based on these rate constants, the enzyme–inhibitor solution recovered from the affinity column is adjusted to contain 100 mM neutral hydroxylamine and 10 mM glutathione disulfide to effect reactivaton. Under these conditions, after a short lag for consumption of excess free inhibitor, the dissociation of E–I complex becomes rate determining with a half-life of ~40 min at 25°. After allowing approximately 6 half-lives for the exchange, the inhibitor oxime and excess reagents are removed by gel filtration using Sephadex G-25. When desired, the enzyme–glutathione conjugate is reduced with 10 mM DTT (half-life < 1 min) to give active enzyme. The purified enzyme is stable indefinitely at −80°.

Results. The affinity purification described above gives a purification of 100,000-fold with at least 50% recovery of activity. As shown in the SDS–PAGE gel of Fig. 2, active enzyme is a heterodimer composed of equimolar amounts of two proteins, determined by mass spectroscopy to have molecular masses of 10,248 Da (p10) and 19,866 Da (p20). Frequently, affinity chromatography also isolates a protein with a molecular mass of 21,456 Da (p22), which has been identified as an alternately processed form of p20.[19]

The specific activity expected for fully active, homogeneous enzyme was determined by titration with an active site directed irreversible inhibi-

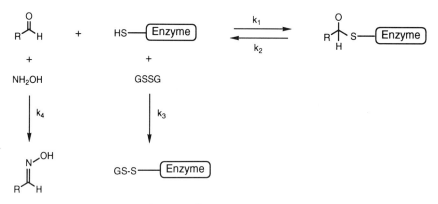

Scheme I.

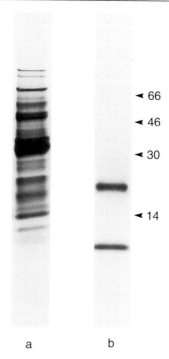

◄ 66

◄ 46

◄ 30

◄ 14

a   b

FIG. 2. Affinity purification. Silver-stained SDS–PAGE: (a) Anion exchange-purified sample applied to the affinity column, and (b) affinity column eluate revealing the two enzyme subunits. Positions of SDS–PAGE molecular weight standards ($\times 10^3$) are shown on the right.

tor, as shown in Fig. 3. The results from several such experiments indicate that 1 unit is equivalent to 18.7 ± 3.1 fmoles enzyme. This corresponds to a theoretical specific activity of 1.8×10^6 units/mg. In our laboratory, the observed specific activity is occasionally as much as 50% lower than this, due to the presence of alternately processed forms of the two subunits, variable degrees of dissociation to catalytically inactive subunits, or variable recovery of activity during removal of the inhibitor.

Precautions. Results of mass spectroscopy indicate that use of 10 mM glutathione to facilitate removal of inhibitor from enzyme frequently results in formation of at least one disulfide bond between the 20 and 10 kDa subunits, which can be reversed by prolonged treatment with DTT (100 mM, 60 min, 37°).

Inhibition

IL-1β converting enzyme meets all chemical and catalytic criteria for a cysteine proteinase. It is not inhibited by standard inhibitors of serine

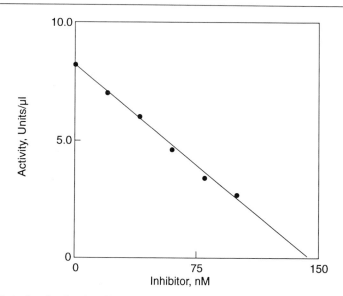

Fig. 3. Active site titration. Enzyme (8.2 U/μl) was incubated with the peptide (acyloxy)-methyl ketone Ac-Tyr-Val-Ala-Asp-CH$_2$-OCO-[2,6-(CF$_3$)$_2$]Ph (0–100 nM) under standard reaction conditions in a total volume of 25 μl. The second-order inactivation rate for this inhibitor is 1.7×10^6 M^{-1} sec^{-1}. Following incubation for 10 min at 25°, residual enzyme activity was measured by diluting 10 μl into a 100-μl assay containing 50 μM Ac-Tyr-Val-Ala-Asp-AMC under standard reaction conditions. After 10 min the reactions were quenched by boiling and addition of 400 μl 0.1% trifluoroacetic acid, and evaluated by HPLC as described. The x-axis intercept (143.5 nM) represents the amount of inhibitor required to inhibit 100% of the enzyme activity. The results of several experiments indicate that 18.7 ± 3.1 nM inhibitor is required to completely inhibit 1 U/μl enzyme, corresponding to 1 U = 0.56 ng enzyme.

(PMSF, diisopropyl fluorophosphate), aspartyl (pepstatin), or metalloproteinases (EDTA). Nonspecific thiol alkylating reagents, such as N-ethylmaleimide and iodoacetate, are competitive inactivators. The enzyme is also inhibited by the metal chelator 1,10-phenanthroline, but this was shown to involve phenanthroline–metal-catalyzed oxidation of the catalytic cysteine.[22] As described below, classic peptide-based reversible and irreversible cysteine proteinase inhibitors, when designed with a peptide sequence that satisfies the specificity requirements for IL-1β converting enzyme, are potent and highly selective, as anticipated from the unusual specificity of the enzyme.

[22] K. Kobashi and B. L. Horecker, *Arch. Biochem. Biophys.* **121,** 178 (1967).

Peptide Aldehydes

Peptide aldehydes are potent, reversible inhibitors of cysteine protein-ases, forming a thiohemiacetal with the active site cysteine.[23] The aldehyde designed with the appropriate peptide recognition sequence for IL-1β converting enzyme, Ac-Tyr-Val-Ala-Asp, is a potent, competitive, revers-ible inhibitor.[7,18] The compound displays slow formation ($k_{on} = 3.8 \times 10^5$ M^{-1} sec^{-1}) and slow dissociation of enzyme–inhibitor complex ($k_{off} = 2.9 \times 10^{-4}$ sec^{-1}, $t_{1/2} = 40$ min), with an overall dissociation constant of 0.76 nM. It is also highly selective, as demonstrated by the affinity chromatography described above.

Peptidyl Diazomethanes

Peptidyl diazomethanes are highly selective, covalent inhibitors of cysteine proteinases although the chemical basis for this selectivity is poorly understood (see [46] in this volume). The tetrapeptidyl diazometh-ane, Ac-Tyr-Val-Ala-Asp-CHN$_2$, is a competitive, irreversible inhibitor of IL-1β converting enzyme, which it inactivates with a second-order rate constant of 1.6×10^4 M^{-1} sec^{-1}.[7]

Peptide (Acyloxy)methyl Ketones

As described in this series, peptide (acyloxy)methyl ketones represent another class of potent, selective inactivators of cysteine proteinases (see [47] this volume). Compounds with the general structure Ac-Tyr-Val-Ala-Asp-CH$_2$-OCOAr are highly selective, covalent inhibitors of IL-1β converting enzyme.[24] Mass spectrometry and sequence analysis indicates that inactivation proceeds through expulsion of the carboxylate leaving group to form a thiomethyl ketone with the active site Cys-285. The second-order inactivation rate is independent of leaving group pK_a with an approximate value of 1×10^6 M^{-1} sec^{-1}. This rate constant is directly proportional to the reaction macroviscosity, indicating that the rate-limit-ing step in inactivation is association of enzyme and inhibitor, rather than any bond-forming reactions.

crmA

The only macromolecular inhibitor of the enzyme that has been identi-fied to date is a 38,000-kDa protein (crmA) from cowpox virus, whose amino acid sequence is similar to those of members of the serpin superfam-ily.[6] This protein, a potent, selective inhibitor of IL-1β converting enzyme,

---

[23] C. A. Lewis and R. Wolfenden, *Biochemistry* **16**(22), 4890 (1977).

[24] N. A. Thornberry, E. P. Peterson, J. J. Zhao, A. D. Howard, P. R. Griffin, and K. T. Chapman, *Biochemistry* **33**, 3934–3940 (1994).

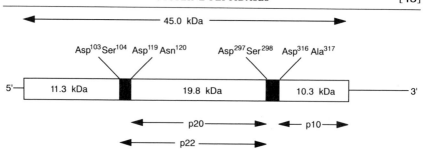

FIG. 4. Structural organization of the proenzyme. Taken from Thornberry *et al.*,[7] with permission.

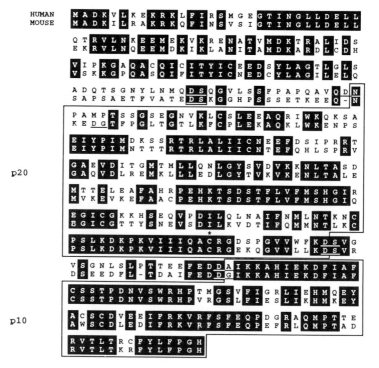

FIG. 5. Sequence comparison of human and murine IL-1β converting enzyme. Alignment of the sequences indicates that the Asp-119 and Pro-308 residues in the human enzyme are deleted in the murine enzyme. Identical amino acids are indicated in black and the active site Cys-285 is denoted with an asterisk. The p20 and p10 are outlined. The putative auto-processing sites are underlined. Taken from Monlineaux *et al.*,[10] with permission.

is the first example of a serpin whose target enzyme is a cysteine proteinase. Through suppression of host IL-1β, this protein is believed to play an important role in inhibition of the host inflammatory response to cowpox infection.

Structure and Composition

Molecular cloning and sequence analysis indicate that p10, p20, and p22 are proteolytically derived from a single 45-kDa proenzyme.[7] As shown in Fig. 4, the proenzyme also encodes an 11.5-kDa N-terminal propeptide, and a 2.0-kDa linker peptide between p20 and p10. The function of these peptides is unknown, but it is clear from the affinity chromatography that they are not required for catalytic activity.

A comparison of the primary sequences of the human[7,8] and murine[10,11] proenzymes, shown in Fig. 5, reveals the following. The catalytic Cys[285], identified by active site labeling with [^{14}C]iodoacetate, is located in a highly conserved stretch of amino acids in p20. Although 6 other Cys residues are conserved in the two subunits, the results of mass spectroscopy indicate that all are reduced in the active protein. Finally, in both species p20 and p10 are flanked by Asp-X bonds in the proenzyme. Because of the unusual specificity of the enzyme for Asp, it seems likely that the mature, catalytically active form of the enzyme is generated, at least in part, by an autocatalytic mechanism.

Three lines of evidence indicate that both subunits are required for catalytic activity. First, as described above, the enzyme undergoes dissociation to catalytically inactive subunits on dilution. Second, deletions of portions of p10 yield proteins without detectable processing activity.[8] Finally, the importance of this subunit is implied by the high degree of sequence identity between the human and murine enzymes (81%), compared to that for p20 (62%), and the propeptide (53%).

Acknowledgments

The author would like to thank Erin P. Peterson, Herbert G. Bull, Kevin T. Chapman, Matthew J. Kostura, Douglas K. Miller, Susan M. Molineaux, John A. Schmidt, and Michael J. Tocci for critical reading of this article.

Note Added in Proof: In the crystal structure of the enzyme, two p20/p10 heterodimers associate to form a tetramer, and it has been argued that this is the catalytically active form of the enzyme.[27]

[27] K. P. Wilson, J. F. Black, J. A. Thomson, E. E. Kim, J. P. Griffith, M. A. Navia, M. A. Murcko, S. P. Chambers, R. A. Aldape, S. A. Raybuck, and D. J. Livingston, *Nature*, **370,** 270–275 (1994).

[44] Isoprenylated Protein Endopeptidase

By ROBERT R. RANDO and YU-TING MA

Introduction

One important class of posttranslational modifications involves protein isoprenylation.[1-7] In this set of modifications, a protein with a carboxyl-terminal CX$_3$[1,3,4,7,8] or, much less frequently, a CXC or a CC(XX) sequence (where C is cysteine, and X is an undefined amino acid)[9-11] is first isoprenylated at the cysteine residue(s) with either all-*trans*-farnesyl (C$_{15}$) or all-*trans*-geranylgeranyl (C$_{20}$) pyrophosphate.[12-14] Soluble isoprenyl transferases that carry out these modifications have already been purified and characterized.[15,16] In the case of modifications at a CX$_3$ motif, proteolysis

[1] P. J. Casey, P. A. Solski, C. J. Der, and J. E. Buss, *Proc. Natl. Acad. Sci. U.S.A.* **86,** 8323 (1989).

[2] C. C. Farnsworth, M. H. Gelb, and J. A. Glomset, *Trends Biochem. Sci.* **15,** 139 (1990).

[3] J. F. Hancock, A. I. Magee, J. E. Childs, and C. J. Marshall, *Cell (Cambridge, Mass.)* **57,** 1167 (1989).

[4] R. K. Lai, D. Perez-Sala, F. J. Cañada, and R. R. Rando, *Proc. Natl. Acad. Sci. U.S.A.* **87,** 7673 (1990).

[5] W. A. Maltese, *FASEB J.* **4,** 3319 (1990).

[6] S. M. Mumby, P. J. Casey, A. G. Gilman, S. Gutowski, and P. C. Sternweis, *Proc. Natl. Acad. Sci. U.S.A.* **87,** 5873 (1990).

[7] W. R. Schafer, R. Kim, R. Sterne, J. Thorner, S.-H. Kim, and J. Rine, *Science* **245,** 379 (1989).

[8] D. R. Lowy and B. M. Willumsen, *Nature (London)* **341,** 384 (1990).

[9] H. Horiuchi, M. Kawata, M. Katayama, Y. Yoshida, T. Musha, S. Ando, and Y. Takai, *J. Biol. Chem.* **266,** 16981 (1991).

[10] C. C. Farnsworth, M. Kawata, Y. Yoshida, Y. Takai, M. H. Gelb, and J. A. Glomset, *Proc. Natl. Acad. Sci. U.S.A.* **88,** 6196 (1991).

[11] R. Khosravi-Far, R. J. Lutz, A. D. Cox, L. Conroy, J. R. Bourne, M. Sinensky, W. E. Balch, J. E. Buss, and C. J. Der, *Proc. Natl. Acad. Sci. U.S.A.* **88,** 6264 (1991).

[12] W. Manne, D. Roberrts, A. Tobin, E. O'Rourke, M. De Virgilio, C. Meyers, N. Ahmed, B. Kurz, M. Resh, H.-F. Kung, and M. Barbacid, *Proc. Natl. Acad. Sci. U.S.A.* **87,** 7541 (1990).

[13] Y. Reiss, J. L. Goldstein, M. C. Seabra, P. J. Casey, and M. S. Brown, *Cell (Cambridge, Mass.)* **62,** 81 (1990).

[14] M. D. Schaber, M. B. O'Hara, W. M. Garsky, S. D. Mosser, J. D. Bergstrom, S. L. Moores, M. S. Marshall, P. A. Friedman, R. A. F. Dixon, and J. B. Gibbs, *J. Biol. Chem.* **265,** 14701 (1990).

[15] Y. Reiss, S. J. Stradley, L. M. Gierasch, M. S. Brown, and J. L. Goldstein, *Proc. Natl. Acad. Sci. U.S.A.* **88,** 732 (1991).

[16] M. C. Seabra, Y. Reiss, P. J. Casey, M. S. Brown, and J. L. Goldstein, *Cell (Cambridge, Mass.)* **65,** 429 (1991).

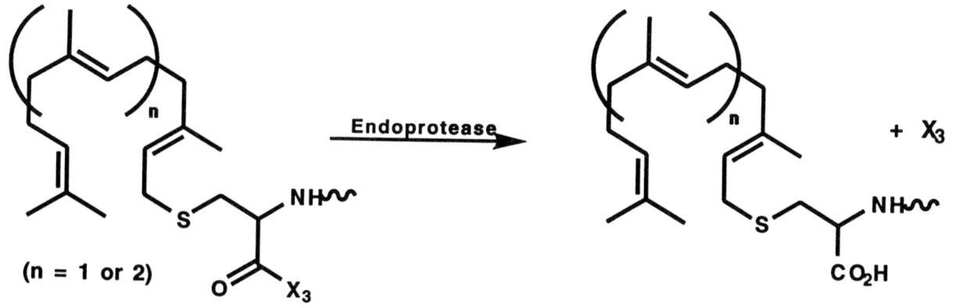

Endoprotease

$+ X_3$

(n = 1 or 2)

NH

S

O

X_3

NH

S

CO_2H

SCHEME I. Endoproteolysis in the isoprenylation pathway.

follows, to generate the isoprenylated cysteine residue as the new carboxyl terminus.[17] This set of modifications is completed by the reversible carboxymethylation of the isoprenylated cysteine residue.[18–25]

Proteolysis occurs in mammals primarily by endoproteolytic cleavage between the modified cysteine residue and the adjacent aliphatic amino acid to liberate the intact X_3 tripeptide, as shown in Scheme I.[26,27] A liver and pancreatic microsomal endoproteolytic activity was identified that produces a single cut between the modified cysteine residue and the X_3 tripeptide, using a synthetic radiolabeled tetrapeptide substrate L-AFC-Val-Ile-Ser[26] (AFC, N-acetyl-S-all-trans-farnesyl-L-cysteine). The protease also specifically cleaves the tripeptide L-AFC-Val-Ile and the dipeptide L-AFC-Val, but not L-AFC amide.[26] Therefore, a dipeptide is minimally required for substrate activity. The enzyme does not cleave substrates containing D-AFC as the first amino acid, demonstrating that the cleavage reaction is stereospecific at the scissile bond.[26] Moreover, the isoprenyl

[17] J. F. Hancock, K. Cadwallader, and C. Marshall, EMBO J. 10, 641 (1991).

[18] S. Clarke, J. P. Vogel, R. J. Deschenes, and J. Stock, Proc. Natl. Acad. Sci. U.S.A. 85, 4643 (1988).

[19] Y. Fukada, T. Takao, H. Ohguro, T. Yoshizawa, T. Akino, and Y. Shimonishi, Nature (London) 346, 658 (1990).

[20] L. Gutierrez, A. I. Magee, C. J. Marshall, and J. F. Hancock, EMBO J. 8, 1093 (1989).

[21] M. Kawata, C. C. Farnsworth, Y. Yoshida, M. M. Gelb, J. A. Glomset, and Y. Takai, Proc. Natl. Acad. Sci. U.S.A. 87, 8960 (1990).

[22] D. Perez-Sala, E. W. Tan, F. J. Cañada, and R. R. Rando, Proc. Natl. Acad. Sci. U.S.A. 88, 3043 (1991).

[23] R. C. Stephenson and S. Clarke, J. Biol. Chem. 265, 16248 (1990).

[24] H. K. Yamane, C. C. Farnsworth, H. Xie, T. Evans, W. N. Howald, M. H. Gelb, J. A. Glomset, S. Clarke, and B. K.-K. Fung, Proc. Natl. Acad. Sci. U.S.A. 88, 286 (1991).

[25] H. K. Yamane, C. C. Farnsworth, H. Xie, W. Howald, B. K.-K. Fung, S. Clarke, M. H. Gelb, and J. A. Glomset, Proc. Natl. Acad. Sci. U.S.A. 87, 5868 (1990).

[26] Y.-T. Ma and R. R. Rando, Proc. Natl. Acad. Sci. U.S.A. 89, 6275 (1992).

[27] M. N. Ashby, D. S. King, and J. Rine, Proc. Natl. Acad. Sci. U.S.A. 89, 4613 (1992).

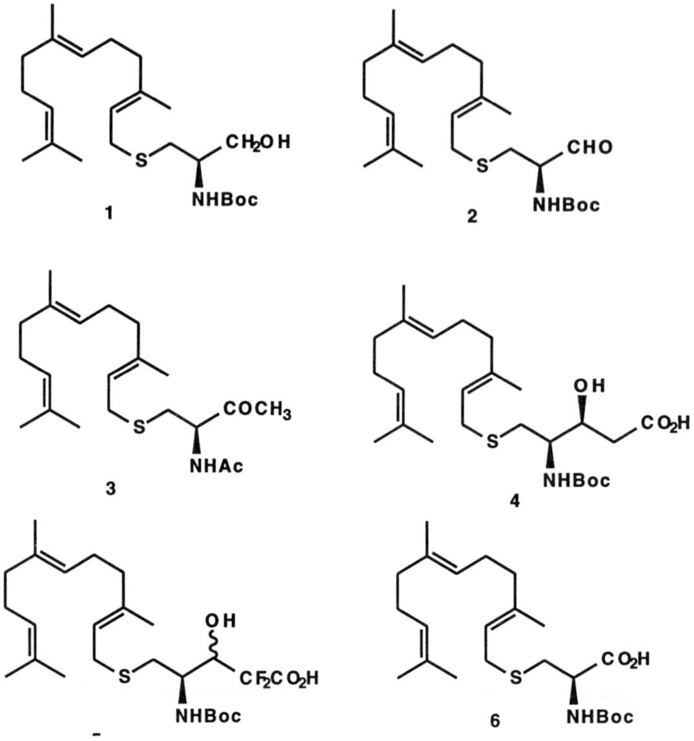

SCHEME II. Farnesyl-L-cysteine-based inhibitors.

group is essential for substrate activity, and either stereospecificity or stereoselectivity is observed at X_3.[28]

Thus far, little information is available on the mechanistic class to which the enzyme belongs. Standard commercially available protease inhibitors designed to inhibit either serine, cysteine, aspartic, or metalloproteases have thus far not been shown capable of inhibiting the endoprotease.[29] However, potent L-AFC-containing peptide analogs have been prepared as inhibitors of the enzyme.[29] In the amino acid series shown in Scheme II (structures 1–6), only the aldehyde analog 2 proved to be a potent competitive inhibitor ($K_I = 1.9 \ \mu M$) of the enzyme. The remaining inhibitors proved to be inactive. The fact that the aldehyde 2 is a potent inhibitor suggests the possibility that the protease is a serine- or cysteine-

[28] Y.-T. Ma, A. Chaudhuri, and R. R. Rando, *Biochemistry* **31**, 11772 (1992).
[29] Y.-T. Ma, B. A. Gilbert, and R. R. Rando, *Biochemistry* **32**, 2386 (1993).

based enzyme.[30] It is noteworthy that neither the statine nor difluorostatine analog (4 and 5) was active as an inhibitor, suggesting that the enzyme is not an aspartic endopeptidase.[31]

Assay Method

Principle

The assay for the endopeptidase is described in detail, as is the synthesis of a substrate to assay the enzyme. Bovine liver microsomes provided the source of enzyme. Various substrates have been utilized, although a tetrapeptide having the sequence AFC-Val-Ile-Ser (all L-amino acids) was first used as a synthetic substrate for the protease.[27] A radiochemical assay was developed in which an [³H]acetyl is incorporated in the substrate. Endoproteolytic cleavage of the substrate leads to the formation of [³H]AFC, which is followed by high-performance liquid chromatography (HPLC) analysis.

The particular peptide sequence chosen (AFC-Val-Ile-Ser) is taken from the carboxyl terminus of the γ subunit of the retinal heterotrimeric G protein transducin.[32–35] The γ subunit of transducin has already been shown to be farnesylated and methylated.[4,19] Because of the real possibility of endopeptidase isoforms, the choice of an individual peptide substrate should be determined by the exact sequence of the isoprenylated protein substrate whose cleavage one is interested in determining. Nevertheless, the assay described here will be with [³H]AFC-Val-Ile-Ser as the substrate. It was shown that this substrate is specifically cleaved by the microsomal enzyme between the isoprenylated cysteine residue and the valine residue to generate [³H]AFC and the intact Val-Ile-Ser.[26] Because [³H]AFC is readily separated from the tetrapeptide substrate by HPLC, the enzymatic activity of the endoprotease can be easily assayed.

[30] A. Vinitsky, C. Michaud, J. C. Powers, and M. Orlowski, *Biochemistry* **31**, 9421 (1992).
[31] A. M. Doherty, I. Sircar, B. E. Kornberg, J. Quin, R. T. Winters, J. S. Kaltenbronn, M. D. Taylor, B. L. Batley, S. R. Rapundalo, M. J. Ryan, and C. A. Painchaud, *J. Med. Chem.* **35**, 214 (1992).
[32] D. C. Medynski, K. Sullivan, D. Smith, C. Van Dop, F.-H. Chang, B. K.-K. Fung, P. H. Seeburg, and H. R. Bourne, *Proc. Natl. Acad. Sci. U.S.A.* **82**, 4311 (1985).
[33] K. Yatsunami and H. G. Khorana, *Proc. Natl. Acad. Sci. U.S.A.* **82**, 4316 (1985).
[34] T. Tanabe, T. Nukada, Y. Nishikawa, K. Sugimoto, H. Suzuki, H. Takahashi, M. Noda, T. Haga, A. Ichiyama, K. Kangawa, N. Minamino, H. Matsuo, and S. Numa, *Nature (London)* **315**, 242 (1985).
[35] J. B. Hurley, H. K. W. Fong, D. B. Teplow, W. J. Dreyer, and M. I. Simon, *Proc. Natl. Acad. Sci. U.S.A.* **81**, 6948 (1984).

Procedure

Kinetic measurements are performed by incubating 0.2 mg/ml protein with peptide substrates in 200 mM 4-(2-hydroxyethyl-1-piperazineethane-sulfonic acid (HEPES) buffer containing 100 mM NaCl and 5 mM MgCl$_2$ at pH 7.4, all in a final volume of 50 μl. The peptide substrates are added in dimethyl sulfoxide (DMSO) to a final concentration of 4% (v/v) at 37°. The enzymatic reaction is quenched at the appropriate time by the addition of 0.5 ml of chloroform/methanol (1/1, v/v). Phase separation is achieved by adding 0.5 ml 1 M citric acid. The chloroform layer is separated and evaporated under nitrogen and the residue is dissolved in *n*-hexane/2-propanol/trifluoroacetic acid [85 : 15 : 0.1 (v/v/v)] containing cold AFC. The formation of the product [³H]AFC is followed by HPLC analysis at 210 nm.[22] The samples are injected onto a normal phase HPLC column (Dynamax 60, Rainin, Woburn, MA) connected to a model LB 506-C on-line radioactivity monitor (Berthold, Nashua, NH). The column is eluted with *n*-hexane/2-propanol/TFA [85 : 15 : 0.1 (v/v/v)[at 1.5 ml/min. As mentioned previously, the enzyme requires an isoprenylated cysteine resi-due in the peptide substrate. Farnesyl, geranylgeranyl, and geranyl moie-ties are sufficient, although *N*-acetyl-*S*-all-*trans*-geranyl-L-cysteine (AGC) derivatives provide the weakest substrates.[28] The retention times for AGC, AFC, and *N*-acetyl-*S*-all-*trans*-geranylgeranyl-L-cysteine (AGGC) are 7.58, 7.26, and 7.00 min, respectively, and, by way of comparison, for AGC-Val-Ile, AFC-Val-Ile, and AGGC-Val-Ile, they are 5.48, 6.54, and 5.08 min, respectively. The limit of efficiency of this assay is 3× the background counts/minute [~150 counts per minute (cpm)], which would readily allow us to assay molecules as substrates whose activities were 0.4% of the V_{max} of AFC-Val-Ile-Ser (1.13 pmol/min/mg).

Source of Enzyme

Although the endopeptidase is likely to be ubiquitously distributed, our own work has centered on the use of bovine liver microsomes as the source of enzyme. Although we have effectively solubilized the enzyme with 3-[(3-cholamidopropyl)dimethylammonio-2-hydroxy-1-propane sul-fonate (CHAPSO), we have yet to achieve its complete purification. There-fore, the studies described here use bovine liver microsomes, which are obtained by the published procedure.[36]

---

[36] P. Walter and G. Blobel, this series, Vol. 96, p. 84.

Syntheses of Peptide Substrates

Synthesis of N-[³H]Acetyl-S-all-trans-farnesyl-L-Val-L-Ile

A mixture of S-all-*trans*-farnesyl-L-cysteine-L-valine-L-isoleucine methyl ester (20 mg, 36 μmol), [³H]acetic anhydride (0.02 mmol, ~10 mCi, 500 mCi/mmol), triethylamine (3.3 μl, 0.024 mmol), and a catalytic amount of 4,4-dimethylaminopyridine in 20 ml of dichloromethane is stirred at room temperature overnight. The mixture is filtered and the solvent is evaporated. The residue is separated by HPLC (Rainin silica, 250 × 4.6 mm, 1.5 ml/min, hexane/2-propanol 85 : 15; UV detection is at 210 nm). Retention time, 4.27 min; yield, 3.57 mCi, 14.28 μmol, 71%.

N-[³H]Acetyl-S-all-*trans*-farnesyl-L-cysteine-L-valine-L-isoleucine methylester(1.19mCi,4.76μmol,250mCi/mmol)isdissolvedin12mlofmethanol.KOH/methanol(5%,12ml)isaddedat0°andthemixtureisstirredat0°for 1hrandthenatroomtemperaturefor18hr.Basichydrolysisundertheseconditions serves to racemize the chiral center specifically at the cysteine residue. AceticacidisaddedtopH7.Themethanolisevaporatedandtheresidueisextracted with ethylacetate twice. The combined organic phase is washed with 10% HCl,thenwithwater,thendriedoveranhydroussodiumsulfate,filtered, and evaporated. Total activity, 1.04 mCi (87%). Purification is carried out by HPLConaRaininsilica250×4.6-mmcolumnataflowrateof1.5ml/min,using hexane/2-propanol/TFA92 : 8 : 0.01aseluant. UVdetectionisat210nm.The retentiontimeoftheDLLisomeris8.1minandfortheLLLisomeritis10.2min.

Synthesis of N-[³H]Acetyl-S-all-trans-farnesyl-L-cysteine-L-valine-L-isoleucine-L-serine

A solution of N-[³H]acetyl-S-all-*trans*-farnesyl-L-cysteine-L-valine-isoleucine (1.12 μmol, 250 mCi/mmol), L-serine hydrochloride methyl ester (20 mg, 0.13 mmol), and 1-hydroxybenzotriazole hydrate (17.4 mg, 0.13 mmol), and 1-hydroxybenzotriazole hydrate (17.4 mg, 0.13 mmol) in 3 ml of dimethylformamide (DMF) is cooled to 0°. 1-(3-Dimethylamino)-propyl-3-ethyl carbodiimide hydrochloride (17.2 mg, 0.09 mmol) and N-methylmorpholine (14.1 μl, 13 mg, 0.13 mmol) are added at 0°. The mixture is stirred at 0° under argon for 2 hr and at room temperature for 24 hr. Water (20 ml) is added and the mixture is extracted with ethyl acetate (4 × 20 ml). The combined organic layer is washed with 10% HCl, saturated sodium bicarbonate, and water, dried over anhydrous magnesium sulfate, filtered, and evaporated. The residue is dissolved in 3 ml of acetonitrile, and 3 ml of 10% aqueous sodium carbonate is added. The mixture is stirred at room temperature for 24 hr; 10% HCl is added to adjust the pH

TABLE I
SUBSTRATE SPECIFICITY STUDIES ON ENDOPROTEASE

Substrate	Isomer	K_m (μM)	V_{max} (pmol/min/mg)
AFC-Val-Ile	LLL	9.2 ± 0.2	57.7 ± 5
AGGC-Val-Ile	LLL	4.01 ± 0.4	26.20 ± 0.97
AGC-Val-Ile	LLL	14.60 ± 2.14	19.35 ± 1.12
Cys-Val-Ile	LLL	Inactive	
tert-Butylthio-Cys-Val-Ile	LLL	Inactive	
AFC-Val-Ile-Met	LLLL	2.96 ± 0.35	126.3 ± 4.7
AFC-Val-Ile-Ser	LLLL	5.76 ± 0.72	251.2 ± 8.5

to 2. The mixture is then extracted with ethyl acetate (3 × 20 ml), and the combined organic layer is washed with water, dried over anhydrous magnesium sulfate, filtered, and evaporated. Purification is carried out by HPLC on a Rainin silica 4.6 × 250-mm column at a flow rate of 1.5 ml/min, using hexane/2-propanol/TFA 85 : 15 : 0.1 as eluant. UV deteection is at 210 nm. Recovery of radioactivity, 28.2 μCi, 1.13 × 10^{-4} mmol; retention time, 6.49 min. ^{1}H-NMR (500 MHz, DMSO-d_6) of the authentic nonradioactive peptide: δ 8.14 (1H, d, J = 8 Hz), 7.97 (1H, d, J = 7 Hz), 7.86 (1H, d, J = 9 Hz), 7.76 (1H, d, J = 8.5 Hz), 5.16 (1H, t, J = 8.5 Hz), 5.05 (2H, m), 4.48 (1H, brdd, J = 6.5, 14 Hz), 4.25 (1H, t, J = 8 Hz), 4.20 (1H, brt, J = 8 Hz), 4.10 (1H, brd, J = 7 Hz), 3.66 (1H, dd, J = 6, 11 Hz), 3.56 (1H, dd, J = 4, 11 Hz), 3.08 (2H, m), 2.73 (1H, dd, J = 5.5, 13.5 Hz), 2.52 (1H, dd, J = 4, 11 Hz), 1.90–2.10 (8H, m), 1.84 (3H, s), 1.73 (2H, m), 1.62 (3H, s), 1.60 (3H, s), 1.54 (6H, s), 1.41 (1H, m), 1.30 (1H, m), 0.80 (12H, m).

Properties

Enzymatic Hydrolysis of AFC-Val-Ile-Ser

In the experiments reported here, a tetrapeptide having the sequence AFC-Val-Ile-Ser (all L-amino acids) was chosen as a synthetic substrate for the endopeptidase. Incubation of the tetrapeptide (labeled with an [^{3}H]acetyl group) with a calf liver microsomal enzyme preparation led to the time-dependent formation of N-acetyl-S-farnesyl-L-cysteine.[26] When the concentration of substrate was varied, saturation was observed, and the K_m and V_{max} were measured to be 5.76 ± 0.72 μM and 251.17 ± 8.51 pmol/min/mg protein, respectively. AFC formation did not occur when the tetrapeptide was D-AFC-Val-Ile-Ser, nor when the carboxyl group of the serine residue was methylated.[26] These data demonstrate that the proteolysis is stereospecific and that a free carboxyl-terminal group is required in the substrate. A pH versus rate profile for the proteolysis reaction showed a broad maximum at approximately pH 7.[26]

It was of special interest to determine where the initial cleavage reaction took place. To accomplish this, the tetrapeptide AFC-Val-Ile-Ser was synthesized with a tritiated serine residue at the carboxyl terminus. This tetrapeptide was subjected to proteolysis by the microsomal enzyme, and the products were treated with *o*-phthalaldehyde and mercaptoethanol to generate the fluorescently tagged peptides.[37] These peptides could be separated easily on reversed-phase HPLC. Incubation of the endoprotease preparation with this substrate initially produced the tripeptide Val-Ile-Ser intact.[27] This shows that the initial proteolytic cut is between the AFC moiety and the adjacent valine residue.

The experiments described here demonstrate that the endopeptidase is stereospecific at the scissile bond; it also requires a free terminal carboxyl group in the tetrapeptide substrate. Substrate specificity was further explored using the substrates in Table I. Clearly, the tripeptide is a weaker substrate for the protease than is the tetrapeptide. Kinetic constants were not obtained for the dipeptide AFC-Val, but it is clearly a substrate, albeit a weak one, for the protease as well. An isoprenyl group is essential for substrate activity. Stereospecificity is also observed with both the di- and tripeptides, because substitution of D-AFC for L-AFC in these molecules led to the complete abolition of substrate activity. Finally, AFC amide is not a substrate for the enzyme, demonstrating that a second amino acid moiety is required, in addition to AFC, in the substrate.

[37] E. Trepman and R. F. Chen, *Arch. Biochem. Biophys.* **204**, 524 (1980).

[45] Affinity Chromatography of Cysteine Peptidases

By DAVID J. BUTTLE

General Considerations

Cysteine peptidases rely on the thiol group of a cysteine residue for catalytic activity. The sulfur atom must be in a reduced, or reducible, form to participate as the essential thiolate anion in catalysis. In order to obtain fully active preparations of cysteine peptidases it is therefore usually necessary to separate active and activatable (reversibly oxidized) forms of the enzyme from the irreversibly oxidized sulfinic acid, as well as from other contaminating proteins. Methods for active site-directed affinity chromatography of cysteine peptidases thus require the use of an immobilized ligand that has selective affinity for the active site of the enzyme in its catalytically competent, reduced state. This is achieved by the use of either chemical groups that have a reversible affinity for thiols, or substrate-like peptides that may act by forming tetrahedral intermediates that do not lead to hydrolysis, or by compounds that combine these two properties. Published methods vary between those that should be generally applicable to any thiol-containing protein, and those that have been developed for the purification of a particular peptidase from a mixture of such enzymes, by the incorporation in the ligand of specificity determinants for the individual enzyme.

Organic Mercurial Ligands

General Principle

A preparation of active cysteine peptidase is passed through a column comprising an immobilized organomercurial compound. Thiol-containing proteins are retained by the column. Advantage can be taken of the very low pK of the active site cysteine of cysteine peptidases (at least those in the papain family) compared to uncomplicated thiols. Thus, if the sample is applied at a mildly acidic pH, most thiol-containing compounds will not bind to the column but cysteine peptidases will. These are subsequently eluted with buffer containing a mercurial salt or a thiol compound. The method is not capable of distinguishing between different cysteine peptidases, so other fractionation procedures are required to do this if a mixture exists.

Examples

The first example of this type of affinity chromatography was described by Sluyterman and Wijdenes.[1,2] p-Aminophenylmercuric acetate was coupled to CNBr-activated Sepharose 4B (Pharmacia, Piscataway, NJ) and the capacity determined with 5,5'-dithiobis(2-nitrobenzoic acid). Application of activated papain (EC 3.4.22.2) at pH 5 and elution with HgCl$_2$ led to the recovery of papain, containing almost exactly 1 atm of Hg/molecule, which could be readily converted to fully active papain with 1.7-fold the specific activity of the applied sample. Elution of active papain rather

---

[1] L. A. Æ. Sluyterman and J. Wijdenes, *Biochim. Biophys. Acta* **200**, 593 (1970).
[2] L. A. Æ. Sluyterman and J. Wijdenes, this series, Vol. 34, p. 544.

than mercuripapain was achieved with 10 mM mercaptoethanol in place of $HgCl_2$.

Kortt and Liu[3] used essentially the same procedure for the purification of streptopain (EC 3.4.22.10), with recovery of 100% activatable enzyme. Raising the pH of the $HgCl_2$-containing buffer to 6.6 was the only modification required for the successful elution of ficain (a mixture of cysteine peptidases obtained from the latex of *Ficus glabrata*; EC 3.4.22.3) with high specific activity when activated with thiols.[4]

Organomercurial-Sepharose has also been used in the purification of human liver cathepsin B (EC 3.4.22.1).[5] Application of pH 5.5 of a partially purified preparation and elution with 10 mM cysteine resulted in a fourfold increase in specific activity and 60% yield. It was not necessary to activate the cathepsin B preparation prior to application, presumably because the freshly prepared enzyme had not oxidized.

Disulfide Exchange

General Principle

A thiol-containing reagent is immobilized on a chromatography matrix and a mixed disulfide is subsequently formed, possibly by use of a thiol-containing spectrophotometric reporter group. The enzyme sample is activated by the addition of a reducing agent and applied to the gel. At low pH the active site thiol reacts preferentially with the immobile mixed disulfide, compared to the low M_r reducing agent, and binds to the matrix, displacing the spectrophotometric reporter thiol. Active enzyme is subsequently eluted from the column by the application of a reducing agent. As with the use of organic mercury, this method is generally incapable of distinguishing between different cysteine peptidases. Glutathione-Sepharose 4B, thiopropyl-Sepharose 6B, and activated thiol-Sepharose 4B are available from Pharmacia LKB Biotechnology.

Examples

Brocklehurst et al.[6] reported the synthesis of immobilized glutathione-2-pyridyl disulfide and its use in the preparation of fully active papain. The capacity of the gel was conveniently determined by measuring A_{343}

---

[3] A. A. Kortt and T. Y. Liu, *Biochemistry* **12**, 320 (1973).
[4] C. D. Anderson and P. L. Hall, *Anal. Biochem.* **60**, 417 (1974).
[5] A. J. Barrett, *Biochem. J.* **131**, 809 (1973).
[6] K. Brocklehurst, J. Carlsson, M. P. J. Kierstan, and E. M. Crook, *Biochem. J.* **133**, 573 (1973).

due to 2-thiopyridone in the eluate, produced as a consequence of the reaction between glutathione and 2,2'-dipyridyl disulfide. The gel was also very easily regenerated by reducing the immobilized glutathione and then allowing it to react again with 2,2'-dipyridyl disulfide. Preparations of active papain are applied to the gel at pH 4 in buffer containing 0.3 M NaCl and 1 mM EDTA. The papain sample may contain up to 5 mM cysteine without any appreciable reaction between the ligand and cysteine.[7] Displacement by papain of 2-thiopyridone on the column is followed by measuring A_{343} of the eluate. After both A_{280} and A_{343} have fallen to zero, the column is reequilibrated to pH 8.0 and fully active papain (containing 1 mol thiol/mol enzyme) is eluted with 50 mM cysteine.

The method was adapted for the purification of larger amounts of papain and ficain by batchwise separation.[8] Partially purified active ficain was mixed batchwise with Sepharose-glutathione-2-pyridyl disulfide at pH 4.5. After washing, the pH was adjusted to 8.0 and fully active ficain was released from the matrix with 20 mM cysteine.[9]

For the purification of actinidain (EC 3.4.22.14), thiopropyl-Sepharose 6B was used in place of immobilized glutathione.[10] Partially purified active enzyme was bound to the gel batchwise at pH 4.4. After washing and equilibration at pH 8.0, the gel was packed into a column and fully active actinidain eluted in 20 mM cysteine and 1 mM dithiothreitol. Very reproducible results have been obtained by this method (A. J. Barrett and M. A. Brown, unpublished results). After recovery of the enzyme, the thiopropyl-Sepharose is very readily reconverted to the 2-pyridyl mixed disulfide, and the greater degree of substitution (compared to glutathione-Sepharose) improves the convenience and performance of the purification.

The same affinity medium was used in the copurification of cathepsins B and H (EC 3.4.22.16) from bovine spleen.[11] The partially purified active enzymes were bound to the column batchwise at pH 4.0. Subsequent recovery of the active enzymes was achieved with 20 mM cysteine at pH 6.8 (due to the instability of the mammalian lysosomal cysteine peptidases above pH 7), and cathepsins B and H were then separated by cation-exchange chromatography. Unfortunately, no data regarding the degree of purification of either enzyme were given. When Sepharose-glutathione-2-pyridyl disulfide was used in place of Sepharose-propyl-2-pyridyl disul-

---

[7] K. Brocklehurst, J. Carlsson, M. P. J. Kierstan, and E. M. Crook, this series, Vol. 34, p. 531.

[8] T. Stuchbury, M. Shipton, R. Norris, J. P. G. Malthouse, K. Brocklehurst, J. A. L. Herbert, and H. Suschitzky, *Biochem. J.* **151**, 417 (1975).

[9] J. P. G. Malthouse and K. Brocklehurst, *Biochem. J.* **159**, 221 (1976).

[10] K. Brocklehurst, B. S. Baines, and P. G. Malthouse, *Biochem. J.* **197**, 739 (1981).

[11] F. Willenbrock and K. Brocklehurst, *Biochem. J.* **227**, 511 (1985).

fide, a large fraction of cathepsin H failed to bind to the column. No information was given about the fate of a third lysosomal cysteine endopeptidase present in bovine spleen, cathepsin S (EC 3.4.22.27).[12,13]

An attempt to build some degree of specificity into disulfide exchange chromatography was made by Evans and Shaw.[14] An immobilized asymmetric mixed disulfide, Affi-Gel 10Gly-Phe-Phe-cystamine, was found to bind partially purified active pig liver cathepsin B at pH 6.5. {Recommended abbreviations [*Eur. J. Biochem.* **138**, 9 (1984)] are used for amino acids and blocking groups.} Unbound material was washed from the column and the enzyme was eluted with 10 mM cysteine at the same pH, resulting in about 50-fold purification. Unfortunately, no account was taken of the other cysteine peptidases that would have been present in the starting material, and the assays used were not specific for cathepsin B. It is therefore not possible to say if the method represents a *specific* purification of cathepsin B, or if other endopeptidases, such as cathepsin L (EC 3.4.22.15), would also bind to and be eluted from the matrix.

Peptide Inhibitors with Unmodified C Termini

General Principle

Cysteine endopeptidases usually have specificity for cleavage of substrates with a bulky hydrophobic residue in P2 (in the accepted terminology[15]). Peptides that have such a residue at the penultimate position to the C terminus, that is, peptides that do not contain the preferred scissile bond, act as reversible inhibitors.[16] The enzyme is bound to such an immobilized ligand under conditions in which the affinity is relatively high, and is subsequently released by changing the conditions such that the affinity is lowered.

Examples

The first example of this type of affinity chromatography for the purification of a cysteine peptidase utilized Sepharose-Gly-Gly-Tyr(Bzl)-Arg in the purification of fully active papain.[17] The K_i for the binding of the

[12] V. Turk, I. Kregar, F. Gubensek, and P. Locnikar, *in* "Protein Turnover and Lysosomal Function" (H. L. Segal and D. J. Doyle, eds.), p. 353. Academic Press, New York, 1978.
[13] H. Kirschke, I. Schmidt, and B. Wiederanders, *Biochem. J.* **240**, 455 (1986).
[14] B. Evans and E. Shaw, *J. Biol. Chem.* **258**, 10227 (1983).
[15] Nomenclature Committee of the IUBMB, *in* "Enzyme Nomenclature," p. 371. Academic Press, London, 1992.
[16] A. Berger and I. Schechter, *Philos. Trans. R. Soc. London B* **257**, 249 (1970).
[17] S. Blumberg, I. Schechter, and A. Berger, *Eur. J. Biochem.* **15**, 97 (1970).

soluble inhibitor to papain was found to be 10 μM at pH 4.3, and 150 μM at pH 6.0. A partially purified preparation of papain was applied to the immobilized ligand in 20 mM EDTA and 10 mM 2-mercaptoethanol, pH 4.3, and fully active enzyme subsequently eluted in water. Burke et al.[18] adapted the method for inclusion in a two-step purification of papain from a crude extract of papaya latex. Such a procedure should, however, be used with caution until the affinity for the ligand of the other cysteine endopeptidases also present in the starting material has been investigated. Two of these enzymes, chymopapain (EC 3.4.22.6) and caricain (EC 3.4.22.30), have specificity very similar to that of papain and are present in the latex in much greater amounts.[19] Burke et al. noticed components in the latex other than papain that bound and were not eluted from the column, thus reducing its binding capacity for papain. It is possible that these components were other cysteine endopeptidases.

Funk et al.[20] were unable to repeat the purification of papain reported by Blumberg et al.[17] They found that, in their hands, both active and nonactivatable papain bound to the column and both forms were eluted from the column, suggesting that the affinity of the irreversibly oxidized form of the enzyme was high enough to favor binding under the chromatographic conditions. Removal of the benzyl group from the tyrosine resulted in a soluble ligand with lower affinity for papain. The immobilized Gly-Gly-Tyr-Arg was then found to bind only active papain, which could be eluted in water.

More recent work vindicates the procedure of Blumberg et al.[17] and Burke et al.[18] Immobilized Gly-Gly-Tyr(Bzl)-Arg, obtained from the Peptide Institute (Osaka, Japan), is more effective than the inhibitor lacking the benzyl group, in the preparation of highly active papain from material from which the other papaya cysteine endopeptidases have already been separated.[21]

Peptides containing D-amino acids sometimes act as weak reversible inhibitors, so that when they are immobilized they may act as affinity ligands. Syu et al.[22] synthesized three ligands containing D-Phe. All bound papain, which could be eluted in 0.5 mM HgCl$_2$. However, there remains some doubt as to whether the compounds were acting as active site-directed affinity ligands. Even the most efficient, ε-aminocaproyl-L-Leu-

[18] D. E. Burke, S. D. Lewis, and J. A. Shafer, Arch. Biochem. Biophys. 164, 30 (1974).
[19] D. J. Buttle, P. M. Dando, P. F. Coe, S. L. Sharp, S. T. Shepherd, and A. J. Barrett, Biol. Chem. Hoppe-Seyler 371, 1083 (1990).
[20] M. O. Funk, Y. Nakagawa, J. Skochdopole, and E. T. Kaiser, Int. J. Pept. Protein Res. 13, 296 (1979).
[21] P. Lindahl, E. Alrikkson, H. Jörnvall, and I. Björk, Biochemistry 27, 5074 (1988).
[22] W.-J. Syu, S. H. Wu, and K.-T. Wang, J. Chromatogr. 262, 346 (1983).

D-Phe, at a concentration of 40 mM gave only 30% inhibition of papain. Because the concentration of ligand attached to agarose was estimated to be 3.2 mM, it is difficult to understand how active enzyme was retained on the column unless nonspecific interactions were also involved. No attempt was made to assay the thiol content of the purified enzyme or to titrate the active site by other means, such as E-64 [*trans*-epoxysuccinyl-L-leucylamido(4-guanidino)butane]. No account was taken of the ability of the other papaya endopeptidases with specificity similar to that of papain to bind to the ligand.

Peptide Aldehydes and Related Compounds

General Principle

The derivatization of the C terminus of a peptide with a moiety that has high affinity for the nucleophilic center of the peptidase can produce tight-binding inhibitors, with specificity for a particular enzyme conferred by the structure of the peptide. Activated enzyme is bound to the gel at its pH optimum. It is subsequently eluted by choosing conditions that weaken the interaction between the enzyme and ligand, in the presence of a reagent that competes for covalent binding of the active site thiol.

Examples

Peptide aldehydes are transition-state analog inhibitors of cysteine and serine endopeptidases. They have been used in the affinity purification of serine endopeptidases, where the extremely tight interaction is reversed by conversion of the aldehyde to the semicarbazone.[23] In some cases affinity for the semicarbazone is sufficient to promote attachment to the column, in which case elution may be achieved with 1 M guanidine hydrochloride.[24] The purification of cathepsin B from human kidney cortex has been achieved by a method that included affinity purification on immobilized Leu-Leu-Argininal.[25] Displacement of active enzyme from the column was accomplished by the use of the 1.5 mM 2,2'-dipyridyl disulfide at pH 4.0. This step in the purification led to a 380-fold increase in specific activity and a 3-fold increase in yield, the latter presumably due to the removal of reversible inhibitors.

[23] A. H. Patel and R. M. Schultz, *Biochem. Biophys. Res. Commun.* **104**, 181 (1982).
[24] A. Basak, X. W. Yuan, N. G. Seidah, M. Chrétien, and C. Lazure, *J. Chromatogr.* **581**, 17 (1992).
[25] P. H. Wang, Y. S. Do, L. Macaulay, T. Shinagawa, P. W. Anderson, J. D. Baxter, and W. A. Hsueh, *J. Biol. Chem.* **266**, 12633 (1991).

The realization that peptide aldehyde semicarbazones are weaker, but still effective inhibitors of cysteine peptidases, compared to the parent aldehydes,[26] prompted Rich *et al.*[27] to investigate their potential as ligands for affinity chromatography. Gly-Phe-glycinal semicarbazone and Phe-glycinal semicarbazone were synthesized and coupled to activated CH Sepharose 4B (Pharmacia, Piscataway, NJ). The interaction with papain was found to be too tight to allow elution of the bound enzyme (K_i of Z-Gly-Phe-glycinal semicarbazone, 9.2 nM)[16] but cathepsin B (K_i 3.4 μM[27]) could be eluted with 2,2'-dipyridyl disulfide. An acetone fraction of human liver was reduced in 2 mM dithiothreitol and mixed batchwise with Sepharose-Ahx-Gly-Phe-glycinal semicarbazone at pH 6.0. Unbound proteins were removed by washing the gel in buffer containing 0.5 M NaCl and the gel was equilibrated to pH 4.0 and packed into a column. Highly active cathepsin B was recovered after incubating the column overnight in 1.5 mM 2,2'-dipyridyl disulfide. A 15-fold increase in specific activity was achieved, with an 80% yield. Importantly, human cathepsin H was shown not to bind to the column, whereas cathepsin L was bound but was not eluted. Sepharose-Ahx-Gly-Phe-glycinal semicarbazone is therefore a specific affinity ligand for the purification of cathepsin B from mammalian sources. On occasion, a M_r 40,000 contaminant, showing no immunological cross-reactivity with cathepsin B and no proteolytic activity, is copurified with cathepsin B. If Sepharose-Ahx-Phe-glycinal semicarbazone is used, cathepsin B is obtained free of this contaminant (D. J. Buttle and A. J. Barrett, unpublished result).

Immobilized Gly-Phe-glycinal and Phe-glycinal semicarbazones do not achieve purification of any of the papaya cysteine endopeptidases, or actinidain from kiwi fruit (D. J. Buttle, unpublished results, and Ref. 27). They have proved valuable, however, in the isolation and characterization of cysteine endopeptidases from the pineapple[28,29] (see [38]). Gly-Phe-glycinal semicarbazone is commercially available from Bachem Feinchemikalien AG (Bubendorf, Switzerland).

Sepharose-Ahx-Phe-phenylalaninal semicarbazone was successfully employed in the purification of a cysteine endopeptidase, histolysain (EC 3.4.22.35) from the protozoan pathogen *Entamoeba histolytica*.[30] The partially purified enzyme was applied to the column at pH 7.5 in buffer containing 4 mM dithiothreitol. Fully active enzyme was subsequently

[26] D. H. Rich, *in* "Proteinase Inhibitors" (A. J. Barrett and G. Salvesen, eds.), p. 153. Elsevier, Amsterdam, 1986.

[27] D. H. Rich, M. A. Brown, and A. J. Barrett, *Biochem. J.* **235,** 731 (1986).

[28] A. D. Rowan, D. J. Buttle, and A. J. Barrett, *Arch. Biochem. Biophys.* **267,** 262 (1988).

[29] A. D. Rowan, D. J. Buttle, and A. J. Barrett, *Biochem. J.* **266,** 869 (1990).

[30] A. L. Luaces and A. J. Barrett, *Biochem. J.* **250,** 903 (1988).

eluted from the column at the same pH in the presence of 50 mM hydroxyethyl disulfide. This single step produced a 100-fold purification with complete recovery of the active enzyme.

In an attempt to find a specific ligand for the purification of chymopapain from papaya latex, a number of dipeptide aldehyde semicarbazones were synthesized and tested as immobilized ligands for the preferential binding and recovery of this enzyme.[31] Sepharose-ECH-Ala-phenylalaninal semicarbazone proved to have the best properties, binding and eluting chymopapain efficiently, and showing some discrimination for this enzyme over others also present in the starting material. Partially purified chymopapain, obtained by acid treatment and cation-exchange chromatography of dried papaya latex, was applied to the column at pH 7.2 in the presence of ethanediol (33%, v/v) and 5 mM cysteine. Ethanediol was found to be effective at preventing nonspecific hydrophobic interactions between components of papaya latex and the matrix. Elution of fully active chymopapain was achieved at pH 4.5 with 10 mM $HgCl_2$, with a 67% yield.[19]

$HgCl_2$ has been found to be a very convenient agent for the recovery of cysteine peptidases from these matrices. The use of disulfides usually requires a period of reequilibration before all of the bound enzyme can be recovered from the matrix. The mercury salt, on the other hand, produces almost instantaneous release of the enzyme so that the flow from the column need not be disrupted. The enzyme is then rapidly recovered in a stable and readily activatable form. Elution with $HgCl_2$ has been successfully achieved with cathepsin B and the pineapple cysteine endopeptidases, from the immobilized Gly-Phe-glycinal and Phe-glycinal semicarbazones (D. J. Buttle, unpublished results; see [38]), and glycyl endopeptidase from the peptide aminoacetonitrile matrix (see [37]).

A method has been described for the quantification of immobilized aldehydes and aldehyde semicarbazones,[32] thus removing one of the unknown factors associated with this form of affinity chromatography. Aldehydes are coupled with 4-phenylazoaniline, which is then displaced with salicylaldehyde and spectrophotometric quantification achieved at A_{405}. It was found that several matrices differed greatly in both the degree and stability of derivatization, and performance as affinity media, with activated CH Sepharose 4B (Pharmacia) proving to be superior to others tested.

Peptide nitriles, like aldehydes, can form covalent complexes with the active site thiol of cysteine peptidases, and probably act as transition state

---

[31] A. J. Barrett, D. J. Buttle and D. H. Rich, *Int. Patent Publ.* WO 90/13561 (1990).
[32] P. M. Dando and A. J. Barrett, *Anal. Biochem.* **204,** 328 (1992).

analogs.[33,34] There is to date only one example of the use of an immobilized peptide nitrile in the affinity purification of cysteine peptidases. Sepharose-Gly-Phe-NHCH$_2$CN is a very effective affinity ligand for the purification of papain and glycyl endopeptidase (see [37]).

Conclusions

It is clear that there are two general methods for the affinity purification of cysteine peptidases, both of which would be expected to function with any member of this class of enzyme, and for which no information is required regarding substrate specificity of the enzyme in question. Thus, organomercurial matrices and methods involving disulfide exchange will continue to be of general usefulness in the purification of existing and newly discovered cysteine peptidases.

There is still a need for the development of more specific ligands that are capable of discriminating between individual enzymes. In many cases more than one cysteine peptidase is present in the starting material, and conventional chromatographic techniques are relied on for the separation of these, either before or after an affinity chromatographic step. The precedent has been set, with the peptide aldehyde semicarbazone ligands for the purification of cathepsin B from the mixture with cathepsins L and H, and the peptide aminoacetonitrile for the efficient separation of papain and glycyl endopeptidase from chymopapain and caricain. There is a real need for the development of affinity chromatographic steps for the purification of physiologically and pathologically important enzymes, such as cathepsins L and S. Current methods are labor intensive and prolonged, and do not result in the preparation of 100% active enzyme. As we learn more about the structure–activity relationships and relative substrate specificities of these enzymes it should be possible to design affinity ligands that produce single-step purification of fully pure and active enzyme. The structures of these ligands may include the mixed disulfide, aldehyde semicarbazone, or nitrile groups that have already shown promise in the development of peptide-conferred specificity in affinity chromatography of the cysteine peptidases.

[33] J. B. Moon, R. S. Coleman, and R. P. Hanzlik, *J. Am. Chem. Soc.* **108,** 1350 (1986).
[34] R. P. Hanzlik, J. Zygmunt, and J. B. Moon, *Biochim. Biophys. Acta* **1035,** 62 (1990).

[46] Peptidyl Diazomethanes as Inhibitors of Cysteine and Serine Proteinases

By ELLIOTT SHAW

Peptidyl diazomethanes are peptide derivatives having the general structure R—C(=O)—CHN$_2$ and thus are ketones, sometimes referred to as peptidyl diazomethyl ketones. Their properties have been summarized.[1] The initial finding that Z—Phe—CHN$_2$ inactivates papain by a stoichiometric alkylation of the active center —SH group[2] has been followed by application to many cysteine proteinases by variations in the peptidyl portion providing high reactivity (i.e., k_{2nd} near 10^6 M^{-1} sec^{-1}) and some degree of specificity.[3–7] Recent developments include demonstration of these properties in the inactivation of cathepsins B, L, S, and H and calpain. Because these inhibitors are cell permeable, they can be used for *in vivo* inactivation of cysteine proteinases. In radioactive form they are valuable for identifying target proteases and for shedding light on their function.

Properties as Affinity-Labeling Reagents

The diazomethyl ketone group has very low reactivity to thiols, but when incorporated within peptides that promote complex formation with a given cysteine proteinase, the catalytic thiol group apparently adds to the ketone, converting it to an unstable adduct that rearranges to the stable thioether, irreversibly inactivating the proteinase. Thus this class of inhibitor is conveniently thought of as enzyme-activated.[3,8] As a result of the low thiol reactivity, these reagents are stable in the presence of mercaptoethanol or dithiothreitol (DTT), which are often essential for cysteine proteinase activity.

[1] E. Shaw and G. D. J. Green, this series, Vol. 80, p. 820.
[2] R. Leary, D. Larsen, H. Watanabe, and E. Shaw, *Biochemistry* **16**, 5857 (1977).
[3] E. Shaw, *Adv. Enzymol.* **63**, 271 (1990).
[4] G. D. J. Green and E. Shaw, *J. Biol. Chem.* **256**, 1923 (1981).
[5] H. Kirschke, P. Wikstrom, and E. Shaw, *FEBS Lett.* **228**, 128 (1988).
[6] C. Crawford, R. W. Mason, P. Wikstrom, and E. Shaw, *Biochem. J.* **253**, 751 (1988).
[7] H. Angliker, P. Wikstrom, H. Kirschke, and E. Shaw, *Biochem. J.* **262**, 63 (1989).
[8] K. Brocklehurst and J. P. G. Malthouse, *Biochem. J.* **175**, 761 (1978).

Inhibitor Structure–Activity Relationships

Cathepsins B, L, S, and H and calpain are susceptible to small peptide diazomethyl ketones with hydrophobic side chains. Examination of some sequence variations has revealed differences in *in vitro* susceptibility that are useful for analytical and functional studies, as summarized in Table I. Z-Tyr-Ala-CHN$_2$ is effective in inactivating cathepsins B, L, and S but does not act on calpain,[6] which apparently does not bind aromatic residues in S2. The activity of this reagent is actually increased on monoiodination without loss of specificity for cathepsins B and L, which, as described below, makes radioiodination studies feasible. Z-Leu-Leu-Tyr-CHN$_2$ does inactivate calpain, which will thus bind an aromatic side chain in S1, as will cathepsins B, L, and S. This reagent may also be iodinated with retention of inhibitory activity.[9] (The iodinatable residue could be placed elsewhere in the sequence if necessary, because of a strict specificity in P1).

If the bulk of the residue in P1 is increased in the general structure Z-Phe-X-CHN$_2$, the ability to inactivate cathepsin L is retained.[5] With Z-Phe-Tyr-(*t*Bu)CHN$_2$, cathepsins B and S are remarkably less sensitive (Table I) and this reagent is essentially specific for cathepsin L under reasonable experimental conditions.

Cystatin C, an endogenous cysteine proteinase inhibitor, undergoes loss of inhibitory activity if cleavage occurs at Gly in the sequence Leu-Val-Gly in the reactive region[10] of the human inhibitor. This suggested Z-Leu-Val-Gly-CHN$_2$ as a cystatin analog with potential for inactivating cysteine proteinases.[11] This inhibitor inactivates a number of cysteine proteinases,[12] including streptopain (streptococcal proteinase). Strain A *Streptococcus* is lethal for mice. It is of interest with respect to the *in vivo* action of peptidyl diazomethanes that mice given a lethal dose of streptococci A could be protected with a single intraperitoneal injection of Z-Leu-Val-Gly-CHN$_2$.[11]

A superior inactivator of streptopain with an unrelated sequence had been described earlier[4] (see Table I), suggesting that the efficacy of cystatin C as a cysteine proteinase inhibitor may not rely merely on the LVG

[9] J. Anagli, J. Hagmann, and E. Shaw, *Biochem. J.* **274**, 497 (1991).

[10] M. Abrahamson, A. Ritonja, M. A. Brown, A. Grubb, W. Machleidt, and A. Barrett, *J. Biol. Chem.* **262**, 9688 (1987).

[11] L. Björck, P. Akesson, M. Bohus, J. Trojnar, M. Abrahamson, I. Olafsson, and A. Grubb, *Nature (London)* **337**, 385 (1989).

[12] M. Abrahamson, R. W. Mason, H. Hansson, D. J. Buttle, A. Grubb, and K. Ohlsson, *Biochem. J.* **273**, 621 (1991).

TABLE I
INACTIVATION OF CYSTEINYL PROTEINASES BY PEPTIDYL DIAZOMETHANES[a]

Reagent	Cysteine Cathepsin k_2 (M^{-1} sec^{-1})				Calpain k_2 (M^{-1} sec^{-1})	Streptopain k_2 (M^{-1} sec^{-1})
	Cathepsin B	Cathepsin L	Cathepsin S	Cathepsin H		
Z-Y-A-CHN$_2$	1800[j]	120,000[j]	1740[j]	<10[j]	<10[b]	
	1180[b]	177,000[b]				
Z-Y(I)-A-CHN$_2$	27,800[b]	1,128,000[b]				
Z-L-L-Y-CHN$_2$	1300[b]	1,500,000[b]	133,000[c]	39[c]	113,000[b]	
Z-F-Y-(tBu)CHN$_2$	10[d]	200,000[d]	30[e]			
H$_2$N-Ser-(Bz)CHN$_2$	9.3[f]			2570[f]		
Z-L-V-G-CHN$_2$	10,640[h]	118,000[h]				102[g]
Z-A-F-A-CHN$_2$	1175[i]					29,500[i]
Z-A-A-CHN$_2$	140[i]					135[i]

[a] Origin of enzymes: Cathepsin S was from bovine spleen in all cases; calpain was form II from chicken gizzard. The other cathepsins are identified from their reference source: b and h, human liver; c, bovine spleen; d, f, and i, rat liver cathepsins L and H, pork liver cathepsin B; j, bovine liver.
[b] C. Crawford, R. W. Mason, P. Wikstrom, and E. Shaw, Biochem. J. 253, 751 (1988).
[c] X.-Q. Xin, B. Gunesekera, and R. W. Mason, Arch. Biochem. Biophys. 299, 334 (1992).
[d] H. Kirschke, P. Wikstrom, and E. Shaw, FEBS Lett. 228, 128 (1988).
[e] E. Shaw, S. Mohanty, A. Colic, V. Stoka, and V. Turk, FEBS Lett. 334, 340 (1993).
[f] H. Angliker, P. Wikstrom, H. Kirschke, and E. Shaw, Biochem. J. 262, 63 (1989).
[g] L. Björck, P. Akesson, M. Bohus, J. Trojnar, M. Abrahamson, I. Olafsson, and A. Grubb, Nature (London) 337, 385 (1989).
[h] M. Abrahamson, R. W. Mason, H. Hansson, D. J. Buttle, A. Grubb, and K. Ohlsson, Biochem. J. 273, 621 (1991).
[i] G. D. J. Green and E. Shaw, J. Biol. Chem. 256, 1923 (1981).
[j] D. J. Buttle and J. Saklatvala, Biochem. J. 287, 657 (1992).

sequence complexed to the active center of the proteinase, but on contacts at other sites.

Peptidase Class Specificity

Peptidyl diazomethanes appeared initially to be specific inactivators of cysteine proteinases because of lack of action on other types of peptidases.[4] Serine proteinases on occasion formed adducts, with eventual destruction of the reagents.[13] However, some exceptions have subsequently been found involving serine peptidases. For the family of chymotrypsin (S1; see [2] in this volume), Z-Phe-Arg-CHN$_2$ is destroyed by trypsin, but it slowly ($k_2 = 250\ M^{-1}\ sec^{-1}$) inactivates human plasma kallikrein.[13] Of special interest is the interaction with prolyl oligopeptidase (the postproline cleaving enzyme). This was initially thought to be a cysteine peptidase, but is now known to be a serine peptidase representing a family (S9) distinct from those of chymotrypsin and subtilisin.[14] Prolyl oligopeptidase is inactivated *in vitro* by Z-Ala-Ala-Pro-CHN$_2$, and the effect is also demonstrable in intact macrophages.[15] However, it was shown by Stone *et al.*[16] that this and related inhibitors are slow binding, and that the initial complex isomerizes to a more tight-binding complex with a K_i of 16 nM for the above inhibitor. Chemical modification of the enzyme was demonstrable only on denaturation of the complex (shown with the acetyl derivative), and the labeled inhibitor was found on the active center serine. For the subtilisin (S8) family, it was shown that Z-Ala-Ala-Phe-CHN$_2$ inactivates subtilisin Carlsberg ($k_2 = 741\ M^{-1}\ sec^{-1}$) and thermitase ($k_2 > 260,000\ M^{-1}\ sec^{-1}$) by alkylation of the active center histidine.[17]

Synthesis and Stability

The peptidyl diazomethanes (diazomethyl ketones) are prepared from a blocked peptide activated as the mixed anhydride followed by reaction with ethereal diazomethane at 0°. Monitored by thin-layer chromatography or high-performance liquid chromatography (HPLC), the reaction is typically completed within 15 min and should be worked up without delay,

[13] A. Zumbrunn, S. Stone, and E. Shaw, *Biochem. J.* **250**, 621 (1988).
[14] D. Rennex, B. A. Hemmings, J. Hofsteenge, and S. R. Stone, *Biochemistry* **30**, 2195 (1991).
[15] G. D. J. Green and E. Shaw, *Arch. Biochem. Biophys.* **225**, 331 (1983).
[16] S. R. Stone, D. Rennex, P. Wikstrom, E. Shaw, and J. Hofsteenge, *Biochem. J.* **283**, 871 (1992).
[17] A. Ermer, H. Baumann, G. Steude, K. Peters, and S. Fittkau, P. Dolaschka, and N. C. Genov, *J. Enzyme Inhib.* **4**, 35 (1990).

if the peptide contains tyrosine, to avoid methyl ether formation. For details, see, for example, Z-Tyr-Ala-CHN$_2$ below. The crystalline solids are stable for years at room temperature, but solutions are much less stable. Even dimethyl sulfoxide (DMSO) stock solutions stored at $-20°$ may lose activity in about a week, depending on the structure. In using HPLC for the assessment of purity, it should be kept in mind, with programs utilizing trifluoroacetic acid (TFA), that the diazomethyl ketone is converted to R—CH$_2$OC(=O)CF$_3$ and the peak that is seen is this derivative. With some peptidyl diazomethanes this reaction is slow, the conversion may be incomplete, and two peaks emerge, although by other criteria the preparation is pure. In such cases, preincubation of a sample in TFA for 15 min before injection will give a single peak. (Programs utilizing ammonium bicarbonate gradients do not exhibit this problem.)

Z-Tyr-Ala-CHN$_2$

Z-Tyr-Ala-OH (1.0 g) in tetrahydrofuran (THF, 10 ml) at $-20°$ is converted to the mixed anhydride with 2.1 mmol of N-methylmorpholine and *tert*-butyl chloroformate. After 15 min, the mixture is poured into 25 ml of ethereal diazomethane (standard preparation) at $0°$. After 15 min the reaction is partitioned between cold aqueous NaHCO$_3$ and ethyl acetate. The organic layer is washed with aqueous NaCl, briefly dried with anhydrous MgSO$_4$, and evaporated under reduced pressure. The solid residue indicates the presence of a major component, R_f 0.2 on TLC in CH$_2$Cl$_2$. This is chromatographed on Si60 (30 g) prepared in CH$_2$Cl$_2$. The sample is applied in 10% CH$_2$Cl$_2$ in methanol (8 ml) (v/v) and eluted with CH$_2$Cl$_2$ containing increasing concentrations of methanol. The eluate is monitored by TLC and fractions are not pooled until examined by HPLC to obtain a preparation of highest purity rather than highest yield, providing 0.6 g, mp 157°–158°.

Z-Leu-Leu-Tyr-CHN$_2$

Z-Leu-Leu-Tyr-OH is obtained via the *O-tert*-butyl ester. For this, the mixed anhydride from Z-Leu-Leu-OH (2.3 g) is prepared and to this is added a solution of Tyr-O-*t*Bu (1.7 g) in dimethyl formamide (DMF, 4 ml) and THF (4 ml). After stirring 1 hr at $-20°$, the suspension is stirred overnight. The THF is removed and the residue stirred with water (50 ml) at $0°$ until a semisolid suspension forms; this can be filtered with suction. The product is taken up in ethyl acetate and CH$_2$Cl$_2$, dried with MgSO$_4$ and, after filtration, the solvent is removed. The yield is quantitative. The ester (1.3 g) is converted to the acid by treatment with dry 2 N HCl in ethyl acetate (70 ml) at room temperature (8 hr). The conversion is followed by HPLC; the ester has a longer retention time. The residue, Z-Leu-Leu-Tyr-OH, is converted to the diazomethyl ketone via the mixed anhydride (see Z-Tyr-Ala-CHN$_2$). The product is purified by silica chromatography as above and crystallized twice from ethyl acetate to yield 0.4 g (mp 166°–167°).

Z-Phe-Tyr(O-tBu)CHN$_2$

Z-Phe-Tyr(O-tBu)-OH is obtained by the condensation of Z-Phe-OSu (4.4 g) with H$_2$N-Tyr(O-tBU)-OH (2.37 g) in DMF (10 ml), triethylamine (3 ml), and dimethoxyethane (10 ml) overnight. The reaction is diluted with water (100 ml) and extracted with ethyl acetate. The blocked dipeptide is liberated from the aqueous layer by acidification and, after drying,[18] a portion is converted to the diazomethyl ketone in the usual manner, chromatographed on silica (mp 137°–138°).[5]

Z-Ala-Ala-Pro-CHN$_2$

The acid is obtained from Z-Ala-Ala-OSu and proline as in the preceding sample and converted to the diazomethyl ketone as described above (mp 136°–138°).[4]

Radioiodination by Affinity Labeling

Z-[^{125}I]Iodo-Tyr-Ala-CHN$_2$

Z-Tyr-Ala-CHN$_2$ (25 μl of 1 mM in 25% ethanolic solution), 10 μl of 50 mM sodium phosphate buffer, pH 7.5, and 10 μl of Na^{125}I (1 mCi) are mixed in an Iodogen-coated (Pierce Chemical Co.) glass tube[19,20] for 10 min at 0°, then diluted with 455 μl of the same buffer and removed from the tube.[19] The specific radioactivity of the inhibitor is determined as TCA-precipitable radioactivity after reaction with a standard papain preparation.

Z-Leu-Leu-[^{125}I]Tyr-CHN$_2$

Z-Leu-Leu-Tyr-CHN$_2$ (25 μl of a 1 mM solution in 50% ethanol), Na^{125}I (10 μl, 1 mCi), and 50 mM phosphate buffer, pH 7.5 (10 μl) are mixed in a reaction vessel at 0°[9] and one Iodo-bead[21] is added. After 10 min, 455 μl of buffer is added and the bead removed.

[18] E. Shaw, P. Wikstrom, and J. Ruscica, Arch. Biochem. Biophys. **222**, 424 (1983).
[19] R. W. Mason, D. Wilcox, P. Wikstrom, and E. Shaw, Biochem. J. **257**, 125 (1989).
[20] P. J. Fraker and J. C. Speck, Biochem. Biophys. Res. Commun. **80**, 849 (1978).
[21] M. A. K. Markwell, Anal. Biochem. **125**, 427 (1982).

Labeling of Cathepsin B and L in Cells with Z-[^{125}I]Tyr-Ala-CHN$_2$

In the case of virally transformed mouse fibroblasts,[19] cells are incubated in medium with or without serum with a low concentration of labeled inhibitor (0.1 μM) for 3 hr followed by TCA precipitation, SDS electrophoresis, and X-ray film exposure. Three bands are revealed with M_r 35,000, 30,000, and 23,000 identified as cathepsin B (35,000) and cathepsin L (30,000 and 23,000) by immunological methods. Precursor forms are not labeled by this procedure. (Labeling is observable as early as 30 min).

When this procedure is applied to other cell types[22] the active cathepsins are more processed. For example, in BALB/c 3T3 cells, in addition to a single-chain form of cathepsin B, a two-chain form is identified in which the label resides in a 5000 M_r form known to contain the active center -SH of cathepsin B. In HT 1080 cells, the cathepsin B is similar, but cathepsin L is found in four forms. Other variations are found, indicating that the population of active forms of cathepsin B and L is dependent on cell type (The assignment of form is supported by immunoprecipitation). The variations might be in the degree of glycosylation, for example.

In tissue extracts prepared at pH 5, only cathepsin B is radiolabeled.[23] If the sample is autolysed at pH 4, 37°, for a few hours, cathepsin L (M_r 25,000) is also found; the identifications are immunologically confirmed. The labeled samples have the characteristic chromatographic properties of the active enzymes on FPLC (fast protein liquid chromatography).[23]

Use of Z-Leu-Leu-[^{125}I]Iodo-Tyr-CHN$_2$ to Label Calpain in Platelets

Washed human platelets (4–5 × 10^8) are incubated in 1 ml, pH 7.4, containing 5 mM CaCl$_2$ with various concentrations of Z-Leu-Leu-Tyr-CHN$_2$ with or without radiolabel (as specified[9]) for 1 hr at 37°, then subjected to a Ca^{2+} ionophore (1 μM 4-bromo-A23187) for 30 min. With unlabeled inhibitor at 10–100 μM, it is possible to show, using SDS–gel electrophoresis, that protection of actin-binding protein and talin, high molecular weight proteins and presumed substrates of calpain, has been achieved. Because the inhibitor also inactivates cathepsin L, a control experiment using Z-Phe-Ala-CHN$_2$ (50 μM), which inactivates cathepsin L but not calpain, is carried out; actin-binding protein and talin are not protected, therefore the protective action of Z-Leu-Leu-Tyr-CHN$_2$ can be ascribed to the inactivation of calpain.[9]

When radiolabeled inhibitor (0.1–0.5 μM) is preincubated with platelets prior to addition of ionophore, and the gels are autoradiographed[9,24]

[22] D. Wilcox and R. W. Mason, *Biochem. J.* **285**, 495 (1992).
[23] R. W. Mason, L. T. Bartholomew, and B. S. Hardwick, *Biochem. J.* **263**, 945 (1989).
[24] J. Anagli, J. Hagmann, and E. Shaw, *Biochem. J.* **289**, 93 (1993).

the Ca^{2+}-dependent labeling of the calpain large subunit is confirmed, along with cathepsin L and some minor bands.

The platelets with inactivated calpain offered the opportunity to examine some other properties ascribed to calpain, for example, release of 5-hydroxytryptamine on thrombin stimulation.[9] This property was not significantly altered.

Comments on Intracellular Inactivation of Proteinases

Irreversible inhibitors other than diazomethyl ketones[3] exist, and it would be difficult to attempt a comparison. An advantage of irreversible inhibitors in general is that, in radioactive form, they permit confirmatory analysis of whether the supposed target enzyme has been reached, indispensible information for an interpretation of mechanism of action. In addition, because they label only active enzyme forms, they may reveal unexpected complexities of processing.[19]

In a group of proteinases with similar substrate specificity, such as the cysteine endopeptidases discussed here, it is difficult to achieve a high degree of selectivity. It should be remembered that when rate differences are found, any selectivity that an inhibitor may possess could disappear on long-term incubation with cells or with use of high concentrations, conditions that will cancel kinetic differences.

When inhibitor-treated cells are transferred to fresh medium a restoration of protease activity may be observed due to new synthesis[25] or activation of inert precursor.

[25] E. Shaw and R. Dean, *Biochem. J.* **186**, 385 (1980).

[47] Peptidyl (Acyloxy)methanes as Quiescent Affinity Labels for Cysteine Proteinases

By ALLEN KRANTZ

Introduction

The design of small-molecule inhibitors of enzymes has evolved from reliance on primarily empirical approaches, to current rational strategies that are mechanism[1] and structure based.[2] A major conceptual advance that spawned the active site-directed reagent involved appending reactive organic chemical functionality to binding determinants targeted for a specific enzyme.[3] This "affinity label" approach led to reagents of improved specificity and potency and avoided the indiscriminant labeling that was characteristic of group modification reagents.

As originally conceived, affinity labels were designed to bring reactive carbon electrophiles in close proximity to active site residues vulnerable to alkylation in S_N2 type reactions. Although such affinity labels have proved to be useful biochemical and pharmacological tools, because of their specificity *in vitro*, their intrinsic chemical reactivity has severely restricted their clinical utility. Clinically useful inhibitors must meet stringent specificity requirements and be stable to nonspecific reactions *in vivo*, to limit potential toxic effects. Thus, to be practical oral agents, affinity labels must possess functional groups that are resistant to nonspecific chemical reactions over the course of hours at physiological temperatures and pH. It is in this sense that they must be *chemically unreactive*. This imperative imposes a severe limitation on design motifs utilizing single atoms as nucleofuges, because single atoms cannot be finely tuned by systematic modification to balance low chemical reactivity versus inhibitory potency.

A characteristic feature of cysteine proteinases of the papain type is their extraordinary nucleophilicity, which is evident even at acid pH with simple carbon electrophiles such as iodoacetamide.[4] The huge rate accelerations (10^8), observed for analogous displacements when an affinity group is appended to a halomethyl function, suggest that high chemical reactivity of the label is not necessarily a prerequisite for the design of an affinity label. In fact, Z-Phe-Ala-CH$_2$F, containing nonoptimal binding determinants for cathepsin B (EC 3.4.22.1) and a modest leaving group, can inactivate this enzyme,[5] with $k/K = 18,000\ M^{-1}\ sec^{-1}$. This implies that the reactivity of such affinity labels can be modulated by judicious choice of affinity and leaving groups, to give reagents that are stable *in vivo* to nonspecific reactions.

To achieve the requisite balance between chemical reactivity and inhibitory potency, we set out to design affinity labels with difficultly displaceable leaving groups, whose reactivity could be controlled by substituent

[1] R. B. Silverman, "Mechanism-Based Enzyme Inactivation: Chemistry and Enzymology," Vol. 1–2. CRC Press, Boca Raton, Florida, 1988.

[2] I. D. Kuntz, *Science* **257**, 1078 (1992).

[3] F. Wold, this series, Vol. 46, p. 3.

[4] P. Halász and L. Polgár, *Eur. J. Biochem.* **71**, 563 (1976).

[5] D. Rasnick, *Anal. Biochem.* **149**, 461 (1985).

effects, and which might undergo rapid displacement in the active site of the target enzyme, by virtue of their proximity to a highly nucleophilic residue.

Design Motifs

In devising reagents that might have the necessary flexibility to serve as lead structures in inhibitor or drug development, benzoates represent appealing design elements for several reasons. First, a broad range of commercially available benzoic acids is available as substrates for synthesis. Further, the use of benzene rings to transmit substituent effects is standard practice. Classically, substituent effects have been used to manipulate the pK_a of benzoic acids by electron withdrawal or donation, which forms the basis of Hammett relationships.[6] Finally, a guiding principle of organic chemical reactivity relevant to S_N2 type displacements holds that within a closely related series of leaving groups (i.e., where the atom type of the scissile bond is kept constant), nucleofugality can be controlled by incremental changes in pK_a. (In general, the lower the pK_a of the leaving group, the greater its nucleofugality or leaving-group ability.)

To be practical, (acyloxy)methanes* must be stable to attack at the benzoyl carbonyl and this can be accomplished by deploying ortho substituents such as alkyl groups to shield the potentially reactive benzoyl carbonyl from nucleophilic attack.[7] To achieve this end, both methyl and trifluoromethyl substituents were commonly employed as design elements in our studies. In experiments with phenyl thiolate in dimethyl sulfoxide (DMSO) at 25°, 2,4,6-trimethylbenzoyloxy (mesitoyloxy) linked to the primary carbon of acetophenone possesses nucleofugality intermediate between the corresponding chloride and fluoride.[8] Strikingly, under these conditions, mesitoate is less reactive by at least an order of magnitude than fluoride, which possesses an identical pK_a. Mesitoate therefore represents a practical leaving group, potentially useful in the design of powerful inactivators, as well as being stable to nonspecific displacement reactions.

A useful measure of the vulnerability of an affinity label to intracellular nucleophiles is afforded by rate measurements of glutathione with the reagent at 37° in buffer of pH 1–7.4. A more rigorous test involves treating

[6] C. D. Johnson, "The Hammett Equation." Cambridge Univ. Press, Cambridge, 1973.
* The peptidyl (acyloxy)methanes have often been termed (acyloxy)methyl ketones in the past, in recognition of the mechanistic importance of the ketonic nature of the peptidyl carbonyl group.
[7] M. S. Newman (ed.), in "Steric Effects in Organic Chemistry," p. 224. Wiley, New York, 19■■.
[8] A. Krantz, L. J. Copp, P. J. Coles, R. A. Smith, and S. B. Heard, *Biochemistry* **30**, 4678 (1991).

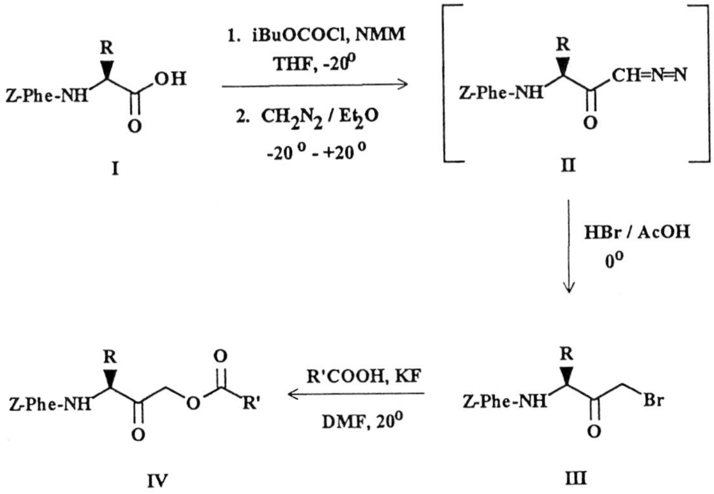

SCHEME I. NMM, N-methylmorpholine; Et$_2$O, diethyl ether; AcOH, acetic acid.

cells with (radio)labeled reagent and then demonstrating, with separation techniques that the label resides exclusively on the target protein. The most severe test involves oral administration of a radiolabeled agent in animals, followed by analysis of tissue distribution and localization to cysteine proteinases.[9]

Synthesis

A diverse series of compounds can be generated from the very large number of commercially available and synthetically accessible carboxylic acids by condensing the latter with activated methyl ketones. Thus, peptidyl (acyloxy)methanes (**IV**) are synthesized via the analogous bromomethane (**III**), as shown in Scheme I. In this procedure, N-protected peptides are prepared by standard procedures and then converted to the corresponding bromomethanes (**III**) via the intermediate diazomethanes (**II**).[10] Displacement of bromide by carboxylate, mediated by potassium fluoride,[11] gives the desired product **IV** in good yield. Although this sequence allows ready access to a variety of (acyloxy)methanes, it suffers

[9] K. Fukishima, M. Arai, Y. Kohno, T. Suwa, and T. Suwa, *Toxicol. Appl. Pharmacol.* **105**, 1 (1990).

[10] G. D. J. Greene and E. Shaw, *J. Biol. Chem.* **256**, 1923 (1981).

[11] J. H. Clark and J. M. Miller, *Tetrahedron Lett.* 599 (1977).

TABLE I
RATES OF CATHEPSIN B INACTIVATION BY PEPTIDYL (ACYLOXY)METHANES:
VARIATION OF LEAVING GROUP

Number	Compound R	k/K (M^{-1} sec^{-1})	K_{inact} (nM)	k_{inact} (sec^{-1})	pK_a^b
Z-Phe-Ala-CH$_2$OCO-R (benzoyloxy)methanes					
1	2,6-(CF$_3$)$_2$-Ph	1,600,000	28 ± 7	0.045 ± 0.008	0.58^c
2	2,6-Cl$_2$Ph	690,000	36 ± 11	0.025 ± 0.004	1.59
3	C$_6$F$_5$	520,000			1.48^d
4	2,6-Me$_2$-4-COOMe-Ph	58,000	610 ± 150	0.036 ± 0.005	2.67^c
5	2,5-(CF$_3$)$_2$-Ph	38,000^c	880 ± 580	0.033 ± 0.013	2.63^c
6	2,6-F$_2$-Ph	26,000	2100 ± 900	0.056 ± 0.017	2.24^f
7	3,5-(CF$_3$)$_2$-Ph	22,000			3.18^c
8	3,5-(NO$_2$)$_2$-Ph	~19,000^g			2.79
9	2-CF$_3$-Ph	17,000^h	770 ± 220	0.013 ± 0.002	2.49^c
10	2,4,6-Me$_3$-Ph	14,000^{h,j}	510 ± 110	0.0073 ± 0.0005	3.45
11	2,6-Me$_2$-Ph	14,000^i			3.35
12	2,4,6-iPr$_3$-Ph	3800^j			
13	3,4-F$_2$-Ph	630^j			3.67^c 3.79^d
14	4-NO$_2$-Ph	610			3.43
15	3-CF$_3$-Ph	420^j			3.76^c 3.75^k
16	2,6-(OMe)$_2$-Ph	300			3.44
17	4-F-Ph	290^k			4.15
18	4-CN-Ph	280^j			3.50^k
19	4-CH$_3$-Ph	260^j			4.37
20	3,5-(OH)$_2$-Ph	140^j			4.04
21	Ph	90^j			4.20
22	4-CF$_3$-Ph	80^j			3.67^l
23	3,5-Me$_2$-Ph	80^j			4.30
24	4-OMe-Ph	ntd^m			4.50
Z-Phe-Ala-CH$_2$OCO-R					
25	CH$_3$	140			4.76
26	CMe$_3$	330^j			5.03
27	CH(CH$_2$CH$_3$)$_2$	70			4.73
28	CH$_2$OCH$_3$	240			3.57
Z-Phe-Lys-CH$_2$OCO-R					
29	2,6-(CF$_3$)$_2$-Ph (HCl salt)	2,000,000^{a,j}			0.58^c
30	2,4,6-Me$_3$-Ph (TFA salt)	230,000^{a,j}	170 ± 50	0.037 ± 0.008	3.45
31	Ph (TFA salt)	9200^{a,j}	2400 ± 800	0.022 ± 0.005	4.20
32	4-OMe-Ph (TFA salt)	660^{a,j}			4.50

[a] Conditions: bovine spleen cathepsin B, 100 mM potassium phosphate, 1.25 mM EDTA, and 1 mM dithiothreitol, pH 6.0, 25°, under argon. The parameter k/K is the second-order rate constant (k_{inact}/K_{inact}), except as noted. Standard errors for $k/K \leq 15\%$, except as noted. In those cases in which saturation kinetics were observed, k_{inact}/K_{inact} and the individual parameters K_{inact} and k_{inact} are given, as determined from a hyperbolic fit. In the other cases, k_{inact}/K_{inact} was obtained from a linear fit.
[b] pK_a (aqueous) of the acyloxy group (RCOOH); values are from E. P. Serjeant and B. Dempsey, "Ionization of Organic Acids in Aqueous Solution." Pergamon, New York, 1979, except as noted.
[c] pK_a determined by UV measurement (or, in the case of 4, by HPLC).
[d] pK_a from L. E. Strong, C. L. Brummel, and P. Lindower, J. Solution Chem. 16, 105 (1987).
[e] Standard error 20–30%.
[f] pK_a from L. E. Strong, C. Van Waes, and K. H. Doolittle, J. Solution Chem. 11, 237 (1982).
[g] Compound instability was evident.
[h] Standard error 15–20%.

from the limitation that intermediates therein, i.e., the bromomethane (III), often possess considerably greater activity than the (acyloxy)methane product. Therefore, in experiments in which a large excess of inhibitor is utilized, it is important that the (acyloxy)methane be rigorously purified. In instances in which low inhibitory activities are observed as well, stoichiometric experiments provide the clearest indication that inactivation of the enzyme is related to the action of the peptidyl (acyloxy)-methane.

Peptidyl (Acyloxy)methanes (IV)

In general, isolated yields for the conversion of bromomethanes to (acyloxy)methanes are in the range 50–80%. Product purification is carried out by recrystallization (usually from ethyl acetate/hexane) and/or by silica gel column chromatography with ethyl acetate/hexane as eluant. Excess (5–10%) carboxylic acid is used in all cases to promote the consumption of all traces of bromomethane. HPLC analyses [Perkin-Elmer (Norwalk, CT) Pecosphere 3X3C C_8 or C_{18} column, 0.46 × 3.3 cm, with an acetonitrile/H_2O gradient at 3.0 ml/min] at 250 nm indicate product purities of >90% and in most cases >96%. HPLC analyses at 220 nm permit the detection of bromo- and chloromethanes to a level equivalent to 0.2% contamination; all (acyloxy)methanes are thereby confirmed to have no significant (i.e., <2.0%) halomethane contamination. Samples of inhibitors 1 and 10 (Table I) are also further purified by preparative HPLC [Whatman (Clifton, NJ) Magnum 20 silica gel column, 30% ethyl acetate/hexane, 11 ml/min] to exclude rigorously any impurities; these samples are then reassayed against cathepsin B to confirm the integrity of the inhibition kinetic data. It should be noted that the very low inactivation potency of the Z-L-Phe-D-Ala derivatives 37 and 49 (Table II), relative to the Z-L-Phe-L-Ala diastereomers 1 and 10, indicates that no significant diastereomeric impurities (epimeric at the P_1 α-carbon) exist in these

---

[i] This value has been revised from a previous report by R. A. Smith, L. J. Copp, P. J. Coles, H. W. Pauls, V. J. Robinson, R. W. Spencer, S. B. Heard, and A. Krantz, *J. Am. Chem. Soc.* **110**, 4429 (1988).

[j] Second-order rate constant (k/[I]) determined at one inhibitor concentration near the solubility limit.

[k] pK_a from M. Ludwig, V. Baron, K. Kalfus, O. Pytela, and M. Vecera, *Collect. Czech. Chem. Commun.* **51**, 2135 (1986).

[l] Calculated pK_a from D. D. Perrin, B. Dempsey, and E. P. Serjeant, "pK_a Prediction for Organic Acids and Bases," p. 129. Chapman & Hall, New York, 1981.

[m] ntd, No time dependence or significant inhibition observed near the compound solubility limit, over a 10- to 20-min period.

TABLE II
RATES OF CATHEPSIN B INACTIVATION BY PEPTIDYL (ACYLOXY)METHANES:
VARIATION OF AFFINITY GROUP[a]

Number	Compound R'	k/K (M^{-1} sec^{-1})	k_{inact} (nM)	k_{inact} (sec^{-1})
R'-CH$_2$OCO-[2,6-(CF$_3$)$_2$Ph				
33	Z-Phe-Gly	4,000,000		
34	Z-Phe-Cys(SBn)	2,900,000	13 ± 2	0.039 ± 0.003
35	Z-Phe-Ser(OBn)	2,600,000	13 ± 15	0.033 ± 0.019
29	Z-Phe-Lys (HCl salt)	>2,000,000		
1	Z-Phe-Ala	1,600,000	28 ± 7	0.045 ± 0.008
36	Z-Phe-Thr(OBn)	>100,000[b]		
37	Z-Phe-D-Ala (diastereomer of 1)	7100[c]		
R'-CH$_2$OCO-[2,4,6-(CH$_3$)]Ph				
30	Z-Phe-Lys (TFA salt)	230,000	170 ± 50	0.037 ± 0.008
38	Z-Phe-Thr(OBn)	30,000		
39	Z-Tyr(OMe)-Ala	19,000[d]	280 ± 60	0.005 ± 0.0003
40	Z-Phe-Ser(OBn)	17,000[d]	1300 ± 400	0.022 ± 0.0002
41	Z-Phe-Ala	14,000[d,e]	510 ± 110	0.0073 ± 0.0005
42	Z-Phe-Phe	9900[e]	1900 ± 900	0.019 ± 0.007
43	Z-Phe-Cys(SBn)	4300		
44	H-Phe-Ala (HCl salt)	4100		
45	Z-Phe	ntd[f]		
46	Z-Phe-β-Ala	ntd		
47	Z-Phe-GABA	ntd		
48	Z-Phe-Sar	ntd		
49	Z-Phe-D-Ala (diastereomer of 10)	ntd		
Isomer of 10				
50	Z-Phe-Ala-OCH$_2$CO-[2,4,6-(CH$_3$)$_3$]Ph	ntd		

[a] Conditions: bovine spleen cathepsin B, 100 mM potassium phosphate, 1.25 mM EDTA, and 1 mM dithiothreitol, pH 6.0, 25°, under argon. The parameter k/K is the second-order rate constant (k_{inact}/K_{inact}), except as noted. Standard errors for $k/K \leq 15\%$, except as noted. In those cases in which saturation kinetics were observed, k_{inact}/K_{inact} and the individual parameters K_{inact} and k_{inact} are given, as determined from a hyperbolic fit. In the other cases, k_{inact}/K_{inact} was obtained from a linear fit.
[b] Compound instability was evident.
[c] Second-order rate constant ($k/[I]$) determined at one concentration near the inhibitor solubility limit. This value represents an upper limit of $k/[I]$ for 37, because it may be accounted for completely or in part by a slight amount (<0.5%) of the diastereomeric compound 1 (a potential contaminant).
[d] Standard error 15–20%.
[e] This value has been revised from a previous report [R. A. Smith, L. J. Copp, P. J. Coles, H. W. Pauls, V. J. Robinson, R. W. Spencer, S. B. Heard, and A. Krantz, J. Am. Chem. Soc. 110, 4429 (1988)].
[f] ntd, No time dependence or significant inhibition observed near the compound solubility limit, over a 10- to 20-min period.

(and presumably other) (acyloxy)methane samples. As well, these results indicate that epimerization at this position does not occur under our cathepsin B assay conditions. The following procedure is representative.

N-(Benzyloxycarbonyl)-L-phenylalanyl-L-alanyl[(2,4,6-trimethylbenzoyl)oxy]methane (10)

Anhydrous potassium fluoride (30 mmol, 1.75 g) is added to a solution of Z-L-phenylalanyl-L-alanine bromomethane (10 mmol, 4.48 g) in 100 ml of anhydrous dimethylformamide (DMF). The mixture is stirred 3 min at room temperature, 2,4,6-trimethylbenzoic acid [Aldrich (Milwaukee, WI) 11 mmol, 1.81 g] is added, and the mixture is stirred for 3 hr at room temperature. The mixture is diluted with ethyl ether and then washed with water (5×) and saturated aqueous $NaHCO_3$ (2×). Ethyl acetate is added to redissolve organic material, and the solution is washed with brine (2×), dried ($MgSO_4$), and rotary evaporated to give a slightly yellow solid. Recrystallization from ethyl acetate/hexane provides 3.42 g (65%) of the product (**10**) as a white powder, mp 171–172°; $[\alpha]_D^{21}$ −35.0° (c = 1.13, acetone); IR (KBr) 1735, 1720, 1685, 1650 cm^{-1}; ^{1}H NMR (CDCl$_3$) δ 7.4–7.1 (m, 10 H, 2Ph), 6.9 [s, 2 H, (CH$_3$)$_3$C$_6$H$_2$CO], 6.4 (br d, J = 6.7 Hz, NH), 5.2 (br d, J = 8.0 Hz, NH), 5.1 (s, PhCH$_2$CO), 4.8 (d, app J = 1.6 Hz, COCH$_2$CN), 4.8–4.2 (m, 2 H, 2NHCHCH), 3.2–3.0 (m, Ph-CH$_2$CH), 2.4 and 2.3 [2 s, (CH$_3$)$_3$Ph], 1.3 (d, J = 7.1 Hz, CH$_3$CH); analyzed for C, H, and N.

Kinetics

Continuous fluorometric assays for cathepsin cysteine proteinases have previously been described along with optimal conditions for determining pseudo-first-order rate constants.[8] The series of (acyloxy)methanes herein are time-dependent inactivators of cathepsin-type cysteine proteinases and the gross inactivation kinetics should parallel that of suicide inhibitors and classical affinity labels.[1]

Continuous Enzyme Assay

An aliquot of assay buffer [2 ml of 0.1 M potassium phosphate, 1.25 mM EDTA, and 1 mM dithiothreitol (DTT), pH 6.0] is placed in a fluorometer cuvette, which is thermostatted at 25° and kept under an argon atmosphere. Cathepsin B, in storage buffer, is then added to give a concentration of approximately 0.4 nM. After 2–5 min of incubation to allow enzyme activation, during which time a steady baseline level of fluorescence is measured, substrate (5 or 50 μl of 1 mM stock solution in DMSO) is added and the resulting increase in fluorescence is monitored continuously. Enzyme activity is monitored by measuring the enzyme-catalyzed hydrolysis of one of two fluorogenic substrates: 7-(benzyloxycarbonylphenylalanylarginyl)-4-methylcoumarylamide (Peninsula Laboratories, San Carlos, CA) (fluorescence λ_{ex}, 370 nm; λ_{em}, 460 nm) or 7-(benzoylvalyllysyllysylarginyl)-4-trifluoromethylcoumarylamide (Enzyme Systems Products, Livermore, CA) (fluorescence λ_{ex}, 400 nm; λ_{em}, 505 nm).

Inhibitors are added to the assay solution as 0.5–20 μl of a stock solution in DMSO or CH_3CN and the fluorescence is monitored. When inhibitors are tested at very low concentrations, the enzyme concentration is reduced to maintain a minimum of a 10-fold excess of inhibitor over enzyme. In the absence of inhibitor, a linear increase in fluorescence with time is observed. On addition of inhibitor, the rate of fluorescence increase is observed to decrease exponentially, to give a rate essentially equal to zero (<2% of initial rate of fluorescence increase). A series of data points (generally 10–20 sets of fluorescence versus time) from the inhibition curve is analyzed by nonlinear regression to the exponential Eq. (1).

$$\text{Fluorescence} = Ae^{-(k_{obs}t)} + B \qquad (1)$$

The pseudo-first order rate constant, k_{obs}, is thereby obtained for each concentration of inhibitor.

Accordingly, the expectation is that the inhibitor binds noncovalently to the target enzyme in a diffusion-controlled process followed by chemical events leading to adduct formation. Progress curves of the inhibition show pseudo-first-order kinetics in a concentration-dependent manner. The second-order rate constant (k_{inact}/K_{inact}) was obtained from the dependence of pseudo-first-order inhibition rates (k_{obs}) on inhibitor concentration [I]. The data were fit to either Eq. (2) or (3), by nonlinear or linear regression,[12] depending on whether saturation kinetics were observed.

$$k_{obs} = k_{inact}[I]/(K_{inact} + [I]) \qquad \text{(hyperbolic fit)} \qquad (2)$$
$$k_{obs} = k_{inact}/K_{inact})[I] \qquad \text{(linear fit)} \qquad (3)$$

A minimal kinetic mechanism describing the general features of the inactivation is indicated in Eq. (4), where E is free enzyme, I is inhibitor, E·I is a reversible complex, and E–I is a stable adduct.[11]

$$\text{E} + \text{I} \underset{k_{-1}}{\overset{k_1}{\rightleftharpoons}} \text{E} \cdot \text{I} \xrightarrow{k_2} \text{E–I} \qquad (4)$$

The physical significance of E·I can, in principle, vary with the (acyloxy)-methane inhibitor: E·I may correspond to a noncovalent (Michaelis–Men-

---

[12] W. W. Cleland, this series, Vol. 63, p. 103.

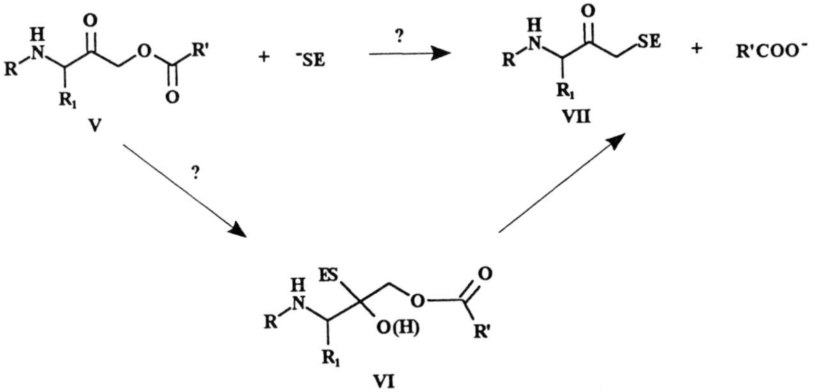

SCHEME II

ten) complex or a tetrahedral intermediate (hemithioketal **VI**, Scheme II), or it may be a minimum along the reaction pathway to E–I (i.e., E + I → E–I). Although a tetrahedral intermediate has been detected at subzero temperatures in the case of a serine proteinase, this type of experiment has posed some difficulty in our series because of solubility considerations.

The irreversibility of the inhibition has been established by both dialysis and dilution assays, as well as by NMR characterization of the stable enzyme adduct as a thiomethane **VII** (Scheme II).[13] (Dilution assays are performed with enzyme at approximately 0.4 nM, which is preincubated with at least a 10-fold excess of inhibitor in assay buffer, followed by removal of aliquots that are diluted 1 : 40 into buffer, and then assayed against substrate. (This type of assay provides a means of ascertaining whether there are potentially reversible inactive enzyme–inhibitor complexes that contribute to the observed time-dependent kinetics.) In addition, the stoichiometry of inactivation was determined to be ~1 : 1 by titration of enzyme activity with **10**. Leupeptin, a competitive inhibitor of cathepsin B, protected the enzyme from inactivation by **1** [an extremely potent peptidyl (acyloxy)methane], with a K_i of 5 nM, providing confirmatory evidence for the active site-directed nature of the inactivation. With the advent of electrospray mass spectrometry, the determination of adduct formation, stoichiometry, and the site of labeling can now be performed directly.[14]

[13] V. J. Robinson, P. J. Coles, H. W. Pauls, R. A Smith, and A. Krantz, *Biorg. Chem.* **20**, 42 (1992).

[14] P. R. Griffin, S. A. Coffman, L. E. Hood, and J. R. Yates, *Int. J. Mass Spectrom. Ion Phys.* **111**, 131 (1991).

Potency: Dependence on the Affinity and Leaving Group

(Acyloxy)methanes with peptide recognition elements that satisfy the specificity requirements of cathepsin B span a range of six orders of magnitude in their ability to inactivate this enzyme (Table I). (Acyloxy)-methanes linked to only a single amino acid residue exhibit poor activity. Proper registry of affinity groups at S1 and S2 subsites is critical for inactivating cathepsin B, as neither amino acids of D configuration, nor β- and γ-amino acids at the P1 position, confer time-dependent activity on dipeptidyl (acyloxy)methanes.

The effect of the leaving group was evaluated in Z-Phe-Ala (acyloxy)-methanes using nucleofuges spanning a range of pK_a values from 0.6 to 5.0 (Table II). The activity of these inhibitors was found to be exquisitely sensitive to the nature of the carboxylate leaving group, as a strong correlation of the logarithm of the second-order inactivation rate with carboxylate pK_a was uncovered [Eq. (5)].

$$\log(k/K) = -1.1(\pm 0.1)pK_a + 7.2(\pm 0.4 \qquad r^2 = 0.82, \quad n = 26 \quad (5)$$

For Z-Phe-Ala (acyloxy)methanes the second-order rate of inactivation falls off at low values when leaving group $pK_a > 4$, yet this threshold can be overcome dramatically by using a tighter binding peptidyl moiety. For example, the Z-Phe-Ala-anisate **24** does not exhibit time-dependent activity in our assay, but the anisate linked to the Z-Phe-Lys framework (**32**) inactivates cathepsin B at a rate essentially identical to that observed for the 4-nitrobenzoate in the Z-Phe-Ala series (**14**). The ability to exploit a tighter binding affinity group, to compensate for the effect of reducing the nucleofugality of the leaving group, should be a useful design principle for fine-tuning the properties of an inhibitor.

Specific Inactivation of Papain-Type Cysteine Proteinases

The selectivity of this series of (acyloxy)methanes for cathepsin B is impressive. They exhibit no time-dependent activity versus other classes of proteinases, including the mechanistically related serine proteinases. For example, even when binding determinants are tailored for serine proteinases, as in the case of Z-Ala-Ala-Pro-Val-CH$_2$OCO[2,6(CF$_3$)$_2$]Ph, no time-dependent activity for human leukocyte elastase (EC 3.4.21.37) is observed versus this enzyme. Also, Z-Phe-Lys-CH$_2$OCO[2,4,6(CH$_3$)$_3$]Ph (**30**) binds to trypsin (EC 3.4.21.4) with a K_i of ~20 μM, but no significant time dependence is observed. Peptidyl (acyloxy)methanes are very weak inhibitors of smooth muscle calpain (EC 3.4.22.17), a Ca^{2+}-dependent

TABLE III

INACTIVATION OF CATHEPSINS L AND S BY PEPTIDYL (ACYLOXY)METHANES
Z-PHE-ALA-CH$_2$OCO-R: VARIATION OF LEAVING GROUP

Number	pK_a+	R	k_2 (sec^{-1})	K_i (μM)	K_2/K_i (M^{-1} sec^{-1})
		Cathepsin L			
1	0.58	2,6-(CF$_3$)$_2$-Ph	0.083 ± 0.037	0.25 ± 0.12	332,000
2	1.59	2,6-Cl$_2$-Ph	0.01 ± 0.001	0.07 ± 0.001	143,000
5	2.63	2,5-(CF$_3$)$_2$-Ph			2400[a]
4	2.67	2,6-(CH$_3$)-4-COOCH$_3$-Ph	0.015 ± 0.003	4.18 ± 1.52	3600
14	3.43	4-NO$_2$-Ph	0.007 ± 0.0003	0.16 ± 0.02	44,000
10	3.45	2,4,6-(CH$_3$)-Ph			4200[a]
18	3.50	4-CN-Ph			1000[a]
28	3.57	CH$_2$OCH$_3$	0.01 ± 0.001	2.26 ± 0.21	4400
22	3.67	4-CF$_3$-Ph			700[a]
24	4.50	4-OCH$_3$-Ph			1000[a]
27	4.73	CH(CH$_2$CH$_3$)$_2$			900[a]
26	5.03	C(CH$_3$)$_3$			7800[a]
		Cathepsin S			
1	0.58	2.6-(CF$_3$)$_2$-Ph	0.040 ± 0.009	0.12 ± 0.04	364,000
2	1.59	2,6-Cl$_2$-Ph	0.096 ± 0.01	0.14 ± 0.02	686,000
5	2.63	2,5-(CF$_3$)$_2$-Ph	0.014 ± 0.002	3.00 ± 0.03	4700
4	2.67	2,6-(CH$_3$)$_2$-4-COOCH$_3$-Ph	0.005 ± 0.0005	0.12 ± 0.02	42,000
14	3.43	4-NO$_2$-Ph	0.066 ± 0.021	19.7 ± 8.9	3300
10	3.45	2,4,6-(CH$_3$)$_3$-Ph			500[a]
18	3.50	4-CN-Ph			200[a]
28	3.57	CH$_2$OCH$_3$	0.01 ± 0.003	20.6 ± 9.2	500
22	3.67	4-CF$_3$-Ph	0.033 ± 0.014	57 ± 25	600
24	4.50	4-OCH$_3$-Ph			100[a]
27	4.73	CH(CH$_2$CH$_3$)$_2$			200[a]
26	5.03	C(CH$_3$)$_3$	0.024 ± 0.002	8.8 ± 0.6	2700

[a] Saturation kinetics were not achieved.

cysteine proteinase.[15] However, they do exhibit activity against other papain-type cysteine proteinases such as rat liver cathepsin L and human cathepsin S (Tables III and IV). Albeit for the latter enzymes, a different inhibitory pattern is observed from that of cathepsin B, which suggests that poorer leaving groups could be used to discriminate in favor of cathepsin L or S, because cathepsin B is the least robust enzyme among the three.

Although this set of inhibitors does not discriminate effectively between cathepsin L and S, the kinetic parameters reveal preferences of these enzymes for specific structural determinants. For example, inhibi-

[15] D. H. Pliura, B. J. Bonaventura, R. A. Smith, P. J. Coles, and A. Krantz, *Biochem. J.* **288,** 759 (1992).

TABLE IV

INACTIVATION OF CATHEPSINS L AND S BY PEPTIDYL (ACYLOXY)METHANES Z-PHE-X-CH$_2$OCO-R: VARIATION OF P1 RESIDUE

Compound number	X	R	K_2 (sec$_{-1}$)	K_i (μM)	K_2/K_i (M^{-1} sec^{-1})
			Cathepsin L		
30	Lys	2,4,6-Me$_3$-Ph	0.002 ± 0.0002	0.028 ± 0.003	71,000
34	Cys(SBn)	2,6-(CF$_3$)$_2$-Ph	0.015 ± 0.002	0.0014 ± 0.0004	10,700,000
35	Ser(OBn)	2,6-(CF$_3$)$_2$-Ph	0.003 ± 0.0012	0.007 ± 0.0003	4,290,000
42	Phe	2,4,6-(CH$_3$)-Ph			100[a]
			Cathepsin S		
30	Lys	2,4,6-(CH$_3$)-Ph	0.04 ± 0.01	0.33 ± 0.69	120,000
34	Cys(SBn)	2,6-(CF$_3$)$_2$-Ph			1,550,000[a]
35	Ser(OBn)	2,6-(CF$_3$)$_2$-Ph	0.033 ± 0.004	0.64 ± 0.15	52,000
42	Phe	2,4,6-(CH$_3$)$_3$-Ph			400[a]

[a] Saturation kinetics were not achieved.

tors that contain an ortho-substituted benzoate leaving group (e.g., **1, 2, 4, 5, 10, 30**) are generally observed to be more effective against cathepsins B or S, than against cathepsin L. This phenomenon contrasts with the usual finding that cysteine proteinase inhibitors react most rapidly with cathepsin L, indicating differences in the S1′ sites of these enzymes, with perhaps a limitation in the size or shape of the S1′ region of cathepsin L.

Future Prospects

From a chemical point of view, two general approaches govern the design of inhibitors, dividing them into two broad subclasses. One approach is based on the idea of generating complements to enzyme active sites, where the goal is to produce tight-binding enzyme inhibitors without resort to chemically reactive functionality, either latent or intrinsic. In this regard, contemporary efforts are devoted to structure-based design in which structural information from NMR spectroscopy and X-ray spectroscopy is used to guide the construction of potentially tight-binding inhibitors.[2]

The other broad approach embraces strategies that exploit principles of organic chemical reactivity designed to produce stable covalent bonds between enzyme and inhibitor and generate essentially irreversible adducts during the course of catalysis (Table V). Mechanism-based inhibitors are one manifestation of such an approach, where the idea is to induce enzyme to generate either a stable enzyme–inhibitor adduct during the

TABLE V
STRATEGIES FOR IRREVERSIBLE COVALENT ENZYME INHIBITION

Reaction	Rationale
Mechanism-based inhibitors	
$E + S' \rightleftharpoons E \cdot S' \rightarrow E \cdot I' \rightarrow E-I'$	I' is a reactive intermediate
$E + S' \rightleftharpoons E \cdot S' \rightleftharpoons E-S' \nrightarrow E \cdot P$	$E-S'$ is trapped in a potential energy well
Affinity labels	
$E + S_{rx} \rightleftharpoons E \cdot S_{rx} \rightarrow E-S'' + X^-$	High effective concentration of a reactive group at active site
$E + S_x \rightleftharpoons E \cdot S_x \rightarrow E-S'' + X^-$	High enzymatic reactivity vs. low chemical reactivity

normal course of catalysis, or a highly reactive form of the inhibitor that can combine rapidly with enzyme to form a new covalent bond, irreversibly. Alternatively, enzyme may be inactivated as a consequence of the intrinsic reactivity of an active site-directed reagent such as a classical affinity label.

Enzymes have the ability to accelerate reaction rates by huge factors. Yet, very few efforts to test the limits of the ability an enzyme to combine irreversibly with reagents of low intrinsic chemical reactivity, in a reaction type distinct from that which the enzyme has evolved to catalyze, have been documented. The demonstration of powerful inactivation of cysteine proteinases by affinity labels containing chemically unreactive groups such as (mesitoyloxy)methanes opens the door to more extensive studies of this concept. An unlimited number of potential leaving groups can be envisaged that conform to the general formula:

$$PCH_2X-\text{(substituted phenyl)}-Y$$

where P is a peptidyl group, and X is an atom or function that, combined with (substituted) phenyl, is a potential leaving group of low chemical reactivity. The enormous variety of substituents and ring substitutive patterns offers reasonable prospects for subtly activating weak electrophilic reagents to attack by nucleophilic enzymes, without compromising their chemical stability.

The classification of inhibitors according to whether they are mechanism based or pure affinity labels hinges on whether an intermediate is formed along the inactivation pathway paralleling the normal course of catalysis, or whether direct displacement occurs without prior bond-making or bond-breaking events. If the direct displacement mechanism is

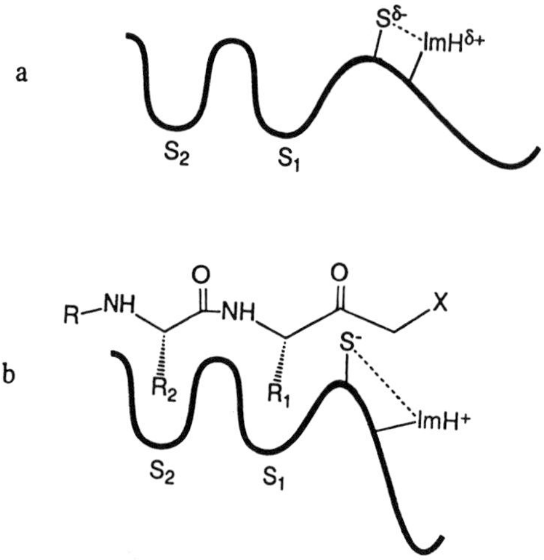

SCHEME III. Schematic representation of a hypothesis to explain enhanced active-site thiolate reactivity resulting from tight-binding inhibitors.

operative, it is of exceptional interest, because the huge inactivation rate observed would imply that the enzyme employs intrinsic binding to lower the overall free energy of activation of an "aberrant" chemical path leading to facile inactivation. It is then quite likely that reactions of this genre could be exploited to inactivate other target enzymes as well, without utilizing a part of their normal catalytic pathway. Relevant inactivators could be designed to react facilely only on binding to the target enzyme and ideally would be "quiescent" in the presence of other nonnucleophiles lacking a proper complementary surface.

Amidomethylation of active site cysteine by iodoacetamide may be an example of a reaction that proceeds by direct displacement of a leaving group. The peptide recognition group in affinity labels then serves to increase the residence time of the electrophilic moiety and properly orient it for displacement, presumably using intrinsic binding to power the displacement reaction. Perhaps even more subtle factors could be operative as well, if the enzyme acts like a chemical machine.[16] To account for enhanced reactivity with tighter binding, Krantz et al. have proposed that tighter binding (achieved by improvements in complementarity between enzyme and affinity group) may lead to increased separation of charge in the thiolate–imidazolium ion pair (Scheme III).[8] The greater the charge separation, the freer is the thiolate; the freer the thiolate, the greater is the nucleophilicity of the enzyme, and the more rapidly displaced is a

[16] R. J. P. Williams, *Trends Biochem. Sci.* **18**, 115 (1993).

given leaving group within a series of inactivators. In the limit, with perfect active site complements, chemistry may not be rate determining and such affinity groups may be used to drive reactions that show little variation with leaving-group structure.

It should be noted that even if enzyme inactivation by α-substituted ketones proceeds through a common tetrahedral intermediate, in all cases parallelling catalytic hydrolysis of substrates, the distinction between "quiescent" acyloxymethanes and chemically reactive halomethanes is still of central importance to the design of clinically relevant inactivators and has, in a sense, revolutionized the affinity label concept. Using a highly electrophilic reagent to inactivate a powerful nucleophilic enzyme, such as a papain-type cysteine proteinase, is tantamount to using a massive hammer to swat a fly. A kinder and gentler reagent will suffice, especially *in vivo*. As more robust enzyme targets are identified, one can anticipate the discovery of reactions at ambient conditions in aqueous media that are unprecedented in the annals of chemistry, and distinct from those which the enzyme has evolved to catalyze.

Acknowledgment

It is a pleasure for me to acknowledge Dr. Roger A. Smith's and Mr. Peter Coles' efforts in organic chemistry and the contributions of Ms. Leslie Copp in the domain of enzymology.

[48] N,O-Diacyl Hydroxamates as Selective and Irreversible Inhibitors of Cysteine Proteinases

By DIETER BRÖMME and HANS-ULRICH DEMUTH

Introduction

Cysteine proteases represent attractive targets for the design of inhibitors because the proteinases have been shown to play a role in degenerative diseases and in malignant tumor and parasite development.[1-4] A variety of low molecular weight inhibitors of cysteine proteases have been devel-

[1] W. H. Baricos, S. L. Cortez, Q. C. Le, L. T. Wu, E. Shaw, K. Hanada, and S. V. Shah, *Arch. Biochem. Biophys.* **288,** 468 (1991).
[2] M. Schmitt, F. Jänicke, and H. Graeff, *Fibrinolysis* **6,** 3 (1992).
[3] M. J. Duffy, *Clin. Exp. Metastasis* **10,** 145 (1992).
[4] M. J. North, J. C. Mottram, and G. H. Coombs, *Parasitol. Today* **6,** 27 (1990).

oped.[5,6] Generally these inhibitors consist of a peptide part binding to the S specificity subsites of the target proteinase and a reactive group interacting with a catalytic active site residue. Well-known cysteine protease inhibitors such as peptide aldehydes,[7] halomethanes,[8] diazomethanes,[9] (see also [46] in this volume), and epoxides[10] allow only a variation of the peptide moiety to modulate their specificity. The chemically reactive function in these inhibitors is not affected greatly by the inhibitor structure and can lead to nonspecific side reactions. In addition, these inhibitors do not allow an exploitation of the S' specificity of the enzyme and developments in inhibitor design for cysteine proteases have overcome these disadvantages. A new class of peptide-derived inhibitors, N,O-diacyl hydroxamates, permits variations of the N-acyl and O-acyl residues and, thus, a selective control of their affinity and reactivity toward the enzymes. Residue R in the general structure N-peptidyl-NHO-CO-R occupies the S' binding region and can be any aliphatic, aromatic, amino acid, or peptidyl residue. These residues can be utilized for the modification of the specificity, stability, and solubility of the inhibitor. Furthermore, extending the inhibitor with amino acids or peptidyl residues allows the investigation of the S' specificity of cysteine proteinases.

N,O-Diacyl hydroxamates are highly effective irreversible inactivators of cysteine proteinases and display second-order rate constants of up to $10^6 \ M^{-1} \ \text{sec}^{-1}$.[11,12] The inhibitors were initially developed as potential mechanism-based inhibitors of dipeptidyl peptidase IV.[13-15] The hydroxamate inhibitors generally show only weak activity, with rate constants of inactivation in the range of $10-1000 \ M^{-1} \ \text{sec}^{-1}$ toward this serine peptidase and other subtilisin-like[16] and trypsinlike[17] proteinases. No inactivation

[5] D. H. Rich, in "Proteinase Inhibitors" (A. J. Barrett and G. Salvesen, eds.), p. 153. Elsevier, Amsterdam, 1986.

[6] H.-U. Demuth, *J. Enzyme Inhib.* **3**, 249 (1990).

[7] R. C. Thompson, this series, Vol. 46, p. 220.

[8] D. Rasnick, *Anal. Biochem.* **149**, 461 (1985).

[9] E. Shaw and G. D. J. Green, this series, Vol. 80, p. 820.

[10] B. J. Gour-Salin, P. Lachance, C. Plouffe, A. C. Storer, and R. Ménard, *J. Med. Chem.* **36**, 720 (1993).

[11] R. A. Smith, P. J. Coles, R. W. Spencer, L. J. Copp, C. S. Jones, and A. Krantz, *Biochem. Biophys. Res. Commun.* **155**, 1201 (1988).

[12] D. Brömme, A. Schierhorn, H. Kirschke, B. Wiederanders, A. Barth, S. Fittkau, and H.-U. Demuth. *Biochem. J.* **263**, 861 (1989).

[13] G. Fischer, H.-U. Demuth, and A. Barth, *Pharmazie* **38**, 249 (1983).

[14] H.-U. Demuth, R. Baumgrass, C. Schaper, G. Fischer, and A. Barth, *J. Enzyme Inhibition* **2**, 129 (1988).

[15] H.-U. Demuth, U. Neumann, and A. Barth, *J. Enzyme Inhib.* **2**, 239 (1989).

[16] H.-U. Demuth, C. Schönlein, and A. Barth, *Biochim. Biophys. Acta* **996**, 19 (1989).

[17] D. Brömme, U. Neumann, H. Kirschke, and H.-U. Demuth, *Biochim. Biophys. Acta* **1202**, 271 (1993).

was detected with aspartic proteinases and metalloproteinases. Therefore, N,O-diacyl hydroxamates can be regarded as a new class of specific and powerful cysteine proteinase inhibitors.

Synthesis and Stability of N,O-Diacyl Hydroxamates

N,O-Diacyl hydroxamates are easily accessible synthetically. Boc- and Z-protected peptidyl methyl esters have been synthesized according to standard methods. The methyl esters are converted into the appropriate hydroxamic acids by treatment with hydroxylamine in sodium methanolate solution.[13,14] Acylation with alkyl or aryl acid chlorides in dry tetrahydrofuran or pyridine as a base gives the diacyl hydroxylamines. The final products are usually crystallized from ethanol/ethyl acetate, ethyl acetate/petrol ether, or acetone/diisopropyl ether and give correct elemental analyses, correct masses determined by mass spectroscopy, and single-peak high-performance liquid chromatography (HPLC) profiles or single spots on thin-layer chromatography (TLC). The structures of the compounds are verified by ^{13}C NMR or ^{1}H NMR. The yield of the last chemical step is between 60 and 85%.

*Synthesis of Z-(Aa)$_n$-NHOH**

A 5-ml volume of a 3.5 M sodium methylate solution is combined with 16 mmol of hydroxylamine hydrochloride dissolved in 10 ml of dry methanol and stirred for 10 min at room temperature. The precipitated sodium chloride is removed by filtration and the hydroxylamine solution is allowed to react with 4 mmol of a dry N-protected peptide methyl ester. The reaction mixture is stored overnight at 4° and then adjusted to pH 4.0 with HCl. The peptidyl hydroxamic acid is extracted with ethyl acetate and the solvent is evaporated. The compound is recrystallized from ethyl acetate/petrol ether with a yield of 60–80%.

Synthesis of Z-(Aa)$_n$-NHO-Bz(4-NO$_2$)

Z-(Aa)$_n$-NHOH (1.5 mmol) is dissolved in 40 ml pyridine and is allowed to react with 1.55 mmol of 4-nitrobenzoyl chloride dissolved in 6 ml tetrahydrofuran at −10°. After stirring for 15 min at −10° and 45 min at 0°, the reaction mixture is concentrated and treated with 250 ml ice-cold 5% KHSO$_4$. The precipitate is washed with water, filtered, and dried over P$_2$O$_5$. The solid product is recrystallized using the appropriate solvents.

---

* Aa, Amino acid; ONp, (4-nitro)phenyl ester.

Synthesis of NH₂-Aa-NHO-Bz(4-NO₂)

Boc-Aa-NHO-BzNO₂ is prepared via the hydroxamic acid derivative as described in the preceding procedures. The N-terminal deprotected *N*-aminoacyl hydroxamates are obtained by removing the Boc group in 1.75 *N* HCl/glacial acetic acid for 30 min. The product is precipitated and washed with diethyl ether and recrystallized from methanol/diethyl ether.

Synthesis of Z-Phe-Gly-NHO-CO-ONp

To 4 mmol of 4-nitrophenyl chloroformate dissolved in 20 ml tetrahydrofuran, 20 ml of Z-Phe-Gly-NHOH (2 mmol) and 4 mmol (triethylamine in tetrahydrofuran are added dropwise at −10°. After stirring for 30 min at −10° and 1 hr at room temperature, the reaction mixture is filtered and the solvent evaporated. The product is recrystallized from ethyl acetate/ petrol ether. The yield is 85% (mp 140°–144°).

Synthesis of Z-Phe-Gly-NHO-CO-Aa (Aa: Gly, Ala, Val, Leu, Phe)

Z-Phe-Gly-NHO-CO-ONp (1 mmol) is dissolved in 6 ml cold dimethylformamide and stirred with 1.25 mmol of a sodium salt of a neutral amino acid at −10° for 10 min and at room temperature for 2 hr. The reaction mixture is acidified with 5% KHSO₄ and the precipitate is dissolved in ethyl acetate, washed 3 times with water, and dried over anhydrous Na₂SO₄. The organic phase is concentrated and treated with petrol ether to give a white precipitate (mp 129°–131° for Aa Gly; mp 84°–85° for Aa Ala; mp 84°–87° for Aa Val; mp 78°–79° for Aa Leu; mp 78°–83° for Aa Phe).

Stability and Nonenzymatic Degradation

As solid compounds, the inhibitors are very stable if stored under anhydrous conditions at 4° or at room temperature. Exposure to light should be avoided, especially for the inhibitors containing the 4-nitrobenzoyl moiety as the leaving group. *N,O*-Diacyl hydroxamates are stable under acidic conditions but less stable toward strong alkali.[18] In aqueous solution *N,O*-diacyl hydroxamates can decompose into the hydroxamic acid and the appropriate acid of the *O*-acyl residue. The decomposition rate is strongly dependent on the nature of the leaving group. The more electron-withdrawing the *O*-acyl residue is or the lower its pK_a value, the

---

[18] H.-U. Demuth, G. Fischer, A. Barth, and R. L. Schowen, *J. Org. Chem.* **54**, 5880 (1989).

TABLE I
HALF-LIFE TIMES OF SPONTANEOUS DECOMPOSITION IN AQUEOUS
SOLUTION FOR N,O-DIACYL HYDROXAMATES

Inactivator	$t_{1/2}$ (min)	pK_a^c	Ref.
Z-Gly-Phe-NHO-Bz(4-OCH$_3$)	504	4.50	d
Z-Gly-Phe-NHO-Bz(4-CH$_3$)	390	4.37	d
Z-Gly-Phe-NHO-Bz	240	4.20	d
Z-Gly-Phe-NHO-Bz(4-NO$_2$)	98	3.43	d
Boc-Phe-Gly-NHO-Bz(4-NO$_2$)^a	900		e
Boc-Phe-Ala-NHO-Bz(4-NO$_2$)^a	365		e
Boc-Phe-Pro-NHO-Bz(4-NO$_2$)^a	209		e
Boc-Ala-Phe-Leu-NHO-Bz(4-NO$_2$)^a	248		e
H-Gly-NHO-Bz(4-NO$_2$)^b	204		f
H-Val-NHO-Bz(4-NO$_2$)^b	30		f
H-Phe-NHO-Bz(4-NO$_2$)^b	24		f
Z-Phe-Phe-NHO-Bz(4-NO$_2$)^a	16		e
Boc-Gly-Phe-Phe-Bz(4-NO$_2$)^a	45		e
Z-Phe-Lys-NHO-Bz(4-NO$_2$)^a	25		g
Z-Val-Val-Lys-NHO-Bz(4-NO$_2$)^a	15		g

^{a} In 40 mM Tricine buffer, pH 7.6 (I = 0.125) at 30°.
^{b} In 40 mM sodium phosphate buffer, pH 6.5 (I = 0.125) at 30°.
The kinetic measurements of nonenzymatic decomposition were
recorded in the range of 225–300 nm by using a Carl-Zeiss spectro-
photometer (M-40).
^{c} The pK_a values of the leaving groups were taken from A. Krantz,
L. J. Copp, P. J. Coles, R. A. Smith, and S. B. Heard, *Biochemistry*
30, 4678 (1991).
^{d} H.-U. Demuth, A. Schierhorn, R. Höfke, H. Kirschke, and D.
Brömme, submitted for publication.
^{e} D. Brömme, A. Schierhorn, H. Kirschke, B. Wiederanders, A.
Barth, S. Fittkau, and H.-U. Demuth, *Biochem. J.* **263**, 861 (1989).
^{f} H.-U. Demuth, A. Stöckel, A. Schierhorn, S. Fittkau, H. Kirschke,
and D. Brömme, *Biochim. Biophys. Acta* **1202**, 265–270 (1993).
^{g} D. Brömme, U. Neumann, H. Kirschke, and H.-U. Demuth, *Bio-
chim. Biophys. Acta* **1202**, 271–276 (1993).

faster is the nonenzymatic degradation (Table I).[14,19] However, the P1
amino acid residue seems to influence the stability of the inhibitors. Comparing all the available stability data of *N,O*-diacyl hydroxamates,[12,14,17,19,20] the following empirical range of decomposition-promoting
ability is obvious (Table I): Gly < Pro,Ala < Leu < Phe,Val < Lys. The

[19] H.-U. Demuth, A. Schierhorn, R. Höfke, H. Kirschke, and D. Brömme, submitted (1993).
[20] H.-U. Demuth, A. Stöckel, A. Schierhorn, S. Fittkau, H. Kirschke, and D. Brömme,
Biochim. Biophys. Acta **1202**, 265 (1993).

half-life of nonenzymatic decomposition varies from several minutes to more than 15 hr (Table I).

The data derived from the nonenzymatic degradation studies may be used in the design of specific inhibitors that, from a pharmacokinetic point of view, would be more stable and efficient in aqueous solutions.

Inhibitor Assays and Mechanistic Studies

Enzyme Assays with Methylcoumarylamide Substrates

Progress curves for the inactivation of the proteinases using methylcoumarylamide (MCA) substrates are monitored in 1-cm cuvettes at 22° (at 30° for the inactivation with N-aminoacyl-O-4-nitrobenzoyl hydroxamates) in a Shimadzu spectrophotometer (UV-300) equipped with a fluorescence detection unit or in a KONTRON SFM 25 fluorimeter at an excitation wavelength of 380 nm and an emission wavelength of 450 nm. The kinetic experiments are carried out with a constant enzyme concentration in 50 mM sodium acetate buffer, pH 5.5, for cathepsin L (0.7 nM); in 50 mM potassium phosphate buffer, pH 6.0, for cathepsin B (0.9 nM); in 50 mM potassium buffer, pH 6.5, for cathepsin H (2.4 to 9.7 nM), and in 50 mM potassium phosphate buffer, pH 6.5, containing 0.01% Triton X-100 for cathepsin S (2.3 nM). The mercury derivatives of rat cathepsins B, L, and H are activated by incubation for 5 min at 25° with 2.5 mM dithioerythritol and 2.5 mM EDTA in the assay buffer. Bovine cathepsin S is activated for 15 min at 25° in its assay buffer containing 5 mM dithioerythritol and 5 mM EDTA. The substrates used are Z-Phe-Arg-MCA (10 and 50 μM for cathepsin B), Z-Phe-Arg-MCA (2.5 and 8 μM for cathepsin L), Arg-MCA (8 and 25 μM for cathepsin H), and Z-Val-Val-Arg-MCA (10 and 50 μM for cathepsin S). The inhibitor stock solutions are prepared either in dimethylformamide or in acetonitrile. The final concentration of the organic solvent is normally kept below 1% (v/v) in the assay mixture.

Irreversibility of Inhibition

N,O-Diacyl hydroxamates react irreversibly with their target cysteine proteinase. No recovery of activity of the enzyme is achieved after the removal of the excess inhibitor. The irreversibility of inactivation can be confirmed by gel-filtration or dilution experiments. In gel-filtration experiments the proteinase and an appropriate inhibitor are incubated for 1–3 hr are in the assay buffer at room temperature. Then, the low molecular weight inhibitors are separated on a Sephadex G-10 column and activity assays are performed. Reference solutions without the inhibitor are tested in the same way. No recovery of activity is observed within 15 hr after removal of excess inhibitor. Dilution experiments are carried out as described by Krantz and co-workers.[21] The enzyme is incubated with a 10- to 20-fold excess of inhibitor and, at time intervals, 50-μl aliquots are taken and the residual activity is monitored as described above. The values obtained from plotting the residual activity versus time and the values from progress curve analysis of the continuously monitored assay are identical, if the inactivation is irreversible.

Kinetics of Enzyme Inactivation

The inactivation rates (k_{obs}) for 8–11 different inhibitor concentrations in the presence of the substrate were determined according to Tian and Tsou.[22] To rule out the competing effect of the substrate toward the inhibitor binding site in the enzyme, the rate constant at zero substrate concentration has to be determined. This can be achieved by fitting the k_{obs} values obtained at one substrate concentration against the inhibitor concentration using Eq. (1). The K_i^{app} must be corrected to zero substrate concentration by the term $1 + [S]/K_m$ in Eq. (2). The same result will be obtained by using different substrate concentrations. Extrapolation of k_{obs} values for the different substrate concentrations to zero substrate concentration gives the rate of inactivation (k_0) in the absence of substrate. Using the k_0 rate constants obtained for different inhibitor concentrations, Eq. (1) can be transformed into $k_0 = k_2/[I]/K_i + [I]$, giving the inactivation parameters K_i and k_2 directly.

$$k_{obs} = k_2[I]/(K_i^{app} + [I]) \tag{1}$$
$$K_i = K_i^{app}/(1 + [S]/K_m) \tag{2}$$

Proposed Mechanism of Inactivation

Several organic reactions proceed via formation of reactive nitrogen intermediates such as nitrenes and nitrenium ions. The classic reactions are nucleophilic migrations from a carbon to a nitrogen atom releasing molecular nitrogen (carbonyl azides), halogen (*n*-halogen amides), or acid anions (N,O-diacyl hydroxamates). The migration and the nucleofuge elimination that yield isocyanates occur either concerted or stepwise via

[21] A. Krantz, L. J. Copp, P. J. Coles, R. A. Smith, and S. B. Heard, *Biochemistry* **30**, 4678 (1991).
[22] W. X. Tian and C. L. Tsou, *Biochemistry* **21**, 1028 (1982).

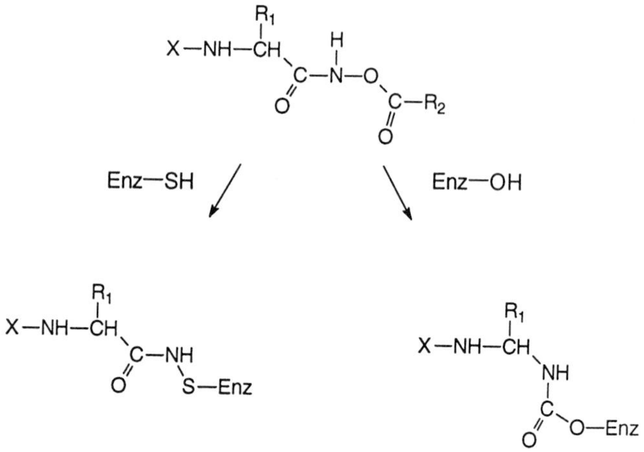

$$X-NH-\underset{\underset{O}{\|}}{\overset{R_1}{\underset{|}{C}H}}\!\!\!\!\!\!\!\underset{O}{\overset{H}{C-N-O}}\!\!\!\!\!\!\underset{O}{\overset{\|}{C-R_2}}$$

Enz—SH   /   \   Enz—OH

$$X-NH-\overset{R_1}{\underset{|}{C}H}$$
$$\underset{O}{\overset{\|}{C}}-NH$$
$$S-Enz$$

$$X-NH-\overset{R_1}{\underset{|}{C}H}$$
$$NH$$
$$\underset{O}{\overset{\|}{C}}-O-Enz$$

SCHEME I. Structure of peptidyl-*O*-acyl hydroxamates and possible mechanisms with cysteine and serine proteinases.

a nitrene intermediate.[23] Such alkyl isocyanates are inhibitors of serine proteinases.[24,25] Because migration of a peptide residue held within the active site of an enzyme might be hindered, nucleophilic attack on the scissile carbonyl carbonamide linkage could promote α-elimination, resulting in short-lived highly electrophilic nitrenes. Incubation of several cysteine and serine proteinases with *N,O*-diacyl hydroxamates resulted in the formation of covalently modified, inactivated enzymes.[11–17,19,20] Using ^{13}C NMR[26] and electrospray mass spectroscopy,[27,28] sulfenamidation was established as the main route to inactivate cysteine proteinases by *N,O*-diacylhydroxamates. In contrast, X-ray structure analysis of a complex of subtilisin Carlsberg inhibited by Boc-Ala-Pro-Phe-NHO-Bz(4-NO$_2$) showed carbamoylsubtilisin formed during a covalent modification (Scheme I).[29]

[23] L. Bauer and O. Exner, *Angew. Chem.* **13**, 376 (1974).
[24] W. F. Brown and F. Wold, *Biochemistry* **12**, 828 (1973).
[25] W. E. Brown and F. Wold, *Biochemistry* **12**, 835 (1973).
[26] V. J. Robinson, P. J. Coles, R. A. Smith, and A. Krantz, *J. Am. Chem. Soc.* **113**, 7760 (1991).
[27] R. Ménard, R. Feng, A. C. Storer, V. J. Robinson, R. A. Smith, and A. Krantz, *FEBS Lett.* **295**, 27 (1991).
[28] D. Brömme and H. Kirschke, *FEBS Lett.* **322**, 211 (1993).
[29] A. Steinmetz and D. Ringe, submitted (1993).

Inactivation of Lysosomal Cysteine Proteinases

N-Peptidyl-O-acyl Hydroxamates

α-NH$_2$-protected *N*-peptidyl-*O*-acyl hydroxamates are powerful inactivators of cysteine endopeptidases of the papainlike proteinases such as the lysosomal cathepsins B, L, and S. All three enzymes have been implicated in a range of inflammatory and tumorigenic processes as well as in protein processing. The highest second-order rate constants of inactivation are in the range of 10^5 to 10^6 M^{-1} sec^{-1}.[11,12,17] A comparison of rate constants for the tested cysteine proteinases reveals generally higher inactivation rates for cathepsin L, followed by cathepsin S and cathepsin B. Cathepsin H, a lysosomal aminoendopeptidase, is only weakly inhibited by peptide-derived hydroxamates (Table II). The inhibitor specificity of *N,O*-diacyl hydroxamates toward the cysteine proteinases is comparable to the results obtained with peptide substrates or other classes of peptide-derived inhibitors. Bulky aromatic residues in P2 and P1 (phenylalanine) are especially favorable for cathepsin L whereas cathepsin S prefers smaller residues in P2 and P3.[12,17] Positively charged residues (lysine) are well accepted in the S1 subsites of papainlike cathepsins whereas proline in P1 and P2 decreases the efficiency of the inhibitors by some orders of magnitude.

N-Aminoacyl-O-4-nitrobenzoyl Hydroxamates

N-Aminoacyl-*O*-4-nitrobenzoyl hydroxamates inhibit cathepsin H irreversibly in a time-dependent reaction.[19] H-Phe-NHO-Bz(4-NO$_2$) exhibits a second-order rate constant of 32,000 M^{-1} sec^{-1} for the inactivation of the enzyme, which represents one of the most effective and specific inhibitors of cathepsin H described so far (Table III). Only a minor inhibition of endopeptidase cathepsins is observed with this inhibitor. The good acceptance of aromatic residues in P1 is in accordance with specificity studies of cathepsin H.[30] Metal-dependent aminopeptidases such as leucyl aminopeptidase and aminopeptidases M and P show only weak, reversible inhibition, or none at all, with the tested inhibitors (Table III).[20] The measured K_i values in the range of 10^{-5} to 10^{-4} M are comparable to values reported for the inhibition of these enzymes with amino acid hydroxamic acids.[31] Because a cleavage of the -NHO-CO bond and the release of

[30] H. Koga, N. Mori, H. Yamada, Y. Nishimura, K. Tokuda, K. Kato, and T. Imoto, *J. Biochem. (Tokyo)* **110,** 939 (1991).
[31] J. C. Powers and J. W. Harper, *in* "Proteinase Inhibitors" (A. J. Barrett and G. Salvesen, eds.), p. 219. Elsevier, Amsterdam, 1986.

TABLE II
INACTIVATION PARAMETERS FOR CATHEPSINS B, L, S, AND H WITH N-PEPTIDYL-O-ACYL HYDROXAMATES

Inactivator	k_2/K_i (M^{-1} sec^{-1})				Ref.
	Cathepsin B[a]	Cathepsin L[b]	Cathepsin S[c]	Cathepsin H[d]	
Z-Phe-Phe-NHO-Ma	2800	1,222,000	21,000	19	g
Boc-Ala-Phe-Leu-NHO-Bz(4-NO₂)	12,000	696,000	229,000	32	g
Z-Phe-Lys-NHO-Bz(4-NO₂)	35,000	3,538,000	471,000	760	h
Z-Val-Val-Lys-NHO-Bz(4-NO₂)	15,000	443,000	606,000	650	h
Z-Lys-Lys-NHO-Bz(4-NO₂)	30,000	47,000	5600	230	h
Z-Phe-Ala-NHO-Mes	640,000	n.d.[e]	n.d.	n.d.	i
Boc-Phe-Ala-NHO-Bz(4-NO₂)	14,000	437,000	42,000	21	g
Boc-Ala-Phe-NHO-Bz(4-NO₂)	ntd[f]	600	30	n.d.	g

[a] In 50 mM potassium phosphate buffer, pH 6.0; source, rat.
[b] In 50 mM sodium acetate buffer, pH 5.5; source, rat.
[c] In 50 mM potassium phosphate buffer, pH 6.5; source, cow.
[d] In 50 mM potassium phosphate buffer, pH 6.5; source, rat.
[e] Not determined.
[f] No time-dependent inhibition.
[g] D. Brömme, A. Schierhorn, H. Kirschke, B. Wiederanders, A. Barth, S. Fittkau, and H.-U. Demuth, Biochem. J. 263, 861 (1989).
[h] D. Brömme, U. Neumann, H. Kirschke, and H.-U. Demuth, Biochim. Biophys. Acta 1202, 271 (1993).
[i] R. A. Smith, P. J. Coles, R. W. Spencer, L. J. Copp, C. S. Jones, and A. Krantz, Biochem. Biophys. Res. Commun. 155, 1201 (1988).

TABLE III

KINETIC PARAMETERS FOR INHIBITION OF AMINOPEPTIDASES BY *N*-AMINOACYL-O-4-
NITROBENZOYL HYDROXAMATES

| | Cathepsin H[a] | | Competitive inhibition, K_i (μM) | |
Inhibitor	k_2/K_i (M^{-1} sec^{-1})	K_i (μM)	Aminopeptidase M[b]	Leucyl aminopeptidase[c]
H-Gly-NHO-Bz(4-NO$_2$)[e]	301	29	124	67
H-Val-NHO-Bz(4-NO$_2$)[e]	116	293	506	413
H-Pro-NHO-Bz(4-NO$_2$)[e]	n.i.[d]	n.i.	473	n.i.
H-Phe-NHO-Bz(4-NO$_2$)[e]	31,800	0.8	102	40

[a] In 50 mM sodium phosphate buffer, pH 6.5, at 30°.
[b] In 40 mM MES buffer, pH 6.5, at 30°.
[c] In 40 mM HEPES buffer, pH 7.0, at 30°.
[d] No inhibition.
[e] H.-U. Demuth, A. Stöckel, A. Schierhorn, S. Fittkau, H. Kirschke, and D. Brömme, *Biochim. Biophys. Acta* **1202**, 265 (1993).

4-nitrobenzoic acid would also generate the appropriate hydroxamic acid, it cannot be excluded that the observed reversible inhibition of the metal-loaminopeptidases by the aminoacyl-*O*-4-nitrobenzoyl hydroxamates is due to an inhibition by the liberated hydroxamic acid derivative.

Influence of Leaving (O-Acyl) Group

The linear free-energy (LFE) relationship of the decomposition reactions shows a modest positive dependence of the reaction rate on the electron-withdrawing forces of the substituents. The Hammett correlation coefficients ρ are estimated in a range between 0.708 and 0.860 for a variety of peptidyl and nonpeptidyl hydroxamates.[14,19,32] The inactivation rate constants k_2 of cysteine proteinases tested are not much dependent on the electronic nature of the substituent of the benzoyl leaving group of the inhibitors.[19] Spans of half-life for the inactivation reaction are from 3.7 to 7.7 sec and from 52 to 67 sec for cathepsin B and papain, respectively. This indicates that the rate-determining step of the inactivation is, at best, only poorly influenced by the nature of the *O*-benzoyl residue of the hydroxamates. Thus the departure of the *O*-benzoyl residue must take place at a distinct step before the chemical modification of the target enzyme by the inhibitor occurs. This is reflected in the stronger dependence of the bimolecular reaction constant (k_2/K_i) on the electron negativ-

[32] W. B. Renfrow and C. R. Hauser, *J. Am. Chem. Soc.* **59**, 2308 (1937).

ity of the benzoyl substituents. Although the reactions of cathepsins B and S are not so sensitive to the leaving group, the inactivation of papain and cathepsin L is similarly controlled by the leaving group of the inhibitors as the rate of the spontaneous decomposition of the compounds. This difference might be interpreted by different impacts of single steps in the catalytic reaction path on the overall rate constant of inactivation.

Considering the H_2O concentration to be 55 M, the second-order rate constants for the hydrolytic decomposition of the Z-Phe-Gly-O-benzoyl hydroxamates are between 10^{-5} and $10^{-6} M^{-1} \sec^{-1}$. Because in both types of reaction, enzyme inhibition and inhibitor decomposition, the N–O bond fission is the key chemical step, the cysteine proteinases enhance this reaction by a factor of up to 10 orders of magnitude, in a process that can be termed "enzyme-activated" inhibition.

The linear free relationships of the inhibiton of cysteine proteinases reveal that, depending on the individual target proteinase, the variation of the leaving O-acyl residue might influence the overall inactivation reaction. It has been shown by Krantz and co-workers[11] that, especially in the inactivation of cathepsin B, steric elements of the departing group greatly influence the specificity constant of the inactivation of the enzyme.

S' Specificity Studies with N-Peptidyl-O-carbamoyl Amino Acid Hydroxamates

N,O-Diacyl hydroxamates may serve as useful tools to study the S' specificity of cysteine proteinases. The introduction of an active carbonyl group allows an extension at the P' site of the hydroxamate inhibitor with amino acids or peptides. The -NH-O-CO- moiety mimics a glycine residue occupying the S1' site in the protease.

Substitutions in the P2' position of the inhibitor have a maximal 10- to 30-fold effect on the second-order rate constants of inactivation for the residues tested so far. The closely related cysteine proteinases papain and the cathepsins L and S display a preference toward aromatic residues in this position, whereas cathepsin B better accommodates branched hydrophobic residues (Fig. 1).[28] The different specificity of cathepsin B may be related to the unique occluding loop that closes the S' binding site in this enzyme and determines its peptidyl-dipeptidase activity.[33]

The second-order rate constants for N-peptidyl-O-carbonyl amino acid hydroxamates[28] are at least one order of magnitude lower than that of the O-nitrobenzoyl hydroxamates.[12,17,19] This lower reactivity can be explained by the weaker electron-withdrawing effect of an amino acid residue

[33] D. Musil, D. Zucic, D. Turk, R. A. Engh, I. Mayr, R. Huber, T. Popovic, V. Turk, T. Towatari, N. Katunuma, and W. Bode, *EMBO J.* **10**, 2321 (1991).

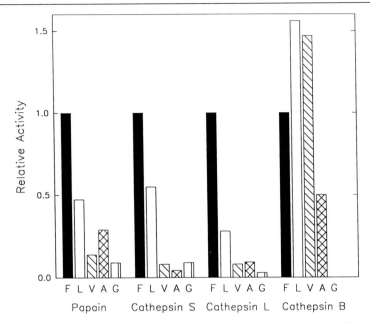

FIG. 1. Comparison of k_2/K_i values of the inactivation of papain and the cathepsins S, L, and B by Z-Phe-Gly-NHO-CO-Aa (Aa: F, Phe; L, Leu; V, Val; A, Ala; G, Gly). The values are normalized to 1 for Z-Phe-Gly-NHO-CO-Phe. From D. Brömme and H. Kirschke, *FEBS Lett.* **322,** 211 (1993).

on the carbonyl function. However, an extension with a longer peptide part on the P' site may lead to a significant improvement of the binding parameters of this class of inactivators.

Selectivity of N,O-Diacyl Hydroxamates toward Cysteine Proteinases

When comparing the second-order rate constants of inactivation of the tested cysteine and serine proteinases, a difference in efficiency of these inhibitors becomes evident. In fact, the results resemble those obtained for the inhibition of cysteine and serine proteinases by diazomethanes (Table IV).

Depending on their specificity and the electronegativity of the leaving group O-acyl residue, hydroxamates with short-chain peptide parts inactivate cysteine proteinases with inactivation constants up to 3,500,000 M^{-1} sec^{-1} (Table II). Inactivation constants for serine proteinases of up to 36,000 M^{-1} sec^{-1} are observed depending on the length of the substrate-analog peptide moiety of the inhibitors and the electronegativity of the

TABLE IV
SECOND-ORDER RATE CONSTANTS OF INACTIVATION OF CYSTEINE AND SERINE PROTEINASES BY
N,O-DIACYL HYDROXAMATES

Inactivator	k_2/K_i $(M^{-1} \sec^{-1})$	Enzyme	Enzyme class	Ref.
Z-Gly-Phe-NHO-Bz(4-NO$_2$)	128	Subtilisin	Serine	a
	1170	Thermitase	Serine	a
Boc-Ala-Leu-Phe-NHO-Bz	36,000	Subtilisin	Serine	g
Boc-Ala-Ala-Bz(4-NO$_2$)	12	Elastase	—	b
Boc-Ala-Pro-NHO-Bz(4-NO$_2$)	2.5	PSE^h	Serine	b
H-Ala-Pro-NHO-Bz(4-NO$_2$)	1.9	DPIV^h	Serine	c
Z-Ala-Ala-Pro-Lys-NHO-Bz(4-NO$_2$)	930	Trypsin	Serine	d
	48	Thrombin	Serine	d
	2800	Cathepsin S	Cysteine	d
Z-Phe-Phe-NHO-MA	1,222,000	Cathepsin L	Cysteine	e
Z-Phe-Lys-NHO-Bz(4-NO$_2$)	3,538,000	Cathepsin L	Cysteine	d
	35,000	Cathepsin B	Cysteine	d
Z-Val-Val-Lys-NHO-Bz(4-NO$_2$)	606,000	Cathepsin S	Cysteine	d
H-Phe-NHO-Bz(4-NO$_2$)	32,000	Cathepsin H	Cysteine	f

^{a} H.-U. Demuth, A. Schierhorn, R. Höfke, H. Kirschke, and D. Brömme, submitted (1993).
^{b} H.-U. Demuth, C. Schönlein, and A. Barth, *Biochim. Biophys. Acta* **996**, 19–22 (1989).
^{c} H.-U. Demuth, R. Baumgrass, C. Schaper, G. Fischer, and A. Barth, *J. Enzyme Inhib.* **2**, 129 (1988).
^{d} D. Brömme, U. Neumann, H. Kirschke, and H.-U. Demuth, *Biochim. Biophys. Acta* **1202**, 271 (1993).
^{e} D. Brömme, A. Schierhorn, H. Kirschke, B. Wiederanders, A. Barth, S. Fittkau, and H.-U. Demuth, *Biochem. J.* **263**, 861 (1989).
^{f} H.-U. Demuth, A. Stöckel, A. Schierhorn, S. Fittkau, H. Kirschke, and D. Brömme, *Biochim. Biophys. Acta* **1202**, 265 (1993).
^{g} H.-U. Demuth, A. Schierhorn, and J. Schultz, submitted (1993).
^{h} PSE, Proline-specific endopeptidase; DPIV, dipeptidyl-peptidase IV.

departing O-acyl residue (Table IV). The results reflect the mechanistic differences of action of serine and cysteine proteinases. Serine proteinases are fully activated at the catalytic site by additional, optimal interactions between the recognition site and the substrate molecule, whereas cysteine proteinases recruit a great deal of their catalytic power by their highly reactive nucleophile.[34] These mechanistic differences can be exploited to discriminate further between cysteine and serine proteinases. Compounds containing O-acyl residues that only weakly stabilize the acid anions in solution (such as the acetyl and propyl residues) only competitively inhibit their target proteinase subtilisin, whereas the cysteine proteinase papain is completely inactivated by the same structures. That means that only

[34] E. Shaw, *Adv. Enzymol.* **63**, 271 (1990).

the fully activated, nucleophilic-attacking target enzyme is capable of inducing the decomposition of the hydroxamic moiety, leading to its suicide inactivation. Additionally, the electronic nature of the leaving O-acyl residue can modulate the efficacy of this reaction in three different ways.[35] (1) Depending on the electron-withdrawing nature of the departing O-acyl residue, the reaction rate can be altered. (2) Depending on the stability of resonance stabilization of the departing anion, an inactivation reaction with serine proteinases of similar specificity can be completely excluded. (3) Depending on the nature (aliphatic, aromatic, or amino acid residues) of the O-acyl residue, additional binding specificity can be added.

Enzymes exhibiting a completely different mechanism, such as aspartic proteinases or metalloproteinases, are not irreversibly inactivated (D. Brömme and H.-U. Demuth, unpublished results). N,O-Diacyl hydroxamates can therefore be considered a new class of potent and specific cysteine proteinase inhibitors. In contrast to other specific cysteine proteinase inhibitors, such as diazomethanes or epoxides, they have the advantage of a higher flexibility for chemical modifications. The access of the P' site for modifications can be exploited in the design of more specific, stable, and bioviable inhibitors of cysteine proteinases.

Acknowledgment

This work has been supported in part by a grant of the Deutsche Forschungsgemeinschaft, Grant-Nr. De 471/1-1.

[35] H.-U. Demuth, A. Schierhorn, and C. Schmidt, submitted (1993).

[49] Cystatins

By MAGNUS ABRAHAMSON

Protein inhibitors of cysteine peptidases in mammalian tissues were first reported in the late 1950s,[1] and have been studied extensively for the past few years. The name "cystatin" was originally coined by Barrett for an inhibitor of papainlike cysteine peptidases that was isolated from chicken egg white,[2] and this name has been used in the nomenclature of evolutionary and functionally related proteins, including those found in

[1] J. T. Finkenstaedt, *Proc. Soc. Exp. Biol. Med.* **95,** 302 (1957).
[2] A. J. Barrett, this series, Vol. 80, p. 771.

mammalian tissues and body fluids. In this article, "cystatin" will be used in the general sense, to designate a functionally active cysteine peptidase inhibitor evolutionarily related to chicken cystatin.

Three types or families of cystatins have been recognized.[3] Families 1 and 2 contain low molecular weight cystatins, proteins of about 100–120 amino acid residues. The proteins in family 1 are normally intracellular, being synthesized without signal peptides, and contain no disulfide bonds. In contrast, those in family 2 are synthesized with signal peptides, and are secreted proteins that contain disulfide bonds. The family 3 cystatins are the kininogens, much larger proteins that contain three cystatin domains, two of which are functional, as well as unrelated domains.

Cystatins from several higher animals, including man, mouse, rat, dog, and cow, have been isolated. Low molecular weight cystatins are also found in many lower organisms, as exemplified by inhibitors isolated from rice seeds, African puff adder venom, and flesh fly hemolymph.[4-6] This chapter will deal with the human cystatins known to date, but the protocols given should be applicable also to the isolation and assay of the counterparts in other mammals. Reviews covering historical aspects of cystatin research, as well as the numerous studies dealing with biological activity of the inhibitors, have been written by Barrett et al.[7] and by Turk and Bode.[8]

Assay Methods

Assay for Cystatin Activity

Because all cystatins known to date inhibit papain by formation of tight, reversible complexes, assays for cystatin activity are most often based on the inhibition of the peptidase activity of this enzyme, which results from preincubation with the inhibitor. The assay we use routinely for human cystatins in our laboratory is essentially that described by Barrett,[2] using benzoyl-DL-arginine 4-nitroanilide as substrate. To obtain a reasonable level of hydrolysis of this substrate, the papain solution to be added in the assay needs to have a concentration of around 5 μM (assay concentration, 0.5 μM). Consequently, the sample to be assayed

[3] N. D. Rawlings and A. J. Barrett, J. Mol. Evol. 30, 60 (1990).
[4] K. Abe, Y. Emori, H. Kondo, K. Suzuki, and S. Arai, J. Biol. Chem. 262, 16793 (1987).
[5] A. Ritonja, H. J. Evans, W. Machleidt, and A. J. Barrett, Biochem. J. 246, 799 (1987).
[6] H. Saito, T. Suzuki, K. Ueno, T. Kubo, and S. Natori, Biochemistry 28, 35 (1989).
[7] A. J. Barrett, N. D. Rawlings, M. E. Davies, W. Machleidt, G. Salvesen, and V. Turk, in "Proteinase Inhibitors" (A. J. Barrett and G. Salvesen, eds.), p. 515. Elsevier, Amsterdam, 1986.
[8] V. Turk and W. Bode, FEBS Lett. 285, 213 (1992).

must have a cystatin concentration of at least 0.5 μM to result in significant inhibition of the papain activity. Given known association rate and equilibrium constants for cystatin–papain interactions (see end of chapter), this concentration is both sufficiently high to result in equilibrium between enzyme and inhibitor within minutes normally, and to render the reaction practically irreversible. Hence, there should be a linear relationship between amount of cystatin added and reduction of papain activity in the assay. Because cystatins seem to bind also to inactive forms of cysteine peptidases, the papain used should preferentially be fully active to allow reliable quantitation. Fully active papain can be prepared by affinity chromatography, e.g., on glycylglycyltyrosylarginine (available from Bachem, Bubendorf, Switzerland) bound to CNBr-activated Sepharose.[9] The molar concentration of active papain to be used should be determined by titration, by substituting known amounts of the irreversible 1 : 1 binding inhibitor L-3-carboxy-2,3-*trans*-epoxypropionylleucylamido(4-guanidino)butane (E-64, from Sigma, St. Louis, MO) for cystatin in the assay described below.[10,11]

Reagents

Substrate: benzoyl-DL-arginine 4-nitroanilide hydrochloride (Bachem). A stock solution is made by dissolution to 100 mM in dimethyl sulfoxide (DMSO; 435 mg/10 ml). Store at 4°.

Assay buffer concentrate: 0.4 M sodium phosphate buffer, pH 6.5, containing 4 mM EDTA. Powdered dithiothreitol (DTT) is added to 4 mM (15.4 mg to 25 ml buffer) before use.

Stopping reagent: 0.1 M sodium acetate buffer, pH 4.3, containing 0.1 M sodium monochloroacetate.

Solution for dilutions: 0.01% (w/v) Brij 35 (Sigma) in water.

Papain solution: preferentially affinity purified papain, in an approximately 5 μM solution. Concentrated enzyme, purified by chromatography on glycylglycyltyrosylarginine-Sepharose,[9] can be stored at −20° in elution buffer without significant loss of activity.[12]

Procedure. In microfuge tubes, add in successive order: 125 μl of assay buffer concentrate, 260 μl of 0.01% Brij 35 solution, 50 μl of papain solution, and 50 μl of cystatin sample. Mix and incubate at room temperature for at least 10 min. Start the reaction by addition of 15 μl substrate solution and incubate for 15 min at 37°. The reaction is stopped by addition

[9] M. O. Funk, Y. Nakagawa, J. Skochdopole, and E. T. Kaiser, *Int. J. Pept. Protein Res.* **13,** 296 (1979).

[10] A. J. Barrett, A. A. Kembhavi, M. A. Brown, H. Kirschke, C. G. Knight, M. Tamai, and K. Hanada, *Biochem. J.* **201,** 189 (1982).

[11] A. J. Barrett and H. Kirschke, this series, Vol. 80, p. 535.

[12] A. Hall, unpublished result (1993).

of 500 μl of the stopping reagent, and liberated nitroaniline is quantified by A_{410} measurement.

Evaluation of Results. Under the conditions described, a standard curve for assay tubes in which the amount of cystatin is varied should be linear for all human cystatins. Thus, and because equal volumes of papain solution and cystatin sample are used, the active cystatin concentration in the sample analyzed can be calculated as fractional reduction of papain activity multiplied by concentration of active papain in the solution added in the assay (known from E-64 titration). If kininogens are analyzed, the molar concentration obtained equals that of active cystatin domains in the sample, which under normal conditions should be twice the molar concentration of kininogen.

Comment. The linearity of the cystatin assay described holds for all native human cystatins. For cystatins modified by limited proteolysis or amino acid substitutions of critical residues, however, this is not necessarily the case. If K_i is of the same order as the assay enzyme concentration, the assay will not be useful for accurate determination of the sample cystatin concentration (but will still be useful to determine if any cystatin is present in the sample analyzed). It can therefore be advised that, in work with cystatins of unknown origin or quality, a series of dilutions of the cystatin sample under study are analyzed to prove or disprove that a linear relation between amount of cystatin and degree of inhibition holds.

Assays for Determination of K_i Values

The equilibrium constants for dissociation, K_i, of most cystatin–cysteine peptidase complexes are in the order of 10^{-9} M (see later). To be able to measure dissociation from such complexes experimentally, assays based on hydrolysis of sensitive fluorogenic aminomethylcoumarin substrates are widely used. A continuous-rate assay as described here for determination of K_i values for cystatin interactions was originally used by Nicklin and Barrett[13] to study dissociation of chicken cystatin from complexes with cathepsin B. The same assay can also be used for studies of cystatin interactions with the other three well-known mammalian cysteine peptidases, cathepsin L, H, and S, as well as for the model cysteine peptidase, papain. For simplicity, an assay buffer of pH 6.0 can be used for all of these five enzymes, and benzyloxycarbonylphenylalanylarginylaminomethylcoumarin can be used as substrate for all enzymes except

---

[13] M. J. H. Nicklin and A. J. Barrett, *Biochem. J.* **223**, 245 (1984).

cathepsin H (for which unblocked arginylaminomethylcoumarin instead is used).[11] The fluorimeter used should preferentially be under computer control, to facilitate calculations from collected data. To be able to determine K_i values at defined temperatures (most reference values have been determined at 37°), it also needs to allow temperature control of the cuvette holder. In addition, a cuvette holder equipped with a magnetic stirrer, if available, facilitates rapid mixing.

Reagents

Substrates: benzyloxycarbonylphenylalanylarginylaminomethylcoumarin and arginylaminomethylcoumarin (both available from Bachem). Stock solutions are made by dissolution to 1.0 mM (6.49 mg/10 ml water and 4.26 mg/10 ml dimethyl sulfoxide, respectively). Store at 4°. The stock solutions are diluted to 200 μM with water before use.

Assay buffer concentrate: 0.4 M sodium phosphate buffer, pH 6.0, containing 4 mM EDTA. Powdered dithiothreitol is added to 4 mM (15.4 mg to 25 ml buffer) before use.

Solution for dilutions: 0.01% (v/v) Brij 35 in water.

Solution for fluorimeter calibration: a stock solution of 0.25 mM aminomethylcoumarin dissolved in dimethyl sulfoxide is kept at 4°. The stock solution is diluted to 0.2 μM with water to obtain the calibration solution.

Enzyme solutions: purified enzymes, diluted to approximately 10 nM solutions. The exact enzyme concentration is not critical for the evaluation of results. The enzyme concentration needed in the assay to give a useful rate of substrate hydrolysis varies (due to K_m differences for the enzymes) between approximately 0.05 nM for papain to around 0.25 nM for cathepsin B.

Cystatin sample: preferentially an approximately 1 mg/ml (75 μM) solution of the cystatin under study, for which the exact concentration of active cystatin has been determined in a cystatin assay (procedure above).

Procedure. Set fluorimeter excitation and emission wavelengths to 360 and 460 nm, respectively. Calibrate with 0.2 μM aminomethylcoumarin to let the maximal response of the fluorimeter correspond to 2% hydrolysis of the assay substrate. Add to the fluorimeter cuvette, fitted into the cuvette holder, 2040 μl 0.01% Brij 35 solution, 750 μl assay buffer concentrate, and 30 μl enzyme solution. Let the temperature adjust to the chosen assay temperature for 5 min, during which time the enzyme also activates. Add 150 μl of substrate solution and start data collection. When a steady-state rate of substrate hydrolysis (v_0) is recorded, 30 μl of cystatin sample is added and the progress curve is recorded until a new steady-state rate (v_i) can be determined.

Calculations. The linear equation derived by Henderson[14] for tight-binding inhibitors can be used to evaluate the results from experiments with different dilutions of the cystatin sample. The relative steady-state rates of substrate hydrolysis before and after addition of cystatin sample are then used to determine the apparent K_i value [$K_{i(app)}$] as the slope from the plot of [I]/(1 − v_i/v_0) against v_0/v_i. This value is then used to calculate the substrate independent K_i from Eq. (1).

$$K_{i(app)} = K_i(1 + [S]/K_m) \tag{1}$$

Comments. The equation used is valid when the assay enzyme concentration is in the same order as K_i and the inhibitor concentration is so high that no significant depletion of free inhibitor in the system is caused by the formation of enzyme–inhibitor complexes. In practice, the procedure described is useful to determine K_i values for most cystatin–cysteine peptidase interactions. However, the very tight complexes between some cystatins and target peptidases, e.g., human cystatin C and papain or cathepsin L, will not significantly dissociate under the given conditions. The K_i for such interactions must be calculated from the individual associa-tion and dissociation rate constants.[15] Evaluation of assay data by the Henderson equation has the advantage that it can also be carried out without a computer-controlled fluorimeter. Steady-state velocities of sub-strate hydrolysis before and after addition of inhibitor, v_0 and v_i, can be assessed on graph paper with a ruler, in case a program to do linear regression of part of the progress curve (such as FLUSYS by Rawlings and Barrett[16]) is not available to prove that linearity has been achieved. Alternative ways to evaluate data collected in assays similar to the one described here include computer-aided fitting of progress curve data to obtain v_0 and v_i values from Eq. (2),[17] and then fitting of these parameters to Eq. (3)[18] to obtain the apparent K_i (which can be corrected for substrate-induced dissociation as described above). The advantage of this procedure is that the experimental rate constant for the reaction, k_{obs}, also is obtained from the data fit to Eq. (2), which for some interactions can be used to

[14] P. J. F. Henderson, *Biochem. J.* **127**, 321 (1972).
[15] P. Lindahl, M. Abrahamson, and I. Björk, *Biochem. J.* **281**, 49 (1992).
[16] N. D. Rawlings and A. J. Barrett, *Comput. Appl. Biosci.* **6**, 118 (1990).
[17] S. Cha, *Biochem. Pharmacol.* **24**, 2177 (1976).
[18] J. F. Morrison, *Trends Biochem. Sci.* **7**, 102 (1982).

calculate both association and dissociation rate constants for the reaction from Eq. (4).

$$[P] = v_i t + (v_0 - v_i)(1 - e^{-k_{obs}t})/k_{obs} \qquad (2)$$
$$v_0/v_i - 1 = [I]/K_i \qquad (3)$$
$$k_{obs} = k_{-1} + k_1[I] \qquad (4)$$

Immunoassays

All human cystatins elicit good antibody responses in rabbits. Immunization is done by subcutaneous injection of the isolated and desalted cystatin (0.1–0.5 mg), emulsified in Freund's complete adjuvant. The injection is repeated after 3 weeks, and the rabbits can then be bled every third week. Enzyme-amplified single radial immunodiffusion[19,20] in agarose gels containing 0.4–0.8% antiserum is a rapid procedure that has been successfully used to quantitate all human family 1 and 2 cystatins, with a sensitivity of approximately 0.3 μg/ml.[21,22] The cross-reactivity between the two differently sized forms of human kininogen, L- and H-kininogen, make it more difficult to set up an accurate, simple immunoassay. The enzyme-amplified single radial immunodiffusion procedure, using 0.4% of an antiserum raised against L-kininogen, works satisfactorily, however.[21]

Electrophoresis of Cystatins

The size difference between human family 1 cystatins (A and B) and the family 2 cystatins (C, D, S, SA, and SN) can readily be detected by SDS–polyacrylamide gel electrophoresis using a discontinuous system as described by Laemmli[23] and separation gels containing 20% acrylamide.[21] Better resolution in the low molecular weight range is obtained by use of 16.5% acrylamide separation gels in the Laemmli-based buffer system described by Schägger and von Jagow,[24] which may be useful in the study of proteolytically modified cystatins.[25] In these electrophoresis systems, a slight overestimation of the cystatin molecular weight will be apparent in comparisons with commonly used size marker proteins. If, e.g., the low molecular weight calibration kit from Pharmacia (Piscataway, NJ;

---

[19] G. Mancini, A. O. Carbonara, and J. F. Heremans, *Immunochemistry* **2**, 235 (1965).

[20] H. Löfberg and A. O. Grubb, *Scand. J. Clin. Lab. Invest.* **39**, 619 (1979).

[21] M. Abrahamson, G. Salvesen, A. J. Barrett, and A. Grubb, *J. Biol. Chem.* **261**, 11282 (1986).

[22] J. P. Freije, M. Balbín, M. Abrahamson, G. Velasco, H. Dalbøge, A. Grubb, and C. López-Otín, *J. Biol. Chem.* **268**, 15737 (1993).

[23] U. K. Laemmli, *Nature (London)* **227**, 680 (1970).

[24] H. Schägger and G. von Jagow, *Anal. Biochem.* **166**, 368 (1987).

[25] A. Hall, H. Dalbøge, A. Grubb, and M. Abrahamson, *Biochem. J.* **291**, 123 (1993).

containing α-lactalbumin and soybean trypsin inhibitor as markers in the cystatin size range) is used to construct a calibration curve, typical apparent molecular weights obtained are around 13,000 and 15,000–16,000 for family 1 and 2 cystatins, respectively.[21] For size electrophoresis under nondenaturing conditions, continuous 16.5% acrylamide gels and the pH 4.0 alanine/acetate buffer system described by Jovin[26] work well also for the very basic cystatin, cystatin C.

To distinguish between the equally sized low molecular weight cystatins, e.g., in column fractions at purification, we have found charge-separating agarose gel electrophoresis as described by Jeppsson et al.[27] very useful. With the pH 8.6 barbital buffer and 1% (w/v) agarose gels routinely used, optimal separation is achieved in less than 45 min and the entire procedure, from casting of the gel to visualization of protein bands by Coomassie blue staining, can be completed in 3 hr.

Purification of Human Cystatins

Human low molecular weight cystatins of families 1 and 2 can be isolated from a number of different tissues and body fluids, whereas the primary starting material for isolation of family 3 cystatins, kininogens, should be blood plasma. The purification protocol described below has been used in the simultaneous purification of cystatins A, B, C, S, and SN as well as L-kininogen from human urine,[21] and should therefore be useful as a general guideline for cystatin isolation. However, depending on which cystatins are present in the starting material for purification, and to what extent the inhibitors are proteolytically modified, further purification steps may be needed to isolate the individual inhibitors. Affinity chromatography on immobilized inactive papain[28] is the key purification step, a procedure that has been widely used for purification of cystatins.[7,8] To obtain good yields of native cystatins in any purification procedure, the single most important step has turned out to be protection against proteolytic breakdown during sample handling and the purification procedure. This is especially important for the family 2 cystatins, because their N-terminal segments seem to be well exposed on the surface of the molecules and hence susceptible to proteolytic attack by bacterial, as well as some endogenous, peptidases.

[26] T. M. Jovin, Ann. N.Y. Acad. Sci. 209, 477 (1973).
[27] J.-O. Jeppsson, C.-B. Laurell, and B. Franzén, Clin. Chem. 25, 629 (1979).
[28] M. Järvinen, J. Invest. Dermatol. 71, 114 (1978).

Starting Materials

Urine. If available, urine from individuals with mixed glomerular–tubular proteinuria will be a good source for most human cystatins. Such urine has been successfully used as starting material for isolation of cystatins A, B, C, S, and SN and L-kininogen.[21] Because of its low cystatin concentration, the urine should preferentially be concentrated before the first chromatography step. This can be achieved by pressure ultrafiltration using YM2 membranes in Diaflo cells (Amicon, Danvers, MA) or, more efficiently, by help of an artificial kidney (e.g., C-DAK, available from Cordia Dow Corp., Miami, FL, with a retention limit of approximately 1500 Da).

Saliva. Whole saliva is the recommended source for purification of human cystatins D, S, SN, and SA, and also contains cystatin C. It can be used directly, after addition of the inhibitor cocktail mentioned below, for carboxymethylpapain (Cm-papain) chromatography.

Blood. Blood plasma will be the best source for L- and H-kininogen, and also contains cystatin C. Granulocytes from blood can be used for purification of cystatin A. The cytosolic fraction from the cells is then used for Cm-papain chromatography.[29]

Tissue Homogenates. If autopsy material is available, liver or spleen has proved to be good starting material for isolation of family 1 cystatins (A and B).[30,31] To obtain a cellular fraction for Cm-papain chromatography, the tissue is homogenized and the resulting supernatant is treated with alkali (pH 11.0) to inactivate cellular cysteine peptidases, which otherwise will form cystatin complexes.[31]

Procedure

To all body fluids or tissue homogenates to be used as starting material, add, as soon as possible, 5% of an inhibitor cocktail concentrate containing 0.1 M benzamidinium chloride, 0.2 M EDTA, and 2% (w/v) sodium azide (final concentrations of 5, 10, and 15 mM, respectively). Cm-papain is prepared as described by Barrett,[2] and coupled to CNBr-activated Sepharose 4B (Pharmacia) according to the manufacturer's instructions. The starting material is mixed with an equal volume of 0.1 M Tris-HCl buffer, pH 7.4, containing 1.0 M NaCl and 5% inhibitor cocktail concentrate, and

[29] J. Brzin, M. Kopitar, V. Turk, and W. Machleidt, *Hoppe-Seyler's Z. Physiol. Chem.* **364,** 1475 (1983).

[30] M. Järvinen and A. Rinne, *Biochim. Biophys. Acta* **708,** 210 (1982).

[31] G. D. J. Green, A. A. Kembhavi, M. E. Davies, and A. J. Barrett, *Biochem. J.* **218,** 939 (1984).

applied to a column packed with the Sepharose-coupled Cm-papain. After thorough washing with 50 mM Tris-HCl buffer, pH 7.4, containing 0.5 M NaCl and 5% inhibitor cocktail concentrate, material with affinity for Cm-papain is eluted with 0.20 M trisodium phosphate, pH 12.1, containing 0.5 M NaCl and 5% inhibitor cocktail concentrate. Eluted material is immediately neutralized with 2 M Tris-HCl buffer, pH 7.4, and concentrated by pressure ultrafiltration using a Diaflo YM2 membrane.

The concentrated eluate is fractionated by ion-exchange chromatography using the Mono Q HR 10/10 column of the FPLC (fast protein liquid chromatography) system (Pharmacia). After dialysis against 20 mM ethanolamine buffer, pH 9.5 (3500-Da cut-off dialysis tubing should be used, e.g., Spectra/Por, available from Spectrum, Houston, TX), the sample is applied and elution is performed with a gradient of the same buffer containing 1.0 M NaCl, as shown in Fig. 1. Most human cystatins will separate under these conditions (Fig. 1) but, depending on the starting material used, additional purification steps such as immunosorption or chromatofocusing may be needed to isolate the individual cystatins. To achieve separation of L- from H-kininogen, both of which are recovered late in the salt gradient from the Mono Q column, the pooled kininogen-containing fractions should be rechromatographed. The same column is used, but

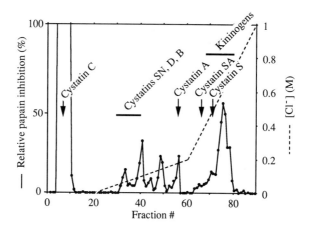

FIG. 1. Ion-exchange chromatography of eluate from a Cm-papain column. The chromatogram shows the elution profile from a Mono Q HR 10/10 column, on chromatography of material with Cm-papain affinity from 2 liters of urine.[21] The sample was applied in 20 mM ethanolamine buffer, pH 9.5, and developed with a gradient of the same buffer containing 1.0 M NaCl. Fractions of 2.0 ml were collected at a flow rate of 4 ml/min. Cystatin activity was measured as described in the text. Typical elution positions for human cystatins are indicated above the diagram, with arrows or, when elution peaks overlap, with bars.

now equilibrated in 10 mM sodium acetate buffer, pH 5.5, and developed with a gradient of the same buffer containing from 0 to 0.5 M NaCl.[32]

Pooled cystatin fractions are concentrated to 0.5 ml and then gel filtered using the Pharmacia FPLC Superdex HR 75 or Superose 12 column, in 50 mM ammonium bicarbonate buffer, pH 8.3, to desalt the fractions and remove minor impurities. The cystatin solutions can be stored at $-20°$, but should not be frozen and thawed repeatedly.

Comments

The purification yields of cystatins in the three-step procedure above are generally high. By use of the pH 12.1 buffer for elution from the Cm-papain column as described, virtually all of the cysteine peptidase inhibitory activity in urine could be recovered in isolated cystatin fractions.[21] However, the Sepharose used to immobilize Cm-papain will hydrolyze to some extent at this high pH, resulting in leakage of Cm-papain if the gel is exposed for longer periods and, eventually, destruction of the column. For purification of cystatins that bind less tightly than cystatin C to Cm-papain, a 50 mM phosphate buffer, pH 11.5, containing 0.5 M NaCl may be used instead of the pH 12.1 elution buffer.

For large-scale purification of cystatins, it is advised that the Cm-papain binding step be done in a bottle by overnight incubation at 4° with gentle agitation, and all successive washes as well as the elution be performed batch-wise on a glass filter funnel.

Characteristics of Human Cystatins

Structures and Physicochemical Properties

Two different groups of low molecular weight cystatins can be distinguished by physicochemical differences. Those inhibitors belonging to family 1 of the cystatin superfamily are single-chain polypeptides of approximately 100 amino acid residues (M_r 11,000–12,000); these lack intrachain disulfide bridges and carbohydrate. The human family 1 cystatins with known primary structures, called cystatins (or by some authors, stefins) A and B, have defined counterparts in other mammals.[7] A third family 1 cystatin, called stefin C, has in addition been isolated from bovine thymus,[33] but no human homolog has been reported yet. The other group

[32] W. Müller-Esterl, D. A. Johnson, G. Salvesen, and A. J. Barrett, this series, Vol. 163, p. 240.

[33] B. Turk, I. Krizaj, B. Kralj, I. Dolenc, T. Popovic, J. G. Bieth, and V. Turk, *J. Biol. Chem.* **268,** 7323 (1993).

of low molecular weight cystatins has defined members in humans; these are cystatins C, D, S, SN, and SA. They are all single-chain polypeptides of 120–122 amino acid residues (M_r 13,000–14,000), and are members of the cystatin family 2. They lack carbohydrate, but have two disulfide bridges, in contrast to the low molecular weight cystatins belonging to family 1. Of the five human family 2 cystatins, the three S-type cystatins (S, SN, and SA) have strikingly similar primary structures, with only approximately 10% amino acid differences when their sequences are aligned. They also display immunochemical cross-reactivity. Presence of at least one S-type cystatin in rat saliva has been documented.[34] Homologs to human cystatin C have been found in several mammals, including mouse, rat, cow, and dog.[35–38]

The physicochemical characteristics of the low molecular weight cystatins are summarized in Table I. If, at purification of cystatins, specific antisera are not available to identify individual inhibitors, N-terminal sequencing needs to be carried out. This is especially important for the family 2 cystatins, because they are frequently (depending on starting material used and whether efficient control of proteolytic breakdown during purification can be achieved) isolated in multiple forms of varying p*I* because of proteolytic cleavage of bonds in the N-terminal segment. The N-terminal sequences of the full-length human low molecular weight cystatins are given in Table I. Commonly seen forms of cystatin C isolated from urine have N-terminal sequences starting at residue Gly-4, Lys-5, Arg-8, Leu-9, or Val-10.[25] Cystatin D isolated from saliva, with maximal protection against proteolysis during sample handling, is a mixture of forms with N-terminal sequences starting at residue Ala-5, Ser-7, Thr-9, or Ala-11.[22] The originally reported forms of cystatins S, SN, and SA isolated from saliva had N-terminal residues Ile-9, Ile-9, and Glu-5, respectively,[39] and cystatin SN (which is identical to the cystatin we originally called cystatin SU[21,40]) is in addition to the full-length molecule also found in a form with Glu-4 as N-terminal residue in saliva. Differences in p*I* values also result from phosphorylation of residues Ser-3 and/or Ser-1 in

[34] P. A. Shaw, J. L. Cox, T. Barka, and Y. Naito, *J. Biol. Chem.* **263**, 18133 (1988).
[35] M. Solem, C. Rawson, K. Lindburg, and D. Barnes, *Biochem. Biophys. Res. Commun.* **172**, 945 (1990).
[36] A. Esnard, F. Esnard, D. Faucher, and F. Gauthier, *FEBS Lett.* **236**, 475 (1988).
[37] M. Hirado, S. Tsunasawa, F. Sakiyama, M. Niinobe, and S. Fujii, *FEBS Lett.* **186**, 41 (1985).
[38] M. D. Poulik, C. S. Shinnick, and O. Smithies, *Mol. Immunol.* **18**, 569 (1981).
[39] S. Isemura, E. Saitoh, and K. Sanada, *J. Biochem.* (*Tokyo*) **102**, 693 (1987).
[40] H. Hansson, A. Hall, M. Abrahamson, and A. Grubb, unpublished results (1993).

TABLE I
MOLECULAR CHARACTERISTICS OF HUMAN LOW MOLECULAR WEIGHT CYSTATINS

Cystatin	Other names	Family	Amino acids $(n)^a$	$M_r{}^a$	N-Terminal sequence^{a,b}	pI
A	Stefin, stefin A, ACPI	1	98	11,006	MIPG-	4.5–4.7
B	Stefin B, NCPI	1	98	11,175	acMMCG-	5.6–6.3
C	γ-Trace, post-γ-globulin	2	120	13,343^c	SSPGKPPRLVG-	9.3
D	—	2	122	13,885^d	GSASAQSRTLAG-	6.8–7.0
S	SAP-1	2	121	14,189	SSSKEENRIIPG-	4.4–4.6
SN	Cystatin SU, Cystatin SA-I	2	121	14,316	WSPKEEDRIIPG-	6.6–6.8
SA	—	2	121	14,351	WSPQEEDRIIEG-	4.6

^{a} From protein sequencing of full-length cystatins, or deduced from their DNA sequences, reported in W. Machleidt, U. Borchart, H. Fritz, J. Brzin, A. Ritonja, and V. Turk, *Hoppe-Seyler's Z. Physiol. Chem.* **364**, 1481 (1983) (cystatin A); A. Ritonja, W. Machleidt, and A. J. Barrett, *Biochem. Biophys. Res. Commun.* **131**, 1187 (1985) (cystatin B); M. Abrahamson, A. Grubb, I. Olafsson, and Å. Lundwall, *FEBS Lett.* **216**, 229 (1987) (cystatin C); J. P. Freije, M. Abrahamson, I. Olafsson, G. Velasco, A. Grubb, and C. López-Otín, *J. Biol. Chem.* **266**, 20538 (1991) (cystatin D); and S. Isemura, E. Saitoh, K. Sanada, and K. Minakata, *J. Biochem. (Tokyo)* **110**, 648 (1991) (cystatins S, SN, and SA). acM, N-acetyl-(blocked) Mot residue.
^{b} Aligned after the evolutionarily conserved Gly residue present in the N-terminal segment of all cystatins (N. D. Rawlings and A. J. Barrett, *J. Mol. Evol.* **30**, 60 (1990)).
^{c} Cystatin C from urine contains a hydroxylated Pro residue, Hyp-3, giving a M_r of 13,359 [A. Grubb and H. Löfberg, *Proc. Natl. Acad. Sci. U.S.A.* **79**, 3024 (1982)].
^{d} Cystatin D also exists in a form with Arg replacing Cys-26, giving a M_r of 13,938 [J. P. Freije, M. Balbín, M. Abrahamson, G. Velasco, H. Dalbøge, A. Grubb, and C. López-Otín, *J. Biol. Chem.* **268**, 15737 (1993)].

salivary cystatin S,[41] and for cystatin D because of a genetic polymorphism resulting in either Arg or Cys as residue 26.[42]

A different group of mammalian cystatins is constituted by the high molecular weight glycoproteins originally named α-cysteine (or thiol) proteinase inhibitors (α-CPI), which were isolated initially from blood plasma

[41] S. Isemura, E. Saitoh, K. Sanada, and K. Minakata, *J. Biochem. (Tokyo)* **110**, 648 (1991).
[42] M. Balbín, J. P. Freije, M. Abrahamson, G. Velasco, A. Grubb, and C. López-Otín, *Hum. Genet.* **90**, 668 (1993).

TABLE II
DISSOCIATION CONSTANTS OF COMPLEXES BETWEEN HUMAN CYSTATINS
AND CYSTEINE PEPTIDASES

	K_i (nM)				
Cystatin	Papain	Cathepsin B	Cathepsin H	Cathepsin L	Cathepsin S
A	0.019[a]	8.2[a]	0.31[a]	1.3[a]	0.05[b]
B	0.12[a]	73[a]	0.58[a]	0.23[a]	0.07[b]
C	0.000011[c]	0.25[d]	0.28[d]	<0.005[d]	0.008[b]
D	1.2[e]	>1000[e]	8.5[f]	25[f]	0.24[f]
S	108[g]	—	—	—	—
SN	0.016[h]	19[h]	—	—	—
SA	0.32[i]	—	—	—	—
L-Kininogen[j]	0.015[k]	600[j]	0.72[j]	0.017[k]	—

[a] G. D. J. Green, A. A. Kembhavi, M. E. Davies, and A. J. Barrett, *Biochem. J.* **218**, 939 (1984).

[b] D. Brömme, R. Rinne, and H. Kirschke, *Biomed. Biochim. Acta* **50**, 631 (1991).

[c] Calculated from association and dissociation rate constants [P. Lindahl, M. Abrahamson, and I. Björk, *Biochem. J.* **281**, 49 (1992)].

[d] A. J. Barrett, M. E. Davies, and A. Grubb, *Biochem. Biophys. Res. Commun.* **120**, 631 (1984).

[e] J. P. Freije, M. Balbín, M. Abrahamson, G. Velasco, H. Dalbøge, A. Grubb, and C. López-Otín, *J. Biol. Chem.* **268**, 15737 (1993).

[f] M. Balbín, A. Hall, A. Grubb, R. W. Mason, C. López-Otín, and M. Abrahamson, *J. Biol. Chem.* (1994), in press.

[g] S. Isemura, E. Saitoh, and K. Sanada, *FEBS Lett.* **198**, 145 (1986).

[h] M. Abrahamson, G. Salvesen, A. J. Barrett, and A. Grubb, *J. Biol. Chem.* **261**, 11282 (1986).

[i] S. Isemura, E. Saitoh, and K. Sanada, *J. Biochem.* (*Tokyo*) **102**, 693 (1987).

[j] Human H-kininogen displays K_i values almost identical to those for L-kininogen [W. Machleidt, A. Ritonja, T. Popovic, M. Kotnik, J. Brzin, V. Turk, I. Machleidt, and W. Müller-Esterl, *in* "Cysteine Proteinases and Their Inhibitors" (V. Turk, ed.), p. 3. de Gruyter, Berlin and New York, 1986].

[k] G. Salvesen, C. Parkes, M. Abrahamson, A. Grubb, and A. J. Barrett, *Biochem. J.* **234**, 429 (1986).

in several forms with M_r 60,000–120,000.[43,44] When a cDNA encoding one of these forms was sequenced,[45] it was discovered that the amino acid sequence deduced from the cDNA was very similar to that of bovine L-kininogen.[46] It was later confirmed that the different α-CPI forms occur-

[43] M. Järvinen, *Acta Chem. Scand.* **B30**, 933 (1976).

[44] M. Sasaki, K. Minakata, H. Yamamoto, M. Niwa, T. Kato, and N. Ito, *Biochem. Biophys. Res. Commun.* **76**, 917 (1977).

[45] I. Ohkubo, K. Kurachi, T. Takasawa, H. Shiokawa, and M. Sasaki, *Biochemistry* **23**, 5691 (1984).

[46] H. Nawa, N. Kitamura, T. Hirose, M. Asai, S. Inayama, and S. Nakanishi, *Proc. Natl. Acad. Sci. U.S.A.* **80**, 90 (1983).

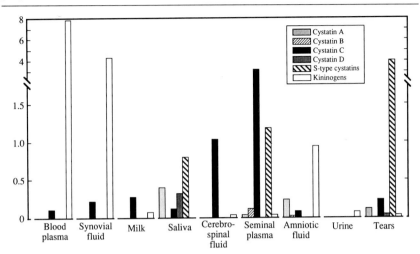

FIG. 2. Molar concentrations of cystatins in human body fluids. The bars denote mean cystatin concentrations (μM) from immunochemical quantitation of 10 samples of each fluid. Data from Refs. 21 and 22.

ring in human blood plasma were identical to differently processed forms of L- and H-kininogens, and that kininogens had cysteine peptidase inhibitory activity (reviewed in Ref. 47). Thus, kininogens are, in addition to precursor molecules for vasoactive kinins, also cysteine peptidase inhibitors, and are by evolutionary relationship members of family 3 in the cystatin superfamily.[3] Besides the L- and H-kininogens generally found in mammals, a third kininogen, T-kininogen, is present in rat blood.[45] Human L-kininogen is composed of 409 amino acid residues and has carbohydrate attached (M_r 68,000 from SDS–polyacrylamide electrophoresis), whereas H-kininogen contains 628 residues and is more extensively glycosylated (M_r 114,000).[32] Both L- and H-kininogen have pI values around 4.5.[32]

The three-dimensional structures of one member from each of cystatin families 1 and 2, recombinant human cystatin B (called stefin B)[48] and chicken cystatin,[49] respectively, have been elucidated by X-ray crystallographic analysis. From this work, and supported by functional studies of

[47] W. Müller-Esterl, S. Iwanaga, and S. Nakanishi, *Trends Biochem. Sci.* **11,** 336 (1986).
[48] M. T. Stubbs, B. Laber, W. Bode, R. Huber, R. Jerala, B. Lenarcic, and V. Turk, *EMBO J.* **9,** 1939 (1990).
[49] W. Bode, R. Engh, D. Musil, U. Thiele, R. Huber, A. Karshikov, J. Brzin, J. Kos, and V. Turk, *EMBO J.* **7,** 2593 (1988).

modified family 2 cystatins, the mechanism of cystatin interaction with target peptidases has been clarified (reviewed in Refs. 8 and 25).

Inhibitory Properties

The cystatins are competitive reversible inhibitors, generally displaying broad specificity and high-affinity binding to target peptidases (Table II). All mammalian cysteine peptidases belonging to the papain superfamily are inhibited to some extent by the human cystatins, as is also the case for a number of cysteine peptidases of plant origin.[7] Cystatin C is the most general inhibitor, in that it binds tightly to all investigated enzymes, whereas cystatin D seems to be more specific based on its poor inhibition of cathepsin B. One of the two functional cystatin domains of the kininogens has inhibitory activity against calpains,[50] a feature not shared by the family 1 and 2 cystatins.

Distribution of Human Cystatins

The family 1 cystatins, A and B, are present in the cytosol of many investigated cell types, although cystatin B seems to have a more general distribution and cystatin A is found primarily in skin and in some white blood cells (reviewed in Ref. 7). The human family 1 cystatins can also be detected in several body fluids (Fig. 2). Measurements of the concentrations of cystatins in human biological fluids reveal that all such fluids contain inhibitors, but the total molar amounts and ratios of different inhibitors vary markedly (Fig. 2). The general pattern seems to be that secretions such as saliva and seminal plasma contain the highest concentrations of family 2 cystatins, whereas the glycoproteins, L- and H-kininogen, are restricted to blood plasma and synovial fluid. Cystatin C has the broadest distribution and is found in all fluids analyzed, cystatin D is found only in saliva and tears, and the S-type cystatins are intermediate in distribution pattern and can be called secretory fluid cystatins.

Acknowledgment

This work was supported by M. Bergvall's Foundation, the Medical Faculty of the University of Lund, and the Swedish Medical Research Council (Project Nos. 9291 and 9915).

[50] G. Salvesen, C. Parkes, M. Abrahamson, A. Grubb, and A. J. Barrett, *Biochem. J.* **234,** 429 (1986).

Author Index

Numbers in parentheses are footnote reference numbers and indicate that an author's work is referred to although the name is not cited in the text.

Subject Index